# MIDDLE EAST MODELS OF JURASSIC/CRETACEOUS CARBONATE SYSTEMS

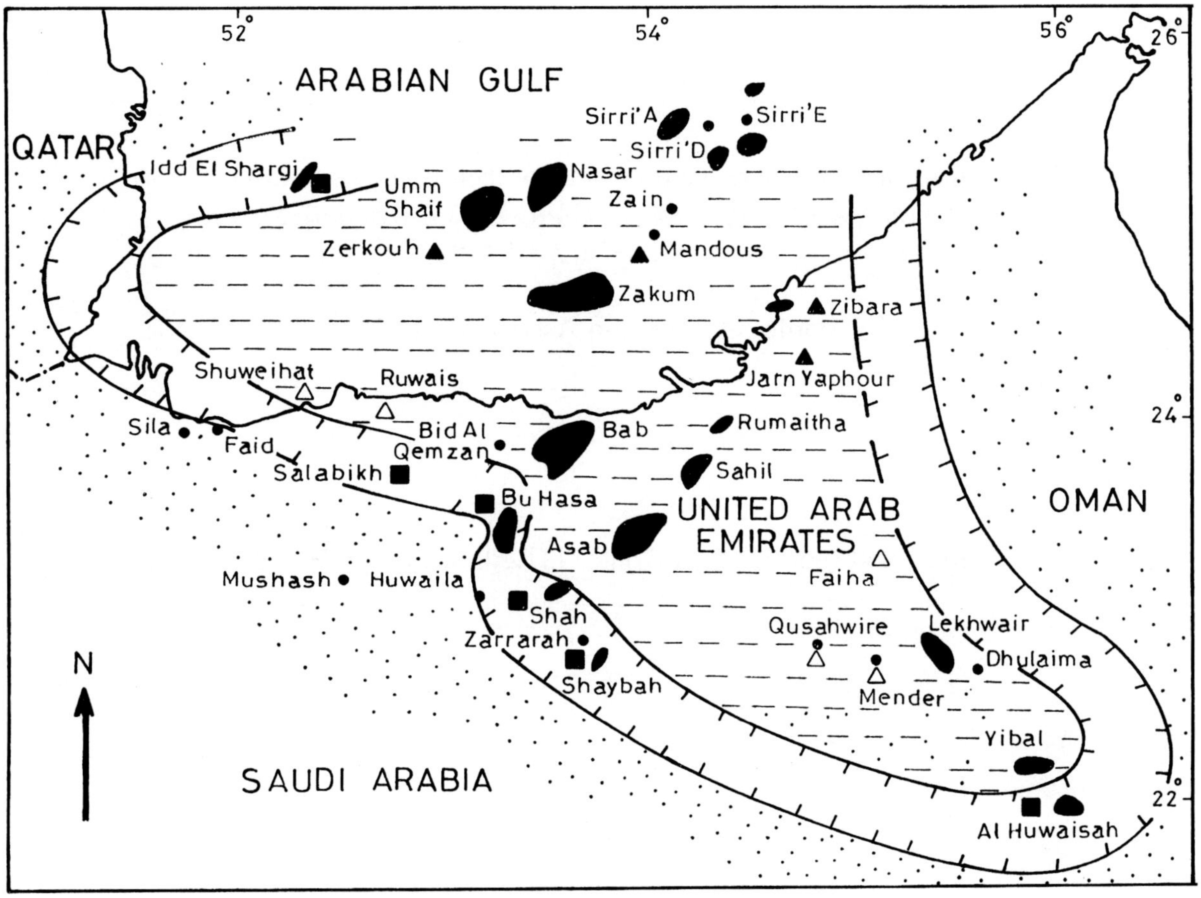


*Edited by:*


ABDULRAHMAN S. ALSHARHAN

*Faculty of Science, United Arab Emirates University, P.O. Box 17551, Al Ain, U.A.E.*

AND

ROBERT W. SCOTT

*Precision Stratigraphy Associates, RR3 Box 103-3, Cleveland, OK 74020, U.S.A.*





*Robert W. Dalrymple, Editor of Special Publications*
*SEPM Special Publication Number 69*

*Tulsa, Oklahoma, U.S.A.* *November, 2000*

SEPM greatly acknowledges the following for their generous contributions to
*Middle East Models of Jurassic/Cretaceous Carbonate Systems*

Badley Ashton and Associates, Ltd.

BP Kuwait

Vrije Universiteit Industrial Associates in Sedimentology Research Grant

ISBN 1-56576-075-1


1741 E. 71st Street
Tulsa, OK 74136-5108

Printed in the United States of America

# CONTENTS

## Introduction



## Part 1: Sequence Stratigraphy, Chronostratigraphy, and Tectonic Influences



## Part 2: Depositional and Diagenetic Models of Carbonate Platforms

## Part 3: Hydrocarbon Habitat and Exploration/Development Case Studies

# Introduction

# INTRODUCTION: MIDDLE EAST MODELS OF JURASSIC/CRETACEOUS CARBONATE SYSTEMS

ABDULRAHMAN S. ALSHARHAN
*Faculty of Science, United Arab Emirates University, P.O. Box 17551, Al Ain, U.A.E.*
AND
ROBERT W. SCOTT
*Precision Stratigraphy Associates, RR3 Box 103-3, Cleveland, OK 74020, U.S.A.*

This volume is the product of the international conference on "Jurassic/Cretaceous Carbonate Platform–Basin Systems, Middle East Models" that was convened in December 1997 jointly by SEPM (Society for Sedimentary Geology) and the United Arab Emirates University in the oasis city of Al Ain, United Arab Emirates (U.A.E.). The concept of this meeting had its roots in the Carbonate Platform Working Group of the Cretaceous Resources, Events and Rhythms Project (CRER), which first convened in 1988 (Ginsburg and Beaudoin, 1990). In order to pursue the CRER goals, the 1997 conference objectives were: (1) to promote and disseminate research on geological and geophysical studies of carbonate platforms in the Tethyan Realm; (2) to provide access to the superb outcrops of the famous Jurassic–Cretaceous platforms and the classic Holocene coastal sabkha of the Arabian Gulf; (3) to promote the interaction between research and exploration/production geoscientists in the region and worldwide; and (4) to demonstrate the interaction of SEPM, academia, and industry personnel as a global community of geoscientists.

The 125 attendees came from four continents; just over 50% were from the Middle East region. Nearly 70% were from the petroleum industry and 30% were from universities or research and consulting groups. Abstracts of the oral and poster presentations were published in the December, 1997, issue of *GeoArabia*, v. 2, no. 4. The conference was summarized by Scott et al. (1998a, b).

The geologic development of the Arabian Platform was closely related to tectonic and sedimentologic processes in the adjacent areas, so that knowledge of the Arabian region informs and constrains hypotheses of the adjacent areas. Today the Arabian Platform is an independent plate locked between the Zagros Mountains suture zone to the northeast in Iran, the Anatolia plate to the north in Turkey, the Red Sea rift to the southwest, and the spreading center in the Gulf of Aden to the southeast (Glennie, 1995). During the Jurassic and Cretaceous Periods, however, the Arabian Platform was attached to the northeastern margin of Africa and straddled the Equator (Smith et al., 1994). In this position it recorded depositional processes upon a low-relief shelf in the tropical zone where mixed clastics and carbonates–evaporites were deposited both laterally and iteratively through time. This geographic position between the Mediterranean–Atlantic oceans to the west, the Indian plate to the south, and the Pacific Ocean to the east allowed this area to experience the migration of diverse biota from several centers of evolution.

During much of the Mesozoic Era shallow shelf deposystems were the norm upon the Arabian Platform, and slope deposition rimmed the passive margin located to the present-day northeast (Murris, 1980). Sometime between the Turonian and Campanian obduction of oceanic crust began along the modern eastern margin (Glennie, 1995). By the latest Campanian–earliest Maastrichtian, slabs of ophiolitic oceanic crust were emplaced upon the eastern margin of the Arabian platform, forming the base of the Oman Mountains. During the Maastrichtian a foreland basin developed along this margin and carbonates buried the ophiolites. Shallow marine deposition resumed during Paleogene time, and during the Oligocene the Oman Mountains were uplifted. Possibly as recently as ten million years ago subduction closed the post-Triassic seaway, Neo-Tethys 2, between the Sirjan–Sanandaj Zone and Central Iran/Lut microcontinent (Glennie, 1995). This brought together along the Zagros Crush zone the Zagros Mountains, which were the leading margin of the Arabian platform, and the Sirjan–Sanandaj microcontinent, thus welding together Arabia and Asia. The Arabian Gulf was flooded during the Pliocene following this collision.

The Arabian platform is significant also because it contains approximately two-thirds of the world's recoverable oil reserves and about one-third of the recoverable gas reserves (Beydoun, 1998). Most of these resources are produced from seven nations: Saudi Arabia, Iraq, Kuwait, Iran, Qatar, the United Arab Emirates (U.A.E.), and Oman. Several factors have converged in this region that resulted in the concentration of hydrocarbons (Alsharhan and Nairn, 1997; Beydoun, 1998). (1) Giant volumes of high-quality source rocks are illustrated by the Upper Jurassic Diyab Formation in the paper by Al-Suwaidi et al. (this volume). (2) The geographic continuity of the northeast Arabian shelf margin for a distance of over 2500 km and its great width up to 2000 km created conditions favorable to widespread deposition of source rocks and reservoirs. (3) The history of sedimentation was uninterrupted for much of the Mesozoic during the passive-margin stage of the Arabian platform; several papers in this volume describe sedimentation from Late Jurassic to end of the Cretaceous. (4) Source rocks are juxtaposed with reservoir rocks, such as the Upper Jurassic Arab Formation and the Aptian Shuaiba Formation described by several papers in this volume. (5) Tertiary salt movement and Neogene tectonics during development of the Zagros foreland basin formed folds and produced fractures and stylolites that enhanced permeability. Fractured and stylolitic reservoirs are described in this volume by Alsharhan and Sadd and by Saotome et al. (6) Finally, the paleolatitudinal position of the Arabian platform changed from the Cambrian to the present, thereby influencing the several factors essential to development of diverse petroleum systems (Beydoun, 1998).

The conference was a dynamic exchange of ideas as well as of data, which has been the mode of meetings of the CRER Carbonate Platform Working Group. The present set of papers is the third publication resulting from these meetings. The two previous publications were edited by Simo, Scott, and Masse (1993) and Philip and Skelton (1995). The papers for this book were selected from the conference oral and poster presentations to document new data and present interpretations on the sequence stratigraphy, diagenesis, depositional models, chronostratigraphy, and reservoir characterization of Jurassic and

Middle East Models of Jurassic/Cretaceous Carbonate Systems
SEPM Special Publication No. 69, Copyright © 2000
SEPM (Society for Sedimentary Geology), ISBN 1-56576-075-1, p. 3–6.

Cretaceous carbonates of the Middle East. A number of key themes were emphasized during the meeting and also continue through this volume.

Sequence stratigraphers are cautious of circular reasoning. Uncritical "correlations" with a global cycle chart may result in unwarranted age assignments, as well as lending false credibility to the theory of eustatic control. Practitioners in this volume recommend the separation of sequence stratigraphic correlations from the global cycle chart, and the compilation of local sequence stratigraphy into a local cycle chart before attempting global correlations. The role of seismic data in sequence stratigraphy is clearly critical, although it must be tied to core, outcrop, and/or log data. Evaluation of "surfaces", such as paleosols caused by subaerial exposure, or hardgrounds resulting from marine drowning, or stylolites by diagenesis is also essential to sequence stratigraphic analysis.

The classic image of stratigraphic dating is that of the lone biostratigrapher working at a microscope; however, this image has been replaced by the "chronostratigraphic team", which integrates multiple technologies and data sets. The challenge now is how to define time lines with greater precision. High-resolution correlations must utilize chemostratigraphy, paleoecology, and cyclostratigraphy together with biostratigraphy to test whether sequence surfaces indeed are time lines. Biostratigraphy must be used to calibrate chemostratigraphy and cyclostratigraphy, because the sedimentary cycles are not uniquely identified. Biostratigraphic analyses may be facilitated in the near future by automated computerized intelligent systems that achieve preliminary microfossil identifications.

The papers in this volume highlight the need for a regional correlation project of Middle Eastern Jurassic and Cretaceous carbonates. Such a project has the potential to contribute to the measurement of tectonic subsidence rates and the scales and durations of hiatuses, factors that must be known for any kind of modeling. For example, the amplitude of relative sea-level cycles is pertinent to reservoir modeling.

Understanding diagenesis is also critical to estimating hydrocarbon reserves and recovery. Even when our predictions of the location of porous grainstones, organic-rich facies, and effective regional seals and traps are accurate, diagenesis can still make or break a prospect. Seals may be breached, source rocks may not be mature or may have matured before traps were developed, and primary porosity may have been destroyed by cementation or compaction. Alternatively, primary porosity may be enhanced by dissolution, dolomitization, or fracturing, and porosity zones may be isolated by cementation. Many diagenetic trends are controlled by depositional facies or structural trends. Once we understand the diagenetic effects, the relationships between controls of primary facies and structure and the geophysical or log responses can be understood, then predictions will be more effective. Exploration and production teams and research teams must be brought together to generate new ideas and improve their practical application to the carbonate reservoirs and to enhance recovery in various Middle East fields.

The papers were selected for this volume in order to illustrate some of these generalizations and are organized under three themes: sequence stratigraphy, chronostratigraphy, and tectonic influences; depositional and diagenetic models of carbonate platforms; and hydrocarbon habitat and exploration/development case studies. A synopsis of each paper follows and serves to highlight the contributions of each.

Section 1 shows that sequence stratigraphic and chronostratigraphic models can be tested by Middle East carbonate examples. It begins with a review of the tectonic evolution of the Middle East (Glennie, this volume), which shows that the eastern margin of the Arabian platform was passive from Late Permian until earliest Cretaceous time, when the Late Permian spreading center, Neo-Tethys 1, evolved into a subduction zone and a second spreading center, Neo-Tethys 2 formed, separating central Iran from the Arabian platform. Mid-Cretaceous plate tectonic movement of the Arabian platform established regional tectonic features that affected structural development and sedimentation in the region.

The 800-meter-thick shallow-water Middle Jurassic section in the Musandam Peninsula, U.A.E., and Oman can be directly correlated into the subsurface of the U.A.E. de Matos and Walkden (this volume) use the ammonites and foraminifers to identify the standard stage boundaries in the region, and they clarify the lithostratigraphy. Walkden and de Matos (this volume) show that subaerial exposure contacts separate the Lower Jurassic section of the Musandam Peninsula into meter-scale cyclothems that reflect sea-level changes at the scale of Milankovitch climatic frequencies. The facies are clues to the depositional processes, and the diagenetic changes significantly affect reservoir character.

The uppermost Jurassic of the Arabian platform is well exposed in central Lebanon, where the sequence stratigraphic succession serves to document long-term depositional cycles. Toland (this volume) defines two unconformity-bounded sequences in the Tithonian section and suggests a correlation with part of the Arab Formation in the subsurface of the U.A.E. and Saudi Arabia. The top of the second sequence is a prominent paleo-karst contact overlain by Cretaceous clastics.

Both the Lower Aptian Shuaiba Formation carbonate in the U.A.E. and a Lower Aptian clastic unit in Alaska consist of ten shoaling-up cycles (Kendall et al., this volume). Their onlapping shelf geometries are similar to geometries of the Bahamian Neogene sections, which record high-frequency eustasy. By dynamic forward modeling of sedimentary fill of the Shuaiba Basin, U.A.E., these authors show that the seismic sequence geometry can be used as input to a published sea-level curve. The modeling tests the timing and magnitude of the processes.

Using the quantitative stratigraphic technique of graphic correlation, Scott et al. (this volume) test the synchroneity of depositional cycles in Oman relative to those defined elsewhere in the Tethyan Realm. Nine cycles are recognized in each of the Albian Nahr Umr and Cenomanian Natih Formations, Oman. These may indeed be eustatic, because most cycles in the Oman section are the same age as cycles elsewhere. Sediment accumulation rates increased progressively from the Albian to the Cenomanian from 1 to 5 cm/1000 years.

Skelton and Masse (this volume) describe Lower Cretaceous rudists of the region and for the first time relate them to Tethyan assemblages. They present the data in a way that enables all stratigraphers to identify and utilize rudists for correlation and environmental analyses of cores and outcrops.

Section 2 includes eight papers about depositional and diagenetic models. The Middle East Mesozoic section exhibits a wide variety of depositional and diagenetic models of carbonate platforms. Middle Jurassic oncoids in the northern Sinai, Egypt, were formed as Mg-calcite in a shallow shelf environment (Holail, this volume). Diverse geochemical data, including depletion of oxygen isotopes, suggest subaerial exposure and meteoric diagenesis.

Berriasian–Hauterivian carbonates are exposed in central Saudi Arabia (Shebl and Alsharhan, this volume). Microfacies of the Sulaiy, Yamama, and Buwaib formations represent open-platform and shelf-lagoon conditions. Regional unconformities separate Valanginian, Hauterivian, and Barremian strata. These disconformities separate mudstone and wackestone below from coarse siliciclastic rocks above.

Microfacies of the Maastrichtian Simsima Formation in northwestern Oman represent the final Cretaceous carbonate platform stage with rudists, corals, algae, and larger foraminifers (Alsharhan et al., this volume). The Simsima Formation is divided into two longer-term depositional cycles that have numerous subaerial exposure contacts at bedding planes within it. Multiple diagenetic processes modified the porosity greatly.

The spectacular Upper Cretaceous rudist biostromes in Oman represent widespread platform paleocommunities that Schumann (this volume) traces across outcrops separated by many kilometers. The rudist communities are monospecific associations of constratal *Vaccinites* that developed in restricted conditions during regional transgressions and did not form reefs because of abiotic factors. Colonial corals with clinger radiolitids are common below and above the *Vaccinities* beds and formed biostromes in a slightly deeper environment than the *Vaccinites* beds.

The Middle to Upper Jurassic shallow marine carbonates in Yemen experienced diverse stages of diagenesis (synsedimentary, shallow burial, deep burial, and uplift diagenesis) that resulted in its reservoir potential (Al-Thour, this volume). Dolomite formed during the latter three stages and dedolomitization operated during weathering oxidation. The Aptian Shuaiba Formation consists of two major reservoirs in the Ghaba North Field, Oman (Al-Awar and Humphrey, this volume). The lower reservoir zone consists of facies deposited in a marine environment with restricted circulation, whereas the upper zone represents open shallow marine conditions. Together they constitute a longer-term, shoaling-up cycle with a regional unconformity at the top. These heterogeneous reservoirs produce from secondary macroporosity, fracture porosity, and intercrystalline microporosity. Meteoric dissolution as indicated by stable isotopes was facilitated by repeated subaerial exposure. Three stylolite types affect the performance of Lower Cretaceous carbonate reservoirs in the U.A.E. (Alsharhan and Sadd, this volume). Stylolites are more abundant toward the flanks of producing fields, and most are subparallel to bedding, indicating burial compaction.

Ten repetitive hardground surfaces in the mainly Albian Nahr Umr Formation developed by meteoric exposure and subsequent marine overprinting (Immenhauser et al., this volume). They are reliable correlation markers in Oman, and they are recognized by the integration of stratigraphic, petrographic, and geochemical evidence.

In Section 3, Middle East hydrocarbon habitat and reservoir case studies provide models for other carbonate basins. Lateral variations in thickness and lithology of Upper Jurassic rocks resulted from sea-level changes, local structure, and spreading of the Neo-Tethys (Ayoub and Ennadi, this volume). About 75% of the present structural configuration in onshore U.A.E. is the result of Late Cretaceous compression produced by overthrusting in Oman. Gas was generated in the Upper Jurassic Hanifa Formation during subsequent subsidence in the U.A.E.

The Upper Jurassic Diyab Formation in the U.A.E. has a basinal lime mudstone facies that sourced reservoirs in the Arab Formation and in the Thamama Group (Al-Suwaidi et al., this volume). The facies of the Diyab change westward, landward in Abu Dhabi to oolitic packstone and grainstone. The authors present new geochemical data to define oil types and propose timing and migration routes from several source rocks in Abu Dhabi. Generation began in the Late Cretaceous.

Davies et al. (this volume) use sequence stratigraphic geometries and stacking patterns to develop reservoir models of the Berriasian–Valanginian Minagish Formation in the Umm Gudair Field, Kuwait. Ten lithofacies suggest deposition on a homoclinal carbonate ramp. Reservoir facies with micritic microporosity are present in most systems tracts. Parasequences are separated by flooding contacts, and stratal stacking influenced the development of intraformational flow barriers and reservoir compartmentalization.

Pore geometry of the Barremian upper Kharaib reservoir in offshore U.A.E. is a result of original depositional fabric and subsequent diagenesis (Saotome et al., this volume). Producing facies are rudist–peloidal grainstone/packstone, algal–peloidal grainstone/packstone, and bioclastic wackestone/mudstone deposited on a carbonate shelf in a shoaling-up succession. Unique pore-throat distributions resulted from depositional textures and dolomitization. Facies and pore-throat distributions were the basis for a detailed reservoir simulation.

The Shuaiba Formation in the U.A.E. is a major reservoir and also records Early Aptian depositional processes. Alsharhan et al. (this volume) demonstrate by means of well logs, petrography, and geochemical analyses that porosity was mainly controlled by diagenesis during early shallowing and later deeper burial. The basinal facies of the Bab Member of the Shuaiba Formation is demonstrated to be a mature source rock that has generated hydrocarbons in the basin. Stable-isotope values of the cement are more depleted than those of the matrix.

Cretaceous sequence stratigraphy in the Mesopotamian basin of southern Iraq was controlled by the culmination of the Neo-Tethys rifting, the movement of the Arabian Platform into the tropical and subtropical climatic zones, and the contemporaneous growth of salt structures (Sadooni and Aqrawi, this volume). Reservoirs were formed within clastic deposits that are intercalated with carbonate platform deposits.

The final paper, by Alsharhan and Scott (this volume), presents a summary of the basins in the eastern Arabian platform and describes many of the reservoir types. The U.A.E. is bounded by structurally positive features on the northwest and on the east. Since the Permian two passive-platform-margin basins developed. Sedimentation in these basins has consisted of epeiric shelf carbonates and minor evaporites and clastics, reflecting long-term and short-term depositional cycles resulting from the interplay of tectonics and eustasy. Oil and gas reserves are in reservoirs of shallow-water Jurassic and Cretaceous carbonates. The source rocks are deep-water shale and argillaceous limestone sealed by evaporites, dense limestone, and shale.

The editors express their appreciation to the SEPM editor of Special Publications, Robert W. Dalrymple, for his guidance and support of this publication. We thank our colleagues for affording us part of their valuable time reviewing manuscripts to make this set of papers more technically sound. They include I. Al-Aasm, I.A. Al Jallal, D. Bosence, R. Bray, D. Budd, M. Cassidy, J.M. Connan, J. de Matos, J.A.D. Dickson, M. Esteban, A.J. Everts, P.C. Franks, E. Gili, K.W. Glennie, B. Granier, R. Handford, P.M. Harris, R. Hoffling, J. Hopkins, G. Hughes, R. Hulstrand, A. Immenhauser, R.W. Jones, C.G.St.C. Kendall, C. Kerans, A. Kirkham, R. Koepnick, J. Kuss, L.S. Land, A. Lomando, R. Loucks, J.-P. Masse, S.J. Mazzullo, P. Nederlof, E. Oswald, J. Philip, J.M. Pons, B. Pratt, D. Prezbindowski, A. Saller, D. Schumann, A.J. Simo, P. Skelton, J. Smewing, A. Smith, A. Strasser, V. Vahrenkamp, F. Van Buchem, and G. Walkden.

The editors express their profound thanks to all contributors to this volume for their support, cooperation, promptness, and patience improving the manuscripts. Special thanks are due to SEPM and the United Arab Emirates University for co-sponsoring the conference from which this volume arose. The University has allowed our participation in this undertaking and supported its successful completion in many ways. For processing the manuscripts, we are particularly grateful to the essential administrative assistance of Ms. Bonita Valdez-Cruzada and Ms. Girija Bhaskaran.

We would like to express our gratitude and appreciation to His Highness Sheikh Nahyan Mubarak Al Nahyan, Minister of

Higher Education and Scientific Research and Chancellor of the U.A.E. University, for his guidance and support during the planning and running of the conference and for his encouragement to create this publication.

As editors, it is our privilege to present this volume to all of you. It is a worthy publication for geoscientists working in the Middle East and comparable areas around the world.

Abdulrahman S. Alsharhan, Al Ain, U.A.E.
Robert W. Scott, Cleveland, Oklahoma, U.S.A.

## REFERENCES

ALSHARHAN, A.S., AND NAIRN, A.E.M., 1997, Sedimentary basins and petroleum geology of the Middle East: Amsterdam, Elsevier Science, 843 p.

BEYDOUN, Z.R., 1998, Arabian Plate oil and gas: Why so rich and so prolific?: Episodes, v. 21, p. 74–81.

GINSBURG, R.N., AND BEAUDOIN, B., 1990, Cretaceous Resources, Events and Rhythms: Background and Plans for Research: NATO ASI series C, Mathematical and Physical Sciences, v. 304: Dordrecht, The Netherlands, Kluwer Academic Publishers, 352 p.

GLENNIE, K.W., 1995, The Geology of the Oman Mountains: An Outline of Their Origin: Beaconsfield, Bucks, UK, Scientific Press Ltd., 92 p.

MURRIS, R.J., 1980, Middle East: Stratigraphic evolution and oil habitat: American Association of Petroleum Geologists, Bulletin, v. 64, p. 597–618.

PHILIP, J., AND SKELTON, P.W., 1995, Palaeonenvironmental models for the benthic associations of Cretaceous carbonate platforms in the Tethyan Realm: Palaeogeography, Palaeoclimatology, Palaeoecology, v. 119, p. 1–199.

SCOTT, R.W., ALSHARHAN, A.S., AND ENOS, P., 1998a, Jurassic/Cretaceous carbonate platform–basin systems — Middle East models: Episodes, v. 21, p. 39.

SCOTT, R.W., ALSHARHAN, A.S., AND ENOS, P., 1998b, Jurassic/Cretaceous carbonate platform–basin systems — Middle East models: Episodes, v. 21, p. 361–362.

SIMO, J.A., SCOTT, R.W., AND MASSE, J.-P., 1993, Cretaceous Carbonate Platforms: American Association of Petroleum Geologists, Memoir 56, 479 p.

SMITH, A.G., SMITH, D.G., AND FUNNELL, B.M., 1994, Atlas of Mesozoic and Cenozoic Coastlines: Cambridge, UK, Cambridge University Press, 99 p.

# Part 1:
# Sequence Stratigraphy, Chronostratigraphy, and Tectonic Influences

# CRETACEOUS TECTONIC EVOLUTION OF ARABIA'S EASTERN PLATE MARGIN: A TALE OF TWO OCEANS

KENNETH W. GLENNIE

*Department of Geology and Petroleum Geology, University of Aberdeen, Meston Building, King's College, Old Aberdeen AB9 2UE, U.K.*

ABSTRACT: Plate-tectonic movements made the Cretaceous a time of major change in the area of the modern Oman and Zagros Mountains. Neo-Tethys 1 had been created in the Late Permian by the calving of a microcontinent (Anatolia, Sanandaj–Sirjan/Central Iran) along the NE margin of Arabia. In the Late Triassic, a second spreading axis, Neo-Tethys 2 (more readily recognizable in Iran than in Oman) replaced that of Neo-Tethys 1 by the separation of the Central Iran and Sanandaj–Sirjan–Kawr microcontinents.

Neo-Tethys 1 had a passive continental margin during the Triassic and Jurassic as the Afro–Arabian portion of Gondwana moved westward away from the actively spreading oceanic ridge of Neo-Tethys 2. Shallow-marine sediments along the continental margin were the source of carbonate turbidity currents that flowed basinward to the abyssal plain of Neo-Tethys 1 until the early Late Cretaceous, whereas the floor of Neo-Tethys 2 seems to have been starved of coarse sediment in its Oman sector.

Early in the Cretaceous, the South American and Afro–Arabian portions of Gondwana began to separate to create the South Atlantic Ocean. South America continued to move to the west, but Afro–Arabia reversed its sense of motion. The ensuing buildup of horizontal compressional stresses led to an eastward-dipping subduction zone within the Oman sector of Neo-Tethys 2, leading to obduction of the Late Permian to mid–Cretaceous Hawasina Series (deposited in Neo-Tethys 1) and the Semail Nappe, which was generated by back-arc spreading. North of the Dibba Line, subduction also took place within Neo-Tethys 1.

The latest Cretaceous was a time of tectonic adjustment and shallow-marine carbonate sedimentation across the area of the present Oman Mountains and southern Zagros, but the effects of late Maastrichtian subduction in Neo-Tethys 2 are visible in the Inner Makran. Evidence of subduction beneath the northern half of the Gulf of Oman suggests that this process has been more or less continuous over the Makran area until today. Uplift of the Oman Mountains began in the Mio–Pliocene, about the same time as the Zagros Mountains began to form.

## INTRODUCTION: GEOLOGICAL SETTING

The tectonic evolution of the northeastern margin of Arabia (Oman and Zagros mountains) involved the successive development of two belts of spreading oceanic crust, referred to as Neo-Tethys 1 and Neo-Tethys 2 (Glennie, 1995). The first formed in the mid Permian when a large microcontinent (Anatolia?–Sanandaj–Sirjan/Central Iran) was calved from the northeastern margin of Arabia; the sedimentary fill of that basin in Oman is known as the Hawasina Series. The second oceanic belt formed in the mid to late Triassic when the axis of spreading "jumped" to the east, where it split the former microcontinent, separating Central Iran/Lut from the Sanandaj–Sirjan Zone (Figs. 1, 2A). The ensuing history of these two oceanic areas had a profound effect not only on the structural development of the Arabian continental margin as seen in the Oman and Zagros mountain belts but also over a much wider area of the Middle East.

The northeast continental margin of Arabia extends from the Arabian Sea/Gulf of Oman to the Mediterranean Sea. The Cretaceous was a period of profound change along this margin and in the two adjacent marine basins (Neo-Tethys 1 and 2). A reversal in the sense of plate-tectonic movement converted a long-lasting passive-margin relationship between continent and ocean into one involving successive ocean–ocean subduction in the two oceanic basins, leading eventually to continent–continent collision in Iran. The history of events during and preceding the Cretaceous in the first of these basins is more clearly discerned in the Oman Mountains, whereas the second is better defined in Iran, the two areas with which this paper is mostly concerned. But how did all this come about? The scene needs to be explained.

## PALEOZOIC HISTORY

Throughout the Paleozoic Era, the site of modern Arabia was located within the Southern Hemisphere (Scotese and McKerrow, 1990; Scotese and Barrett, 1990; Schandelmeier and Reynolds, 1997) near the northeastern edge of the supercontinent Gondwana, where it was flanked by the ocean Paleo-Tethys. For most of that long time span, Arabia lay about midway between the South Pole and the Equator, with siliciclastics dominating sedimentation (Schandelmeier and Reynolds, 1997). Towards the end of the Paleozoic Era, however, there was a relatively rapid change in Oman from the climatic conditions needed for a Late Carboniferous to Early Permian glaciation (Braakman et al., 1982; Love, 1994) to the warm waters of Late Permian coral reefs and other carbonate environments (Blendinger et al., 1990).

The glaciation in SE Oman occurred while the area was probably no farther from the Equator than about 45° (Scotese and Langford, 1995) and possibly closer. Al-Beloushi et al. (1996) suggest that the glaciation was of mountain type, and connected with uplift prior to the creation of an Early Permian triple junction adjacent to the SE end of the present Oman Mountains; Le Métour et al. (1994) reported associated erosion of up to 4000 m at this time in northern Saih Hatat. The NW–SE trending arms were associated with the calving of microcontinents (Anatolia, Iran, Tibet) from the northeast margin of Arabia (present north) and the development of Neo-Tethys 1. The other arm involved the coeval separation of SE Arabia and the Helmand block (now southern Afghanistan) and creation of an intervening Batin–Helmand seaway (Al-Beloushi et al., 1996), which, on the basis of mid-Permian marine faunas, possibly extended as far as Madagascar (Flores, 1970); this separation developed fully only during the final Late Jurassic breakup of eastern Gondwana (see Immenhauser, 1996; Gnos et al., 1997).

The early introduction of warm-water faunas is attributed to the rapid collapse of the glaciated mountains and the development of nearby newly formed oceanic areas (Al-Beloushi et al., 1996). This event was coupled with a postglacial rise in sea level and the lowering and perhaps partial collapse of the flanking land

Middle East Models of Jurassic/Cretaceous Carbonate Systems
SEPM Special Publication No. 69, Copyright © 2000
SEPM (Society for Sedimentary Geology), ISBN 1-56576-075-1, p. 9–20.

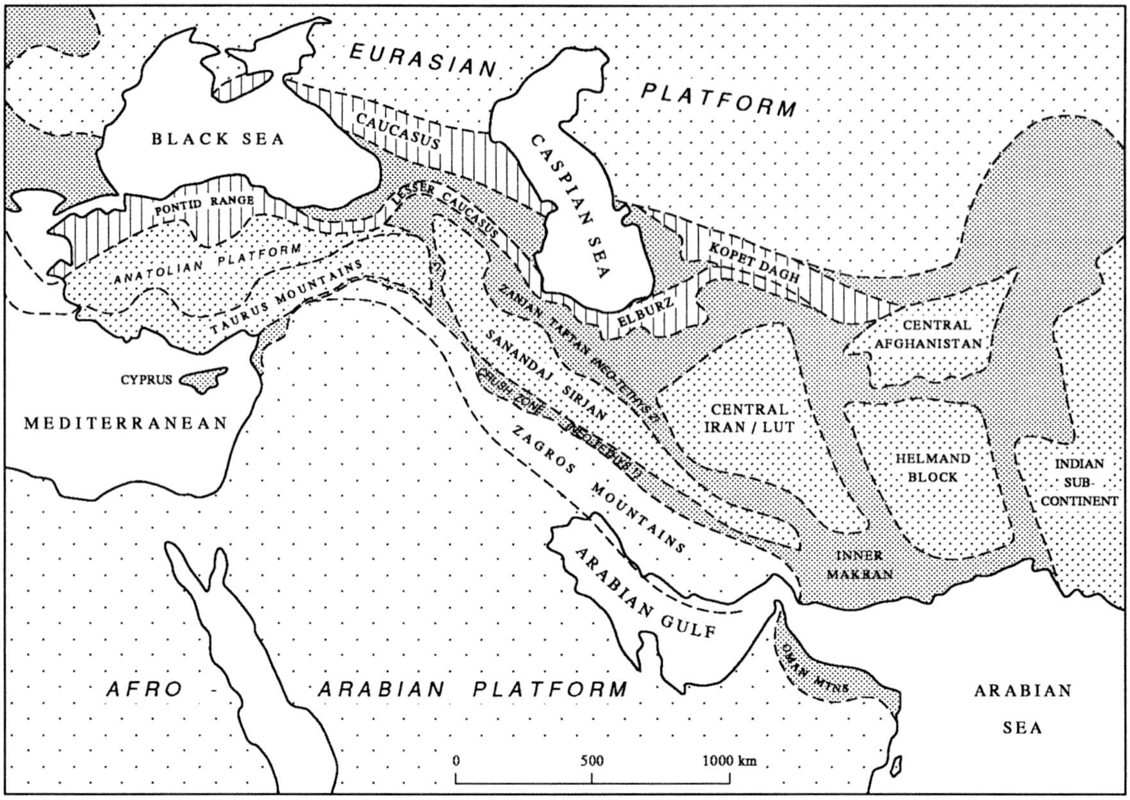


FIG. 1.—Plate-tectonic distribution of microcontinents (e.g., Anatolia/Taurus, Sanandaj–Sirjan, Central Iran/Lut, Helmand Block, Central Afghanistan) north of a line from the Arabian Sea to the Mediterranean, which are thought to have existed between the Arabian and Eurasian platforms during much of the Mesozoic. The Indian microcontinent finally separated from Gondwana only in the Late Cretaceous. The intervening areas (dark stipple) are interpreted to contain relics of the oceanic crust that formerly separated the microcontinents: Crush Zone = Neo-Tethys 1; Zanjan–Taftan Zone = Neo-Tethys 2. The Lake Van area between Anatolia and Sanandaj–Sirjan may represent early separation between the two microcontinents. Modified from figure 15 in Glennie (1995).

areas as they moved away from the new axes of spreading, resulting in a mid to late Permian transgression over much of eastern Arabia (fig. 6 in Murris, 1980; Blendinger, 1990; Le Métour et al., 1994; Glennie, 1995). A slow northward drift of Arabia towards and across the Equator added to the climatic change and ensured that carbonates were the dominant marine sediment throughout the Mesozoic Era except where deposited below the carbonate compensation depth (e.g., figs. 1 and 8 in Bernoulli et al., 1990).

## THE VARYING CHARACTER OF NEO-TETHYS 1 AND 2

The former existence of Neo-Tethys (Stöcklin, 1974), or the same ocean under other names (e.g., Hawasina Ocean of Glennie et al., 1973, 1974; Fig. 4A; or the Southern Tethys of Stoneley, 1990) has been inferred by many workers, whereas two oceanic areas (Fig. 1) subparallel to the continental margin of Arabia have been proposed by Glennie et al., 1990) and Glennie (1995), and are indicated by the ophiolite distribution in figure 2 of Sengör (1990). Evidence for inferring their existence is given in these and other publications, the main points only being summarized here. Because the supporting evidence differs in Oman and Iran, each will be treated separately.

### *Neo-Tethys 1 in Oman*

The best evidence for the existence of Neo-Tethys 1 is found in the Oman Mountains. Le Métour et al. (1994) cited considerable data to support an early Late Permian origin for Neo-Tethys 1 (e.g., in northern Saih Hatat, continental separation is indicated by sedimentary sequences dated by Permian faunas that are interbedded with basaltic pillow lavas and are associated with sills and dikes, and even Early Permian volcanics (Le Métour et al., 1994).

Palinspastic reconstructions of the Hawasina sequences indicate that northeast of the Arabian platform margin, a continental slope developed, down which carbonate-rich turbidity currents flowed towards the abyssal plain from the mid or late Permian until the mid Cretaceous. Paleocurrent data show the source of turbidity currents to have been in the west and southwest (Glennie et al., 1974; Watts, 1987; Blendinger, 1988; Blendinger et al., 1990; Cooper, 1990; Glennie, 1995); many of the carbonate grains found in the resulting turbidites consist of microfossils that confirm the shallow-marine platform-margin origin of the sediments that form them. For example, Late Triassic fore-reef talus of the Sumeini Group (Glennie et al., 1974; Watts, 1990) was later removed tectonically from proximity to its original reef; its faunas have been described in some detail by Bernecker (1996). In the

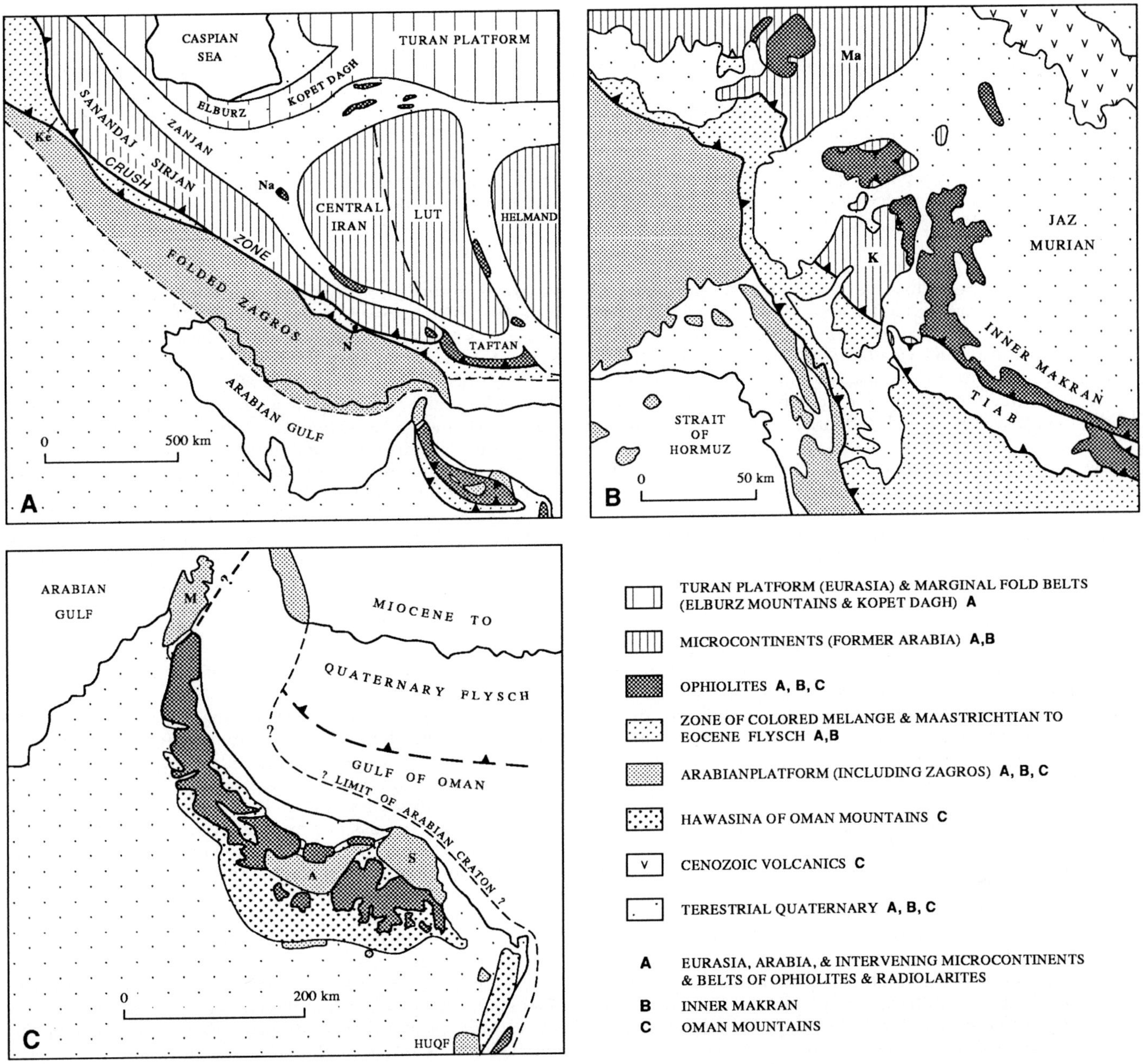


FIG. 2.—Three tectonic sketches. (A) Distribution of microcontinents of Iran and intervening areas of oceanic sediments and ophiolites. (B) Local detail in the western Inner Makran. The Tiab Zone is a klippe comprising Cretaceous limestones and Eocene flysch overlying Upper Jurassic pillow lavas dated by interpillow calpionellids. The ophiolites to its north, which have been similarly dated, tectonically overlie metamorphic rocks of the Sanandaj–Sirjan Zone. (C) Autochthonous rocks of the Oman Mountains overlain by the Hawasina and Semail Nappes. Note extent of Miocene to Recent flysch of Neo-Tethys 2 over northern floor of Gulf of Oman. A, Jebel Akhdar; Ke, Kermanshah; N, Neyriz; Na, Nain; M, Musandam Peninsula; Ma, Mardin Metamorphics; S, Saih Hatat.

deepest parts of the depositional realm, middle to Late Permian radiolarians occur near the base of the allochthonous Hawasina (De Wever et al., 1990).

From the order of superposition of Hawasina nappes in the Oman Mountains, a fairly regular depositional pattern can be deduced across Neo-Tethys 1 (Hamrat Duru Basin) from southwest to northeast: up to 1600 m of continental-slope reef talus, slumps, slides, and minor turbidites (Sumeini Group); base-of-slope turbidites up to 1000 m thick (Hamrat Duru Group) grading into about 300 m of finer-grained turbidites of a more siliceous nature (e.g., Wahrah Formation) beneath the deep waters of the abyssal plain (Glennie et al., 1974).

Neither continental nor confirmed oceanic crust has been identified from below the oldest basinal units of the Hawasina. There is commonly an association with mostly alkaline volcanic rocks, which have a MORB tholeiitic character in the Hawasina

Window. Béchennec et al. (1990) interpret the alkaline volcanics as the product of Late Permian extensional thinning of a continental crust; this crust possibly became fully disrupted over northeastern parts of the Hamrat Duru Basin, leading to the emplacement of pseudo-oceanic crust. A submarine volcanic component of such crust might be present in submarine volcanics within the basal Zulla Formation (see Bernoulli et al., 1990). In the most distal depositional area of the Hawasina, shallow-marine limestones (Oman Exotics; Glennie et al., 1974) are underlain by the Haybi Volcanics, which include pillow lavas that also have a MORB character. Lippard et al. (1986) suggest that the Haybi Volcanics indicate an origin in a small ocean basin like the Red Sea. This interpretation is supported by El-Shazly et al. (1994), who note that greenschists, blueschists, and eclogites in the Saih Hatat area were originally enriched MORB erupted axially in a small oceanic basin in the Late Permian. The pressure required for blueschist–eclogite metamorphism was provided within the subduction trench, which Searle et al. (1994) suggest was achieved at depths of up to 80 km during Late Cretaceous subduction.

Glennie et al. (1974), Glennie et al. (1990), and Glennie (1995) considered that the sub-Hawasina volcanics and the lack of evidence of underlying continental crust indicate the ease of basal Hawasina décollement during later nappe emplacement; this, they believed, also favored an interpretation of Hawasina deposition over an oceanic crust (Neo-Tethys 1 or Hawasina Ocean) that developed from about the mid Permian to the Middle Triassic.

Obduction of the Hawasina and Semail nappes has been shown to have taken place predominantly within the Campanian to Maastrichtian time span. While the Hawasina sediments all predate the time of obduction, the age of the overlying Semail ophiolite is approximately coeval with the time of obduction (i.e., Late Cretaceous) (Glennie et al., 1974; Lippard et al., 1986; Béchennec et al., 1990).

### *Neo-Tethys 1 in Iran*

The arcuate but discontinuous distribution of ophiolites and associated "radiolarites" (see below) northwest of Oman, along the Crush Zone of Iran, and south of Anatolia to Syria and Cyprus (see Fig. 1 and, e.g., fig. 2 in Sengör, 1990) is taken as evidence that a relatively narrow seaway formerly connected the Hawasina Basin to the present Mediterranean area. In Iran, the former basin fill and overlying ophiolites are displayed only in two widely separated areas, near Kermanshah in the northwest (Fig. 2A) and near Neyriz in the southeast (Fig. 2A, 3). As in Oman, the oceanic crust of the Crush Zone is also of Late Cretaceous age, and similarities in basin fill suggest a similar history of development. This distribution was named by Ricou (1971) "Le Croissant ophiolitique peri-Arabe", an interpretation supported by Biju-Duval et al. (1974) and Coleman and Irwin (1974).

The term "radiolarites" is used as an abbreviation in the sense of Stöcklin (1974); it includes a range of sediments from the characteristic reddish radiolarian chert to carbonate turbidites and masses of white marble. Contained fossils indicate a time span of deposition from Permian to mid Cretaceous, although the fauna in many sequences has been destroyed by recrystallization. Because of late Cenozoic transpression, some large slabs of the ophiolites and radiolarites are now both tectonically and sedimentologically interleaved with younger sedimentary sequences ranging in age from late Maastrichtian to Miocene, and much additional evidence is now presumed to lie beneath the overthrust Sanandaj–Sirjan zone.

As in Oman, the imbricated sedimentary fill of radiolarites was thrusted over shallow-marine carbonates of the continental margin during the Late Cretaceous (Fig. 3, NW of Neyriz). This sequence, in turn, is overlain tectonically by slabs of ophiolitic material rich in clearly recognizable peridotite as well as extrusive rocks. A schematic cross section of these relationships near Neyriz is given in figure 4 of Stöcklin (1974).

The broad similarity with the Hawasina and Semail of the Oman Mountains is striking. The ophiolites, like those in Oman, are considered by Hooper et al. (1994) to result from a Late Cretaceous "supra-subduction zone (i.e., back arc) spreading". These authors believe that the Arabian plate margin comprised a series of promontories and embayments, which may explain the presence of only two areas where ophiolites and radiolarites are exposed (K and N in Figure 2A). In the Neyriz area, Ricou (1971) considered the sedimentary basin to have had a minimum width of 50 to 100 km, with another 100 km for the ophiolite nappe. This compares with a minimum deduced width for the Hawasina Basin of the order of 400 km (Glennie et al., 1974).

### *Neo-Tethys 2 in Oman*

The northeast side of the Hawasina Basin was the site of deposition of the Oman Exotics (Kawr Group of Béchennec et al., 1990) on the Kawr (or Misfar) Ridge (Fig. 4B). The mainly shallow-marine carbonates of the Exotics were deposited over two distinct time spans, early Late to latest Permian and Late Triassic to Early Jurassic; many smaller Exotics have been fully recrystallized to marble. The Liassic top to the 1000 m-thick Late Triassic limestone of Jebel Kawr (Figs. 3, 4) is truncated and capped by thin Cretaceous argillaceous and radiolaria-rich deep-water limestones that imply sediment starvation. The Late Permian and especially Late Triassic Exotics are commonly underlain by pillow lavas and other volcanics, which, as mentioned above, are suggestive of an oceanic location. In the case of Jebel Qamar (just south of the Dibba Line; Fig. 3), however, which is underlain by the Ordovician Rann quartzites, a continental substrate is indicated.

The sub-Exotic alkaline to transitional tholeiitic lavas of Triassic age (Haybi Volcanics, Lippard et al., 1986) are believed by Searle and Graham (1982) to have been erupted during the early stages of continental rifting. As already implied, evidence from Iran indicates that there were probably two events of continental separation, which in Oman may have been recorded by the distinct Permian and Triassic ages of the Oman Exotics. The Exotics were flanked by deep water into which flows and slides

→

FIG. 3.—Map showing how the Oman and Zagros mountains on the eastern edge of Arabia relate to the Sanandaj–Sirjan and Central Iran/Lut microcontinents and the intervening areas of ophiolites and ocean-floor sediments. Cenozoic volcanics mask much of the ophiolite–radiolarite sequence surrounding Central Iran/Lut. The southern Makran was the site of late Cenozoic subduction of Neo-Tethys 2, and subduction of its floor is still occurring beneath the northern half of the Gulf of Oman. The Musandam Peninsula, north of the Dibba Line, is probably more closely related to the Zagros than the rest of the Oman Mountains; this highlights the origin of the Dibba Line as a probable major transform fault separating the two areas. A, Jebel Akhdar; B, Nappes of Batain Coast; K, Kushmandar; Kw, Jebel Kawr; M, Musandam Peninsula; Ma, Mardin Metamorphics; N, Nain; Q, Jebel Qamar; S, Saih Hatat; SA, Salakh Arch.

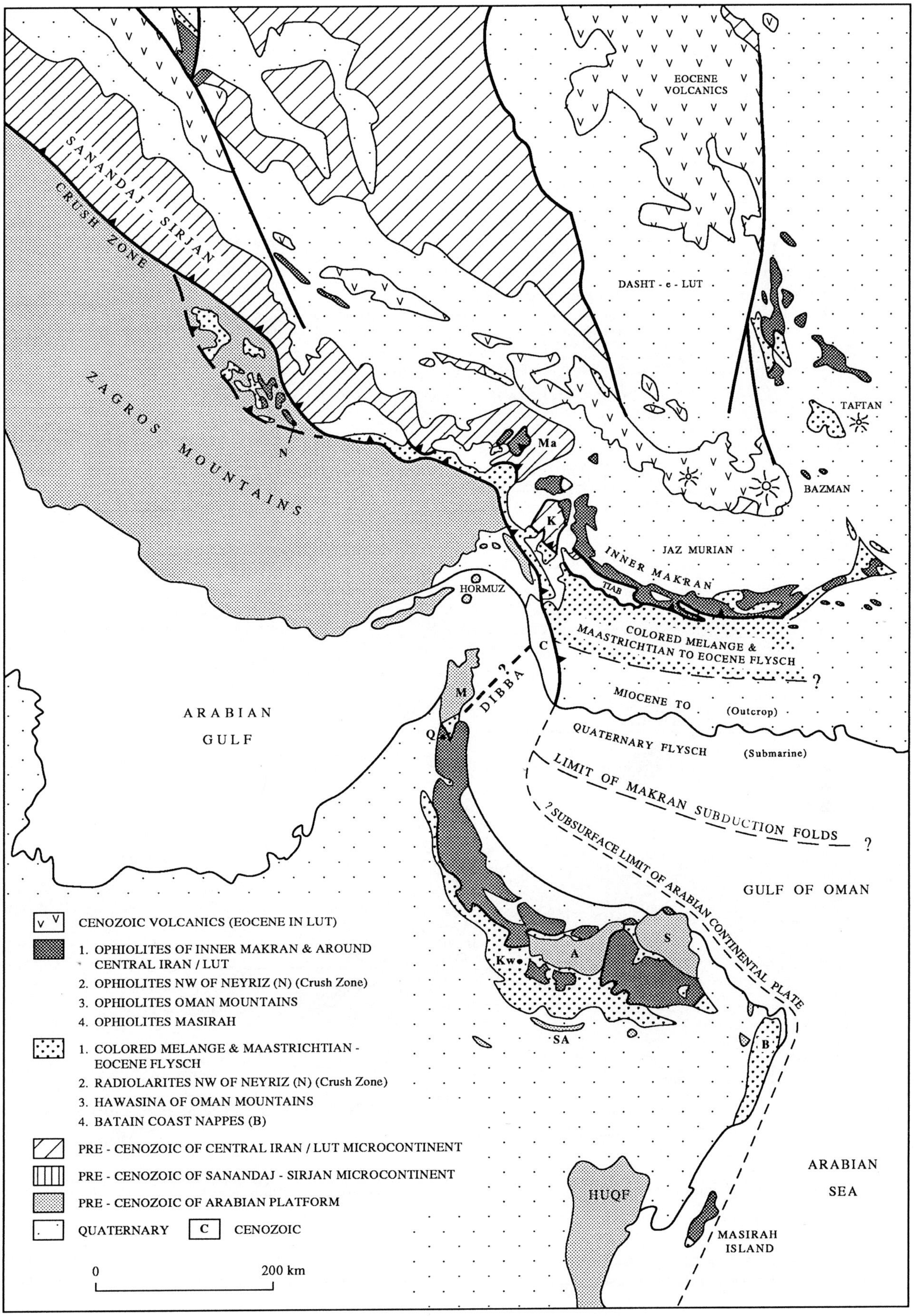
EOCENE VOLCANICS
SANANDAJ - SIRJAN
CRUSH ZONE
DASHT - e - LUT
ZAGROS MOUNTAINS
N
Ma
TAFTAN
BAZMAN
K
JAZ MURIAN
INNER MAKRAN
TIAB
HORMUZ
COLORED MELANGE &
MAASTRICHTIAN TO EOCENE FLYSCH
C
DIBBA
M
Q
MIOCENE TO
(Outcrop)
QUATERNARY FLYSCH
(Submarine)
ARABIAN
GULF
LIMIT OF MAKRAN SUBDUCTION FOLDS
? SUBSURFACE LIMIT OF ARABIAN CONTINENTAL PLATE
GULF OF OMAN
S
A
Kw
SA
B
HUQF
MASIRAH
ISLAND
ARABIAN
SEA
CENOZOIC VOLCANICS (EOCENE IN LUT)
1. OPHIOLITES OF INNER MAKRAN & AROUND CENTRAL IRAN / LUT
2. OPHIOLITES NW OF NEYRIZ (N) (Crush Zone)
3. OPHIOLITES OMAN MOUNTAINS
4. OPHIOLITES MASIRAH
1. COLORED MELANGE & MAASTRICHTIAN - EOCENE FLYSCH
2. RADIOLARITES NW OF NEYRIZ (N) (Crush Zone)
3. HAWASINA OF OMAN MOUNTAINS
4. BATAIN COAST NAPPES (B)
PRE - CENOZOIC OF CENTRAL IRAN / LUT MICROCONTINENT
PRE - CENOZOIC OF SANANDAJ - SIRJAN MICROCONTINENT
PRE - CENOZOIC OF ARABIAN PLATFORM
QUATERNARY
C
CENOZOIC
0
200 km

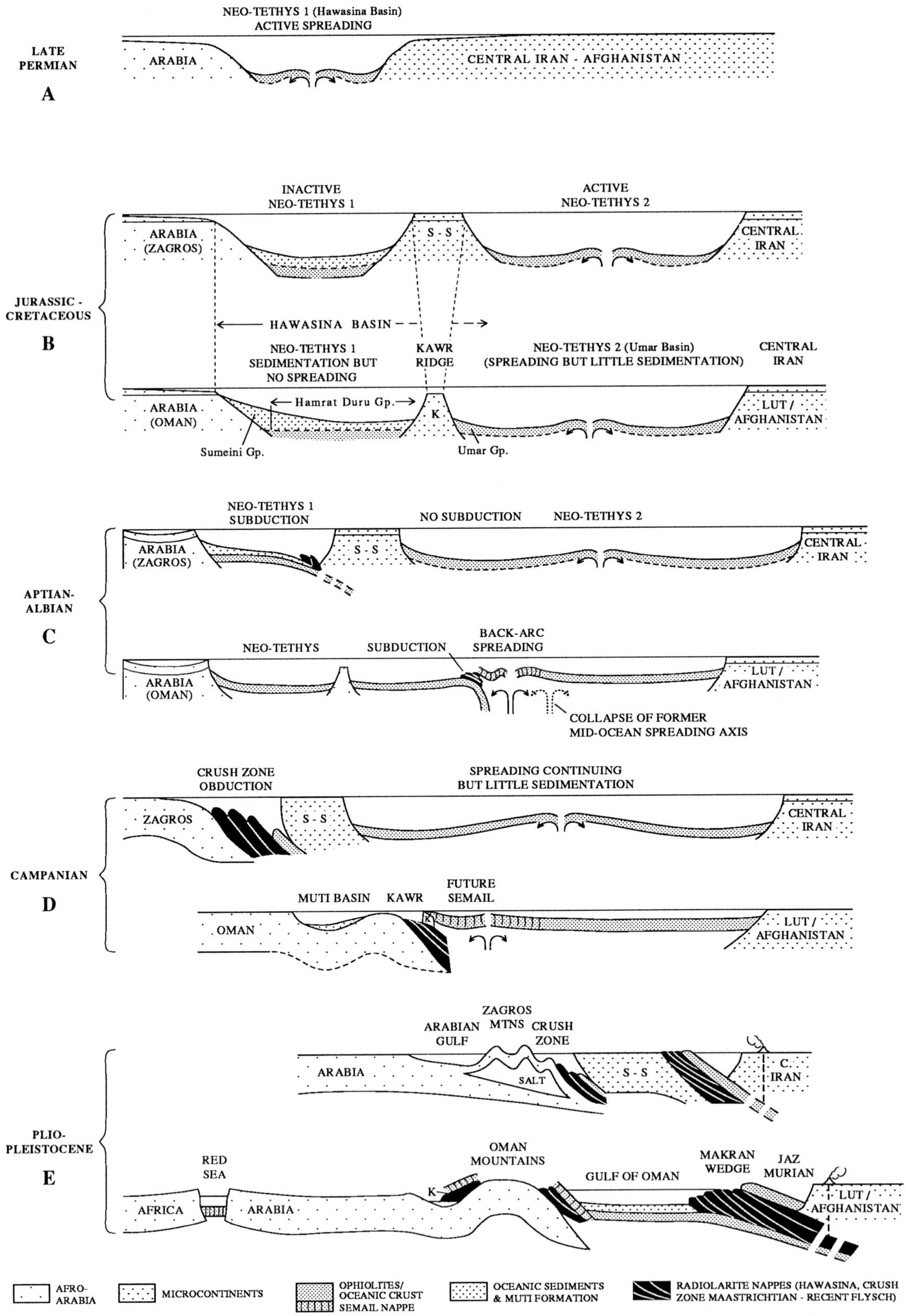


FIG. 4.—A series of cartoons illustrating similarities and differences between the history of Neo-Tethys 1 and 2 in Iran and Oman. (A) Late Permian. (B) Jurassic–Cretaceous boundary. (C) Aptian–Albian. (D) Campanian. (E) Plio-Pleistocene. No fixed scale is implied.

of derived coarser shallow-marine carbonate material were deposited (Fig. 4B).

The sediments deposited in the Umar Basin (Fig. 4B) (Béchennec et al., 1990), northeast of the Kawr Ridge, are very fine grained, with only limited coarser detritus thought to have been derived from the Triassic Exotics of the ridge itself. Like Searle and Graham (1982), Béchennec et al. (1990) associated the Haybi Volcanics with continental separation. Separation in this case involved the creation of Neo-Tethys 2. Because the narrow Kawr Ridge intervened between the depositional areas of Neo-Tethys 1 and 2, the Umar Group had no sedimentary input from the continental margin of Arabia.

### *Neo-Tethys 2 in Iran*

The foregoing geometrical evidence related to the Oman Exotics and overlying Umar Group supports the hypothesis that Neo-Tethys had two parts to it. The older Neo-Tethys 1 is presumed to have stopped spreading in about mid Triassic time, when the axis of crustal spreading "jumped" eastward to separate the Sanandaj–Kawr Ridge from Central Iran/Lut. Neo-Tethys 2 possibly continued to spread from about the mid Triassic until the mid or late Cenozoic (Glennie, 1995), but at least a part of this basin was involved in subduction in the latest Cretaceous. In Oman, sedimentary evidence for the existence of the Umar Basin (Neo-Tethys 2) is poorly displayed in small outcrops between the Oman Exotics and the sheared base of the flanking Semail Nappe. In Iran, the margins of Neo-Tethys 2 are sharply defined by its flanking continental areas, but its sedimentary basin fill and associated ophiolites are, in general, not nearly so well exposed as in Oman, and are further complicated by intense tectono-sedimentary reworking from the Maastrichtian to late in the Cenozoic (Figs. 2B, 3).

In the NW Makran, the exposed metamorphic rocks of the Mardin and Kushmandar areas (Ma and K in Figure 3) at the SE end of the Sanandaj–Sirjan Zone plunge to the southeast beneath the ophiolites of the Inner Makran and adjacent Tiab klippe (Figs. 2B, 3; see Glennie et al., 1990; Glennie, 1995). The metamorphic rocks are capped by mountain-size blocks of white marble that are reminiscent of the Oman Exotics. These mountains of marble are tentatively correlated with the Oman Exotics of the Kawr Ridge, recrystallization having possibly occurred during a Carnian metamorphic event that affected this area (see Hooper et al., 1994); if so, metamorphism would have coincided with the initiation of Neo-Tethys 2. If this correlation is correct, then the Kawr Ridge plays an important role in relating the geology of Oman to the Inner Makran, and for reconstructing the Mesozoic history of the Arabian continental margin in these areas of the Middle East.

The Sanandaj–Sirjan Zone is separated from the continental crust of Central Iran by a low-lying belt that is typified by Tertiary extrusive rocks (Fig. 3) but through which an ophiolitic suite of rocks is locally exposed (e.g., near Nain; Na in Figure 2A). Using the Tertiary volcanics as a guide, this second zone of ophiolites and radiolarites can be traced from between the Lesser Caucasus and the NE corner of Anatolia (Fig. 1), southeastward via the Zanjan area of NW Iran (Fig. 2A) to the Makran and on to the vicinity of the Quaternary Bazman and Taftan volcanoes south and southeast of the Lut Block (Fig. 3; see also distribution of "Colored Mélange" (Gansser, 1955) in figure 1 of Stöcklin, 1974). The Colored Mélange can be separated into its constituent sedimentary and ophiolitic parts. As individual blocks within a wildflysch, some of the sedimentary sequences are similar to those found in the Crush Zone (grainstone turbidites, radiolarian cherts, and blocks of marble or shallow-marine limestone that are associated with basic igneous rocks). The basic and ultrabasic ophiolite components are dominantly of peridotite but include dike swarms overlain by pillow lavas.

The Mardin and Kushmandar metamorphics (Fig. 3) are overlain by Alpine-type peridotite masses that are highly sheared and serpentinized along their basal contacts. To the east and southeast, less metamorphosed peridotites grade up into gabbros similar to those mapped in the Oman Mountains (e.g., Reinhardt, 1969, Lippard et al., 1986). Dike swarms are overlain by, and imbricated with, mudstones bearing Tithonian to Berriasian calpionellids and radiolaria. These ages for ophiolite generation are in strong contrast to the Late Cretaceous age recorded by inter-pillow microfossils of the Semail Nappe in Oman. Along the margin of the Inner Makran, the gabbros grade up into dike swarms that form the feeders to massive lava flows, tuffs, and pillow lavas, which are overlain by Eocene lithic sandstones. This ophiolite slab is in high-angle thrust contact with the Tiab Klippe (Figs. 2B, 3; see also figure 3B of Glennie et al., 1990). Beneath a cover of Eocene conglomerates, sandstones, and shales, the klippe comprises Maastrichtian to Paleocene shallow-marine limestones unconformably overlying Lower Cretaceous limestones and Upper Jurassic to Lower Cretaceous sandstones and shales; these, in turn, can grade down into sheared serpentinite or pillow lavas that again enclose Tithonian to Berriasian calpionellid-bearing limestone and shale. The Late Jurassic age of the ophiolites in the Inner Makran contrasts strongly with the Late Cretaceous age (also on interpillow fossils) of the Semail ophiolite in Oman.

### *Events Leading to Cretaceous Nappe Emplacement along the Arabian Margin*

To summarize the foregoing events, it is clear that crustal separation took place along the northeastern margin of Arabia in two distinct phases: (1) From the early Late Permian until the middle or Late Triassic (Carnian), Neo-Tethys 1 was actively spreading in the Hawasina–Crush Zone–Cyprus ocean. (2) Spreading then "jumped" to Neo-Tethys 2, which began to spread during the Middle to Late Triassic, and in the Sabsavar Zone (Fig. 2A) continued to spread until possibly the mid Cenozoic.

Passive-margin conditions of sedimentation prevailed within Neo-Tethys 1 and along the SW margin of Neo-Tethys 2 as Gondwana moved southwestward away from the axis of spreading. The early Barremian start of spreading in the South Atlantic Ocean (Milner et al., 1995) is thought to have been responsible for the Aptian (?) onset of subduction in Neo-Tethys. As Afro-Arabia moved away from the Atlantic axis of spreading, the resulting increase in crustal compression in Neo-Tethys 1 and 2, intensified by continued spreading in Neo-Tethys 2, is thought to have been relieved by the initiation of subduction. By the late Maastrichtian, the sedimentary fill of Neo-Tethys 1 had been obducted onto the continental margin of Arabia.

Whereas obduction in the Iranian sector of Neo-Tethys 2 may have begun only in the latest Cretaceous or Paleocene, subduction of its remaining oceanic crust seems to be continuing today beneath the northern Gulf of Oman (White and Klitgord, 1976; White, 1982).

The following is an attempt to rationalize the above developments into a coherent history of events (Fig. 4; Table 1).

1. Mid to Late Permian: Creation of Neo-Tethys 1 by the calving of a major microcontinent (or linear series of them) from the northeastern margin of Arabian Gondwana (see, e.g., figure 9 in Al Beloushi et al., 1996). By the Mid to Late Triassic, Neo-Tethys 1 was some 400 km wide in the Oman sector (Fig. 4A).

TABLE 1.—Generalized timing of major events in the structural history of the Oman and Iranian sectors of the Arabian continental margin, and the associated Neo-Tethys 1 and still extant Neo-Tethys 2 (Gulf of Oman).

| AGE | ARABIAN CONTINENTAL MARGIN | NEO-TETHYS 1 (HAWASINA/CRUSH ZONE) | NEO-TETHYS 2 (MAKRAN) |
|---|---|---|---|
| Plio-Pleistocene | Active folding of post–Hormuz Zagros sequences | Crush-Zone transpression | Northward subduction of floor of Gulf of Oman |
| Late Miocene | Flooding of Arabian Gulf | | Closure of Neo-Tethys 2 in Iran? Deep-marine sediments in Gulf of Oman |
| Oligocene | Uplift of Oman Mountains | | Right-lateral transpression |
| Eocene | Shallow-marine deposition | | Flysch deposition |
| Paleocene | Shallow-marine & turbidite deposition | | Wildflysch deposition |
| Maastrichtian | Shallow-marine Simsima deposition End Semail/Hawasina obduction Juweiza/Amiran deposition | Closure of Hawasina/Crush-Zone ocean | Start of active subduction S and W of Lut Block |
| Campanian | Start of Muti/Faraghan deposition & Hawasina obduction | | |
| Cenomanian | Youngest Hajar Supergroup | | |
| Albian | Second up-warping of continental edge | Start of Semail back-arc spreading | |
| Aptian | First up-warping of continental edge | Start of Hawasina subduction | |
| | | | |
| Portlandian | Deep-marine continental margin | | Dated oceanic crust |
| | | | |
| Triassic (late) | | End of spreading in Neo-Tethys 1 | Start of Zanjan–Taftan (Neo-Tethys 2) spreading (& Umar Group of Hawasina) |
| Triassic (early) | | Neo-Tethys 1 actively spreading | |
| Permian | Marine transgression by Hajar Supergroup Triple-junction ice cap | Neo-Tethys 1 starts to spread | |

2. Approximately Middle to Late Triassic: Crustal spreading ceased to be active in Neo-Tethys 1 and the axis "jumped" to NE to initiate spreading between the Sanandaj–Sirjan Zone and Central Iran/Lut. Geometric evidence of the Oman Exotics and the Kawr Ridge suggests correlation with Sanandaj–Sirjan Zone, which could be crucial in relating Neo-Tethys 2 in Oman and Iran.
3. The Permian to Early Cretaceous basin fill of Neo-Tethys 1 in Iran is not well enough known to reconstruct its width and sedimentary history in such detail as in Oman. In Oman, palinspastic reconstructions indicate a wedge of oceanic sediment that thins to the northeast towards the Kawr Ridge.
4. No sediment from the Arabian continental margin reached the Umar Basin of Neo-Tethys 2, and none is known to have been derived from the more easterly microcontinents (Fig. 4B). A palinspastic reconstruction has not been attempted on the sedimentary fill of Neo-Tethys 2 in either Oman or Iran; this would be possible but difficult to assemble in the Inner Makran (see, e.g., figure 3B in Glennie et al., 1990).
5. The Umar Group forms the uppermost part of the Hawasina Series of Oman. If sedimentation of this Group followed initiation of Neo-Tethys 2, as implied by its relationship to the Haybi Volcanics and the Kawr Ridge, then the subduction that led to obduction of the Hawasina and Semail must have begun in Neo-Tethys 2 (Fig. 4C) and then involved Neo-Tethys 1 with little change of pace because the narrow Kawr Ridge provided almost no resistance to the subduction/obduction process.
6. In Iran, the Sanandaj–Sirjan Zone was much broader than the Kawr Ridge. Furthermore, subduction seems to have begun within the Crush Zone sector of Neo-Tethys 1 (approx. Campanian) before that of the Inner Makran area of Neo-Tethys 2 (late Maastrichtian). The difference in timing and geometry in the Oman sector can be explained if the Dibba Line (Figs. 2C, 3) is considered as a major transform fault of long standing (since the mid Triassic or even mid Permian), on either side of which both crustal spreading and later subduction acted independently of each other.
7. In the Oman sector of Neo-Tethys 2, the Semail ophiolites were generated during the Late Cretaceous by back-arc extension associated with subduction (Lippard et al., 1986; Glennie, 1995, fig. 6). Although less well constrained, a similar origin is believed for its Iranian sector. No clear representatives of an underlying Triassic or older oceanic crust is known from Neo-Tethys 1, although the sub-Exotics Haybi Volcanics and the volcanics beneath and interbedded with the basal Zulla Formation in the Hawasina Window in Oman come close. In the Makran sector of Neo-Tethys 2, the known oceanic crust was generated in the Late Jurassic–Early Cretaceous. As field evidence indicates a time span of generation from Mid Triassic to Late Maastrichtian (150 Ma or more) before subduction began, the width of Neo-Tethys 2 could have reached 7500 km by the Late Maastrichtian if the spreading rate averaged 5 cm/year.
8. The start of subduction during the Aptian (Fig 4C) was accompanied by gentle crustal warping of the Arabian continental platform (see later). Warping was repeated at the start of obduction over the continental platform edge during the Campanian (Fig. 4D). This latter event coincided with development of the Muti (Oman) and Faraghun (Iran) fore-arc basins, which were partly covered by the advancing nappes before obduction ceased.
9. Subduction led to closure of Neo-Tethys 1 and obduction of the Hawasina, Semail, and Crush Zone nappes by the mid

Maastrichtian. Subduction then ceased because Arabia's thick and relatively buoyant continental margin could not be consumed into the mantle. The South Atlantic Ocean and Neo-Tethys 2 were still spreading, however, and the resulting increase in crustal compression led in Iran to the creation of a new subduction zone in Neo-Tethys 2 before the end of the Maastrichtian.

10. The Late Maastrichtian release of crustal compression is possibly the reason why the obducted nappes remained close to sea level rather than forming a high mountain chain. The youngest Cretaceous sediments in the vicinity of the Oman Mountains are the mainly shallow-marine post-obduction carbonates of the Simsima Formation.

    The areal and volumetric differences in ophiolite exposure between Iran and Oman might be explained as follows:
    a. In Iran, subduction resulted in continent–continent collision, the ophiolites remaining buried beneath the relatively broad but rigid Sanandaj–Sirjan Zone microcontinent.
    b. In Oman, however, the collision was of continent–ocean type; subduction ceased when the second subduction zone became active (as in Iran), thereby releasing the compressive forces, and allowed isostatic uplift of the combined volume of narrow Kawr Group continental crust and a large volume of hanging-wall future Semail. With no further destruction caused by horizontal compression, the Semail Nappe remains the largest and best preserved ophiolite in the world.
11. The start of mountain building in Oman coincided with the opening of the Red Sea and with a major plate-tectonic collision to the northeast between India and Eurasia (see Gnos et al., 1996). In Iran, the Zagros Mountains began to form in the Mio-Pliocene. There, the Eo-Cambrian Hormuz salt acted as a plane of décollement between basement and overburden, above which long anticlinal folds of younger strata are pierced by many diapirs (Fig. 4E). The Oman Mountains lack halite, and where it is present beneath interior Oman, salt piercements, under the influence of plate-tectonic movements in the Indian Ocean, occur through the overburden rather than causing major folds; the Salakh Arch may be an exception
12. Right-lateral continent–continent collision between Arabia and the Sanandaj–Sirjan Zone makes the Crush Zone an area of strong seismicity. In the Oman sector, the Arabian margin is flanked by the remnants of Neo-Tethys 2, the underlying oceanic crust of which is being subducted south of the Makran coast (Figs. 3, 4E).

## SOME CRETACEOUS EVENTS ON THE ARABIAN PLATFORM

Deposition of the earliest Cretaceous Kahmah (Thamama) Group saw a change from deep-marine conditions along the continental margin to a shallow-marine environment farther west (Fig. 5A). The eastward progradation of shallow-marine reservoir rocks over potential source rocks of the Habshan Formation (Aziz and El Sattar, 1997) adds to the economic interest of SE Arabia.

The initiation of subduction caused warping of the Arabian continental margin (Figs. 4C, 5B) to produce both basins and erosional highs (e.g., Murris, 1980, Masse et al., 1997), the depositional details of which can now be deduced using high-resolution sequence stratigraphy (e.g., Van Buchem et al., 1996). The warping was sufficient for the development of grainstone reservoir rocks (e.g., Shuaiba) around erosional highs that were separated by euxinic basins in which hydrocarbon source rocks were deposited (Van Buchem et al., 1996). During the Albian–Cenomanian, the siliciclastic Wasia Sandstone prograded eastward from the Arabian Shield over the western part of the carbonate shelf (Murris, 1980). At the onset of obduction during the Turonian or earliest Campanian, this episode was accompanied by the widespread "post-Wasia break" and was succeeded by the creation of the Muti foreland basin in Oman and the less well known Faraghun Basin along the Crush-Zone edge of the Zagros Mountains. Ophiolite components occur in the younger Juweiza (Oman) and Amiran (Zagros) formations of the Late Cretaceous. In interior Oman, dextral transtension and sinistral transpression occurred along fault zones under the influence of plate-tectonic movements that were taking place in the Indian Ocean (Loosveld et al., 1996).

After subduction ceased in the Oman sector of Neo Tethys in the Late Maastrichtian, the obducted Hawasina and Semail did not form a mountain range but lay close to sea level. The erosion of exposed highs led to the deposition of local fluvial conglomerates (e.g., Qahlah Formation in the SE Oman Mountains; Glennie et al., 1974) while widespread shallow-marine grainstones of the Simsima Formation covered most of the remaining area of obducted rocks and spread far onto the continental platform, although it is rather patchy in southern Oman. A deeper-water facies was deposited in the Suneinah Trough west of the northern Oman Mountains (Hughes Clarke, 1988). In the southeastern Oman Mountains, local Paleocene to Eocene deep-water conglomerates and turbidites (Filbrandt et al., 1990) possibly mark the true southeastern fault-scarp limit of the Oman Mountains; the adjacent Batain Coast deposits are associated with obduction of the Arabian Sea's Masirah ophiolite at the Cretaceous–Tertiary boundary (Immenhauser, 1996).

## CONCLUSIONS

The northeastern margin of the Arabian platform is deformed into the Zagros (Iran) and Oman Mountains, which achieved their current stature only during the Neogene. The forerunners to these mountain ranges comprise platform-margin rocks over which the Permian to mid Cretaceous sedimentary fill of a narrow oceanic basin (Neo-Tethys 1), together with Late Cretaceous back-arc ophiolites, were obducted between the Campanian and Early Maastrichtian.

Neo-Tethys 1 came into being when the Anatolia–Central Iran–Afghanistan microcontinent calved from the northeastern margin of Arabia (Gondwana) during the Permian; it had a short history of spreading from about mid Permian until the Mid Triassic, when the spreading axis jumped eastward to separate Central Iran/Lut from the Sanandaj–Sirjan Zone of Iran, thereby creating Neo-Tethys 2.

Neo-Tethys 1 received a basin fill of sediments (e.g., Hawasina Series of Oman) derived from the Arabian platform's passive margin. The spreading axis of Neo-Tethys 2 was probably active until the end of the Mesozoic, although its only known oceanic crust dates as Tithonian to Berriasian.

Gondwana migrated passively westward until the Aptian, when the effects of the newly opening South Atlantic Ocean caused Afro-Arabia to move towards the spreading axis in Neo-Tethys 2. The resulting crustal compression was relieved when a subduction zone formed within Neo-Tethys 1 north of the Dibba transform fault and in Neo-Tethys 2 south of that fault, leading to obduction of the basin fill and associated back-arc oceanic crust over Arabia's platform margin. Obduction ceased because the Arabian margin was too thick and buoyant to be consumed down the subduction trench, but because the South Atlantic was still opening and Neo-Tethys 2 was still spreading, another trench had to form in Neo-Tethys 2. Subduction has continued within and to the south of the Makran coast until today, while the Musandam Peninsula is in process of being subducted beneath the Strait of Hormuz.

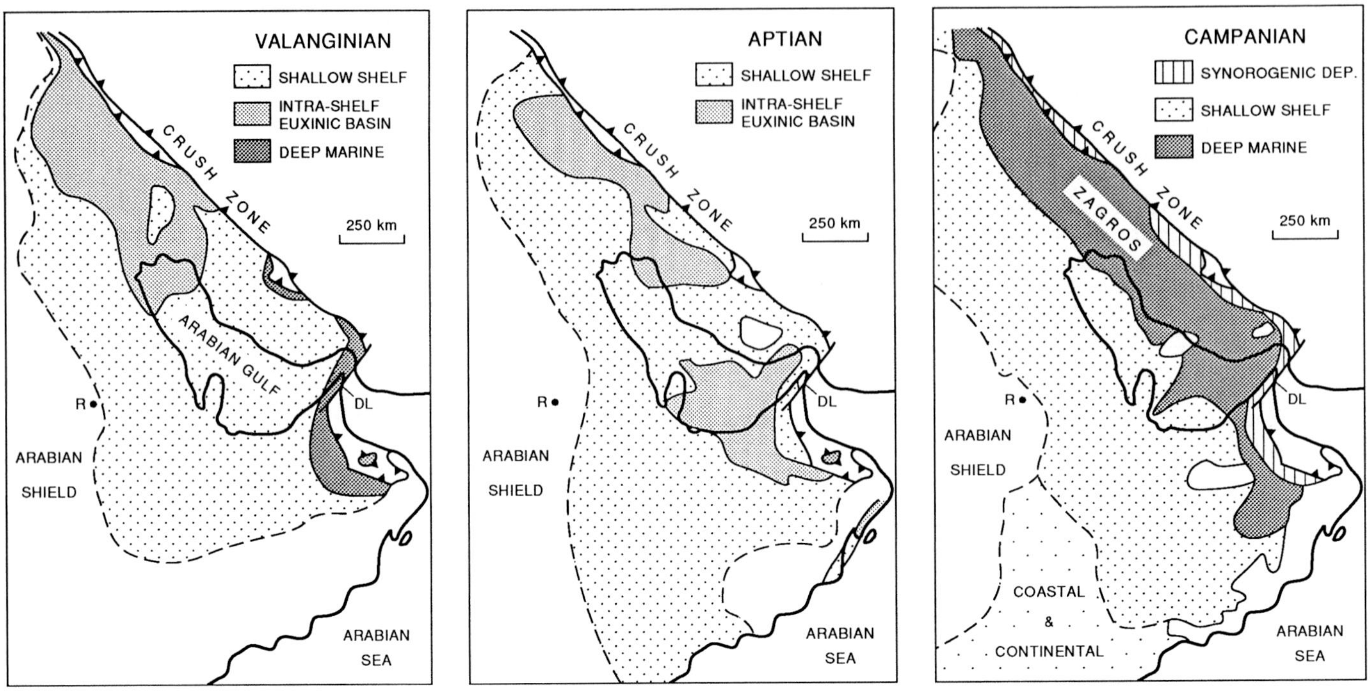


FIG. 5.—Sedimentation over eastern Arabia at three Cretaceous times. (A) Valanginian: most of the area covered by shallow-marine carbonates. Euxinic intra-shelf deposition in the Luristan Basin in the north and deeper open-marine conditions along the continental margin in NE Oman and in the Neyriz region of the Zagros. (B) Aptian: warping of the Arabian platform created both reservoir grainstones and intra-shelf euxinic basins west of the continental margin. (C) Campanian: outer edge of the Arabian platform depressed as it starts to be subducted. Erosion of advancing Hawasina led to deposition of Muti and Faraghun-type sediments west of the nappe front in Oman and the Zagros thrust front in Iran. Later erosion of the ophiolite thrust front led to Juweiza and Amiran deposition in northern Oman and Iran. Note the effect of the Dibba Line (DL) on deposition. R, Riyadh. Simplified from Murris (1980).

## ACKNOWLEDGMENTS

Heiko Oterdoom (PD Oman) reviewed the final draft. Reviewers Brian R. Pratt and John D. Smewing improved the readability of the text. Barry Fulton, University of Aberdeen, is thanked for drafting the figures.

## REFERENCES

AL-BELOUSHHI, J.D., GLENNIE, K.W., AND WILLIAMS, B.J.P., 1996, Permo-Carboniferous glaciogenic Al-Khlata Formation, Oman: a new hypothesis for origin of its glaciation: GeoArabia, v. 1, p. 389–404.

AZIZ, S.K., AND ABD EL-SATTAR, M.M., 1997, Sequence Stratigraphic modeling of the Lower Thamama Group, east onshore Abu Dhabi, United Arab Emirates: GeoArabia, v. 2, p. 179–202.

BÉCHENNEC, F., LE MÉTOUR, J., RABU, D., BOURDILLON-DE-GRISSAC, CH., DE WEVER, P, BEURRIER, M., AND VILLEY, M., 1990, The Hawasina Nappes: stratigraphy, palaeogeography and structural evolution of a fragment of the south-Tethyan passive continental margin, *in* Robertson, A.H.F., Searle, M.P., and Ries, A.C., eds., The Geology and Tectonics of the Oman Region: Geological Society of London, Special Publication 49, p. 213–223.

BÉCHENNEC, F., LE MÉTOUR, J., PLATEL, J-P., AND ROGER, J., 1994, Doming and down-warping of the Arabian Platform in Oman in relation to eoalpine tectonics, *in* Al-Husseini, M.I., ed., Middle East Petroleum Geoscience Conference: Geo'94, Manama, Bahrain, Gulf PetroLink, p. 167–178.

BERNECKER, M., 1996, Upper Triassic reefs of the Oman Mountains: data from the South Tethyan margin: Erlangen, Facies, v. 34, p. 41–76.

BERNOULLI, D., WEISSERT, H., AND BLOME, C.D., 1990, Evolution of the Triassic Hawasina Basin, Central Oman Mountains, *in* Robertson, A.H.F., Searle, M.P., and Ries, A.C., eds., The Geology and Tectonics of the Oman Region: Geological Society of London, Special Publication 49, p. 189–202.

BIJU-DUVAL, B., LETOUZEY, J., MANTADERT, L., COURRIER, P., MUGNIOT, J.F., AND SANCHO, J., 1974, Geology of the Mediterranean Sea Basins, *in* Burk, C.A., and Drake, C.L., eds., The Geology of Continental Margins: New York, Springer-Verlag, p. 695–721.

BLENDINGER, W., 1988, Permian to Jurassic deep water sediments of the eastern Oman Mountains: their significance for the evolution of the Arabian Margin of South Tethys: Erlangen, Facies, v. 19, p. 1–32.

BLENDINGER, W., VAN VLIET, A., AND HUGHES CLARKE, M.W., 1990, Updoming, rifting and continental margin development during the Late Palaeozoic in northern Oman, *in* Robertson, A.H.F., Searle, M.P., and Ries, A.C., eds., The Geology and Tectonics of the Oman Region: Geological Society of London, Special Publication 49, p. 27–37.

BRAAKMAN, J.H., LEVELL, B.K., MARTIN, J.H., POTTER, T.L., AND VAN VLIET, A, 1982, Late Palaeozoic Gondwana glaciation in Oman: Nature, v. 229, p. 48–50.

CHRISTIAN, L., 1997, Cretaceous subsurface geology of the Middle East region: GeoArabia, v. 2, p. 239–256

COLEMAN, R.G., AND IRWIN, W.P., 1974, Ophiolites and Ancient Continental Margins, *in* Burk, C.A., and Drake, C.L., eds., The Geology of Continental Margins: New York, Springer-Verlag, p. 921–931.

Cooper, D.J.W., 1990, Sedimentary evolution and palaeogeographical reconstruction of the Mesozoic continental rise in Oman: evidence from the Hamrat Duru Group, *in* Robertson, A.H.F., Searle, M.P., and Ries, A.C., eds., The Geology and Tectonics of the Oman Region: Geological Society of London, Special Publication 49, p. 161–187.

De Wever, P., Bourdillon-De-Grissac, Ch., and Béchennec, F., 1990, Permian to Cretaceous radiolarian biostratigraphic data from the Hawasina Complex, Oman Mountains, *in* Robertson, A.H.F., Searle, M.P., and Ries, A.C., eds., The Geology and Tectonics of the Oman Region: Geological Society of London, Special Publication 49, p. 225–238.

El-Shazly, A.K., Worthing, M.A., Jayawardane, J., and Varne, R., 1994, Geochemistry of metamorphosed mafic rocks from Saih Hatat: pre-obduction history of NE Oman: Geological Society of London, Journal, v. 151, p. 999–1016.

Filbrandt, J.B., Nolan, S.C., and Ries, A.C., 1990, Late Cretaceous and early Tertiary evolution of Jebel Ja'alan and adjacent areas, NE Oman, *in* Robertson, A.H.F., Searle, M.P., and Ries, A.C., eds., The Geology and Tectonics of the Oman Region: Geological Society of London, Special Publication 49, p. 697–714.

Flores, G., 1970, Suggested origin of the Mozambique Channel: Geological Society of South Africa, Transactions, v. 73, p. 1–16.

Gansser, A., 1955, New aspects of the geology of central Iran: 4th World Petroleum Congress, Rome, Proceedings, section I/A/5, p. 280–300

Glennie, K.W., 1995, The Geology of the Oman Mountains: an outline of their origin: Beaconsfield, U.K., Scientific Press, 92 p.

Glennie, K.W., Boeuf, M.G.A., Hughes Clarke, M.W., Moody-Stuart, M., Pilaar, W.F.H., and Reinhardt, B.M., 1973, Late Cretaceous nappes in the Oman Mountains and their geologic evolution: American Association of Petroleum Geologists, Bulletin, v. 57, p. 5–27.

Glennie, K.W., Boeuf, M.G.A., Hughes Clarke, M.W., Moody-Stuart, M., Pilaar, W.F.H., and Reinhardt, B.M., 1974, Geology of the Oman Mountains: Koninklijk Nederlands Geologisch Mijnbouwkundig Genootschaap, Verhandelingen, v. 31, 423 p.

Glennie, K.W., Hughes Clarke, M.W., Bouef, M.G.A., Pilaar, W.F.H., and Reinhardt, B.M, 1990, Inter-relationship of the Makran Oman Mountains belts of convergence, *in* Robertson, A.H.F., Searle, M.P. and Ries, A.C., eds., The Geology and Tectonics of the Oman Region: Geological Society of London, Special Publication 49, p. 773–786.

Gnos, E., Immenhauser, A., and Peters, T.J., 1997, Late Cretaceous/early Tertiary convergence between the Indian and Arabian plates recorded in ophiolites and related sediments: Tectonophysics, v. 271, p. 1–19.

Hooper, R.J., Baron, I.R., Agah, S., and Hatcher, R.D., 1994, The Cenomanian to Recent development of the southern Tethyan margin in Iran, *in* Al-Husseini, M.I., ed., Middle East Petroleum Geoscience Conference: Geo'94, Manama, Bahrain, Gulf PetroLink, p. 505–516.

Hughes Clarke, M.W., 1988, Stratigraphy and rock-unit nomenclature in the oil-producing areas of Interior Oman: Journal of Petroleum Geology, v. 11, p. 5–60.

Immenhauser, A., 1996, Cretaceous sedimentary rocks on the Masirah Ophiolite (Sultanate of Oman): evidence for an unusual bathymetric history: Geological Society of London, Journal, v. 153, p. 539–551.

Le Métour, J., Béchennec, F., Platel, J-P., and Roger, J., 1994, Late Permian birth of the Neo-Tethys and development of its southern continental margin in Oman, *in* Al-Husseini, M.I., ed., Middle East Petroleum Geoscience Conference: Geo'94, Manama, Bahrain, Gulf PetroLink, p. 643–654.

Le Métour, J., Michel, J.C., Béchennec, F., Platel, J-P., and Roger, J., 1995, Geology and Mineral Wealth of the Sultanate of Oman: Muscat, Sultanate of Oman, Ministry of Petroleum and Minerals, 285 p.

Lippard, S.J., Shelton, A.W., and Gass, I.G., 1986, The ophiolite of northern Oman: Geological Society of London, Memoir 11, 178 p.

Loosveldt, R.J.H., Bell, A., and Terken, J.J.M., 1996, The tectonic evolution of Interior Oman: GeoArabia, v. 1, p. 28–51.

Love, C.F., 1994, The palynostratigraphy of the Haushi Group (Westphalian–Artinskian) in Oman, *in* Simmons, M.D., ed., Micropalaeontology and Hydrocarbon Exploration in the Middle East: London, Chapman & Hall, p. 23–39.

Masse, J-P., Borgomano, J., and Al Maskiry, S., 1997, Stratigraphy and tectonosedimentary evolution of the late Aptian–Albian carbonate margin: the northeastern Jebel Akhdar (Sultanate of Oman): Sedimentary Geology, v. 113, p. 269–280.

Milner, S.C., Le Roex, A.P., and O'Conner, J.M., 1995, Age of Mesozoic igneous rocks in northwestern Namibia, and their relationship to continental breakup: Geological Society of London, Journal, v. 152, p. 97–104.

Murris, R.J., 1980, The Middle East: stratigraphic evolution and oil habitat: American Association of Petroleum Geologists, Bulletin, v. 64, p. 597–618.

Reinhardt, B.M., 1969, On the genesis and emplacement of ophiolites in the Oman Mountains geosyncline: Schweitzerische Mineralogische und Petrographische Mitteilungen, v. 49, p. 1–30.

Ricou, L.E., 1971, Le Croissant ophiolitique peri-Arabe. Une ceinture de nappes mise en place au Crétacé Supérieur: Révue de Geographie Physique et de Géologie Dynamique (2) VIII Fasc. 4, p. 327–350.

Robertson, A.H.F., Blome, C.D., Cooper, D.W.J., Kemp, A.E.S., and Searle, M.P., 1990, Evolution of the Arabian continental margin in the Dibba Zone, Northern Oman Mountains, *in* Robertson, A.H.F., Searle, M.P., and Ries, A.C., eds., The Geology and Tectonics of the Oman Region: Geological Society of London, Special Publication 49, p. 251–284

Schandelmeier, H., and Reynolds, P.O., 1997, Palaeogeographic–Palaeotectonic Atlas of North-Eastern Africa, Arabia, and adjacent areas: Late Neoproterozoic to Holocene: Rotterdam, The Netherlands, Balkema, 160 p.

Scotese, C.R., and Barrett, S.F., 1990, Gondwana's movement over the South Pole during the Palaeozoic: evidence from lithological indicators of climate, *in* McKerrow, W.S., and Scotese, C.R., eds., Palaeozoic Palaeogeography and Biogeography: Geological Society of London, Memoir 12, p. 75–85.

Scotese, C.R., and Langford, R.P., 1995, Pangea and the Paleogeography of the Permian, *in* Scholle, P.A., Peryt, T.M., and Ulmer-Scholle, D.S., eds., The Permian of Northern Pangea, v. 1: Berlin, Springer-Verlag, p. 3–19.

Scotese, C.R., and McKerrow, W.S., 1990, Revised World Maps and Introduction, *in* McKerrow, W.S., and Scotese, C.R., eds., Palaeozoic Palaeogeography and Biogeography: Geological Society of London, Memoir 12, p. 1–21.

Searle, M.P., and Graham, G.M., 1982, Oman Exotics: Oceanic carbonate build-ups associated with the early stages of continental rifting: Geology, v. 19, p. 43–49.

Searle, M.P., Waters, D.J., Martin, H.N., and Rex, D.C., 1994, Structure and metamorphism of blueschist–eclogite facies rocks from the northeastern Oman Mountains: Geological Society of London, Journal, v. 151, p. 555–576.

Sengör, A.M.C., 1990, A new model for the late Palaeozoic–Mesozoic tectonic evolution of Iran and implications for Oman, *in* Robertson A.H.F., Searle, M.P., and Ries, A.C., eds., The Geology and Tectonics of the Oman Region: Geological Society of London, Special Publication 49, p. 797–831.

Stöcklin, J., 1968, Structural history and tectonics of Iran: American Association of Petroleum Geologists, Bulletin, v. 52, p. 1229–1258.

Stöcklin, J., 1974, Possible ancient continental margins in Iran, *in* Burk, C.A., and Drake, C.L., eds., The Geology of Continental Margins: New York, Springer-Verlag, p. 873–887.

Stoneley, R., 1990, The Arabian continental margin in Iran during the Late Cretaceous, *in* Robertson, A.H.F., Searle, M.P., and Ries, A.C., eds., The Geology and Tectonics of the Oman Region: Geological Society of London, Special Publication 49, p. 787–795.

VAN BUCHEM, S.P., RAZIN, P., HOMEWOOD, P.W., PHILIP, J.M., EBERLI, G.P., PLATEL, J.P., ROGER, J., ESCHARD, R., DESAUBLIAUX, G.M.J., BOISSEAU, T., LEDUC, J-P., LABOURDETTE, R., AND CANTALOUBE, S., 1996, High resolution sequence stratigraphy of the Natih Formation (Cenomanian/Turonian) in northern Oman: distribution of source rocks and reservoir facies: GeoArabia, v. 1, p. 65–89.

WATTS, K.F., 1987, Triassic carbonate submarine fans along the Arabian platform margin, Sumeini Group, Oman: Sedimentology, v. 34, p. 43–71.

WHITE, R.S., AND KLITGORD, K.D., 1976, Sediment deformation and plate tectonics in the Gulf of Oman: Earth and Planetary Science Letters, v. 32, p. 199–209.

WHITE, R.S., 1982, Deformation of the Makran accretionary sediment prism in the Gulf of Oman (north-west Indian Ocean), *in* Leggett, J.K., ed., Trench-Forearc Geology: Sedimentation and Tectonics on Modern and Ancient Active Plate Margins: Geological Society of London, Special Publication 10, p. 357–372.

# STRATIGRAPHY AND SEDIMENTATION OF THE MIDDLE JURASSIC, U.A.E.

JOSE E. DE MATOS

*Petroprimo, Av. da Republica, 50-4°, 1050 Lisbon, Portugal*

AND

GORDON M. WALKDEN

*Department of Geology and Petroleum Geology, University of Aberdeen, Meston Bldg, King's College, Old Aberdeen AB9 2UE, U.K.*

ABSTRACT: Wadi Naqab, SE of Ras Al Khaimah in the U.A.E., exposes an 800 m thick, shallow-water, Middle Jurassic succession. The base of the Bajocian is placed at a sequence boundary at ca. 460 m above the top Triassic and 45 m beneath the horizon that yielded the ammonite *Poecilomorphus* sp. The Bajocian/Bathonian contact is marked by the highest occurrence of *Haurania deserta*. The questionable Bathonian/Callovian boundary either coincides with the last occurrence of *Alzonella cuvillieri* or is near the base of the *Trocholina palastiniensis* zone. The last occurrence of *Kilianina blancheti* indicates the top of the Callovian. A new Bajocian foraminifer, *Pseudodictyopsella jurassica*, has been identified.

In Wadi Naqab, the Middle Jurassic corresponds to a broadly shallowing succession comprising multiple, meter-scale, fifth-order cycles. The Bajocian is predominantly subtidal, and cycles are commonly terminated by thick, massive oncoidal/peloidal packstones or grainstones. Most of the Bathonian and Callovian cycles start with spicular wackestones and end with cross-bedded peloidal/oolithic grainstones and/or stromatoporoid/coral rudstones. A direct comparison is established between the Middle Jurassic of Wadi Naqab and the subsurface of the Emirates. In the subsurface of Abu Dhabi, the Bajocian Izhara and Bathonian–Callovian Araej formations are the rock units that make up the Middle Jurassic. As a result of regional comparisons performed in this study, the Izhara Formation is redefined and a new type section proposed.

## INTRODUCTION

The Mesozoic section of the Musandam, studied and described by Hudson and Chatton (1959), is a composite of small segments from different wadis of the Ras Al Khaimah area. Since then and especially in the last ten years, the Jurassic of the Arabian Peninsula has been studied in particular detail (Moshrif, 1987; Hassan, 1989; Droste, 1990 and 1993; Manivit et al., 1990; Koepnick, 1993; McGuire et al., 1993; Toland et al., 1993; de Matos, 1997; de Matos and Hulstrand, 1994; de Matos et al., 1994; Alsharhan and Nairn, 1994; Alsharhan and Magara, 1994, 1995; Alsharhan and Whittle, 1995a, 1995b; and de Matos and Walkden, 1995). The Arabian Jurassic picture is therefore now clearer, but there are stratigraphic uncertainties regarding the Jurassic of U.A.E. that still need to be addressed. Published biostratigraphic studies of the Jurassic of onshore and offshore Abu Dhabi are scarce.

Therefore, and in order to understand Jurassic stratigraphy and sedimentation and to evaluate reservoir characteristics of the subsurface of Abu Dhabi, a surface section of ca. 1310 m was measured, sampled (at ca. 1–2 m intervals), and described (on 1/100 scale) in Wadi Naqab, 16 km southeast of Ras Al Khaimah (Figs. 1, 2). This section is located between ca. 25° 42' N, 56° 03' E and 25° 44' N, 56° 09' E, within the Rub'al-Khali basin. The field work was undertaken between November 1992 and May 1995, and the Middle Jurassic section turns out to be 800 m thick.

This study also incorporates onshore and offshore subsurface data for the Jurassic. A total of 75 wells were included in the study, over 1300 thin sections and more than 150 cores were either reviewed or described for the first time to better define the sedimentology and microfossil content of the Jurassic in the subsurface of Abu Dhabi. In addition, palynological analyses of selected cores, cuttings, and field samples were undertaken to establish chronostratigraphic control and a better definition of sedimentology. Whenever possible, seismic data were used to study stratal patterns and to check and/or complement log correlations.

The purpose of this study was to describe in detail the biostratigraphy, cyclicity, and sequence stratigraphy, to analyze and to establish the relationship between the stratigraphic units of the Middle Jurassic section of the subsurface of Abu Dhabi with the section of the Musandam Peninsula. The Middle Jurassic formations of the subsurface are the Izhara and Araej formations, but only the Izhara Formation is discussed in this study.

The Oman Mountains have a complex geology. Studies by Glennie et al. (1974), Glennie (1992, 1995), Hughes Clarke (1988), and those included in Robertson et al. (1990) have contributed substantially to the understanding of it. Hudson and Chatton (1959) were the first to compile a comprehensive stratigraphy of the Mesozoic of the Musandam Peninsula, and Middle Jurassic to Lower Cretaceous source rock–reservoir–seal relationships in the Middle East were outlined by Murris (1980). Alsharhan and Kendall (1986) reviewed and interpreted the occurrence of oil and gas in the Precambrian to Jurassic rocks from the Arabian Peninsula, Iraq, and Iran. Middle Jurassic stratigraphy and hydrocarbon occurrences in the central and western parts of offshore Abu Dhabi were further discussed by Hassan (1989).

Until 1990 little attention had been paid to the Middle Jurassic of the Emirates in general and Musandam Peninsula in particular. Recently, however, the Middle Jurassic Araej Formation of U.A.E. and Qatar has been the subject of more focused research (Al-Saad et al., 1992; Alsharhan and Magara, 1994, 1995; Alsharhan and Whittle, 1995a, 1995b).

## BIOSTRATIGRAPHY

The present study provides new biostratigraphic data on Middle Jurassic foraminifera, algae, and stromatoporoids. The stratigraphic ranges of most genera and species of algae, foraminifera, and stromatoporoids recognized in the Jurassic of Wadi Naqab, are illustrated in de Matos (1997, fig. 4.1). These data confirm the previous stratigraphic ranges of most of the foraminifera (Septfontaine et al., 1991, fig. 2) and algae (Granier and Deloffre, 1993) of the southern margin of Tethys. However, the

Middle East Models of Jurassic/Cretaceous Carbonate Systems
SEPM Special Publication No. 69, Copyright © 2000
SEPM (Society for Sedimentary Geology), ISBN 1-56576-075-1, p. 21–35.

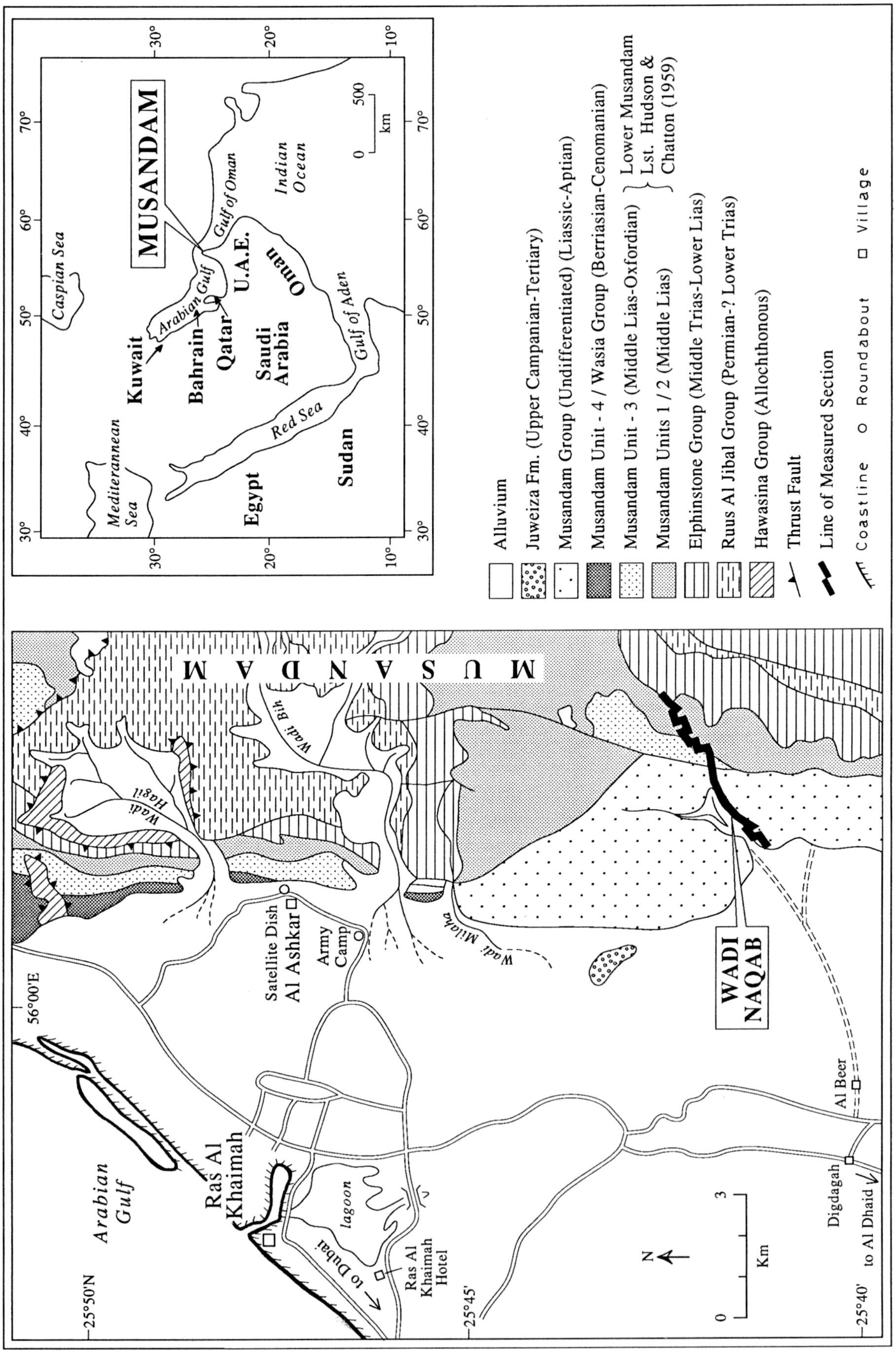


FIG. 1.—Geographical location of Wadi Naqab, SW Musandam Peninsula (U.A.E.) with schematic geological map.

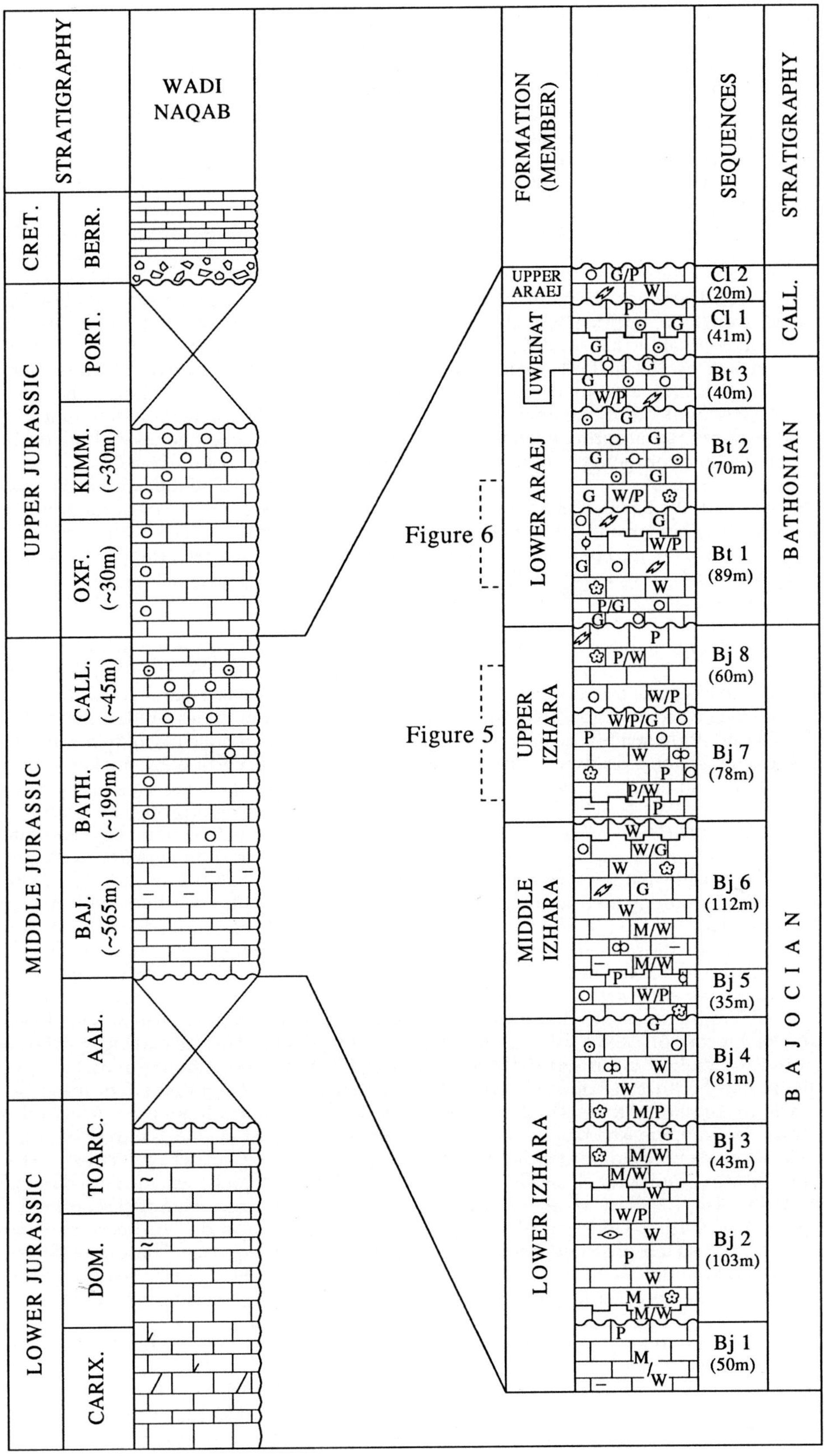


FIG. 2.—Schematic lithological log of the Jurassic of Wadi Naqab. Detailed Middle Jurassic section showing the sequences and the formations of the subsurface is included. Also indicated are the intervals illustrated in Figures 5 and 6.

local stratigraphic ranges of some stromatoporoids (de Matos, 1997) do not conform with Wood (1987) or Toland (1994).

### *Bajocian*

In Wadi Naqab, the limits of the Bajocian are placed on the basis of a combination of sequence stratigraphic and biostratigraphic data. The concentration of omission surfaces (cf. Walkden and de Matos, this volume) at ca. 460 m above top Triassic and ca. 45 m below the horizon yielding the sole Lower Bajocian ammonite *Poecilomorphus* sp. is interpreted as the regional hiatus of a large part of the Toarcian and all the Aalenian in the Arabian Peninsula (Manivit et al., 1990). Therefore, we place the base of the first sequence of the Bajocian (Bj1) coincident with the uppermost omission surface of that interval (de Matos, 1997, fig. 4.15).

The top Bajocian is marked by the highest occurrence of *Haurania deserta* Henson, the first certain occurrence of *Protopeneroplis striata* Weynschenk, and the inceptions of *Pseudomarssonella* sp. and *Pfenderella* sp. at approximately 1025 m in the section. The joint appearance of these foraminifera indicates early Bathonian.

The microfossil biozonation of the Bajocian is not as clear as the rest of the Middle Jurassic (Bathonian and Callovian stages) because of several intervals with a sparse biota. Four biozones were nevertheless defined as: zone 1: Dasycladacean BjA; zone 2: *Thaumatoporella parvovesiculifera*; zone 3: *Amijiella amiji*; and zone 4: *Pseudodictyopsella jurassica*. These are separated by four intervals with a nondiagnostic fossil content (designated A, B, C, and D, Fig. 3).

A new foraminifer, *Pseudodictyopsella jurassica* (Septfontaine and de Matos, 1998), of biozone 4 occurs higher than the Early Middle Jurassic *Timidonella*-rich storm layers. The new genus, in association with *Spiraloconulus* sp. and *Selliporella donzellii*, is dated as Bajocian (Septfontaine and de Matos, 1998).

### *Bathonian*

In Wadi Naqab the thickness of the Bathonian is between 175 and 200 m, depending of the definitions used for both boundaries (de Matos, 1997, appendix 4.3) (Fig. 2). The base is constrained by the highest occurrence of *Haurania deserta* (Fig. 3). Normally, the base of the Bathonian is defined by the highest occurrence of the association *Haurania deserta–Amijiella amiji* (Athersuch et al., 1992). In Wadi Naqab the two forms have different ranges. The highest record of *Haurania deserta* is at the base of the Bathonian but the highest occurrence of *Amijiella amiji* is in the late Bathonian because it coincides with the inception of *Trochamijiella gollesstanehi* Athersuch, Banner and Simmons, a late Bathonian foraminifer (Athersuch et al., 1992). The Bathonian/Callovian boundary could be at either the last questionable occurrence of the foraminifer *Alzonella cuvillieri* (Bajocian to Bathonian) or near the base of the *Trocholina palastiniensis* zone (Fig. 3). According to Gollesstaneh (1965) in South Iran *Trocholina palastiniensis* Henson occurs sporadically near the Bathonian–Callovian boundary.

A regional break of sedimentation between the early Bathonian and the middle Callovian was recognized in Saudi Arabia (Manivit et al., 1990). Comparable hiatuses undoubtedly occurred in Wadi Naqab and are probably present in other areas of the Arabian Peninsula.

### *Callovian*

The Callovian is approximately 45 m thick (de Matos, 1997, appendix 4.3). Its top (top Middle Jurassic) is placed directly above the grainstones containing the last occurrence of *Kilianina blancheti* Pfender in association with *Kurnubia palastiniensis* Henson (late Callovian to Kimmeridgian), *Trocholina palastiniensis* (?Bathonian to lower Kimmeridgian), *Trocholina elongata* (Callovian to Kimmeridgian) and *Praekurnubia crusei* (late Bajocian to middle Oxfordian).

## CYCLICITY, SEDIMENTATION, AND SEQUENCE STRATIGRAPHY

The Middle Jurassic of the Musandam Peninsula is a succession of shallowing-upward cycles arranged in sequences of different orders, meters to decameters thick, in a lagoonal to locally intertidal environment. Meter-scale shallowing-upward, fifth-order cycles are the building blocks of larger depositional sequences.

The Middle Jurassic of Wadi Naqab displays a remarkably consistent shallowing-up trend. The lower part of the Bajocian is dominated by lime mudstones and the middle Bajocian by wackestones. Packstones predominate higher up in the upper Bajocian, and the grainstones with subordinate rudstones are the major constituents of the remainder.

Cycle definition, identification of parasequence sets, and analyses of the omission surfaces of the Middle Jurassic indicate the existence of eight sequences in the Bajocian, three in the Bathonian and two in the Callovian (Fig. 2; de Matos, 1997, appendix 4.3).

### *Bajocian*

A total of at least 146 small shallowing-upward cycles were identified in the ca. 565 m of the Bajocian of Wadi Naqab. The characteristics of Bajocian cycles vary throughout the section, with the subtidal cycles predominant. A typical subtidal cycle starts with thin-bedded nodular lime mudstones to wackestones and commonly ends with thicker (decimeter-thick) massive oncoidal/peloidal wackestones, packstones, or grainstones (Figs. 4, 5). Rarely, the cycle is capped by intertidal/tidal-flat limestones with birdseye structures and algal laminites.

The pattern of sedimentation indicates a stable balance between subsidence and sediment accumulation, which maintained broadly open shelf conditions with increasing energy towards the top of the Bajocian. Assuming a period of 7.3 million years for the Bajocian (Hardenbol et al., 1998), then each small cycle took ca. 50,000 years to be formed, which means that the Bajocian would be composed of fifth-order cycles and the average rate of subsidence would have been 7.5 cm/1000 years.

More than half of the sequences of the Bajocian (sequences Bj2 to Bj5 and Bj8) end with irregular, burrowed iron-stained omission surfaces representing either firmgrounds or hardgrounds. Only a few exhibit limonitic crusts and iron-coated molds of burrows suggesting the removal of the mineralized crust by mechanical erosion. These cemented surfaces are more resistant to weathering than the enclosing sediments and therefore stand out clearly in outcrop. A high degree of bioturbation or hardground development may indicate a decrease of sedimentation at the top of a regressive sequence. These omission surfaces probably indicate a rapid deepening of the sea after the break of sedimentation at the top of the Bajocian sequences.

### *Bathonian*

A total of 34 cycles was identified in the 199 m of the Bathonian of Wadi Naqab. Most of the cycles of the Bathonian consist of spicular wackestones, locally filament-rich at the base and cross-bedded peloidal/oolithic grainstones and/or stromatoporoid/coral rudstones at the top (Fig. 6). Inclusion of a rudstone as the start of a cycle, in the middle or at the top, depends on the cycle constituents and the cycle stacked facies.

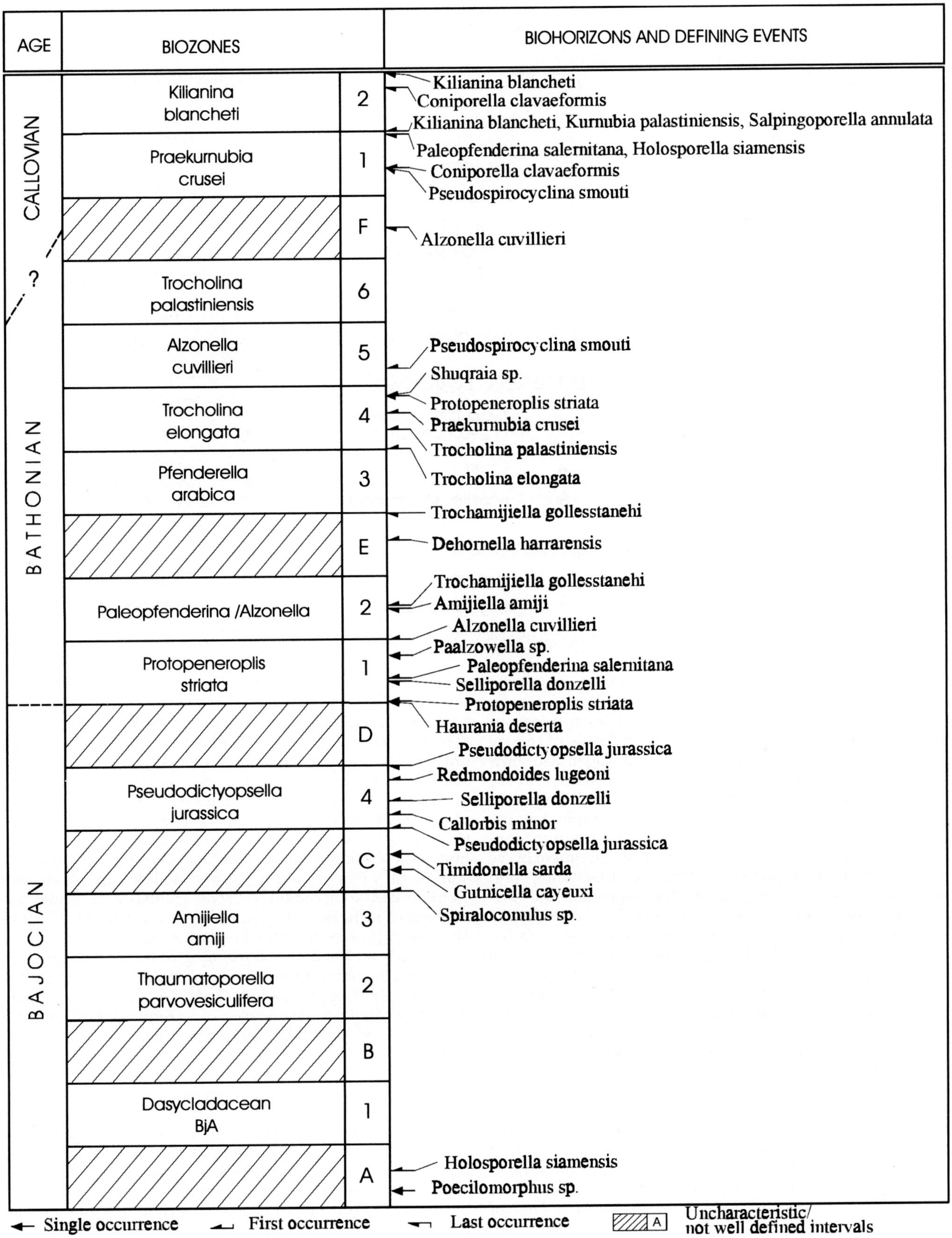


FIG. 3.—Micropaleontological biozonation of the Middle Jurassic of Wadi Naqab.

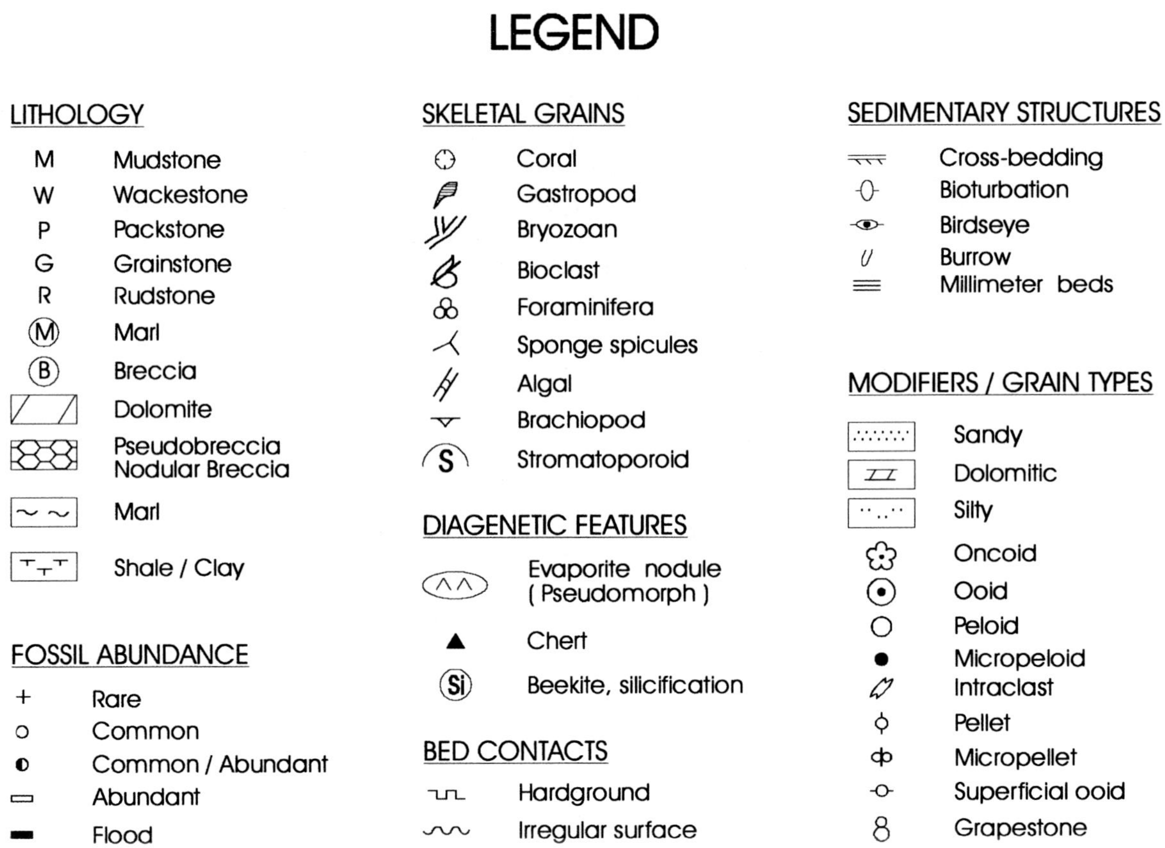


FIG. 4.—Legend for Figures 5, 6, 9, and 10.

Rudstones occur in tidal flats and channels. They also occur as condensed horizons. Similar to the reworked fossils of the lag deposits, the bioclasts of the rudstones of Wadi Naqab also show breakage, coatings, borings, and locally geopetal structures. The intraclasts may consist either of rip-up clasts and/or of fragments of rocks not actually represented in the local stratigraphic column because of total erosion. Consequently, a large part of the time of the Bathonian and Callovian (i.e., time of the development of the stacked coral/stromatoporoid rudstones, floatstones, and boundstones) was consumed during the deposition of those rocks. This may explain why the thickness of the Bathonian and Callovian of Wadi Naqab is so small in comparison with the thickness of the Bajocian. The differences in the time spans of each stage would not justify such a great variation in thickness.

Assuming a period of 4.8 million years for the Bathonian (Hardenbol et al., 1998), then the average rate of subsidence would be ca. 4 cm/1000 years and each cycle would have taken ca. 141,000 years to be formed. There is, however, a strong possibility that each cycle was deposited in a much shorter period. If the hiatus in sedimentation defined in Saudi Arabia between the early Bathonian and the base of the middle Callovian (Manivit et al., 1990; Le Nindre et al., 1990) also exists in the Musandam Peninsula, then the 34 cycles of the Bathonian of Wadi Naqab needed only ca. 1.3 million years (the time span of the Lower Bathonian) to be deposited. The presence of this hiatus is not evident in the studied section. The concentration of the rudstone lag beds might, however, explain the discrepancies.

The first horizon with silica replacement occurs in the transgressive systems tract (TST) of the first sequence of the Bathonian (de Matos, 1997, fig. 4.45). Silicification (replacement by beekite) of the stromatoporoids, corals, and other bioclasts is common throughout the Bathonian, Callovian, and Upper Jurassic. Similar large silicified stromatoporoid colonies also occur in the Early Oxfordian of Ethiopia, Tigre Province (Bosellini et al., 1997) and in the Jurassic of Yemen (Bosellini and Al Thour, personal communication). The presence of beekite suggests a low rate of subsidence with discontinuous deposition and major time breaks in sedimentation (Kazanci and Varol, 1993). In addition, chert nodules are abundant and concentrated along certain horizons. The existence of silicification can be used in the field as a guide to distinguish the Middle Jurassic and Upper Jurassic from the Lower Jurassic.

There is a clear upward thinning of the sequences and TST (i.e., from the oldest to the youngest sequence) of the Bathonian (de Matos, 1997, appendix 4.3). This, together with the occurrence of beekite, suggests the decrease of accommodation space favoring erosion and/or nondeposition towards the top of the Bathonian succession.

### *Callovian*

A total of 11 cycles, distributed over two sequences, were recognized in the 45 m of Callovian sediments of Wadi Naqab (de Matos, 1997, figs. 4.65 and 4.80). Composition and characteristics of the Callovian cycles are very similar to those of the Bathonian.

In conclusion, the durations of the time span of each stage of the Middle Jurassic (Bajocian, 7.3 million years; Bathonian, 4.8 million years; and Callovian, 5 million years) are not proportional to the respective thickness differences (Bajocian, 565 m; Bathonian, 199 m; and Callovian, 45 m), so the preserved sedi-

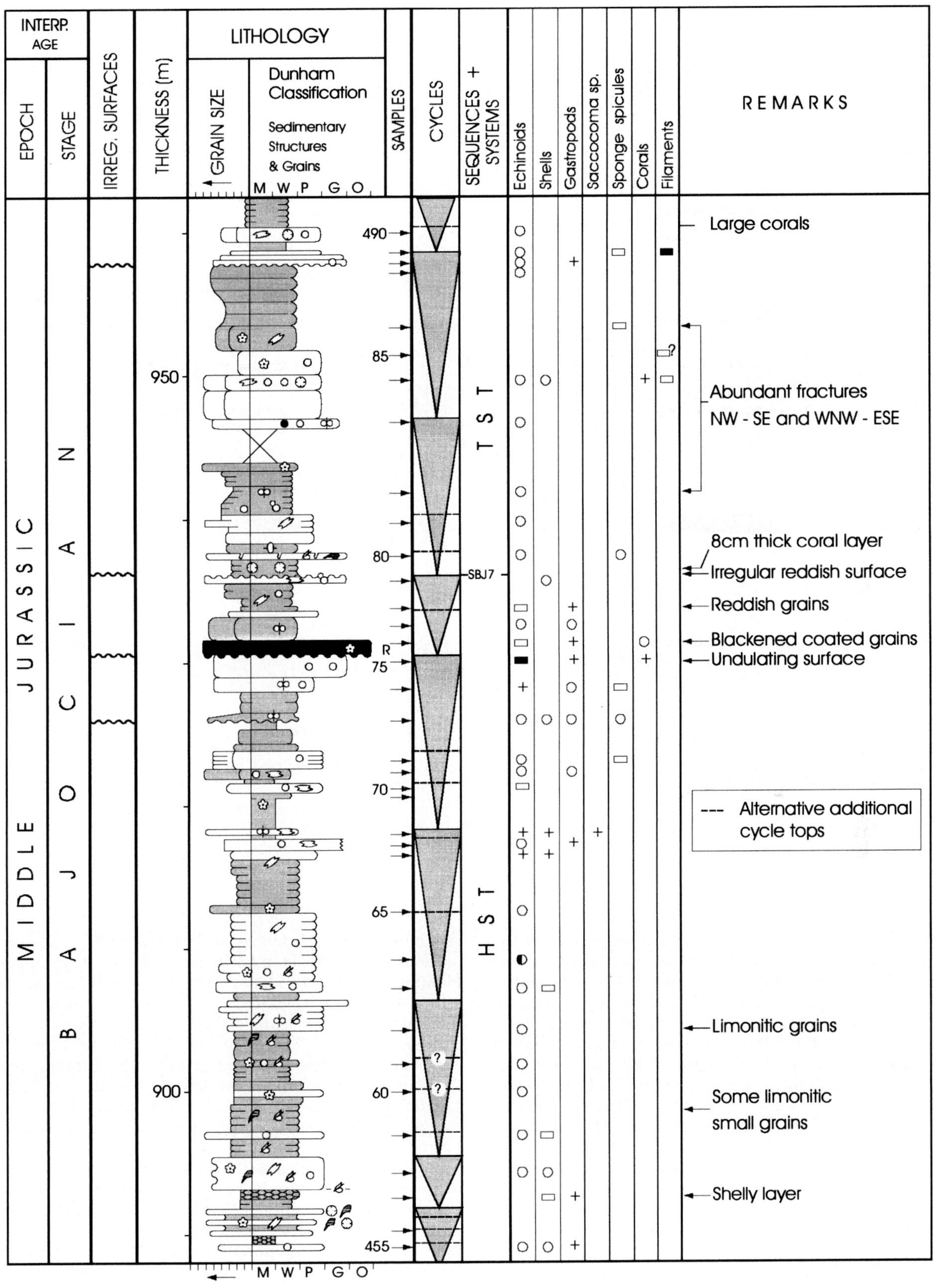


FIG. 5.—Bajocian of Wadi Naqab. Lithology, relative abundance of major groups of organisms, and cycle composition of the HST of sequence Bj7 and the TST of sequence Bj8. The column designated "thickness (m)" indicates the thickness of the section above top Triassic.

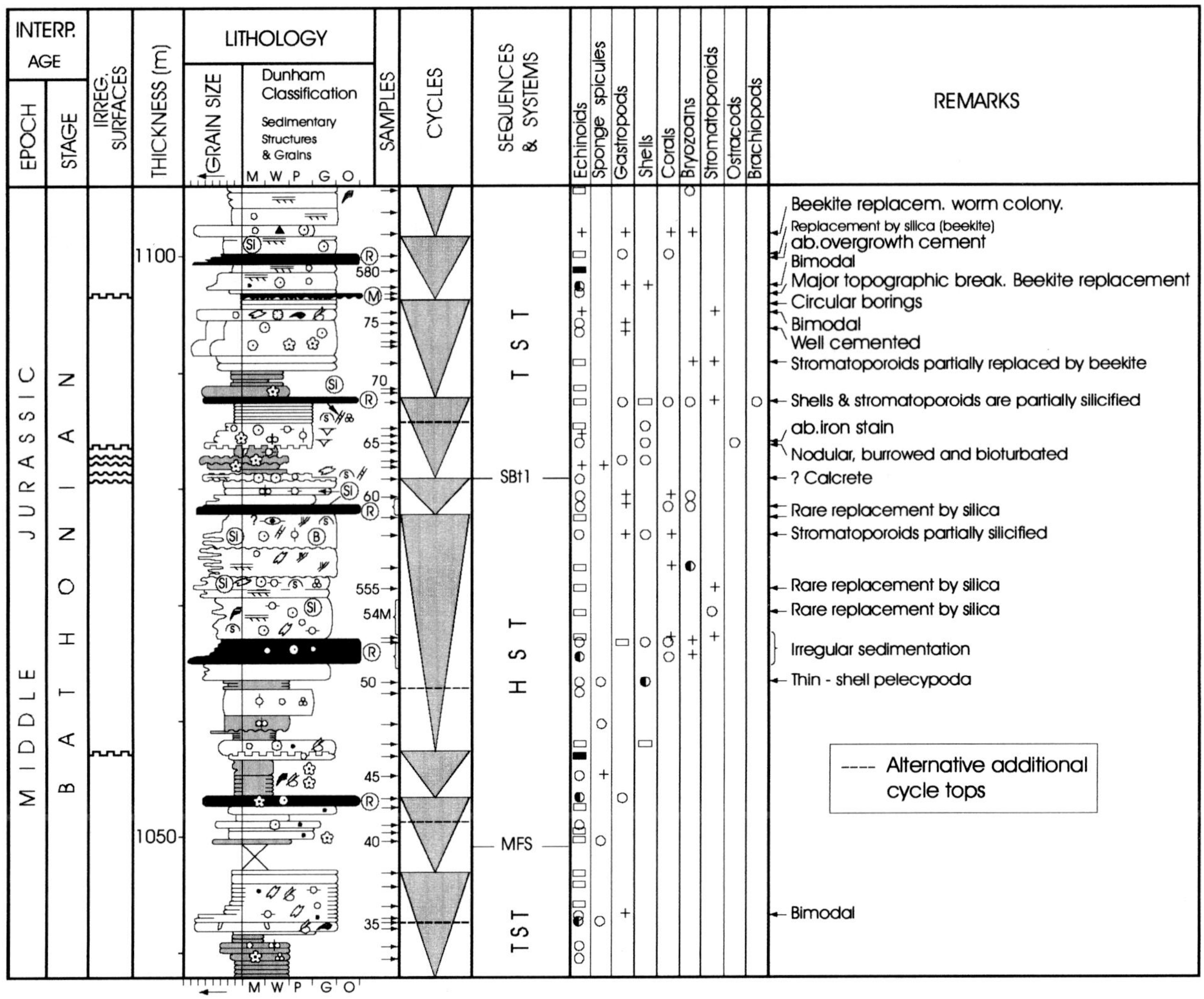


FIG. 6.—Bathonian of Wadi Naqab. Lithology, relative abundance of major groups of organisms, and cycle composition of the upper part of the TST of sequence Bt1 and part of the TST of sequence Bt2. The column designated "thickness (m)" indicates the thickness of the section above top Triassic.

ments must therefore represent just a small proportion of total elapsed time. Large parts of the Bathonian and Callovian must have been taken up by condensed deposition and reworking leading to the development of the stacked lag deposits (coral/stromatoporoid rudstones, floatstones, and boundstones). One should also bear in mind the late Bathonian–late Callovian unconformity identified in the Middle East by Lewy (1981) and the break in sedimentation recognized in Saudi Arabia (Le Nindre et al., 1990) of at least 7 million years between the early Bathonian and the middle Callovian. There are also hiatuses in the late Lias, Bajocian, and Bathonian in Jebel Akhdar in Oman (Le Nindre et al., 1990). It is clear, therefore, that during parts of the Middle Jurassic there were regional breaks of sedimentation in the Arabian Peninsula, and similar important hiatuses clearly occurred in Wadi Naqab.

## THE MIDDLE JURASSIC OF THE ABU DHABI SUBSURFACE

The Middle Jurassic of the subsurface of the Emirates comprises the Izhara and Araej formations (Fig. 7). In Abu Dhabi, the section between the Izhara and Marrat formations, previously correlated with the Hamlah Formation, is not equivalent to the Hamlah as defined in Qatar. The Hamlah Formation of Qatar, which has been considered to be Lower Jurassic since 1975 (Sugden and Strandring, 1975), was recently dated as upper Triassic (Hassan, 1989; de Matos et al., 1994; de Matos, 1997). Log correlation has demonstrated that the Abu Dhabi equivalent of the Hamlah Formation lies beneath the Marrat and Minjur formations (Fig. 8). It was necessary, therefore, to decide how to allocate that part of the subsurface sequence formerly designated Hamlah Formation by ADCO geologists.

Palynologic analyses from onshore and offshore Abu Dhabi wells have indicated a Bajocian age for that part of the section between the Marrat and Izhara formations. In addition, sedimentation seems to have been continuous between the Marrat and the top of the Araej Formation (the formation directly overlying the Izhara Formation, Fig. 7). These factors favored the inclusion of that sequence in the Izhara Formation.

Here we offer a detailed description of the redefined Izhara Formation, and its regional setting and distribution in the Arabian Peninsula in general and in the Emirates in particular.

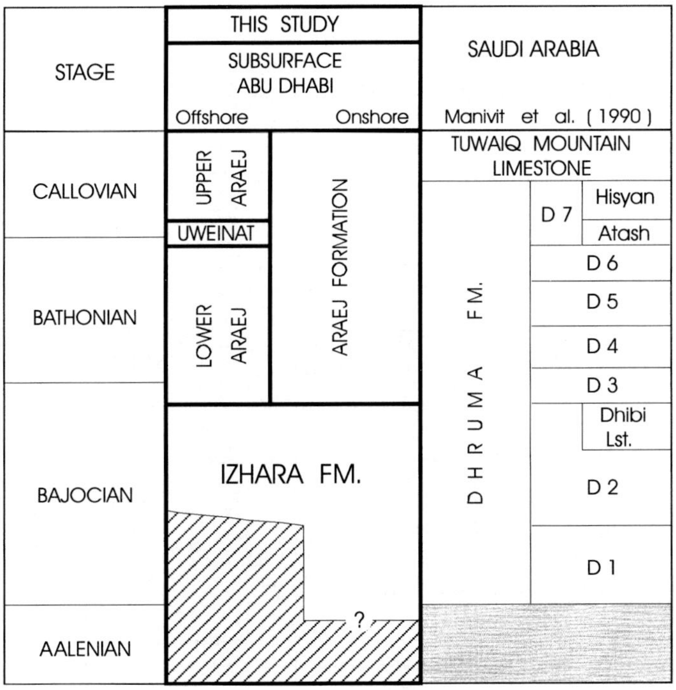


FIG. 7.—Middle Jurassic stratigraphy of the subsurface of Abu Dhabi and equivalent formations in Saudi Arabia.

## *Stratigraphic and Sedimentological Revision of the Izhara Formation*

*New Type Section.—*

The section of the ADCO well Jarn Yaphour-7 (Jy-7), previously interpreted as Izhara plus Hamlah formations (362 m thick), was chosen as the type section of the newly redefined Izhara Formation (Fig. 9). The well Jy-7 of onshore Abu Dhabi is located at 24° 15' 50.6" N, 54° 38' 58.9" E. The previous type section was in Qatar Petroleum Company well Kharaib-1, where this formation is ca. 137 m thick (Sugden and Strandring, 1975). The offshore Abu Dhabi reference section is ca. 208 m thick in the ADMA well Ghasha-7 (Gh-7) (Hassan, 1989).

*Lithology.—*

In Abu Dhabi, the Izhara Formation can be divided into three members, lower, middle and upper (Fig. 9) based on both lithology and gamma-ray log characteristics. The Lower and Middle Members are equivalent to what has been considered the Hamlah Formation. The Upper Member corresponds to the former definition of the Izhara Formation in the "Lexique Stratigraphique of Qatar" (Sugden and Strandring, 1975).

The top of the Lower Izhara Member is located at the top of a clean low-gamma-ray section composed mainly of lime mudstones and wackestones (Figs. 8, 9). The boundary between the Middle and Upper Izhara Members correspond to a clear log break between a clean low-gamma-ray interval, composed of peloidal wackestones and packstones, and the overlying siltstones and shales with high gamma-ray response (Fig. 9).

The Lower Izhara Member in Jy-7 is about 98 m thick and is dominated by gray, brown, buff, hard and dense bioclastic lime mudstones and wackestones with a variable degree of dolomitization. It is locally chalky with pyrite and minor shales, and packstone streaks are also present. The biota consist of echinoids, ostracodes, gastropods, benthonic foraminifera, and recrystallized, thin-shelled molluscan fragments. Syntaxial overgrowths are present on echinoid fragments. The uppermost 16 m of the Lower Izhara in Jy-7 contains abundant anhydrite laths and nodules at several horizons. Cryptalgal laminites associated with meter-scale peritidal cycles are evident.

The Middle Izhara Member is 142 m thick in Jy-7 and comprises a lower clean limestone section (91 m thick), with lithologies similar to the Lower Izhara Member, and an upper more argillaceous section, 51 m thick, where limestones predominate but gray platy hard fissile shales and siltstones also occur.

The Upper Izhara Member is ca. 122 m thick in Jy-7 and was divided into two units based on gamma-ray log response (Steele et al., 1983), a lower unit with a high gamma-ray trace (Unit A) and an upper unit (Unit B) with a clean low-gamma-ray pattern (Fig. 9). The lower unit (Unit A = Hassan's Unit 4 or Unit 2 in QPC well Kharaib-1) is 89 m thick in Jy-7 and consists of interbedded shelly, silty (locally sandy) lime mudstones, calcareous siltstones, and shales. The uppermost Izhara Unit B (equivalent to Dhibi Limestone of the Dhruma Formation in Saudi Arabia, or to Unit 1 of Kharaib-1 well of Qatar or Hassan's Unit 5 in Abu Dhabi) is 33 m thick in Jy-7 and is muddy at the base, changing to packstones and grainstones with some lime mudstones and bioclastic wackestones at the top.

A detailed study of the Upper Izhara Member has been conducted in an offshore well (Fig. 10). The lower unit of the Upper Izhara in this well is mostly of argillaceous lime mudstones to wackestones. Several pyritic, shelly packstone, condensed layers occur locally with subrounded coarse quartz sand grains or silt-size grains. The shales are calcareous and pyritic, and subrounded phosphate particles are abundant in the condensed shelly and packstone layers. The upper unit (equivalent to Dhibi Limestone) consists mainly of lime mudstones to wackestones.

*Biostratigraphy and Age.—*

The age of the Izhara Formation in the previous type locality was considered to be Bajocian (Sugden and Strandring, 1975), although Hassan (1989) attributed a Liassic age to the entire Izhara Formation of the offshore Abu Dhabi. More recently, Manivit et al. (1990), in a comprehensive study of the entire Jurassic of Saudi Arabia, consider that the Lower Dhruma Formation (equivalent to the Izhara Formation) is also Bajocian. Because of this controversy and owing to a lack of microfossil content, particularly in the Lower and Middle Members of the Izhara Formation, palynological analyses were undertaken in several onshore and offshore wells (Figure 8 shows some of these). The miospores and palynomorphs support an age not older than late Bajocian for the Middle Izhara in the studied wells. The microfossil association recognized in the upper unit of the Izhara Formation (Fig. 10) also supports a late Bajocian age for the topmost Izhara. Therefore, all the Izhara Formation belongs to the Middle Jurassic–Bajocian.

*Environment of Deposition.—*

The occurrence of the miospores *Classopolis* spp. and *Araucariacites australis* Cookson in the Izhara Formation indicate that arid climatic conditions were dominant. The occurrence of fine-grained and bioturbated limestones in the Lower and Middle Izhara indicates low-energy conditions, while the accompanying fairly diverse biota suggests deposition in an open marine-shelf environment. The limestones of the uppermost Middle Izhara Member in southeast Abu Dhabi are locally rich in pyrite and

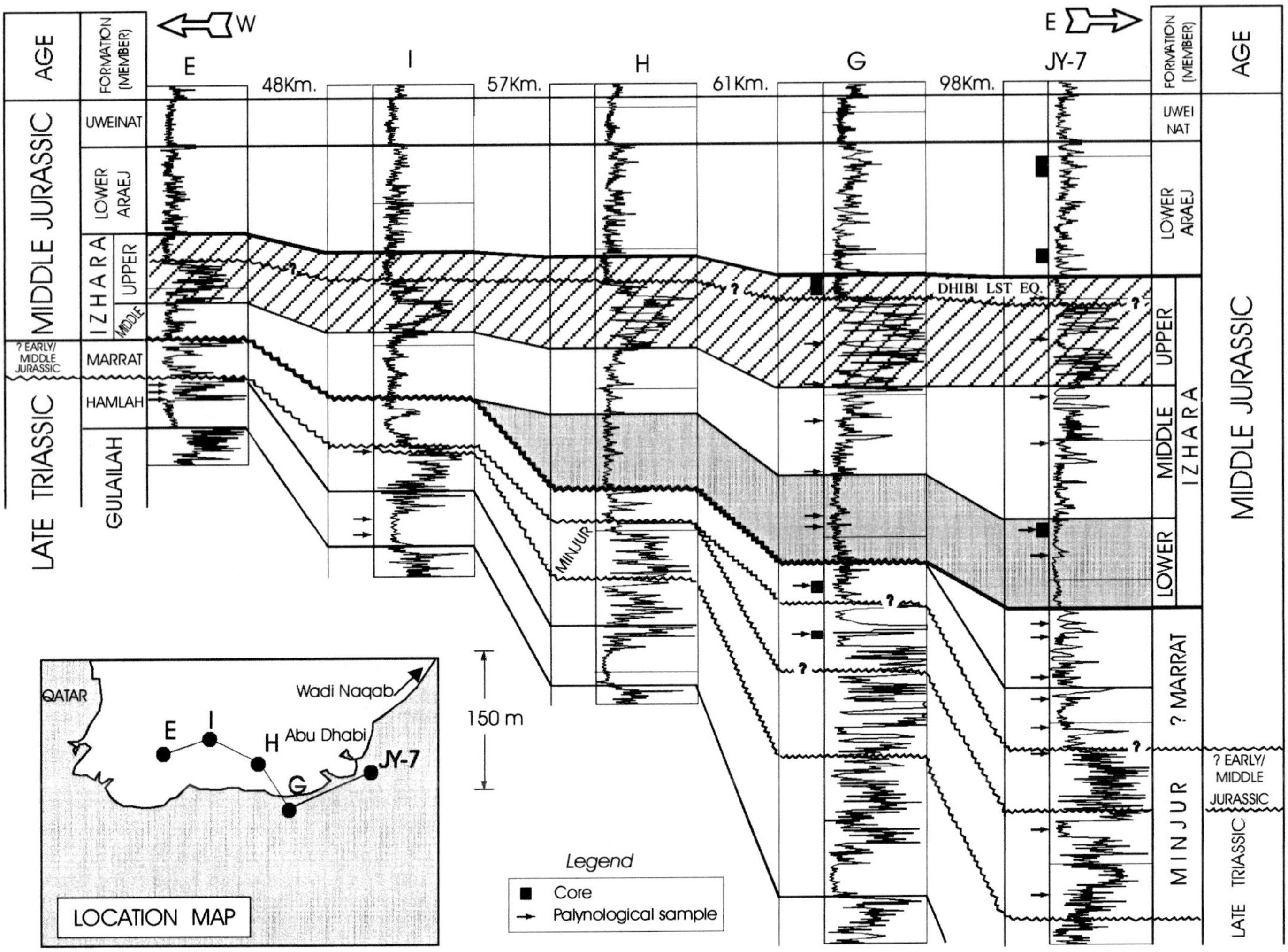


FIG. 8.—East–west gamma-ray log correlation, Late Triassic to the Middle Jurassic Izhara Formation is highlighted.

contain abundant pelagic bivalves (filaments), sponge spicules, and shell debris, suggesting deposition under deeper open-marine conditions.

The lithology and bioclasts of the lower part of the Upper Izhara (Fig. 10) indicate deposition in a marine euxinic environment with periodic siliciclastic influxes. Phosphate particles are abundant in the condensed shaly and packstone layers closer to the maximum flooding surface (MFS), which is marked by a very high gamma-ray peak (Figure 10). Shallow, low-energy-shelf conditions probably prevailed during deposition of the topmost Izhara.

*Boundaries.—*

The upper contact with the Araej Formation is conformable (Fig. 8). The upper boundary coincides with a lithologic change and a log break, and is placed at the top of a low-gamma-ray sequence of limestones beneath the argillaceous limestones of the Lower Araej Member (Fig. 8). The underlying contact is predominantly with the Marrat Formation, and is unconformable in offshore and in large areas of onshore Abu Dhabi (Fig. 8). Also, in most areas of the Qatar Arch, the Izhara Formation directly overlies the Triassic Gulailah or Jilh formations (de Matos, 1997, fig. 4.92). The lower contact, however, is conformable in eastern onshore Abu Dhabi, where lowstand deposits occur at the top of the Marrat Formation (Fig. 8; de Matos et al., 1994).

*Regional Setting and Distribution.—*

In Abu Dhabi, the thickness of the Izhara Formation decreases from east to west, with a maximum of 362 m onshore (type locality) and ca. 120 m in the offshore around Dalma Island (Fig. 11). This thickness variation is mainly the result of: 1) depositional onlap of the Lower Izhara Member to the west, over the irregularly eroded top surface of the Marrat Formation, and 2) the gradual thinning of the Middle and Upper Izhara Members westward (Fig. 8). No Lower Izhara sediments were deposited to the west of the 200 m contour on the Izhara isopach map (compare Figs. 8 and 11).

The Upper Izhara Member maintains similar lithologic characteristics all over Abu Dhabi. The section of the Upper Izhara of onshore and offshore Abu Dhabi, equivalent to the Dhibi Limestone Member, as in Saudi Arabia (Powers, 1968), has a remarkably constant thickness of ca. 30 m. The "Pre-Dhibi" unconformity identified at the base of the Dhibi Limestone in Saudi Arabia (Enay et al., 1987) and noted by Banner et al. (1991), has not been recognized so far in Abu Dhabi. However, the

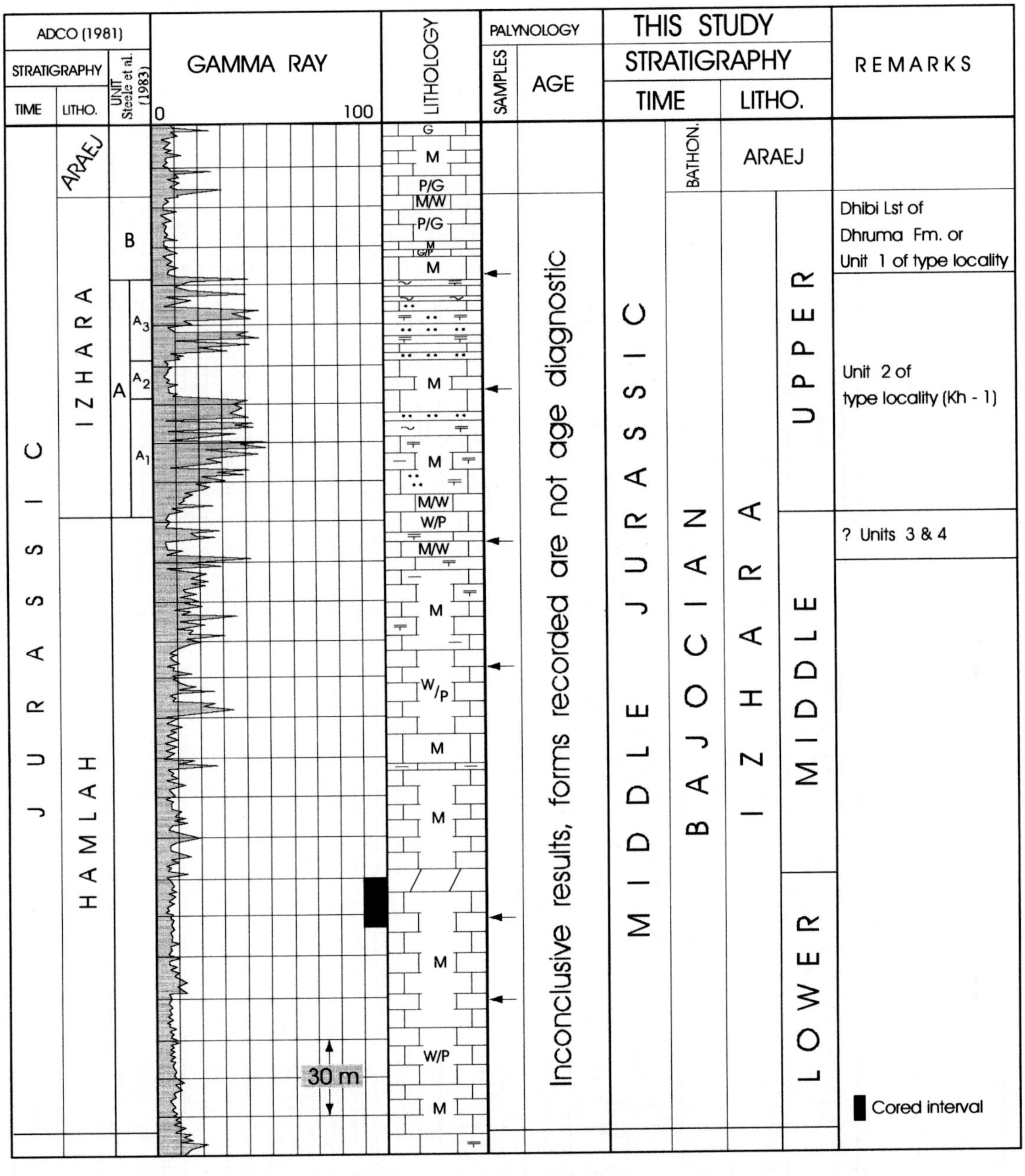


FIG. 9.—New type section for Izhara Formation (Well-Jarn Yaphour-7). Comparison between previous and present interpretation.

thinning of the lower part of the Upper Izhara westward (Fig. 8) suggests missing section at that level. Future additional data (cores, logs, and seismic) might confirm the existence of that hiatus.

### *Comparison of the Bajocian Section of Wadi Naqab with the Izhara Formation in the Subsurface of Abu Dhabi*

The section of Wadi Naqab equivalent to the Lower Izhara Member of the subsurface of Abu Dhabi comprises sequences Bj2–Bj4 (i.e., approximately between samples MUS-300 and 387, (de Matos, 1997, appendix 4.3). Like the succession of Jy-7, this interval of Wadi Naqab is composed mainly of lime mudstones to wackestones, and includes some horizons with birdseye structures and cryptalgal laminites. If this interpretation is correct, then there is a large change in thickness between the subsurface (98 m in the well Jy-7, Fig. 9) and the outcrop (ca. 230 m in Wadi Naqab, de Matos, 1997, appendix 4.3). The difference in thickness suggests that the increased subsidence rate during the Liassic of Wadi Naqab as compared to the rest of the Arabian Peninsula continued into the Middle Jurassic. The subsidence rates became essentially uniform only during the time of deposition of the Middle Izhara Formation and later.

As far as the Middle Izhara Member is concerned, the equivalent section of Wadi Naqab includes two sequences Bj5 and Bj6 and corresponds to the interval between samples MUS-388 and 442b (de Matos, 1997, appendix 4.3). This section is represented by subtidal cycles. Each cycle starts with nodular argillaceous lime mudstones to wackestones (which can produce a high gamma-ray reading) and ends with oncoidal, peloidal packstones

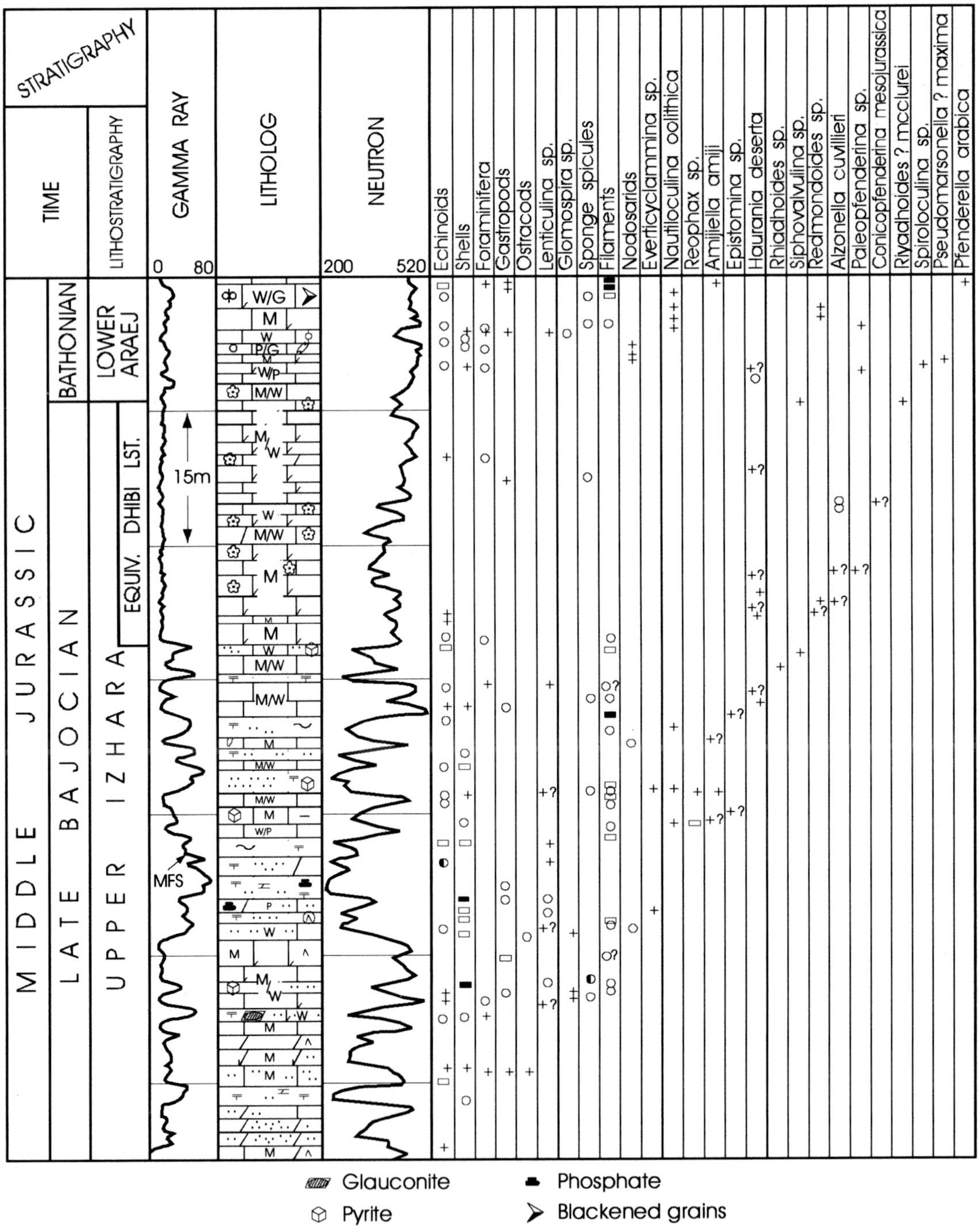


FIG. 10.—Upper Izhara lithology and fossil distribution for western offshore Abu Dhabi.

(which would have a cleaner and lower gamma-ray value). There is, therefore, likely correspondence in gamma-ray response between the Middle Izhara in the subsurface and the type section. Although sponge spicules and *Saccocoma* sp. have been recognized in the uppermost Middle Izhara of Wadi Naqab, it seems that the Middle Izhara of the subsurface is more argillaceous and silty, probably reflecting deposition under deeper conditions. The 140 m thickness of the Middle Izhara in Wadi Naqab (de Matos, 1997, appendix 4.3) matches well with the 142 m of equivalent section in the type locality (Fig. 9).

The Upper Izhara Member of Wadi Naqab and subsurface of Abu Dhabi have similar thickness, and both show a predominance of oncoidal packstones and larger concentration of oncoids towards the top. They differ, however, particularly in the lower part of the Upper Izhara. In the subsurface there is a pyritic, shelly, silty and shaly interval locally with phosphatic particles, characterized by a high gamma-ray response (Fig. 10). Filaments are particularly abundant, indicating deposition under pelagic conditions. Filaments are also abundant locally in Wadi Naqab. There are also some marly layers, and

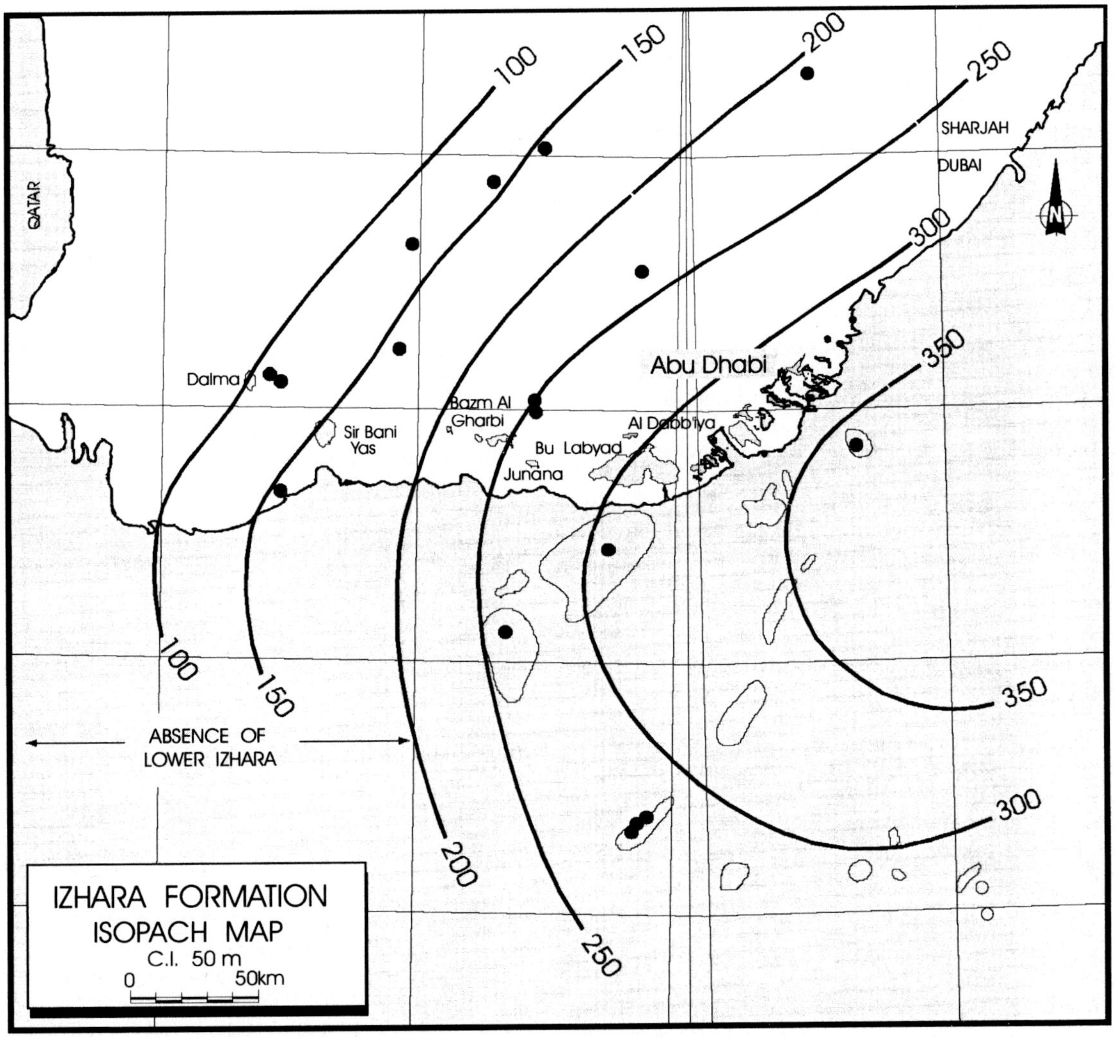


FIG. 11.—Izhara Formation isopach map in Abu Dhabi.

locally shell-rich horizons, but no silt, sand, phosphate, or glauconite were found in this part of the section of Wadi Naqab. A gamma-ray log through this part of the Wadi Naqab succession, which is thin-bedded, might show spiky higher values than would be expected in the sections above and below. Thus, the gamma-ray expression of the Upper Izhara in Wadi Naqab and in the subsurface of Abu Dhabi might be very similar. The upper part of the Upper Izhara (the Dhibi Limestone) shows thicker beds and oncoid-rich layers, and is also comparable in both localities.

## SUMMARY AND CONCLUSIONS

The defining biostratigraphic events of the Bajocian, Bathonian, and Callovian of Wadi Naqab are as follows:

1. The Toarcian/Bajocian boundary is marked by an omission surface located ca. 45 m beneath the layer that yielded the Lower Bajocian ammonite *Poecilomorphus* sp.
2. The Bajocian/Bathonian contact is defined by the highest occurrence of *Haurania deserta* and the inceptions of *Protopeneroplis striata*, *Pseudomarssonella* sp., and *Pfenderella* sp.
3. A new Bajocian foraminifer, *Pseudodictyopsella jurassica,* has been identified, in association with *Spiraloconulus* sp. and *Selliporella donzelli.*
4. The boundary between the Bathonian and Callovian could be either at the last questionable occurrence of the foraminifer *Alzonella cuvillieri* or near the base of the *Trocholina palastiniensis* zone.
5. Top Callovian (i.e., top of the Middle Jurassic) is placed coincident with the top of the grainstones that contain the last occurrence of *Kilianina blancheti* in association with *Kurnubia palastiniensis*, *Trocholina palastiniensis*, *Trocholina elongata,* and *Praekurnubia crusei.*
6. As far as the dasyclad algae are concerned, although *Holosporella siamensis* has a stratigraphic range (Bajocian to Callovian) that fits with the worldwide range, the occurrence of *Coniporella clavaeformis* (Callovian, Fig. 3) is younger than previously recorded (middle Bathonian).

This study confirms the stratigraphic range of most of the foraminifera (Septfontaine et al., 1991) and algae (Granier and Deloffre, 1993) of the South Tethyan realm.

The Middle Jurassic consists of a generally shallowing-upward succession of multiple meter-scale fifth-order cycles.

Subtidal cycles commonly terminated by thick massive oncoidal and peloidal packstones and grainstones are dominant in the Bajocian. Most Bathonian and Callovian cycles start with spicular wackestones and end with cross-bedded peloidal and oolitic grainstones and/or stromatoporoid and coral rudstones. The lags are more common at the top of the Middle Jurassic, where the section becomes condensed.

A total of 13 sequences were defined in the Middle Jurassic: eight in the Bajocian, three in the Bathonian, and two in the Callovian.

Silicification (replacement by beekite) of the stromatoporoids, corals, and other bioclasts is common throughout the Bathonian, Callovian, and Upper Jurassic. In the Musandam Peninsula, the existence of silicification can be used as a guide to distinguish the Middle and Upper Jurassic from the Lower Jurassic.

The type section for the redefined Izhara Formation is the section between top Marrat and top of former Izhara Formation in well Jy-7. The age of the Izhara Formation is Bajocian, supported by palynomorphs and foraminifera. The Izhara Formation comprises three members. The Upper Member corresponds to the former definition of the Izhara Formation (Sugden and Strandring, 1975). In the subsurface of Abu Dhabi, the Izhara formation typically lies unconformably over the Marrat Formation. The lower contact, however, is conformable in some parts of eastern onshore Abu Dhabi where lowstand deposits exist at the top of the Marrat. Above the Qatar Arch, the Izhara Formation directly overlies the Triassic. The upper contact of the Izhara with the Araej Formation is conformable. In Abu Dhabi, there is a westward thinning of the Izhara as a consequence of the onlap of the Lower Izhara Member over the Marrat, plus the gradual thinning of the Middle and Upper Izhara Members to the west.

## ACKNOWLEDGMENTS

This paper resulted from the principal author's doctoral work at the University of Aberdeen. This research project received joint financial support from PARTEX SERVICES CORPORATION and ADCO, to whom we wish to express our gratitude. ADNOC and ADMA are also thanked for allowing access to study seismic lines, cores, and additional lithological well data of offshore Abu Dhabi. We are most grateful to R. Hulstrand and Maria J. E. de Matos for their assistance during the field work. We are also very grateful to M. Septfontaine for his assistance in the identification of the foraminifera and M. Conrad and B. Granier for their help with the identification of the dasycladaceans. Thanks are also due to Justino Carvalho for his assistance in the preparation of the figures.

## REFERENCES

AL-SAAD, H., HEWAIDY, A.G., AND KENDALL, C.G.ST.C., 1992, Lithofacies and depositional setting of the Middle Jurassic Araej Formation in the Dukhan oil field, western Qatar, Arabian Gulf: Neues Jahrbuch für Geologie und Paläontologie, Monatshefte, v. 10, p. 614–630.

ALSHARHAN, A.S., AND KENDALL, C.G.ST.C., 1986, Precambrian to Jurassic rocks of Arabian Gulf and adjacent areas: Their facies, depositional setting and hydrocarbon habitat: American Association of Petroleum Geologists, Bulletin, v. 70, p. 977–1002.

ALSHARHAN, A.S., AND MAGARA, K., 1994, The Jurassic of the Arabian Gulf Basin: facies, depositional setting and hydrocarbon habitat, *in* Embry, A.F., Beauchamp, B., and Glass, D.J., eds., Pangea; Global Environments and Resources: Canadian Society of Petroleum Geologists, Memoir 17, p. 397–412.

ALSHARHAN, A.S., AND MAGARA, K., 1995, Nature and distribution of porosity and permeability in Jurassic carbonate reservoirs of the Arabian Gulf Basin: Facies, v. 32, p. 237–254.

ALSHARHAN, A.S., AND NAIRN, A.E.M., 1994, Geology and hydrocarbon habitat in the Arabian Basin: The Mesozoic of the State of Qatar: Geologie en Mijnbouw, v. 72, p. 265–294.

ALSHARHAN, A.S., AND WHITTLE, G.L., 1995a, Chronostratigraphy and hydrocarbon habitat associated with the Jurassic carbonates of Abu Dhabi, United Arab Emirates (abstract): American Association of Petroleum Geologists, International Conference, Nice.

ALSHARHAN, A.S., AND WHITTLE, G.L., 1995b, Sedimentary-diagenetic interpretation and reservoir characteristics of the Middle Jurassic (Araej Formation) in the Southern Arabian Gulf: Marine and Petroleum Geology, v. 12, p. 615–628.

ATHERSUCH, J., BANNER, F.T., AND SIMMONS, M.D., 1992, On *Trochamijiella gollesstanehi*, n. gen. n. sp (Foraminiferida, Loftusiacea), an index for the Middle Eastern Marine Late Bathonian: Journal of Micropalaeontology, v. 11, p. 7–12.

BANNER, F.T., SIMMONS, M.D., AND WHITTAKER, J.E., 1991, The Mesozoic Chrysalidinidae (Foraminifera, Textulariacea) of the Middle East: the Redmond taxa and their relatives: British Museum Natural History, Bulletin (Geology), v. 47, p. 101–152.

BOSELLINI, A., RUSSO, A., FANTOZZI, P.L., ASSEFA, G., AND SOLOMON, T., 1997, The Mesozoic succession of the Mekele Outlier (Tigre Province, Ethiopia): Memorie di Scienze Geologiche, v. 49, p. 95–116.

DE MATOS, J.E., 1997, Stratigraphy, sedimentation and oil potential of the Lower Jurassic to Kimmeridgian of the United Arab Emirates; outcrop and subsurface compared: Ph.D. thesis, University of Aberdeen, 469 p.

DE MATOS, J.E., AND HULSTRAND, R.F., 1994, Regional characteristics and depositional sequences of the Oxfordian and Kimmeridgian, Abu Dhabi, *in* Al-Husseini, M.I., ed., GEO'94: The Middle East Petroleum Geosciences: Gulf PetroLink, v. 1, p. 346–356.

DE MATOS, J.E., HULSTRAND, R.F., AND WALKDEN, G.M., 1994, The biostratigraphy of the Lower Jurassic of the U.A.E.: outcrop and subsurface compared: 6th Abu Dhabi International Petroleum Exhibition and Conference, ADSPE 51, Proceedings, p. 449–459.

DE MATOS, J.E., AND WALKDEN, G.M., 1995, Biostratigraphy and sequence stratigraphy of a Lower Jurassic shallow carbonate Platform, Musandam Peninsula (abstract): International Meeting "Fauna, Flora and Sequence Stratigraphy", Paris, December 14–15.

DROSTE, H., 1990, Depositional cycles and source rock development in an epeiric intraplatform basin: The Hanifa Formation of the Arabian Peninsula: Sedimentary Geology, v. 69, p. 281–296.

DROSTE, H., 1993, Source rock development and relative sealevel changes in an intraplatform basin: The Upper Jurassic Hanifa Formation of the Arabian Peninsula (abstract): American Association of Petroleum Geologists, Annual Convention, New Orleans, April 1993.

ENAY, R., LE NINDRE, Y.M., MANGOLD, C., MANIVIT, J., AND VASLET, D., 1987, Le Jurassique d'Arabie Saoudite Centrale: Nouvelles données sur la lithostratigraphie, les paléoenvironments, les faunes d'ammonites, les ages et les correlations, *in* Enay, R., ed., Le Jurassique d'Arabie Saoudite Centrale: Géobios, Mémoire Spécial, v. 9, p. 13–65.

GLENNIE, K.W., 1992, Plate tectonics and the Oman Mountains: Tribulus, v. 2(2), p. 11–21.

GLENNIE, K.W., 1995, The geology of the Oman Mountains: an outline of their origin: Beaconsfield, U.K., Scientific Press, 92 p.

GLENNIE, K.W., BOEUF, M.G.A., HUGHES CLARKE, M.W., MOODY-STUART, M., PILAAR, W.F.H., AND REINHARDT, B.M., 1974, Geology of the Oman Mountains: Koninklijk Nederlands Geologisch Mijnbouwkundig Genootschap, Verhandelingen, v. 31, 423 p.

GOLLESSTANEH, A., 1965, A micropaleontological study of the Upper Jurassic and Lower Cretaceous of Southern Iran: University of London, Ph.D. thesis (unpublished).

GRANIER, B., AND DELOFFRE, R., 1993, Inventaire critique des Algues Dasycladales Fossiles II Partie — Les Algues Dasycladales du Jurassique et du Cretacé: Revue de Paléobiologie, v. 12, p. 19–65.

HARDENBOL, J., THIERRY, J., FARLEY, M.B., JACQUIN, TH., DE GRACIANSKY, P.-C., AND VAIL, P.R., 1998, Mesozoic and Cenozoic sequence chronostratigraphic framework of European basins, *in* de Graciansky, P.-C., Hardenbol, J., Jacquin, Th., and Vail, P.R., eds., Mesozoic and Cenozoic Sequence Stratigraphy of European Basins: SEPM, Special Publication 60, p. 3–13.

HASSAN, T.H., 1989, The Lower and Middle Jurassic in Offshore Abu Dhabi: Stratigraphy and hydrocarbon occurrence: 6th Middle East Oil Show, Society of Petroleum Engineers, SPE # 18011, Bahrain, p. 847–858.

HUDSON, R.G.S., AND CHATTON, M., 1959, The Musandam Limestone (Jurassic to Lower Cretaceous) of Oman, Arabia: Museum National d'Histoire Naturelle, Paris, Notes et Mémoirs, Moyen-orient v. 7, p. 69–93.

HUGHES CLARKE, M.W., 1988, Stratigraphy and rock unit nomenclature in the oilproducing area of Interior Oman: Journal of Petroleum Gcology, v. 11, p. 5–60.

KAZANCI, N., AND VAROL, B., 1993, The occurrence and significance of beekite in Palaeocene alluvial-fan deposits in Central Anatolia, Turkey: Terra Nova, v. 5, p. 36–39.

KOEPNICK, R.B., 1993, Reservoir architecture in Late Jurassic outerramp carbonates, Hadriya and Hanifa Formations, Saudi Arabia: Abstracts of invited lecture given for Society of the Explorationists of the Emirates (SEE) at Abu Dhabi, U.A.E.

LE NINDRE, Y.M., MANIVIT, J., AND VASLET, D., 1990, Géodynamique et paléogéographie de la plate-forme Arabe du Permien au Jurassique: histoire géologique de la bordure occidentale de la plate-forme Arabe, 2, "France", Bureau de Recherches Géologiques et Minière, Document 192, 287 p.

LEWY, Z., 1981, A Late Bathonian–Late Callovian unconformity in the Middle East: Newsletter on Stratigraphy, v. 10, p. 27–33.

MANIVIT, J., LE NINDRE, Y.M., AND VASLET, D., 1990, Le Jurassique d'Arabie Centrale: histoire géologique de la bordure occidentale de la plateforme Arabe, 4, "France", Bureau de Recherches Géologiques et Minière, Document 194, 559 p.

MCGUIRE, M.D., KOEPNICK, R.B., MARKELLO, J.R., STOCKTON, M.L., WAITE, L.E., KOMPANIK, G.S., AL-SHAMMERY, M.J., AND AL-AMONDI, M.O., 1993, Importance of sequence stratigraphic concepts in development of reservoir architecture in Upper Jurassic Grainstones, Hadriya and Hanifa reservoirs, Saudi Arabia: 6th Middle East Oil Show, Society of Petroleum Engineers, SPE # 25578, Bahrain, p. 489–500.

MOSHRIF, M.A., 1987, Sedimentary history and paleogeography of Lower and Middle Jurassic Rocks, Central Saudi Arabia: Journal of Petroleum Geology, v. 10, p. 335–350.

MURRIS, R.J., 1980, Middle East stratigraphy evolution and oil habitat: American Association of Petroleum Geologists, Bulletin, v. 64, p. 597–618.

POWER, R.W., 1968, Saudi Arabia: Lexique Stratigraphique International; Asie: Centre National de Recherches Scientifiques, Paris, v. III, fasc. 10bl, 177 p.

ROBERTSON, A.H.F., SEARLE, M.P., AND RIES, A.C., eds., 1990, The Geology and Tectonics of the Oman Region: Geological Society of London, Special Publication 49, 845 p.

SEPTFONTAINE, M., ARNAUD-VANNEAU, A., BASSOULLET, J.-P., GUSIC, Y., RAMALHO, M., AND VELIC, I., 1991, Les foraminifères imperforés des plate-formes carbonatées Jurassiques: état des connaissances et perspectives d'avenir: Société Vaudoise des Sciences Naturelles, Bulletin, v. 80, p. 255–277.

SEPTFONTAINE, M., AND DE MATOS, J.E., 1998, *Pseudodictyopsella jurassica* n. gen., n. sp., a new foraminifera from the Early Dogger of the Musandam Peninsula, N-Oman Mountains, and its sedimentological and stratigraphical context: Revue de Micropaléontologie, v. 41, p. 71–87.

STEELE, R.P., ET AL., 1983, Biostratigraphical, sedimentological and geochemical study of the PreCretaceous section in ADCO well Bu Hasa 175 and a comparison with Jarn Yaphour 7 and Menderl: ADCO in-house report.

SUGDEN, W., AND STRANDRING, A.J., 1975, Qatar Peninsula, Lexique Stratigraphique International: Centre National de la Recherche Scientifique, Vol. III, fasc. 10b3, Paris, 120 p.

TOLAND, C., 1994, Late Mesozoic Stromatoporoids: Their use as stratigraphic tools and paleoenvironmental indicators, *in* Simmons, M.D., ed., Micropaleontology and Hydrocarbon Exploration in the Middle East: London, Chapman & Hall, p. 113–126.

TOLAND, C., PEEBLES, R.G., AND WALKDEN, G.M., 1993, Upper Jurassic and basal Cretaceous outcrop sequence stratigraphy of Wadi Hagil, Ras Al Khaimah: 6th Middle East Oil Show, Society of Petroleum Engineers, SPE # 25581, Bahrain, p. 533–543.

WOOD, R., 1987, Biology and revised systematics of some Late Mesozoic stromatoporoids: The Paleontological Association, London, Special Papers in Paleontology, no. 37, 89 p.

# "TUNING" HIGH-FREQUENCY CYCLIC CARBONATE PLATFORM SUCCESSIONS USING OMISSION SURFACES: LOWER JURASSIC OF THE U.A.E. AND OMAN

GORDON M. WALKDEN

*Department of Geology and Petroleum Geology, University of Aberdeen, Meston Bldg, King's College, Old Aberdeen AB24 3UE, U.K.*

AND

JOSE DE MATOS

*Petroprimo, Av. da Republica, 50-4°, 1050 Lisbon, Portugal*

ABSTRACT: Meter-scale cyclicity is developed on a number of Lower Jurassic peri-Tethyan carbonate platforms, and comparable cyclicity is known from basinal successions in parts of Europe. Analysis of the 234-m-thick Hettangian to Early Pliensbachian section of a recently constructed reference section through inner- to mid-shelf Jurassic carbonates in Wadi Naqab, U.A.E./Oman border, shows that facies patterns alone give an inaccurate and even misleading record of this cyclicity. Omission surfaces, however, can provide far more reliable evidence for sea-level change and enable the "tuning" of cyclicity data derived from facies studies. Clear distinctions need to be made, in the field, between a number of different marine and subaerial omission surfaces, including firmgrounds, hardgrounds, karsts, corrosion surfaces, and paleosols. Some of these surfaces display similar or ambiguous characteristics in the field, and difficulties can be compounded by the re-texturing and remodeling effects of burial diagenesis. High-resolution data, incorporating systematically observed and logged omission surfaces, have potentially high value in correlation and greatly improve the reliability of interpretations. In the Wadi Naqab succession the incorporation of carefully observed cycle-boundary data more than doubles the number of cycles seen. The data show that average duration of the high-frequency cycles was ca. 90,000 years, well within the range of Milankovitch forcing factors but outside the 40,000 years obtained from basinal Milankovitch successions of the same age in Europe. The difference reflects the incompleteness of the shallow-water succession, and it is likely that both types of cyclic successions in the Lower Jurassic share the same climatic and/or eustatic causes. Fischer-plot analysis of the 149 cycles reveals the additional operation of two strong third-order sea-level fluctuations, and possibly two further smaller ones, between late Hettangian and early Pliensbachian. Subaerial omission surfaces are best developed and most numerous on the falling limbs of these, where cycles are thin. These points would correspond to the falling stage of third-order sea-level change when there was limited creation of new accommodation and maximum subaerial exposure. Omission surface data therefore support the case for both fifth-order and third-order sea-level change in the Jurassic. A particular characteristic of this and other shallow-water Milankovitch successions is their thin-bedded nature. This is likely to impart a strong flow anisotropy in reservoir settings, enhanced by stratiform weathering and diagenetic features, but many of the definitive features are likely to be missed in seismic sections and well logs. Actual omission surfaces can be hard to detect even in core.

## INTRODUCTION

Cyclicity within the range of fourth- to sixth-order frequency (10–100 thousand years) is a mark of Jurassic sedimentary patterns world-wide (e.g., Valdes et al., 1995; House, 1985). In the Lower Jurassic such cyclicity is noted in both basinal successions (House, 1985) and shallow carbonate platform successions (Crevello, 1991; Bassoulet and Bergougnan, 1981; de Matos et al., 1994). It has been attributed to orbital forcing of climate (e.g., Crevello, 1991; Weedon, 1993), although tectonic causes have also been invoked (Moses, 1995). In the peri-Tethyan carbonate successions of Morocco and the Arabian Peninsula the key evidence for a causative mechanism is the meter scale of the cycles, their widespread development, and the presence of subaerial discontinuity surfaces (including karsts, calcrete, and soils) separating normal marine facies. Eustatic change of at least a few meters is, thus, the prime candidate (de Matos et al., 1994), providing a causative link with the very similar but much better documented eustatic cyclicity of the Middle to Late Triassic in the same province (e.g., Goldhammer et al., 1987, 1990; Fischer and Bottjer, 1991).

Cyclic changes with a 10–100 thousand years periodicity that are synchronous and worldwide provide exciting potential for high-resolution correlation, not only locally but globally, and not only between platforms but also between these and their basinal equivalents. This would be particularly valuable in shallow carbonate successions in which some of the best biostratigraphic markers, such as ammonoids, are persistently absent. The likelihood that the great majority of cycle boundaries in high-frequency cyclic successions might represent precise, unique, and potentially correlatable ticks of the great geological clock has ensured that such successions throughout the geological column have received much attention in the last decade. Despite this, there remain many problems in the use of high-frequency cycles in correlation (e.g., Cotillon, 1995). These include:

- separating eustatic and tectonic effects;
- recognizing gaps in the shallow-water record caused by prolonged sea-level falls;
- homogenization of the record in deep-water sequences;
- no systematic means of reliably calculating cycle magnitude or period;
- cycles may be incomplete or hard to distinguish.

The purpose of this paper is to address the last of these problems, that of improving discrimination within cyclic sequences. In many shallow-water cyclic successions the cycles (more appropriately, but seldom, termed cyclothems; e.g., Walkden, 1987) are commonly incomplete, lacking crucial evidence of a succession of depth-related facies. In some cases very similar facies, but representing separate cycles, are built directly on top of one another (cycle amalgamation). Reliable discrimination thus breaks down, reducing potential for correlation and for identifying causative mechanisms. This is particularly inhibiting where statistical techniques are being attempted, for example power spectral analysis in cases where astronomical forcing

Middle East Models of Jurassic/Cretaceous Carbonate Systems
SEPM Special Publication No. 69, Copyright © 2000
SEPM (Society for Sedimentary Geology), ISBN 1-56576-075-1, p. 37–52.

factors may be involved (e.g., Weedon, 1993). In sequence stratigraphic terms, the cycles produced by fifth-order cycles of sea-level change could be regarded as parasequences, but it takes confidence to designate a thin cycle, perhaps even comprising a single bed, as a parasequence.

Our study is based in the Liassic of the Musandam Peninsula (U.A.E./ Oman, southern Arabian Gulf) (Fig. 1). Here we have established that careful study of omission surfaces is an independent tool in defining cycles, determining the nature of early Jurassic sea-level change, recognizing parasequence sets, and identifying lower-order (longer-term) patterns of relative sea-level change. This use of omission surfaces depends upon reliably distinguishing between a number of depositional, erosional, and diagenetic effects at bedding discontinuities that can create similar and sometimes easily confused phenomena (Fig. 2). Commonly these effects are difficult to resolve even under ideal outcrop conditions, and their recognition in core is particularly difficult.

## THE LOWER JURASSIC OF THE MUSANDAM PENINSULA

The Jurassic of the Musandam Peninsula has been studied in detail in Wadi Naqab, ca. 16 km SE of Ras Al Kaimah, U.A.E. (Fig. 1). The first reference section through the Jurassic in the region has recently been established here, comprising 1310 m of inner- to mid-shelf carbonates (de Matos, 1997; de Matos and Walkden, this volume), of which the Lower Jurassic is at least 470 m thick. This is built of meter-scale peritidal cycles, and we have logged in particular detail the first 234 m, up to a prominent and widespread lithological marker, a 10 m thick white micritic peritidal laminite. The Lower Jurassic here is poorly constrained biostratigraphically, especially towards the base, where the macrofauna is sparse and microfauna and flora lack diversity. The Hettangian definition is unsatisfactory, and although we have tried to constrain it paleontologically no forms were found that reliably prove its extent. The Triassic–Jurassic boundary is taken at a condensed horizon comprising a crinoidal phosphatic grainstone with an identifiable vertebrate and invertebrate fauna (de Matos et al., 1994; de Matos, 1997). The Sinemurian is more reliably constrained on the basis of foraminifera and algae, and the "white band" (peritidal micrite) described above is mid-Pliensbachian (de Matos, 1997).

Carbonate sediments in the Lower Jurassic of Wadi Naqab have been assigned to 23 microfacies types, which are described in detail in de Matos (1997). Facies recognition is informed by a combination of criteria, including grain types, grain size and texture, rock texture, rock color, biota (microfossils, macrofossils and algae), and sedimentary structures. The contribution from biota is variable, reflecting the patchy distribution of reliable faunas, but in the higher parts of the logged section foraminifera, bivalves, gastropods, sponges, echinoderms, ostracods, and dasycladaceans are increasingly common. The first identifiable corals appear in the Pliensbachian but ammonites remain absent.

For the purposes of cycle recognition, microfacies can be grouped into three marine facies associations, viz: low-energy subtidal, high-energy subtidal/shoal, and intertidal. In addition, a number of supratidal subfacies and phenomena have been recognized. These are discussed in detail below. Commonly occurring and characteristic representatives of the three facies associations are: nodular clay/carbonate pseudobreccia and spicular lime mudstones (low-energy subtidal); oncoidal wackestones, *Lithiotis*-rich (bivalve) foraminiferal packstones, *Orbitopsella* (foraminifera) wackestones and oolitic packstones to grainstones (high-energy subtidal); stromatolitic and cryptalgal laminites, birdseye mudstones to packstones and ostracod-rich grainstones to mudstones (mainly intertidal). The supratidal environment is characterized by karst, rubbly soils, rhizolith horizons, collapse breccias, and rare calcretes. In the Wadi Naqab succession the majority of carbonates relate to low intertidal to shallow subtidal environment, with high intertidal algal mat and low-energy subtidal facies less well represented.

## CYCLIC PATTERNS AND CYCLE BOUNDARIES

Asymmetry of facies stacking patterns, with a dominant shallowing-up regressive trend, is normal in meter-scale to decameter-scale cycles in the geological record (e.g. Goodwin and Anderson, 1985; Walkden and Walkden, 1990). Many authors prefer to attribute such facies asymmetry to a rapid generative sea-level rise that outpaces sedimentation, although others consider that carbonate "startup" is slow, leading to a distinctive transgressive lag (Ginsburg 1971; Read et al., 1986; Goldhammer et al., 1987; Walkden and Walkden, 1990) (Fig. 3A). No doubt both are important, but another possible cause of asymmetry is storm reworking. Probability ensures that the earliest deposited facies of any developing cycle are automatically at greatest risk of removal or severe modification because they are in place for the greatest period of time. Most inner-shelf to mid-shelf sites are probably most prone to storm reworking at the deepest, most open-water phase, so that replacement facies are more likely to be regressive than transgressive. The effects of such reworking upon a hypothetical cycle are shown in Figure 3B. Early lithification through subaberial exposure might further reinforce such a bias by protecting regressive facies from erosion during the next transgressive phase.

Facies patterns in shallow-water carbonate settings are therefore unlikely to deliver a full record of the sea-level changes that have built them. Similar facies deposited directly above one another, but from differing cycles, might be taken to represent a single cycle, whilst a couplet of regressive facies, separated only by partial reworking through the effects of a single storm, might be taken as separate cycles. In Wadi Naqab only three cycles out of 149 have been found to contain representatives of all three depth-defined carbonate facies associations plus a subaerial cap. Distinction between one cycle and the next based only upon facies stacking pattern is therefore hazardous. Cycles are easily misidentified at the outset, cycle numbers and thicknesses will consequently be wrong, and overall cyclic stacking patterns will be misconstrued. All these are crucial in the application of high-frequency cyclic successions to correlation and in their interpretation in terms of autocyclic or allocyclic processes. In such incomplete successions proper study of omission surfaces, starting with the accurate identification and logging of these in the field, provides an invaluable aid to cycle definition.

## OMISSION SURFACES AS CYCLE BOUNDARY MARKERS

An omission surface is any sedimentary interface (discontinuity; diastem; time-elapsed surface) where there is evidence for the effects of one or more nondepositional processes that may account for a substantial period of time. In shallow marine environments a very high proportion of cycle time may be represented by omission surfaces (Koerschner and Read, 1989; Wilkinson et al., 1991; Sadler, 1994), so that features they display may reflect conditions representative of a very significant but otherwise unrecorded part of geological history (e.g., Figs. 3A, 3B). They can also reveal important geochemical, sedimentological, biological, or climatic information. Examples are subaerial dissolution surfaces (including paleokarsts, pebbly soils, and intertidal corrosion zones), diagenetically modified surfaces (including dolomite caps, calcretes, hardgrounds, and mineralized crusts), or-

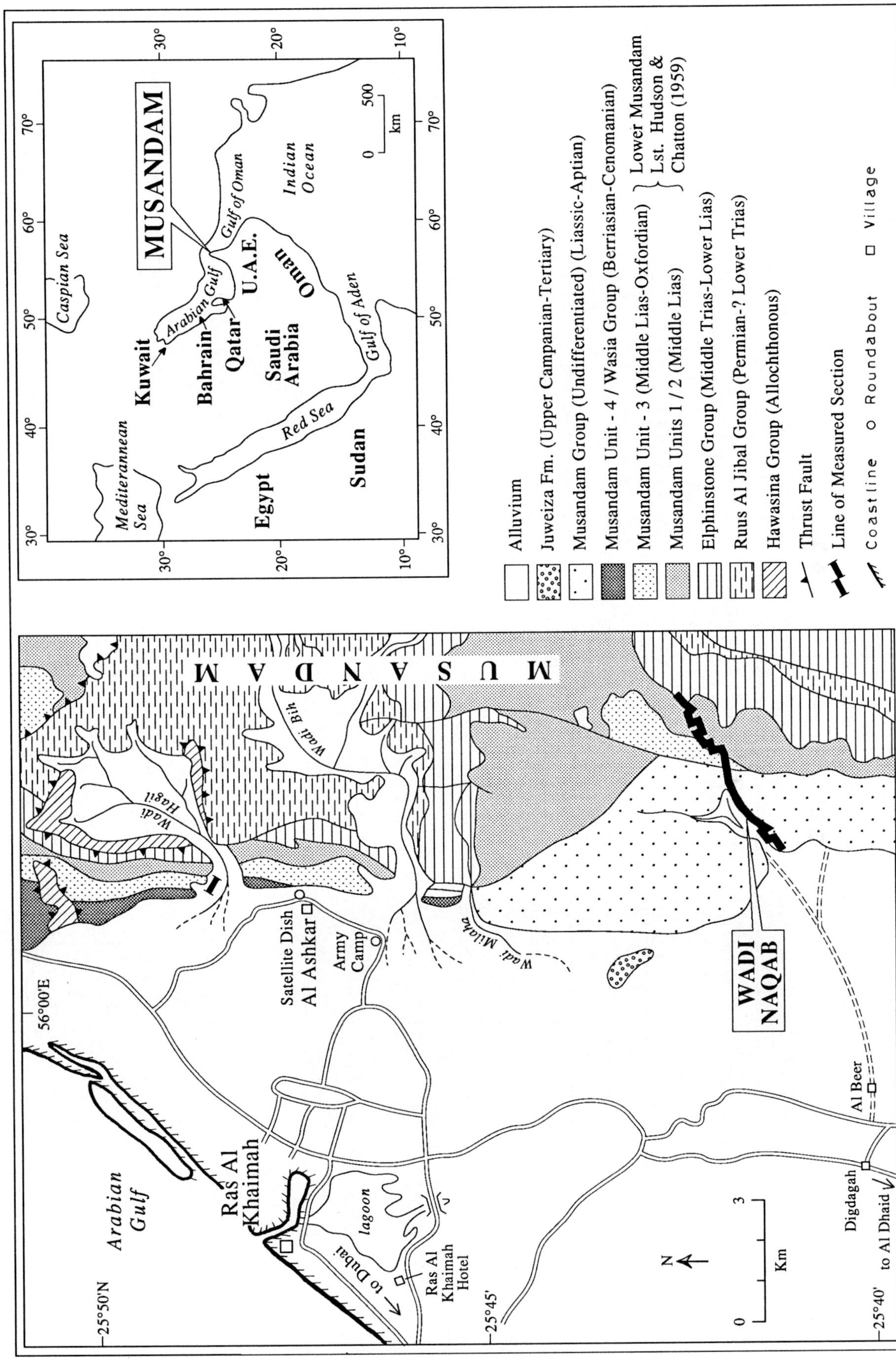


FIG. 1.—Geographical location of Wadi Naqab, SW Musandam Peninsula (U.A.E.) with schematic geological map.

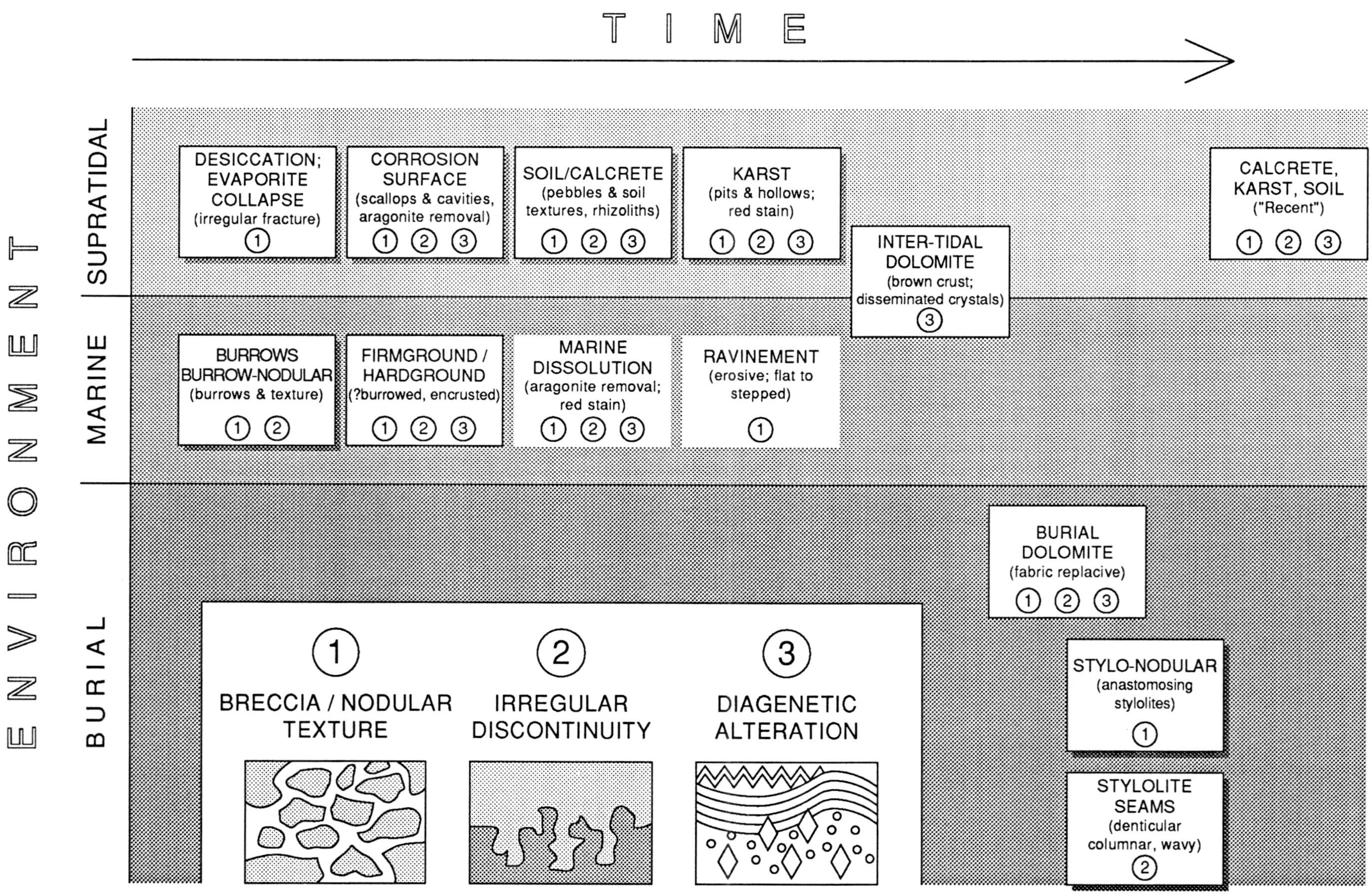


FIG. 2.—Summary of the main categories of omission surface in carbonate environments and their associated main physical/diagenetic characteristics. The depth of frame (three types, strong to none) indicates the relative importance of each type of omission surface in the Lower Jurassic reference section. The numbers relate to the inset boxes. These show three general textural categories of omission surface, emphasizing the ambiguity that can arise through the effects of different geological processes. The horizontal timeframe is approximate.

ganically colonized surfaces (including rhizolith mats, bored or encrusted surfaces, and burrowed firmgrounds), and submarine erosion or nondeposition surfaces (including lag surfaces, ravinements, and sea-floor corrosion surfaces). In theory, any bedding plane is a candidate omission surface, but evidence of at least one of the above effects is needed.

Any one of the surface types listed might mark a cycle boundary, but some surfaces can be formed at any point within a cycle and so provide equivocal evidence. For example, a bored hardground, a marine dissolution surface, or a glauconite and/or phosphate crust could form at or around a maximum flooding surface, but a karst and paleosol can form only during emergence. Some omission surfaces may have a compound history, such as a hardground exhumed by marine reworking and subsequently subaerially exposed (e.g., Immenhauser et al., this volume). Furthermore, the effects of diagenesis may exaggerate features (a stylolitized hardground might be impossible to distinguish from a karst) or may destroy distinguishing textures (a stylobreccia might look like a fossil soil). Many omission surfaces have probably been subjected to more than one type of contemporaneous modification, and most are affected to a greater or lesser extent by shallow to deep diagenesis during burial.

The Wadi Naqab succession presents an ideal case study for developing "ground rules" for recognizing the varied characteristics of omission surfaces and in particular for identifying omission surfaces that are also cycle boundaries. The succession is extensive, the cycles are well exposed, and the inner- to mid-shelf setting ensured that only relatively small sea-level changes were sufficient to exert a substantial effect on depositional environment, in most cases leading to emergence of the shelf. Set against this advantageous shallow shelf setting is the near certainty that superimposed longer-term sea-level changes, such as third-order sea-level lows, could have caused nondepositional gaps in the record with the loss of cycles or groups of cycles. Such lost cycles (e.g., Fig. 3A) might be picked up downdip, upon as yet unrecognized more distal parts of the shelf, or they may be present only in the basin beyond. Nonetheless, closer attention to omission surfaces in the delineation of cycles inevitably leads to improved data and the refinement of our knowledge of the overall cycle stacking pattern. A more accurate picture of the distribution of subaerial omission surfaces could reveal the location of longer-term sea-level lows through the vertical clustering and/or relative maturity of subaerial surfaces.

### *Characteristics of Subaerial Omission Surfaces*

Subaerial effects in carbonate depositional environments are reviewed in Esteban and Klappa (1983) and Demicco and Hardie (1994), and good case studies are in Wright (1986) and James and Choquette (1988). Reviews of the sequence stratigraphic context of emergent surfaces are in Tucker (1993), Loucks and Sarg (1993), Read and Horbury (1993), and Budd et al. (1995). Three principal types of emergent surface have been recognized in the Lower Jurassic of the Musandam Peninsula, viz.: paleokarsts, corrosion surfaces, and paleosols. The features of these surfaces are intergradational, and the distinction especially between karsts and corrosion surfaces is probably genetically artificial. Calcrete textures have been noted, although surfaces extensively altered to calcrete are absent. Surfaces commonly display red staining through precipitation of iron oxide, probably at a soil/rock interface. Geopetal silts and contemporaneous speleothem-like calcite cements also appear directly beneath some surfaces but are not easily identified in the field. Rhizoliths are commonly associated with each of these types of surface. Eolian facies have not been recognized.

*Paleokarsts.—*

These are here recognized as small-scale dissolution surfaces with rolling, rounded, hummocky, or pitted surfaces that are essentially smooth rather than fretted. Their appearance is very similar to paleokarsts described by Walkden (1974), and surface relief up to 0.5 m has been observed. These surfaces clearly incise underlying sediments, and they are normally filled by horizontally bedded sediments that onlap the karstic irregularities (Fig. 4). Associated irregular cracks or fissures are also filled with carbonate (Fig. 5). Paleokarsts of this type probably formed in porous carbonates beneath a thin layer of mineral or organic soil, locally assisted by plants (e.g., Herwitz, 1993). If so, any direct evidence of soil or substantial plant material is usually lost.

*Corrosion Surfaces.—*

The distinguishing features of corrosion surfaces are a scalloped or fretted surface morphology, usually underlain by sediment-filled cavities. Corrosion surfaces sometimes underlie a pebbly paleosol (see below), but they can also be buried by carbonate mud and show up as a result of contrasting color or lithology. In either case randomly shaped dissolution pipes and cavities may extend well beneath the corrosion surface, commonly opening into geopetally filled microcaves up to tens of centimeters across (Fig. 6). Key evidence of a dissolutional origin (as opposed to a burrowed origin) is that sediment removal was mineral-selective and texture-selective, leaving, for example, sediment-filled mollusc molds (Fig. 7) or algal thalli. More resistant material such as oncoids and large forams survive, although the sediment–fossil interface has become the focus of dissolution, with the gap later filled by infiltered mud. Demicco and Hardie (1994) note that such surfaces could result from bioerosion of rocky surfaces, but our clear evidence for dissolution discounts this. Nonetheless some Lower Jurassic corrosion surfaces can display burrow-like features. In many cases these are probably root casts, but vertical bioturbation (and even desiccation) could well have opened up some sediment surfaces to penetrative dissolution, thereby contributing to the creation of this distinct category of subaerial omission surface.

Corrosion surfaces can be formed in the intertidal region (e.g., Read and Grover, 1977; Esteban and Klappa, 1983), but the broad, shallow, dissolution hollows of Recent intertidal surfaces are in marked contrast to the penetrative and microcave-riddled features noted in the Lower Jurassic of Wadi Naqab. In numerous examples, a corrosion surface has evidently finally evolved, after prolonged dissolution, into a corroded breccia, but diagenetic effects may also contribute to this texture (see below).

*Paleosols.—*

These are structureless and unfossiliferous clay horizons with a gray, ochre, or red color (original color retention is unlikely in this modern arid environment). They may incorporate pebbles (made of recognizable carbonate sediment) or nodules (formed in randomly shaped micritic carbonate). They are seldom more than a few centimeters thick, but thicken appreciably where they overlie and fill a karst or corrosion surface (Fig. 8). X-ray diffractograms reveal poorly ordered mixed-layer clays with dominant smectite/illite and kaolinite.

### *Features Providing Key Confirmation of Emergence*

Subaerial surfaces can appear in unexpected places, capping incomplete cycles or separating otherwise very similar facies. This means that they can provide key but unique evidence of

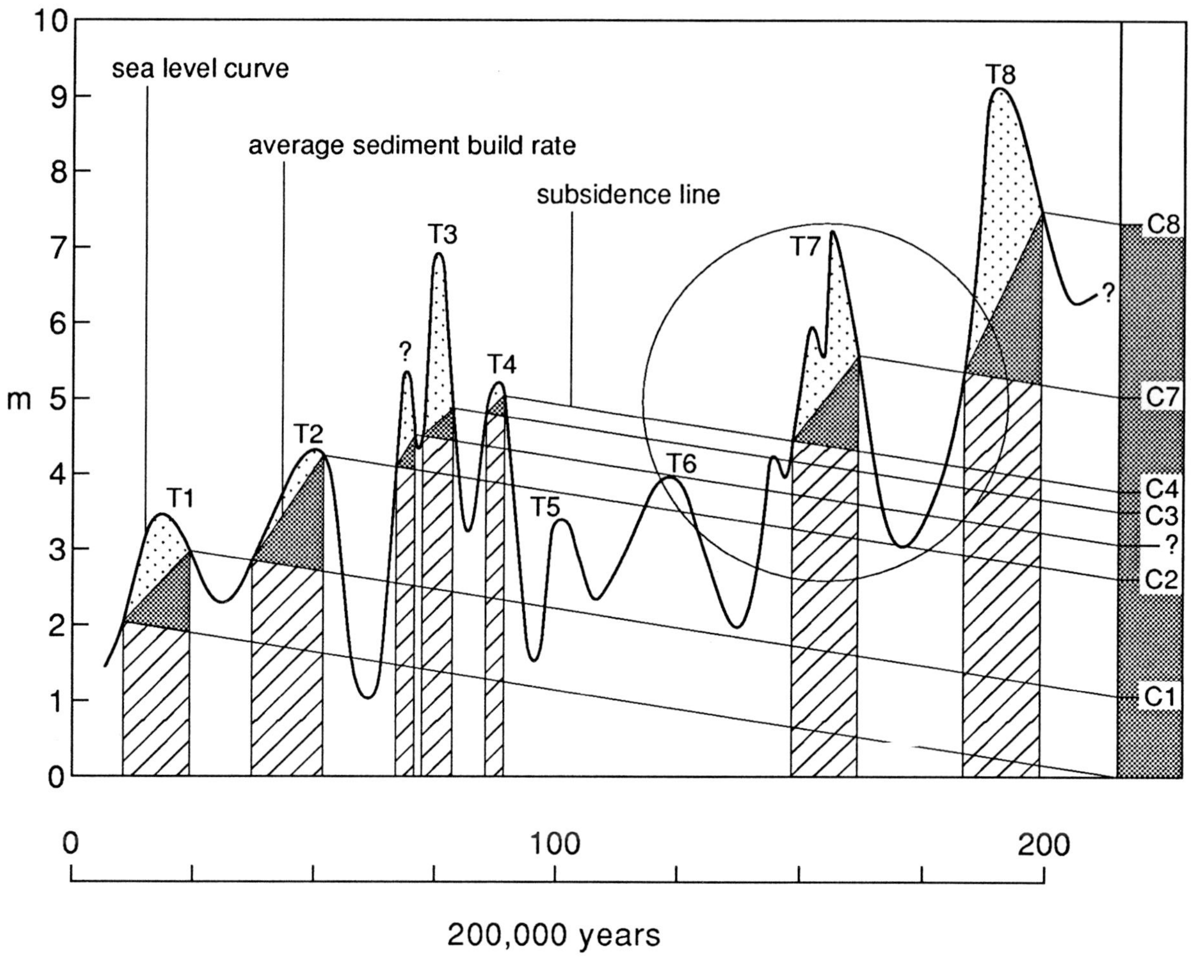


FIG. 3A.—Synthetic fifth-order cycles simulating an inner shelf setting. The sea-level curve (labeled) is deliberately irregular in both timing and magnitude, modeling the combined effects of several Milankovitch forcing factors, but with the strongest signal at ca. 100,000 years. Cyclicity of only a few meters of relative sea-level change leads to short periods of deposition (toned columns) separated by long periods of exposure. The column on the right is the synthetic rock column showing the positions of the resultant omission surfaces. T1–T8 = time points for each fifth-order upwave; C1–C8 = the respective cycle boundaries created by each downwave in the rock column. The tielines between equivalent T and C points show subsidence with time. The average sediment build rate (labeled) varies, and sediments never fill all the accommodation created during the short-lived peaks of sea level (stippled). Upwaves 5 and 6 fail to inundate the shelf in this proximal position; no sediment is deposited, and the respective cycles are consequently missing in the rock record. The question mark indicates uncertainty over what should, and should not, be regarded as a distinct cycle. The ringed cycle is enlarged and explained in more detail in Figure 3B.

sudden and probably very brief sea-level falls. In these instances, where surface geometry is the only criterion available for emergence, their identification in the field needs supporting evidence. In a few cases, probable karst surfaces have had to be ignored because they lack much surface relief and display no other evidence of identity. In other cases breccia horizons have provided ambiguous evidence for soil-forming processes. Associated features such as calcrete textures, rhizoliths, mineral crusts, or early meteoric cements therefore provide vital corroboration.

*Rhizoliths.—*

Many subaerial omission surfaces show locally concentrated festoons of micron- to millimeter-sized, branching and mainly vertically arranged tubules, as well as less common larger (up to 10 mm) meandering vertical tubes. These extend tens of centimeters beneath the surface and are normally obvious, being filled mainly with mottled brown calcite. Where the texture is particularly dense there may be abundant millimeter-scale fracturing of the sediment. In thin section the tubes may be centered within, or contain concentric laminae of brown micrite and may show geopetal blockages of clear equant micrite or silt (vadose silts). Larger tubes have blockages of sediment that commonly show mottled and anastomosing networks of apparently collapsed tubules. Crystal silt, comprising evidently corroded calcite fragments in a matrix of marine carbonate, is also present within some of the larger tubes. The tubules and tubes might be burrows, but the association with karsts and corrosion surfaces, together with crystal silt, brown laminated calcite fills, and fracturing suggests that they are root molds and rhizoliths.

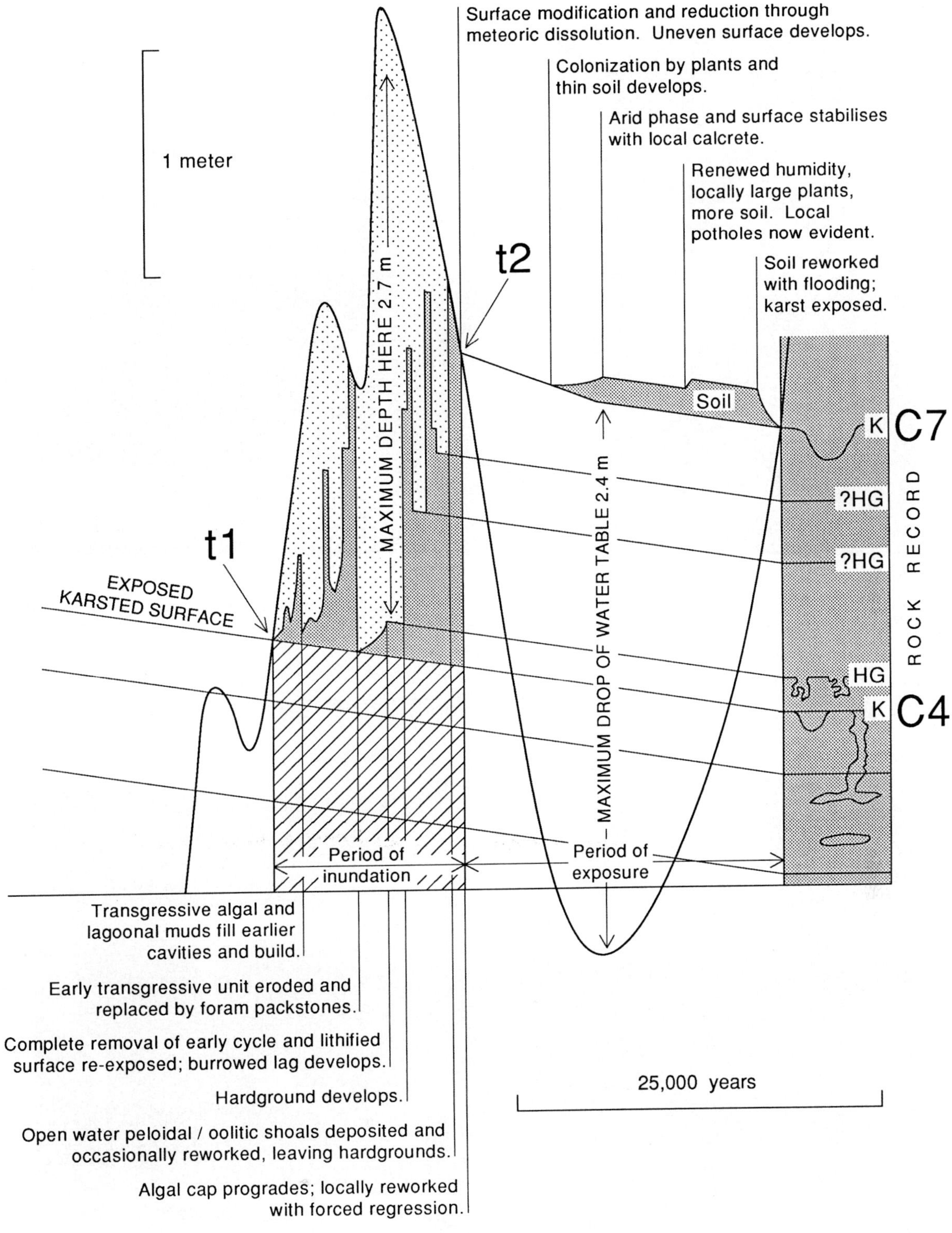


FIG. 3B.—Enlargement of the synthetic fifth-order upwave at T7 in Figure 3A, together with the relevant extract of the rock record. Rising sea-level inundates the shelf at t1 (inundating surface C4, created in Figure 3A), depositing the cycle C7 (recall that surface C4 also encompases cycles C5 and C6, which were not recorded). Falling sea level then reexposes the shelf at t2 and subaerial modification of C7 begins. The inset comments trace a theoretical history of deposition and exposure during cycle T7/C7. This involved repeated deposition and reworking of marine and subaerial sediment and the creation of marine and subaerial omission surfaces. Note how transgressive facies are deposited, but regressive facies are the most likely to be preserved in the rock record. This diagram illustrates a method of displaying and testing detailed analysis and interpretation of intracyclic depositional environments and erosive events.

FIG. 4.—A ca. 250-mm-magnitude paleokarst at cycle 85, near the top of the Sinemurian. The sediment fill is mainly mudstone, showing onlap and compaction effects where it fills an erosional hollow. A hammer to the left (ca. 300 mm long) rests on the karstified surface itself.

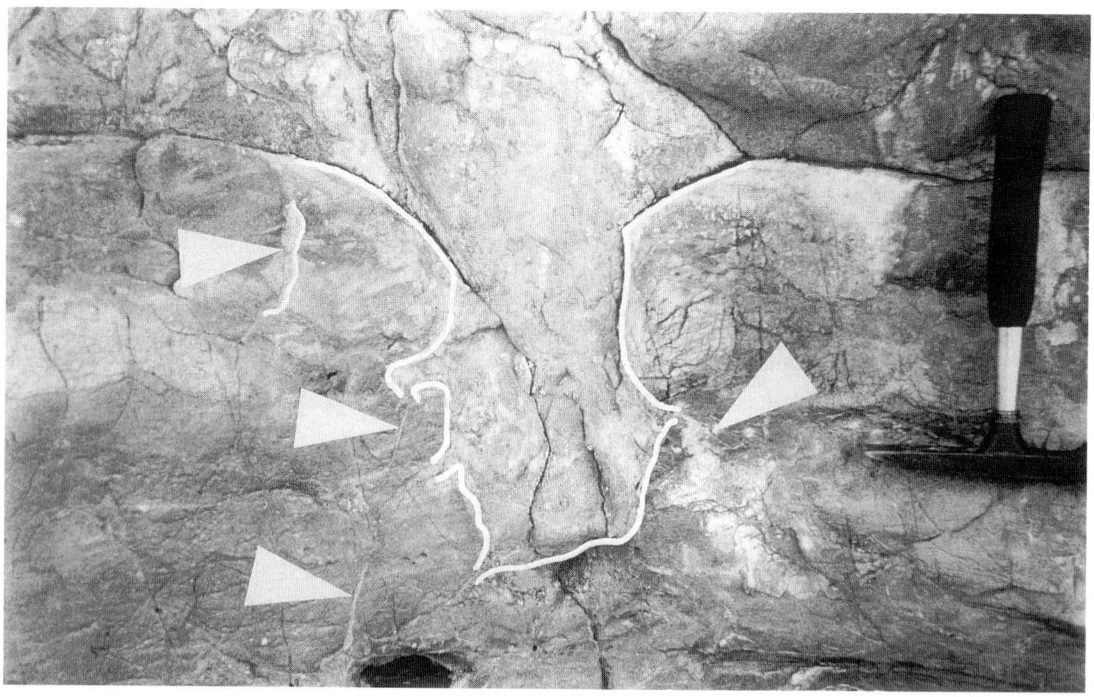

FIG. 5.—A karstic pit at the top of cycle 95, near the Sinemurian–Pliensbachian boundary. Undolomitized gray packstone top-fills a dolomitized karst surface. The interface is marked in white, and an intricate mud-filled crack system extends from the karstic pit into the surrounding dolomite (arrowed). Hammer is ca. 300 mm long.

FIG. 6.—One of several corrosion zones concentrated around cycles 137–144. The rubbly layer is formed in dark gray packstone. This is topped by funnel-shaped holes (A) that extend into cavities and fracture systems with irregular, scalloped margins. Interstices are filled by light gray lithoclastic mud (B). The fills are geopetal, and some of the detail of cavity margins (e.g., at B) is determined by mud that has intruded the molds left by dissolved millimeter-sized algal thalli. A similar phenomenon is seen in Figure 7. Hammer is ca. 300 mm long.

FIG. 7.—A mud-replaced gastropod in the corrosion zone topping cycle 49 (mid-Sinemurian). The gastropod lies exposed and intersected at the top of the omission surface. It has experienced meteoric dissolution of the aragonitic shell followed by intrusion of the mold by soft mud. The same mud also occurs within nearby millimeter-sized molds of algal thalli and extends into underlying centimeter-sized microcaves (not shown). Width of pencil is ca. 8 mm.

FIG. 8.—A prominent pebbly paleosol close to the top of the Hettangian at cycle 9. The clay component of the paleosol is structureless, and pebbles comprise irregular, scalloped lithoclasts up to 50 mm. These are derived from muddy lithologies not necessarily recognizable within the beds immediately above or below. Rarely, such pebbles show calcrete textures.

*Calcrete.—*

On exposed surfaces in arid environments care needs to be taken when identifying calcrete, because it may be Recent. Unequivocally Jurassic calcretes are rare, and a full caliche profile has not been seen. Evidence is confined to patchy brown-stained laminated, brecciated, and tubule-rich alteration fabrics and superficial crusts. These coat surfaces and line cavities and may relate to rhizolith formation rather than meteoric alteration. Pebbles within soil horizons locally show calcrete-like brown staining.

*Mineral Crusts.—*

Red and ochre (iron oxide and hydroxide) crusts on discontinuities are common. Those present on dissolution surfaces beneath probable paleosols may be an original feature, representing, for example, a redox interface beneath a gleyed soil. Such crusts could survive reworking and removal of a soil during transgression and, therefore, merely lack a confirmatory soil cover. However, oxide crusts appear in pelagic successions (Tucker and Wright, 1990) and are also common beneath pyritic horizons where shallow diagenetic or recent weathering reactions cannot be ruled out. They are, therefore, not a reliable diagnostic feature of emergence.

Dolomite pervades some karst and corrosion surfaces, weathering to a characteristic buff or brown color rather than the normal gray (lost on fresh surfaces). Cracks in such surfaces have been observed filled with normal undolomitized sediment (Fig. 5), evidently confirming early alteration of the host rock. Supratidal dolomitization (e.g., Shinn, 1983) was probably responsible, although later selective alteration cannot be discounted. Sediments infilling the distal parts of dissolution tubes can also be dolomitic, in contrast to the host sediment. These may demonstrate that the sediment was already (at least partly) dolomitic when it was emplaced.

*Meteoric Cement.—*

This is a poor criterion, because meteoric effects can be pervasive for some distance below emergent surfaces, spanning several cycles (e.g., Walkden, 1987; Read and Horbury, 1993). They are also hard to recognize in the field. Vadose silts are found below some Musandam omission surfaces, including within fenestral peritidal deposits. Together with rare dark brown cements (full of presumed organic inclusions) they occur as the earliest obviously crystalline precipitates within rhizoliths and calcretes. More easily recognized in the field are brown cements (but clear in thin section) that line all types of cavity, but these are common throughout the Lower Jurassic of Wadi Naqab and are unlikely to relate to local subaerial surfaces.

## *Characteristics of Submarine Omission Surfaces*

Submarine omission surfaces include hardgrounds, firmgrounds, and ravinements. Hardgrounds and firmgrounds are described in Scholle et al. (1983) and reviewed in James and Choquette (1990) and Demicco and Hardie (1994). They differ only in original degree of cementation and so are difficult to distinguish in lithified sediments. Marine corrosion of hardgrounds is common in deeper-water settings (Scholle et al., 1983), but this is not further considered here. Ravinements have not been identified as such in the Lower Jurassic, although intracyclic erosion is common (Fig. 3A). Submarine omission surfaces are characteristically intracyclic, potentially occurring anywhere within a cycle, so that they are not necessarily useful cycle boundaries. However their proper identification is essential because they have characteristics that might be confused with, or even overstamp, those of subaerial surfaces. For example, irregular hardgrounds and burrow-scalloped firmgrounds, given appropriate fill and later modification by pressure dissolution, might look like corrosion surfaces, and subtidal burrow systems might be taken for supratidal root molds. Key evidence of marine diagenesis comes from features such as marine cements, organic encrustation or borings, mineral coatings, erosion-intersected fossils, derived pebbles, and fractured overhangs in cavernous hardgrounds. All these are accessory features and may be absent. Hardgrounds form at any water depth, but they commonly occur with lag deposits and may be used to define transgressive surfaces and maximum flooding surfaces within parasequence sets (e.g., Emery and Myers, 1996). Modern examples are seen in sand shoals of the Bahamas (Demicco and Hardie, 1994) and in shallow subtidal settings in the Arabian Gulf (e.g., Walkden and Williams, 1998). Firmgrounds (e.g., Bromley, 1996) lack early lithification, but candidates must display evidence of resistance to erosion and self-supporting characteristics.

*Hardgrounds.—*

Only one unequivocal biologically encrusted hardground has been recognized in the Lower Jurassic of Musandam, comprising small compound corals on a burrow-modified discontinuity (Fig. 9). No examples of bored hardgrounds have been seen, but they are common in the Middle Jurassic of Musandam (de Matos and Walkden, this volume). Their absence is probably linked with the general impoverishment of the Lower Jurassic fauna. Otherwise, hardgrounds have been identified only on the basis of irregular, eroded discontinuities, sometimes burrowed, showing sharp lithological contrasts. These tend to be grouped in multiples of two or three, a feature we have also seen in Recent hardgrounds beneath the Arabian Gulf and previously noted by Shinn (1969).

*Firmgrounds.—*

Top-filled or empty burrow systems truncated by planar erosion surfaces, and burrow-scalloped surfaces with overhangs might qualify as firmgrounds. These are common in the Musandam succession, but in the absence of positive evidence for lithification, many surfaces dismissed as probable firmgrounds may actually be hardgrounds. Similarly, many surfaces such as sharply defined bedding planes that lack any other features might have originated either as firmgrounds or as hardgrounds.

### *Omission Surfaces Modified by Diagenesis*

A number of diagenetic processes are capable of destroying, mimicking, or enhancing key characteristics of omission surfaces, thereby removing evidence or creating ambiguity (Fig. 2). For example, meteoric alteration of Pleistocene limestones presently beneath the Arabian Gulf has removed aragonitic cements from marine hardgrounds, rendering them almost indistinguishable within subsea cores (personal observations). On the other hand, pressure dissolution acting at a textural interface can enhance unimportant discontinuities in more deeply buried limestones, giving them undue prominence. Pressure-dissolution seams (stylolites) create sutured interfaces that might be taken for corrosion zones, and burrowed intervals are reduced to rubble by pressure dissolution so that they resemble soil or corrosion zones (Fig. 2). This diagenetic retexturing or masking is a major barrier to the recognition of omission surfaces and highlights the danger of overreliance on surface morphology alone.

Breccias, conglomerates, and pebbly deposits can provide important corroboration for omission surfaces where they accumulated as transgressive lags, soils, and solution breccias. Confusion arises, though, in trying to distinguish between these and so-called "pseudobreccias", which are diagenetically produced nodu-

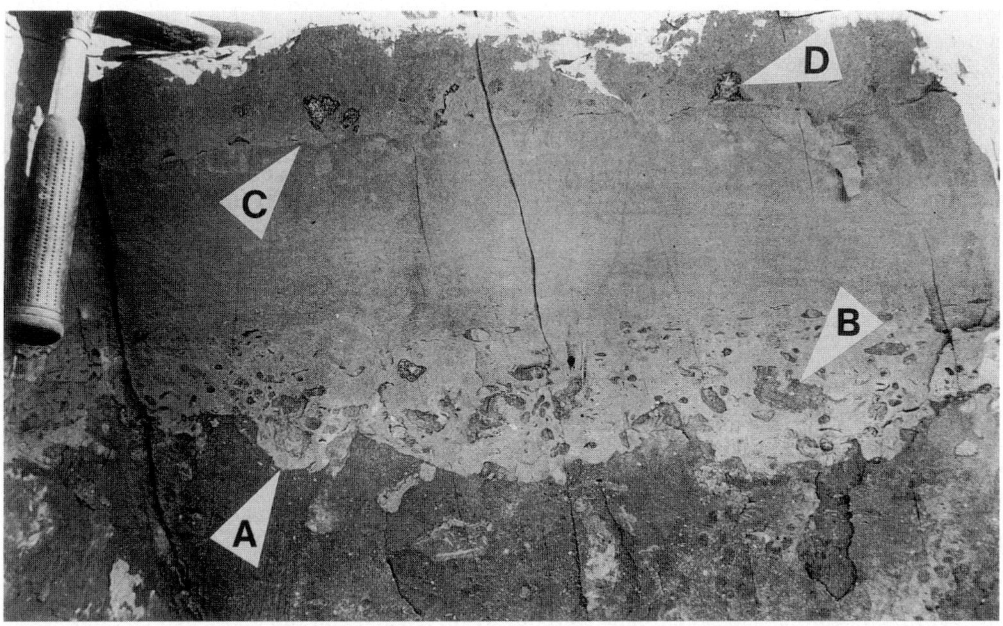


FIG. 9.—Two probable marine hardgrounds within mid-Sinemurian cycle 54. The lower of the two hardgrounds (A) divides burrowed packstones from lithoclastic rudstone. The rudstone may be a lag deposit, and contains eroded fragments of the underlying bed (B). The upper discontinuity (C) is also burrowed and tops coarse grainstones. This is colonized by small cerioid coral colonies in life position (D).

lar fabrics, including those created by localized cementation (e.g., Scholle et al., 1983), and those created or enhanced by pressure dissolution (e.g., Logan and Semeniuk, 1976). Nodular fabrics can be formed through selective lithification of burrowed carbonates both at the surface as successive hardgrounds and in the subsurface accompanying differential compaction. Petrocalcic soils can also create nodular horizons. Stylobreccias and stylonodular fabrics form through pressure dissolution acting along texturally defined or randomly distributed interfaces and can modify any of the primary and secondary textures already noted. Such textural convergence means that any nodular or brecciated horizon showing evidence for pressure dissolution might have originated as a cycle boundary, and it demands careful attention. This problem is common in the Musandam succession, and several criteria are needed to distinguish between "stylonodular", "burrownodular" and "pedonodular" categories. These include the morphology and associated features of the underlying discontinuity, the lithology and internal texture of nodules (e.g., calcreted clasts, burrow, fabrics, growth fabrics), the morphology of nodules (e.g., randomness, fitted fabric), the lithology, volume, texture, and the color of matrix, and the presence or absence of fossils within it.

### *Omission Surfaces and Climate*

Climate plays a major role in determining the type and nature of omission surfaces that can develop. In the marine environment, for example, hardground development is encouraged by aridity and evaporation (Tucker, 1993; Walkden and Williams, 1998). In the case of subaerial surfaces, aridity inhibits dissolution and encourages the development of calcrete (e.g., Hird and Tucker, 1988; Read and Horbury, 1993). Aridity also promotes the development of dolomite crusts (Shinn, 1983). A more humid regime is more likely to lead to a dense plant cover, enhancing soil development and encouraging the growth of pitted karsts. These are broad generalizations, though, and a strongly seasonal climate is likely to encourage a range of contradictory phenomena.

In the Musandam succession, however, there are few contradictions. Dissolution features at cycle tops are common throughout the Hettangian, Sinemurian, and Lower Pliensbachian whilst calcretes and dolomite crusts are uncommon. In the Upper Pliensbachian dissolution features become less prominent at cycle tops, and collapse breccias and desiccation features become common. Dolomitization is also more common in the higher parts of the succession, although not confined to cycle tops. These lines of evidence point strongly to the prevalence of moderately humid conditions for most of the early Jurassic, giving way to more general aridity in the Late Pliensbachian.

## OMISSION-SURFACE ANALYSIS APPLIED TO THE CYCLIC LOWER JURASSIC

The use of omission surfaces, particularly those of subaerial origin, has provided a valuable tool in the recognition and placing of cycle boundaries, leading to a substantial increase in resolution of cycles and, hence, creating a more accurate cycle database. This increases the value of the section for statistical analysis of cycles and of cycle stacking patterns, for example through the use of Fischer plots. The relative completeness of the database also permits accurate determination of the average duration of cycles, but in such a stratigraphically incomplete proximal shelf setting, where erosion and nondeposition account for more than half of the sedimentary record, the results of methods such as power spectral analysis are less likely to be meaningful and have not been attempted. Of 149 cycles recognized in the 234 m succession, more than half are topped by subaerial omission surfaces. Paleokarsts or corrosion surfaces account for 77 cycle tops whilst prominent soils, with or without breccia or pebbles and sometimes lacking an obvious underlying dissolution surface, cap 10 cycles. At least 17 of the subaerial surfaces display penetrative dissolution features between 20 cm and 50 cm deep. An additional nine surfaces have been logged as cycle boundaries but lack clear evidence of a subaerial origin. As regards marine omission surfaces, only 15 probable hardgrounds have been recognized, mostly in clusters, and few provide sufficient evidence (e.g., sharply backstepping facies) to invoke a cycle boundary.

Many cycle boundaries produce an outcrop bench, so that little excavation was necessary to identify crucial features and very few could not be examined in detail owing to poor exposure. It is worth noting, however, that in a less ideal setting such as a freshly quarried outcrop, a more fully vegetated area, or a less accessible setting, many cycle boundaries would have gone unnoticed and fewer cycles would have been identified. Furthermore, many of the characteristics used to define cycle tops in this outcrop study are on a scale or pattern of distribution that would render them difficult to recognize in core. Even in core from the floor of the present-day Arabian Gulf, Pleistocene marine and subaerial omission surfaces are difficult to recognize owing to subsequent and pervasive meteoric alteration (personal observation and in preparation).

## INTERPRETATION OF LOWER JURASSIC CYCLES AND CYCLE STACKING PATTERNS

The probable time span of Jurassic stages Hettangian–Sinemurian–mid-Pliensbachian was about 14 million years (Hardenbol et al., 1998; Gradstein et al., 1994). This gives an average duration of around 90,000 years each for the 149 lower Jurassic cycles covering this time span. The figure is comfortably within the range of Milankovitch orbital forcing periods (e.g., House, 1995) and close to the strongest of these in the present day, the 95,000 year eccentricity cycle. With an unknown number of cycles missing through nondeposition or erosion in this proximal shelf setting, however, it is predictable that the calculated figure for cycle duration should be greater than the true average during the Early Jurassic. In certain European Lower Jurassic hemipelagic mud successions, where breaks in sedimentation are less deleterious to the sedimentary record, Milankovitch orbital forcing is regarded as the main cause of minor cyclicity (e.g., House, 1985; Weedon, 1993; Weedon and Jenkyns, 1990). Average durations of around 40,000 years have been obtained for such minor cycles by House (1985) and Weedon et al. (2000), and it is likely that basins and platforms share the same orbitally forced influences. Continental massing in the Jurassic may have favored limited ice buildup (Crowley et al., 1992; Valdes et al., 1995) and the abundant omission surfaces in the Wadi Naqab succession could be the first direct evidence of glacioeustatic sea-level change in the Early Jurassic. However, other possible orbitally forced links with sea-level change have been discussed, such as varying storage of groundwater on the continents (Jacobs and Sahagian, 1993) and thermal expansion of the oceans (Schulz and Schafer-Neth, 1997).

Thickness of cycles ranges from less than 1 m to around 10 m, with an average of 1.5 m. In Figure 10, thicknesses of all 149 cycles are plotted in histograms against the corresponding Fischer plot. Fischer plots use thickness data to show the cumulative departure from cycle mean thickness (e.g., Sadler et al., 1993). They are a method of smoothing raw thickness data and, by assuming constant cycle periodicity, are an analog of changes in the rate of creation of accommodation. The plot reveals some clear trends in cycle thickness and stacking patterns that may reflect at least two distinct third-order sea-level variations superimposed upon the fifth-order variations represented by the cycles themselves. The first of these, a peak around cycles 8–12, may be incomplete, the initial cycles lying beneath our measured section and not included in our database. The second major Fischer peak, around cycles 82–88, reflects a steady buildup in cycle thickness followed by an abrupt reduction to near average thickness.

Because each of the cycles can be regarded as a parasequence, the pattern produced is a classic example of the variations in parasequence stacking pattern caused by third-order relative sea-level change noted by numerous authors (e.g., papers in Loucks and Sarg, 1993; Tucker, 1993). Assuming that accommodation was always quickly filled, the thicker parasequences marked by the rising Fischer curve show where background tectonic subsidence and third-order sea-level rise worked together in the creation of accommodation, so that each superimposed fifth-order eustatic rise had maximum effect. This was followed during third-order sea-level fall by the canceling out of tectonic subsidence so that superimposed fifth-order sea-level rises could create little to no additional accommodation. Parasequences are correspondingly thin and some may be missing. Tucker (1993) notes how dominant facies change in tandem with the changing parasequence stacking patterns.

Amongst others, Read et al. (1986) and Walkden and Walkden (1990) have used original computer simulations to generate variations in cycle stacking patterns. They show how, during the transition from highstand to lowstand, when parasequences are thin, much time is taken up with exposure rather than deposition, providing an extended opportunity for subaerial modification at cycle boundaries. This prediction from computer modeling points to an important application for data on omission surfaces derived from outcrop. Given that subaerial dissolution features are found throughout the Lower Jurassic, their variable development, particularly the depth of dissolutional penetration, could provide a proxy for the duration of exposure. Clearly there are numerous factors that can influence the development of karsts and corrosion surfaces, including lithology, vegetation cover, rainfall, prevailing wind direction, and depth of the meteoric vadose zone. Neverthless, length of time exposed and elevation during sea-level falls, both of which are directly related to sea-level change, must be amongst the most important influences on the amount of dissolution that can take place.

In Figure 10B all cycles with firmly identified subaerial omission surfaces are plotted, together with a general indication of the depth of dissolution at each of these. Clearly, subaerial omission surfaces, especially the best developed of these, are not evenly distributed. In particular there is an out-of-phase relationship between periods of maximum cycle thickness and periods of maximum subaerial modification. The main Fischer peaks in Figure 10A, showing where third order sea-level rise had its maximum additive effect upon cycle thickness, are around cycles 8 and 84. These are directly followed by thinner cycles, where the reverse effect was felt, and it is in these positions, in the regions of cycles 20 and 96 (trends 1 and 3 in Figure 10B), that omission surfaces tend to be best developed or most numerous. Evidently the best developed omission surfaces, and to some extent also their clustered distribution, reveal the parts of the succession where third-order sea-level fall caused a reduction in the period of deposition for each minor cycle and a consequent lengthening of the intervening periods of exposure. This pattern is consistent with the predicted one, and provides independent support for the interpretation of the Fischer peaks as reflecting long-term secular sea-level change. Of added interest, therefore, is the similar trend between cycles 38 and 60 (trend 2 in Figure 10B), and the less distinct example between cycles 112 and 140 (trend 4 in Figure 10B). Evidently the relatively minor mid-Sinemurian rise in the Fischer plot around cycle 40, which might otherwise be dismissed as a statistical artefact, was a real third-order sea-level event. A similar case can be made for the indistinct "hump" close to cycle 112, which lies on the back of the main Pliensbachian third-order fall.

## CONCLUSIONS

1. High-frequency (fifth-order) cyclicity can be recognized in both basinal and carbonate platform successions in the peri-Tethyan Lower Jurassic, providing exciting potential for high-resolution correlation and detailed characterization of sea-level change. However, many problems remain in ensuring accuracy in the correlation of such successions. One of the most serious of these relates to the use of facies patterns alone for defining cycles in high-frequency shallow-water carbonate successions. These are characteristically incomplete, and the diversity and forced rate of change in facies are so great that ideal regressive patterns are rarely developed. We are commonly uncertain about where cycle boundaries lie and about how many cycles are contained within a given interval.
2. Closer study and systematic logging of omission surfaces offers a solution. Omission surfaces are likely to be abundant in shallow-water carbonate successions, and subaerial surfaces, in particular, are valuable direct indicators of sea-level change. Consequently these have considerable potential for improving cycle discrimination, and they provide us with a means of "tuning" high-frequency cyclic successions.
3. A recently described, biostratigraphically constrained, Lower Jurassic Hettangian to lower Pliensbachian reference section close to the Oman/U.A.E. border includes marine hardgrounds, burrowed firmgrounds, karsts, corrosion surfaces, and paleosols. These account for a substantial proportion of elapsed time, but they share certain key physical characteristics and their recognition is not always straightforward. In particular, identification of omission surface types is hampered by diagenetic changes, which render some of their characteristics ambiguous.
4. Systematic study and logging of subaerial omission surfaces has substantially improved discrimination of cycles in our reference section and has increased known numbers by more than 100%. This improved database enables high-confidence analysis of the Lower Jurassic cyclic record and reveals an average cycle period within the Milankovitch range of around 90,000 years. This is longer than the 40,000 years obtained from equivalent basinal successions in Europe, and reflects the inherent incompleteness of shallow-water successions in which some cycles fail to be recorded. The recognition of a Milankovitch periodicity strengthens the likelihood that the sea-level change responsible for the fifth-order platform cycles was climatically generated, and a small-scale glacioeustatic mechanism might have been the cause.
5. Fischer analysis of cycle data reveals two major accommodation peaks that may relate to third-order sea-level change, one at the Hettangian–Sinemurian boundary and one at the Sinemurian–Pliensbachian boundary. Subaerial omission surfaces are most common and best developed on the falling limbs of the Fischer plot, which corresponds to a time when the third-order sea-level "downwave" would maximize subaerial exposure at fifth-order cycle boundaries. Extending this analysis to two smaller fluctuations on the Fischer plot, one mid-Sinemurian and the other mid-Pliensbachian, the distribution and intensity of omission surfaces suggests that these fluctuations are also likely also to relate to longer-term sea-level change. The inclusion of reliable omission surface data

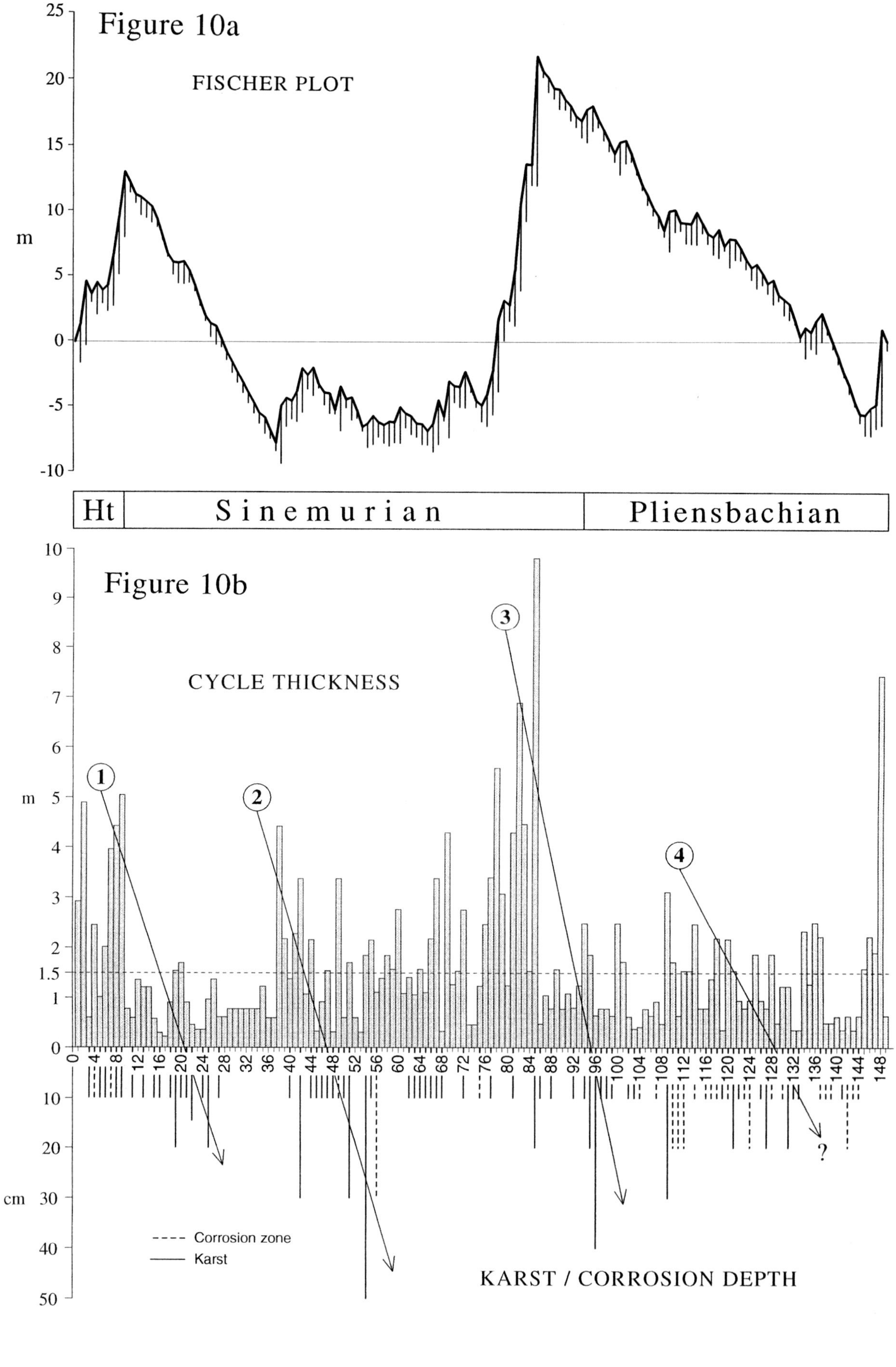
Figure 10a
FISCHER PLOT
m
Ht
Sinemurian
Pliensbachian
Figure 10b
CYCLE THICKNESS
m
cm
Corrosion zone
Karst
KARST / CORROSION DEPTH
1
2
3
4
?

therefore not only improves discrimination of cycles but it increases confidence in the use of Fischer analysis both as a means of detecting secular change in cylicity and as a tool in the interpretation of cycle stacking patterns. Evidently the pattern of sea-level change on this Early Jurassic Tethyan carbonate platform was complex, but not random.

6. Milankovitch-related cyclicity exerts a strong control on facies development and formation architecture in shallow-water carbonates and was pervasive during the early Mesozoic. It is likely that similar patterns will be found globally and in other parts of the Mesozoic where there was an appropriate combination of climatic and palaeogeographic conditions. In the subsurface, the resultant characteristics, including submeter-scale bedding, weathered zones, soils, and early diagenetic caps, can strongly influence water and hydrocarbon movement, creating potentially strong anisotropy, particularly in nonfractured reservoirs. Nevertheless, many of the features by which such formations could be characterized are very unlikely to be revealed in seismic, will be missed in most well logs, and can be difficult to recognize in core.

## ACKNOWLEDGMENTS

The authors would like to thank ADCO and PARTEX, who supported the original field work upon which this paper is based. Parexpro and the University of Aberdeen have provided facilities for its completion. In particular, we gratefully acknowledge Barry Fulton (UoA), who finalized the diagrams. Thanks also go to Charlie Kerans and Dennis Prezbindowski, who reviewed the original manuscript and made helpful suggestions that undoubtedly led to some improvements.

## REFERENCES

Bassoulet, J.P., and Bergougnan, H., 1981, Fauna et facies typique du domaine sud-tethysien: lias du Munzur Dag (Anatolie orientale): Société Geologique de France, Bulletin, v. 7, p. 83–93.

Bromley, R.G., 1996, Trace Fossils; Biology, Taphonomy and Applications, 2nd Edition: London, Chapman & Hall, 361 p.

Budd, D.A., Sailer, A.H., and Harris, P.M., eds., 1995, Unconformities and Porosity in Carbonate Strata: American Association of Petroleum Geologists, Memoir 63, 313 p.

Cotillon, P., 1995, Constraints for using high-frequency sedimentary cycles in cyclostratigraphy, *in* House, M.R., and Gale, A.S., eds., Orbital Forcing Time Scales and Cyclostratigraphy: Geological Society of London, Special Publication 85, p. 133–141.

Crowley, T.J., Baum, S.K., and Hyde, W.T., 1992, Milankovitch fluctuations on supercontinents: Geophysical Research Letters, v. 9, p. 793–796.

Crevello, P.D., 1991, High-frequency carbonate cycles and stacking patterns: Interplay of orbital forcing and subsidence on Lower Jurassic rift platforms, High Atlas, Morocco, *in* Franseen, E.K., Watney, W.L., Kendall, C.G.St.C., and Ross, W.H., eds. Sedimentary Modeling — Computer Simulations and Methods for Improved Parameter Definition: Kansas Geological Survey, Bulletin 233, p. 207–230.

de Matos, J.E., 1997, Stratigraphy, Sedimentation and Oil Potential of the Lower Jurassic to Kimmeridgian of the United Arab Emirates, outcrop and subsurface compared: PhD. thesis, University of Aberdeen, 469 p.

de Matos, J.E., Hulstrand, R.F., and Walkden, G.M., 1994, Biostratigraphy of the Lower Jurassic of the U.A.E.: Outcrop and subsurface compared: 6th Abu Dhabi International Petroleum Exhibition and Conference, Proceedings, p. 449–459.

Demicco, R.V., and Hardie, L.A., 1994, Sedimentary Structures and Early Diagenetic Features of Shallow Marine Carbonate Deposits: SEPM, Atlas Series 1, 265 p.

Emery, D., and Myers, K., 1996, Sequence Stratigraphy: Oxford, U.K., Blackwell Science, 297 p.

Esteban, M., and Klappa, C.F., 1983, Subaerial exposure environment, *in* Scholle, P.A., Bebout, D.G., and Moore, C.H., eds., Carbonate Depositional Environments: American Association of Petroleum Geologists, Memoir 33, p. 1–54.

Fischer, A., and Bottjer, D.J., 1991, Orbital forcing and sedimentary sequences: Journal of Sedimentary Petrology, v. 61, p. 1063–1069.

Ginsburg, R.N., 1971, Landward movement of carbonate mud: new model for regressive cycles in carbonates (abst.): American Association of Petroleum Geologists, Bulletin, v. 55, p. 340.

Goldhammer, R.K., Dunn, P.A., and Hardie, L.A., 1987, high-frequency glacioeustatic sea-level oscillations with Milankovitch characteristics recorded in middle Triassic platform carbonates in northern Italy: American Journal of Science, v. 287, p. 853–892.

Goldhammer, R.K., Dunn, P.A., and Hardie, L.A., 1990, Depositional cycles, composite sea-level changes, cycle stacking patterns, and the hierarchy of stratigraphic forcing: Examples from Alpine Triassic carbonates: Geological Society of America, Bulletin, v. 102, p. 535–562.

Goodwin, P.W., and Anderson, E.J., 1985, Punctuated aggradational cycles: a general hypothesis of episodic stratigraphic accumulation: Journal of Geology, v. 93, p. 515–533.

Gradstein, F.M., Agterberg, F.P., Ogg, J.G., Hardenbol, J., Van Veen, P., Thierry, J., and Huang, Z., 1994, A Mesozoic time scale: Journal of Geophysical Research, v. 99, p. 24,051–24,074.

Hardenbol, J., Thierry, J., Farley, M.B., Jacquin, Th., de Graciansky, P.-C., and Vail, P.R., 1998, Mesozoic and Cenozoic sequence chronostratigraphic framework of European basins, *in* de Graciansky, P.-C., Hardenbol, J., Jacquin, Th., and Vail, P.R., eds., Mesozoic and Cenozoic Sequence Stratigraphy of European Basins: SEPM, Special Publication 60, p. 3–13.

←

Fig. 10.—Bed thicknesses, omission surface data, and Fischer plots for the Lower Jurassic of Wadi Naqab. A) Fischer plot of cycle thickness data presented in B. The plot shows cumulative deviation from average cycle thickness, and indicates two major peaks in the creation of accommodation, one close to the Hettangian–Sinemurian boundary and one near the Sinemurian–Pliensbachian boundary. Only the early Pliensbachian is represented in the data set, and the late upwave on the right of the Fisher plot is part of a peak marking the start of the late Pliensbachian. Two further minor peaks seem possible, one in the mid-Sinemurian and one in the early Pliensbachian, superimposed on the back of the much larger peak. B) Histogram of the cycle thicknesses used in the Fischer plot, and average depths of penetration of subaerial dissolution features at respective cycle tops taken from field records and photographs. The numbered arrows (1–4) indicate positions of trends in the combined data between thick cycles with few subaerial omission surfaces and thinner cycles with numerous and well developed (deeply pitted or corroded) subaerial omission surfaces. The fourth is the least distinct, but each coincides with the "downwave" side of a Fischer peak. The combined data show that the times of maximum creation of accommodation and the times of maximum cycle-top modification are out of phase. This is exactly as predicted by computer modeling and cycle stacking pattern theory. The inclusion of data on fifth-order omission surfaces has substantially enhanced recognition of cycles and provides independent confirmation of secular third-order sea-level change.

HERWITZ, S.R., 1993, Stemflow influences on the formation of solution pipes in Bermuda eolianite: Geomorphology, v. 6, p. 253–271.

HIRD, K., AND TUCKER, M.E., 1988, Contrasting diagenesis of two Carboniferous oolites from South Wales: a tale of climatic influence: Sedimentology, v. 35, p. 587–602.

HOUSE, M.R., 1985, A new approach to an absolute timescale from measurements of orbital cycles and sedimentary microrhythms: Nature, v. 93, p. 721–725.

HOUSE, M.R., 1995, Orbital forcing timescales: an introduction, *in* House, M.R., and Gale, A.S., eds., Orbital Forcing Timescales and Cyclostratigraphy: Geological Society of London, Special Publication 85, p. 1–18.

JACOBS, D.K., AND SAHAGIAN, D.L., 1993, Climate-induced fluctuations in sea-level during non-glacial times: Nature, v. 361, p. 710–712.

JAMES, N.P., AND CHOQUETTE, P.W., eds., 1988, Paleokarst: New York, Springer-Verlag, 416 p.

JAMES, N.P., AND CHOQUETTE, P.W., 1990, Limestones — The Sea Floor Diagenetic Environment, *in* McIlreath, I.A., and Morrow, D.W., eds., Diagenesis: Geoscience Canada, Reprint Series 4, p. 13–34.

KOERSCHNER, W.F., AND READ, J.F., 1989, Field and modeling studies of Cambrian carbonate cycles, Virginia Appalachians: Journal of Sedimentary Petrology, v. 59, p. 654–687.

LOGAN, B.W., AND SEMENIUK, V., 1976, Dynamic metamorphism: processes and products in Devonian carbonate rocks, Canning Basin, Western Australia: Geological Society of Australia, Special Publication 16, p. 1–138.

LOUCKS, R.G., AND SARG, J.F., EDS., 1993, Carbonate Sequence Stratigraphy: American Association of Petroleum Geologists, Memoir 57, 545 p.

MOSES, G.P.G., 1995, Calibration analysis and interpretation of depositional cycles in the Early Toarcian of Yorkshire, UK, *in* House, M.R., and Gale, A.S., eds. Orbital Forcing Timescales and Cyclostratigraphy: Geological Society of London, Special Publication 85, p. 67–74.

READ, J.F., GROTZINGER, J.P., BOVA, J.A., AND KOERSCHNER, W.F., 1986, Models for generation of carbonate cycles: Geology, v. 14, p. 107–110.

READ, J.F., AND GROVER, G.A., 1977, Scalloped and planar erosion surfaces, Middle Ordovician limestones, Virginia: analogues of Holocene exposed karst or tidal rock platforms: Journal of Sedimentary Petrology, v. 47, p. 956–972.

READ, J.F., AND HORBURY, A.D., 1993, Eustatic and tectonic controls on porosity evolution beneath sequence-bounding unconformities, *in* Horbury, A.D., and Robinson, A.G., eds., Diagenesis and Basin Development: American Association of Petroleum Geologists, Studies in Geology, no. 36, p. 155–197.

SADLER, P.M., 1994, The expected duration of upward-shallowing peritidal carbonate cycles and their terminal hiatuses: Geological Society of America, Bulletin, v. 106, p. 791–802.

SADLER, P.M., OSLEGER, D.A., AND MONTAÑEZ, P., 1993, On the labeling, length, and objective basis of Fischer plots: Journal of Sedimentary Petrology, v. 63, p. 360–368.

SCHOLLE, P.A., ARTHUR, M.A., AND EKDALE, A.A., 1983, Pelagic environments, *in* Scholle, P.A., Bebout, D.G., and Moore, C.H., eds., Carbonate Depositional Environments: American Association of Petroleum Geologists, Memoir 33, p. 620–691.

SCHULZ, M., AND SCHÄFER-NETH, C., 1997, Translating Milankovitch climate forcing into eustatic fluctuations via thermal deep water expansion — a conceptual link: Terra Nova, v. 9, p. 228–231.

SHINN, K.A., 1969, Submarine lithification of Holocene carbonate sediments in the Persian Gulf: Sedimentology, v. 121, p. 109–144.

SHINN, K.A., 1983, Tidal Flat, *in* Scholle, P.A., Bebout, D.G., and Moore, C.H., eds., Carbonate Depositional Environments: American Association of Petroleum Geologists, Memoir 33, p. 171–210.

TUCKER, M.E., 1993, Carbonate diagenesis and sequence stratigraphy, *in* Wright, V.P., ed., Sedimentology Review 1: Oxford, U.K., Blackwell Scientific Publications, p. 51–72.

TUCKER, M.E., AND WRIGHT, V.P., 1990, Carbonate Sedimentology: Oxford, U.K., Blackwell Scientific Publications, 482 p.

VALDES, P.J., SELLWOOD, B.W., AND PRICE, G.D., 1995, Modeling Late Jurassic Milankovitch climate variations, *in* House, M.R., and Gale, A.S., eds., Orbital Forcing Timescales and Cyclostratigraphy: Geological Society of London, Special Publication 85, p. 115–132.

WALKDEN, G.M., 1974, Paleokarstic surfaces in Upper Viséan (Carboniferous) limestones of the Derbyshire block, England: Journal of Sedimentary Petrology, v. 44, p. 1232–1247.

WALKDEN, G.M., 1987, Sedimentary and diagenetic styles in late Dinantian carbonates of Britain, *in* Miller, J., Adams, A.E., and Wright, V.P., eds., European Dinantian Environments: Chichester, U.K., Wiley, p. 131–155.

WALKDEN, G.M., AND WALKDEN, G.D., 1990, Cyclic sedimentation in mixed carbonate–clastic environments: four simulation programs for a desktop computer, *in* Tucker, M.E., Wilson, J.L., Crevello, P.D., Sarg, J.F., and Read, J.F., eds., Carbonate Platforms, Facies Sequences and Evolution: International Association of Sedimentologists, Special Publication 9, p. 55–78.

WALKDEN, G.M., AND WILLIAMS, A., 1998, Carbonate ramps and the Pleistocene–Recent depositional systems of the Arabian Gulf, *in* Wright, V.P., and Burchette, T.P., eds., Carbonate Ramps: Geological Society of London, Special Publication 149, p. 43–53.

WALKDEN, G.M., AND WILLIAMS, D.O., 1991, The diagenesis of the late Dinantian Derbyshire–east Midland carbonate shelf, central England: Sedimentology, v. 38, p. 643–670.

WEEDON, G.P., 1993, The recognition and stratigraphic implications of orbital forcing of climate and sedimentary cycles, *in* Wright, V.P., ed., Sedimentary Review 1: Oxford, U.K., Blackwell, p. 31–50.

WEEDON, G.P., AND JENKYNS, H.C., 1990, Regular and irregular climatic cycles and the Belemnite Marls (Pliensbachian, Lower Jurassic, Wessex Basin): Geological Society of London, Journal, v. 147, p. 915–918.

WEEDON, G.P., JENKYNS, H.C., COE, A.L., AND HESSELBO, S.P., 2000, Astronomical calibration of the Jurassic time-scale from cyclostratigraphy in British formations: Philosophical Transactions of the Royal Society of London, series A, v. 357, p. 1787–1813.

WILKINSON, B.H., OPDYKE, B.N., AND ALGEO, T.J., 1991, Time partitioning in cratonic carbonate rocks: Geology, v. 19, p. 1093–1096.

WRIGHT, V.P., ED., 1986, Paleosols; Their Recognition and Interpretation: Oxford, U.K., Blackwell Scientific Publications, 315 p.

# A SEQUENCE STRATIGRAPHIC REFERENCE SECTION FOR THE TITHONIAN OF LEBANON

CHRISTOPHER TOLAND
*Oolithica Geoscience Ltd., 489 Union Street, Aberdeen AB11 6DB, Scotland, United Kingdom*

ABSTRACT: Central Lebanon provides some of the best exposed and most readily accessible Upper Jurassic (Kimmeridgian–Tithonian) sections in the Middle East, and is one of the few places where lateral equivalents of the prolific Arab Formation (Kimmeridgian–Tithonian) reservoirs of Peninsular Arabia can be studied at outcrop.

At the Bikfaya outcrop section (35 km ENE of Beirut), the uppermost Jurassic comprises at least two disconformity-bounded third-order depositional sequences. Sequence 1 ("Falaise de Bikfaya") is 61 m+ thick and comprises a progradational succession (highstand systems tract) of foreshoal micropeloid packstones, shoal-crest stromatoporoid floatstones, back-shoal *Permocalculus* wackestones, and (?attached mainland-) shoreface facies culminating in tidal-flat deposits. This interval is of Early to Middle Tithonian age.

Sequence 2 ("Calcaire de Salima") is ca. 63 m thick, and commences with an abrupt transgressive surface and an associated influx of calcareous dinocysts. The lower part of this sequence comprises strongly argillaceous micropeloidal packstones and occasional peloid–intraclast packstones, interpreted as an offshore transition zone facies association. The initial marine flooding event is of late Middle Tithonian (upper *fallauxi-ponti* Zone) age. A candidate maximum flooding surface (MFS) is recognized within the late Middle Tithonian *ponti* Zone, coincident with calcareous dinocyst species and abundance maximum. (A *ponti* Zone MFS has also been identified elsewhere in the Middle East.) The recessive middle part of Sequence 2 is largely unexposed, whereas the upper cliff-forming part comprises ca. 22 m of Upper Tithonian ooid–skeletal grainstones that coarsen and thicken up-section. These grainstones are characterized by pronounced planar cross-stratification with set heights of up to 3.6 m, and are interpreted as a wave-dominated shoal complex culminating in emergent foreshore facies. Sequence 2 is terminated by a prominent paleo–karst (Type 1 sequence boundary) corresponding to the Jurassic–Cretaceous boundary, and is overlain by basal Cretaceous wacke–ironstones that form the lower part of the Chouf Sandstone Formation.

The lower part of the Chouf Sandstone Formation lacks age-diagnostic fossils. The timing of initial Cretaceous onlap is thus only poorly constrained by the presence of Late Tithonian taxa in the underlying "Calcaire de Salima" and the presence of Barremian spores in the upper part of the Chouf Sandstone Formation.

## INTRODUCTION

The Bikfaya reference section is located in central Lebanon, ca. 35 km due ENE of Beirut, at 33°55'20"N, 35°42'40"E (Fig. 1). The section is reached by traveling from Beirut to Antilyas on the Beirut–Juniyah coast road, and then eastward on the Antilyas–Bikfaya road. The base of section is located ca. 2 km east of Bikfaya town center, in the floor of a partially infilled roadside quarry on the south side of the Bikfaya–Zahle road, a short distance east of the Hamlaya turnoff. The top of the section is located directly below the hilltop monastery of Mar Elias (Figs. 1, 2A).

There have been relatively few recent studies of the Upper Jurassic in Lebanon. Given that the succession is well exposed and more accessible than almost any other in the Middle East, Lebanon may be an ideal place to establish a sequence stratigraphic reference section for the Oxfordian–Tithonian succession, which hosts vast petroleum resources across the adjacent Arabian Peninsula.

The macropaleontology of the Upper Jurassic succession has attracted most previous attention, with early studies by Felix (1904), Krumbeck (1905), Renz (1930), Pfender (1937), Delpey (1940), and Hudson (1954). The micropaleontology is less well documented, with significant publications by Bischoff (1964) and Basson and Edgell (1971).

Summaries of the Upper Jurassic stratigraphy of Lebanon are provided by Zumoffen (1926), Heybroek (1942), Renouard (1951), and Dubertret (1950a, 1963, 1966, 1975), and Picard and Hirsch (1987) present an excellent up-to-date review and synthesis of previous work. The stratigraphy of the basal Cretaceous has been formalized by Walley (1983). The study area is represented by the 1:50,000 Beirut geological map sheet and accompanying explanatory memoir (Dubertret, 1951), and adjacent Upper Jurassic outcrops in the northern Anti-Lebanon (Zebdani–Rayak region) and Israeli-occupied Syria (Mount Hermon) have been described by Dubertret (1949, 1950b), Ponikarov (1967), and Toland (1998).

Details of Lebanese Upper Jurassic subsurface sections are provided in Renouard (1955) and Beydoun (1977, 1981). The paleogeographic reconstructions of May (1991) suggest a mid-shelf to outer-shelf paleo-position for the Bikfaya area, ca.10 km east (i.e., landward) of a N–S trending Jurassic shelf break. The correlation of Tethyan and Boreal Tithonian biozones as used in this study is based on Geyssant (1997) (Fig. 3).

## STRATIGRAPHY

The measured section at Bikfaya comprises at least two uppermost Jurassic depositional sequences and the lower part of a basal Cretaceous sequence (Fig. 4):

- Sequence 3: basal Cretaceous (123.85–151.1 m)
- Sequence 2: upper Middle–Upper Tithonian (61.1–123.85 m)
- Sequence 1: Lower to lower Middle Tithonian [pre-*Chitinoidella* Zone] (0–61.1 m)

Sequence 1 corresponds to the "Falaise de Bikfaya" of Dubertret (1951), the "Falaise de Djisr el Qadi" of Heybroek (1942) and the "Niveau 7" of Renouard (1951). Sequence 2 corresponds to the "Calcaire de Salima" of Dubertret (1951), the "Couches Jaunes Superieur" of Heybroek (1942), and the "Niveau 8" of Renouard (1951). Sequence 3 corresponds to the lower part of the basal Cretaceous "Grés de Base" of earlier authors, and the "Chouf Sandstone Formation" of Walley (1983). In the present study thin-section biostratigraphic analysis (benthic forams, calcareous algae, calcareous dinocysts, and stromatoporoids) was undertaken at ca. 2 m intervals (Fig. 5).

Middle East Models of Jurassic/Cretaceous Carbonate Systems
SEPM Special Publication No. 69, Copyright © 2000
SEPM (Society for Sedimentary Geology), ISBN 1-56576-075-1, p. 53–64.

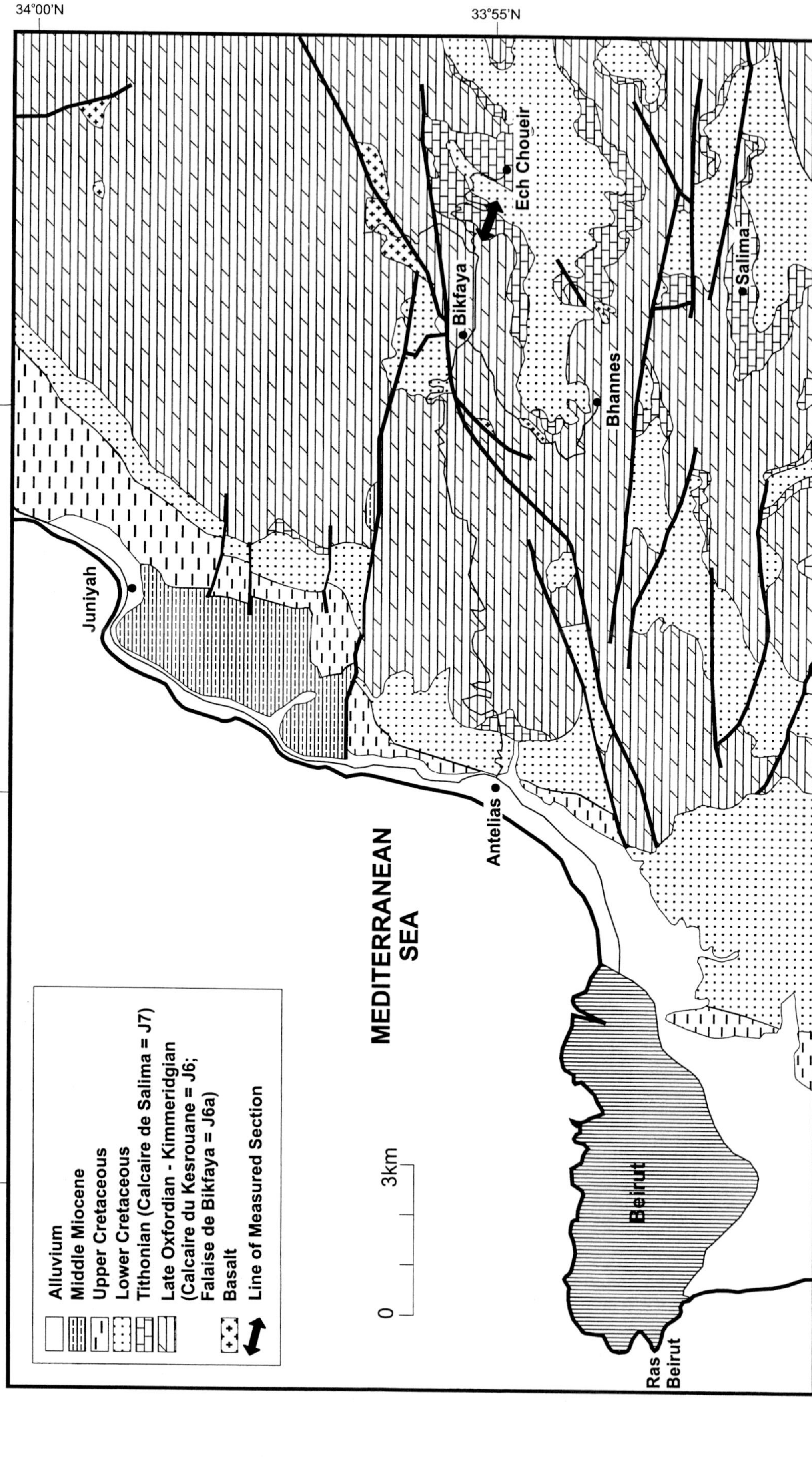


FIG. 1.—Simplified geologic map of central Lebanon showing location of Bikfaya studied section (modified after Dubertret, 1951).

FIG. 2—A. General eastward view of the Bikfaya reference section. The base of section is located in the floor of the quarry at bottom right, with the top of the quarry face corresponding to an elevation of 18 m above base of section. The measured section continues along an ephemeral river bed beyond the quarry face to an elevation of 67 m, and is then offset to the scree-covered slopes behind the three houses at center left. The long cliff face below the monastery of Mar Elias (top left) corresponds to the upper part of Sequence 2 (the "Calcaire de Salima"). B. Closer view of the upper part of the "Calcaire de Salima", here forming a cliff below Mar Elias (center left to center right). The cliff is almost 1 km in length, and is dominated by a single 3.6-m-thick tabular-planar cross-set, with foreset laminae dipping consistently at ca. 20° towards the ESE. The scree-covered slopes above the cliff correspond to the basal part of the "Chouf Sandstone Formation" (Sequence 3).

### *Sequence 1: Lower to lower Middle Tithonian "Falaise de Bikfaya"*

*Description.*—

Sequence 1 comprises three informal members (0–61.1 m) (Fig. 4): the cliff-forming "lower member" (0–32 m), comprises medium-bedded, silt-grade peloidal packstones and stromatoporoid floatstones; a distinct 2-m-thick coral–stromatoporoid rudstone unit occurs at 15.8–17.8 m. Discrete beds of coalesced decimeter-scale chert nodules occur sporadically throughout. The sparse, moderately diverse biota recorded in this study (Fig. 5) is dominated by agglutinated foraminifera (*Ammobaculites* sp., *Pfenderina neocomiensis-trochoidea*, *Pseudocyclammina lituus* Yokoyama) and stromatoporoids (*Actinostromarianina lecompti* Hudson, *Burgundia trinorchii* Dehorne, *Parastromatopora libani* Hudson, *Shuqraia zuffardi* Wells), with sparse calcareous algae (*Actinoporella podolica* Alth, *Permocalculus* sp. cf. *P. inopinatus* Elliott, *Salpingoporella annulata* Carozzi, *Rajkaella* (=*Likanella*) *bartheli* Bernier, *Campbeliella* (= "*Vaginella*") *striata* Carozzi, *Lithocodium aggregatum* Elliott, *Terquemella* sp.) and gastropods (*Nerinea salinensis* d'Orbigny = *N. maroni* Krumbeck). The associated coral assemblage described by Felix (1904) includes *Actinarea sponginoides* Blanckenhorn, *Baryhelia hexaconena* Felix, *Centrastrea leptomeres* Felix, *C. polystila* Felix, *Diplocoenia punica* Blanckenhorn, *Latimaeandra zumoffeni* Felix, *L. amphitrites* Felix, *L. sulcata* Fromentel, *Metatsrea cf. delmontana* Koby, *Stephanocoenia pentagonalis* Becker, *S. helmopotis* Felix, *S. trochiformis* d'Orbigny, *Stylina girodi* Etallon, and *S. bullosa* Blanckenhorn. The molluscan fauna recorded by Krumbeck (1905), Delpey (1940), Heybroek (1942), and Renouard (1951) includes *Ampullina vautrini* Delpey, "*Cardium*" *corallium* Leymerie, *Harpagodes oceani* Brogniart, *Lima acutorostris* Krumbeck, *L. densistriata* Krumbeck, *L. libanensis* Krumbeck, *Natica (= Ampullina) dido* Krumbeck, *N. (= Ampullina) mulitta* Krumbeck, *N.* sp. (= *Ampullina krumbecki* Delpey), *Nerinea maroni* Krumbeck (= *N. salinensis* d'Orbigny), *N. desvoidyi pauciplicata* Krumbeck, *N. sesostris* (?= *N. contorta* var. *sesostris* Krumbeck), *Nerinella* sp. cf. *N. elatior* d'Orbigny, *Nerita littoralis* Krumbeck, *Ostrea* sp. cf. *O. expansa* Sowerby, *O. matronensis* De Loriol, *Pachyerosma blanckenhorni* Krumbeck, *Pecten palmyrensis* Krumbeck, and *Phylobrissus thevenini* Etallon.

The "lower member" is interpreted as a lower-shoreface succession, with the stromatoporoid floatstone units representing episodic storm deposits. The 2-m-thick coral-stromatoporoid rudstone unit at 15.8–17.8 m is interpreted as a subtidal biostrome. The co-occurrence of stromatoporoids and the codiacean alga *L. aggregatum* at 16.5 m suggests paleo-water depths on the order of 10–15 m (Banner and Simmons, 1994; Toland, 1994).

The "lower member" is conformably overlain by a "middle member" (32–47.5 m), comprising recessive, slightly argillaceous lime mudstones (grading to silt-grade peloidal packstones) and peloid–algal wackestones / packstones. This interval is characterized by abundant ramifying *Thalassinoides* burrows, which result in a distinct pseudo-nodular weathering fabric. The low-diversity biota is dominated by calcareous algae (*Permocalculus* sp. cf. *P. inopinatus* in flood abundance, *Cylindroporella arabica* Elliott, and *Actinoporella podolica*), ostracodes, and indeterminate ex-aragonitic molluscan debris. There is no precise indication of paleo–water depth, *Permocalculus* having a wide depth range of ca. 5–50 m (Banner and Simmons, 1994). Elsewhere in the Middle East *Permocalculus* is known to form important backshoal "lagoonal" buildups, as in the Rumailah Field of Southern Iraq (Gaddo, 1971). It may, thus, have occupied a niche similar to present-day halimedid-form algae, which form extensive lagoonal buildups leeward of the Great Barrier Reef (e.g., Orme and Salama, 1988).

The base of the "upper member" (47.5–61.1 m) is marked by a 0.25 m skeletal packstone rib with prominent *Rhizocorallium* burrows, yielding abundant *P. inopinatus* and occasional *Rectocyclammina chouberti* Hottinger. This is overlain by interbedded recessive lime mudstones and cortoid–ooid–dasyclad

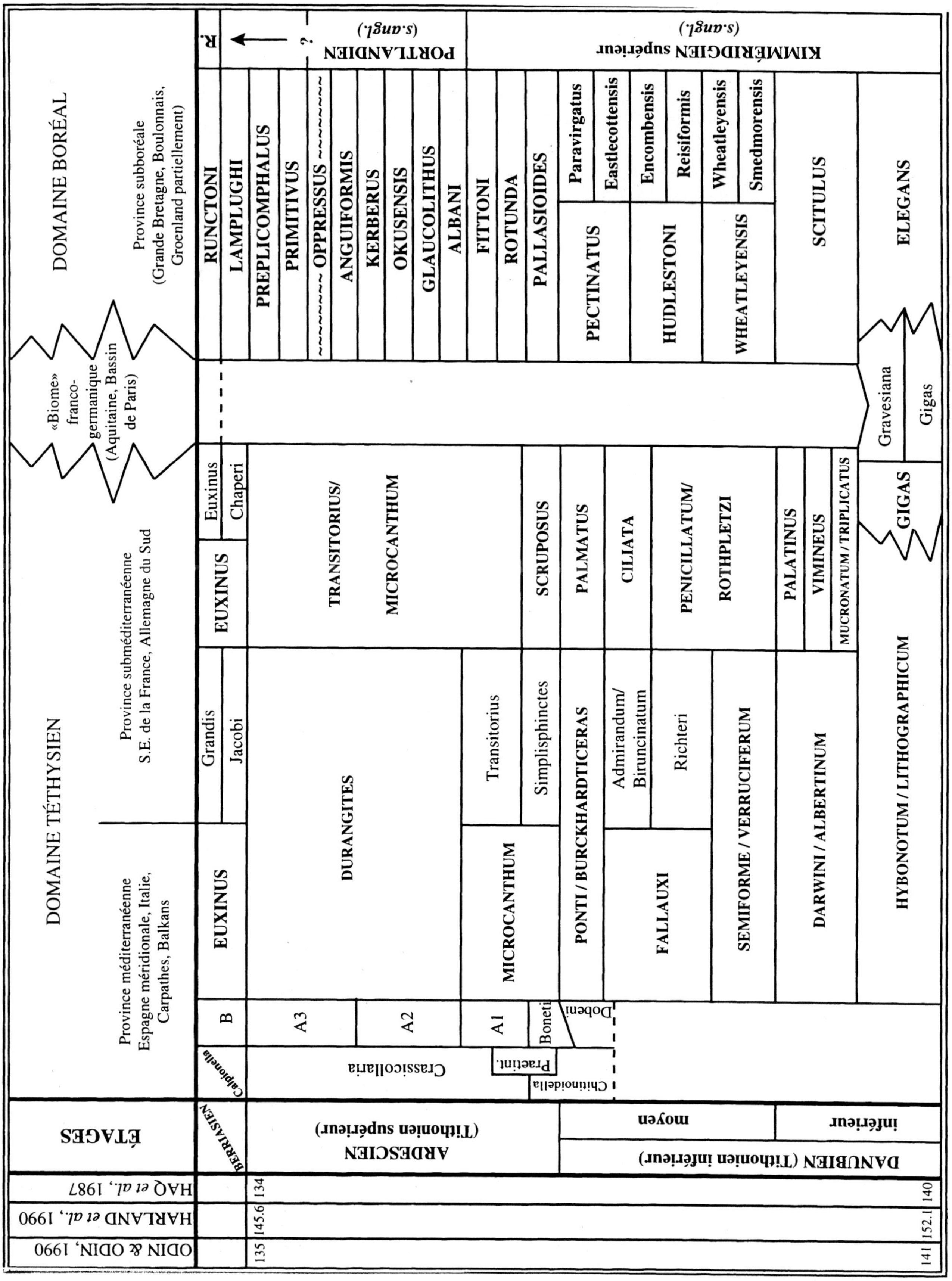


Fig. 3.—Tithonian biozonation scheme in Tethyan and Boreal seas (after Geyssant, 1997).

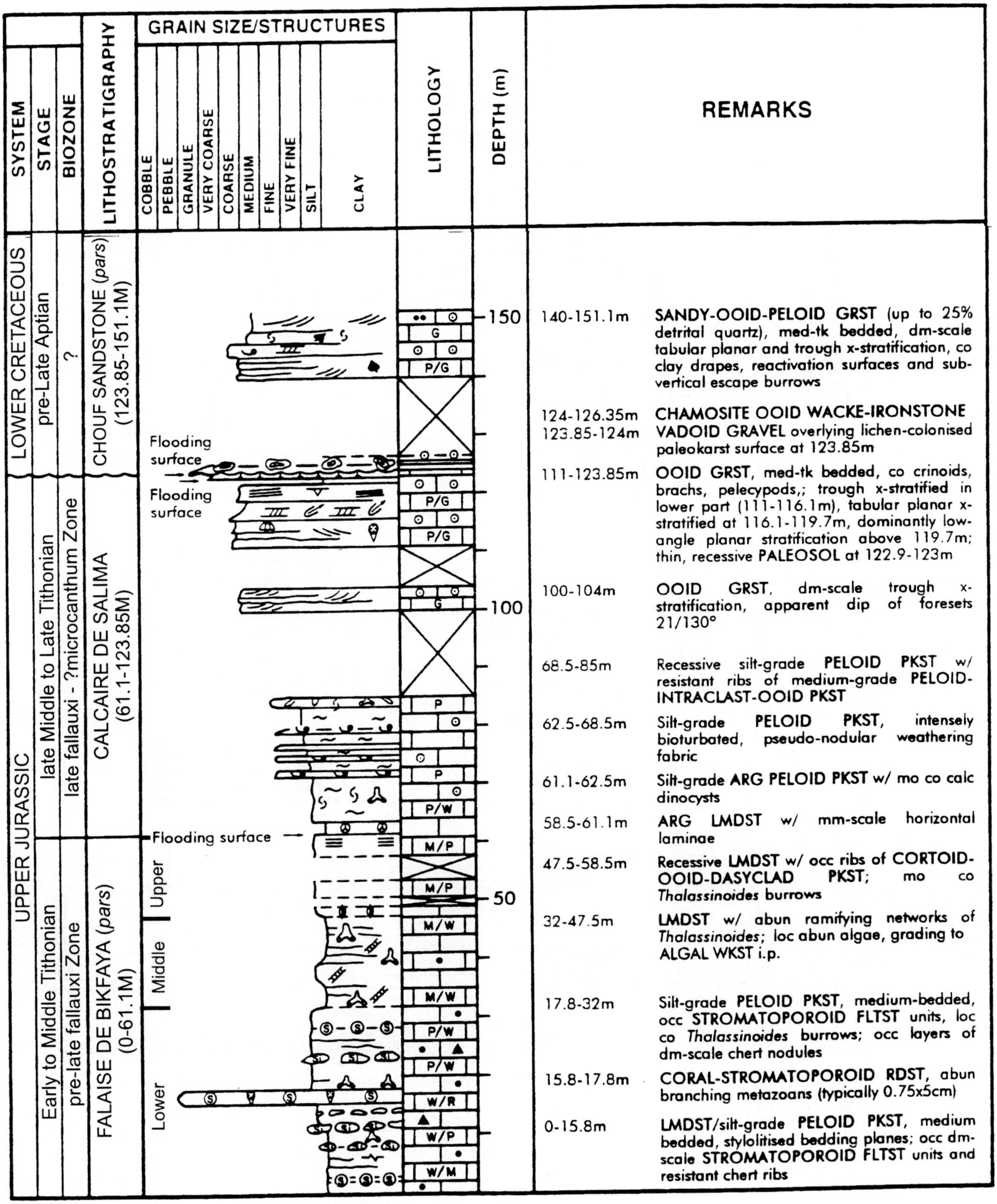


FIG. 4.—Lithologic succession of Upper Jurassic–Lower Cretaceous strata at Bikfaya reference section.

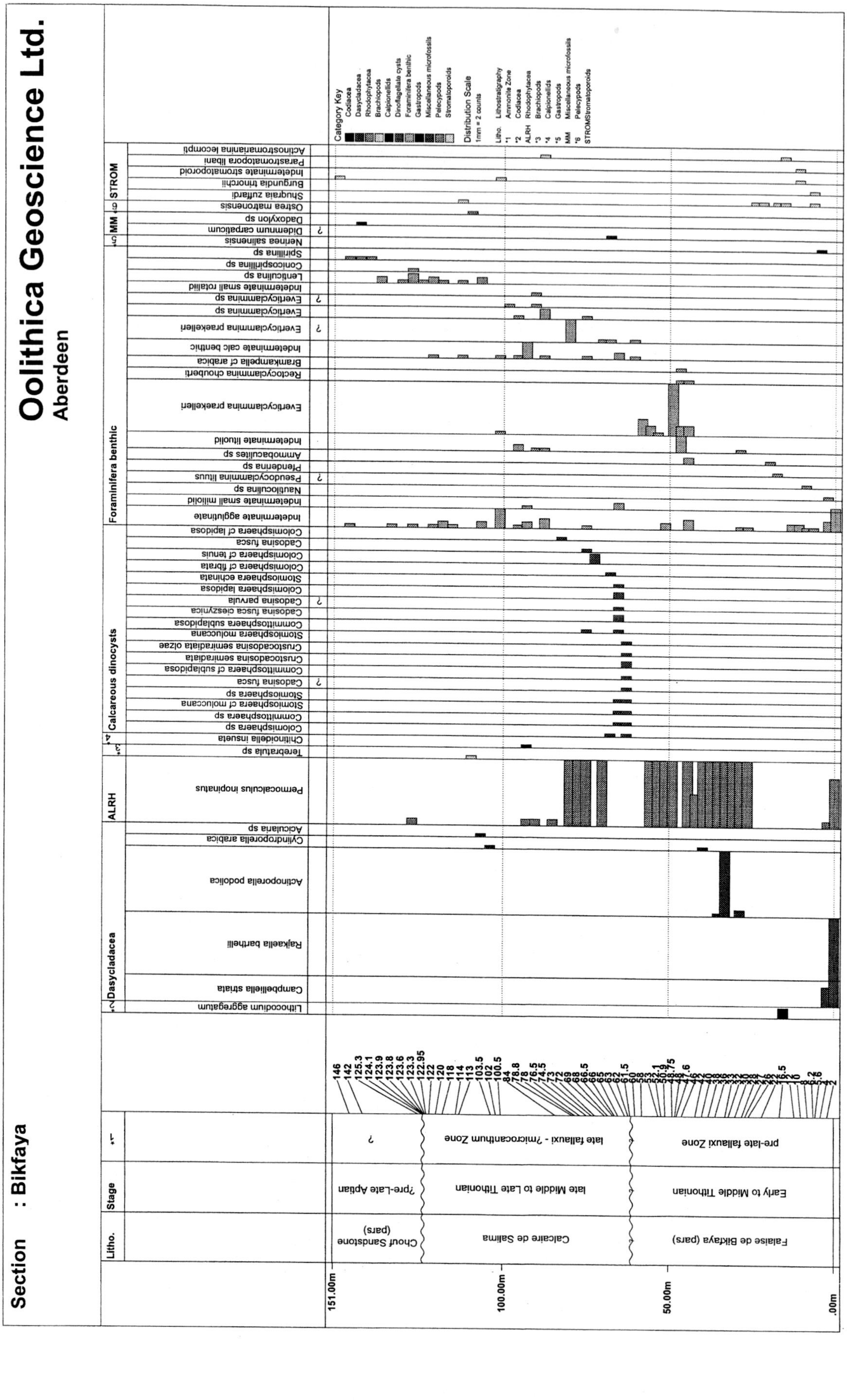


FIG. 5.—Paleontological distribution chart in Bikfaya reference section.

packstones. The lime mudstones are characterized by *Thalassinoides* burrow systems, and become increasingly argillaceous upsection, grading to calcareous claystones above 58.5 m. The uppermost part of this interval (58.5–61.1 m) is characterized by fine, millimeter-scale (?tidal) parallel laminae. The "upper member" biota is dominated by abundant *Everticyclammina praekelleri* Banner and Highton (1960) and *Permocalculus inopinatus*. The presence of *E. praekelleri* in the lower part of the "upper member" suggests a shallow neritic (?upper shoreface) paleoenvironment (Banner and Highton, 1990). The strongly argillaceous upper part of the "upper member" is tentatively interpreted as a (?brackish) intertidal mudflat facies, based on the millimeter-scale (?tidal) lamination, the absence of bioturbation, the restricted biota (ostracodes and bivalves), and the complete absence of stenohaline faunal elements.

Sequence 1 is dated as no older than Early Tithonian, on the basis of the presence of *P. inopinatus* and *A. podolica* (Simmons and Al-Thour, 1994; Bassoullet, 1997). This is consistent with the presence of *N. salinensis* (= *N. maroni*) in the "lower member", a species known from the basal Tithonian *gigas* Zone in Europe (Delpey, 1940). The upper part of the sequence is no younger than lowermost Upper Tithonian on the basis of a single specimen of *Chitinoidella insueta* Rehanek in the basal part of the overlying interval. The range of *C. insueta* is uppermost *fallauxi* to lower *microcanthum* Zone (Geyssant, 1997; Remane, 1997).

*Sequence Stratigraphy.—*

The interval 0–61.1 m is tentatively interpreted as a single, large-scale shoaling-up cycle, commencing with storm-influenced lower-shoreface facies ("lower member"), passing upward through subtidal *Permocalculus* algal-bank facies ("middle member"), culminating in intertidal facies ("upper member"). This is consistent with an aggradational–progradational highstand systems tract (HST) interpretation. The implication is that the measured section represents only the HST and not a complete depositional sequence. The TST and basal bounding surface are presumed to occur below base of section.

### *Sequence 2: Upper Middle to Upper Tithonian "Calcaire de Salima"*

*Description.—*

Sequence 2 (61.1–123.85 m) commences with an abrupt marine flooding surface and an associated influx of calcareous dinocysts, including *Colomisphaera lapidosa* Vogler, *C. tenuis* Nagy, *Crustocadosina semiradiata* Wanner, *C. semiradiata olzae* Nowak, *Stomiosphaera* sp. cf. *S. moluccana* Wanner, *S. echinata* Nowak, *Cadosina fusca* Wanner, *C. fusca cieszynica* Nowak, and *Commitosphaera sublapidosa* Vogler (Figs. 4, 5). This is the first record of calcareous dinocysts from Lebanon, though a very similar dinocyst assemblage occurs at about the same stratigraphic level in southern Yemen (Toland et al., 1995; Toland, unpublished data.).

The lower part of Sequence 2 (61.1–85 m) comprises recessive, strongly argillaceous silt-grade peloidal packstones and lime mudstones, with local resistant decimeter-scale interbeds of peloid–intraclast–superficial ooid packstone (Fig. 6A). The recessive silt-grade peloidal packstones and lime mudstones are characterized by local simple horizontal burrows (*Planolites* sp). The peloid–intraclast–ooid packstone interbeds are, in contrast, intensely bioturbated with abundant, small vertical and oblique escape burrows, and with bed tops heavily reworked by *Thalassinoides* networks. This is interpreted as an offshore transition zone facies association, with the resistant interbeds representing episodic storm deposits that have been biogenically reworked during fair-weather interludes. In addition to the rich dinocyst assemblage noted above, this interval has yielded agglutinated foraminifera (*Everticyclammina ?praekelleri* Banner and Highton, *E. ?virguliana* Koechlin, *?Ammobaculites* sp) and, at 78 m above base of measured section, a single specimen of *Chitinoidella insueta* Rehánek (Fig. 5) (Bischoff, 1964) also records the ostracodes *Cytherelloidea bikfayaensis* Bischoff and *C. salimaensis* Bischoff from this interval. The age of this lower interval (61.1–85 m) can be constrained to the upper Middle Tithonian (*ponti* Zone)–lowermost Upper Tithonian (lower *microcanthum* Zone), based on the presence of *C. tenuis* (*ponti*–lower *microcanthum* Zone), *C. sublapidosa* (upper *darwini* to lower *microcanthum* Zone), and *C. insueta* (uppermost *fallauxi*–lower *microcanthum* Zone).

The middle part of Sequence 2 (85–100 m) is poorly exposed at this locality, being covered by scree and scrub vegetation. This recessive (?marly) unit may be the source of rare cephalopods (*Phylloceras salima* Krumbeck, *Nautilus turcicus* Krumbeck, *Aspidoceras* sp., *Berriasella richteri* Oppel) described by Felix (1904), Krumbeck (1905), and Zumoffen (1926). Dr. Michael Howarth of the Natural History Museum (London) (personal communication, 1994). has suggested that the specimen referred to as "*B. richteri*" may, in fact, belong to the Upper Tithonian genus, *Pseudosubplanites* (Zumoffen's specimens were removed from St. Joseph's University to the Ecole Superieure d'Ingenieure de Beyrouth (ESIB) during the recent civil war, but an attempt to trace them there in 1994 proved unsuccessful and they may now be lost.)

The upper part of Sequence 2 (100–123.85 m) is, in contrast, well exposed in the low cliffs and abandoned quarries below Mar Elias, where it comprises medium- to thick-bedded, ooid–skeletal grainstones (Fig. 6B) which coarsen and thicken upsection. These grainstones are characterized by decimeter-scale unidirectional trough and tabular-planar cross-stratification, with foreset laminae dipping consistently towards the ESE (90–130°), i.e., onshore-directed. A prominent 3.6-m-thick tabular planar cross-set at 116.1–119.7 m can be traced laterally for almost 1 km along the quarry face and adjacent hillside (Fig. 2B). This cross-set represents a straight-crested subtidal sandwave, with the set height of 3.6 m, indicating minimum paleo-water depth. This sandwave has its long axis oriented NNE–SSW, i.e., normal to the foreset dip azimuth and broadly parallel to the regional platform-margin trend. This orientation suggests a wave-dominated setting (in a tide-dominated setting, by contrast, the major bedforms would tend to be oriented normal to the margin).

Directly overlying the sandwave unit at 116.1–119.7 m above base of section is a 3.2-m-thick interval characterized by indistinct, very-low-angle, planar bedding, with high-angle trough cross-sets (119.7–122.9 m). This is very tentatively interpreted as a ridge-and-runnell foreshore facies association, with the low-angle sets perhaps representing swash laminae. Diagnostic foreshore criteria such as keystone vugs were, however, not observed. This (?)foreshore member is overlain by a thin, recessive paleosol at 122.9–123.0 m (Fig. 6C), and a thin, terminal, trough cross-stratified oolitic grainstone unit (123.0–123.85 m). The sequence is abruptly terminated at 123.85 m by a prominent, irregular paleokarst surface with up to 0.2 m vertical relief (Fig. 6C). This former exposure surface was colonized to a depth of 5 mm by iron-stained, fossil endolithic lichen colonies, representing a proto-paleosol (Fig. 6D).

The sparse microfauna in the upper part of Sequence 2 (100–123.85 m) is dominated by calcareous benthic foraminifera (*Conicospirillina* sp; *Lenticulina* sp.). An abundant and diverse macrofauna includes bivalves (*Ostrea matronensis*), terebratulid

FIG. 6.—A. Burrowed lime mudstones at 62 m, forming the basal part of the "Calcaire de Salima" (Sequence 2). Samples from this locality yielded abundant calcareous dinocysts of late Middle to earliest Late Tithonian age (*ponti* to lower *microcanthum* Zone). B. Close-up view of the upper "Calcaire de Salima" at 113 m, comprising ooid grainstone with abundant bivalve, brachiopod, and crinoid debris. C. General view of the upper "Calcaire de Salima" (lower half of frame) disconformably overlain by brown, recessive ooid wacke–ironstones of the basal "Chouf Sandstone Formation". The thin recessive notch within the upper "Calcaire de Salima" is a paleosol. The upper bounding surface of the "Calcaire de Salima" is a paleokarst, shown in more detail in Figure 6D. Close-up view of the paleokarstified upper surface of the "Calcaire de Salima" at 123.85 m. The pale orange-brown areas are vadoid gravels infilling paleokarst cavities. The thin dark brown laminae are interpreted as limonitized paleo–lichen colonies (i.e., a protosol).

brachiopods, corals, crinoids, and echinoids (Fig. 6B). Krumbeck (1905) records *Terebratula asiatica* Krumbeck, *T. beyroutiana* Krumbeck, *T. curtirostris* Krumbeck, *T. logistimata* Krumbeck, *T. bauhini* Etallon, *T. bisuffaricinata* Schloth., *Kingenia gutta* Quenstedt, *K. cubica* Quenstedt, *K. latifrons* Krumbeck, *Rhynchonella drusorum* Krumbeck, *Alectryonia hastellata* Schloth, *Mytilus alatus* Krumbeck, *Ceratomya excentrica* Agas., *Pecten lycosensis* Krumbeck, *Lima (= Plagiostoma) sublaeviuscula* Krumbeck, *Lithodomus zumoffeni* Krumbeck, *Turbo (Chilodonta) antonini* Krumbeck, *Hemipedina eliasensis* Krumbeck, and *Trigonia libanensis* Krumbeck from this interval. The associated coral assemblage, described by Felix (1904), includes *Dimorphastrea kobyi* Felix, *Protoseris* cf. *foliosa* Beckmann, *Stephanocoenia trochiformis* d'Orbigny, *Centrastrea blanckenhorni* Felix, and *Actinarea* cf. *granulata* d'Orbigny. Delpey (1940) records the Tithonian gastropod *Nerinea* cf. *zeuschneri* Peters from the terminal part of the Jurassic at Bikfaya.

### *Sequence Stratigraphy.—*

The significant flooding surface at 61.1 m, which marks the base of Sequence 2, is interpreted as a cryptic "shale-on-shale" sequence boundary, with open-marine transgressive mudstones resting directly on paleontologically barren (?intertidal) mudstones of the underlying sequence. This initial marine flooding event is of inferred upper Middle Tithonian age (upper *fallauxi-ponti* Zone) given the presence of *C. insueta* 0.9 m up section. Inferred maximum flooding occurs within the *Chitinoidella* Zone at ca. 62 m, coincident with calcareous dinocyst species and abundance maxima. This suggests a potential correlation with a *ponti* Zone MFS, which has been identified at about this same stratigraphic level elsewhere in the Middle East (Toland, 1998), and a broadly coincident *pectinatus* Zone MFS is recognized in the northern North Sea (Partington et al., 1993). The interval 62–85 m is interpreted as a sediment-starved distal early HST with sedimentation occurring in the offshore transition zone, above storm wave base but below fair-weather wave base.

The overlying recessive unexposed unit at 85–100 m above base of section may be the source of previous ammonite finds. Alternatively, at least some of the ammonites may originate from the basal Cretaceous. Further fieldwork is required to resolve this uncertainty.

The ooid grainstones that comprise the uppermost part of the Jurassic section (100–122.9 m) could be interpreted as an aggradational HST, genetically linked to the underlying interval. Alternatively they could represent a forced regressive wedge systems tract, bounded below by a basal surface of forced regression and above by a sequence boundary. Further examination of the poorly exposed lower bounding surface is required to determine which of these two interpretations is most likely.

### *Sequence 3:* *Lower Cretaceous "Chouf Sandstone Formation" [pars]*

*Description.—*

Sequence 3 (123.85–151.1m) commences with a thin vadoid gravel (123.85–124.0 m) that infills solution hollows, pipes, and cavities in the underlying paleokarst surface; this is perhaps the same "pisolite bed" as documented by Ponikarov (1967) from the "Upper Tithonian" of Jebel Sheikh Mansour, NW Syria. This vadoid gravel represents a paleosol that may have supported conifer stands, judging by the presence of coniferous wood fragments a short distance upsection.

The basal vadoid gravel is overlain by a recessive, strongly argillaceous, oolitic wacke-ironstone bed (124–126.35 m), yielding a sparse, non-age-diagnostic biota including echinoid debris, coniferous wood fragments (*Dadoxylon* (= *Araucarioxylon*) sp.), and calcareous benthic foraminifera ("*Spirillina* sp."). These wacke-ironstones may represent transgressive reworking of contemporaneous lateritic iron-rich paleosols. Basalt lava flows, which are locally present directly beneath the Chouf Sandstone at Bhannes ca. 3 km WSW of the measured section, are a potential source of iron. The Bhannes basalt has recently yielded a late Middle to Late Tithonian radiometric age of 146.2 ± 2.5 Ma (Mike Simmons, personal communication), i.e., significantly older than the 139 ± 3 Ma Late Berriasian age recorded previously by Saint-Marc (1980).

The overlying interval 126.35–140 m is not exposed at this locality, though its recessive weathering profile suggests that it is continuous with the underlying argillaceous wacke-ironstone facies.

The uppermost part of the measured section (140–151.1 m) comprises cliff-forming, medium- to thick-bedded, sandy–ooid–peloid grainstones (Fig. 7A, B). Content of detrital quartz increases progressively upsection to a maximum of 24% bulk volume at the top of measured section. Glauconitized and phosphatized grains are also moderately common, accounting for up to 5% and 1% of bulk volume, respectively. This interval is characterized by prominent tabular-planar and trough cross-stratification, with unidirectional foresets dipping towards the east (60–110°). This is only slightly different from the ENE mean dip azimuth recorded in the underlying sequence. The presence of reactivation surfaces, clay-draped foreset laminae, and a *Skolithos–Ophiomorpha–Diplocraterion* ichnofauna all suggest a tide-influenced setting, with the unidirectional landward-facing foresets indicating a floodtide-dominated regime. The strong

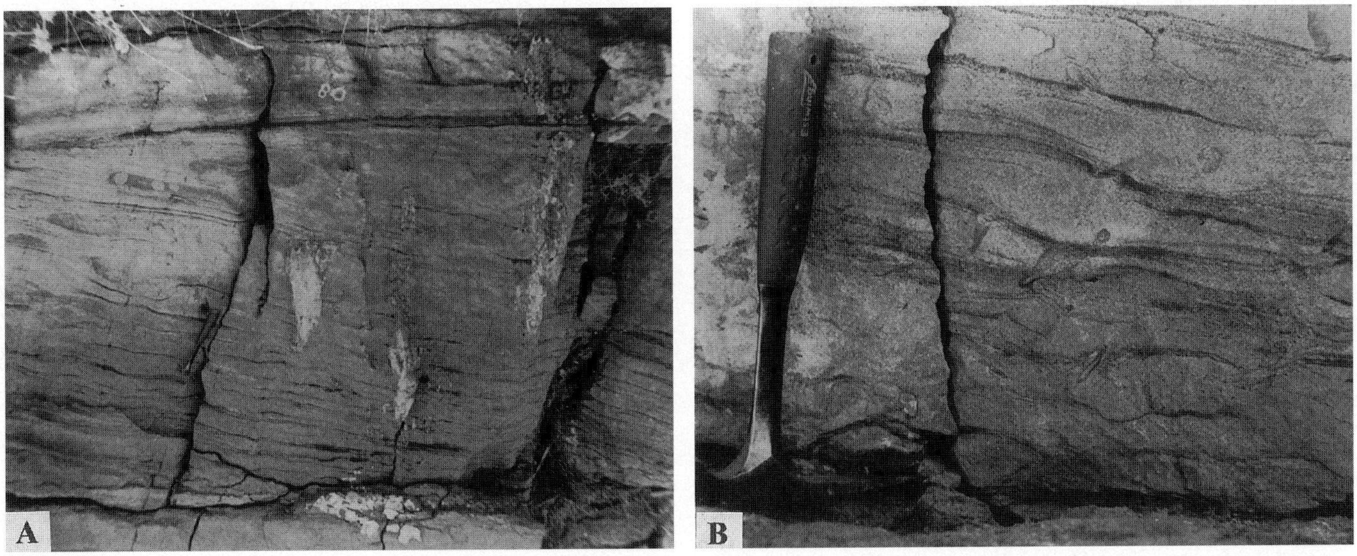


FIG. 7.—A. Tabular-planar cross-stratified ooid grainstone unit at 144.5 m ("Chouf Sandstone Formation"). The foresets are highlighted by thin, slightly recessive clay drapes suggesting a tidal regime. B. Close-up view of clay-draped foresets within the "Chouf Sandstone Formation" at 144.5 m. Note also the numerous clay-lined subvertical escape burrows, which suggest tidal sedimentation.

tidal signature is consistent with minimal frictional damping on a relatively narrow shelf, and supports the full separation of Apulia from the Levant by Early Cretaceous times, as suggested by Dercourt et al. (1986).

The sparse, non-age-diagnostic biota in the upper part of Sequence 3 (140–151.1 m) is dominated by echinoid debris and lituolid foraminifera (?*Ammobaculites* sp.), together with minor bivalve, bryozoan, and calcareous sponge debris. This is perhaps the same fauna of *Haplophragmoides* sp., *Ammobaculites subaequalis*, *Pseudocyclammina* sp., and *Globulina* sp. documented by Ponikarov (1967) from a 6-m-thick "Upper Tithonian" black marl unit at Ain Qaya, northern Anti-Lebanon. In the absence of an age-diagnostic biota, perhaps the best way of dating this interval would be Ar–Ar single-grain laser ablation radiometric dating of the glauconite and phosphate grains and/or single-grain laser ablation Sr/Sr dating of low-Mg calcite bivalve fragments.

Dubertret (1951) records a total thickness of 100–220 m for the "Grès de Base" and implies a Berriasian–Barremian age, on the basis of definitive Early Aptian echinoids and orbitolinid foraminifera in the directly overlying "Couches Orbitolina". Heybroek (1942) assigns an Early Cretaceous age to the "Grès de Base", and notes the occurrence of the Aptian bivalve, *Trigonia syriaca*, in the overlying succession cited previously by Fraas (1878) and Zummoffen (1926). A pre-Aptian Early Cretaceous age is further substantiated by the presence of the Early Aptian charophyte *Atopochara trivolvis* Peck in the overlying interval (Grambast and Lorsch, 1968; Tixier, 1972). The only fossils recorded previously from the "Grès de Base" itself are wood fragments (Dubertret, 1975), the Early Cretaceous (?Barremian) pteridophyte taxon *Weichselia reticulata* Stokes and Webb (Koert, 1924), and Barremian spores (Mroueh, 1980).

### *Sequence Stratigraphy.—*

The karstified upper surface of the "Calcaire de Salima" (Sequence 2) is an abrupt, disconformable, Type 1 sequence boundary. The contact with the overlying "Grès de Base" is clearly not gradational, as has been suggested previously by numerous authors. The mature paleokarst at 123.85 m represents a prolonged period of subaerial exposure, i.e., a relative sea-level lowstand. The timing and duration of this lowstand event are poorly constrained, though it is clearly no older than (late?) Late Tithonian, and probably spans the Jurassic–Cretaceous boundary. A broadly coincident end-Jurassic erosional hiatus is evident across much of the Arabian Plate (Toland, 1998), with the Upper Tithonian being generally absent except in some of the intra-shelf grabens of Yemen and at Chia Gara in northern Iraq (Howarth, 1992).

The thin vadoid gravel at 123.85–124.0 m represents transgressive reworking of a paleosol developed on the karstified upper surface of the "Calcaire de Salima". This basal lag is overlain by an argillaceous oolitic wacke-ironstone bed which represents a sediment-starved drowned shelf in "catch-up" mode. The winnowing and concentration of ooids to form packstone ribs suggests water depths above storm wave base during inferred maximum flooding at ca. 124 m. The inferred relatively shallow water depths are consistent with May's (1991) paleogeographic reconstruction for the Early Cretaceous of the Eastern Mediterranean, which depicts a paleoshoreline located a short distance east of Bikfaya.

The timing of base-Cretaceous onlap is only poorly constrained by the presence of Late Tithonian taxa in the underlying section and the presence of Barremian spores in the upper part of the Chouf Sandstone Formation. Onlap may be broadly coincident with a pronounced lower Middle Berriasian (*subalpina* Subzone in the upper part of *Calpionella* Zone B) transgression evident throughout much of Arabia (Toland, 1998; Toland et al., 1993) and farther afield in NW Europe (Jan du Chene et al., 1993; Coe, 1996), though there is at present no hard evidence to substantiate this possibility.

The stacked tidal facies that form the uppermost part of the Bikfaya section (140–151.1 m) are tentatively interpreted as an Early Cretaceous HST, with the presence of glauconite and phosphate suggesting starved sedimentation. This condensed interval may represent sediment bypassing during a relative stillstand, with abundant sediment supply but no accommodation space creation. The upper sequence boundary was not observed at this locality, and is assumed to occur some distance upsection.

## CONCLUSIONS

1. The upper part of the "Falaise de Bikfaya" extends into the Lower (and perhaps lower Middle) Tithonian, i.e., considerably younger than the Kimmeridgian age suggested previously.
2. The lower part of the "Calcaire de Salima" is precisely dated for the first time as upper Middle Tithonian (*ponti* Zone) on the basis of a previously undescribed, rich assemblage of calcareous dinocysts.
3. The Tithonian of the Bikfaya section comprises parts of at least two third-order depositional sequences and their component systems tracts.
4. The upper part of the "Falaise de Bikfaya" is interpreted as an aggradational–progradational highstand systems tract.
5. The base of the overlying "Calcaire de Salima" is a significant flooding surface of inferred upper Middle Tithonian age (upper *fallauxi-ponti* Zone).
6. A candidate maximum flooding surface (MFS) occurs ca. 1 m above the base of the "Calcaire de Salima", coincident with planktic species and abundance maximum; this MFS is of upper Middle Tithonian *ponti* Zone age, i.e., broadly coincident with a *ponti* Zone MFS recognized elsewhere in the Middle East.
7. The top of the "Calcaire de Salima" is marked by a prominent paleokarst (sequence boundary) and superimposed transgressive surface corresponding to the Jurassic–Cretaceous boundary.
8. The timing of base-Cretaceous onlap is uncertain; it may be broadly coincident with a pronounced lowermost Middle Berriasian (*subalpina* Zone in the upper *Calpionella* Zone B) transgression evident throughout much of Arabia, though there is no hard evidence to substantiate this at present.

## ACKNOWLEDGMENTS

It is my pleasure to thank Dr. Mike Simmons (Aberdeen University) and Dr. Bruno Granier (Total) for assistance with foram and algal identifications. The late Dr. Ziad Beydoun provided generous logistical assistance in the field. Financial and logistical support for this study was provided by Halliburton Reservoir Description Services and Oolithica Geoscience.

## REFERENCES

Banner, F.T., and Highton, J., 1990, On *Everticyclammina* Redmond (Foraminifera), especially *E. kelleri* (Henson): Journal of Micropalaeontology, v. 9, p. 1–14.

Banner, F.T., and Simmons, M.D., 1994, Calcareous algae and foraminifera as water-depth indicators: an example from the Early Cretaceous carbonates of northeast Arabia, *in* Simmons, M.D., ed., Micropalaeontology and Hydrocarbon Exploration in the Middle East: London, Chapman & Hall, p. 243–252.

BASSON, P.W., AND EDGELL, H.S., 1971, Calcareous algae from the Jurassic and Cretaceous of Lebanon: Micropaleontology, v. 17, p. 411–433.

BASSOULLET, J.-P., 1997, Algues Dasycladales, *in* Cariou, E., and Hantzpergue, P., eds., Biostratigraphie du Jurassique Ouest-Européen et Méditerranéen: Bulletin des Centres de Recherches Exploration-Production Elf-Aquitaine, Mémoir 17, p. 339–342.

BEYDOUN, Z.R., 1977, Petroleum prospects of Lebanon: reevaluation: American Association of Petroleum Geologists, Bulletin, v. 61, p. 43–64.

BEYDOUN, Z.R., 1981, Some open questions relating to the petroleum prospects of Lebanon: Journal of Petroleum Geology, v. 3, p. 303–314.

BISCHOFF, G., 1964, Die Gattung Cytherelloidea im Oberen Jura und in der Unterkreide: Senckenbergiana Lethaea [Frankfurt], v. 45, p. 1–27.

COE, A.L., 1996, Unconformities within the Portlandian Stage of the Wessex Basin and their sequence-stratigraphic significance, *in* Hesselbo, S.P., and Parkinson, D.N., eds., Sequence Stratigraphy in British Geology: Geological Society of London, Special Publication 103, p. 109–143.

DELPEY, G., 1940, Les Gastéropodes Mésozoïques de la région Libanaise: Notes et Mémoires Moyen-Orient, v. 3, p. 5–323.

DERCOURT, J., ZONENSHAIN, C.P., RICO, L.E., KAZMIN, V.G., LE PICHON, X., KNIPPER, A.L., GRANDJAQUET, C., SBURTSHIKOV, I.M., GEYSSANT, J., LEPURIER, C., PECHERSKY, D.H., BOULIN, J., SIBUET, J.C., SAVOSTIN, L.A., SOROKHTIN, O., WESTPHAL, M., BAZHENOV, M.L., LAUER, J.P., AND BIJU-DUVAL, B., 1986, Geologic evolution of the Tethys belt from the Atlantic to the Pamirs since the Lias: Tectonophysics, v. 123, p. 241–315.

DUBERTRET, L., 1949, Carte Géologique a 1/50000e du Liban: Feuille de Zebdan: République Libanaise, Ministére des Travaux Publics, Beyrouth.

DUBERTRET, L., 1950a, Sur le Jurassique du Liban et de l'Anti-Liban: Société Géologique de France, Compte Rendu Sommaire des Séances, v. 7, p. 12.

DUBERTRET, L., 1950b, Carte Géologique a 1/50000e du Liban: Feuille de Rayak: République Libanaise, Ministére des Travaux Publics, Beyrouth.

DUBERTRET, L., 1951, Carte Géologique a 1/50000e du Liban: Feuille de Beyrouth: République Libanaise, Ministére des Travaux Publics, Beyrouth.

DUBERTRET, L., 1963, Liban, Syrie: Lexique Stratigraphique International, v. III (Asie), fasc. 10 c 1 (Liban, Syrie, Jordanie): Center National de la Recherche Scientifique, Paris.

DUBERTRET, L., 1966, Liban, Syrie et bordure des pays voisins: Notes et Mémoires Moyen-Orient, v. 8, p. 355–394.

DUBERTRET, L., 1975, Introduction to la carte géologique a 1/50000e du Liban: Notes et Mémoires Moyen-Orient, v. 13, p. 345–403.

FELIX, J., 1904, Die fossile Fauna des Libanesischen Jurakalkes, I: die Anthozoenfauna des Glandarienkalkes: Beiträge zur Paläontologie und Geologie Österreich–Ungarns und des Orients, v. 15, p. 163–184.

FRAAS, O., 1878, Geologisches aus dem Libanon: Verein für Vaterländische Naturkunde in Württemberg, Jahresheft, v. 34, p. 257–391.

GADDO, J.Z.H., 1971, The Mishrif Formation palaeoenvironment in the Rumailah /Tuba/ Zubair region of S. Iraq: Geological Society of Iraq, Journal v. 4, p. 1–12.

GEYSSANT, J., 1997, Tithonien, *in* Cariou, E., and Hantzpergue, P., eds., Biostratigraphie du Jurassique Ouest-Européen et Méditerranéen: Bulletin des Centres de Recherches Exploration-Production Elf-Aquitaine, Mémoir 17, p. 97–102.

GRAMBAST, L., AND LORCH, J., 1968, Une flore de charophytes du Crétacé inférieur du proche-orient: Naturalia Monspeliensia, série Botanica, v. 19, p. 47–56.

HEYBROEK, F., 1942, La géologie d'une partie du Liban sud: Leidsche Geologische Mededeelingen, v. 12, p. 251–470.

HOWARTH, M.K., 1992, Tithonian and Berriasian ammonites from the Chia Gara Formation in northern Iraq: Palaeontology, v. 35, p. 597–655.

HUDSON, R.G.S., 1954, Jurassic stromatoporoids from the Lebanon: Journal of Paleontology, v. 28, p. 657–661.

JAN DU CHENE, R., BUSNARDO, R., CHAROLLAIS, J., CLAVEL, B., DECONINCK, J.-F., EMMANUEL, L., GARDIN, S., GORIN, G., MANIVIT, H., MONTEIL, E., RAYNAUD, J.-F., RENARD, M., STEFFEN, D., STEINHAUSER, N., STRASSER, A., STROHMENGER, C., AND VAIL, P.R., 1993, Sequence-stratigraphic interpretation of Upper Tithonian–Berriasian reference sections in Southeast France: a multidisciplinary approach: Bulletin des Centres de Recherches Exploration-Production Elf-Aquitaine, v. 17, p. 151–181.

KOERT, W., 1924, Geologische Beobachtungen in Syrien und Palästina während des Feldzuges 1917/18.: IV. Tektonik und Wealden auf der Nordwestseite des Kneysseh im Libanon: Deutsche Geologische Gesellschaft, Zeitschrift, v. 76, p. 23–29.

KRUMBECK, L., 1905, Die Brachiopoden und Molluskenfauna des Glandarienkalkes: Beiträge zur Paläontologie und Geologie Österreich–Ungarns und des Orients, v. 18, p. 65–162.

MAY, P.R., 1991, The eastern Mediterranean Mesozoic basin: evolution and oil habitat: American Association of Petroleum Geologists, Bulletin, v. 75, p. 1215–1232.

MROUEH, M., 1980, Spores and pollen from Lower Cretaceous sediments, Central Lebanon (abst.): 26th International Geological Congress, Paris, France.

ORME, G.R., AND SALAMA, M.S., 1988, Form and seismic stratigraphy of Halimeda banks in part of the northern Great Barrier Reef Province: Coral Reefs, v. 6, p. 131–137.

PARTINGTON, M.A., COPESTAKE, P., MITCHENER, B.C., AND UNDERHILL, J.R., 1993, Biostratigraphic calibration of genetic sequences in the Jurassic–lowermost Cretaceous (Hettangian–Ryazanian) of the North sea and adjacent areas, *in* Parker, J.R., ed., Petroleum Geology of Northwest Europe: Proceedings of the 4th Conference, Geological Society of London, p. 371–386.

PFENDER, J., 1937, Quelques Hydrozoaires de la Syrie septentrionale: Haut-Commissariat de la République Française en Syrie et au Liban, Notes et Mémoires, v. 2, p. 125–136.

PICARD, L., AND HIRSCH, F., 1987, The Jurassic Stratigraphy in Israel and the Adjacent Countries: Jerusalem, The Israel Academy of Sciences and Humanities, 106 p.

PONIKAROV, V.P., ed., 1967, The Geology of Syria. Explanatory Notes on the Geological Map of Syria, Scale 1:500000, Part I, Stratigraphy, Igneous Rocks and Tectonics: Syrian Arab Republic, Ministry of Industry, 229 p.

REMANE, J., 1997, Calpionelles, *in* Cariou, E., and Hantzpergue, P., eds., Biostratigraphie du Jurassique Ouest-Européen et Méditerranéen: Bulletin des Centres de Recherches Exploration-Production Elf-Aquitaine, Mémoir 17, p. 243–247.

RENOUARD, G., 1951, Sur la dècouverte de Jurassique inférieur (?) et du Jurassique moyen au Liban: Acadèmie des Sciences [Paris], Comptes Rendus, v. 232, p. 992–994.

RENOUARD, G., 1955, Oil prospects of Lebanon: American Association of Petroleum Geologists, Bulletin, v. 39, p. 2125–2169.

RENZ, C., 1930, Neue Korallengattungen im Libanon und Antilibanon in Syrien: Schweizere Paläontologische Gesellschaft, Abhandlungen, v. 50, p. 1–4.

SAINT-MARC, P., 1980, Le passage Jurassique–Crétacé et le Crétacé inférieur de la région de Ghazir (Liban central): Géologie Méditerranéene, v. 7, p. 237–245.

SIMMONS, M.D., AND AL-THOUR, K., 1994, Micropalaeontological biozonation of the Amran Series (Jurassic) in the Sana'a region, Yemen Republic: *in* Simmons, M.D., ed., Micropalaeontology and Hydrocarbon Exploration in the Middle East: London, Chapman & Hall, p. 43–80.

TIXIER, B., 1972, Le 'Grès de Base' Crétacé du Liban: étude stratigraphique et sédimentologique: Notes et Mémoires Moyen-Orient, v. 12, p. 187–215.

Toland, C., 1994, Late Mesozoic stromatoporoids: their use as stratigraphic tools and paleoenvironmental indicators, *in* Simmons, M.D., ed., Micropalaeontology and Hydrocarbon Exploration in the Middle East: London, Chapman & Hall, p. 113–125.

Toland, C., 1998, A Sequence Stratigraphic Framework for the Upper Callovian–Lower Valanginian of the Arabian Plate, 2nd edition: Oolithica Geoscience Ltd non-proprietary study. 2 vols.

Toland, C., unpublished., Calcareous dinocyst assemblages from the Upper Kimmeridgian–Lower Valanginian of the Arabian Plate: systematics and stratigraphic significance.

Toland, C., Peebles, R.G., and Walkden, G.M., 1993, Upper Jurassic and basal Cretaceous sequence stratigraphy of Wadi Hagil, Oman Mountains: 8th Middle East Oil Show and Conference, Bahrain, SPE Paper 25581, Proceedings, v. 2, p. 533–544.

Toland, C., Simmons, M.D., and Walkden, G.M., 1995, A new sequence stratigraphic reference section for the Upper Jurassic of Yemen, *in* Al-Husseini, M.I., ed., Geo '94 Middle East Geoscience Conference: Gulf PetroLink, Manama, Bahrain, p. 891–899.

Walley, C.D., 1983, A revision of the Lower Cretaceous stratigraphy of Lebanon: Geologische Rundschau, v. 72, p. 377–388.

Zumoffen, G., 1926, Géologie du Liban, avec carte géologique au 1:200 000: Paris, Barrere, 165 p.

# CAN THE SEDIMENTARY RECORD BE DATED FROM A SEA-LEVEL CHART? EXAMPLES FROM THE APTIAN OF THE U.A.E. AND ALASKA

CHRISTOPHER G. ST. C. KENDALL
*Department of Geological Sciences, University of South Carolina, Columbia, SC 29208, U.S.A.*
ABDULRAHMAN S. ALSHARHAN
*Faculty of Science, United Arab Emirates, P.O. Box 17551, Al Ain, U.A.E.*
AND
KURT JOHNSTON AND SEAN R. RYAN
*Department of Geological Sciences, University of South Carolina, Columbia, SC 29208, U.S.A.*

ABSTRACT: Ten depositional shoaling-upward cycles have been identified in the Aptian Shuaiba Formation of the United Arab Emirates (U.A.E.). A similar number of cycles have been recognized in the National Petroleum Reserve of Alaska (NPRA). Similarities of these cycles with the onlapping shelf geometries of the Neogene of the Bahamas suggest that the sequence geometries of Aptian strata of the NPRA and the U.A.E. are a response to high-frequency changes in eustatic sea-level position. Because the Aptian cycles of the NPRA match similarly dated, events in the U.A.E., it is suggested that where biostratigraphic data are poor the sedimentary section can be tentatively dated by relating the geometries of the shelf margin to the character of the coastal onlap curve and its coincident sea-level chart. Thus, a sea-level chart might be used at locations for which biostratigraphic data is sparse to determine and constrain preliminary depositional models for specific time intervals.

With this in mind, two biostratigraphic models for dating the Shuaiba Formation and the Bab Member were tested against the sea-level curve of Haq et al. (1987) using a sedimentary simulation. The results were ambiguous because both biostratigraphic models could not be matched. Also in both cases, the simulation suggested that just prior to the deposition of the Bab Member the basin margin was uplifted and then subsided, causing a local relative sea-level fall followed by a rise, an event not found on the sea-level chart of Haq et al. (1987). Additionally, the sedimentary simulation supports the position that the Aptian in the U.A.E. is bounded by erosional unconformity surfaces and contains higher-frequency cycles.

## INTRODUCTION

When making sequence stratigraphic interpretations, sequences are identified on the basis of bounding unconformities and the onlapping and offlapping relationship of sedimentary geometries (Vail et al. 1977; Vail and Todd, 1981; Van Wagoner et al., 1987; Vail, 1988; Posamentier and Vail, 1988; Posamentier et al., 1988). For stratigraphic sections in which fossil evidence is sparse, the hypothesis is proposed that the ages of these sequences can be inferred by matching the size, shape, and position of the onlapping sequence geometries to the relative magnitudes and geometries of the sea-level events on an eustatic chart (Kendall et al., 1997a) (Fig. 1).

In other words, the events on the sea-level chart reported by Haq et al. (1987) have produced unique responses in the sedimentary section through time. Each sea-level cycle transgressed the basin margins by different amounts. This produced onlapping sequences of sedimentary fill that should have a unique onlap position on the shelf margin that, when compared to the position of the adjacent underlying and overlying sequences, can be directly related to the shape of the sea-level chart. We tested this working hypothesis using the Haq et al. (1987) sea-level chart with a simulation, and we have used the results of this work to date the geologic section of the Neogene of the Bahamas. We then confirmed these dates with biostratigraphic data (Lidz and McNeil, 1998; Kendall and Sen, 1998).

A major problem with this test is that, realistically, eustatic sea-level does not alone control the position of the sea surface with respect to the adjacent sediment surface. Both sediment supply and accumulation rates as well as tectonics affect this accommodation too! So, contrary to the findings described above, if a series of chronostratigraphic charts is created from seismic profiles at different positions across a shelf margin, the position and onlapping character of sequences with similar ages can be expected to vary from one geographic locale to another, reflecting the different tectonic and sedimentary histories at each location. Recognizing that there is some variability in the onlapping geometries produced by variations in local tectonics and sedimentation, we selected the profiles for this study because they had simple tectonic and sedimentary histories and because their sequence geometries appeared to be largely a response to the major sea-level events. Simulations developed to recreate these sections used the sea-level values of Haq et al. (1987) as a known variable, and rates of tectonic movement were uniform for a simulation run. The sedimentation accumulation rates were changed only as a last resort to achieve a better geometric match to the section. The resulting onlapping sequence geometries of the simulation reflected the effects of the eustatic events used in the simulation and, we believe, provided realistic solutions.

To set the stage for application of the hypothesis to the Lower Cretaceous of the National Petroleum Reserve of Alaska (NPRA) and the United Arab Emirates (U.A.E.), this paper begins by describing the test of the hypothesis made by Kendall and Sen (1998). For this purpose, they dated the Neogene section of the Bahamas by matching the shapes of the onlapping sequence geometries to the sea-level chart of Haq et al. (1987). These dates were confirmed with biostratigraphic data. We then report on our studies of the Aptian–Albian sections of the NPRA of Alaska and of the U.A.E.

The rationale for picking these latter areas and profiles was that though they represent a less well defined part of the geological column than the Neogene, the strata are of similar age and are associated with sufficient data to enable further examination of our hypothesis. Both data sets have seismic profiles that contain evidence of onlapping and prograding basin-margin sequences.

Middle East Models of Jurassic/Cretaceous Carbonate Systems
SEPM Special Publication No. 69, Copyright © 2000
SEPM (Society for Sedimentary Geology), ISBN 1-56576-075-1, p. 65–75.

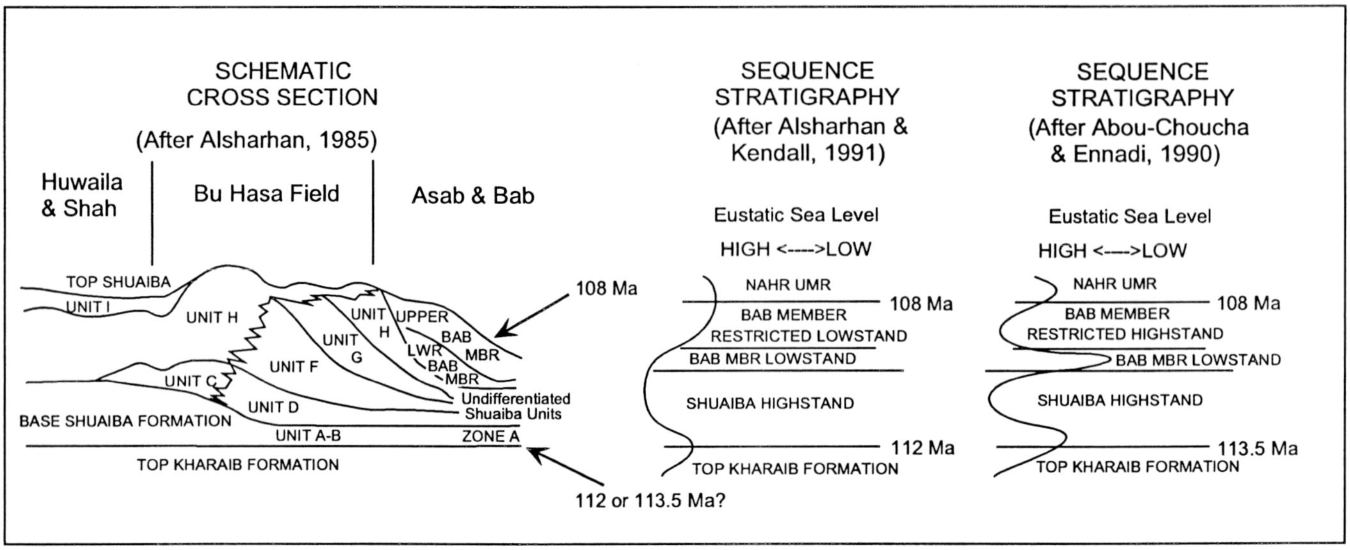


FIG. 1.—Schematic cross section of the Shuaiba Formation and Bab Member at Bu Hasa field in the U.A.E. with two alternative time frames. The upper part of the Shuaiba Formation forms a thick porous shelf-margin buildup trend with stratigraphic relief equivalent to the less porous carbonates shelfwards and basinwards. The overlying Bab Member pinches out basinward and shelf ward (Alsharhan, 1985; Abou-Choucha and Ennadi, 1990; Alsharhan and Kendall, 1991; Simmons 1994, Witt and Gökdag, 1994; Fischer et al., 1995). Sea-level chart based on Haq et al. (1987).

The NPRA seismic data set shows a clearer set of high-frequency sequences but has imprecise biostratigraphic control, while the U.A.E. seismic section shows poorer resolution of the high-frequency sequences but is associated with more detailed biostratigraphy. This paper focuses more on the U.A.E. because this was the area for which we had the most data.

For both areas we were able to demonstrate, through the medium of a sedimentary simulation, that eustatic events produced distinct sedimentary responses. The products of these eustatic events were used to provide preliminary dates for parts of the sections, suggesting that in a region where biostratigraphic data are sparse, a sea-level chart coupled with sedimentary simulations could be used as a powerful tool for interpreting the sedimentary section. The major assumption to the hypothesis is that subsidence and sedimentation are uniform so that for the duration of the simulation the eustatic sea-level cycles produce distinct and identifiable sequence geometries.

## INITIAL TEST OF THE HYPOTHESIS

The above hypothesis was tested with a study of the Neogene section of the Bahamian Banks (Eberli et al., 1994; Kendall and Sen, 1998). To achieve this, a seismic line from the Bahamas (Fig. 2) was interpreted by first identifying the second-order type 1 unconformities (Fig. 3). These second-order unconformities were then related to the second-order events on the sea-level chart of Haq et al. (1987). Then, counting the third-order unconformities between the second-order ones, the shapes of these third-order cycles were matched to events on the sea-level chart. The ages of the sea-level events were then given to the third order sequence boundaries that had been interpreted on the seismic profile.

A sedimentary simulation, Sedpak (Eberli et al., 1994), was then used to test the match between the ages of the cycles and the interpretation of the sequences on the seismic (Kendall and Sen, 1998). Kendall and Sen (1998) iteratively changed the inputs to the simulation, including trying a variety of different rates of tectonic subsidence and sedimentation, and sea-level models. The sedimentary simulation that provided the best match to the seismic interpretation eventually used as inputs constant rates of tectonic subsidence and carbonate accumulation, taking the timing and magnitude of the sea-level history from the Haq et al. (1987) chart.

The resulting position of the onlapping geometries of the simulation (Fig. 4) and those of the interpreted seismic data (Fig. 3) are similar. The sequence boundaries were tied to biostratigraphic data (Lidz and McNeil, 1998; Kendall and Sen, 1998) which suggests that the Haq et al. (1987), sea-level chart for the Neogene was accurate. The authors were able to identify individual eustatic sea-level events, determine their magnitude, and establish the timing of sequences identified on the seismic profile. In fact they argued that, if the rates of subsidence are low in an area and the rates of sedimentation were high, the stratigraphic signal produced by sea-level change should be clear. This model assumed that the rate of carbonate sedimentation was such that shoreward accommodation filled to sea level and that the sediment surface on the shelf could be taken as a proxy for sea level. Not only that, but if the rates of subsidence and carbonate accumulation were constant for several cycles of eustatic sea level, then the frequency and amplitude of the geometries of the onlapping carbonate bodies of the shelf were the product of the frequency and size of the cycles of eustatic sea level. So, if a simulation with a low rate of constant subsidence and a high rate of constant carbonate accumulation is used with the Haq et al. (1987) curve and produces cross sections that match that of the interpreted seismic, then the height of the eustatic events could be correct and their ages might be read from the eustatic chart. Nevertheless, this conclusion is dependent on the models proposed for the seismic and sedimentary geometries. Essentially, use of the sea-level chart cannot be divorced from confirmatory data from biostratigraphy, and should be considered only as a first pass until better biostratigraphic data become available.

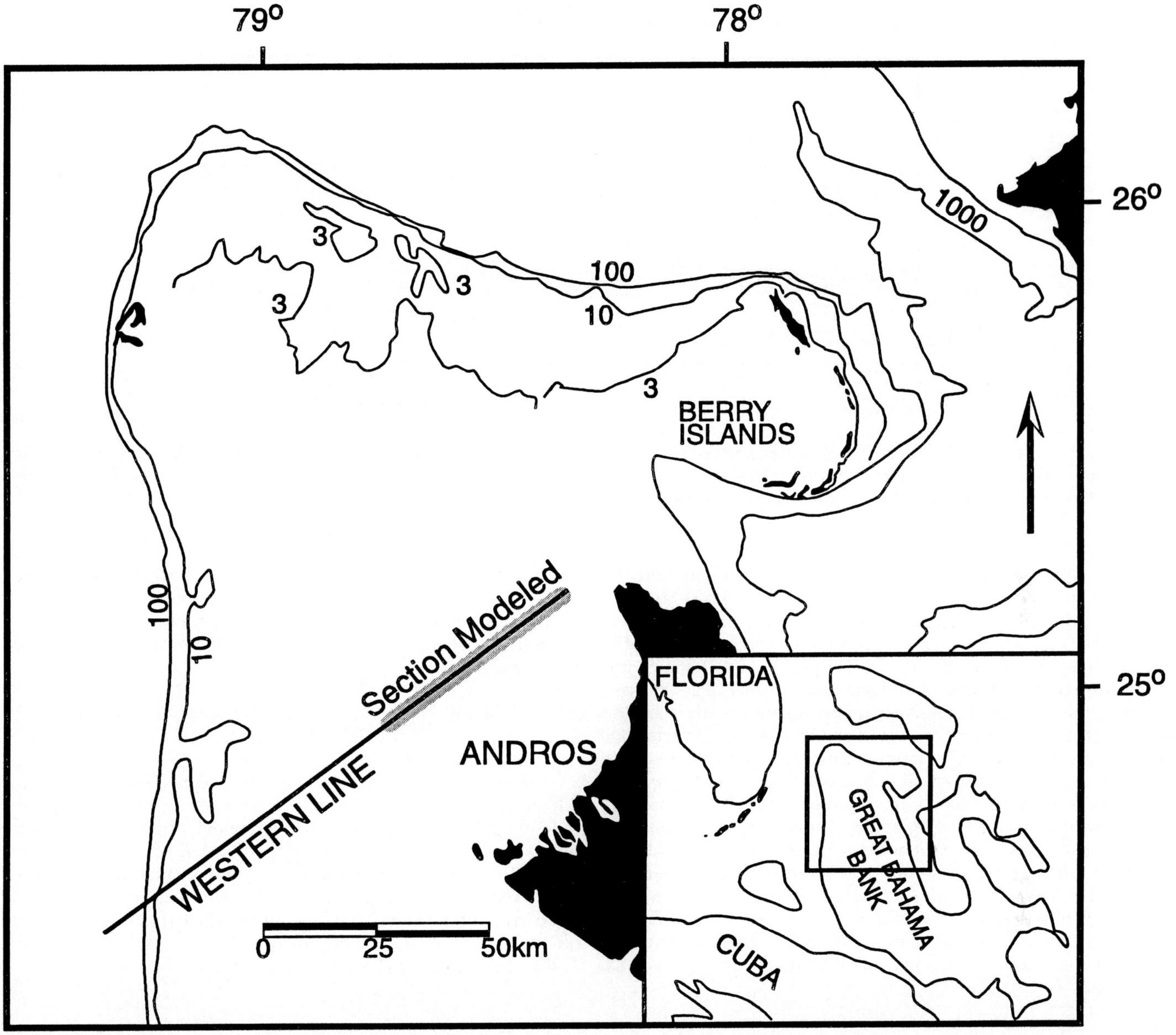


FIG. 2.—Map of Great Bahamas Bank showing the location of the Western line and the highlighted part of this line modeled in this paper. Four depth contours (1000, 100, 10, and 3 meters) are shown (after Kendall and Sen, 1998).

## DATING THE APTIAN OF ALASKA BY SEA-LEVEL CHART

Using the hypothesis described above we examined and modeled two Aptian to Albian geological sections. We first examined the Lower Cretaceous Torok and Nannashuk Formations in the NPRA area north of the Brooks Mountain Range (Kendall et al., 1997b; Ryan, 1997). Here a large wedge of deltaic sediments has prograded northwestward across the northern cratonic edge of Alaska (Bird and Molenaar, 1992). The Nanushuk Formation consists of the proximal sandier facies of the Torok Formation prograding complex. The depositional sequence framework for the Torok and Nanushuk Formations was established using regional seismic (Fig. 5) and well data, and reveals an eastward-prograding shelf edge that contains a series of type 1 unconformities (Fig. 6). It is our interpretation that high-amplitude third-order eustatic events were superimposed on the longer-term second-order sea-level events. Collectively these eustatic events created the characteristic geometric patterns of the depositional margin, including packages of onlapping highstand sigmoidal reflectors and downlapping lowstand systems tracts. These depositional geometries were simulated in order to determine the relationship between the sequences and the variables affecting relative sea level (sea-level position, clastic supply, and subsidence). Though the sections thin onto the cratonic margin to the north, the general lack of faulting and the uniformity of the onlapping sediment thicknesses suggest that changes in base level were driven by eustasy.

Ryan (1997) identified the second-order, type 1 unconformities on seismic profiles within the Aptian–Albian part of the section and correlated these to the second-order, type 1 sea-level falls on

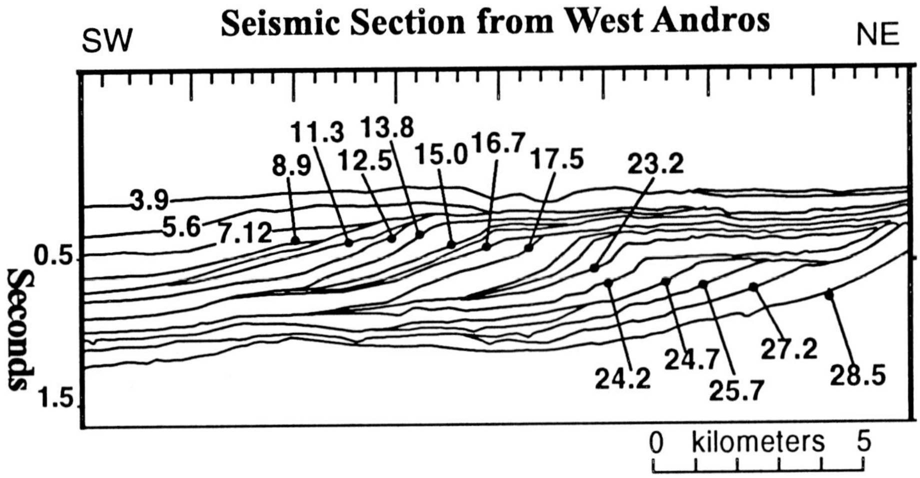


FIG. 3.—Line drawing of the interpreted seismic section for the Neogene carbonate section of the Andros Channel, Bahamas, showing the major sequence boundaries identified in the section with their interpreted ages (after Kendall and Sen, 1998). The numbers linked by lines to the surfaces are ages in millions of years.

the sea-level curve of Haq et al. (1987). He proposed that the unconformities were of the same ages as the sea-level falls. He then identified unconformities in the third-order sequences between these second-order, type 1 events and matched these to the third-order sea-level events on the Haq et al. (1987) chart, and assumed that the ages of these intermediate cycles, or sequences, were those seen on the chart. After using a variety of inputs to the simulation, he iteratively converged on using sea-level variations from the Haq et al. (1987) chart, eventually assuming a simple model of constant sedimentation and constant subsidence. This latter model of tectonic behavior was in part based on the uniformity of the thickness of the onlapping sequence geometries. The simulation output was matched to the seismic by iteratively tweaking rates of constant sedimentation and of constant subsidence until the simulation output matched the seismic (Fig. 6). The biostratigraphic ages for the National Petroleum Reserve of Alaska were used to bracket portions of the Aptian–Albian. The modeling of the intermediate section confirmed the probable existence of major second-order and third-order eustatic sea-level variations in the Aptian–Albian. It demonstrated that these major eustatic changes were probably the primary control for the development of the stratal geometries in the Torok and Nanushuk Formations.

## USE OF THE SEA-LEVEL CHART FOR DATING THE APTIAN OF THE U.A.E.

We next tested the hypothesis in the U.A.E., where the Aptian–Albian sedimentary section has a more detailed biostratigraphy than in the NPRA. In the U.A.E. the Lower Cretaceous is made up of major cycles of carbonate deposition (Fig. 7). These were terminated with an unconformity in the Turonian that coincided with a major marine regression. The Lower Cretaceous accumulated in an intercratonic shallow basin (Fig. 8) that was developed on the Arabian Shield back from the southern margin of the Tethys Sea. Its final character was imposed by the onset of the

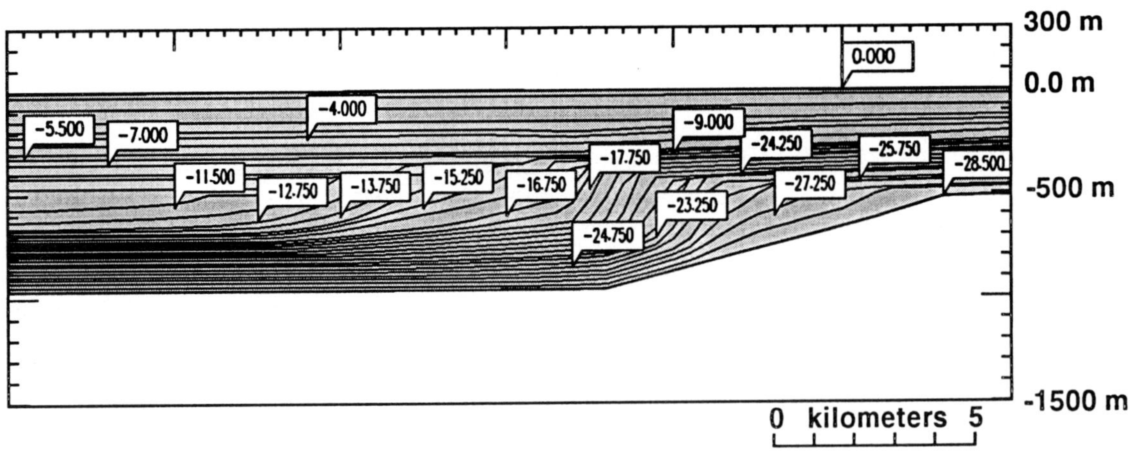


FIG. 4.—Simulation output showing the modeled Neogene sequences separately along with the resultant sequence-boundary ages for the Neogene carbonate section of the Andros Channel, Bahamas. Ages except for 28.5 Ma refer to top of sediment packages (after Kendall and Sen, 1998). The numbers in the arrowed boxes are ages in millions of years.

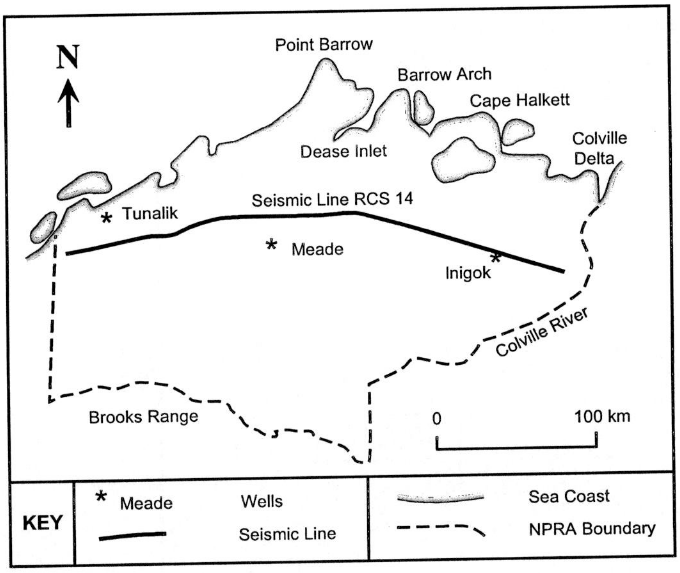


FIG. 5.—Map of National Petroleum Reserve of Alaska showing the seismic line simulated in Figure 6 (after Ryan, 1997).

collision of the Asian Plate with the Arabian Shield, which began to crumple in the Turonian (Alsharhan and Nairn, 1988).

Initially a basin with a relatively flat bottom was formed. It was rimmed by shallow-water carbonates that climaxed with rudistid reefs (Fig. 8). These prograded into the basin, producing large variations in the thickness between the sediments of the basin and those of the shelf (Aldabal and Alsharhan, 1989) (Fig. 1). By the end of the deposition of the Aptian Shuaiba Formation, the thickness of the shelf deposits was approximately 400 meters, and at the center of the basin sediment thicknesses were on the order of 100 to 150 meters (Azzam and Taher, 1995). It is proposed that the progradational shelf margin formed as a series of lowstand, transgressive, and highstand carbonate wedges that extended out from the edge of the basin. These eventually filled the margins of this basin with a rudistid facies. The depositional model for the Shuaiba Formation suggests that shortly after the deposition of the Kharaib Formation a transgression occurred, the water deepened a little, and a shallow-water shelfal limestone with *Lithocodium* sp. accumulated (Alsharhan, 1985). Progressively through the Aptian these slightly deeper shelf sediments shoaled upward and eventually acquired a cap of rudistid reefs (Calavan et al., 1992). Finally the buildups were exposed and then onlapped by the Albian Nahr Umr Formation (Alsharhan, 1991). On the basis of lithology, diagenesis, and fauna, we interpret these lower Shuaiba carbonate bodies as not having reached sea level. We propose that they were prevented from reaching the air/water interface by the onset of the anoxic conditions produced when the basin became isolated. Later, by the end of Shuaiba Formation deposition, the carbonate platform reached sea level. The platform margin is represented by an enormous buildup created by the accumulation of rudistids and their debris. The seismic line (Fig. 9) shows the progradational character of this margin.

The characterization of the biostratigraphy by Simmons (1994), Witt and Gökdag (1994), Calavan et al. (1992), Fischer et al. (1994), and Kendall et al. (1997b), suggests that at least two chronostratigraphic models constrain the interpretation of the timing of the Shuaiba carbonate wedges. The first model was that of Alsharhan and Kendall (1991), Simmons (1994), and Witt and Gökdag (1994), in which the base of the Shuaiba Formation was correlated with the lower part of the Aptian and matched with the 112 Ma transgression of the Haq et al. (1987). The rest of the Shuaiba Formation was modeled as a series of progradational units that terminated at 108 Ma just before the major sea-level drop at 107.5 Ma. This drop was followed by the deposition of the onlapping Nahr Umr Formation (Figs. 2, 10).

The second group of chronostratigraphic interpretations that we modeled were those of Calavan et al. (1992) and Fischer et al. (1994). They interpreted that deposition of the Shuaiba Formation spanned from approximately 113.5 Ma to 108 Ma (Fig. 1). Thus, in contrast to Alsharhan and Kendall (1991), the base of the Shuaiba was placed at 113.5 Ma, and the lower Shuaiba was interpreted to represent deposition during a slightly higher sea level, followed by a drop in sea level at 112 Ma. During deposition of the rest of the Shuaiba Formation two further progradational units were identified and dated from 112 Ma to 109.5 Ma and from 109.5 Ma, to approximately 108 Ma, after which a major sea-level drop occurred. This drop was followed by the deposition of the onlapping Nahr Umr Formation (Figs. 1, 9).

We tested both biostratigraphic models against the Haq et al. (1987) curve. In both group one and group two models (112 Ma to 108 Ma, and 113.5 Ma to 108Ma), to simulate the geometric shapes of the carbonate bodies responding to sea level, we counted the number of well-log lithologic cycles that we saw in the Shuaiba Formation. We then assumed that each carbonate cycle accumulated over a period of around 40,000 to 100,000 years and that the coincident sea-level variations had a sinusoidal character. For the time intervals involved in both models, we superimposed these fourth-order sinusoidal high-frequency events onto the chart of Haq et al. (1987), producing what we considered to be the sea-level curve responsible for the various geometries that can be seen in the Shuaiba Formation. We then ran the Sedpak simulation using a basin with a flat bottom. This shape was assumed on the basis of the changes in thickness in the shelf margin and basin, and on a water depth of no more than 50 meters. Carbonate sedimentation and subsidence were varied but kept as constant as possible, so the character of the geometries responded most to the variations in sea level. The results were a match to the geometries that can be seen on both the seismic and stratigraphic cross sections that have been developed for the Shuaiba Formation (Figs. 10, 11).

We assumed that eustasy, carbonate production rates, and tectonics were the major variables controlling sequence geometry in the Shuaiba Formation. The sea-level curve used in the simulation was modified from Haq et al. (1987). We added a fourth-order, 100,000 year cycle to it. The timing of the resulting sequences is assumed to be equal to that of this sea-level event. We then assumed that the sea-level curve that we developed for this study was reasonable and needed no further adjustments. Carbonate production and tectonics in the simulation were iteratively varied through time to acquire the best match between the interpretation and the simulation. When we kept these parameters constant, the resulting simulations did not convincingly match the seismic interpretation.

The most accurate simulation of the Shuaiba Formation sequence stratigraphy during the interval from 112 Ma to 108 Ma were achieved with the following input variables. From 112 to 110 Ma, the carbonate production rate was set at 35 cm/1000 years close the air/sea surface. This rate declined quickly with depth. During the same interval, the subsidence rates was at 20 cm/1000 years. The results of the simulation (Fig. 10) show a fairly rapid change from a ramp margin to a well defined shelf edge by the end of Shuaiba Formation deposition. The depositional sequences of the Shuaiba Formation aggrade with minor progradation at the shelf edge.

At 110 Ma the shelf was uplifted at a rate of 5 cm/1000 years to maintain the position of sea level below the shelf edge throughout

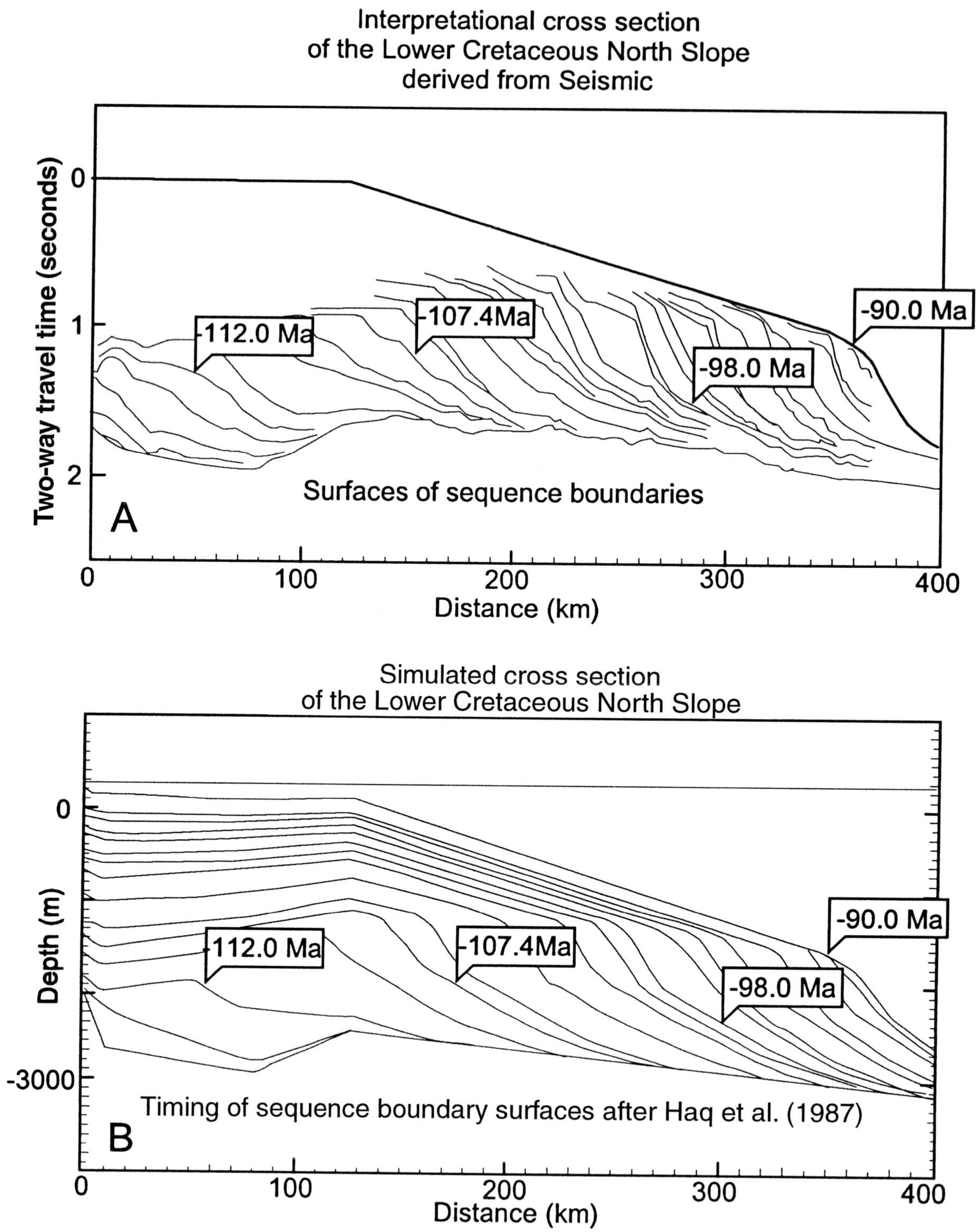


FIG. 6.—A) Seismic cross section of the Torok and Nanushuk Formations shelf margin and B) the associated simulation created with the program Sedpak (after Ryan, 1997). The numbers in the arrowed boxes are ages in millions of years.

the deposition of the Bab Member. Because of the lowered sea level, Bab carbonate production was reduced to 3 cm/1000 years at the surface and declined slowly to near zero at 100 meters water depth. This simulation resulted in two progradational sequences. The first progradational sequence ranges in age from 112 to 110 Ma and onlaps the Shuaiba slope. Following the tectonic uplift and a resulting relative sea-level fall at 110 Ma, the Bab Member onlaps onto the lower Shuaiba Formation margin and progrades out into the basin.

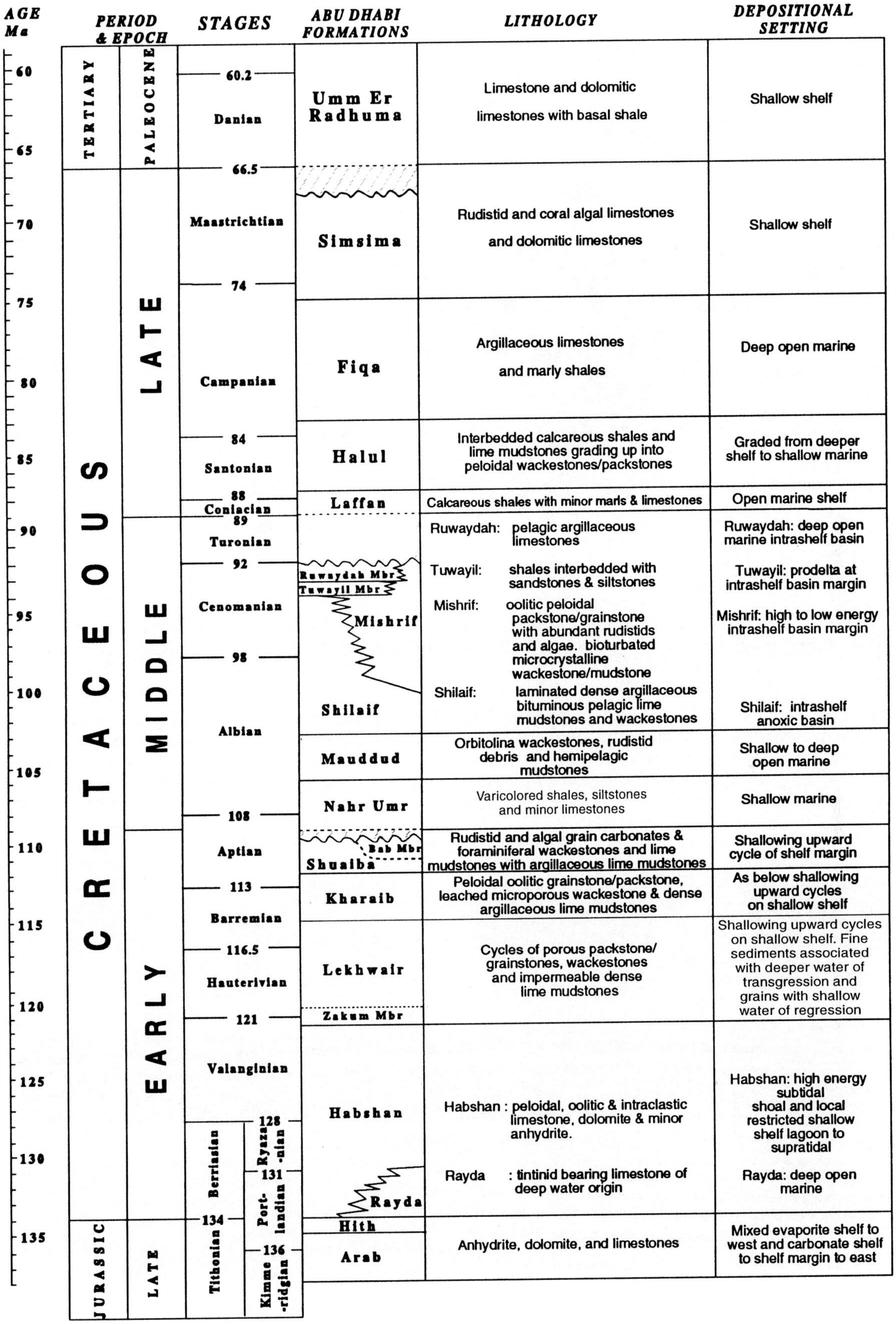


FIG. 7.—Lower Cretaceous lithostratigraphy of Abu Dhabi (after Alsharhan and Nairn, 1988).

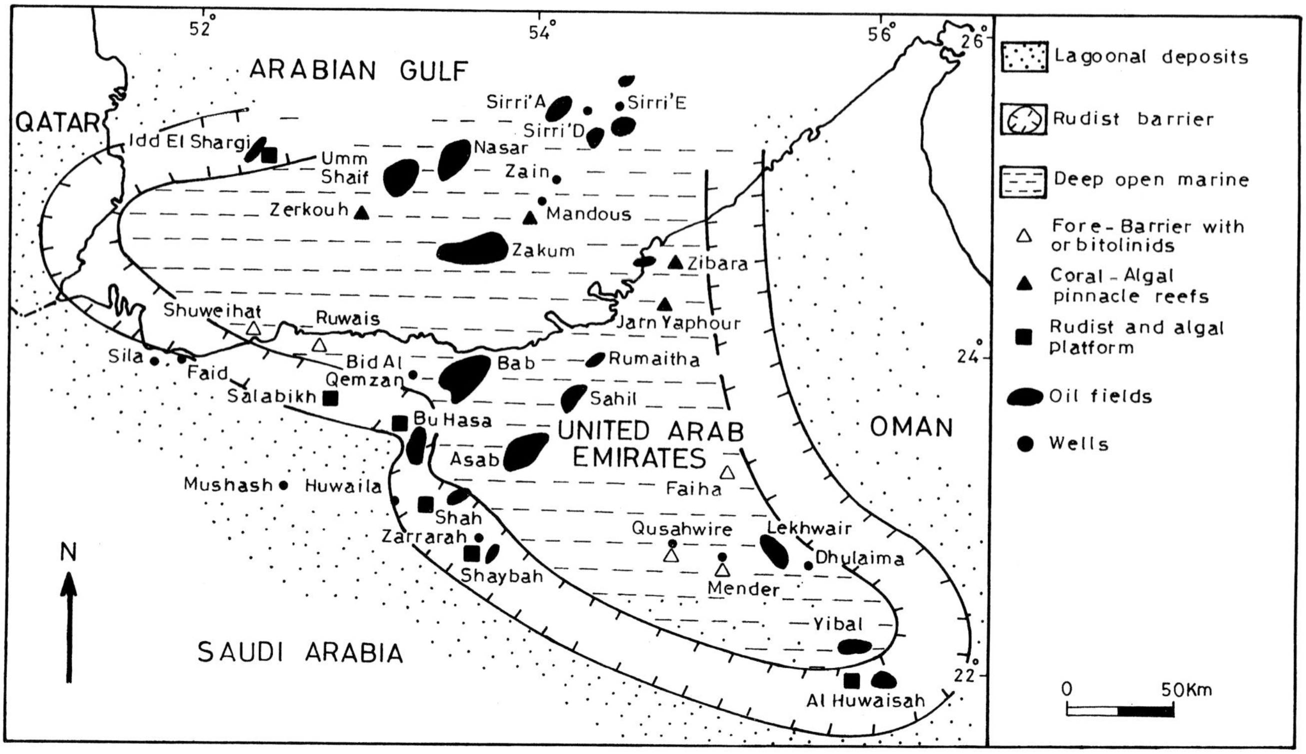


FIG. 8.—Location map of the Aptian Basin formed during the deposition of the Shuaiba Formation. Note the intrashelf basin and shelf-margin trend of the Shuaiba Formation rudist barrier formed during the Aptian. (After Al Dabal and Alsharhan, 1989.)

The change from subsidence to uplift occurred fairly rapidly. A coincident uplift was required to maintain the geometries of the Bab sequences, which were restricted to deposition below the Shuaiba shelf edge. Without uplift, Bab deposition would have onlapped onto the shelf, and the stratigraphic evidence shows no evidence of this.

In contrast, for the 113.5 Ma to 108 Ma Shuaiba Formation model, the best simulation inputs and results were as follows (Fig.11). From 113.5 to 112 Ma a carbonate production rate was set at 90 cm/1000 years at the sea surface. This rate declined sharply with water depth. During the same interval, the subsidence rates began at 40 cm/1000 years, and near the end of this interval it declined abruptly to 1 cm/1000 years. The results of the simulation (Fig. 11) show a fairly rapid change from a ramp margin to a well defined shelf edge by the end of Shuaiba deposition. The Shuaiba depositional sequences aggrade with minor progradation of the shelf edge.

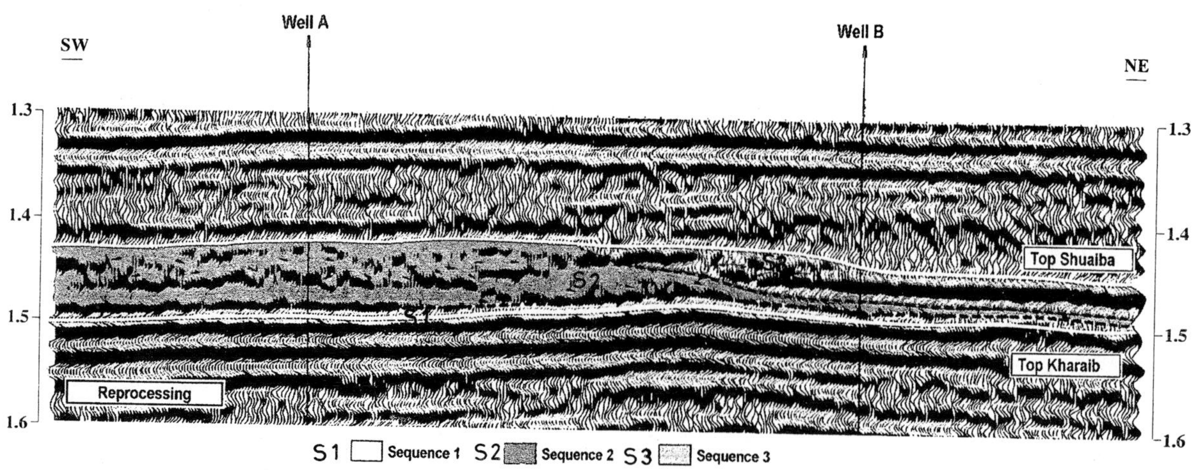


FIG. 9.—Seismic cross section of the Shuaiba Formation shelf margin after (Fischer et al., 1994). The white lines identify the major sequence boundaries.

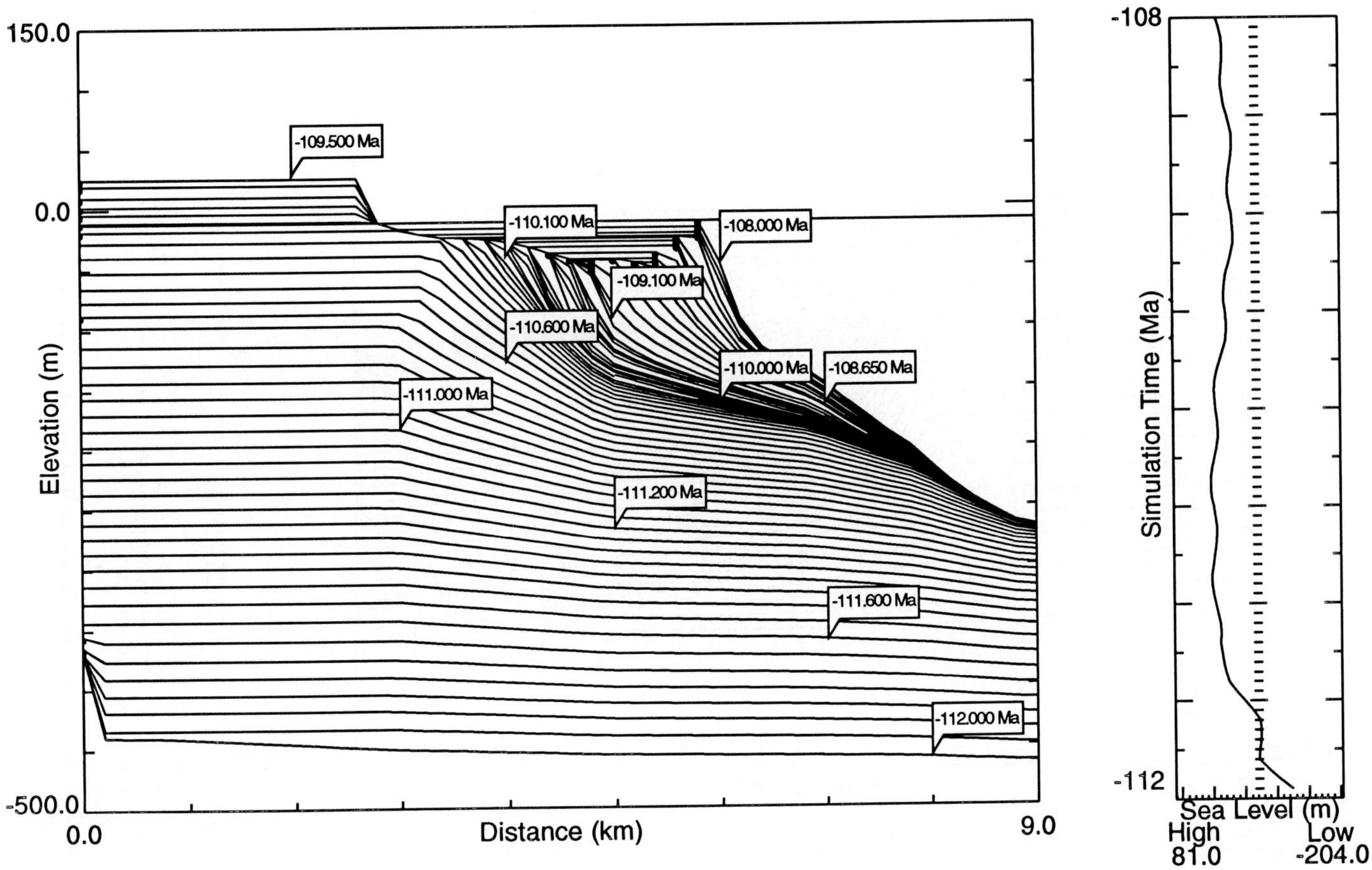


FIG. 10.—Simulation of the Shuaiba Formation cross section of Figures 1 and 9 with the program Sedpak from 112 Ma to 108 Ma. The numbers in the arrowed boxes are ages in millions of years.

During the second simulation run following a major sea-level drop at 112 Ma, the shelf was uplifted at a rate of 3 cm/1000 years to maintain the position of sea level below the shelf edge throughout the deposition of the Bab Member. Because of the restriction of the sea surface during Bab deposition, carbonate production was dramatically less than it was during Shuaiba deposition. Carbonate production was 6 cm/1000 years at the sea surface, decreasing slowly to near zero at 100 meters water depth. The simulation produced two progradational sequences. The first ranged in age from 112 to 109.3 Ma and onlapped the Shuaiba slope. Following a smaller sea-level fall at 109.3 Ma, a second sequence, the upper part of the Bab Member, onlapped onto the lower sequence before prograding farther out into the basin.

The change from subsidence to uplift occurred fairly rapidly, and coincided with the 112 Ma sea-level fall. A coincident uplift was required to maintain the geometries of the Bab sequences, which were restricted to deposition below the Shuaiba shelf edge. Without uplift, Bab deposition would have onlapped onto the shelf.

From the above discussion it can seen that simulation of the section using two chronostratigraphic models of the Aptian section in the U.A.E. does not immediately resolve which model is more reasonable. In both simulations, however, the Shuaiba Formation was bounded by an erosional unconformity and contained higher-frequency cycles that were the function of the time steps used. The shape and stacking patterns of these higher-frequency cycles appears to be influenced by their position within the associated third-order cycles. Thus at the base of the first third-order cycle, the higher-frequency cycles were more like those in a lowstand sequence. Upward in the third-order sequence these higher-frequency cycles became more transgressive in character. In the upper part of the third-order cycles the higher-frequency cycles had the character of a highstand systems tract. In the adjacent intershelf basin, both cycles contain lowstand and highstand systems tracts that accumulated in a deeper and more open marine setting dominated by deposition of argillaceous limestone.

From the seismic data, we know that the Bab Member is restricted to the basin. Two explanations for the restricted nature of the Bab sediments are proposed:

1. The tectonic uplift was required to maintain depositional integrity. The cause of the uplift is not known but may be related to the collision of the Arabian plate with the Asian continent.
2. The amount of coastal onlap after 112 Ma, and hence, the scale of the transgression, used in both the simulations was not correct. If this latter is true, then a reduction in the magnitude of post–112 Ma coastal onlap should produce the same geometries with a subsidence rate that changed very little.

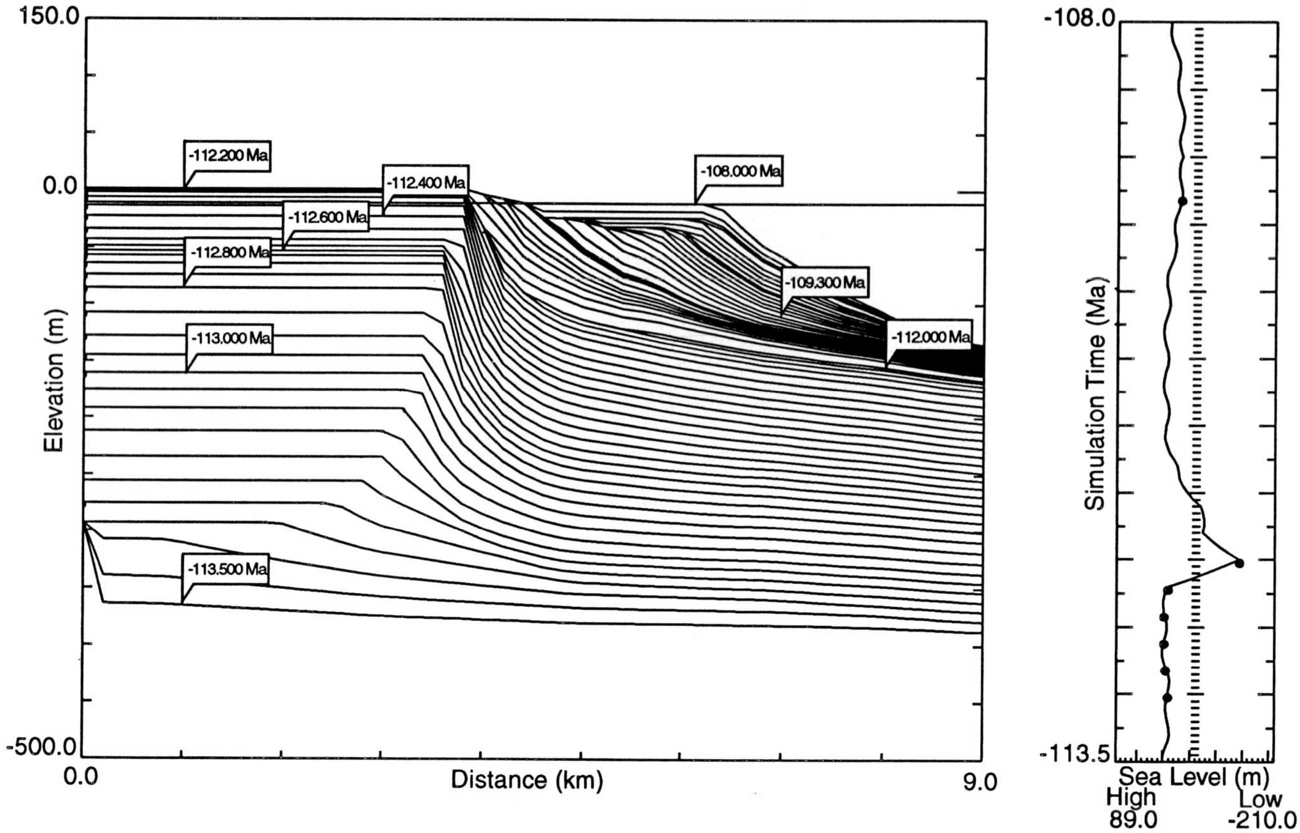


FIG. 11.—Simulation of the Shuaiba Formation cross section of Figures 1 and 9 with the program Sedpak from 113.5 Ma to 108 Ma. The numbers in the arrowed boxes are ages in millions of years.

Both explanations are equally valid from a standpoint of the geologic evidence, and neither is affected by the two different biostratigraphic models. If anything, the dating of the Shuaiba Formation and Bab Member as between 112 Ma and 108 Ma, or between 113.5 Ma and 108 Ma, supports the proposed tectonic uplift to explain the character of the downdip but onlapping Bab geometries.

## CONCLUSIONS

1. It was demonstrated that the hypothesis that events on the sea-level chart of Haq et al. (1987) produced unique responses in the sedimentary section which could potentially be used to date that section. To this end a sea-level chart was used, in conjunction with a simulation, to date the geologic section of the Neogene of the Bahamas. These dates were confirmed with biostratigraphic data (Kendall and Sen 1998).
2. Unconformities bounding the third-order sequences of the seismic sections of the Aptian–Albian of the National Petroleum Reserve of Alaska were matched to the third-order sea-level events on the Haq et al. (1987) chart. A sedimentary simulation using sea-level variations from this chart, and a simple model of constant sedimentation and constant subsidence, was matched to this seismic section. The modeling of the section confirmed the existence of major second-order and third-order eustatic sea-level variations in the Aptian–Albian and demonstrated that these major eustatic changes might be the primary control for the development of the stratal geometries in the Torok and Nanushuk Formations.
3. The Aptian of the U.A.E. has been dated by two biostratigraphic schemes. The use of the Haq et al. (1987) sea-level chart to determine which of these two models was more likely was unsuccessful, but the simulation used in the test suggests that the basin margin was uplifted just before the deposition of the Bab Member.

It has been argued in this paper that by combining stratigraphic interpretations with a sedimentary computer simulation, the ages of eustatic sea-level events, and their sizes, can be tentatively coupled to confirm the ages of the sequence boundaries in interpreted sections, but the results are ambiguous. Not only that, but further tests should be made on the effect of eustasy, tectonics, and sedimentation on the distribution of sequence geometries. Further it is also possible to hypothesize on the initial basin shape, and its accommodation history (the latter by combining tectonic and eustatic history), and the sediment accumulation history of the basin. Finally it is also possible to identify and predict geometries and lithofacies variations within the sequences and to test current and future sequence stratigraphic interpretations and models built from well cross sections and seismic. The problem is that any model that simulates the sedimentary geometry from the geologic record is unquestionably non-unique, and dependent upon data resolution to establish the best interpretation for the geology. So for the techniques outlined in the paper to have merit there is a need for further study if the hypothesis is to be more widely accepted.

ACKNOWLEDGMENTS

We would like to express our appreciation for the help and constructive criticism of Colin Plank, Animikh Sen, James Knapp, Michael Simmons, Robert Scott, and Dennis Kerr, who have all read and, in many cases, offered extensive suggestions for the improvement of the original manuscript of this paper. We take responsibility for the contents of the current paper, and none of these people can be blamed for the mistakes we may have made.

## REFERENCES

ABOU-CHOUCHA, M.K.A., AND ENNADI, I.M., 1990, Possible stratigraphic traps in the Shuaiba Formation, Onshore Abu Dhabi: Society of Petroleum Engineers, 4th Abu Dhabi Petroleum Conference, paper no. 303, 7 p.

ALDABAL, M.A., AND ALSHARHAN, A.S., 1989, Geological model and reservoir evaluation of the Lower Cretaceous Bab Member in the Zakum field, Abu Dhabi, U.A.E.: Society of Petroleum Engineers, Proceeding 6th Middle East Oil Show, Bahrain, 8 p.

ALSHARHAN, A.S., AND NAIRN, A.E.M., 1988, A review of the Cretaceous Formations in the Arabian Peninsula and Gulf: Part II. Mid-Cretaceous (Wasia Group) stratigraphy and paleogeography: Journal of Petroleum Geology, v. 11, p. 89–112.

ALSHARHAN, A.S., 1991, Sedimentological interpretation of the Albian Nahr Umr Formation in the United Arab Emirates: Sedimentary Geology, v. 80, p. 317–327.

ALSHARHAN, A.S., AND KENDALL, C.G.ST.C., 1991, Cretaceous chronostratigraphy, unconformities and eustatic sea-level changes in the sediments of Abu Dhabi, U.A.E.: Cretaceous Research, v. 12, p. 379–401.

ALSHARHAN, A.S., 1985, Depositional environment reservoir units evolution and hydrocarbon habitat of Shuaiba Formation, Lower Cretaceous, Abu Dhabi, U.A.E.: American Association Petroleum Geologists, Bulletin, v. 69, p. 899–912.

AZZAM, I.N., AND TAHER, A.K., 1995, Sequence stratigraphy and source rock potential of the Aptian (Bab Member) in east onshore Abu Dhabi: A model approach to oil exploration: Society of Petroleum Engineers, 9th Middle East Oil Show, Bahrain, Proceedings, p. 305– 317.

BIRD, K.J., AND MOLENAAR, M.C., 1992, The North Slope Foreland Basin, Alaska, *in* Macqueen, R.W., and Leckie, D.A., eds., Foreland Basins and Fold Belts: American Association of Petroleum Geologists, Memoir 55, p. 363–393.

CALAVAN, C.W., HAGERTY, R.M., MITCHEL, J.C., AND SCHUTTER, R.C., 1992, Integrated reservoir study and modeling of the Shuaiba Reservoir, Bu Haas, Abu Dhabi, United Arab Emirates: ADNOC/SPE 5th Abu Dhabi Petroleum Conference 18–20 May 1992, p. 299–312.

EBERLI, G.P., WHITTLE, G.L., KENDALL, C.G.ST.C., CANNON, R.L., AND MOORE, P.D., 1994, Testing a seismic interpretation of the great Bahamas bank with a computer simulation: American Association of Petroleum Geologists, Bulletin, v. 78, p. 981–1004.

FISCHER, K.C., MOELLER, U., AND MARSCHALL, R., 1994, Development of an exploration concept for the Shuiaiba Formation (Thamama Group) using seismic sequence and facies analysis in combination with forward modeling, *in* Al-Husseini, M.I., ed., 'Geo 94, The Middle East Petroleum Geoscience: Manama, Bahrain, Gulf Petrolink, v. 1, p. 377–386.

HAQ, B.U., HARDENBOL, J., AND VAIL, P.R., 1987, Chronology of fluctuating sea levels since the Triassic (250 million years ago to present): Science, v. 235, p. 1156–1167.

KENDALL, C.G.ST.C., SIREGAR, P., LEVINE, P., SEN, A., CRESCINI, C., JOHNSTON, K., AND EBERLI, G., 1997a, Confirmation of the magnitude of eustatic events of the Haq and others curve for the Neogene from a sequence stratigraphic analysis of the Great Bahamas Bank (abst.), American Association Petroleum Geologists, Annual Convention, Dallas, Official Program, p. 18.

KENDALL, C.G.ST.C., ALSHARHAN A.S., AND JOHNSTON K., 1997b, Aptian depositional cycles of the U.A.E.–A response to eustatic events which match those of South Africa and Alaska: SEPM, International Conference "Jurassic/Cretaceous Carbonate Platform–Basin Systems", Dec 7–9, 1997, Convention Abstracts: GeoArabia, v. 2, p. 487.

KENDALL, C.G.ST.C., AND SEN, A., 1998, Use of sedimentary simulations for dating sequence boundaries and measuring the size of eustatic sea-level changes: an example from the Neogene of the Bahamas, *in* Harff, J., Lemke, W., and Stattegger, K. eds., Computerized Modeling of Sedimentary Systems: Berlin, Springer-Verlag, p. 291–306.

LIDZ, B.H., AND MCNEIL, D.F., 1998, New allocyclic dimensions in a prograding carbonate bank: evidence for eustatic, tectonic, and paleoceanographic control (Late Neogene, Bahamas): Journal of Sedimentary Research, v. 68, p. 269–282.

POSAMENTIER, H.W., AND VAIL, P.R., 1988, Eustatic controls on clastic deposition II. Sequence and system tract models, *in* Wilgus, C.K., Hastings, B., Kendall, C.G.St.C., Posamentier, H.W., Ross, C.A., and Van Wagoner, J.C., eds., Sea-Level Changes: An Integrated approach: SEPM, Special Publication 42, p. 125–154.

POSAMENTIER, H.W., JERVEY, M.T., AND VAIL, P.R., 1988, Eustatic controls on clastic deposition I. Conceptual framework, *in* Wilgus, C.K., Hastings, B., Kendall, C.G.St.C., Posamentier, H.W., Ross, C.A., and Van Wagoner, J.C., eds., Sea-Level Changes: An Integrated Approach: SEPM, Special Publication 42, p. 109–124

RYAN, S.R., 1997, Age determination from depositional sequence analysis and stratigraphic simulation of Lower Cretaceous Torok and Nanushuk Formation, North Slope of Alaska: M. Sc. thesis, University of South Carolina, 90 p.

SIMMONS M.D., 1994, Micropalaeontological biozonation of the Kahmah Group (Early Cretaceous), Central Oman Mountains, *in* Simmons, M.D., ed., Microplaeontology and Hydrocarbon Exploration in the Middle East: London, Chapman & Hall, p. 177–205.

VAIL, P.R., 1988, Seismic stratigraphy interpretation using sequence stratigraphy, Pt 1: Seismic stratigraphy interpretation procedure, *in* Bally, A.W., ed., Atlas of Seismic Stratigraphy: American Association of Petroleum Geologists, Studies in Geology, No. 27, v. 1, p. 1–10.

VAIL, P.R., MITCHUM, R.M., JR., TODD, R.G., WIDMIER, J.M., THOMPSON, S., III, SANGREE, J.B., BUBB, J.N., AND HATLELID, W.G., 1977, Seismic stratigraphy and global changes of sea level, *in* Payton, C.E., ed., Seismic Stratigraphy; Applications to Hydrocarbon Exploration: American Association of Petroleum Geologists, Memoir 26, p. 49–212.

VAIL, P.R., AND TODD, R.G., 1981, Northern North Sea Jurassic unconformities, chronostratigraphy and sea-level changes from seismic stratigraphy, *in* Illing, L.V., and Hobson, G.D., eds., Petroleum Geology of the Continental Shelf of North–West Europe: London, Institute of Petroleum, Heyden, p. 216–235.

VAN WAGONER, J.C., MITCHUM, R.M., POSAMENTIER, H.W., AND VAIL, P.R., 1987 Key definitions of seismic stratigraphy, *in* Bally, A.W., ed., Atlas of Seismic Stratigraphy: American Association of Petroleum Geologists, Studies in Geology, No. 27, v. 1, p. 11–14.

WITT, W., AND GÖKDAG, H., 1994, Orbitolinid biostratigraphy of the Shuaibah Formation (Aptian), Oman—Implications for reservoir, *in* Simmons, M.D., ed., Microplaeontology and Hydrocarbon Exploration in the Middle East: London, Chapman & Hall, p. 221–233.

# ARE MID-CRETACEOUS EUSTATIC EVENTS RECORDED IN MIDDLE EAST CARBONATE PLATFORMS?

ROBERT W. SCOTT

*Precision Stratigraphy Associates, RR3 Box 103-3, Cleveland, OK 74020, U.S.A.*

WOLFGANG SCHLAGER

*Aardetenschappen, Vrije Universiteit, 1081 HV Amsterdam, The Netherlands*

BRUCE FOUKE

*Department of Geology, University of Illinois, Urbana, IL 61801, U.S.A.*

AND

SANDRA A. NEDERBRAGT

*Department of Geological Sciences, University College London, Gower St., London WC1E 6BT, U.K.*

ABSTRACT: The mid-Cretaceous is an informal term that, in the Middle East, includes the Cretaceous stages of Hauterivian through Cenomanian. During this 37 to 45 million years time interval twenty cycles of relative change of coastal onlap were proposed by Haq et al. (1988), and since that time, some of these cycles have been subdivided further. The average duration of these so-called second-order cycles is 1.8 to 2.2 million years, depending upon which time scale is used.

The challenge is to identify accurately depositional cycles in the stratigraphic record of the Middle East platform and to correlate them precisely with a reference section of cycles. The quantitative stratigraphic technique of *graphic correlation* achieves this goal. Graphic correlation is a numerical correlation technique that is simple to apply and gives precise and reproducible correlations. Graphic correlation is based on an integrated database of fossil tops and bases and other geologic events. The technique creates hypotheses of correlation that make no assumptions about the completeness of each fossil range.

A composite standard database from the Aptian through the Turonian has been constructed using 42 geologic sections in the Tethyan Realm. More than 1000 bioevents of ammonites, inoceramids, planktic foraminifers, selected rudists, benthic foraminifers, nannofossils, and dinoflagellates have been integrated with nearly 100 depositional and geochemical events. Among these fossils are many of the zonal indicators. The stage boundaries are defined by key taxa in generally accepted reference sections in France, Tunisia, and Texas. The scale is calibrated to the Harland time scale.

Platform exposures of the middle Cretaceous in Oman record six or more global cycle boundaries at the Habshan/Lekhwair contact, at the Shuaiba/Nahr Umr contact, several within the Nahr Umr, possibly at the base and top of the Natih "F" Member, at the top of the "E" Member, and at the contact between the "C" and "B" Members. Other cycles may be recognized as more detailed stratigraphic information is collected.

Sediment accumulation rates progressively increased during deposition of the Albian–Cenomanian carbonate platform. Depositional rate of the Nahr Umr was approximately 1.00 cm/1000 years. Depositional rate of the Natih was about 5.00 cm/1000 years. Clearly, many submarine hiatuses developed during deposition of the Nahr Umr.

## INTRODUCTION

In the eastern Arabian region of the Tethyan Realm a carbonate platform developed during the middle stages of the Cretaceous from the Hauterivian through the Cenomanian. This study focuses on the latest Aptian through the latest Cenomanian and earliest Turonian cycle of platform deposition, represented by the Wasia Group in Oman and the United Arab Emirates. The duration of this interval ranges from 17.5 to 23 million years, depending on which of four time scales is chosen. The ages vary from 109 to 91.5 Ma on the Haq et al. (1988) scale, 113–90 Ma on the Harland et al. (1990) scale, and 113–93 Ma on the scales of Obradovich (1993) and Gradstein et al. (1995). Consequently, the duration of depositional cycles will vary accordingly. Haq et al. (1988) delineated eleven cycles from the latest Aptian to the earliest Turonian that ranged in duration from 1.59 to 2.09 million years. Hardenbol et al. (1998) identified seventeen sequence cycles in the European basins during this 20 million year interval. Our evaluation of sequences in Tethyan reference sections in Tunisia and the North American Gulf Coast recognizes sixteen cycles in the same interval (Fig. 1). In the Harland scale this time interval is from 114.6 to 90.1 Ma and the mean cycle duration is 1.53 million years.

Depositional cycles are sedimentary deposits that record fluctuating environmental conditions regardless of duration and genesis (Scott et al., 1994). Longer-term depositional cycles may be bounded by unconformities across the shelf and shelf break, which may be depositional sequences. Shorter-term cycles may be bounded by transgressive contacts that represent a progressive change in depositional conditions. The duration of cycle hierarchy and their genetic mechanisms cannot be assumed from the scale of the cycle only. Some depositional cycles record a depositional episode of shoreline advance and retreat (Galloway, 1989) or of climatic fluctuations. Regardless of their duration or origins, depositional cycles must be defined by physical and biotic data found in the rocks so they can be identified uniquely and consistently from one section to another. A second challenge is to date the cycle duration represented by both the rock and the contact. Generally zonal schemes are unable to measure the duration of the hiatus represented by the contact. A precise, objective, and testable technique must be used to date the cycles. Such a method of quantitative stratigraphy is graphic correlation.

The succession of facies of the Wasia Group has been interpreted to be the result of tectono-eustatic processes (Scott et al., 1988; Scott, 1990a; Alsharhan and Kendall, 1991; Burchette, 1993; Philip et al., 1995) or of eustatic processes primarily (van

Middle East Models of Jurassic/Cretaceous Carbonate Systems
SEPM Special Publication No. 69, Copyright © 2000
SEPM (Society for Sedimentary Geology), ISBN 1-56576-075-1, p. 77–88.

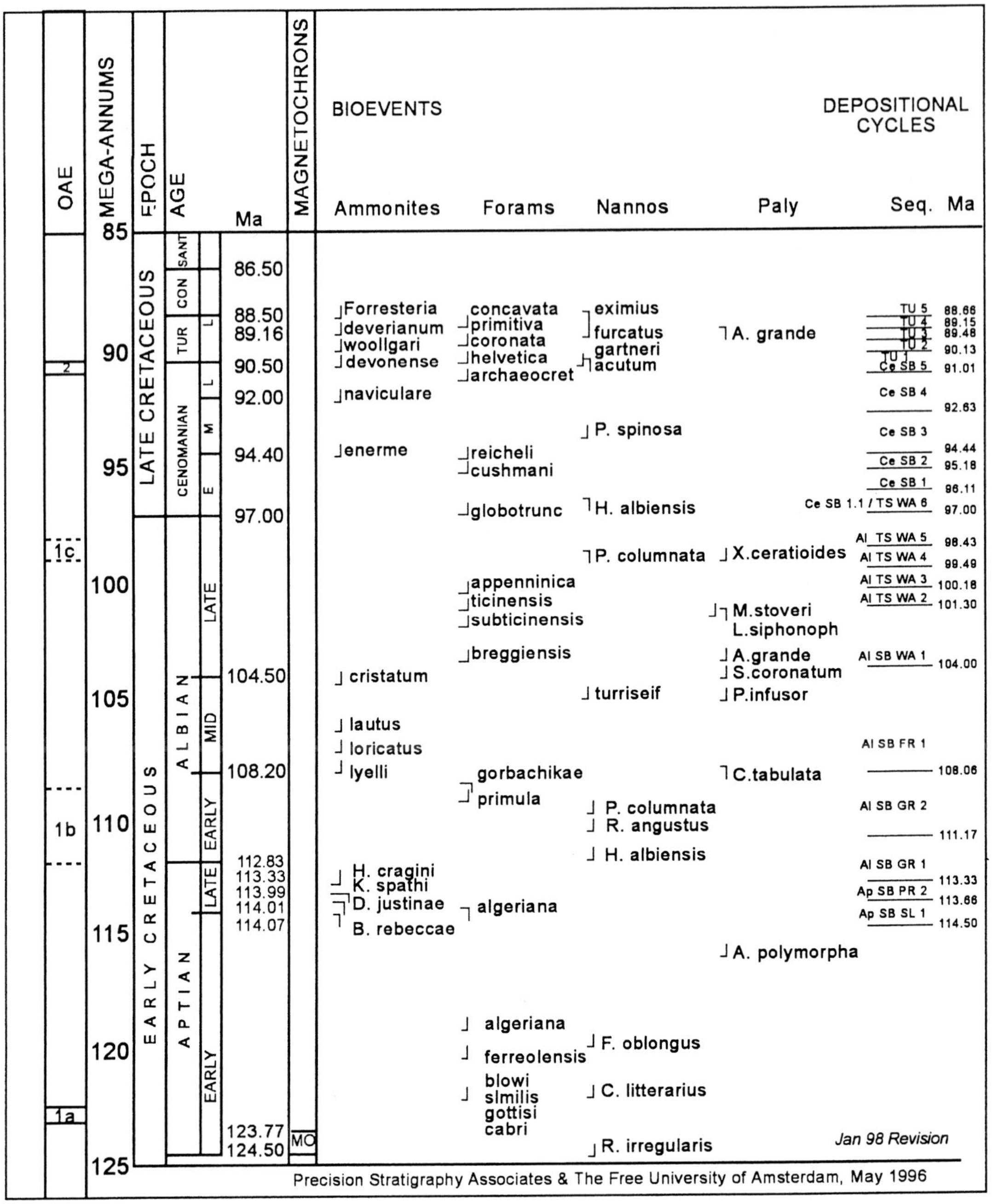

FIG. 1.—Chart of depositional and biotic events of the Aptian–Turonian defined in the Midk3 composite standard.

Buchem et al., 1996; Immenhauser et al., 1997) or of tectonic processes mainly (Rabu et al., 1990; Pratt and Smewing, 1993). The eastern margin of the Arabian platform has been tectonically active and was subjected to intense deformation during the Cretaceous and Tertiary. Dating these events is dependent upon dating the unconformities. However, separating the effects of tectonics from eustasy is difficult. The null hypothesis that unconformities are not the same age as inferred eustatic events could be tested by means of high-precision chronostratigraphy. First, a technique must be available to date precisely the stratal contacts, and second, the contacts must be accurately defined and correlated. Finally, the depositional environments of beds above and below the contacts must be fully understood. New data defining, correlating, and interpreting environments of stratal contacts in the Nahr Umr and Natih Formations are available in Immenhauser et al. (1997), Immenhauser et al. (1999), and Immenhauser et al. (this volume), by Philip et al. (1995), and van Buchem et al. (1996).

## GRAPHIC CORRELATION METHODOLOGY

Graphic correlation (GC) of sections to a composite standard database is a proven technique (Carney and Pierce 1995; Scott et al., 1994). GC is a quantitative but nonstatistical technique that determines the coeval relationships between two sections by comparing the ranges of taxa in both sections. A graph of any pair of sections is an $x$–$y$ plot of the tops and bases of taxa found in both sections; it compares the rate of sediment accumulation in one section with that in the other (Carney and Pierce 1995). The GC technique enables the stratigrapher to consider sedimentological events, such as depositional cycles, along with biotic events so that conclusions based on one can be tested by the other. Also, the addition of event beds, such as ash beds, adds to the precision and accuracy of the correlation.

The elements of a graph are the scaled $x$–$y$ axes, the tops and bases of the fossils and event beds, and the line of correlation (LOC). The scale of the $x$ axis is in thickness, relative units, or million years. The $y$ axis is scaled in thickness of the measured section or well. Fossil tops are plotted as +'s and bases are □'s. The LOC is positioned by the stratigrapher to indicate the most constrained hypothesis of synchroneity between the two sections. The LOC is placed through the maximum number of tops and bases, and depositional hiatuses indicated by the lithostratigraphic record are shown by horizontal terraces. The position of the LOC is defined by the equation for a regression line. By iteratively graphing numerous sections, a mature composite standard database is constructed of the tops and bases of all taxa, geochemical markers, and key sedimentological strata.

## DATABASE

The composite standard (CS) database used in this study consists of data from 42 published outcrop and core-hole data primarily from the Tethyan Realm (Table 1). The ranges of more than 1100 bioevents, planktic and benthic foraminifers, nannofossils, dinoflagellates, ammonites, bivalves, echinoids, calcareous algae, magnetochrons, geochemical events, and sequence stratigraphic markers are defined in this database. This project was initially conceived by W. Schlager and funded at the Free University of Amsterdam with the collaboration of the staff. A. Nederbragt provided verification of foraminiferal taxonomy and range data in selected sections, and B. Fouke provided lithostratigraphy and geochemistry of selected sections.

The following protocols and conventions were followed in the construction of the MIDK3 Composite Standard:

1. Published data were utilized with foraminiferal taxonomy evaluated by Nederbragt. Authorship of the nannofossil and palynomorph data is indicated in each of the section files.
2. Where the sample position is given as a depth interval, the base was defined by the shallow end of the interval, and the top by the deeper end of the interval in order to minimize range extensions by sampling.
3. The scale of the MIDK3 Composite Standard is in megaannums, and the time scale of Harland et al. (1990) was selected by the Free University group as being the most thoroughly researched at the time. To integrate this scale into the CS, the Kalaat Senan, Tunisia section (1; number refers to section list in Table 1) was taken as a complete record of uniform deposition from the beginning to the end of the Turonian (Fig. 2) (Robaszynski et al., 1993). The positions in meters of the bases of the Turonian and Coniacian as defined by ammonites were plotted to the Harland scale and graphed. The Upper Albian through Lower Turonian section in Santa Rosa Canyon (Fig. 3) (Ice and McNulty, 1980; Ross and McNulty, 1980), Mexico has been shown by unpublished graphic correlation experiments with the Amoco composite standard to represent continuous and uniform deposition. This interval is also represented in the sections of DSDP 547 (4) and DSDP 545 (5) offshore Morocco. The Cenomanian substages were defined in the Cenomanian part of the Kalaat Senan section (Fig. 2) (Robaszynski et al., 1990). The depositional cycles in the Texas sections have been described by Young (1986), Scott (1990b, 1993) Scott et al. (1994), and Yurewicz et al. (1993). The Aptian interval is represented by the Gorgo a Cerbara (43) (Bralower et al., 1994; Cecca et al., 1994) and Pie' del Dosso, Italy (26) sections (Erba and Quadro, 1987).
4. Geochemical marker beds defined by the $\delta^{13}C$ isotopic ratio were included in the database by selecting the inflection points as the base/top of the peak. Sequence boundaries and other types of depositionally significant contacts are defined in reference sections by the specialists working the sections. They were not used to control the LOC in the reference

TABLE 1.—List of sections that contributed data to the Midk3 composite standard.

1 Kalaat Senan, Turonian, Tunisia.
2 Harland Geologic Time-scale, 1990.
3 Santa Rosa Canyon, Mexico.
4 DSDP 547 Offshore Morocco.
5 DSDP 545 Offshore Morocco.
6 Type Cenomanian, France.
7 Djebel Mrhila, Tunisia.
8 Djebel Bireno, Tunisia.
9 Amoco No. 1 Bounds, Western Kansas.
10 Kalaat Senan, Cenomanian, Tunisia.
11 DSDP 386 Offshore USA.
12 Boulonnais, Coast of France.
13 Piobbico Core, Italy.
14 DSDP 369A, Offshore Morocco.
15 Pueblo, Colorado Reference Section.
16 Wadi Mi'aidin Section, Oman (Scott, 1990a).
17 Wadi Mi'aidin Section (Simmons and Hart, 1987).
18 Shell Chapman Core, South Texas.
19 Shell Tomasek Core, South Texas.
20 Trinity River Comp. Section, North Texas.
21 Austin Comp. Section, Central Texas.
22 Nahr Ibrahim Outcrop, Lebanon.
23 Djebel Chenan Aair Outcrop, Lebanon.
24 Mt. Risou, Rosans, SE France.
25 Cap Blanc-Nez Outcrop, Northern France.
26 Cismon Outcrop, NE Italy.
27 Pie' del Dosso Outcrop, North-central Italy.
28 Estella Outcrop, North-central Spain.
29 Eastbourne Composite Section, South England.
30 Selbukhra Section, Crimea, Russia.
31 Kef el Azreg, Tunisia.
32 Sopeira Section, Pyrenees, Spain.
33 Flamicell Section, Pyrenees, Spain.
34 Montsec Section, Pyrenees, Spain.
35 Peace River Composite, Alberta, Canada.
36 Anderson Husky Roros Well, Alberta.
37 Type Shell Creek Fm., Wyoming.
38 Ida ou Tanane Outcrop, Morocco.
39 Chichaoua I Borehole, Morocco.
40 Timinoun Outcrop, Morocco.
41 ODP 641C Offshore Portugal.
42 Gorgo a Cebara section, Italy.

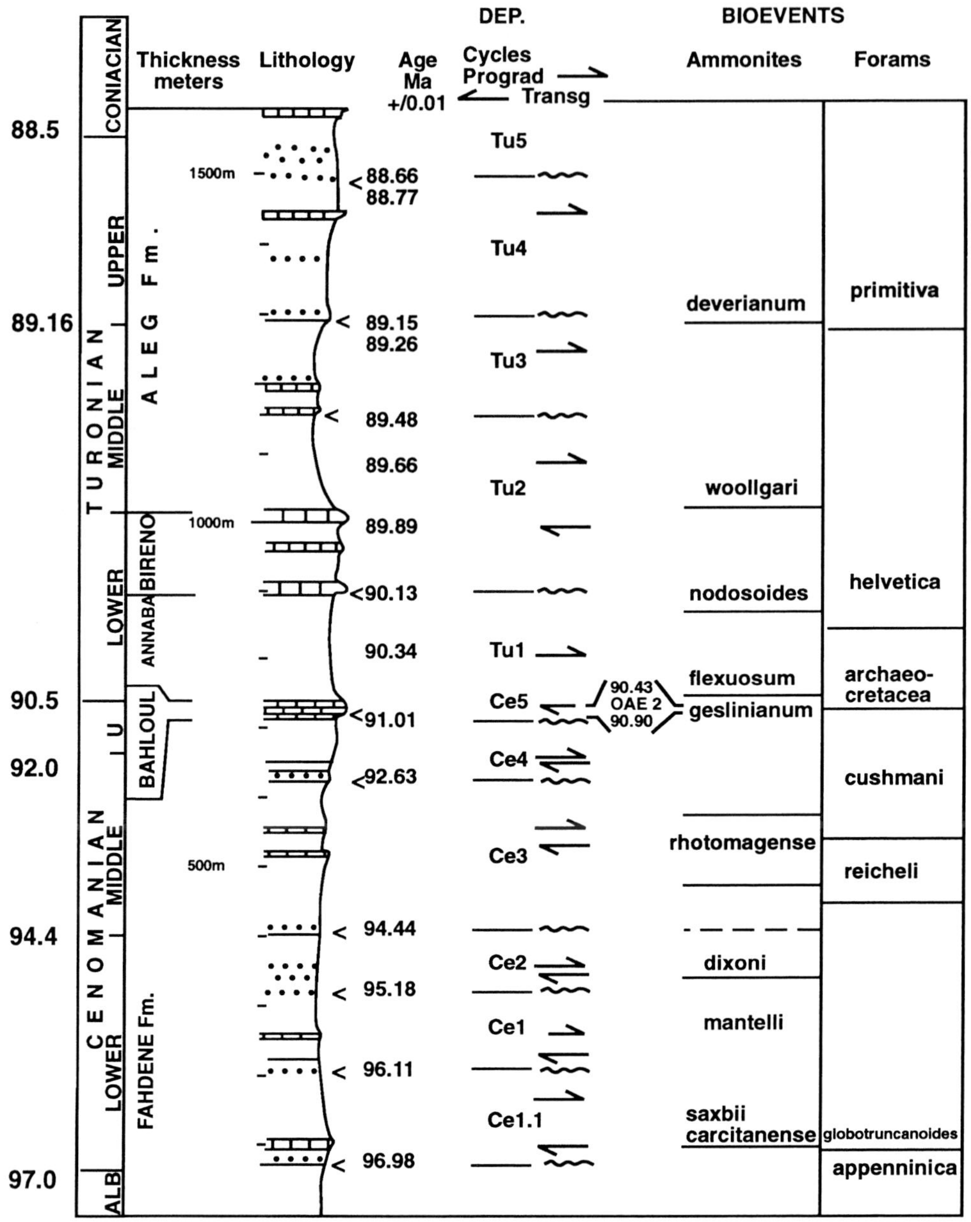

FIG. 2.—Cenomanian–Turonian sequence stratigraphy of Kalaat Senan, Tunisia, reference section based on data and analyses by Robaszynski et al. (1990, 1993).

sections; however, in sections where the paleontological data were wide-ranging they have been used together with biotic events. Milankovitch-scale sedimentary cycles in the U.S. Western Interior Cenomanian were defined in a cored reference section, the Amoco Bounds (9) and in an outcrop at Pueblo, Colorado (15).

5. Stage/Substage boundary definitions (Table 2) follow, where possible, recommendations of the Second International Symposium on Cretaceous Stage Boundaries in 1995 in Brussels (Rawson et al., 1996).

## ALBIAN–CENOMANIAN OF OMAN

The Wasia Group in Oman consists of the Nahr Umr and Natih Formations (Fig. 4) (Hughes Clarke, 1988; Alsharhan and Nairn, 1988; Alsharhan, 1989). The Wasia is bounded below and above by regional unconformities; it overlies the Aptian Shuaiba Formation and underlies the Coniacian and younger Aruma Group. The Nahr Umr in Oman is an interbedded shale–limestone unit 150 to 200 m thick. This unit represents a regional clastic depositional phase. The contact with the overlying Natih is conformable. The Natih is a limestone unit about 430 m thick that is divided into seven informal subunits designated "a" to "g". The type section of the Natih is a well in Oman, and the outcrop reference section is in Wadi Mi'aidin.

### *Basal Nahr Umr Unconformity*

The basal contact of the Nahr Umr Formation overlying the Shuaiba Formation is a regional unconformity having a complex

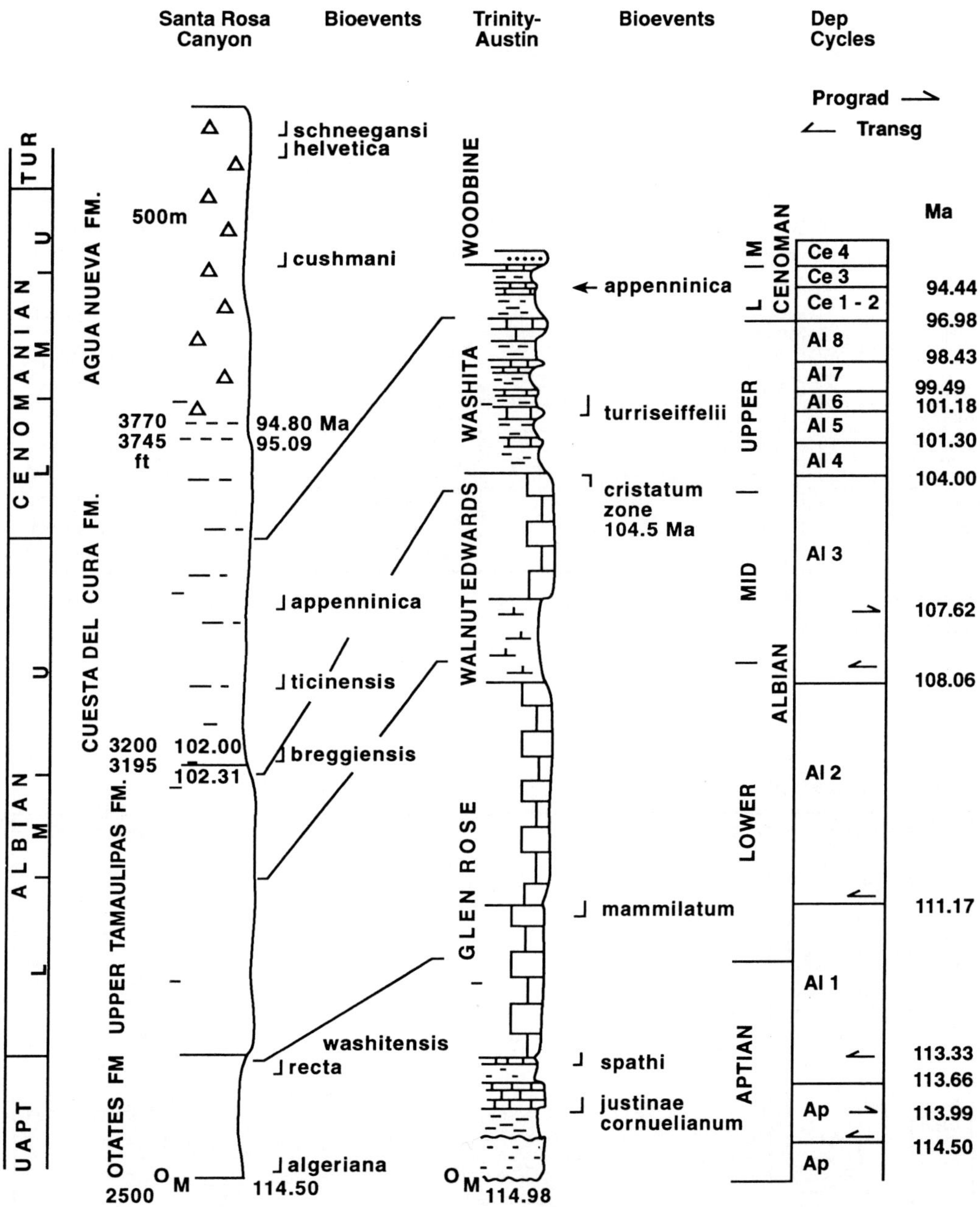


FIG. 3.—Aptian–Albian sequence stratigraphy of U. S. Gulf Coast and Mexico based on data and analyses by Young (1986), Scott et al. (1988), Scott (1990b, 1993), and Yurewicz et al. (1993).

history that in outcrop may span up to 2.2 million years or more (Scott et al., 1988). On the platform exposed in the Oman Mountains the Shuaiba is a shallow-water limestone less than 100 m thick (Hughes Clarke, 1988) of Lower Aptian age (Simmons and Hart, 1987; Scott 1990a). To the west in the Bab Basin the Shuaiba is up to 120 m thick, and the uppermost 30–40 m is a dark gray shale and lime mudstone unit called the Bab Member (Alsharhan and Nairn, 1986). These authors report Upper Aptian ammonites in the Bab. A paleontological analysis of diverse fossil groups indicates that the lower part of the Shuaiba is Lower Aptian and the upper part is Upper Aptian (Witt and Gökdag, 1994), although the lower–upper division is ill-defined. The Upper Aptian age of the Bab is supported by the carbon-isotope profiles from wells in the Bab Basin (Vahrenkamp, 1996; Grötsch et al., 1998). Thus, it seems likely that the hiatus at the Shuaiba–Nahr Umr unconformity is greater near the outer platform in the Oman Mountains than in the Bab Basin.

In the basal part of the Shuaiba Formation in the Bab Basin the carbon-isotope curve has a distinct positive inflection from about 2‰ to nearly 5‰ that is interpreted to be a global signal, OAE 1a (Wagner, 1990; Vahrenkamp, 1996; Grötsch et al., 1998). This inflection is dated at about 122–120 Ma (Bralower et al., 1994; Erba, 1994; Föllmi et al., 1994); Bralower et al. (1994) estimated its duration to be about 0.5 million years. In the Midk3 composite standard database OAE 1a ranges from 123.05 to 122.32 Ma, 0.73 million years duration, on the basis of

TABLE 2.—Biomarkers of stage and substage boundaries, the positions of the fossil bases (B) or tops (T) in the reference sections (section number), and their ages by graphic correlation.

| Published Position | Bioevent | Reference Section | Position/age in section |
|---|---|---|---|
| B Aptian | B CMOR | Pie'del Dosso (27) | 17.58 m = 124.50 Ma |
| B Late Aptian | B *Eprolithus floralis* | DSDP 545 (5) | 528.4 m = 116.61 Ma |
| | alternative: B *Dufrenoyia justinae* | Austin, TX (21) | 120 M = 113.99 Ma |
| B Albian | T *Hypacanthoplites cragini* | Austin, TX (21) | 395 ft = 112.00 Ma |
| B Mid Albian | B *Ceratostreon texanum* (proxy for *L. lyelli)* | Trinity River, TX (20) | 230 ft = 108.20 Ma |
| B Late Albian | B *Dipoloceras cristatum* | Boulonnais (12) | 4.2 m = 104.50 Ma |
| B Cenomanian | B *Rotalipora globotruncanoides* | Risou (24) | 136 m = 97.00 Ma |
| B Mid Cenomanian | B *Cunningtoniceras inerme* | Cap Blanc Nez (25) | 27 m = 94.26 Ma |
| B Late Cenomanian | B *Calycoceras naviculare* | Type Cenoman (6) | 88 m = 92.00 Ma |
| B Turonian | B *Watinoceras devonense* | Pueblo, Colo. (15) | 133.9 m = 90.50 Ma |
| B Mid Turonian | B *Romaniceras kallesi* | Estella Basin (28) | 110 m = 90.12 Ma |
| B Late Turonian | B *Romaniceras deverianum* | Kalaat Senan (1) | 678 m = 89.16 Ma |
| B Coniacian | B *Forresteria* sp. | Kalaat Senan, (1) | 947.5 m = 88.50 Ma |

The age of the base of CMOR is from Rawson et al. (1996) and of the Albian, Cenomanian, Turonian, and Coniacian is from Harland et al. (1990).

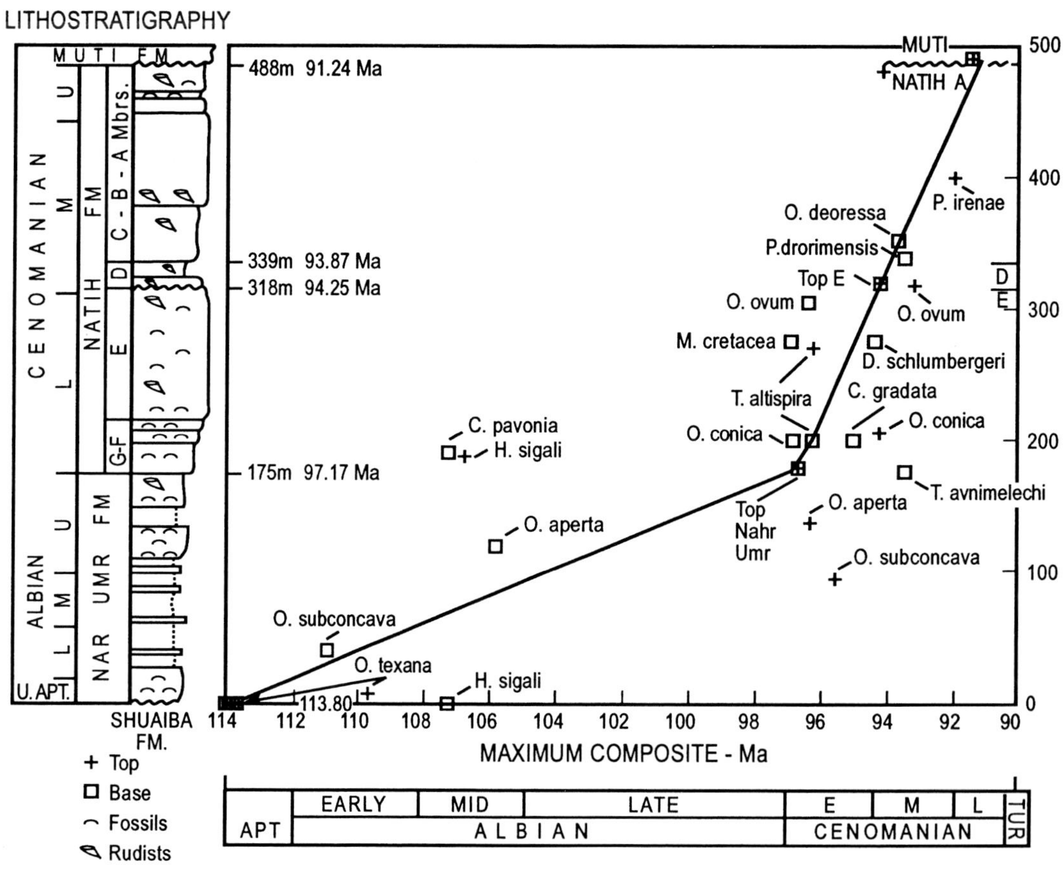


FIG. 4.—Graphic correlation of the Wasia Group in the Wadi Mi'aidin, Oman section and the Midk3 composite standard 9-98. Section measured and biostratigraphic data analyzed by Simmons and Hart (1987).

the carbon-isotope curve from the section at Cismon, Italy (*in* Erba, 1994). The positive value of the carbon-isotope curve is maintained to the top of the Shuaiba near the platform margin (Wagner, 1990); no diagenetic negative shift is recorded. However, in the Bab Basin up to three negative shifts of 0.5 to 2‰ are present in the lower part of the Shuaiba and in the Bab Member (Vahrenkamp, 1996; Grötsch et al., 1998).

The processes of formation of this Shuaiba–Nahr Umr unconformity are equivocal and include subaerial exposure (Frost et al., 1983; Immenhauser et al., 1999) and transgressive drowning by sea-level rise (Scott et al., 1988). A regional study of carbon and oxygen isotopes suggested that "little or no evidence exists for a paleo-exposure of the top of the Shuaiba Formation" (Wagner, 1990, p. 130). Likewise, little or no petrographic evidence of exposure is present in Shuaiba cores in the Sajaa Field, Sharjah, United Arab Emirates (Moshier et al., 1988). However, in the Omani Ghaba North Field, central Oman, petrographic data indicate that the uppermost Shuaiba has experienced meteoric diagenesis (Al-Awar and Humphrey, 1997 and this volume).

### *Intra–Nahr Umr Diastems*

The Nahr Umr Formation in Oman is composed of repetitive bed-sets of mudstone, wackestone, and orbitolinid packstone bounded at the top by bored, ironstone hardground or firmground surfaces that represent diastems (Simmons and Hart, 1987; Scott, 1990a; Immenhauser et al., 1997; Immenhauser et al., 1999; and Immenhauser et al., this volume). In any given section from one to ten such contacts are recognized by local phosphate nodules and burrows filled with dolomitic, iron-rich micrite (Immenhauser et al., 1999). These surfaces tend to cap incomplete shoaling-up successions of subtidal orbitolinid packstone. Immenhauser et al. (1999) correlated many of the surfaces in the Oman Mountains. The hardground surfaces seem to represent a complex sea-level history that began during sea-level fall. A period of subaerial exposure is suggested by distinct negative shifts in carbon and oxygen isotope curves, by brackish-water fluid inclusions, and by soil-forming features (Immenhauser et al., 1999). The final stage was marine flooding and resumption of subtidal deposition.

### *Natih Disconformities*

Ten hardground disconformities have been identified within and at the top of the Natih Formation (Fig. 5) (van Buchem et al., 1996). We have identified these contacts in the Wadi Mi'aidin section of Philip et al. (1995) and determined their positions in the section so that their ages can be projected by means of the line of correlation in the graph of biostratigraphic data from Philip et al. (1995) (Fig. 5). Van Buchem et al. (1996) did not publish biostratigraphic data for their section.

The top of the "e" member of the Natih Formation in Jebel Akhdar is a bored, iron-stained, phosphatic hardground traceable in outcrop and into the subsurface (Scott, 1990a; Wagner, 1990; Philip et al., 1995; van Buchem et al., 1996). Graphic correlation analysis suggested that this disconformity was virtually synchronous in Oman and was coeval with the global mid-Cenomanian sequence boundary (Scott et al., 1988; Scott, 1990a). The marked negative shift in the carbon-isotope signal at the top of the "e" member is evidence of meteoric diagenesis and subaerial exposure (Wagner, 1990). At places as much as seven meters of channelized erosion is evident (van Buchem et al., 1996).

A hardground also is present at the top of the Natih "b" member in some outcrops at Jebel Akhdar (Philip et al., 1995). It is a sharp but non-erosional contact between praealveolinid limestone overlain by basinal lime mudstone. Van Buchem et al. (1996) recognized hardgrounds by the presence of boring, bioturbation, and iron encrustation in the Natih "c" and "d" members. Four of these surfaces demarcate the bases of their accommodation cycles 5, 6, 7, and 8 and can be traced throughout their sections in the Oman Foothills. They use surfaces 5a, 6a, and 8a as contacts for the tops of members "e", "d", and "c".

The top of the Natih Formation, top "a" member, is a regional unconformity that represents termination of the carbonate platform prior to development of the foreland basin filled with clastics of the Coniacian–Maastrichtian Aruma Group (Rabu et al., 1990; Scott, 1990a; Philip et al., 1995; van Buchem et al., 1996). The surface in many places is thickly encrusted by ironstone and phosphate. The age of the Natih at this surface varies from Late Cenomanian to Early Turonian (Scott, 1990a; Philip et al., 1995); however, at any one exposure erosional relief is not evident. Subaerial exposure is evident from the isotopic signal (Wagner, 1990), and subsequently the surface was drowned and buried by mainly Coniacian and younger marine sediments. This unconformity was formed by tectonic processes.

### *Wasia Group Depositional Cycles*

Four longer-term depositional cycles were recognized in the Wasia Group: (1) Nahr Umr–Natih members "g" and "f", (2) Natih member "e", (3) Natih members "d" and "c", and (4) Natih members "b" and "a" (Scott, 1990a) (Fig. 4). Philip et al. (1995) used hardground surfaces to define five longer-term cycles: (1) members "g", "f", and basal "e", (2) member "e", (3) member "d", (4) member "c", and (5) members "b" and "a" (Fig. 5). They further subdivided these cycles into shorter-term shoaling-up subcycles. Most recently, van Buchem et al. (1996) and van Buchem et al. (1997) divided the Natih members "e" through "a" into ten medium-term accommodation cycles using the hardground surfaces 1a at the base of the "e", 2a, 3a, and 4a within the "e", 5a at the base of the "d", 6a at the base of the "c", 7a within the "c", 8a at the base of the "b", and 9a and 10a within the "a" member (Fig. 5). On the basis of paleobathymetry data, they interpreted these as deepening-up to shoaling-up cycles. The contacts of these medium-term cycles were correlated throughout the Oman outcrop belt. These cycles were grouped into three large-scale or third-order cycles using surfaces 1a, 5a, 7a, and top Natih. Van Buchem et al. (1996) inferred that these were eustatically controlled. The origin of the shorter-term cycles was more complex and uncertain. The correlation of the cycles defined by van Buchem et al. with those of Philip et al. is shown in Figure 5. It appears that placement in the section of the member boundaries is not consistent between these two groups because the positions of these surfaces relative to the member contacts is not the same. However, we matched the cycle surfaces in van Buchem's unpublished section at Wadi Miáidin with the lithologies in the section by Philip et al. (1995).

## TIMING OF DISCONTINUITY CONTACTS

The timing of the several sets of depositional cycles in the Natih Formation in the Wadi Mi'aidin section is tested by the technique of graphic correlation. The biostratigraphic data published for this section by Simmons and Hart (1987), Scott (1990a), and Philip et al. (1995) were treated as three separate sections to avoid the problems of integrating three different sets of measurements. The three measured sections can be correlated by means of the lithostratigraphic contacts because they were measured at the same site. The line of correlation (LOC) is constrained by the base and top of the Nahr Umr Formation and

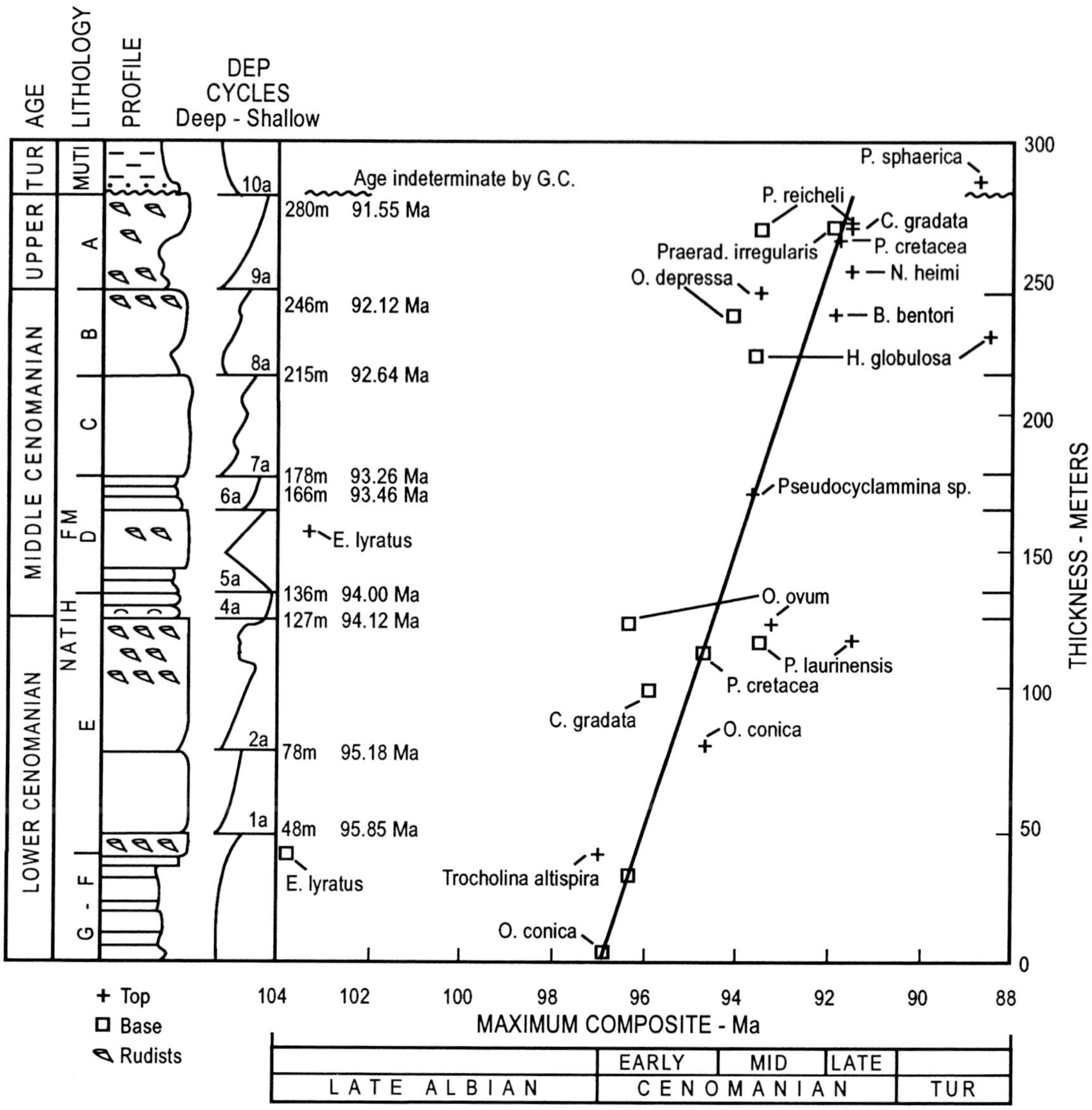


FIG. 5.—Graphic correlation of the Natih Formation in the Wadi Mi'aidin, Oman section and the Midk3 composite standard 9-98. Section measured by Philip et al. (1995) and sequence stratigraphic analysis by van Buchem et al. (1996, 1997).

the top of the "e" member of the Natih (Fig. 4). Note that a number of fossil bases also constrain the LOC. This graph shows a remarkable change in slope of the LOC, which indicates that the rate of sediment accumulation increased from 1.05 cm/1000 years during Nahr Umr deposition to 5.28 cm/1000 years for the Natih, a five times increase. Previous estimates of this differential rate using a different composite standard database developed by Amoco Corporation were 1.43 and 4.84 cm/1000 years, respectively (Scott, 1990a).

The graph of the Wadi Mi'aidin section (Fig. 4) dates key lithostratigraphic contacts, but the numerous hardgrounds in the Nahr Umr and Natih Formations are not identifiable in the data of Simmons or Scott. The ten hardground horizons in the Nahr Umr are best seen in the Wadi Bani Kharus section of Immenhauser et al. (1997), Immenhauser et al. (1999), and Immenhauser et al. (this volume), where the graph is presented. The base of the LOC in Wadi Bani Kharus is constrained by the bases of *Mesorbitolina texana* (Roemer) and *Mesorbitolina subconcava* (Leymerie). The middle segment is bracketed by the bases of *Eoradiolites lyratus* (Conrad) and *Mesorbitolina aperta* (Erman). The top of the LOC is constrained by the top of *Mesorbitolina subconcava* and the top of the Nahr Umr Formation. Such a correlation suggests that this lithostratigraphic contact is synchronous between Wadi Bani Kharus and Wadi Mi'aidin, which is not unreasonable and was proposed by the correlation of Scott (1990a, his fig. 7). The graphic interpretation proposes that Immenhauser's hardgrounds 1, 2, 5, 6, 8, and 10 correlate with transgressive–regressive depositional cycles in the Upper Aptian through Upper Albian section in north-central Texas (Fig. 6). Also, a maximum-flooding hardground in Wadi Bani Kharus is dated at 108.2 Ma, which correlates with the Al SB FR 1, the base of the transgressive Paluxy Sandstone of the Fredericksburg Group. This correlation suggests that some of the Nahr Umr hardgrounds formed synchronously with relative sea-level changes in the U.S. Gulf Coast that may be the response to global eustatic processes. The other hardgrounds, 3, 4, 7, and 9 may reflect hiatuses formed during maximum flooding, or local sea-level changes responding to tectonism, or eustatic events not recognized or recorded in Texas.

The ten hardgrounds in the Natih Formation defined by van Buchem et al. (1996) and van Buchem et al. (1997) are dated by the graph of biostratigraphic data of Philip et al. (1995) (Fig. 5). The LOC is well constrained by tops and bases of eight taxa, and the rate of sediment accumulation is 5.14 cm / 1000 years, close to that measured in the data of Simmons and Hart (1987) (Fig. 4). Hardgrounds 1a, 2a, 4a, and 8a correlate with depositional cycles defined in the Tunisian section. Hardgrounds 5a, 6a, 7a, and 9a may correlate with transgressive or downlap surfaces in the Tunisian section (Table 3).

## CONCLUSIONS

In order to recognize global eustatic events in the stratigraphic record, hypotheses of synchroneity of depositional surfaces must be tested quantitatively. Graphic correlation is an effective tool, as shown by numerous previous studies, to test such hypotheses. A global, multidisciplinary, quantitative stratigraphic database is necessary to conduct such tests. A database of forty-two middle Cretaceous sections yielding foraminifers, nannofossils, dinoflagellates, megafossils, magnetochrons, and geochemical shifts was constructed by a team of stratigraphers at the Free University, Amsterdam and Precision Stratigraphy Associates. This database is here applied to testing the synchroneity of hardground surfaces in the Wasia Group, Oman Mountains, with depositional cycles recognized in reference sections (Fig. 6).

In the latest Aptian–Albian of Texas nine cycles are recognized by shelf-to-basin facies successions and biotic assemblage changes that are separated by locally mappable contacts. The average duration of these cycles is 1.94 million years. In the Cenomanian of Tunisia similar criteria were used in basinal facies to define six cycles that average 1.08 million years in duration. The ages of the defining contacts were constrained in the composite standard database by graphic correlation of diverse fossils.

In Oman, depositional cycles in the Nahr Umr Formation, the basal unit in the Wasia Group, were defined by means of hardgrounds. Petrographic, geochemical, and biofacies data demonstrated the complex history of sea-level fall, exposure, drowning, and resumption of deposition. These contacts were dated by projecting them into a line of correlation constrained by fossil bases and tops and by a stratal contact, between the Nahr Umr and Natih Formations, that is synchronous between two reference sections. Six of ten of the hardgrounds and one maximum flooding contact are of the same age within ± 0.7 million years or less as latest Aptian–Albian cycles in the U.S. Gulf Coast (Table 3).

Depositional cycles of the Natih Formation were defined by petrographic and facies criteria, and their contacts mapped throughout the Oman Mountains. They represent relative sea-level changes. They were dated by projection into the composite standard constrained by fossil tops and bases. Four of ten Natih cycle boundaries correlate closely with Cenomanian sequence boundaries in Tunisia; one correlates with a transgressive surface and two correlate with downlap surfaces. The cycle boundary at the top of the formation is a regional tectonically controlled unconformity that does not correlate with cycles outside of the region. It is reasonable to consider that seven cycles represent the effects of eustatic processes upon the Natih carbonate shelf in Oman.

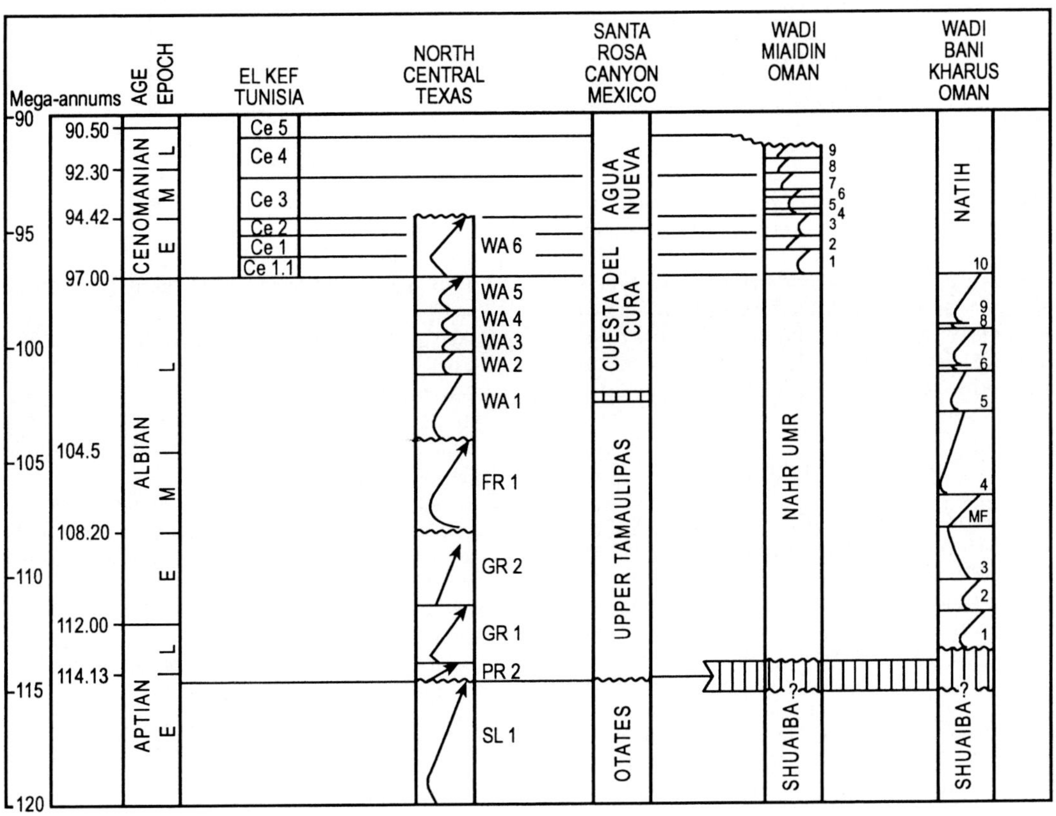


FIG. 6.—Comparison of deposition cycle ages defined in three Tethyan reference sections and in the Albian–Cenomanian sections of Oman. Arrows indicate deepening–shoaling changes.

TABLE 3.—Comparison of ages in mega-annums of surfaces separating Albian–Cenomanian depositional cycles in global reference sections and in Oman.

| Reference Sections | Wadi Bani Kharus | Wadi Miáidin |
|---|---|---|
| Ce DL 5 - 90.3 | | |
| Ce TS 5 - 90.8 –90.7 | | |
| Ce SB 5 - 91.0 –90.9 | | |
| Ce DL 4 - 92.3 –92.0 | | Hg 9a - 92.12 |
| Ce TS 4 - 92.6 –92.0 | | |
| Ce SB 4 - 92.7 –92.6 | | Hg 8a - 92.64 |
| Ce DL 3 - 93.24 –93.16 | | Hg 7a - 93.26 |
| Ce TS 3 - 93.50 | | Hg 6a - 93.36 |
| Ce SB 3 - 94.46 –94.33 | | Hg 4a - 94.12 |
| Ce DL 2 - 94.86 | | |
| Ce TS 2 - 95.01 | | Hg 2a - 95.18 |
| Ce SB 2 - 95.33 –95.16 | | |
| Ce DL 1 - 95.54 | | |
| Ce TS 1 - 95.98 –95.78 | | Hg 1a - 95.85 |
| Ce SB 1 - 96.08 –95.89 | | |
| Ce TS 1.1 - 96.64 | | |
| Ce SB 1.1 - 97.0–96.80 | Hg 10 - 97.12 | |
| Al TS WA 5 - 98.43 | | |
| | Hg 9 - 99.12 | |
| Al TS WA 4 - 99.78 | Hg 8 - 99.33 | |
| Al TS WA 3 - 100.57 | Hg 7 - 100.85 | |
| Al TS WA 2 - 101.85 | Hg 6 - 101.16 | |
| Al SB WA 1 - 104.00–103.50 | Hg 5 - 103.03 | |
| Al MF FR 1 - 106.9 | Hg 4 - 106.78 | |
| Al SB FR 1 - 108.18 | MF - 108.24 | |
| | Hg 3 - 110.26 | |
| Al SB GR 2 - 111.26 | Hg 2 - 111.82 | |
| Al SB GR 1 - 113.40 | Hg 1 - 113.37 | |
| Ap SB PR 2 - 113.72 | | |
| Ap SB SL 1 - 114.57 | | |

These hypotheses of correlation of depositional cycles in the Wasia Group with cycles in global reference sections need to be tested further by incorporating more fossil groups from the Oman sections to constrain the lines of correlation more precisely. Further, geochemical analyses of the hardground surfaces in the Natih Formation would evaluate the subaerial exposure origin of these contacts. Finally, correlation of these cycles in sections of other basins is needed to test for their global, and thus eustatic, origins.

## ACKNOWLEDGMENTS

Amoco Corporation is credited with the pioneering application of quantitative stratigraphic techniques to practical stratigraphic problems and with the development of procedures to construct composite standard databases. The Free University at Amsterdam provided financial support to begin the construction of a database to test the synchroneity of Cretaceous depositional events. Encouragement and advice on taxonomic questions were provided by J. A. Bergen, R. W. Hedlund, B. A. Masters, and J. A. Stein. Stimulating and fruitful discussions about Cretaceous cycles in the Middle East were shared with A.S. Alsharhan, E.J. Kauffman, and C. G. St. C. Kendall.

## REFERENCES

AL-AWAR, A.A., AND HUMPHREY, J.D., 1997, Reservoir characterization of the Shu'aiba Formation, Ghaba North Field, Oman: GeoArabia, v. 2, p. 476.

ALSHARHAN, A.S., 1989, Petroleum geology of the United Arab Emirates: Journal of Petroleum Geology, v. 12, p. 253–288.

ALSHARHAN, A.S., AND KENDALL, C.G.ST.C., 1991, Cretaceous chronostratigraphy, unconformities and eustatic sea-level changes in the sediments of Abu Dhabi, United Arab Emirates: Cretaceous Research, v. 12, p. 379–401.

ALSHARHAN, A.S., AND NAIRN, A.E.M, 1986, A review of the Cretaceous formations in the Arabian Peninsula and Gulf: Part I. Lower Cretaceous (Thamama Group) stratigraphy and paleogeography: Journal of Petroleum Geology, v. 9, p. 365–392.

ALSHARHAN, A.S., AND NAIRN, A.E.M., 1988, A review of the Cretaceous formations in the Arabian Peninsula and Gulf: Part II. Mid-Cretaceous (Wasia Group) stratigraphy and paleogeography: Journal of Petroleum Geology, v. 11, p. 89–112.

BRALOWER, T.J., ARTHUR, M.A, LECKIE, R.M., SLITER, W.V., ALLARD, D.J., AND SCHLANGER, S., 1994, Timing and paleoceanography of oceanic dysoxia/anoxia in the Late Barremian to Early Aptian (Early Cretaceous): Palaios, v. 9, p. 335–369.

BURCHETTE, T.P., 1993, Mishrif Formation (Cenomanian–Turonian), southern Arabian Gulf: carbonate platform growth along a cratonic basin margin, *in* Simo, J.A., Scott, R.W., and Masse, J.-P., eds., Cretaceous Carbonate Platforms: American Association of Petroleum Geologists, Memoir 56, p. 185–199.

CARNEY, J.L., AND PIERCE, R.W., 1995, Graphic correlation and composite standard databases as tools for the exploration biostratigrapher, *in* Mann, K.O., and Lane, H.R., eds., Graphic Correlation: SEPM, Special Publication 53, p. 23–43.

CECCA, F., PALLINI, G., ERBA, E., PREMOLI-SILVA, I., AND COCCIONI, R., 1994, Hauterivian–Barremian chronostratigraphy based on ammonites, nannofossils, planktonic foraminifera and magnetic chrons from the Mediterranean domain: Cretaceous Research, v. 15, p. 457–467.

ERBA, E., 1994, Nannofossils and superplumes: the Early Aptian "nannoconid crisis": Paleoceanography, v. 9, p. 483–501.

ERBA, E., AND QUADRO, B., 1987, Biostratigrafia a nannofossili calcarei calpionellidi e formaniferi planktonici della Maiolica (Titoniano superiore–Aptiano) nelle prealpi Bresciane (Italia Settentrionale): Rivista Italiana di Paleontologia e Stratigrafia, v. 93, p. 3–108.

FÖLLMI, K.B, WEISSERT, H., BISPING, M., AND FUNK, H., 1994, Phosphogenesis, carbon-isotope stratigraphy, and carbonate-platform evolution along the Lower Cretaceous northern Tethyan margin: Geological Society of America, Bulletin, v. 106, p. 729–746.

FROST, S.H., BLIEFNICK, D.M., AND HARRIS, P.M., 1983, Deposition and porosity evolution of a Lower Cretaceous rudist buildup, Shuaiba Formation of eastern Arabian Peninsula: Society of Economic Paleontologists and Mineralogists, Core Workshop 4, p. 381–410.

GALLOWAY, W.E., 1989, Genetic stratigraphic sequences in basin analysis I: Architecture and genesis of flooding-surface bounded depositional units: American Association of Petroleum Geologists, Bulletin, v. 73, p. 125–142.

GRADSTEIN, F.M., AGTERBERG, F.P., OGG, J.G., HARDENBOL, J., VAN VEEN, P., THIERRY, J., AND HUANG, Z., 1995, A Triassic, Jurassic and Cretaceous time scale, *in* Berggren, W.A., Kent, D.V., Aubry, M.P., and Hardenbol, J., eds., Geochronology, Time Scales, and Global Stratigraphic Correlation: SEPM, Special Publication 54, p. 95–126.

GRÖTSCH, J., BILLING, I., AND VAHRENKAMP, V., 1998, Carbon-isotope stratigraphy in shallow-water carbonates: implications for Cretaceous black-shale deposition: Sedimentology, v. 45, p. 623–634.

HAQ, B.U., HARDENBOL, J., AND VAIL, P.R., 1988, Mesozoic and Cenozoic chronostratigraphy and eustatic cycles, *in* Wilgus, C.K., Hastings, B.S., Posamentier, H., Van Wagoner, J., Ross, C.A., and Kendall, C.G.St.C., eds., Sea-Level Changes: An Integrated Approach: SEPM , Special Publication 42, p. 71–108.

HARDENBOL, J., THIERRY, J., FARLEY, M.B., JACQUIN, TH., DE GRACIANSKY, P.-C., AND VAIL, P.R., 1998, Mesozoic and Cenozoic sequence chronostratigraphic framework of European Basins, *in* de Graciansky, P.-C., Hardenbol, J., Jacquin, Th., and Vail, P.R., eds., Mesozoic and Cenozoic sequence stratigraphy of European Basins: SEPM, Special Publication 60, p. 3–13.

HARLAND, W.B., ARMSTRONG, R.L., COX, A.V., CRAIG, L.E., SMITH, A.G., AND SMITH, D.G., 1990, A Geologic Time Scale 1989: Cambridge, U.K., Cambridge University Press, 263 p.

HUGHES CLARK, M.W., 1988, Stratigraphy and rock-unit nomenclature in the oil-producing area of interior Oman: Journal of Petroleum Geology, v. 11, p. 5–60.

ICE, R.G., AND MCNULTY, C.L., 1980, Foraminifers and calcispheres from the Cuesta del Cura and lower Agua Nueva(?) Formations (Cretaceous) in east-central Mexico: Gulf Coast Association of Geological Societies, Transactions, v. 30, p. 403–425.

IMMENHAUSER, A., SCHLAGER, W., BURNS, S.J., AND SCOTT, R.W., 1997, Albian sea-level fluctuations constrained by stable isotopes, fluid inclusions and the morphology of benthic foraminifera (Nahr Umr Formation ) Oman (abstract): GeoArabia, v. 2, p. 486–487.

IMMENHAUSER, A., SCHLAGER, W., BURNS, S.J., SCOTT, R.W., GEEL, T., LEHMANN, J., VAN DER GAAST, S., AND BOLDER-SCHRIJVER, L.J.A., 1999, Late Aptian to Late Albian sea-level fluctuations constrained by geochemical and biological evidence (Nahr Umr Formation, Oman): Journal of Sedimentary Research, v. 69, p. 434–446.

MOSHIER, S.O., HANDFORD, C.R., SCOTT, R.W., AND BOUTELL, R.D., 1988, Giant gas accumulation in a "chalky"-textured micritic limestone, Lower Cretaceous Shuaiba Formation, Eastern United Arab Emirates: SEPM, Core Workshop 12, p. 229–272.

OBRADOVICH, J.D., 1993, A Cretaceous time scale, *in* Caldwell, W.G.E., and Kauffman, E.G., eds., Evolution of the Western Interior Basin: Geological Association of Canada, Special Paper 39, p. 379–396.

PHILIP, J., BORGOMANO, J., AND AL-MASKIRY, S., 1995, Cenomanian–Early Turonian carbonate platform of northern Oman: stratigraphy and palaeo-environments: Palaeogeography, Palaeoclimatology, Palaeoecology, v. 119, p. 77–92.

PRATT, B.R., AND SMEWING, J.D., 1993, Early Cretaceous platform margin, Oman, eastern Arabian Peninsula, *in* Simo, J.A., Scott, R.W., and Masse, J.-P., eds., Cretaceous Carbonate Platforms: American Association of Petroleum Geologists, Memoir 56, p. 201–212.

RABU, D., LE MÉTOUR, J., BÉCHENNEC, F., BEURRIER, M., VILLEY, M., AND BOURDILLON-JEUDY DE GRISSAC, C., 1990, Sedimentary aspects of the Eo-Alpine cycle on the northeast edge of the Arabian platform (Oman Mountains), *in* Robertson, A.H.F., Searle, M.P., and Ries, A.C., eds., The Geology and Tectonics of the Oman Region: Geological Society of London, Special Publication 49, p. 49–68.

RAWSON, P.F., DHONDT, A.V, HANCOCK, J.M., and KENNEDY, W.J., 1996, Proceedings "Second International Symposium on Cretaceous Stage Boundaries" Brussels 8–16 September 1995: Institut Royal des Sciences Naturelles de Belgique, Bulletin, Sciences de la Terre Aardwetenschappen, v. 66, Supplement, 117 p.

ROBASZYNSKI, F., CARON, M., DUPUIS, C., AMEDRO, F., GONZALEZ DONOSO, J.M., LINARES, D., HARDENBOL, J., GARTNER, S., CALANDRA, F., AND DELOFFRE, R., 1990, A tentative integrated stratigraphy in the Turonian of central Tunisia: formations, zones and sequential stratigraphy in the Kalaat Senan area: Bulletin Centres Recherches Exploration-Production Elf-Aquitaine, v. 14, p. 213–384.

ROBASZYNSKI, F., CARON, M., AMEDRO, F., DUPUIS, C., HARDENBOL, J., GONZALEZ DONOSO, J.M., LINARES, D., and GARTNER, S., 1993, Le Cénomanien de la région de Kalaat Senan (Tunisie centrale): Litho-biostratigraphie et interprétation séquentielle: Revue de Paléobiologie, v. 12, p. 351–505.

ROSS, M.A., AND MCNULTY, C.L., 1980, Some microfossils of the Tamaulipas Limestone (Hauterivian–Lower Albian) in Santa Rose Canyon, Sierra de Santa Rose, Nuevo León, Mexico: Gulf Coast Association of Geological Societies, Supplement to Transactions, v. 31, p. 461–469.

SCOTT, R.W., 1990a, Chronostratigraphy of the Cretaceous carbonate shelf, southeastern Arabia, *in* Robertson, A.H.F., Searle, M.P., and Ries, A.C., eds., The Geology and Tectonics of the Oman Region: Geological Society of London, Special Publication 49, p. 89–108.

SCOTT, R.W., 1990b, Models and Stratigraphy of Mid-Cretaceous Reef Communities, Gulf of Mexico: SEPM, Concepts in Sedimentology and Paleontology, v. 2, 102 p.

SCOTT, R.W., 1993, Cretaceous carbonate platform, U.S. Gulf Coast, *in* Simo, J.A., Scott, R.W., and Masse, J.-P., eds., Cretaceous Carbonate Platforms: American Association of Petroleum Geologists, Memoir 56, p. 97–109.

SCOTT, R.W., FROST, S.H., AND SHAFFER, B.L., 1988, Early Cretaceous sea level curves, Gulf Coast southeastern Arabia, *in* Wilgus, C.K., Hastings, B.S., Posamentier, H., Van Wagoner, J., Ross, C.A., and Kendall, C.G.St.C., eds., Sea-Level Changes: An Integrated Approach: SEPM, Special Publication 42, p. 275–284.

SCOTT, R.W., FRANKS, P.C., STEIN, J.A., BERGEN, J.A., AND EVETTS, M.J., 1994, Graphic correlation tests the synchronous Mid-Cretaceous depositional cycles: Western interior and Gulf Coast, *in* Dolson, J.C., ed.,

Unconformity-Related Hydrocarbons in Sedimentary Sequences: Rocky Mountain Association Geologists, p. 89–98.

SIMMONS, M.D., AND HART, M.B., 1987, The biostratigraphy and microfacies of the Early to mid-Cretaceous carbonates of Wadi Mi'aidin, central Oman Mountains, *in* Hart, M.B., ed., Micropalaeontology of Carbonate Environments: Chichester, UK, Ellis Horwood Ltd., p. 176–207.

VAHRENKAMP, V.C., 1996, Carbon isotope stratigraphy of the upper Kharaib and Shuaiba Formations: implications for the Early Cretaceous evolution of the Arabian Gulf region: American Association Petroleum of Geologists, Bulletin, v. 80, p. 647–662.

VAN BUCHEM, F.S.P., RAZIN, P., HOMEWOOD, P.W., PHILIP, J.M., EBERLI, G.P., PLATEL, J.-P., ROGER, J., ESCHARD, R., DESAUBLIAUX, G.M.J., BOISSEAU, T., LEDUC, J.-P., LABOURDETTE, R., AND CANTALOUBE, S., 1996, High resolution sequence stratigraphy of the Natih Formation (Cenomanian/Turonian) in northern Oman: distribution of source rocks and reservoir rocks: GeoArabia, v. 1, p. 65–91.

VAN BUCHEM, F.S.P., RAZIN, P., CASANOVA, P., HOMEWOOD, P.W., WALGENWITZ, F., SCHWAB, A., OTERDOON, H., AND PHILIP, J.M., 1997, The stratigraphic architecture of the Cenomanian/Turonian carbonate petroleum system in Natih Formation northern Oman (abstract): GeoArabia, v. 2, p. 496.

WAGNER, P.D, 1990, Geochemical stratigraphy and porosity controls in Cretaceous carbonates near the Oman Mountains, *in* Robertson, A.H.F., Searle, M.P., and Ries, A.C., eds., The Geology and Tectonics of the Oman Region: Geological Society of London, Special Publication 49, p. 127–137.

WITT, W., AND GÖKDAG, H., 1994, Orbitolinid biostratigraphyof the Shuaiba Formation (Aptian), Oman—implications for reservoir development, *in* Simmons, M.D., ed., Micropalaeontology and Hydrocarbon Exploration in the Middle East: London, Chapman & Hall, p. 221–241.

YOUNG, K., 1986, Cretaceous, marine inundations of the San Marcos Platform, Texas: Cretaceous Research, v. 7, p. 117–140.

YUREWICZ, D.A., MARLER, T.B., MEYERHOLTZ, K.A., AND SIROKY, F.X., 1993, Early Cretaceous carbonate platform, north rim of the Gulf of Mexico, Mississippi and Louisiana, *in* Simo, J.A., Scott, R.W., and Masse, J.-P., eds., Cretaceous Carbonate Platforms: American Association of Petroleum Geologists, Memoir 56, p. 81–97.

# SYNOPTIC GUIDE TO LOWER CRETACEOUS RUDIST BIVALVES OF ARABIA

PETER W. SKELTON
*Department of Earth Sciences, Open University, Walton, Milton Keynes MK7 6AA, UK*
AND
JEAN-PIERRE MASSE
*Centre de Sédimentologie-Paléontologie, UPRESA 6019 du CNRS, Université de Provence, 13331 Marseille cedex 03, France*

ABSTRACT: The diversity of rudists in the Lower Cretaceous of Arabia offers great potential both for biostratigraphical correlation (particularly of certain key levels, such as the Lower/Upper Aptian boundary) and for facies analysis. With the aim of raising awareness of this potential among geologists working on core and outcrop material, we provide here synoptic diagnoses of the seventeen species (at least) so far recognized, together with comments on their stratigraphical distributions and paleoecology, and a range chart.

## INTRODUCTION

Rudist bivalves are among the most characteristic and abundant macrofaunal elements in Cretaceous Tethyan carbonate-platform deposits. Growing either as upright forms implanted in sediment (*elevators*), or lying prone on the surface as substrate-hugging *clingers*, or free-lying *recumbents*, vast congregations of their shells formed extensive biostromes, interbedded with related bioclastic units, together constituting *rudist formations* (Masse and Philip, 1981; Ross and Skelton, 1993; Gili et al., 1995).

Though rudists are present in pre-Hauterivian units (e.g., from boulders contained in debris flow deposits in Wadi Hagil, Ras al Khaimah; Toland et al., 1993), rudist formations are common in the Arabian Lower Cretaceous only from the Hauterivian (Habshan p.p.) upwards. They reach a distinct acme in the Lower Aptian (Shuaiba) (see Witt and Gökdag, 1994; Masse et al., 1998a), but this development is terminated at the Lower/Upper Aptian boundary, at which many rudist taxa disappear. Upper Aptian–Albian rudists are restricted largely to the northeastern part of the Jebel Akhdar tectonic inlier, in northern Oman (in the Nakhl area, e.g., Wadi El Assyi; Masse et al., 1997). Nevertheless, farther to the west there is also some localized evidence in the subsurface of the northeastern Rub' Al Khali for lowermost Upper Aptian rudist facies succeeding the Shuaiba. Known ranges of the taxa in Arabia are summarized in Figure 1, and brief descriptions and comments on their distributions follow.

## SYNOPTIC DIAGNOSES, WITH COMMENTS ON DISTRIBUTIONS

Major clues to rudist identification are as follows:

1. *Shell structure*. Rudists have an outer shell layer of fibrillar prismatic calcite and an inner shell that was originally aragonitic, though now usually either replaced by sparry calcite or leached. The outer layer is normally cream-colored to pale honey brown but may be blackened following deep burial (as in the tectonic inliers of the Oman Mountains). Such blackening is especially noticeable in those taxa with a relatively thick outer shell (e.g., Figs. 2.1, 5.2). The relative development of the shell layers varies considerably between taxa. In one family (Radiolitidae), the outer shell layer contains distinctive layers of tiny hollow cells (*celluloprismatic structure*). Other taxa (mostly Caprinidae) have blind-ending hollow canals (*pallial canals*) in the inner shell.
2. *Shell shape and valve of attachment*. In the majority of Cretaceous rudists, the shell has a more or less uncoiled shell form, with a conical to cylindrical attached right valve (*RV*) and a capuloid to operculiform left valve (*LV*). A few (Requieniidae) attached by a spirally coiled *LV*, capped by a flattened *RV*.
3. *Myocardinal system*. One of the most important keys to rudist identification is the arrangement of the hinge teeth and the internal shelly supports for the adductor muscles (*myophores*). Tilting of the myophores within the shell sometimes separates off distinct *accessory cavities* from the main *body cavity*, and these may also be diagnostic (see Skelton and Masse, 1998, for further details). Myocardinal arrangements are often difficult to interpret in random sections, however, such that numerous specimens need to be inspected for a secure identification. It is therefore advisable to inspect, draw, and/or photograph several specimens at each site for identification purposes.

### Superfamily HIPPURITOIDEA Gray, 1848 (the group informally known as "rudists")

The family classification used here is based mainly on Skelton (1978) with further modifications from Skelton and Masse (1998). At present, the family-level taxonomy of Early Cretaceous rudists is in a state of flux, as new approaches to classification (especially cladistics) are applied. Genera, on the other hand, are more firmly established, and so are more likely to remain stable. Many species within genera still present the traditional taxonomic problems of possible synonymy, or of being merely successive chronospecies within single evolutionary lineages. Current morphometric studies (as in Skelton and Masse, 1998) are likely to lead to resolution of some of these problems in the next few years, again possibly resulting in a few changes to the specific names given here. Except where otherwise indicated, the accompanying drawings show specimens at natural size.

[*A note on the illustrations*. Though all are based on real specimens, some of the drawings shown here are composites from more than one specimen or have been simplified for clarity. A few are based on non-Arabian material (as indicated), to show particular features. These drawings are intended purely as an aid to identification in the field or in core material: they should *not* be regarded as accurate portrayals of individual specimens for the purposes of descriptive taxonomy.]

Middle East Models of Jurassic/Cretaceous Carbonate Systems
SEPM Special Publication No. 69, Copyright © 2000
SEPM (Society for Sedimentary Geology), ISBN 1-56576-075-1, p. 89–99.

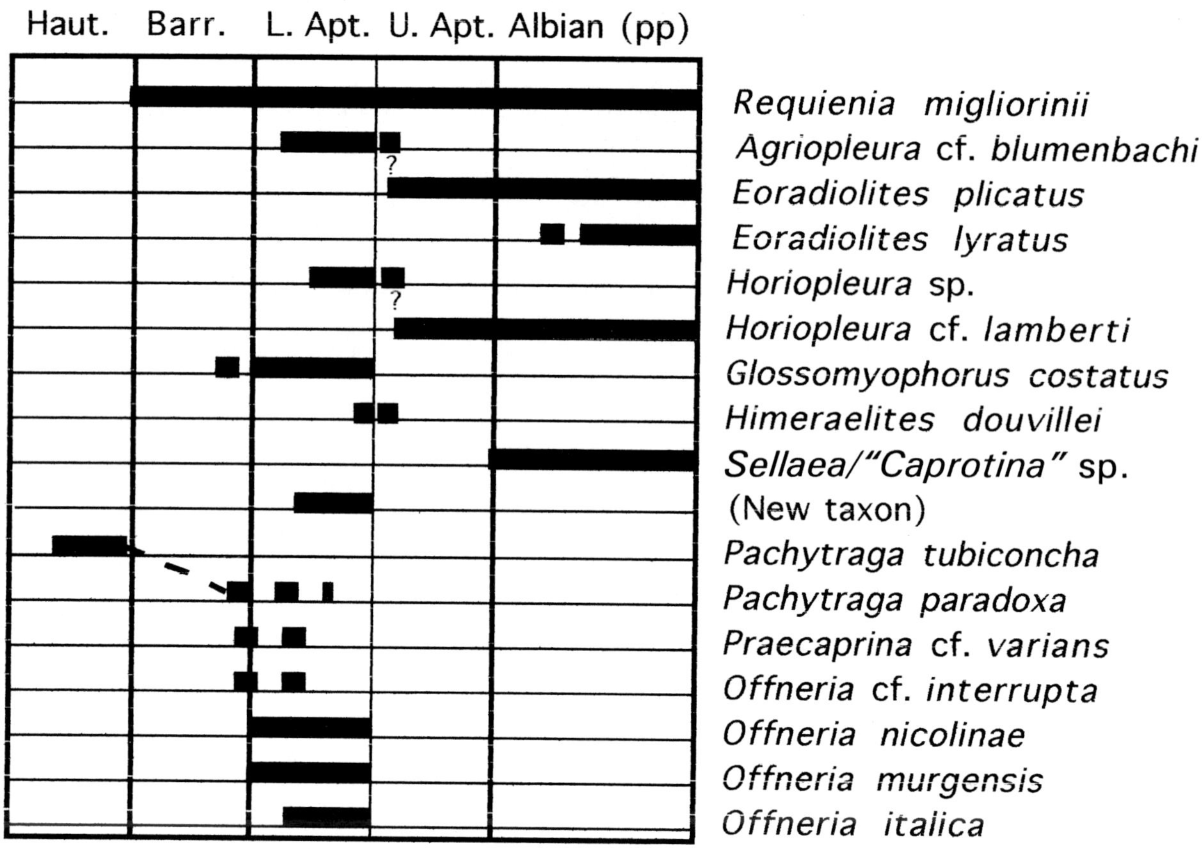


FIG. 1.—Known ranges of Early Cretaceous rudists in Arabia.

Family REQUIENIIDAE Douvillé, 1915

*Main features of family*.—Attached by *LV*. Ligament external, hence shell growth spirogyrate. Outer shell layer relatively thick. Two teeth in the *RV* and one in the *LV*.

*Requienia migliorinii* (Tavani, 1948).
(Fig. 2.1)

*What to look for*.—*LV* has extended corkscrew shape, often giving the appearance in section of a stack of rounded-zigzag or

FIG. 2.—Diagrammatic sketches of Arabian Early Cretaceous rudists. All drawings are approximately natural size, except where indicated. Key to abbreviations: *ab*, anterior radial band; *am*, anterior myophore; *at*, anterior tooth; *ats*, anterior tooth socket; *(b)*, borings; *bc*, body cavity; *cp*, celluloprismatic shell structure; *cs*, central tooth socket; *ct*, central tooth; *dd*, dorsal depression; *lc*, ligamentary crest; *o*, accessory cavity; *pb*, posterior radial band; *pm*, posterior myophore; *pt* posterior tooth; *pts*, posterior tooth socket; *v*, antero-ventral carina.

1. *Requienia migliorinii* Tavani. Longitudinal section through spirally coiled *RV* (outer shell, black; inner shell, white).

2–6. *Agriopleura* sp. cf. *A. blumenbachi* (Stüder). 2, postero-ventral aspect of whole shell; 3, antero-dorsal view of *LV*; 4, interior of quadrate *RV*; 5, section across rounded *RV* (detail shows compact fibrous prismatic structure of outer shell layer); and 6, radial section through both valves (outer shell, black; inner shell, white).

7. *Eoradiolites plicatus* (Conrad). Section across *RV* outer shell (about twice natural size). Note localized development of celluloprismatic structure.

8. *Eoradiolites lyratus* (Conrad). Section across *RV* outer shell (about twice natural size). Note development of celluloprismatic structure throughout outer shell layer.

9–12. *Horiopleura* sp. 9, ventral aspect of whole shell (larger form); 10, section across posterior parts of both valves of the similar *H. distefanoi* (Parona), seen from behind (outer shell, black; inner shell, white; matrix, stippled) [from Italy (Parona, 1909)]; 11, 12, interiors of *LV* (11) and *RV* (12) of larger form.

13. *Horiopleura* sp. cf. *H. lamberti* Douvillé. Section across both valves, viewed from ventral side (outer shell, black; inner shell, white). Reduced; commissural diameter may exceed 10 cms.

14–18. *Glossomyophorus costatus* Masse, Skelton and Sliskovic. 14, ventral aspect of whole shell; 15, 16, interiors of *LV* (15) and *RV* (16); 17, 18, sections across *LV* (17) and *RV* (18) (outer shell, black; inner shell, white; matrix, stippled).

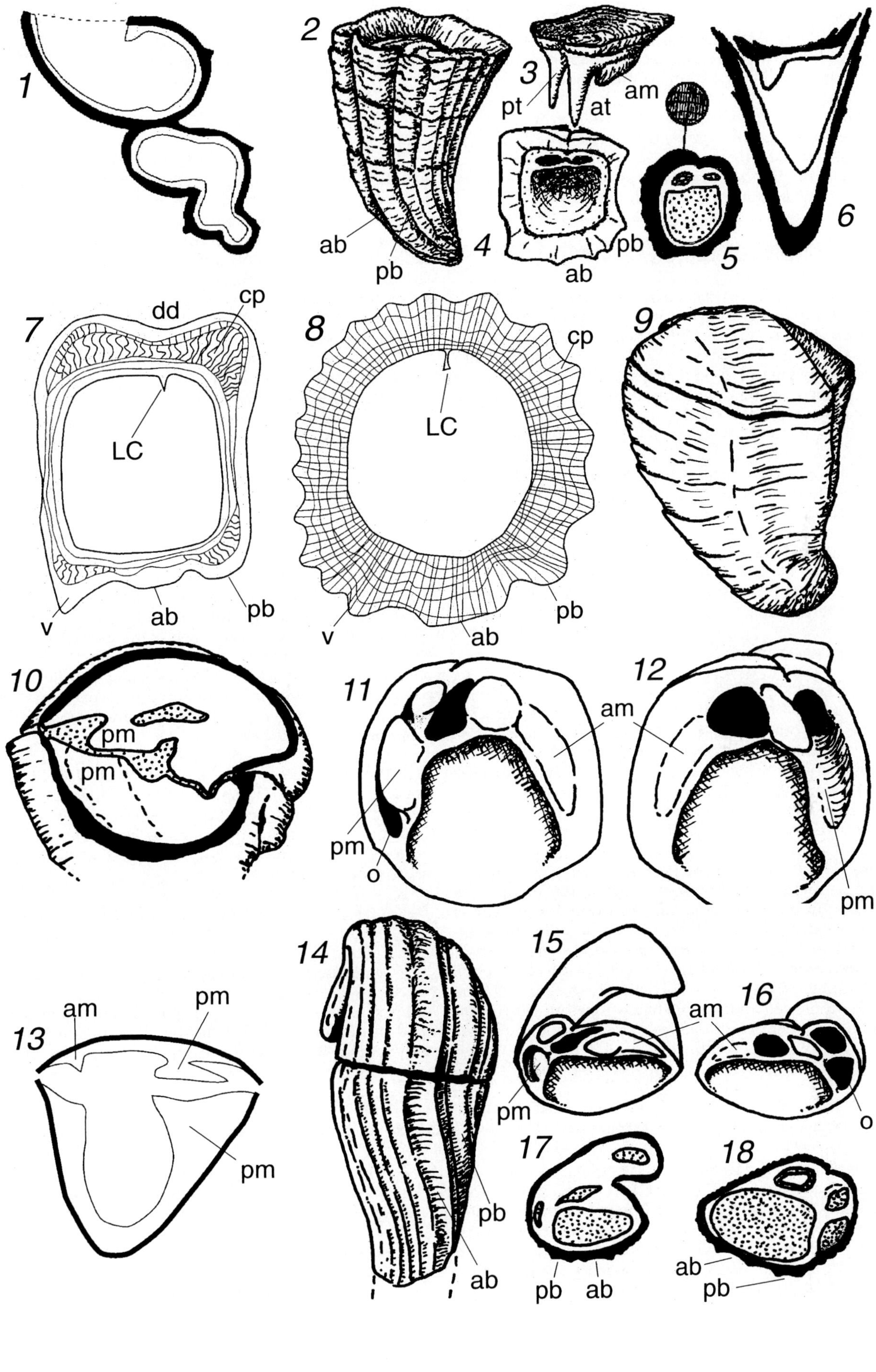
1
2
3
pt
at
am
ab
pb
4
pb
ab
5
6
7
dd
cp
LC
v
ab
pb
8
cp
LC
v
ab
pb
9
10
pm
pm
11
am
pm
o
12
pm
13
am
pm
pm
14
pb
ab
15
am
pm
16
am
o
17
pb
ab
18
ab
pb

kidney-shaped loops of increasing size (Fig. 2.1); *RV* lid-like. Outer shell relatively thick (at least 2 mm), externally smooth (and typically blackened in sections seen at outcrop in the Oman Mountains); inner shell thin, except around hinge area.

*Where found.*—Sporadically present from the Barremian to the Albian, though most common in the Upper Aptian–Albian, e.g., in Wadi El Assyi (Nakhl area). Grew as ascending spiral clinger, semi-immersed in platform-top wackestones to packstones, though often found as transported shells in floatstones.

### Family MONOPLEURIDAE Munier Chalmas, 1873

*Main features of family.*—Attached by *RV*. Ligament shallowly invaginated, hence shell growth uncoiled. *RV* cylindrico-conical; *LV* low-domed to operculiform. Outer shell variable in thickness. Two subequal teeth in *LV* and one in *RV*. Myophores form ledge-like ventral extensions of hinge plates; those of *LV* may be outwardly inclined, facing down onto their depressed counterparts on the inner walls of the *RV*.

Although the genus *Monopleura*, itself, has often been cited from the Arabian Lower Cretaceous, no detailed descriptions have yet been given to confirm its presence. Many citations are evidently misidentifications of other taxa (e.g., the RVs of *Glossomyophorus* or of the new taxon noted below). However, the presence of primitive *Monopleura* sp., has been confirmed by one of us (P.W.S.) in sub-Hauterivian Cretaceous limestones in Wadi Hagil, Ras al Khaimah, U.A.E. (see Toland et al., 1993). The only representative of the family common in the Arabian Lower Cretaceous is *Agriopleura*.

#### *Agriopleura* sp. cf. *A. blumenbachi* (Stüder, 1834)
(Fig. 2.2–2.6)

*What to look for.*—*RV* conical; *LV* concave, lid-like (Figs. 2.2, 2.3). Commissural outline quadrate to rounded (Figs. 2.4, 2.5). Outer shell layer thick (a few to several millimeters), with coarse external ribs and steeply inclined flaring growth lamellae in the *RV* (Fig. 2.6). Two slightly protruding radial bands, externally flat to gently embayed, the anterior one situated ventrally (*ab*) and the other, postero-ventrally (*pb*) (Figs. 2.2, 2.4). Outer shell layer of compact fibrillar prismatic microstructure (Fig. 2.5, detail), entirely lacking cells. *LV* anterior and posterior teeth subequal (*at* > *pt*), projecting straight into narrow sockets in *RV*. Both myophores in *LV* project as low buttresses adjoining teeth (Fig. 2.3). *RV* varies in shape from moderately conical, like that supposedly characterizing "*A. marticensis* (D'Orbigny)", to slender, as in "*A. blumenbachi*", and radial band morphology is like that seen in both of these (cf. Douvillé, 1918). Apart from a possibly slightly thicker shell, the Arabian forms thus seem indistinguishable from those European forms, which may themselves be synonymous (though further work on this question is needed).

*Where found.*—Widespread in the Shuaiba Formation, though most common as scattered individuals and small elevator clusters in inner-platform facies (Hughes, 1997).

### Family RADIOLITIDAE D'Orbigny, 1847

*Main features of family.*—Attached by *RV*. Ligament invaginated on distinct inward projection of outer shell layer (*ligamentary crest*). Shell growth uncoiled, with *RV* conical to cylindrical, and *LV* more or less operculiform. Outer shell layer thick, with celluloprismatic microstructure in *RV*, and with characteristic radial bands on the ventral to posterior flanks of the shell. Two straight teeth in *LV*, flanked by stalked and outward-facing myophores, together projecting as a dorsal arcade into *RV*; central tooth in *RV* reduced, and adductor insertions situated on the inner valve walls.

This general diagnosis applies with rare exceptions in the Lower Cretaceous, though there are more exceptions in the Upper Cretaceous, usually involving secondary modification or loss of features, of no concern here. For further details on Upper Aptian taxa, see Masse and Gallo Maresca (1997).

#### *Eoradiolites plicatus* (Conrad, 1852)
(Fig. 2.7)

*What to look for.*—External form similar to that of *Agriopleura* (with sub-quadrate transverse section); but outer surface relatively smooth rather than ribbed, apart from the radial bands (*ab* and *pb*). Main difference is the presence of celluloprismatic structure (*cp*) in the *RV*, though this is usually limited to a meandriform radial network in the thicker parts of the shell (mainly dorsal, but also at the antero-ventral and postero-ventral corners); thinner parts remain compact (Fig. 2.7; see also Douvillé, 1910). Otherwise, a dorsal depression (*dd*) externally corresponds to a triangular ligamentary crest (*lc*) within, and the antero-ventral corner (*v*) projects as a distinct carina.

*Where found.*—Upper Aptian–Albian of Wadi El Assyi, Wadi Hedeck, and Wadi Hammam (North Oman, Nakhl area), as locally clustered elevators. Probably derived from *Agriopleura*, its first appearance currently seems to lie just above the Lower/Upper Aptian boundary, for which it may thus serve as an approximate marker.

#### *Eoradiolites lyratus* (Conrad, 1852)
(Fig. 2.8)

*What to look for.*—Commissural outline more circular than the previous species, with rounded longitudinal ribs all around shell, and fine rectangular celluloprismatic microstructure (*cp*) developed throughout the *RV* wall (Fig. 2.8). Ligamentary crest (*lc*) with pedicle. Body cavity may contain transverse tabulae.

---

FIG. 3.—Diagrammatic sketches of Arabian Early Cretaceous rudists. See caption to Figure 2 for scale and abbreviations.

1–4. *Himeraelites douvillei* Di Stefano. 1, ventral aspect of whole shell; 2, 3, interiors of *LV* (2, based on Italian specimen of Di Stefano, 1888) and *RV* (3, from upper Shuaiba); 4, section across posterior part of both valves, seen from behind (outer shell, black; inner shell, white; matrix, stippled).

5. *Sellaea*/"*Caprotina*" sp. (*sensu* Di Stefano, 1888). Section across *LV* (reduced to about half size).

6–12. New genus, new species. 6, interior of slender *RV*, viewed from ventral side; 7, section across slender *RV* (outer shell, and tooth sockets, black; inner shell, white; matrix, stippled); 8, antero-ventral aspect of whole shell (more robust example); 9, 10, interiors of *LV* (9) and *RV* (10); 11, 12, sections across *LV* (11) and bored *RV* (12) (outer shell, black; inner shell, white; matrix, stippled).

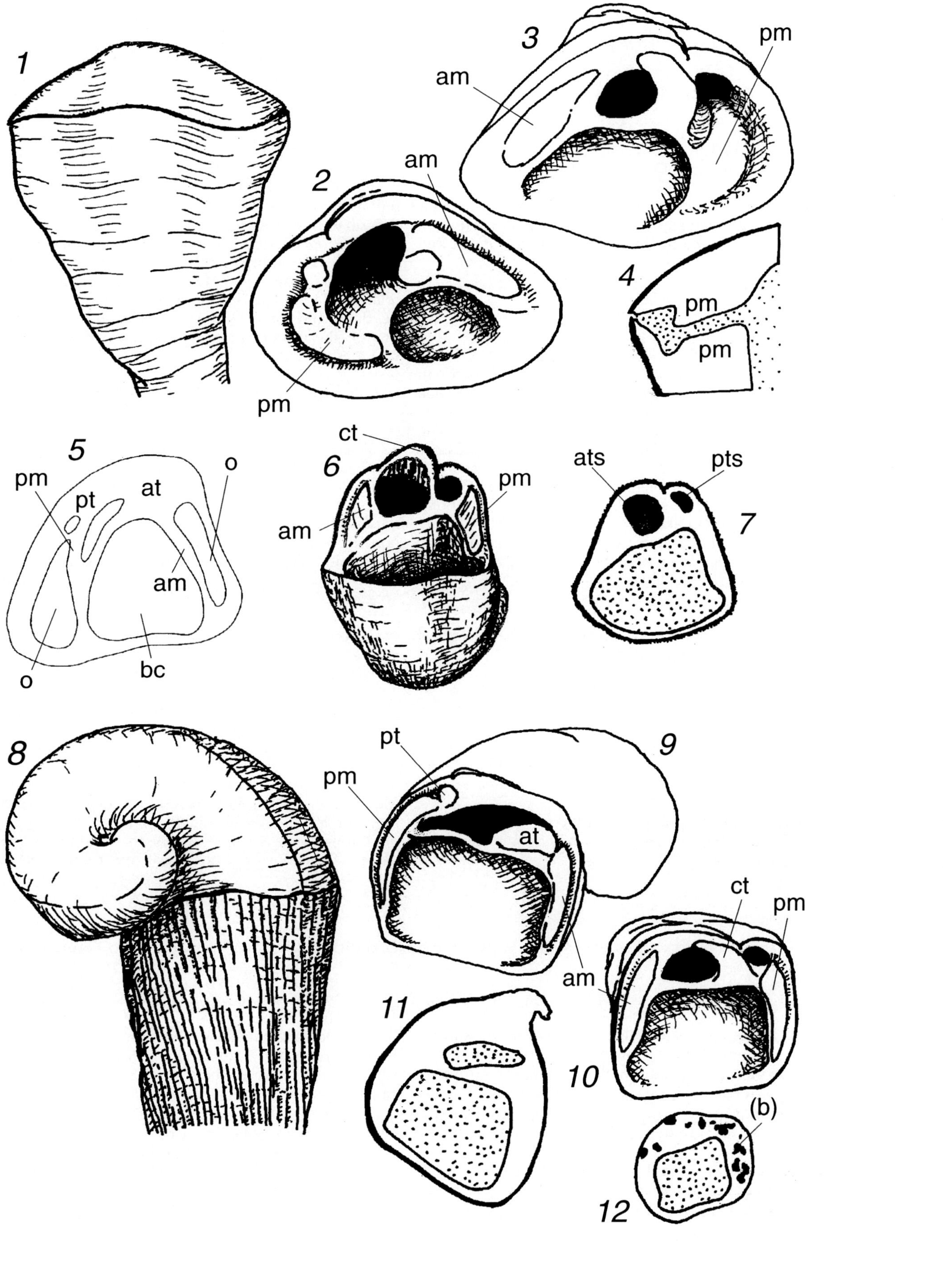
1
3
am
pm
2
am
pm
4
pm
pm
5
pm
pt
at
o
am
o
bc
6
ct
pm
am
ats
pts
7
8
9
pt
pm
at
am
ct
pm
11
10
(b)
12

*Where found.*—Albian of Wadi El Assyi (also continues on into Cenomanian, e.g., Mishrif Formation).

Family POLYCONITIDAE Mac Gillavry, 1937

*Main features of family.*—Attached by *RV*. Ligament shallowly invaginated, hence shell growth uncoiled. *RV* twisted conical; *LV* low-domed to operculiform. Outer shell relatively thick, especially in *RV* (Fig. 5.2). Two subequal teeth in *LV* and one in *RV*. Myophores form ventral extensions of hinge plates; posterior myophore of *LV* pedunculate, projecting back over peripheral accessory cavity and facing down onto ledge-like or depressed myophore in *RV*.

*Horiopleura* sp.
(Fig. 2.9–2.12)

*What to look for.*—*RV* stoutly conical to cylindrical; *LV* low-domed. Both valves fairly smooth externally, except for growth rugae and perhaps some weak ribbing on the *RV*. Outer shell moderately thick, especially in *RV* (at least 2 mm) (Figs. 2.9, 2.10). Main diagnostic feature is pedunculate posterior myophore (*pm*) in *LV* (Fig. 2.11), which flares back over small posterior accessory cavity (*o*), and faces down onto distinct myophoral shelf (*pm*) in *RV* (Figs. 2.10, 2.12). Small specimens in the Shuaiba Formation are similar to the Barremian–Lower Aptian *H. dumortieri* (Matheron) from Europe (illustrated in Skelton and Masse, 1998). A slightly larger, probably descendent, form, similar to that described by Baron-Szabo and Steuber (1996) from the Lower Aptian of central Greece, here characterizes the lowermost Upper Aptian.

*Where found.*—Shuaiba Formation, in the upper part of the Lower Aptian, as well as in overlying limestones of the lowermost Upper Aptian locally encountered in the subsurface.

*Horiopleura* sp. cf. *H. lamberti* (Douvillé, 1889)
(Figs. 2.13, 5.2)

*What to look for.*—Similar in form to the previous species, though larger (commissural diameter may exceed 10 cm), and probably a descendent chronospecies.

*Where found.*—Upper Aptian–Albian succession in NE Jebel Akhdar (Nakhl area) (Fig. 5.2), most typically in outer-platform facies (Masse et al., 1997).

Family CAPROTINIDAE Gray, 1848

*Main features of family.*—Attached by *RV*. Ligament shallowly invaginated, hence shell growth uncoiled (see Skelton 1978 for explanation); *RV* twisted conical to cylindrical; *LV* gently domed to enrolled-capuloid. Outer shell variable in thickness. Two subequal teeth in *LV* and one solid tooth in *RV*. Myophores attached to hinge plates, with differing orientations, sometimes creating various accessory cavities.

Despite removal of the monopleurids and polyconitids from the family as previously defined by Skelton (1978), the "caprotinids" described here are still merely a residual grouping of various primitive uncoiled rudists whose exact relationships remain to be resolved. For example, although *Himeraelites* and the *Sellaea*/ "*Caprotina*" group (*sensu* Di Stefano, 1888) appear to be closely related (Masse et al., 1998b), their evolution was probably independent from that of *Glossomyophorus*, and, for that matter, from that of the Cenomanian type species of *Caprotina*, *C. striata* d'Orbigny.

*Glossomyophorus costatus* (Masse, Skelton, and Sliskovic, 1984)
(Figs. 2.14–2.18, 5.1)

*What to look for.*—*RV* elongate, straight, slightly twisted; *LV* tall enrolled capuloid. Outer shell moderately thick (at least 1 mm), with strongly developed, broad but salient ribs, and two prominent radial bands (anterior, *ab* and posterior, *pb*) on the ventral flanks of both valves (Fig. 2.14). Outer shell layer characteristically honey brown in core material from areas west of the Oman Mountains. Tongue-shaped posterior myophore (*pm*) projects across commissural plane from *LV* (Fig. 2.15) into posterior accessory (myophoral) cavity (*o*) in the *RV* (Fig. 2.16). Hence sections of the *RV* (Fig. 2.18) typically show a row of three holes of similar size, i.e., two sockets for the *LV* teeth and the posterior accessory cavity. See Masse et al. (1984) for further illustrations and discussion.

*Where found.*—Most abundant rudist in Lower Aptian of Arabia, though the relatively recent date of recognition as a distinct taxon means that it was frequently mistaken for other rudist taxa in older literature. Limited to the uppermost Barremian to Lower Aptian (middle Kharaib to Shuaiba formations; see Masse et al., 1998a). Widely found in biostromes ranging from more or less *in situ* elevator congregations (Fig. 5.1), with a packstone to wackestone matrix, to transported drifts of shell debris (floatstones to rudstones); most commonly associated

Fig. 4.—Diagrammatic sketches of Arabian Early Cretaceous rudists (caprinids). See caption to Figure 2 for scale and abbreviations.

1–6. *Pachytraga tubiconcha* Astre [1–4 based on material from Portugal and France (Skelton and Masse, 1998)]. 1, antero-ventral aspect of whole shell; 2, 3, interiors of *LV* and *RV* (slightly magnified); 4–6, sections across *LV* (4) and *RV* (5, 6) (outer shell, black; inner shell, white; matrix, stippled).

7, 8. *Pachytraga paradoxa* (Pictet and Campiche) [based on material from France (Skelton and Masse, 1998)]. Sections across *LV* (7) and *RV* (8) (outer shell, black; inner shell, white; matrix, stippled).

9. *Praecaprina* sp. cf. *varians* Paquier (from Masse and others, 1998a). Section across *LV* (*note*: viewed from apical side; inner shell here shown in black).

10. *Offneria* sp. cf. *O. interrupta* Paquier (from Masse and others, 1998a). Section across *LV* (*note*: viewed from apical side; inner shell here shown in black).

11–13. *Offneria* sp. cf. *O. nicolinae* (Mainelli). 11, antero-ventral aspect of whole shell; 12, 13, sections across *LV* and *RV*.

14–16. *Offneria murgensis* Masse. 14, 15, sections across *LV* and *RV* (some obscuring of features through recrystallization suggested in dorsal part of 15; a typical occurrence); 16, longitudinal section of a shell fragment, showing typical frequency of transverse partitions in pallial canals.

17. *Offneria italica* Masse (from Masse and others, 1998a). Section across *LV* (*note*: viewed towards apex; inner shell here shown in black).

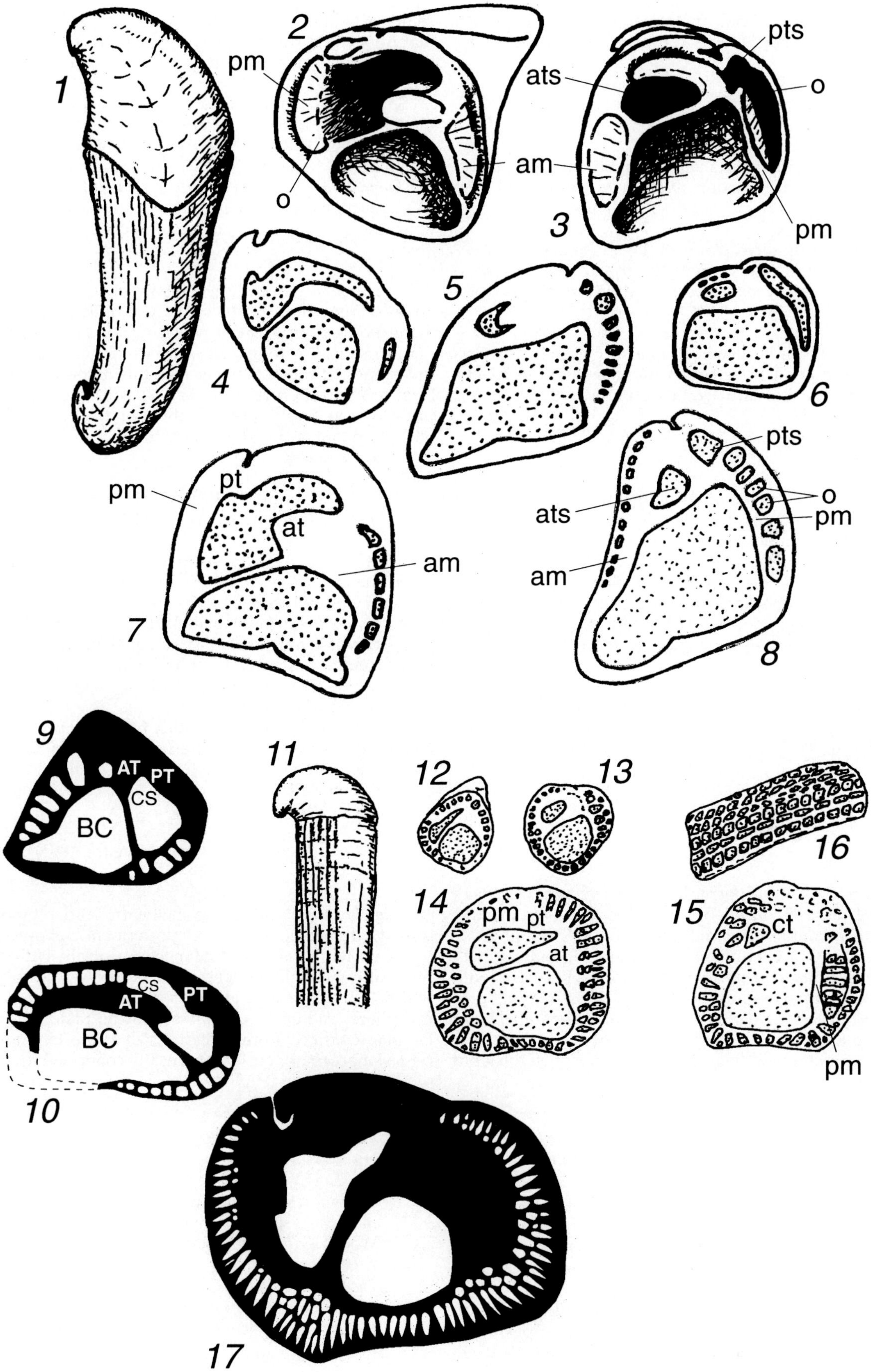
1
2
pm
o
3
pts
ats
o
am
pm
4
5
6
7
pt
pm
at
am
8
pts
ats
o
pm
am
9
AT
PT
CS
BC
10
CS
PT
AT
BC
11
12
13
14
pm
pt
at
15
ct
pm
16
17

with outer platform-top facies, immediately behind caprinid-dominated platform-margin facies (Hughes, 1997; Masse et al., 1998a).

*Himeraelites douvillei* (Di Stefano, 1888)
(Figs. 3.1–3.4)

*What to look for.*—Large solid shell; *RV* slightly twisted conical; *LV* low-domed. Commissure dorso-ventrally compressed, giving broad ventral aspect (Fig. 3.1). Outer shell quite thin (around 1 mm); inner layer thick and compact. *LV* posterior myophore (*pm*) a raised, inwardly tilted platform (Fig. 3.2), lacking a well-marked peripheral accessory cavity (Fig. 3.4), in contrast to *Horiopleura* (see above). *LV pm* faces down onto a broad myophoral shelf (*pm*) in *RV* (Fig. 3.3).

*Where found.*—So far known in Arabia only from the subsurface, around the Lower–Upper Aptian boundary, as rare elevator in inner-platform wackestones (see also Hughes, 1997, Fig. 6a).

*Sellaea / Caprotina* sp. (*sensu* Di Stefano, 1888)
(Figs. 3.5, 5.4)

*What to look for.*—Elongate, robust shell with high capuloid *LV*. Prominent *LV* posterior myophore (*pm*) connects posterior tooth (*pt*) to lamina running from anterior tooth (*at*) to postero-ventral margin (Fig. 3.5). Large accessory cavity (*o*) behind *LV pm*. *LV* anterior myophore (*am*) also separated from margin by accessory cavity (*o*). These myocardinal arrangements are common to the various species assigned to *Sellaea* and *"Caprotina"* by Di Stefano (1888), which is an assemblage of forms in need of further taxonomic revision.

*Where found.*—So far known only from scattered specimens in the Albian inner platform-top facies of NE Jebel Akhdar (Nakhl area) (Masse et al., 1997).

Family Uncertain

New genus, new species
(Figs. 3.6–3.12)

*What to look for.*—*RV* elongate, slightly twisted cylindrico-conical; *LV* tall, spirally enrolled. Outer shell thin (less than 0.5 mm), with fine moderately spaced ribbing on *RV*, though *LV* is smooth (Figs. 3.8, 3.11). Commissural outline sub-quadrate, with blunt antero-ventral and postero-ventral carinae. Posterior tooth (*pt*) much smaller than anterior tooth (*at*) (Fig. 3.9). Myophores extended ventrally as raised platforms oriented subparallel to the commissural plane in both valves or facing outwards somewhat in *LV* and depressed in *RV* (Figs. 3.9, 3.10). Inner shell thick but compact, without any pallial canals (Fig. 3.11) [note, however, that frequent borings (*b*) may be mistaken for canals, although their irregular outlines and distribution rule them out as such (Fig. 3.12)].

Some more slender shells, with the myophores sloping down more steeply into the *RV* interior (Figs. 3.6, 3.7) also belong with this new taxon [in the draft field guide on which this paper is based (Skelton and Masse, 1997), a *RV* of this slender form was attributed to *Monopleura* sp., but new material, including spirally enrolled *LV*s, indicate assignment instead to the new taxon.] Naming of this new taxon awaits a formal published description (Skelton, *in prep.*).

*Where found.*—Locally accompanying *Glossomyophorus* in the Shuaiba Formation, in which it may be quite common, especially in the more external platform facies around the Abu Dhabi intra-platform basin. A robust elevator, usually found transported in floatstones to rudstones. Because this taxon has only recently been recognized, further finds of it will be of particular value prior to formal publication; in the past, thick-shelled specimens with abundant borings have sometimes been erroneously identified as canaliculate caprinids.

Family CAPRINIDAE d'Orbigny
(emend. Skelton and Masse, 1998)

*Main features of family.*—Attached by *RV*. Ligament invaginated, hence valves uncoiled. *RV* slightly twisted, straight to curved, cylindrico-conical; *LV* more or less extended and dorsally curved or enrolled. Outer shell layer thin; inner shell thick, with pallial canals in most taxa. Two unequal teeth in *LV* (*at* > *pt*), and robust central tooth in *RV*. In both valves, myocardinal lamina separates posterior accessory cavity, containing inclined posterior myophore, from body cavity.

A recent revision of this family by Skelton and Masse (1998) has identified the myocardinal system as the key diagnostic character, instead of the presence of pallial canals, because certain other rudist groups may show canals, too, and a few caprinids lack them.

*Pachytraga tubiconcha* (Astre, 1961)
(Figs. 4.1–4.6)

*What to look for.*—Elongate shell with slender, tubular *RV* and somewhat erect capuloid *LV*. *LV* posterior myophore (*pm*) situated on the posterior valve wall, connecting dorsally with small posterior tooth, and inclined steeply into posterior accessory cavity (*o*), which itself adjoins central tooth socket (Figs. 4.2, 4.4). *RV* posterior myophore (*pm*) an erect, outward-facing plate on myocardinal lamina, separated from posterior valve wall by a long, narrow posterior accessory cavity (*o*) dorsally connected to the posterior

FIG. 5.—Rudists at outcrop in the Oman Mountains.

1. Detail of *Glossomyophorus* biostrome, in vertical section, with specimen in life position to right of lens cap, surrounded by displaced valves. Shuaiba Formation (Lower Aptian) of Wadi Mi'aidin, Oman Mountains. Lens cap is 5.5 cms across.
2. Displaced articulated *Horiopleura* in rudist floatstone, in vertical section. Al Hassanat Formation (Albian) of Wadi El Assyi, Oman Mountains. Large specimen is at least 10 cm across.
3. Transported caprinid assemblage, dominated by *Offneria*, in vertical section. Kharaib Formation (Lower Aptian) of Wadi Hedeck, Oman Mountains. Larger specimens up to c. 10 cm across.
4. Displaced *Sellaea* / *"Caprotina"* floatstone in vertical section. All Hassanat Formation (Albian) of Wadi El Assyi, Oman Mountains. Specimens up to ca. 8 cm across.

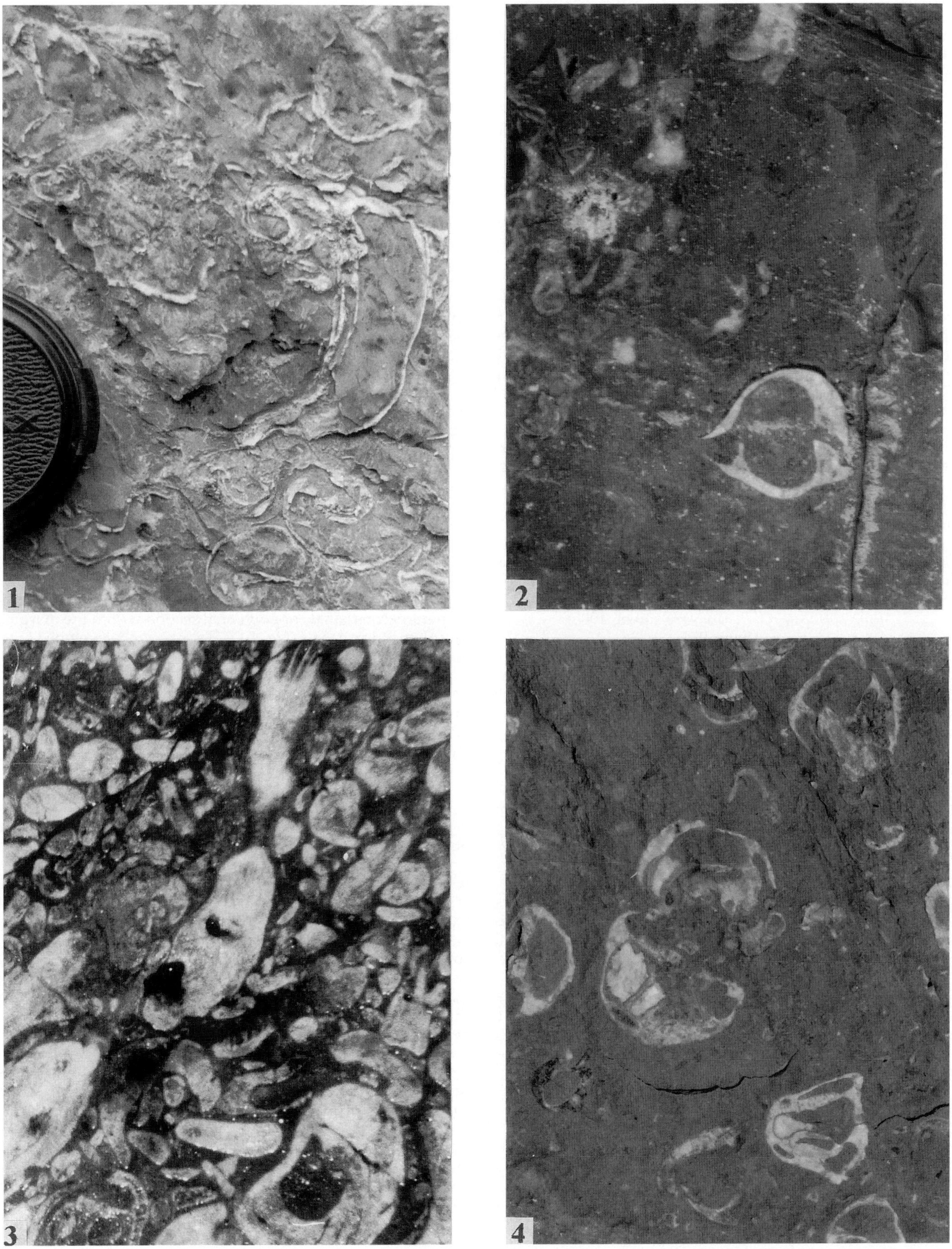

tooth socket (*pts*) (Figs. 4.3, 4.6). Bottom of *RV* posterior accessory cavity may be subdivided by radial laminae, especially in larger specimens (Fig. 4.5). A few small pallial canals may be present around the antero-dorsal margin of the *RV* (Fig. 4.6).

*Where found.*—Hauterivian, Habshan Formation of Wadi Mi'Aidin and basal Lekhwair Formation equivalent of Wadi Bani Kharus (central Jebal Akhdar), as common transported shells in outer platform-top floatstones. In Europe, this species is known to form elevator clusters in platform-top packstones to wackestones (Skelton and Masse, 1998).

*Pachytraga paradoxa* (Pictet and Campiche, 1869)
(Figs. 4.7, 4.8)

*What to look for.*—Morphology essentially similar to that of the previous species, though reaching significantly greater size, with more pronounced antero-ventral and postero-ventral carinae. Also, more distinct accessory cavity in front of anterior myophore (*am*) in *LV*, which may be subdivided to form pallial canals (Fig. 4.7); canals may also develop in front of *am* in *RV* (Fig. 4.8). Small internal ridge common on the mid-ventral wall.

*Where found.*—So far only dubiously noted in Arabia, in the Shuaiba Formation (Wadi Hedeck, Nakhl area). In Europe the species is known predominantly as a robust elevator, but also as a facultative recumbent, in outer platform-top packstones to rudstone/grainstones.

These two species of *Pachytraga* appear to represent a chronospecies lineage, spanning the Hauterivian to Lower Aptian and showing phyletic size increase (Skelton and Masse, 1998).

*Praecaprina* sp. cf. *P. varians* (Paquier, 1905)
(Fig. 4.9)

*What to look for.*—Similar to *Pachytraga*, though broader antero-posteriorly, with pallial canals well developed outside both myophores in the *LV* only (Fig. 4.9).

*Where found.*—Only recently recorded from Arabia (Masse et al., 1998a), from the middle Kharaib Formation equivalent of Wadi Hedeck and adjacent sections (NE Jebel Akhdar, Nakhl area).

*Offneria* sp. cf. *O. interrupta* (Paquier, 1905)
(Fig. 4.10)

*What to look for.*—Pallial canals in the margins of both valves, containing transverse partitions, concave towards the commissure. In this relatively primitive *Offneria* species, the canals do not extend all the way around the ventral margin (Fig. 4.10).

*Where found.*—Again, recently recorded from Arabia (Masse et al., 1998a), from the middle Kharaib Formation equivalent of Wadi Hedeck. Also recorded here was another closely related *Offneria* species, *O. rhodanica* Paquier, 1905, again previously known only from Europe.

*Offneria* sp. cf. *O. nicolinae* (Mainelli, 1983)
(Figs. 4.11–4.13)

*What to look for.*—Shell small, with long slender *RV* and capuloid *LV* (Fig. 4.11). Commissural outline rounded. Row of simple canals extends around entire margins of both valves, with some transverse partitions on the anterior and posterior sides (Figs. 4.12, 4.13). For further details of this and the next two species of *Offneria*, see Masse (1992).

*Where found.*—From the middle Kharaib Formation equivalent of Wadi Hedeck (Masse et al., 1998a); also scattered, though not common, in the Shuaiba Formation of several areas. Elevator, locally clustered in platform-top wackestones to packstones.

*Offneria murgensis* (Masse, 1992)
(Figs. 4.14–4.16, 5.3)

*What to look for.*—Medium-sized shell, also with elongate *RV* and capuloid *LV*. Commissural outline develops blunt angulations around the antero-ventral and postero-ventral corners, with a slight mid-ventral depression. Row of simple canals on the ventral margin of each valve expands to multiple rows around the remaining margins (Figs. 4.14, 4.15), with abundant transverse partitions (Fig. 4.16). Tabulae, concave towards the commissure, are also well developed in the body cavity.

*Where found.*—Most abundant of the *Offneria* species in Arabia. Scattered in the middle Kharaib Formation and equivalents in the Jebel Akhdar region (Masse et al., 1998a), but also widespread in the Shuaiba Formation (e.g., Hughes, 1997). Usually found only as transported specimens in outer platform-top floatstones to rudstones (Fig. 5.3), it appears to have been a facultative elevator to recumbent.

*Offneria italica* (Masse, 1992)
(Fig. 4.17)

*What to look for.*—Larger, again, than the previous species, with numerous canals of polygonal section within the marginal row of pyriform canals (Fig. 4.17). Tabulae well developed in the body cavity.

*Where found.*—So far, positively identified in Arabia only from the Shuaiba Formation of northern Oman (Masse et al., 1998a). Like *O. murgensis*, it appears to have been a facultative elevator to recumbent.

## CONCLUSIONS

The diversity of rudists in the Lower Cretaceous carbonate platform deposits of Arabia offer much potential both for biostratigraphy and for paleoenvironmental analysis, and hence for the development of facies models for production development.

In relation to biostratigraphy (Fig. 1), the uppermost Barremian to Lower Aptian interval is especially well marked by a diverse fauna, amongst which *Glossomyophorus* is a notable and widespread element, together with a variety of caprinids. Few lineages from this fauna survive into the Upper Aptian, which can nevertheless be identified by the appearance of the first radiolitids, showing celluloprismatic structure. The Albian is, in turn, marked by the appearance of the *Sellaea*/ "*Caprotina*" (*sensu* Di Stefano, 1888) group.

With respect to paleoecology, distinct biofacies become especially evident in the Shuaiba Formation, reflecting the diversity of rudists present. Robust, predominantly recumbent caprinids tend to dominate the coarse bioclastic outer platform-margin facies, while elevator *Glossomyophorus*, locally accompanied by the new taxon, form extensive biostromes behind, on the outer platform top (see, for example, Hughes, 1997; Masse et al., 1998a). In more internal platform facies, *Agriopleura* becomes dominant. Less well marked patterns of this kind can also be seen at other levels. In the Habshan Formation, for example, the robust *Pachytraga* tends to be most common in the outer-platform facies. In the Albian, the open platform-margin bioclastic facies is dominated by *Eoradiolites* and *Horiopleura*, together with corals, while the *Sellaea*/"*Caprotina*" assemblage occupies a more internal setting (e.g., Al Hassanat Formation of Jebel Akhdar; Masse et al., 1997).

## ACKNOWLEDGMENTS

We are grateful to numerous colleagues with whom we have worked on Arabian rudists over many years, and to Prof. A. S. Alsharhan and Dr. R. W. Scott for inviting us to submit this paper, originally as a field guide for the Jurassic/Cretaceous platform-basin systems conference in Al Ain, in 1997, and to Prof. J. Philip and Dra. E. Gili for their constructive reviews of the paper. Thanks to Janet Dryden and Ashea Tambe for formatting of the manuscript and to Andrew Lloyd for digitally composing the specimen figures (drawn by PWS).

## REFERENCES

BARON-SZABO, R.C., AND STEUBER, T., 1996, Korallen und rudisten aus dem Apt im tertiären Flysch des Parnass-Gebirges bei Delphi-Arachowa (Mittelgriechenland): Berliner Geowissenschaftliche Abhandlungen, v. E18, p. 3–75.

DI STEFANO, G., 1888, Studi stratigrafici e paleontologici sul sistema Cretaceo della Sicilia. 1. Gli strati con *Caprotina* di Termini-Imerese: Reale Accademia di Scienze, Lettere e Belle Arti, Atti (N. S.), v. 10, 44 p.

DOUVILLÉ, H., 1910, Études sur les Rudistes: Rudistes de Sicile, d'Algérie, d'Égypte, du Liban et de la Perse: Société Géologique de France, Mémoires, Paléontologie, no. 41, 84 p.

DOUVILLÉ, H., 1918, Le Barrémien supérieur de Brouzet—Troisième Partie: les Rudistes: Société Géologique de France, Mémoires, Paléontologie, no. 52, p. 5–19.

GILI, E., MASSE, J.-P., AND SKELTON P.W., 1995, Rudists as gregarious sediment-dwellers, not reef-builders, on Cretaceous carbonate platforms: Palaeogeography, Palaeoclimatology, Palaeoecology, v. 118, p. 245–267.

HUGHES, G.W., 1997, The Great Pearl Bank Barrier of the Arabian Gulf as a possible Shuaiba analogue: GeoArabia, v. 2, p. 279–304.

MASSE, J.-P., 1992, Les rudistes de l'Aptien inférieur d'Italie continentale: aspects systématiques, stratigraphiques et paléobiogéographiques: Geologica Romana, v. 28, p. 243–260.

MASSE, J.-P., AND GALLO MARESCA, M., 1997, Late Aptian Radiolitidae (rudist bivalves) from the Mediterranean and Southwest Asiatic regions: taxonomic, biostratigraphic and palaeobiogeographic aspects: Palaeogeography, Palaeoclimatology, Palaeoecology, v. 128, p. 101–110.

MASSE, J.-P., AND PHILIP, J., 1981, Cretaceous coral–rudistid buildups of France, *in* Toomey, D.F., ed., European Fossil Reef Models: Society of Economic Paleontologists and Mineralogists, Special Publication 30, p. 399–426.

MASSE, J.-P., SKELTON P.W., AND SLISKOVIC, T., 1984, Glossomyophorus costatus nov. gen. nov. sp., rudiste (Caprotinidae) nouveau de l'Aptien du domaine méditerranéen central et oriental: Géobios, v. 17, p. 723–732.

MASSE, J.-P., BORGOMANO, J., AND AL MASKIRY, S., 1997, Stratigraphy and tectonosedimentary evolution of a late Aptian–Albian carbonate margin: the northeastern Jebel Akhdar (Sultanate of Oman): Sedimentary Geology, v. 113, p. 269–280.

MASSE, J.-P., CHARTROUSSE, A., AND BORGOMANO, J., 1998a, The Lower Cretaceous (Upper Barremian–Lower Aptian) caprinid rudists from northern Oman: Géobios, Mémoire Spécial, no. 22, p. 211–223.

MASSE, J.-P., GALLO MARESCA, M., AND LUPERTO SINNI, E., 1998b, Albian Rudist faunas from Southern Italy: taxonomic, biostratigraphic and palaeobiogeographic aspects: Géobios, v. 31, p. 47–59.

PARONA, C.F., 1909, La fauna coralligena del Cretaceo dei Monti d'Ocre nell'Abruzzo Aquilano: Memorie per servire alla Descrizione della Carta geologica d'Italia, no. 5. 233 p.

ROSS, D.J., AND SKELTON, P.W., 1993, Rudist formations of the Cretaceous: a palaeoecological, sedimentological and stratigraphical review: Sedimentology Review, v. 1, p. 73–91.

SKELTON, P.W., 1978, The evolution of functional design in rudists (Hippuritacea) and its taxonomic implications: Royal Society [London], Philosophical Transactions v. B284, p. 305–318.

SKELTON, P.W., AND MASSE, J.-P., 1997, Biostratigraphy and palaeoecology of rudists in the Lower Cretaceous of Arabia, *in* Alsharhan, A.S., compiler, Mesozoic Carbonate Platform–Basin systems, United Arab Emirates; Guidebook prepared for the research conference on Jurassic/Cretaceous carbonate platform–basin systems—Middle East models, 4–12 December, 1997: Unpublished field-guide, U.A.E. University, Al Ain, U.A.E., p. 24–36.

SKELTON, P.W., AND MASSE, J.-P., 1998, Revision of the Lower Cretaceous rudist genera *Pachytraga* Paquier and *Retha* Cox (Bivalvia: Hippuritacea), and the origins of the Caprinidae: Géobios, Mémoire Spécial, no. 22, p. 331–370.

TOLAND, C., PEEBLES, R.G., AND WALKDEN, G.M., 1993, Upper Jurassic and basal Cretaceous outcrop sequence stratigraphy of Wadi Hagil, Ras al Khaimah: Society of Petroleum Engineers, 8th Middle East Oil Show and Conference, Bahrain, Proceedings, SPE paper 025581, p. 533–543.

WITT, W., AND GÖKDAG, H., 1994, Orbitolinid biostratigraphy of the Shuaiba Formation (Aptian), Oman—implications for reservoir development, *in* Simmons, M.D., ed., Micropalaeontology and Hydrocarbon Exploration in the Middle East: London, Chapman & Hall, p. 221–241.

# Part 2:
# Depositional and Diagenetic Models of Carbonate Platforms

# ENVIRONMENTAL SIGNIFICANCE OF MIDDLE JURASSIC ONCOIDS, NORTHERN SINAI, EGYPT

HANAFY HOLAIL

*Geology Department, Faculty of Science, Alexandria University, Alexandria, Egypt*

ABSTRACT: The depositional environment and the diagenetic history of the Middle Jurassic carbonate (Bir Maghara Formation), north Sinai, Egypt, have been evaluated through comprehensive petrographic and geochemical studies of oncoid grains. Petrographically the oncoids are composed of micrite and vary from subspherical to spherical bodies. The abundance of the oncoids and their close association with ooids, pellets, and shelf fauna indicate that the Middle Jurassic carbonate sequence was deposited in a shallow marine environment within the photic zone.

The oncoids are characterized by relatively low Fe and Mn concentrations, normal Sr concentrations, light $\delta^{18}O$ values (from -3.9 to -5.5‰ PDB), and heavy $\delta^{13}C$ values (from +2.2 to +3.1‰ PDB). Enhanced magnesium concentrations (3.6 mol % $MgCO_3$) in these oncoids, relative to carbonate matrix (0.6 mol% $MgCO_3$), suggest their original Mg-calcite mineralogy. The $\delta^{18}O$ and $\delta^{13}C$ values show a narrow range, and the oncoid grains from the same stratigraphic level do not yield any large difference in their isotopic composition. Isotopic comparison of these oncoids (mean $\delta^{18}O$ = -5.4‰; mean $\delta^{13}C$ = +2.1‰ PDB) with the sparry calcite cements (mean $\delta^{18}O$ = -9.8‰; mean $\delta^{13}C$ = +0.6‰ PDB) suggests that meteoric modification was coincident with the development of subaerial exposure and meteoric influx during emergence of the carbonate platform. Such diagenetic modification is probably responsible for $^{18}O$ depletion of the oncoids and the carbonate matrix.

## INTRODUCTION

Oncoid grains are common in the Middle Jurassic shallow marine carbonate deposits of north Sinai, Egypt, and exhibit a wide range of morphologies. Similar oncoids are described worldwide in the Phanerozoic sedimentary column from marine, lacustrine, and fluvial environments (Abell et al., 1978; Nickel, 1983; Andrews,1986; Ratcliffe, 1988; Smith and Mason, 1991; Zamarreno et al., 1997). The examination of oncoids, which are constructed by algae, is of interest in both geological and biological studies, where the algae are important producers of both calcium carbonate and organic matter in many shelf areas from tropics to temperate regions (Milliman, 1974; Morse and Mackenzie, 1990). Therefore, the variation in oncoid morphologies and their excellent preservation in the study area provide the opportunity to examine the occurrence of oncoid grains in relation to their environmental setting.

Despite the wealth of literature relating to oncoids of Mesozoic carbonate shelves (e.g., Bishop, 1968; de Klasz, 1978; Purser, 1978; Becher and Moore, 1979; Peryt, 1981, 1983), few attempts have been made to understand the relationship between palaeoenvironmental and palaeotemperature conditions, oncoid morphology, and the resultant texture of carbonates. Meanwhile, research on the mineralogy and geochemistry (trace elements and stable isotopes) of oncoids is still scanty. The oxygen and carbon isotopic composition of oncoids record paleoenvironmental and paleotemperature conditions during the Jurassic if the effects of diagenesis have not modified the original signal and if the algae did not impart a fractionation of carbon and oxygen isotopes.

The present study reports detailed elemental and isotopic compositions of the Middle Jurassic oncoids from the north Sinai, Egypt. The major goal of this study is to unravel the environmental and diagenetic history of these oncolitic limestones. Diagenetic phases and their environments are identified by detailed petrography and geochemistry. Consequently, the stable-isotope compositions of the oncoids provide a reasonably reliable proxy for Middle Jurassic marine $\delta^{18}O$ and $\delta^{13}C$ values, as well as of paleotemperature of sea water at that time.

## GEOLOGICAL SETTING

The Middle Jurassic Bir Maghara Formation is exposed in the Gebel Maghara area and in the subsurface in bore hole No. H-1 (Fig. 1). It overlies the Shusha Formation and is overlain by the clastic deposits of Safa Formation. The Bir Maghara Formation consists of two members, the Mahl Member and the Mowerib Member (Fig. 2). Stratigraphy and sedimentology of the Middle Jurassic rocks have been studied by a number of authors (e.g., Al Far, 1966; Goldberg et al., 1971; Jenkins, 1990, Said, 1990; Keeley and Wallis, 1991).

In general, the Middle Jurassic deposits of north Sinai comprise a section of predominantly marine shelf sediments dominated by carbonates. Sedimentation of these deposits was controlled by the three competing factors: (1) progressive eustatic sea level rise, (2) intermittent and relatively minor subsidence related to continuing deepening of the depositional basin, and (3) progradation of argillaceous and siliciclastic sediments from the south basement massif. These factors resulted in complex, cyclic deposition in the study area.

Each cycle culminated with either reduced deposition or erosion. Transgressions permitted carbonate deposition over north Sinai (including the study area), whereas shallowing of the depositional environment led to deposition of peritidal carbonate and progradation of nonmarine sediments. As a result, widespread emergence occurred at the tops of individual cycles. In the study area, the carbonate facies comprise oolitic, oncolitic, and peloidal grainstones and packstones facies (Fig. 2). Meanwhile, the faunally restricted subtidal to peritidal wackestone facies are also present. Terrigenous shales and sandstones are interbedded with carbonates throughout the study sequence (Fig. 2).

Tectonically, during the Jurassic the continental margin of the Tethyan Sea was affected by block faulting and tensional movements (Bernoulli, 1972). This extensional regime persisted until the Late Cretaceous and was associated with high heat flow (Spohn and Schubert, 1982). The study area encompasses a transitional area from an extensional regime to a compressional regime (Freund et al., 1975). This compressional regime led to the development of the Syrian Arc system and was also manifested by a

Middle East Models of Jurassic/Cretaceous Carbonate Systems
SEPM Special Publication No. 69, Copyright © 2000
SEPM (Society for Sedimentary Geology), ISBN 1-56576-075-1, p. 103–113.

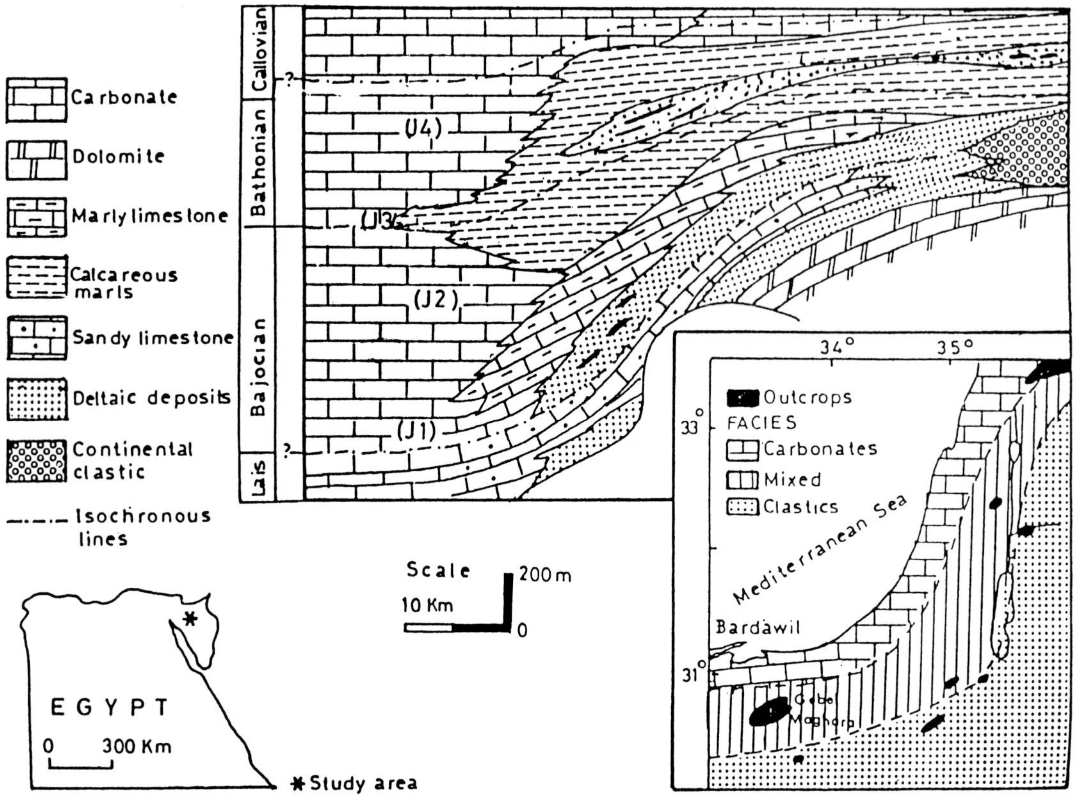


FIG. 1.—Schematic facies distribution of Middle Jurassic in north Sinai (Modified from Schlumberger, 1984).

change from high to low heat flow. In the Cretaceous more gradual regional subsidence took place, which gave rise to post-uplift folding and compaction of the argillaceous Jurassic strata. Finally, the Jurassic sequences were uplifted and subsequently truncated by Early Cretaceous erosion. This erosion has variably reduced the thickness of the Jurassic section.

## SAMPLING AND ANALYTICAL PROCEDURES

This study is based on extensive field work on surface and subsurface Middle Jurassic sections, examination of over 80 polished thin sections, 60 major element and trace-element analyses, and 60 oxygen and carbon isotope analyses from two sections of Bir Maghara Formation (Fig. 2). Approximately 10 samples of oncoid grains and matrix were studied using a JEOL scanning electron microscope (SEM) with an on-line Exl-EDS system. Gold-coated thin sections and broken surfaces were studied either unetched or after slight etching with dilute HCl (1N) for up to 30 seconds. X-ray powder diffraction was used for routine determination of the mineralogic purity of all samples prior to elemental and isotopic analyses.

Documentation of primary textures, and of chemical changes that occurred in the studied marine oncoid grains during diagenesis, requires detailed geochemical analyses of various petrographic components (oncoid grains, matrix, and calcite cement). Accordingly, microsampling is essential. Isotopic analysis of oxygen and carbon was performed on a Finnigan MAT stable-isotope-ratio mass spectrometer at the Stable Isotope Laboratory, University of Michigan. The 0.5 mg powdered carbonate samples of oncoid grains, matrix, and calcite cement were obtained using a dental drill and dissolved in 100% anhydrous phosphoric acid at 70°C. All data have been corrected following procedure modified from Craig (1957). Precision was monitored through daily analyses of NBS-20, which is better than 0.08‰ for oxygen and 0.05‰ for carbon determinations.

Major-element and trace-element compositions for various petrographic components were measured by atomic absorption spectrophotometry (AAS). Powders of 0.1 g were dissolved in concentrated HCl. Despite the low insoluble residue contents in most of the analyzed samples, elements absorbed to clays and organic phases may have contributed to the measured trace-element contents. Analytical precision at the 95% confidence level determined on replicate analyses is better than 5%.

## RESULTS AND DISCUSSION

### *Petrography and Diagenesis*

Oncoids are abundant in some layers of the studied Middle Jurassic carbonate sequences and are associated with ooids, pellets, and skeletal grains. Micrite is the predominant fabric of these oncoids (Fig. 3). In general, the Middle Jurassic carbonate include argillaceous mudstones/wackestones that contain oncoids, passing shoreward into oncoidal–pelodial and peloidal–bioclastic facies and then into oolite-flat grainstones. The studied oncolites formed by the physical movement of intraclasts or fossils (serving as the nuclei) that were coated by microbial mats (Fig. 3). Some oncoids are spherical and have continuous laminae that completely envelop the nucleus (Fig. 3B), which most probably indicates continuous agitation.

The Middle Jurassic oncoids are composed of micrite and meteoric spar that form spongiostromate fabrics (Monty,

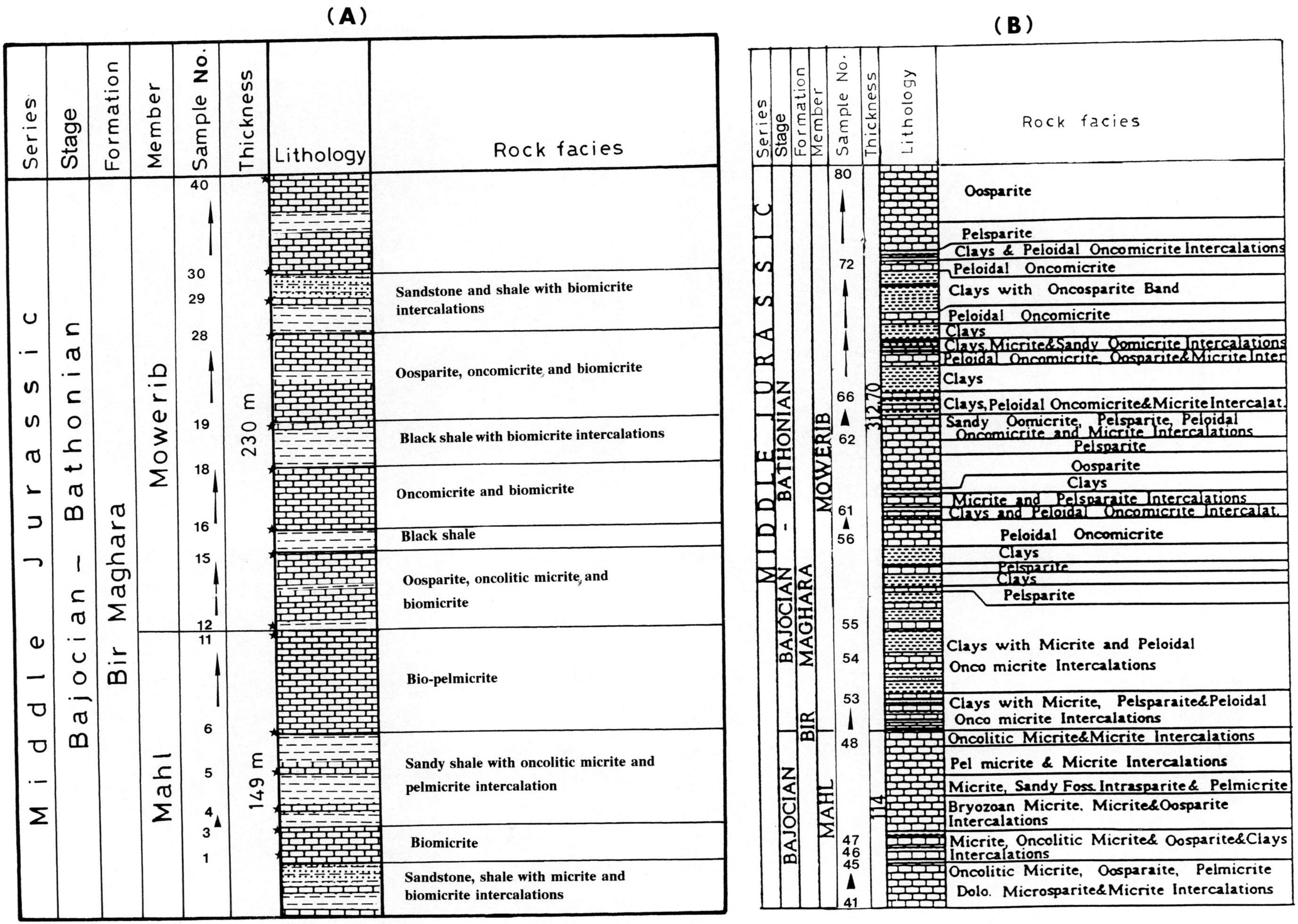


FIG. 2.—Lithostratigraphic sections of Middle Jurassic rocks. A) Bajocian–Bathonian surface section of Gebel Maghara. B) Subsurface section of bore hole No. H-1.

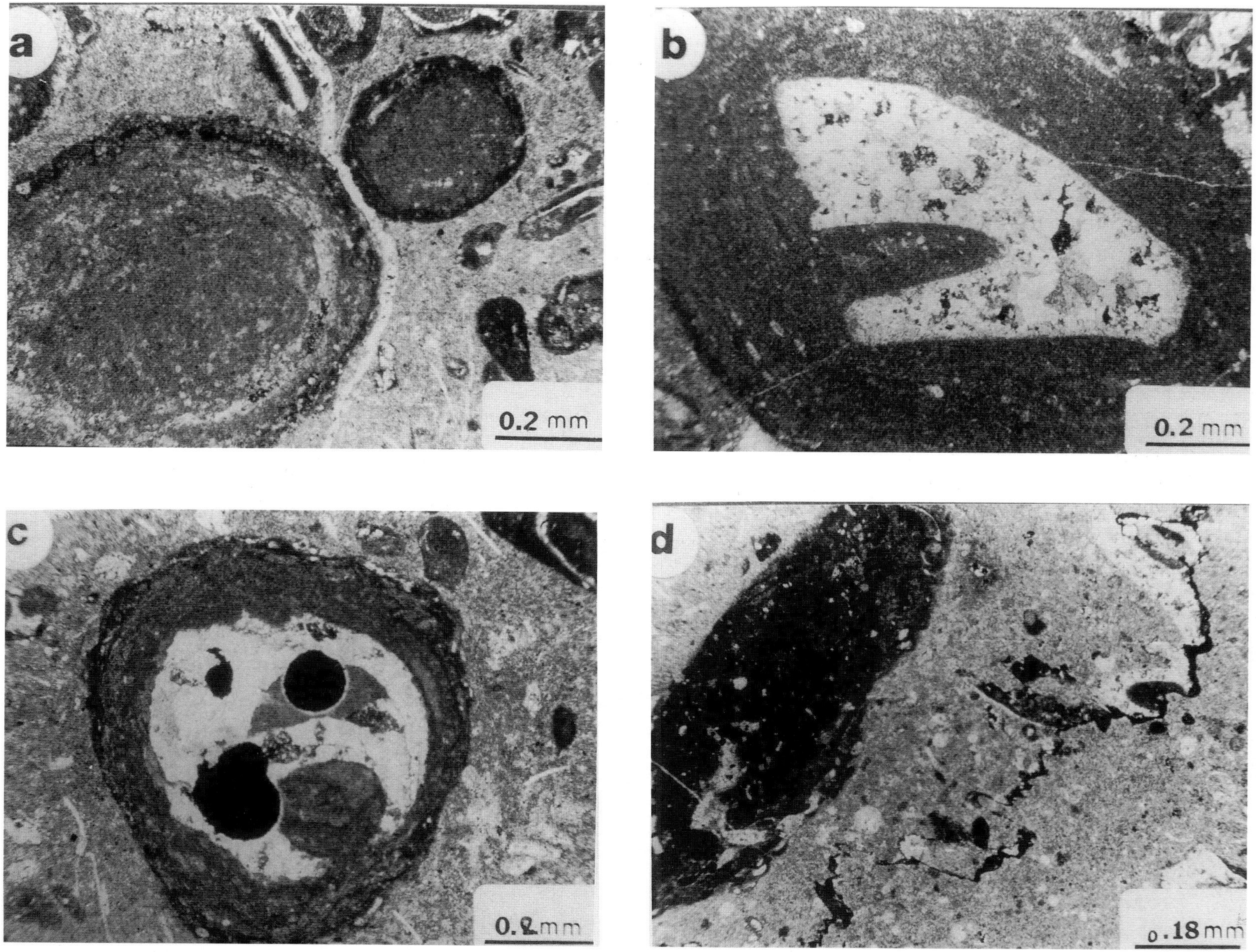


FIG. 3.—Photomicrographs of the studied Middle Jurassic oncoids. A) Micritized subspherical dense oncoids with carbonate intraclasts as nuclei (surface sample). Note the grain-contact suture between oncoid grains. B) Laminated coating with late calcite filling the core of the oncoid (subsurface sample). C) Spherical oncoid displaying symmetrical growth (subsurface sample). Note the partial dissolution and the formation of vugs. D) Oncoid with micritic matrix (subsurface sample). Note the stylolite associated with authigenic calcite and dolomite crystals.

1981). Despite this, the obvious feature of these oncoids is their morphological variations, forming a continuous series from small (9 mm) subspherical–spherical forms to large (up to 1 cm) irregular forms (Fig. 3). The internal structure of the cortex of the subspherical–spherical oncoids is symmetrical, but this cortex is strongly asymmetrical in the irregular oncoids. For descriptive purposes the studied oncoids are divided into three morphological types, A, B, and C. Type A oncoids are small, smooth-surfaced, and spherical to subspherical, and composed dominantly of concentrically laminated fabric (Figs. 3A, 4A). This type dominates in peloidal mudstones to wackestones. The concentric laminae of this type suggest that oncoids were frequently, if not continuously, rolled during formation (Wright, 1983). The associated micritic matrix indicates that the environment was not sufficiently energetic to remove all the fine-grained sediment. Type B oncoids have a greater size range than type A. Their cortex is continuous and composed of irregular laminae. Jones and Wilkinson (1978) suggested that the continuous cortex may grow around allochems with no rolling. Type C oncoids are irregular (Fig. 3D). Types B and C oncoids could not have formed in an environment where rolling occurred. Peryt (1981) concluded that all oncoids are associated with slow sedimentation rates, and that oncoids composed of spongiostromate fabrics typify high-energy conditions.

In general, early marine cementation, dissolution of metastable calcium carbonate components, and dolomitization were the dominant diagenetic processes that affected the studied carbonate sequences (Fig. 5). Marine calcite cementation in interparticle (especially ooids) pore spaces began penecontemporaneously with Middle Jurassic carbonate deposition. Meniscus and pendant meteoric cements were formed prior to marine cementation. In many samples, early marine cementation was followed by leaching of allochems and other mineralogically unstable particles. Oncoids tended to be partially dissolved by the formation of vugs within them (Fig. 3C). Meanwhile, oncoid grains resisted dolomitization. Fossils were commonly dissolved entirely. Early dolomitization and late replacement dolomites are a common but rarely abundant component of Middle Jurassic carbonate.

Other diagenetic processes that affected the studied carbonates include stylolites and other pressure-dissolution features, burial calcite and dolomite cements, and fracturing, both tectonic and caused by collapse of partially dissolved framework allochems. Stylolites crosscut oncoid and ooid grains. Authigenic calcite and dolomite crystals are occasionally associated with stylolites (Fig. 3D). The Middle Jurassic carbonates are characterized by stylolite zones, which developed mainly in the peloncolitic micrite facies (Fig. 3D). This observation is in agreement with the Jurassic Burgogne oncolitic mudstones (Bathurst, 1984).

Grain-contact suturing is common between the oncoid grains (Fig. 3A). This may be because the oncoid grains are more susceptible to dissolution than the other allochems. Stylolites do not crosscut late void-filling calcite and dolomite cements. The coarse calcite crystals and the saddle dolomite cements are generally interpreted as having formed by burial diagenetic fluids at elevated temperature (Choquette and James, 1987). Finally, late sparry equant calcite cements were precipitated after uplift of the Jurassic rocks. The petrographic characteristics suggest that the studied Middle Jurassic carbonate underwent long-lasting, meteoric modification, and dependent on meteoric recharge. These modifications could have been coincident with the development of regional, post-Jurassic unconformity corresponding to emergence of the Jurassic carbonates.

## *Geochemistry*

### *Elemental Compositions.—*

Oncoid grains, matrix, and calcite cements from surface and subsurface sequences did not yield any large difference in their elemental compositions. The mean Mg and Sr concentrations of oncoid grains are more than the corresponding matrix values, but the mean Fe and Mn concentrations are consistently less than the corresponding matrix values (Table 1; Fig. 6). During meteoric stabilization and alteration in a system with relatively high water/rock ratio, a steady state is reached for the oncoid grains. Therefore, the elemental compositions (Sr, Fe, and Mn) of the oncoid grains are approximately consistent with compositions expected for abiotic marine precipitates formed in Jurassic sea water. The relatively high $MgCO_3$ concentrations of the oncoid grains (up to 3.9 mol %) is probably due to their original high-Mg calcite mineral composition. There is no indication of dolomitization of oncoid grains. On the other hand, the relatively low Sr contents of the matrix are quite common in ancient carbonate sequences. Moreover, the petrographic characters suggest that the Fe and Mn entered the matrix during a recrystallization event.

The Fe and Mn contents covary in the matrix and the meteoric sparry calcite cements and show a negative correlation with Sr content (Fig. 6). In general, the geochemical data indicate that the matrix and the calcite cement are richer in Fe and Mn and poorer in Sr and Mg than the oncoid grains. Therefore, the primary geochemical signature of the matrix was reset by later fluids enriched in Fe and Mn. Concentrations of $MgCO_3$ (0.35 to 1.2 mol %) and $SrCO_3$ (0.01 to 0.08 mol %) contents of the meteoric sparry calcite cements are lower than those of the Jurassic marine calcite cement and the studied oncoid grains (Table 1). In general, marine calcite cements formed in shallow depositional settings range from 10 to 21 mol % $MgCO_3$, with a modal composition of 12 mol % (Carpenter and Lohmann, 1992; Burton, 1993). In contrast, calcite cements with less than 2 mol % $MgCO_3$ are typical of cementation from water with low $Mg^{2+}/Ca^{2+}$ ratio (Allan and Matthews, 1982). Sr concentrations are highly variable in modern abiotic calcite (200–2000 ppm; average, 1200 ppm; Burton 1993). In comparison, the studied calcite cement has substantially lower $MgCO_3$ and $SrCO_3$ contents (Table 1). Such low values can be attributed to their formation from meteoric water. Meanwhile, the concentrations of Mn and Fe in the studied calcite cements are higher than those of the oncoid grains (Table 1). Modern calcite cement precipitated from oxic marine water has Mn = 1 ppm and Fe = 39 ppm (Burton, 1993). Consequently, there is compelling evidence that trace quantities of Fe and Mn present in the studied calcite cements were added during the cement formation from meteoric water and the primary mineral composition of these cements was low-Mg calcite.

### *Carbon Isotopes.—*

The mean $\delta^{13}C$ value of +2.6‰ for the oncoid grains in the study Middle Jurassic carbonate sections (Fig. 7) is typical of inorganic precipitates from marine pore fluids. Unfortunately, there is no record of stable-isotope compositions for Middle Jurassic marine carbonate in Egypt. Because marine carbonates normally have $\delta^{13}C$ values of about 0 ± 3‰ (Lundegard and Land, 1986), the $\delta^{13}C$ values of oncoids (between +2.1 and +3.2‰ PDB) suggest that marine carbonate was dominant and isotopically light carbon was not involved in formation of the oncoids. Positive carbon-isotope values also suggest that meteoric water was not involved in the diagenesis of these oncoids, at least not in large proportion, because extensive meteoric water influx would

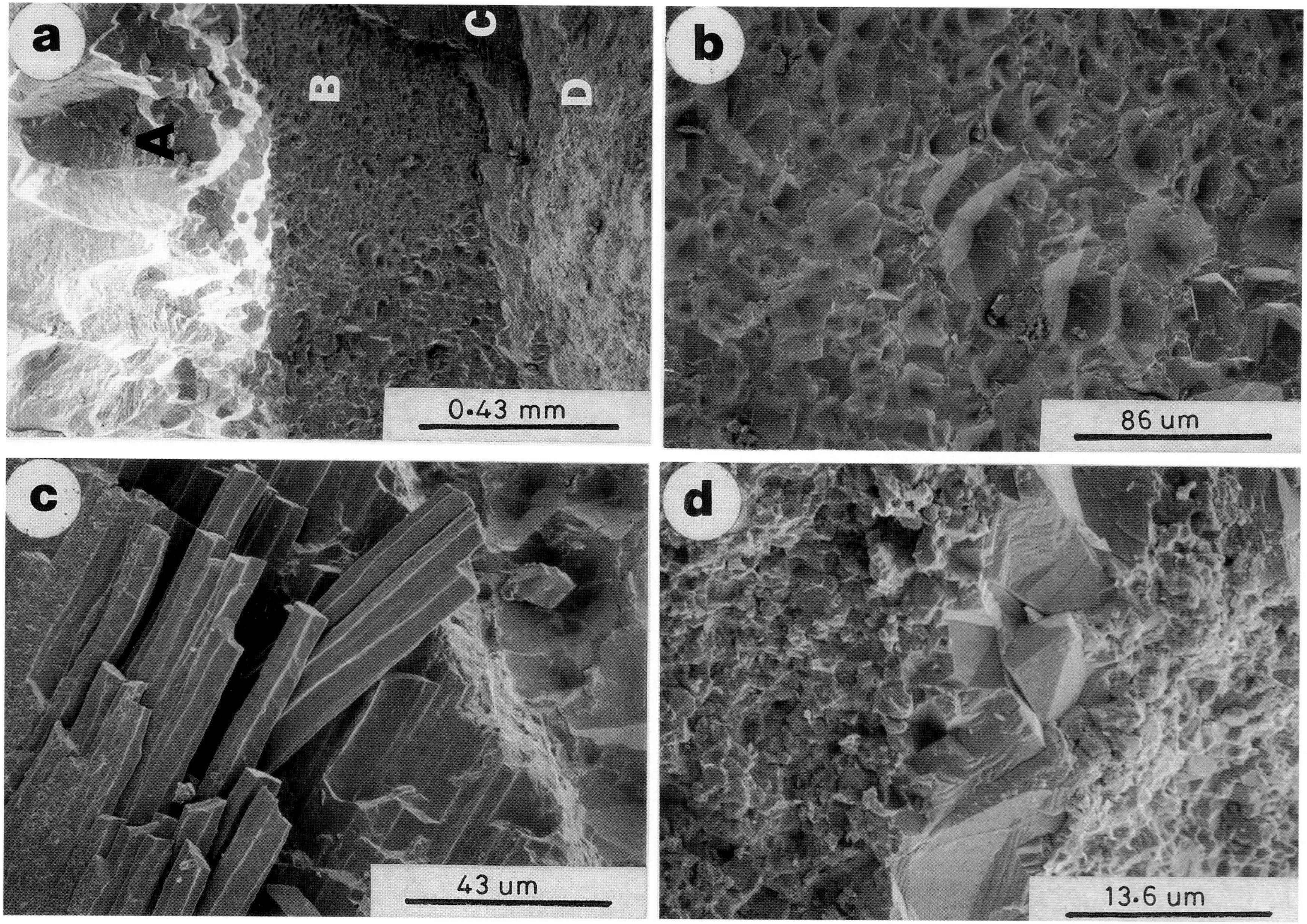


FIG. 4.—SEM photomicrographs of oncoids, matrix, and calcite cement. A) Microstructure of oncoid types (b and c). Note the calcite cement (a) and the matrix (d). B) Enlargement of b. Note the calcite spars which forming the laminated coating. C) Enlargement of c. Note the relic of the original high-Mg calcite mineral. D) Matrix displaying calcite microspar and porosity.

TABLE 1.—Chemical and isotopic compositions of Middle Jurassic carbonate.

(a) Surface samples, Gebel El Maghara

| Sample | $MgCO_3$(mol%) | | | $MnCO_3$(mol%) | | | $FeCO_3$(mol%) | | | $SrCO_3$(mol%) | | | $\delta^{18}O$‰ | | | $\delta^{13}C$‰ | | |
|---|---|---|---|---|---|---|---|---|---|---|---|---|---|---|---|---|---|---|
| No. | O | M | C | O | M | C | O | M | C | O | M | C | O | M | C | O | M | C |
| 4 | 2.61 | 0.67 | 0.68 | 0.26 | 0.60 | 0.71 | 0.60 | 1.40 | 0.90 | 0.14 | 0.10 | 0.03 | -5.4 | -7.5 | -10.5 | +2.3 | -0.02 | -2.9 |
| 5 | 3.42 | 0.81 | 0.92 | 0.14 | 0.80 | 0.63 | 0.20 | 096 | 0.60 | 0.10 | 0.07 | 0.02 | -5.1 | -7.0 | -11.5 | +2.2 | +0.1 | -2.0 |
| 12 | 1.56 | 0.54 | 0.41 | 0.06 | 1.20 | 0.92 | 0.10 | 0.84 | 0.80 | 0.09 | 0.08 | 0.01 | -4.2 | -6.7 | -10.3 | +2.5 | +1.0 | 0.0 |
| 15 | 2.80 | 1.15 | – | 0.10 | 0.90 | – | 0.15 | 0.90 | – | 0.16 | 0.11 | – | -4.6 | -5.6 | – | +2.8 | +1.5 | – |
| 17 | 3.65 | 1.20 | – | 0.42 | 2.10 | – | 0.16 | 1.50 | – | 0.21 | 0.13 | – | -5.6 | -6.5 | – | +2.1 | +0.3 | – |
| 20 | 3.81 | 1.13 | 1.10 | 0.28 | 1.60 | 1.20 | 0.08 | 1.70 | 1.20 | 0.18 | 0.10 | 0.04 | -3.8 | -5.0 | -10.0 | +3.2 | +2.5 | -1.5 |
| 23 | 1.60 | 0.94 | 0.54 | 0.18 | 1.64 | 0.81 | 0.12 | 0.74 | 1.10 | 0.22 | 0.15 | 0.03 | -5.3 | -6.6 | -10.8 | +2.4 | +2.7 | +1.5 |
| 24 | 1.92 | 1.31 | 0.35 | 0.04 | 1.40 | 0.94 | 0.14 | 1.20 | 1.30 | 0.08 | 0.08 | 0.01 | -5.4 | -7.2 | -9.5 | +2.8 | +1.6 | +0.1 |
| 26 | 0.94 | 0.42 | 0.48 | 0.08 | 0.60 | 1.20 | 0.04 | 0.60 | 1.10 | 0.15 | 0.12 | 0.02 | -4.6 | -7.3 | -10.6 | +3.1 | +0.5 | -0.4 |
| 28 | 0.82 | 0.71 | 0.62 | 0.10 | 0.84 | 1.30 | 0.06 | 1.30 | 1.40 | 0.19 | 0.16 | 0.01 | -4.4 | -5.6 | -10.4 | +2.9 | +2.5 | -0.5 |

(b) Subsurface samples, Borehole No. H-1

| Sample | $MgCO_3$(mol%) | | | $MnCO_3$(mol%) | | | $FeCO_3$(mol%) | | | $SrCO_3$(mol%) | | | $\delta^{18}O$‰ | | | $\delta^{13}C$‰ | | |
|---|---|---|---|---|---|---|---|---|---|---|---|---|---|---|---|---|---|---|
| No. | O | M | C | O | M | C | O | M | C | O | M | C | O | M | C | O | M | C |
| 41 | 1.86 | 0.80 | – | 0.34 | 0.56 | – | 0.12 | 1.71 | – | 0.11 | 0.05 | – | -4.0 | -6.1 | – | +2.2 | +1.8 | – |
| 44 | 2.9 | 1.20 | – | 0.08 | 0.40 | – | 0.20 | 0.92 | – | 0.08 | 0.07 | – | -4.0 | -4.2 | – | +2.7 | +2.5 | – |
| 47 | 3.6 | 1.90 | 0.85 | 0.49 | 1.06 | 0.7 | 0.80 | 1.20 | 0.81 | 0.21 | 0.10 | 0.04 | -4.1 | -4.9 | -9.9 | +3.0 | +2.6 | -1.8 |
| 50 | 3.80 | 0.45 | 0.38 | 0.13 | 0.49 | 0.65 | 0.14 | 0.64 | 0.74 | 0.19 | 0.11 | 0.05 | -5.1 | -6.4 | -9.6 | +2.3 | +1.6 | +0.8 |
| 54 | 2.75 | 0.60 | 0.6 | 0.28 | 1.12 | 0.87 | 0.19 | 1.22 | 0.68 | 0.24 | 0.18 | 0.08 | -5.5 | -5.5 | -9.4 | +2.5 | +2.2 | +0.4 |
| 58 | 1.96 | 0.84 | 0.26 | 0.33 | 0.98 | 0.49 | 0.35 | 1.43 | 1.20 | 0.30 | 0.13 | 0.05 | -5.0 | -3.8 | -9.7 | +3.0 | +3.0 | +0.7 |
| 61 | 3.90 | 1.44 | – | 0.64 | 1.27 | – | 0.04 | 0.82 | – | 0.18 | 0.06 | – | -4.5 | -4.8 | – | +3.1 | +2.8 | – |
| 66 | 2.6 | 0.64 | 1.2 | 0.51 | 0.96 | 0.74 | 0.40 | 1.66 | 0.94 | 0.28 | 0.17 | 0.06 | -4.5 | -5.4 | -9.0 | +2.5 | +1.6 | -1.2 |
| 69 | 1.75 | 0.38 | – | 0.22 | 1.06 | – | 0.31 | 0.88 | – | 0.19 | 0.06 | – | -4.3 | -6.4 | – | +2.8 | +1.4 | – |
| 72 | 3.4 | 0.86 | 0.4 | 0.57 | 2.18 | 1.40 | 0.41 | 2.53 | 1.12 | 0.32 | 0.12 | 0.03 | -4.4 | -6.2 | -8.2 | +2.2 | +1.5 | +1.2 |

O = oncoid grains, M = matrix, C = meteoric calcite cement

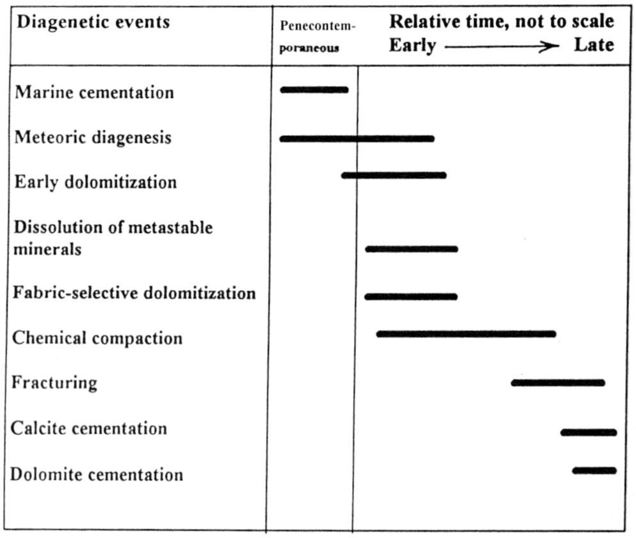


FIG. 5.—Paragenetic sequence for the Middle Jurassic carbonate (Bir Maghara Formation), Sinai.

have led to a relative depletion in $^{13}$C (James and Choquette, 1984; Holail, 1992). This might indicate that the stabilization of the original high-Mg calcite to low-Mg calcite occurred in a marine environment. This would have led to loss of Mg, but the $\delta^{13}$C values of the low-Mg calcite are identical to the marine signal. Moreover, the enriched $^{13}$C of oncoids reflect that the vital effects have not modified the original signal.

The mean $\delta^{13}$C value of +1.7‰ for the studied Middle Jurassic carbonate matrix (Fig. 7) is a characteristic of matrix alteration and recrystallization in systems with high water–rock ratios. These values are quite similar to that reported as typical for Jurassic carbonate matrix (Marshall and Ashton, 1980). The relative depletion in $^{13}$C for the matrix suggests that the carbon-isotope compositions were probably influenced by the $^{13}$C-depleted $CO_2$ inputs from meteoric fluids. Therefore, the relative depletion in $^{13}$C is related to diagenesis.

The $\delta^{13}$C values range widely, from +1.5 to -2.9‰ for the meteoric sparry calcite cements (Figs. 7). Samples from the two studied sections have similar $\delta^{13}$C values. These values are lighter than those for other modern marine cements, which might reflect incorporation of isotopically light carbon from meteoric or subsurface water. This means that these values were strongly influenced by relatively $^{13}$C-enriched $CO_2$ coming from dissolution of the Middle Jurassic carbonates and by $^{13}$C-depleted $CO_2$ coming from recharge of meteoric water through a soil zone. The wide range of $\delta^{13}$C compositions generally follow Lohmann's (1988) meteoric-water line.

### *Oxygen Isotopes.—*

The $\delta^{18}$O value is a sensitive parameter for evaluating water–rock ratio during Middle Jurassic carbonate diagenesis. This is because of the comparatively large temperature-related isotopic fractionation of oxygen isotopes, the large size of the pore-water oxygen reservoir, and the potentially large difference in $\delta^{18}$O values of marine vs. meteoric waters.

The mean $\delta^{18}$O value of -4.6‰ for the studied Middle Jurassic oncoids is significantly shifted from values consistent with precipitation from Jurassic sea water (Fig. 7). The surface and subsurface oncoid samples do not yield any large difference in their oxygen-isotope compositions (from -3.8 to -5.6‰ and from -4.0 to -5.5‰ PDB, respectively). These values show a narrow range, however, and are much heavier than those of the Middle Jurassic meteoric cements and are close to those reported for presumed marine cements (-2.0 to -3.0‰ PDB; Marshall and Ashton, 1980). Assuming the precipitation of oncoids from sea water with $\delta^{18}$O= -1‰ SMOW (1‰ lighter than modern marine sea water), the paleothermometry equation of Friedman and O'Neil (1977) predicts formation of the Middle Jurassic oncoids at temperatures ranging from 29°C ($\delta^{18}$O = -3.8‰ PDB) to 33°C ($\delta^{18}$O = -5.6‰ PDB). These temperatures are quite high and are not in agreement with those predicted for normal sea water and are most probably related to the reseting of the oncoid $\delta^{18}$O values by diagenesis. The paleothermometry equation of Friedman and O'Neil (1977) for low-Mg calcite is used because the mean mineral composition of oncoid is approximately 3.6 mol % $MgCO_3$ (Table 1).

Stable oxygen-isotope compositions also suggest that if the studied oncoids lost their Mg during alteration. Land and Epstein (1970) showed that when high-Mg calcite fossils were converted to low-Mg calcite in meteoric waters, the oxygen-isotope composition of the resulting low-Mg calcite became lighter than the precursor carbonate, approaching isotopic equilibrium with meteoric waters. Oxygen-isotope data (Fig. 7) from the Middle Jurassic oncoids, which were originally of high-Mg calcite composition, show influence of meteoric waters. Therefore, the oncoids have stable-isotope composition consistent with alteration (stabilization) in meteoric waters ($\delta^{18}$O = -3.8 to -5.6‰ PDB; $\delta^{13}$C = +2.2 to +3.2‰ PDB).

Regarding the studied carbonate matrix, its $\delta^{18}$O and $\delta^{13}$C values define a trend distinct from the oncoids, with the subsurface matrix being isotopically heavier ($\delta^{18}$O = -3.8 to -6.4‰; $\delta^{13}$C =+1.4 to +3.6‰) than the surface matrix ($\delta^{18}$O = -5.0 to -7.5‰; $\delta^{13}$C = -0.1 to -2.7‰). There is great danger in just comparing numbers, and the matrix petrography and elemental composition should be considered to place constraints on the isotope interpretations. One particular problem is that there may have been secular fluctuations in sea-water $\delta^{18}$O and $\delta^{13}$C (Lohmann and Walker, 1989; Veizer and Hoefs, 1976), so that there could be variations through time of the isotopic signal of marine carbonate matrix. Therefore, the interpretation of these isotopic data assumes that the matrix had an initial Jurassic marine composition of about +3.0‰ $\delta^{13}$C and -2.0 to -3.0‰ $\delta^{18}$O PDB. Judging from these data the isotopically heaviest matrix samples are interpreted to have been least altered. The isotopically lighter matrix samples are interpreted to have been subjected to different recrystallization episodes.

The isotopic data also constrain the origin of the sparry calcite cements, which are the main occluders of porosity in the studied Middle Jurassic carbonates. It is clear that in the interpretation of the origins of these cements, the isotopic values are ambiguous. Light $\delta^{18}$O values indicate either light (meteoric) water at low temperatures or heavy (marine or subsurface) water at high temperatures. The formation of these cements from marine water ($\delta^{18}$O = -1.0‰ SMOW) would necessitate unrealistically high temperatures (i.e., 80°C ± 10°C; calcite–water geothermometer of Friedman and O'Neil, 1977). Because these calcite cements are isotopically distinct from the oncoids and the carbonate matrix (Fig. 7), it is improbable that isotopic reequilibration at higher temperature occurred. It appears likely that these calcite cements formed from low $^{18}$O pore water (meteoric). Late diagenetic fluids with a $\delta^{18}$O value of -6.0 to -7.0‰ (SMOW) result in a more realistic formation temperature (21°–24°C) for the late calcite. Carbonate–water interactions and mixing with meteoric water can yield low $\delta^{18}$O late diagenetic fluids. Moreover, the relatively low $\delta^{13}$C values of these cements (from -2.9 to +1.5‰ PDB) indicate that $^{13}$C-depleted $CO_2$ coming

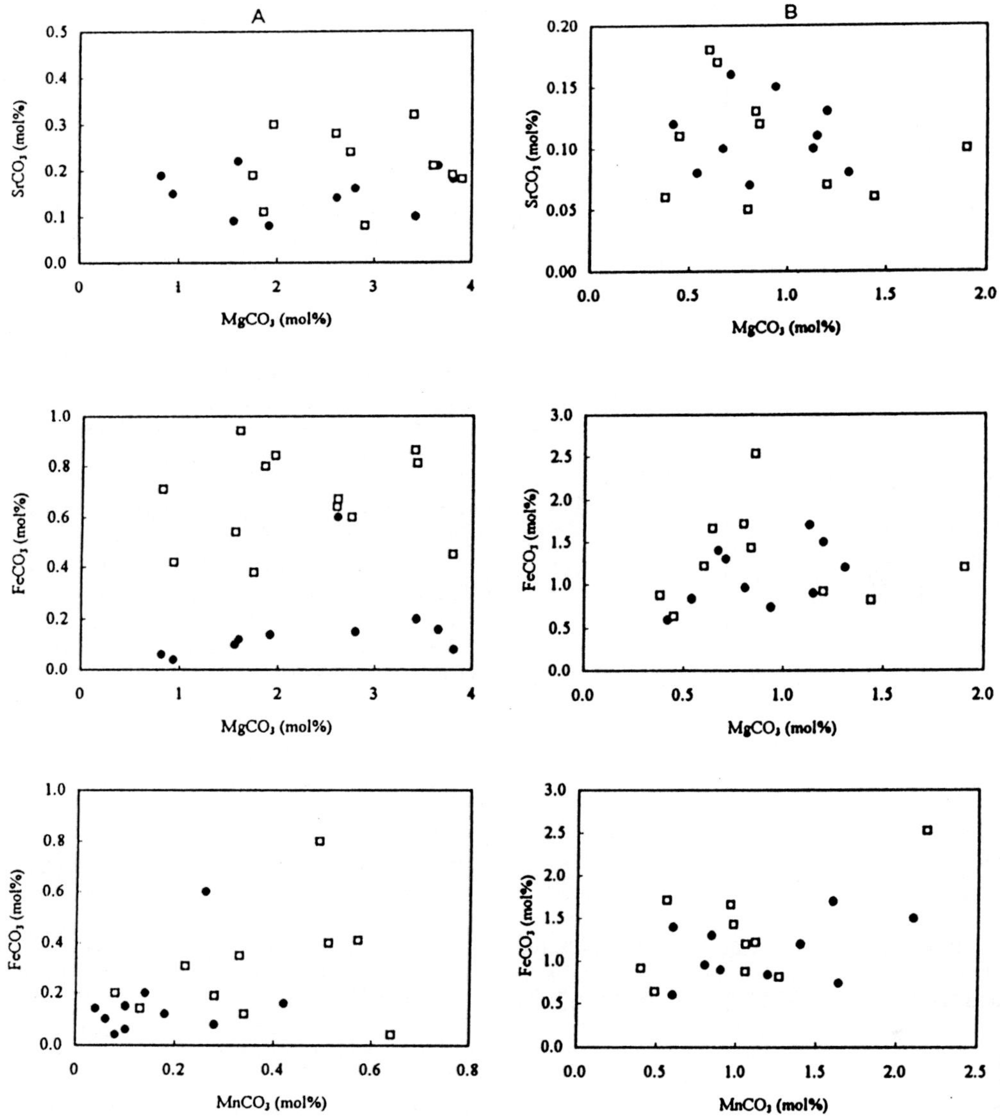


FIG. 6.—Element variations for the studied Middle Jurassic carbonate. A) For oncoids. B) For the matrix. • surface samples; □ subsurface samples.

from soil gas and the meteoric influx was a major source of carbon, in addition to normal marine bicarbonate (0.0‰ PDB).

## CONCLUSIONS

Petrographic and geochemical data complement the study of Middle Jurassic oncoids in north Sinai, Egypt. The following major conclusions can be drawn from this study.

1. Apart from the textural evidence, the geochemical signatures support a shallow marine origin for Middle Jurassic carbonate. Meanwhile, the oncoid distribution and fluctuation are superimposed on the overall regressive pattern, and the influx of meteoric water associated with this pattern had a major influence on diagenesis.
2. $\delta^{18}O$ resetting is least obvious in the oncoid grains. Correlation of various geochemical parameters demonstrates that $\delta^{18}O$ values of these oncoids decrease with decreasing Sr and with increasing Fe and Mn contents. The preserved oncoids are closest to the initial isotopic composition and least altered trace-element composition.
3. The wide range of $\delta^{13}C$ values for sparry calcite cement is related to variable carbon sources. The different values correspond to differences in mixing of $^{13}C$-depleted $CO_2$ from soil gas (meteoric end member) and $^{13}C$-enriched $CO_2$ coming from the dissolution of Middle Jurassic carbonate (marine end member). These changes and the highly depleted $^{18}O$ values are linked to meteoric diagenesis.

## ACKNOWLEDGMENTS

I express my appreciation to Dr. R. Rifai and the Geological Survey of Egypt for providing access to core samples. I would like to thank Profs. K. C. Lohmann and B. Wilkinson for using the

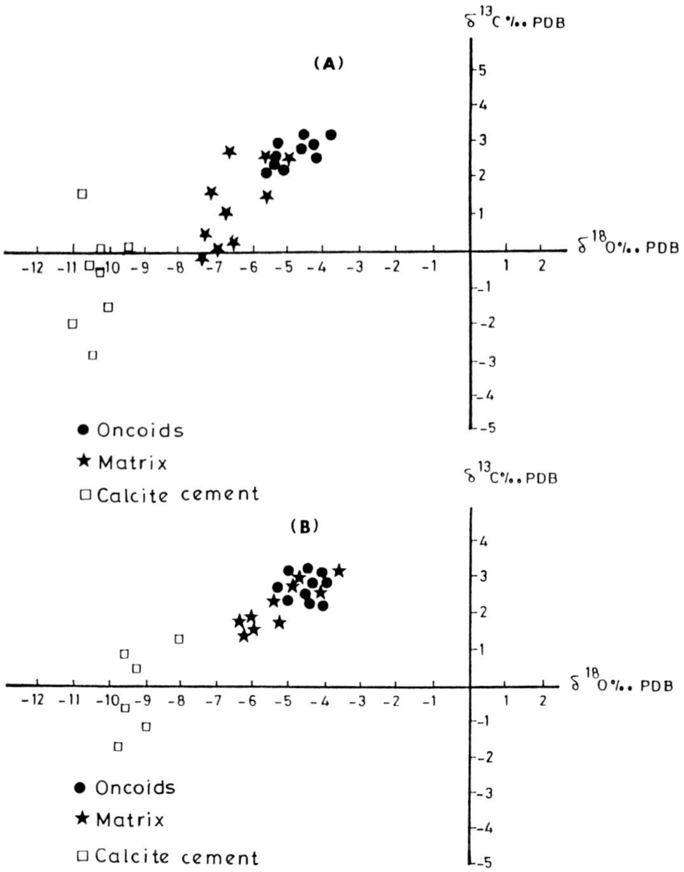


FIG. 7.—Cross-plot showing the carbon and oxygen isotopic compositions of oncoids, matrix, and meteoric calcite cement from Middle Jurassic carbonate, Sinai, Egypt. A) Gebel Maghara section. B) Bore hole No. H-1 section.

Stable Isotope Laboratory at the University of Michigan, USA. The manuscript benefitted from the comments and suggestions of L. S. Land and S. J. Mazzullo. I commend the editors A.S. Alsharhan and R.W. Scott for their efforts on our behalf.

## REFERENCES

ABELL, P.I., AWRAMIK, S.M., AND OSBORNE, R.H., 1978, Oxygen, and carbon isotopic variation in Pleistocene lacustrine stromatolites from Lake Turkana, Kenya: (abstr.) American Association of Petroleum Geologists, Bulletin, v. 62, p. 488.

AL FAR, D.M., 1966, Geology and coal deposits of Gebel El-Maghara, northern Sinai: United Arab Republic Geological Survey, Paper 27, 59 p.

ALLAN, J.R., AND MATTHEWS, R.K., 1982, Isotope signature associated with early meteoric diagenesis: Sedimentology, v. 29, p. 797–817.

ANDREWS, J., 1986, Microfacies and geochemistry of Middle Jurassic algal limestones from Scotland: Sedimentology, v. 33, p. 499–520.

BATHURST, R.G., 1984, The integration of pressure-solution with mechanical composition and cementation, *in* Stylolite and Associated Phenomena Relevance to Hydrocarbon Reservoirs: Abu Dhabi National Reservoir Research Foundation Special Publication, Abu Dhabi, U.A.E., p. 41–55.

BECHER, J.W., AND MOORE, C.H., 1979, The Walker Creek field: A Smackover diagenetic trap, *in* Moore, C.H., ed., Geology of Carbonate Porosity: American Association of Petroleum Geologists, Course Note Series No. 11, p. 24–46.

BERNOULLI, D., 1972, North Atlantic and Mediterranean Mesozoic facies: a comparison: Deep Sea Drilling Project, Initial Reports, Washington, D.C., v. 11, p. 801–822.

BISHOP, W.F., 1968, Petrology of Upper Smackover limestone in North Haynesville field, Clairborne Parish, Louisiana: American Association of Petroleum Geologists, Bulletin, v. 52, p. 92–128.

BURTON, E.A., 1993, Controls on marine carbonate cement mineralogy: review and reassessment: Chemical Geology, v. 105, p. 163–179.

CARPENTER, S.J., AND LOHMANN, K.C., 1992, Sr/Mg ratios of modern marine calcite: empirical indicators of ocean chemistry and precipitation rate: Geochimica et Cosmochimica Acta, v. 56, p. 1837–1849.

CHOQUETTE, P.W., AND JAMES, N.P., 1987, Diagenesis # 12. Diagenesis in Limestone—3. The deep burial environment: Geoscience Canada, v. 14, p. 3–35.

CRAIG, H., 1957, Isotopic standards for carbon and oxygen and correction factors for mass spectrometric analysis of carbon dioxide: Geochimica et Cosmochimica Acta, v. 12, p. 133–149.

DE KLASZ, I., 1978, The West African sedimentary basins, *in* Moullade, M., and Nairn, A., eds., The Phanerozoic Geology of the World II, the Mesozoic: Amsterdam, Elsevier, p. 371–399.

FRIEDMAN, I., AND O'NEIL, J.R., 1977, Compilation of stable isotope fractionation factors of geochemical interest: U.S. Geological Survey, Professional Paper 440-KK, 12 p.

FREUND, R., GOLDBERG, M., WEISSBROD, T., DRUCKMAN, Y., AND DERIN, B., 1975, The Triassic–Jurassic structure of Israel and its relation to the origin of the eastern Mediterranean: Israel Geological Survey, Bulletin, v. 65L, p. 1–26.

GOLDBERG, M., BARZEL, A., COOK, P., AND MIMRAN, Y., 1971, Preliminary columnar section of the Jurassic of Gebel Maghara: Israel Geological Survey, Report, MM/1/71.

HOLAIL, H., 1992, Coordinated petrography–isotopic–chemical investigation of meteoric calcite cement (Jurassic–Pleistocene), Egypt: Carbonates and Evaporites, v. 7, p. 48–55.

JAMES, N.P., AND CHOQUETTE, P.W., 1984, Diagenesis # 9: Limestones—the meteoric diagenetic environment: Geoscience Canada, v. 11, p. 161–194.

JENKINS, D.A., 1990, North and Central Sinai, *in* Said, R., ed., The Geology of Egypt: Rotterdam, A. A. Balkema, p. 361–380.

JONES, F.G., AND WILKINSON, B.H., 1978, The structure and growth of lacustrine pisoliths from recent Michigan marl lakes: Journal of Sedimentary Petrology, v. 48, p. 1103–1110.

KEELEY, M.L., AND WALLIS, R.T., 1991, The Jurassic system in Northern Egypt: II depositional and tectonic regimes: Journal of Petroleum Geology, v. 14, p. 49–64.

LAND, L.S., AND EPSTEIN, S., 1970, Late Pleistocene diagenesis and dolomitization, North Jamaica: Sedimentology, v. 14, p. 187–200.

LOHMANN, K.C., 1988, Geohemical patterns of meteoric diagenetic systems and their application to studies of paleokarst, *in* James, N.P., and Choquette, P.W., eds., Paleokarst: Berlin, Springer-Verlag, p. 58–80.

LOHMANN, K.C., AND WALKER, J.C., 1989, The $\delta^{18}O$ record of Phanerozoic abiotic marine calcite cements: Geophysical Research Letters, v. 16, p. 316–322.

LUNDEGARD, P.D., AND LAND, L.S., 1986, Carbon dioxide and organic acids: their role in porosity enhancement and cementation: Paleogene of the Texas Gulf Coast, *in* Gautier, D.L., ed., The Role of Organic Matter in Sediment Diagenesis: SEPM, Special Publication 38, p. 129–146.

MARSHALL, J.F., AND ASHTON, M., 1980, Isotopic and trace element evidence for submarine lithification hard-grounds in Jurassic of Eastern England: Sedimentology, v. 27, p. 271–289.

MILLIMAN, J.D., 1974, Recent Sedimentary Carbonates 1, Marine Carbonates: Berlin, Springer-Verlag, 375 p.

MONTY, C.L., 1981, Spongiostromate vs. porostromate stromatolites and oncolites, *in* Monty, C.L., ed., Phanerozoic Stromatolites: Berlin, Springer-Verlag, p. 273–275.

MORSE, J.W., AND MACKENZIE, F.T., 1990, Geochemistry of Sedimentary Carbonates: Amsterdam, Elsevier, 707 p.

NICKEL, E., 1983, Environmental significance of freshwater oncoids, Eocene Guarga Formation, Southern Pyrenees, Spain, *in* Peryt, T., ed., Coated Grains: Berlin, Springer-Verlag, p. 308–329.

PERYT, T.M., 1981, Phanerozoic oncoids: a review: Facies, v. 4, p. 197–214.

PERYT, T.M., ED., 1983, Coated Grains: Berlin, Springer-Verlag, 655 p.

PURSER, B.H., 1978, Early diagenesis and preservation of porosity in Jurassic limestones: Journal of Petroleum Geology, v. 1, p. 83–94.

RATCLIFFE, K.T., 1988, Oncoids as environmental indicators in the Much Wenlock limestone Formation of the English Midlands: Geological Society of London, Journal, v. 145, p. 117–124.

SAID, R., 1990, The Geology of Egypt: Rotterdam, A. A. Balkema, 731 p.

Schlumberger, 1984, Well evaluation Conference, Egypt: Schlumberger, Paris, 310 p.

SMITH, A.M., AND MASON, T.R., 1991, Pleistocene, multiple-growth, lacustrine oncoids from the Poacher's Point Formation, Etosha Pan, northern Namibia: Sedimentology, v. 38, p. 591–599.

SPOHN, T., AND SCHUBERT, G., 1982, Convective thinning of the lithosphere: a mechanism for the initiation of continental rifting: Journal of Geophysical Research, v. 87, p. 4669–4681.

VEIZER, J., AND HOEFS, J., 1976, The nature of $^{18}O/^{16}O$ and $^{13}C/^{12}C$ secular trends in sedimentary carbonate rocks: Geochimica et Cosmochimica Acta, v. 40, p. 1387–1395.

WRIGHT, V.P., 1983, Morphogenesis of oncoids in the Lower Carboniferous Llanelley Formation of South Wales, *in* Peryt, T.M., ed., Coated Grains: Berlin, Springer-Verlag, p. 424–436.

ZAMARRENO, I., ANADON, P., AND UTRILLA, R., 1997, Sedimentology and isotopic composition of Upper Palaeocene to Eocene non-marine stromatolites, eastern Ebro Basin, NE Spain: Sedimentology, v. 44, p. 159–176.

# MICROFACIES ANALYSIS OF BERRIASIAN–HAUTERIVIAN CARBONATES, CENTRAL SAUDI ARABIA

HISHAM T. SHEBL
*Core Laboratories, P.O. Box 8503, Abu Dhabi, U.A.E.*
AND
ABDULRAHMAN S. ALSHARHAN
*Faculty of Science, United Arab Emirates University, P.O. Box 17551, Al Ain, U.A.E.*

ABSTRACT: The deposition of the Lower Cretaceous in Saudi Arabia is interpreted within the framework of a model of a simple eastward-dipping carbonate ramp that formed as the result of the extensive flooding of the Arabian Plate during Early Berriasian time. Carbonate deposition was only slightly interrupted by a minor marine regression after the deposition of the Yamama Formation, and it was succeeding by bioclastic limestones of the Buwaib Formation. It was later terminated by a Barremian regression during which clastic deposition dominated most of Central and Eastern Arabia.

The Early Cretaceous carbonates of the Sulaiy, Yamama, and Buwaib formations crop out in central Saudi Arabia. According to sedimentological and paleontological data these formations consist of twenty-two successive microfacies correlated with the published standard microfacies types and belts. The Sulaiy, Yamama, and Buwaib formations were deposited in open-platform and lagoonal settings interrupted by two breaks in sedimentation: (1) a pre-Buwaib disconformity, which is marked by an abrupt change from the fine-grained lime mudstones of the upper Yamama to the sandy wackestones containing cyclamminids of the basal Buwaib, and (2) the more pronounced unconformity between the Buwaib bioclastic wackestone and the overlying Biyadh Sandstone.

## INTRODUCTION

Despite the fact that the Berriasian–Hauterivian carbonate succession in the subsurface of both Eastern Saudi Arabia and the Arabian Gulf region contains a number of hydrocarbon reservoirs, very little has been published on that succession, particularly at the type localities in Saudi Arabia. Published information on the Lower Cretaceous outcrops of Saudi Arabia includes the works of Steineke and Bramkamp (1952), Steineke et al. (1958), Powers et al. (1966), Powers (1968), and Moshrif (1981), who were generally concerned with the subdivision of the outcrop sequence into lithostratigraphic units. A few faunal lists were included, but the results of detailed biostratigraphical or microfacies analysis have never been published except for a description of the lithofacies of the Buwaib Formation by Moshrif (1981). This means that understanding of the stratigraphy is confused by the fact that little subsurface data have been published by the operating oil companies, and the majority of the type sections were chosen in the outcrop belts of Central Arabia, far from the oil fields of the Arabian Gulf. Previous study by Shebl and Alsharhan (1994) shows the relationship between the outcropping sequences of these rocks in Central Arabia and the subsurface sequences of the oil-producing areas in eastern Saudi Arabia. The current study was conducted in an attempt to clarify and revise the stratigraphy of the Lower Cretaceous type sections in Central Arabia by tying these to microfacies analysis of this carbonate succession. Such analysis can be used for regional correlation of the studied sections as well as for their paleoecological and paleogeographical interpretations.

## GEOLOGIC SETTING

The Lower Cretaceous rocks that crop out in Central Arabia occur in the interior homocline of the Arabian Shelf (Fig. 1). The type Sulaiy Formation was designated by Bramkamp and Berger (1938, cited in Powers et al., 1966) in the cliff section above the Dahl Hit escarpment to the east of Al Riyadh City (Fig. 1). Similarly the type section of the Yamama Formation was taken by Steineke et al. (1958) from a number of short exposures on Al Qusay'a plateau to the south of the town of Al Kharj. The Buwaib type section was also measured by the same authors in Khashm Buwaib. However, in 1962 Powers and McClure (cited in Powers et al., 1966) suggested a reference section for the Buwaib Formation about 14 km northeast of Khafs Daghrah. Here the Buwaib Limestone unconformably overlies the Sulaiy Formation without the intervening Yamama Formation. This encouraged Moshrif (1981) to choose another reference section for the Buwaib Formation at Khashm Al Thamamah, where the contacts of the formation with both the underlying Yamama and the overlying Biyadh formations are obvious. This was despite the persistent nature of the pre-Buwaib disconformity and the overstepping character of the lower boundary of the Buwaib Formation (Fig. 2). Moshrif's proposal is accepted here, and the Khashm Al Thamamah section is treated as a reference section for the Buwaib Formation. All these type and reference sections have been visited, studied, and sampled for the present investigation (Fig. 1). For this study, samples were collected at a regular interval of 3 m within uniform lithologic units, and at both lithological changes and boundaries. Sampling was carried out using a Brunton compass, an Abney level, and a measuring tape. Each sample was described in the field and its location was recorded. A summary of the location of each sampled section, its thickness, and the number of samples collected is in Figure 1.

## STRATIGRAPHIC FRAMEWORK

The first synthesis of the stratigraphic terminology in Arabia was given by Steineke and Bramkamp (1952), who raised the Thamama Formation to group status and its constituent members to formation rank. Consequently the Thamama Group included (from bottom to top) the Sulaiy, Yamama, and Buwaib formations. The Thamama Group is characteristically represented by its limestones, which clearly distinguishes it from both the underlying Upper Jurassic Hith Anhydrite and the overlying Barremian and younger Biyadh Sandstone. Aramco

Middle East Models of Jurassic/Cretaceous Carbonate Systems
SEPM Special Publication No. 69, Copyright © 2000
SEPM (Society for Sedimentary Geology), ISBN 1-56576-075-1, p. 115–127.

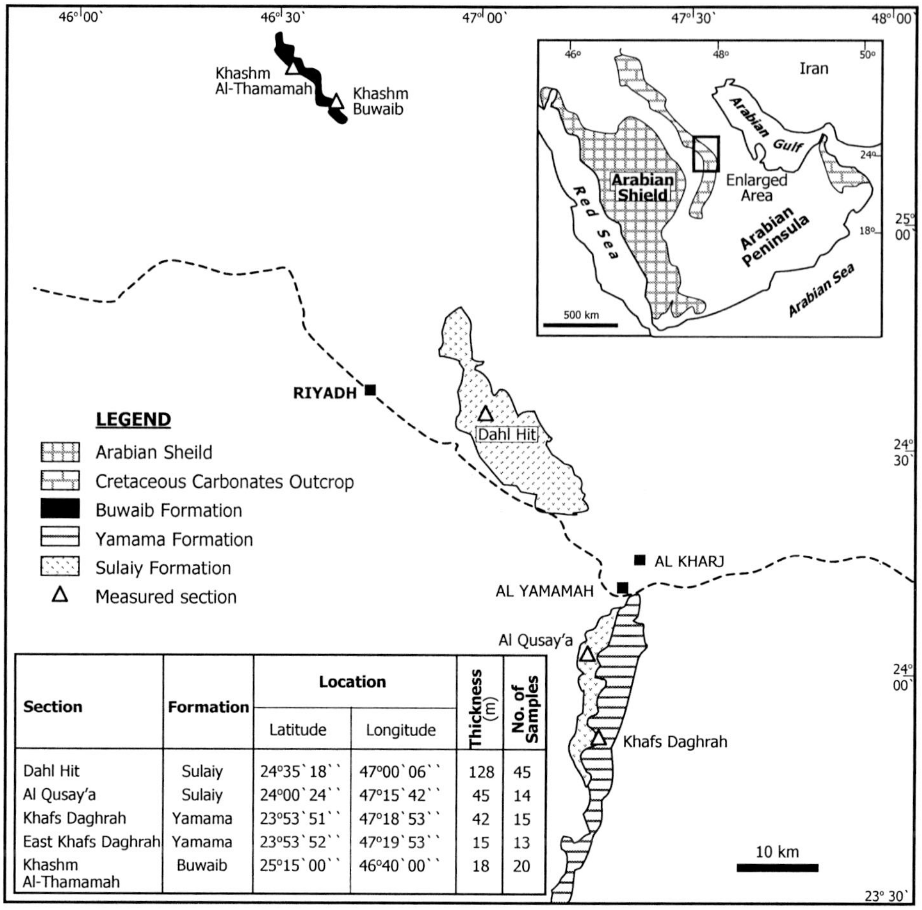


| Section | Formation | Location | | Thickness (m) | No. of Samples |
|---|---|---|---|---|---|
| | | Latitude | Longitude | | |
| Dahl Hit | Sulaiy | 24°35`18`` | 47°00`06`` | 128 | 45 |
| Al Qusay'a | Sulaiy | 24°00`24`` | 47°15`42`` | 45 | 14 |
| Khafs Daghrah | Yamama | 23°53`51`` | 47°18`53`` | 42 | 15 |
| East Khafs Daghrah | Yamama | 23°53`52`` | 47°19`53`` | 15 | 13 |
| Khashm Al-Thamamah | Buwaib | 25°15`00`` | 46°40`00`` | 18 | 20 |

FIG. 1.—Exposed section of the Berriasian–Hauterivian carbonates of central Saudi Arabia, where the type localities were measured and sampled for this study. Simplified location map of outcropping Cretaceous succession of central Saudi Arabia is shown in the right corner of the figure.

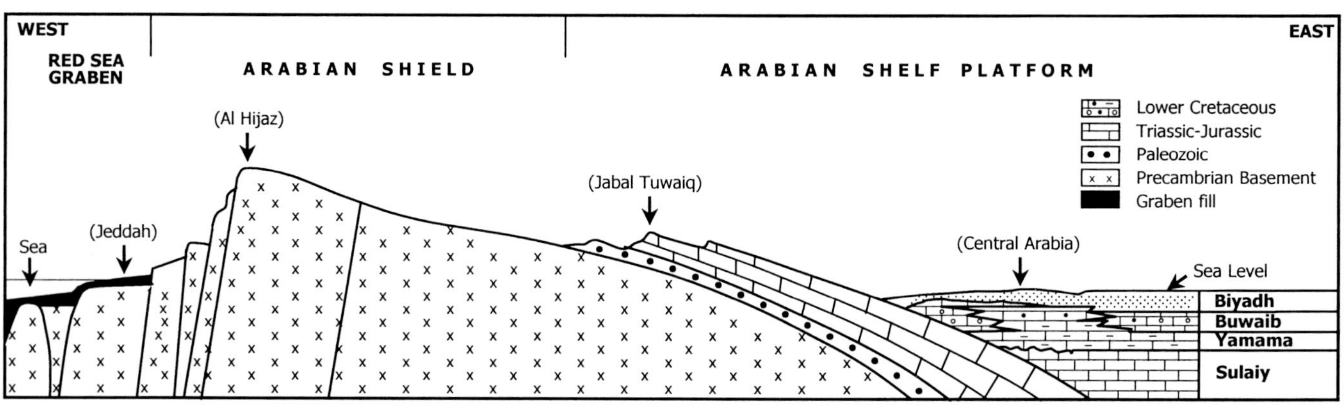


FIG. 2.—Schematic geologic cross section of Precambrian–Lower Cretaceous from western to central Saudi Arabia showing the tilted position with the uplifted western part where the Precambrian basement is cropping out and deepening toward the east (not to scale) (modified from Bayer, 1984; Moshrif and Kelling, 1984).

geologists R.W. Powers and his colleagues (see Powers et al., 1966; Powers 1968) expanded the concept of the Thamama Group (Fig. 3) by including the Biyadh Sandstone and the overlying Shuaiba Limestone, wherever present.

The Lower Cretaceous succession of the Arabian Peninsula was deposited upon an eastward-dipping carbonate ramp similar to that described by Murris (1980). The Early Cretaceous ramp of Arabia is a typical homoclinal ramp. The paleoslope dipped gently and uniformly from Central Arabia (the shoreline) to the east, toward Oman (the basin) (Murris, 1980; Alsharhan and Nairn, 1986). The general paleoenvironmental interpretation and depositional setting of the Lower Cretaceous in Arabia is illustrated in Figure 4.

During the Early Berriasian the basal part of Sulaiy Formation in central Arabia was deposited under shelf conditions with restricted circulation and upon tidal flats and protected isolated lagoons. More consistently open-platform shelf-lagoon sediments of the Yamama Formation accumulated during the Berriasian to Hauterivian and were terminated by the pre-Buwaib disconformity. Again, in the Hauterivian shelf-lagoon sediments resumed, accompanied by an influx of clastic sediments from the western part of Arabia resulting in deposition of the Buwaib Formation.

## *Sulaiy Formation*

The Sulaiy Formation is about 165 to 180 m thick and consists of massively bedded bioclastic wackestones to packstones with interbedded thin layers of mudstones and recrystallized coquina and oolitic grainstones mostly as its base (Alsharhan and Nairn, 1997). The contact between the Upper Jurassic Hith Anhydrite and the brecciated limestone of the Sulaiy is disconformable (Steineke et al., 1958). Powers et al. (1966) considered the boundary to be conformable and the brecciation to be the result of subsurface solution collapse rather than subaerial erosion. The brecciated limestone horizon near the base of the Sulaiy Limestone was once placed by Powers (1968) within the Hith Formation. We reject this interpretation because the brecciated limestone has the same lithology and fossil content as the overlying Sulaiy Limestone. Consequently, the lower boundary of the Sulaiy Formation is drawn at the appearance of the first limestone bed overlying the evaporites of the Hith Formation. The upper boundary is placed in the section where the dense massive limestone of the uppermost Sulaiy Formation is overlain by the detrital limestone of the Yamama Formation.

Powers (1968) listed the following fossils in the Sulaiy Formation: *Bramkampella arabica* Redmond, *Nautiloculina* sp., *Pseudocyclammina sulaiyana* Redmond, *Trocholina* spp., *Everticyclammina* sp., *Anchispirocyclina lusitanica* Egger; *Nerinea* sp. aff. *N. blancheti* Pictet and Campiche, *Nerinea* sp., *Aetostreon latissima* Lamark, *Ostrea* sp., *Milleporidium* sp., *Aporrhais* sp., *Diceras?* sp., *Chaetetes* sp., and miscellaneous cidaroid radioles. In addition, the following forms have been recorded in the present study: *Textularia* sp., *Trochammina* sp., *Palaeomiliolina* sp., *Berthelinella* sp., and *Pseudocyclammina lituus* Yokoyama. On the basis of both its stratigraphic position and fossil content, the Sulaiy Formation is here considered to be of ?Tithonian–Valanginian age.

## *Yamama Formation*

The Yamama Formation, about 25 to 45 m thick, is composed mainly of bioclastic wackestones and mudstones. Away from the type area it contains progressively muddier sediments dominated by micritic limestones with very thin shale interbeds (Alsharhan and Nairn, 1997). In the Al Qusay'a and the Khafs Daghrah sections (Fig. 1) it consists of golden-brown, peloidal, foraminiferal

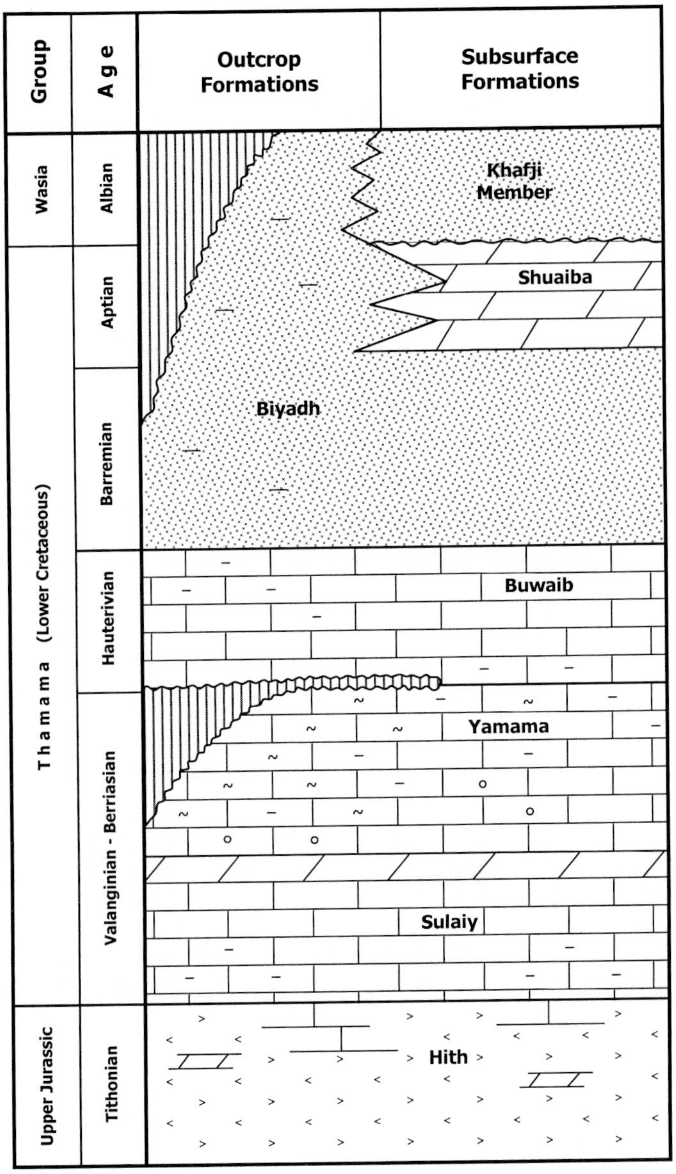


FIG. 3.—Lithostratigraphy of Lower Cretaceous in Saudi Arabia.

wackstones commonly with thin interbeds of bioclastic wackestones and mudstones.

The Yamama Formation rests conformably on the Sulaiy Formation. Steineke et al. (1958) believed that the Yamama Formation is conformably overlain by the Buwaib Formation, whereas Powers et al. (1966) and Shebl (1989) found an unconformity. This angular unconformity at the top of the Yamama is seen only locally in outcrops and in a few oilfields. This upper boundary is mostly conformable in the subsurface of eastern Arabia, except in a few oilfields (Ma'agala, Khurais, and Jawf) where an angular unconformity is observed.

Powers (1968) recorded the following fossils in the reference section of the Yamama Formation: *Everticyclammina eccentrica*, *E. kelleri* Redmond, *E. hellen* Redmond, *Pseudocyclammina cylindrica* Redmond, *P.* spp., *Cyclammina* spp., *Haplophragmoides* sp., and *Trocholina* sp. on the basis of the occurrence of the echinoid fossils, *Pygurus rostratus* Agassiz, *Trematopygus* sp. cf. *T. grasi* d'Orbigny,

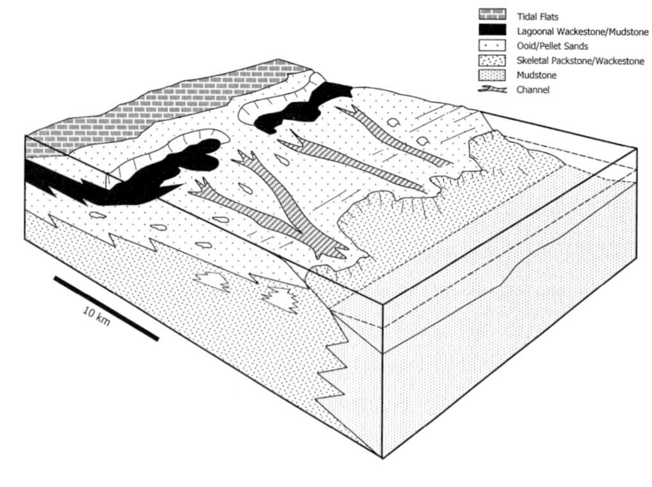


FIG. 4.—Depositional model of Lower Cretaceous carbonate-ramp model in Saudi Arabia, based on model of Read (1985) (reproduced with permission from American Association of Petroleum Geologists).

*Astrocoenia?* sp., *Milliporidium* sp. and cidaroid radioles, and foraminifers, *Bigenerina* sp., *Globuligerina hoterivica* (Subbotina), *Flabellammina* sp., *Pseudonummoloculina* sp., *Nautiloculina oolithica* Mohler, *Trocholina elongata* Leupold, and indeterminate planktonic foraminifera.

Other macrofossils recorded include *Anisocardia* sp., *Aporrhais* sp., *Corbis* sp., *Corbula* sp., *Chione* sp., *Cucullaea?* sp., *Exogyra* spp., *Gyrodes?* sp., *Gryphaea balli* Stefanini, *Homomya* sp., *Lima* sp., *Modiolus?* sp., *Natica* sp., *Nuculana* sp., *Ostrea* sp., *Pholadomya* sp., *P. decussata*, *Pleurotomaria?* sp., *Siliqua* sp., *Tellina* sp., *Trigonia* sp., *Trochus* sp., *Venus* sp., unidentified ammonite genus showing affinities with *Hypengonoceras*, and unidentified small molluscs. The Yamama Formation is considered to be of Valanginian age.

### *Buwaib Formation*

The type section of this formation consists of 11.2 m of micritic and calcarenitic limestones with intercalated sandstone and shale beds. The Buwaib Formation is 17.7 m thick at Khafs Daghrah and consists of interbedded shales, dolomites, and calcarenitic and micritic limestones with occasional very thin beds of quartz sandstone (Powers, 1968). In contrast, the Buwaib Formation at Khashm Al Thamamah is about 17 m thick and is composed of bioclastic wackestones interbedded with argillaceous mudstones (Shebl, 1989).

The upper contact of the Buwaib Formation is disconformable, changing from cyclamminid bioclastic wackestones of the Buwaib Formation to black-weathering, cross-bedded sandstones of the basal Biyadh Formation (Alsharhan and Nairn, 1997).

Fossils recorded in both the studied section and in the type Buwaib include *Ammobaculites?* sp., *Everticyclammina greigi* Henson, *Everticyclammina hensoni* Redmond, orbitolinids, unidentified lituolids, and unidentified small molluscs. On the basis of both its fossil content and stratigraphic position an upper Hauterivian to Lower Barremian age is accepted here for the Buwaib Formation, as proposed by Banner and Highton (1990). In the drilled subsurface sections it was reported to include *Choffatella decipiens* Schlumberger, indeterminate orbitolinids, and unidentified small molluscs in addition to the forms recorded in the surface sections. These were taken to confirm the Hauterivian age of the Buwaib Formation (Powers et al., 1966).

### *Biyadh Formation*

In outcrop, the Biyadh Formation is approximately 300–400 m thick and has an age span from Barremian to Aptian? or younger. This formation consists of dark-weathering, cross-bedded quartzose sandstone with some interbedded variegated shale. Variations in the depositional conditions are indicated by several horizons of conglomerates. The top of the section consists of cross-bedded quartz sandstone with abundant quartz pebbles (Powers et al., 1966; Moshrif, 1980a, 1980b; Moshrif and Kelling, 1984; Alsharhan and Nairn, 1997). The bulk of the Biyadh appears to have been deposited in alluvial channels (mainly as point-bar sands) and in flood-plain settings. In the subsurface the Biyadh is bracketed between the Buwaib Formation below and the Shuaiba Formation above. The section, as much as 625 m thick, appears to be richer in shale and argillaceous mudstone than in outcrop and marks a marginal-marine setting as indicated by two thin beds of orbitolinid limestone within the succession of predominantly green shale.

### *Shuaiba Formation*

The Shuaiba Formation is known only from the subsurface of Saudi Arabia and is restricted to Aptian age. It consists of massive, commonly porous and vuggy dolomite with occasional limestone. Some levels containing orbitolinids have been found. The thickness ranges from 25 to 110 m.

## MICROFACIES ANALYSIS

The type sections and reference sections of the Berriasian–Hauterivian carbonate sequence in Saudi Arabia comprise twenty-two stratigraphically superposed microfacies (Fig. 5). The microfacies are described in the following sections in ascending stratigraphic order. Facies nomenclature and depositional environments follow Flügel (1982) and Wilson (1975). The microfossil remains have been determined using the following publications: Bozorgnia and Banafti (1964), Redmond (1964), Majewske (1969), Sampo (1969), Simmons and Hart (1987), Banner and Highton (1990), and Banner and Whittaker (1991).

## MICROFACIES OF SULAIY FORMATION (A)

### *Microfacies A-1: Textulariid Peloidal Grainstone*

#### *Lithology.—*

This microfacies consists of about 8 m of numerous peloidal grainstones with a few superficial ooids, aggregate grains, and bioclasts cemented by calcite spar (Fig. 6A). The peloids are mostly pseudopeloids and are commonly associated with very small aggregate grains. This microfacies constitutes the basal part of the typical Sulaiy Limestone in the Dahl Hith section. It directly overlies the Hith Anhydrite with an irregular contact, which was probably produced by solution collapse.

#### *Fossils.—*

*Textularia* spp., *Trochammina* sp., miliolids, and algae.

#### *Depositional Setting.—*

This microfacies is correlated to the standard microfacies (SMF) type 17, which was placed in either standard facies belt 8 (restricted platform) or 7 (open-platform shelf lagoon). However, because the pseudopeloids may have been derived from occasional reworking of sediments in an extremely shallow-water

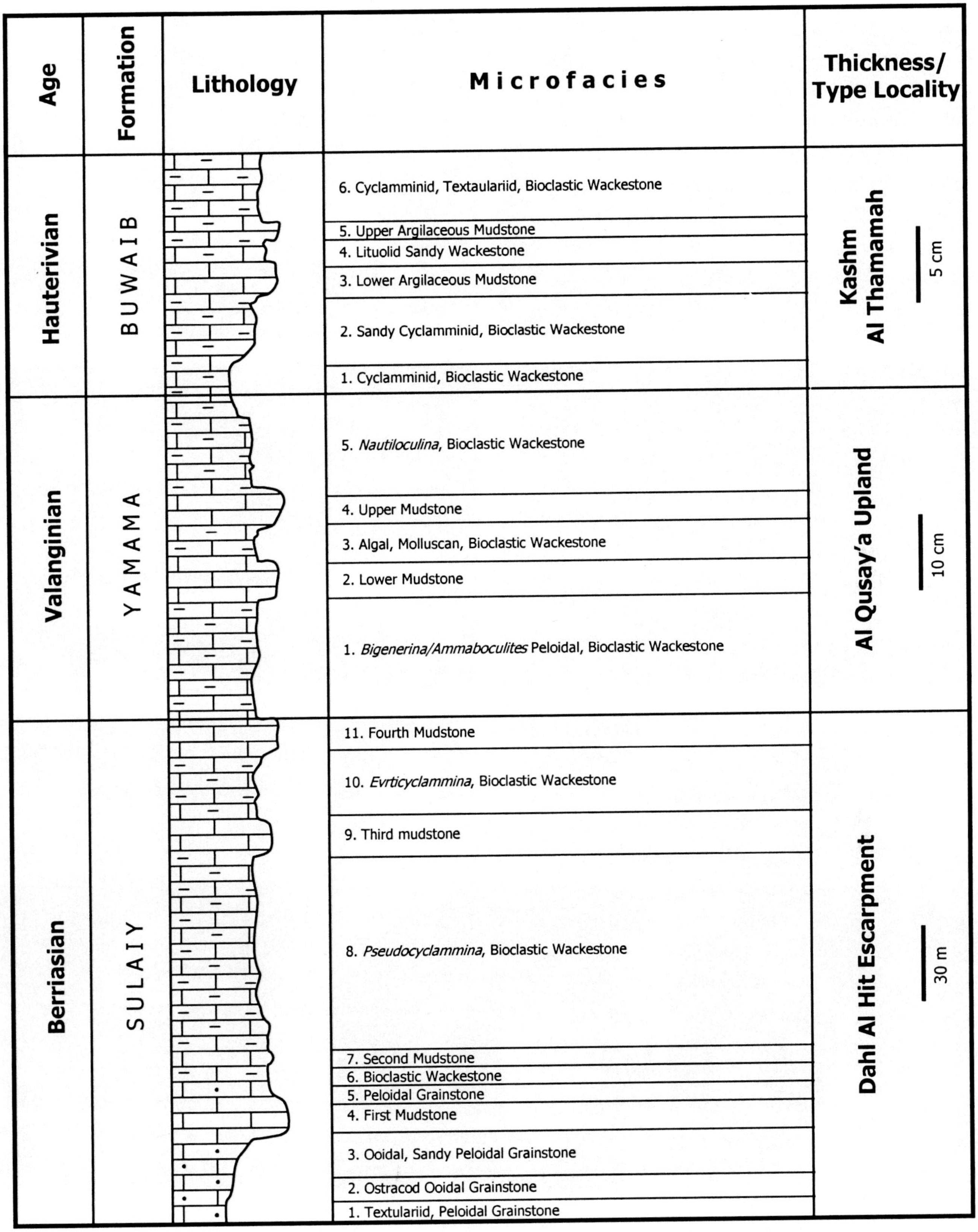


FIG. 5.—Composite section of several outcrops showing the stratigraphic distribution of the microfacies in the Sulaiy, Yamama, and Buwaib formations of central Arabia. Sections taken from Shebl and Alsharhan (1994). Not to scale.

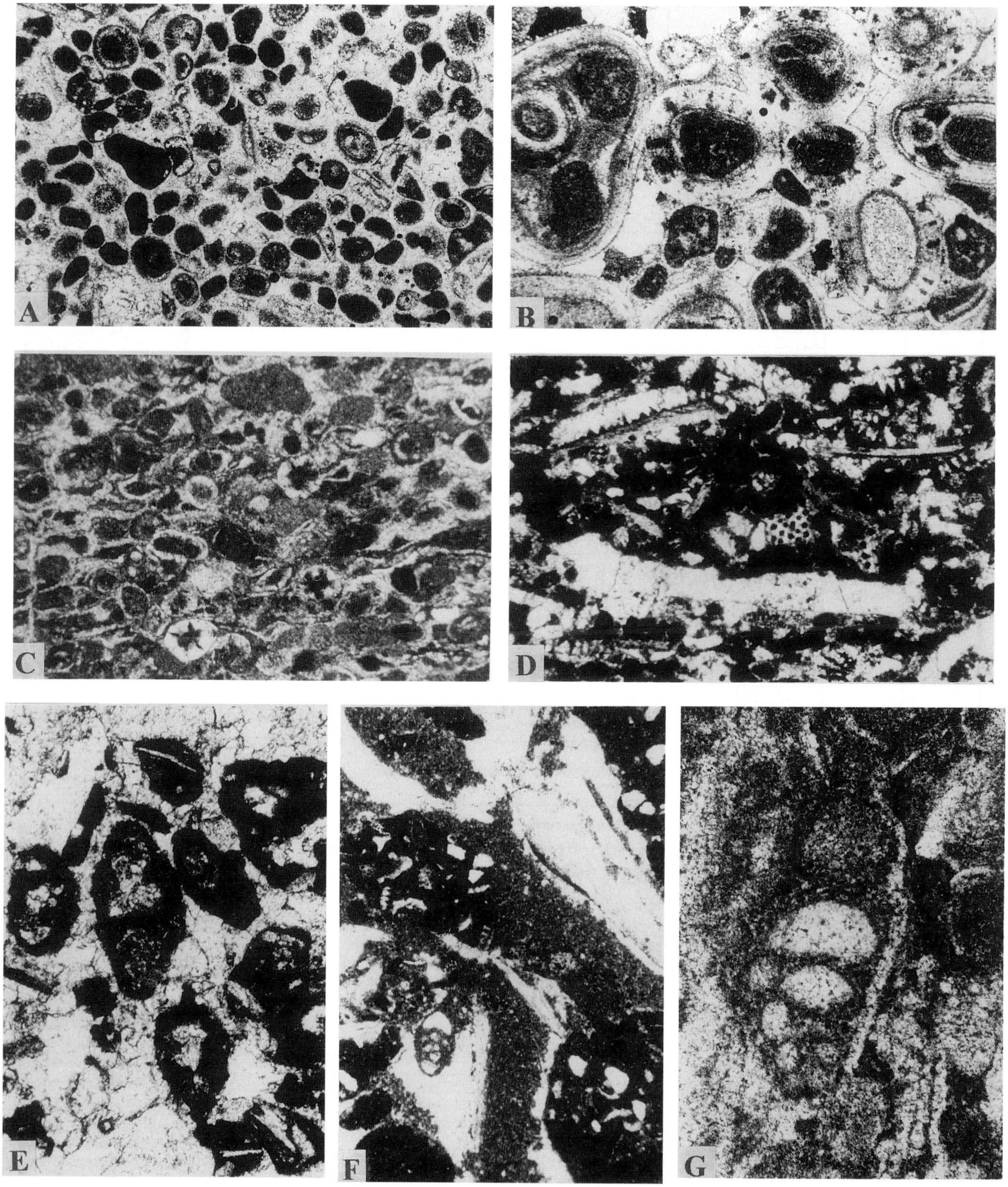

FIG. 6.—Microfacies of Sulaiy Formation: A) Peloids, aggregate grains, superficial ooids, and bioclasts in a sparry matrix (x 22); B) ooids, some multiple ooids, and peloids as well as coated grains in a sparry matrix (x 22); C) oolitic and peloidal grainstones with scattered of quartz grains (x 22); D) recrystallized molluscan, bryozoan, algal, and echinoderm remains (x 22); E) axial and nearly axial section of *Pseudocyclammina sulaiyana* (Redmond) in a recrystallized matrix (x 32); F) recrystallized shell fragments and an axial section of *Nautiloculina* sp. (x 22); G) axial section of textulariid foraminifera (x 32).

setting with restricted sedimentation (Grass and Carozzi, 1967), the interpretation of a restricted platform is preferred here. The presence of aggregate grains supports this interpretation because these grains usually form in shallow subtidal and intertidal shallow-water settings with restricted circulation.

### *Microfacies A-2: Ostracod–Ooidal Grainstone*

*Lithology.—*

This microfacies consists of about 6 m of oolitic grainstone with recrystallized radial ooids and poly-ooids, peloids, coated grains, and bioclasts (Fig. 6B).

*Fossils.—*

Numerous coated ostracode carapaces, small molluscan shells, shell fragments, and foraminifera.

*Depositional Setting.—*

Ooids with radial structures are usually found in abundance in moderate- to low-energy carbonate deposits and are formed in restricted shelf lagoons with elevated salinities. This facies is correlated to the standard microfacies (SMF) type 17, which was placed in facies belt 8 (restricted platform). This interpretation is supported by the greatly reduced diversity of organisms, by the common percentage of ostracode remains, and by broken ooids.

### *Microfacies A-3: Ooid Sandy Peloidal Grainstone*

*Lithology.—*

This microfacies consists of a 1-m-thick peloidal grainstone (Fig. 6C), interrupted by a 3-m-thick horizon of silty, fecal-pellet limestone. Superficial oolites are rare, but both the quartz grains and the pellets are more abundant. The quartz grains are generally subangular to subrounded and highly pitted.

*Fossils.—*

Molluscs (mainly gastropods), algal remains, and agglutinated foraminifera.

*Depositional Setting.—*

The peloidal grainstones of this microfacies correspond to SMF type 17, which was formed in shallow shelf settings where water circulation is restricted in such places as cut-off lagoons, coastal ponds, and tidal flats. The intercalated fecal pellets are correlated with SMF type 16, which was deposited in a very warm, shallow-water setting with only moderate water circulation but with connections to the open sea.

### *Microfacies A-4: First Mudstone*

*Lithology.—*

This microfacies is composed of about 12 m of lime mudstone with rare, scattered, recrystallized shell and bone fragments.

*Depositional Setting.—*

This microfacies corresponds to standard facies belt 7 and SMF Type 17, which is placed in an open-platform shelf lagoon setting. In such lagoonal areas marine grasses can act as sediment trappers and stabilizers, slowing down water currents and stabilizing mud.

### *Microfacies A-5: Peloidal Grainstone*

*Lithology.—*

This microfacies reaches some 3 m in thickness and is composed of peloidal grainstone with abundant detrital grains.

*Fossils.—*

Rare ostracodes, agglutinated foraminifera, and shell fragments.

*Depositional Setting.—*

This microfacies is correlated to SMF type 17 (Standard Facies Belt 7), which is a shelf-lagoon setting with restricted water circulation.

### *Microfacies A-6: Bioclastic Wackestone*

*Lithology.—*

This microfacies consists of a recrystallized bioclastic wackestone (Fig. 6D) about 3 m thick, associated with quartz.

*Fossils.—*

Mainly composed of completely recrystallized bivalve shell fragments.

*Depositional Setting.—*

The bioclastic wackestone microfacies is correlated to SMF type 9, deposited either in standard facies belt 2 (open-sea shelf) or 7 (open-platform shelf lagoon). The latter interpretation is supported by the absence of planktonic foraminifers as well as by the great diversity of the molluscan remains.

### *Microfacies A-7: Second Mudstone*

*Lithology.—*

This microfacies is composed of some 3 m of lime mudstone stained with iron oxide and containing scattered microskeletal fragments and rare recrystallized ostracode carapaces.

*Depositional Setting.—*

The facies was deposited in bays and open lagoons in an open platform and correlated with SMF Type 17 (Standard Facies Belt 7).

### *Microfacies A-8: Pseudocyclammina Bioclastic Wackestone*

*Lithology.—*

This microfacies is composed of about 78 m of bioclastic wackestone interrupted by 3 m of pure lime mudstone, in its lower third, which grades upward into about 3 m of a recrystallized limestone.

*Fossils.—*

A fauna rich in foraminifera: *Nautiloculina* sp. (Fig. 6E), *Pseudocyclammina sulaiyana* Redmond (Fig. 6F), *Pseudocyclammina* sp., *Bramkampella? arabica* Redmond, *Everticyclammina* spp., *Textularia* spp. (Fig. 6G), *Guembelitria* spp., other planktonic and benthonic foraminifera, and abundant brachiopods, gastropods, bivalves, algae, bryozoans, and echinoderms.

*Depositional Setting.—*

This microfacies is correlated to SMF type 9, which is placed either in standard facies belt 2 (open shelf) or 7 (open-platform shelf lagoon). The latter interpretation is more probable because the rock has a very low planktonic/benthonic ratio and contains stenohaline forms including brachiopods and echinoderms. This microfacies consists of abundant fragments of molluscs, bryozoa, brachiopods, echinoderms, and other benthonic remains, whereas only very rare planktonic foraminifers were recorded. It can be further subdivided into five distinctive submicrofacies based on the presence of an intercalating lime mudstone bed in its lower third and a recrystallized limestone bed near the middle.

### *Microfacies A-9: Third Mudstones*

*Lithology.—*

This microfacies consists of 12-m-thick, locally iron-stained lime mudstone with rare foraminifera.

*Fossils.—*

Some *Cyclammina* sp. and *Everticyclammina* sp., together with recrystallized shell fragments.

*Depositional Setting.—*

These sediments accumulated in low-energy, slightly restricted bays and open lagoons on an open platform, and are correlated with SMF Type 17 (Standard Facies Belt 7).

### *Microfacies A-10: Everticyclammina Bioclastic Wackestones*

*Lithology.—*

This microfacies is composed of some 24 m of bioclastic wackestones, with partly recrystallized molluscan and other skeletal fragments as well as abundant benthonic and rare planktonic foraminifers.

*Fossils.—*

*Everticyclammina* spp., *Anchispirocyclina lusitanica* (Egger), *Bramkampella arabica* Redmond, *Pseudocyclammina lituus* Yokoyama, and planktonic foraminifers.

*Depositional Setting.—*

This microfacies correlates to SMF Type 9 and represents an open-platform shelf (standard facies belt 7), as reflected by the more abundant agglutinated benthonic foraminifera and the scarcity of planktonic forms.

### *Microfacies A-11: Fourth Mudstone*

*Lithology.—*

This microfacies consists of some 9 m of lime mudstone with rare, scattered and recrystallized microskeletal fragments.

*Depositional Setting.—*

These sediments were deposited in bays and open lagoons on an open platform and are correlated to SMF Type 17 (Standard Facies Belt 7).

## MICROFACIES OF THE YAMAMA FORMATION (B)

### *Microfacies B-1: Ammobaculites Peloidal Bioclastic Wackestone*

*Lithology.—*

This microfacies is composed of about 9 m of lime mudstone to wackestone that can be divided into three zones: the lower and upper zones consist of peloidal bioclastic wackestones (Fig. 7A), and the middle zone is composed mainly of lime mudstone and wackestone.

*Fossils.—*

*Ammobaculites* sp., *Pseudocyclammina* sp., (Fig. 7B), *Everticyclammina hellen* Redmond, and *Textularia* sp., together with some globotextularids and haplophragmoids.

*Depositional Setting.—*

This microfacies is correlated to SMF type 9, which is placed in either the standard facies belt 2 (open-sea shelf) or 7 (open-platform shelf lagoon). Because of the peloidal nature of the grains an origin in an open-platform shelf lagoon can be assumed. This confirms the occurrence of open-platform shelf lagoons in central Arabia during most of the Early Cretaceous deposition.

### *Microfacies B-2: Lower Mudstone*

*Lithology.—*

This microfacies is composed of about 3 m of lime mudstone with rare recrystallized shell fragments.

*Depositional Setting.—*

Bays and open lagoons on an open platform.

### *Microfacies B-3: Algal, Molluscan, Bioclastic Wackestone*

*Lithology.—*

This microfacies is composed of some 9 m of bioclastic wackestone with recrystallized algal remains, molluscan shell fragments, and together with benthonic and planktonic foraminifera (Fig. 7C).

*Fossils.—*

*Hensonella* sp., bivalves and gastropods, *Cyclammina* sp., *Globuligerina hoterivica* (Fig. 7D).

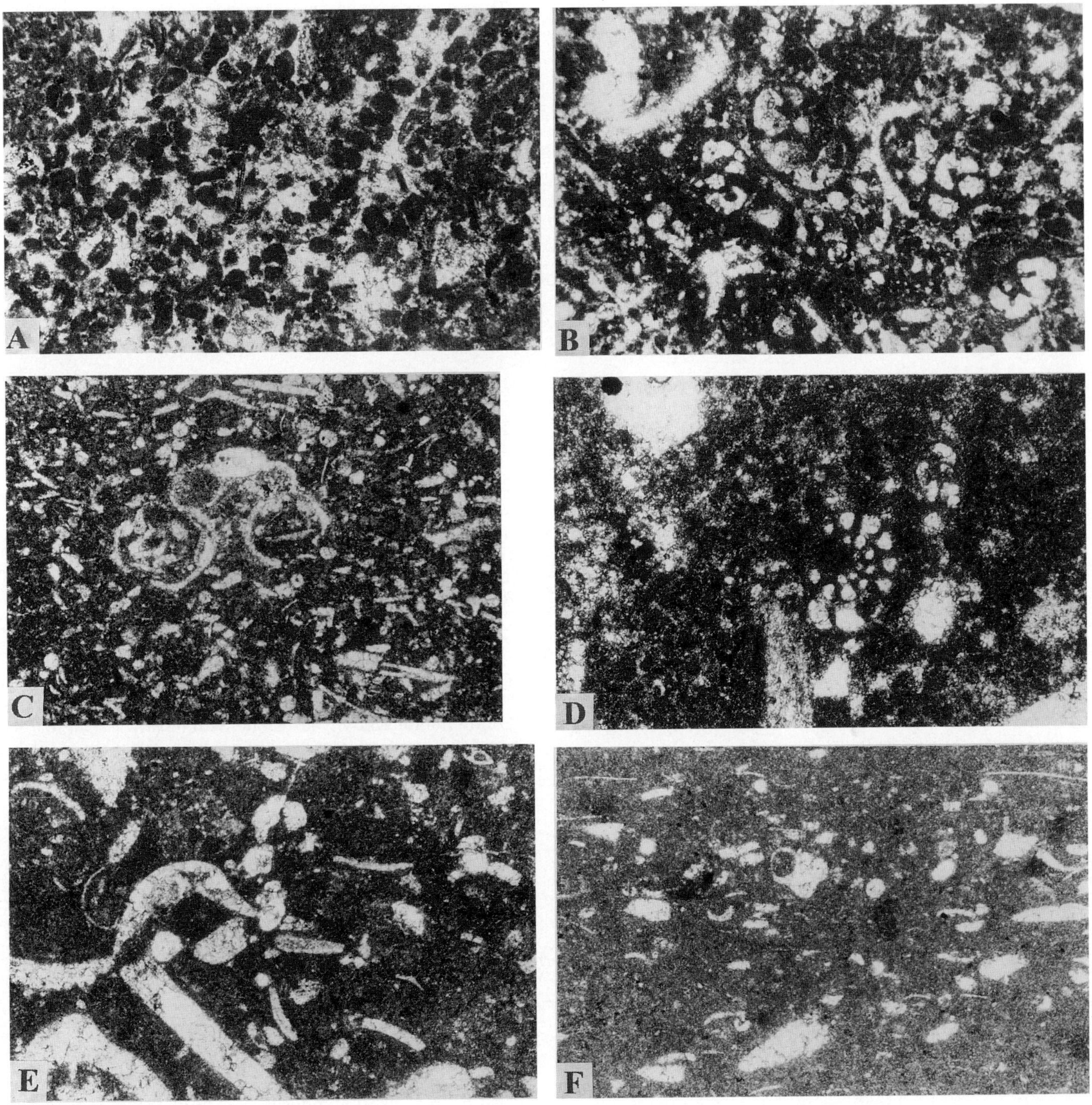


FIG. 7.—Microfacies of Yamama Formation: A) Peloids and bioclast wackestone in microfacies B-1 (x 21); B) peloids, bioclasts, and axial section of *Everticyclammina* spp. ( x 31); C) an axial section of *Globuligerina hoterivica* (Subbotina) (x 82); D) nearly equatorial section of *Nautiloculina oolithica* Mohler (x 31); E) bioclastic wackestone facies with partially recrystallized skeletal fragments (x 21); F) foraminiferan, molluscan, and algal remains in an argillaceous limestone matrix (x 21).

*Depositional Setting.—*

This microfacies is correlated to SMF type 9, and standard facies belt 7, representing an open-platform shelf lagoon.

### *Microfacies B-4: Upper Mudstone*

*Lithology.—*

This microfacies is composed of 3 m of lime mudstone with very rare, recrystallized microskeletal fragments.

*Depositional Setting.—*

This microfacies represents deposition in bays and open lagoons on an open platform.

### *Microfacies B-5: Flabellammina, Nautiloculina Bioclastic Wackestone*

*Lithology.—*

This microfacies is about 18 m thick and constitutes the uppermost part of the Yamama Formation. It changes gradually upward from bioclastic wackestone at the base into mudstones near the top. The fine-grained matrix increases upward.

*Fossils.—*

*Flabellamina* sp., *Cyclammina* sp., "*Nummoloculina*" sp., *Nautiloculina oolithica* Mohler (Fig. 7E), *Trocholina elongata* (Leupold), and planktonic foraminifera, recrystallized ostracods, bryozoans, and echinoderm and molluscan fragments (Fig. 7F).

*Depositional Setting.—*

This microfacies is correlated with SMF type 9, and is interpreted to represent an open-platform shelf lagoon (standard facies belt 7).

## MICROFACIES OF THE BUWAIB FORMATION (C)

### *Microfacies C-1: Cyclamminid Bioclastic Wackestone*

*Lithology.—*

This bioclastic wackestone with a large number of foraminifers is about 2 m in thickness and rests directly above the pre-Buwaib disconformity (Fig. 8A).

*Fossils.—*

*Everticyclammina greigi* Henson (Fig. 8B), *Everticyclammina hensoni* Redmond, *E. contorta* Redmond, *E.* spp., *Ammobaculites?* sp., other benthonic foraminifera remains, and recrystallized skeletal fragments.

*Depositional Setting.—*

This microfacies is correlated to SMF type 9 and was deposited in an open-platform shelf lagoon (standard facies belt 7) as reflected by the bioclastic wackestone composition of the facies and the high percentage of benthonic foraminifers.

### *Microfacies C-2: Sandy Cyclamminid Bioclastic Wackestone*

*Lithology.—*

This microfacies consists of 5 m of bioclastic wackestones with abundant, scattered quartz grains and local iron staining near the base.

*Fossils.—*

Mainly benthonic foraminferids: *Everticyclammina hellen* Redmond (Fig. 8C), *E. greigi* Redmond (Fig. 8D), *Everticyclammina* spp., *Pseudocyclammina* spp., *Ammobaculites?* sp., and orbitolinids.

*Depositional Setting.—*

Both the abundance of quartz grains and the great diversity of the benthonic foraminifera equate this microfacies to SMF type 9 and belt 7 (open-platform shelf lagoon).

### *Microfacies C-3: Lower Argillaceous Mudstone*

*Lithology.—*

This microfacies is about 2 m thick and is composed of argillaceous mudstone with very rare skeletal fragments.

*Depositional Setting.—*

This facies suggests a very shallow environment with minimal wave and current activity such as the landward flanks of bars in a lagoonal region or the lee side of barriers in a broad shallow platform.

### *Microfacies C-4: Lituolid Sandy Wackestone*

*Lithology.—*

Foraminiferal sandy wackestone with mostly subangular to subrounded quartz grains (Fig. 8E), reaching a maximum thickness of 2 m.

*Fossils.—*

*Lituola* sp., and algal and echinodermal remains.

*Depositional Setting.—*

A high-energy, relatively shallow marine setting is suggested, probably located on the inner shelf or contemporaneous bars where most of the current activity was concentrated.

### *Microfacies C-5: Upper Argillaceous Mudstone*

*Lithology.—*

This microfacies is composed of 1 m of argillaceous mudstone (Fig. 8F) with a few scattered microskeletal fragments.

*Depositional Setting.—*

A very shallow-water environment similar to Microfacies C-3 is suggested.

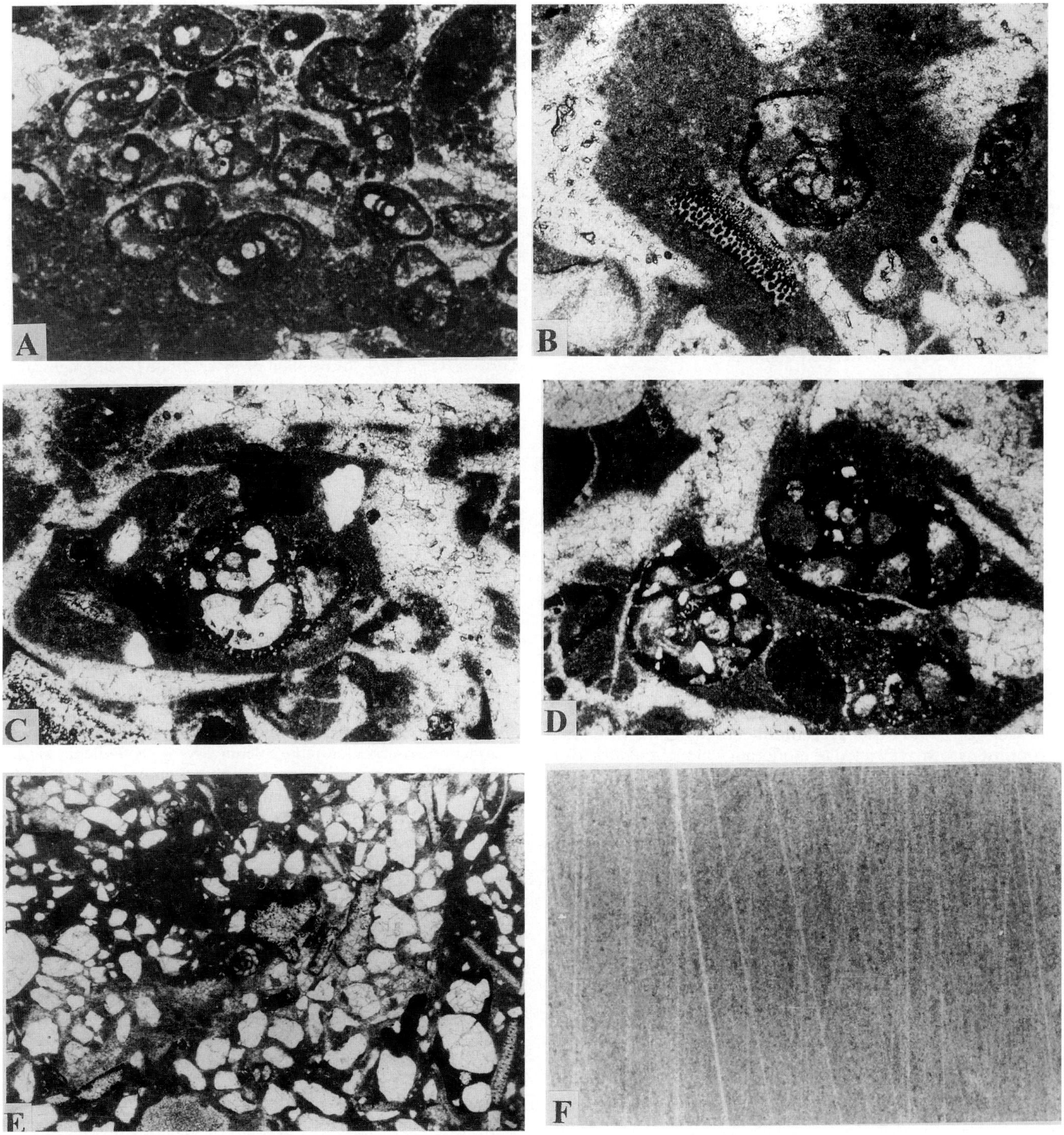


Fig. 8.—Microfacies of Buwaib Formation: A) Random sections of *Cyclammina* sp. and *Everticyclammina* spp. (x 21); B) an equatorial section of *Everticyclammina greigi* (Henson) (x 31); C) an equatorial section of *Everticyclammina hellen* (Redmond) (x 21); D) an equatorial section of *Everticyclammina greigi* (Redmond) (x 31); E) random section in algal and echinoderm remains with abundant quartz grains embedded in an argillaceous limestone matrix (x 21); F) argillaceous lime mudstone (x 21).

### *Microfacies C-6: Cyclamminid Textulariid Bioclastic Wackestone*

*Lithology.—*

This microfacies is composed of about 6 m of bioclastic wackestone recrystallized in the lower and upper parts and iron stained in the middle.

*Fossils.—*

Foraminifera: *Everticyclammina hellen* Redmond, *E. contorta* Redmond, *Cyclammina* sp., *Textularia* spp., *Flabellammina* sp., and some molluscan skeletal remains.

*Depositional Setting.—*

This constitutes the uppermost part of the measured Buwaib section in the Khashm Al Thamamah area, where it underlies the Biyadh Sandstone. It is correlated to SMF type 9, and standard facies belt 7 (open-platform shelf lagoon).

## CONCLUSIONS

1. The Lower Cretaceous strata over the Arabian Peninsula were deposited on an eastward-dipping carbonate ramp characterized by open-shelf carbonate lagoons of central Arabia and by deep shelf-margin conditions in Oman.
2. The Berriasian–Hauterivian rocks of central Saudi Arabia represent a transgressive carbonate phase that began with the deposition of the Sulaiy Limestone during the Berriasian and ended with the deposition of the Buwaib Formation (Hauterivian), with a minor local break (pre-Buwaib disconformity) during the Valanginian or between the Valanginian and the Hauterivian.
3. The Berriasian–Hauterivian section of Central Saudi Arabia contains twenty-two successive microfacies. Paleoecological interpretation of the microfacies indicates that these sediments in Central Saudi Arabia accumulated under marine shelf conditions with restricted to moderate water circulation during the Berriasian. This interpretation is indicated by the peloidal, ooidal grainstones and open-marine fossils which are locally associated with subangular to subrounded quartz grains. Such marine shelf conditions were followed by open-platform shelf lagoons in which the bioclastic wackestones with interbeds lime mudstones accumulated.

## ACKNOWLEDGMENTS

The first author is deeply grateful to Prof. Z. R. El Naggar for his assistance during the field work. The authors would like to thank Prof. C. G. St. C. Kendall, Prof. J. Kuss, Dr. R. W. Jones and Dr. R. W. Scott for reading the manuscript and for offering valuable comments and suggestions which improved the paper.

## REFERENCES

ALSHARHAN, A.S., AND NAIRN, A.E.M., 1986, A review of the Cretaceous formations in the Arabian Peninsula and Gulf; Part 1. Lower Cretaceous (Thamama Group) stratigraphy and paleogeography: Journal of Petroleum Geology, v. 9, p. 365–392.

ALSHARHAN, A.S., AND NAIRN, A.E.M., 1997, Sedimentary Basins and Petroleum Geology of the Middle East: Amsterdam, Elsevier Science, 942 p.

ALSHARHAN, A.S., AND WHITTAKER, J.E., 1991, Redmond's new lituolid foraminifera from the Mesozoic of Saudi Arabia: Micropaleontology, v. 37, p. 41–59.

BANNER, F.T., AND HIGHTON, J., 1990, On *Everticyclammina* Redmond, especially *E. kelleri* (Henson): Journal of Micropalaeontology, v. 9, p. 1–14.

BANNER, F.T., AND WHITTAKER, J.E., 1991, Redmond's new lituolid foraminifera from the Mesozoic of Saudi Arabia: Micropaleontology, v. 37, p. 41–59.

BAYER, H.J., 1984, The Arabian Shield, *in* Jado, A., and Zotl, J.O., eds., Quaternary Period in Saudi Arabia: New York, Springer-Verlag, p. 5–12.

BOZORGNIA, F., AND BANAFTI, S., 1964, Microfacies and micro-organisms of Paleozoic through Tertiary sediments of some parts of Iran: Published by National Iranian Oil Company, Tehran, Iran, 22 p.

FLÜGEL, E., 1982, Microfacies Analysis of Limestone: Berlin, Springer-Verlag, 633 p.

GRASS, B., AND CAROZZI, A.V., 1965, Petrology and Upper Devonian pelletoidal limestones, Arrow Canyon Range, Clark Country, Nevada: Sedimentology, v. 4, p. 197–222.

MAJEWSKE, O.P., 1969, Recognition of Invertebrate Fossil Fragments in Rocks and Thin Sections: Leiden, The Netherlands, E. J. Brill, International Sedimentary Petrology Series XI, 101 p.

MOSHRIF, M.A., 1981, Sedimentation and paleogeography of the Buwaib Formation (Lower Cretaceous) in Central Saudi Arabia: University of Riyadh, College of Science, Journal, v. 12, p. 205–231.

MOSHRIF, M.A., 1980a, Recognition of fluvial environments in the Biyadh–Wasia sandstones (Lower–Middle Cretaceous) as revealed by textural analysis: Journal of Sedimentary Petrology, v. 50, p. 603–612.

MOSHRIF, M.A., 1980b, Cementation, diagenesis and paragenetic sequence in the Biyadh–Wasi sandstones (Lower–Middle Cretaceous) of central Saudi Arabia: University of Kuwait, College of Science, Journal, v. 7, p. 25–261.

MOSHRIF, M.A., AND KELLING, G., 1984, Stratigraphy and sedimentary history of Upper-Lower and Middle Cretaceous rocks, Central Saudi Arabia: Saudi Arabia Deputy Ministry for Mineral Resources, Bulletin, v. 28, p. 1–28.

MURRIS, R.J., 1980, Middle East stratigraphic evolution and oil habitat: American Association of Petroleum Geologists, Bulletin, v. 64, p. 597–618.

POWERS, R.W., 1968, Arabie Saoudite, Lexique Stratigraphic International, vol. III, Asie, Fasc. 10b 1: Geological Congress Commission on Stratigraphy, Centre National del Recherche, Paris, France, 177 p.

POWERS, R.W., RAMIREZ, L.F., REDMOND, C.D., AND ELBERG, E.L., 1966, Geology of the Arabian Peninsula, Sedimentary Geology of Saudi Arabia: U.S. Geological Survey, Professional Paper 560D, 147 p.

READ, J.F., 1985, Carbonate platform facies models: American Association of Petroleum Geologists, Bulletin, v. 69, p. 1–21.

REDMOND, C.D., 1964, Lituoliid foraminifera from the Jurassic and Cretaceous of Saudi Arabia: Micropaleontology, v. 10, p. 405–414.

SAMPO, M., 1969, Microfacies and Microfossils of the Zagros Area, Southwestern Iran (from Pre-Permian to Miocene): Leiden, The Netherlands, E. J. Brill, 102 p.

SHEBL, H.T., 1989, Stratigraphical and microfacies analysis of some Early Cretaceous type sections in Saudi Arabia: unpublished M.Sc. thesis, King Fahd University of Petroleum and Minerals, Dhahran, Saudi Arabia, 116 p.

SHEBL, H.T., AND ALSHARHAN, A.S., 1994, Sedimentary facies and hydrocarbon potential of Berriasian–Hauterivian carbonates in Central Arabia, *in* Simmons, M.D., ed., Micropalaeontology and Hydrocarbon Exploration in the Middle East: London, Chapman & Hall, p. 159–175.

SIMMONS, M.D., AND HART, M.B., 1987, The biostratigraphy and microfacies of the Early Cretaceous to mid-Cretaceous carbonates of Wadi Mi'aidin, Central Oman Mountains, *in* Hart, M.B., ed.,

Micropalaeontology of Carbonate Environments: U.K., Ellis Horwood, p. 176–207.

STEINEKE, M., AND BRAMKAMP, R.A., 1952, Mesozoic rocks of Eastern Saudi Arabia (abstract): American Association of Petroleum Geologists, Bulletin, v. 36, p. 909.

STEINEKE, M., BRAMKAMP, R.A., AND SANDER, N.J., 1958, Stratigraphic relations of Arabian Jurassic oil, *in* Weeks, L.G., ed., Habitat of Oil—a Symposium: American Association of Petroleum Geologists, p. 1294–1329.

WILSON, J.L., 1975, Carbonate Facies in Geologic History: New York, Springer-Verlag, 469 p.

# MICROFACIES AND MICROFABRICS OF MAASTRICHTIAN CARBONATES, NORTHWESTERN OMAN MOUNTAINS, U.A.E.

ABDULRAHMAN S. ALSHARHAN
*Faculty of Science, U.A.E. University, P.O. Box 17551, Al Ain, United Arab Emirates*
ABDELMOHSIN ZIKO
*Faculty of Science, U.A.E. University, P.O. Box 17551, Al Ain, United Arab Emirates*
*Present address: Faculty of Science, Department of Geology, Zagazig University, Egypt*
HISHAM T. SHEBL
*Faculty of Science, U.A.E. University, P.O. Box 17551, Al Ain, United Arab Emirates*
*Present address: Core Laboratories, P.O. Box: 8503, Abu Dhabi, United Arab Emirates*
AND
GREGORY L. WHITTLE
*Faculty of Science, U.A.E. University, P.O. Box 17551, Al Ain, United Arab Emirates*
*Present address: 1201 Elmwood Blvd Park, MS 5112, New Orleans, LA 70123, U.S.A.*

Abstract: The Simsima Formation (lower Upper Maastrichtian) is a shallow-water carbonate that crops out in eastern United Arab Emirates along the northwestern border of the Oman Mountains. Its thickness ranges in this region from 30 to 80 m. From detailed sedimentological and petrographical analysis, various microfabric types and eight sedimentary facies are recognized in the Simsima Formation. These microfacies are arranged according to their frequency and collected into three groups as follows: packstone group (orbitoidal packstone, bioclastic packstone, algal packstone, and planktonic foraminiferal packstone); boundstone group (milleporid, coral, and algae), and dolostone, which consists mainly of dolomitized limestone. The Simsima Formation in this study can be subdivided into two new members, called here the lower member and the upper member, each with a distinct character and considerable vertical and horizontal distribution.

## INTRODUCTION

The Late Cretaceous was one of the most significant periods in the geologic history of the Arabian region. It was a time when major tectonic events affected the Arabian Plate. Subduction at the Zagros geosuture and obduction at the Oman line, together with the consequent tectonic events, left their mark on the region (Murris, 1980). These Late Cretaceous tectonic movements were associated with reactivation of older structures and differential mobilization of ancient sediment deposits.

In the vicinity of the present-day Oman Mountains there is evidence for an oceanic fore-deep basin, which existed during Campanian time. This was filled by the Hawasina sediments and covered by emplacement of the Semail ophiolite (allochthonous complex) (Alsharhan and Nairn, 1990, 1997; Noweir et al., 1998). The ophiolite nappes are overlain, with a marked regional unconformity, by siliciclastic sediments deposited in continental-margin to shallow-marine settings. These, in turn, are overlain with local minor unconformity by the shallow-marine carbonates of the Simsima Formation (Fig. 1). The Simsima Formation in this region is uncomformably overlain by the Muthaymimah Formation (Nolan et al., 1990 and Skelton et al., 1990). Though faulted at many localites, this unconformity is visible on Jebel Thanais (Skelton et al., 1990). The Muthaymimah was originally dated as Paleocene to Eocene by Nolan et al. (1990), but in Sayh Muthaymimah (SE of Buraymi) it ranges from the latest Maastrichtian to the Paleocene. From planktonic foraminifers, they recognize the succession of zones across the Maastrichtian–Danian boundary in its basal part. Both the allochthonous complex and the autochthonous strata (Qahlah and Simsima formations) are exposed along the Oman Mountains as a result of post-Miocene uplift. Other autochthonous strata crop out in peripheral jebels around the mountain chain (Glennie et al., 1974; Skelton et al., 1990; Alsharhan and Nasr, 1996; Smith et al., 1995a).

As a consequence of the tectonic history, the stratigraphic record shows a great diversity in its sedimentologic and paleogeographic patterns and in its fossil content. The regional Wasia / Aruma unconformity separates uppermost Cretaceous rocks from middle Cretaceous strata (Glennie et al., 1974; Harris et al., 1984; Alsharhan and Nairn, 1986, 1993; Scott, 1990). The magnitude of this unconformity varies from place to place throughout the Arabian Peninsula. In the eastern part of the United Arab Emirates, this unconformity represents the early Turonian to the early Campanian (Alsharhan and Nairn, 1988).

This study focuses on the Maastrichtian carbonates of the Simsima Formation that crop out in the northwestern Oman Mountains (United Arab Emirates territory) (Fig. 2A). Sections from various localities in the Al Faiyah region were sampled and measured (Fig. 2B). Both hand specimens and thin sections were described to determine the lithological variations, microfacies, microfabrics, and carbonate petrography, which could help with diagenetic history and the interpretation of the paleoenvironment of the Simsima Formation at these localities. Such analyses can be used in regional correlation of the studied sections to determine their age and make a regional paleoenvironmental interpretation of the Maastrichtian carbonates in NW Oman Mountains. Our study incorporates previous studies by Skelton et al. (1990), who provided the first detailed sedimentological and facies analysis and published lithostratigraphic sections and faunal lists (especially rudist bivalve faunas). Smith et al. (1995a) studied the fossils of this region and concluded that combining the evidence from ammonites, inoceramid bivalves, echinoids, and global sea-level curves most of the Simsima Formation appears to be in the lower Upper Maastrichtian *Anapachydiscus fresvillensis* Zone or higher, with which we agree.

Middle East Models of Jurassic/Cretaceous Carbonate Systems
SEPM Special Publication No. 69, Copyright © 2000
SEPM (Society for Sedimentary Geology), ISBN 1-56576-075-1, p. 129–142.

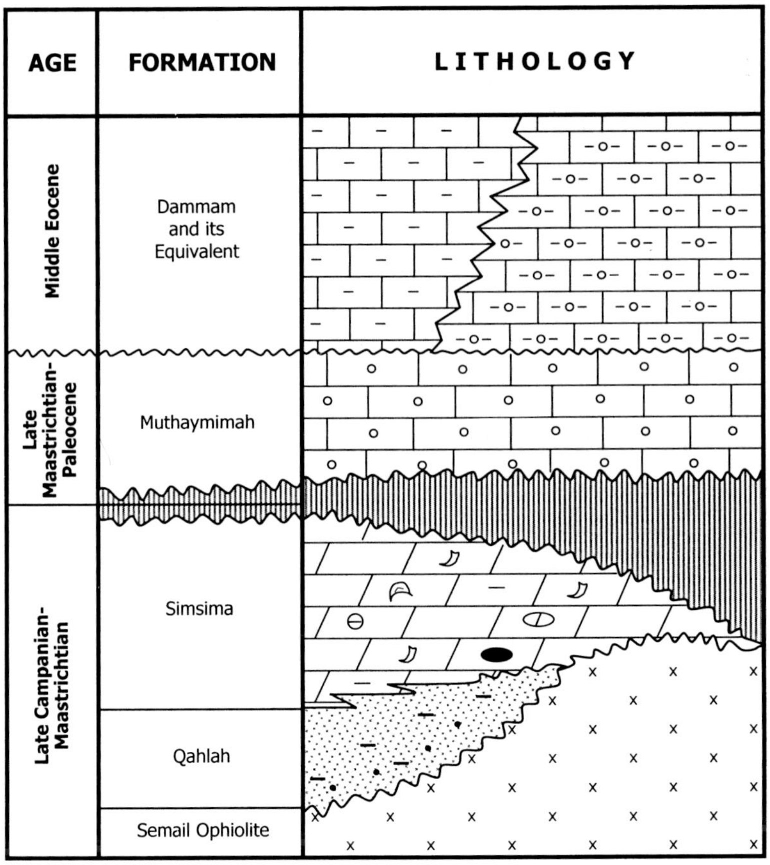


FIG. 1.—Generalized Late Cretaceous–early Tertiary lithostratigraphy in northwestern Oman.

Smith et al. (1995b) established the paleoenvironmental setting of Simsima sediments and concluded that the succession was deposited during a single sea-level cycle and that the basal transgression preserves beach hardground faunas and subtidal shoreface sands. Shallow-water offshore sediments deposited below storm wave base make up the majority of the lower beds. The upper beds are composed of semi-reclining rudists and well-sorted calcarenites deposited on stable shoals above active wave base. Although we support most of their paleoenvironmental conclusions, we disagree with them in some minor points, and our views are presented in the lithostratigraphic section below. The taxonomy of rudists by Morris and Skelton (1995), ammonites by Kennedy (1995), nautiloids and inoceramids by Morris (1995a, 1995b) and echinoids by Smith (1995) have been documented in detail in this area and are not repeated in this study.

## LITHOSTRATIGRAPHY OF THE SIMSIMA FORMATION

The term Simsima was originally introduced by Sugden (in an unpublished report cited in Sugden and Standring, 1975) from the Dukhan-28 well, in western onshore Qatar (lat. 25°24'N; long. 50°45'46"E). These authors designated the Simsima type section in the Dukhan-55 well, between the depths of 353 and 502 m. The first published definition of the Simsima Formation outside this type area was given by Glennie et al. (1974), who erected an outcrop reference section in Oman.

Nolan et al. (1990) questioned the validity of the name Simsima. Although the name was used in unpublished reports by several oil company geologists beginning in 1956, it was first published and validated by Glennie et al. (1974), and a type section was designated. A second section was designated as the type section by Sugden and Standring (1975), who were among the original users of the name. The fact that this section is in a well and that it was not the first well in which Simsima was applied does not invalidate the name. Furthermore, although the Simsima can be correlated with the Aruma Limestone in Saudi Arabia and with the Tayarat Formation in southern Iraq, which were named earlier than the Simsima, Simsima in the U.A.E. is still a valid name that can be used without question. This definition was accepted by Alsharhan and Nairn (1990), who selected a reference section of the Simsima Formation in the Shah Field of onshore Abu Dhabi, the only field to produce from this formation to date. This definition of the Simsima is followed here. We are in strong support of the use of the term Simsima because it is well defined in the subsurface (see Sugden and Standring, 1975; Alsharhan and Nairn, 1990; Alsharhan, 1995), has a wide extent, and is the important Upper Cretaceous producing formation in the United Arab Emirates.

The reference outcrop section of the Simsima Formation in the Al Faiyah region in northwestern Oman Mountains used in this study was also used by Nolan et al. (1990). The Simsima can be subdivided into two informal members of distinct character and considerable vertical and horizontal extent. The following is a detailed description of these members.

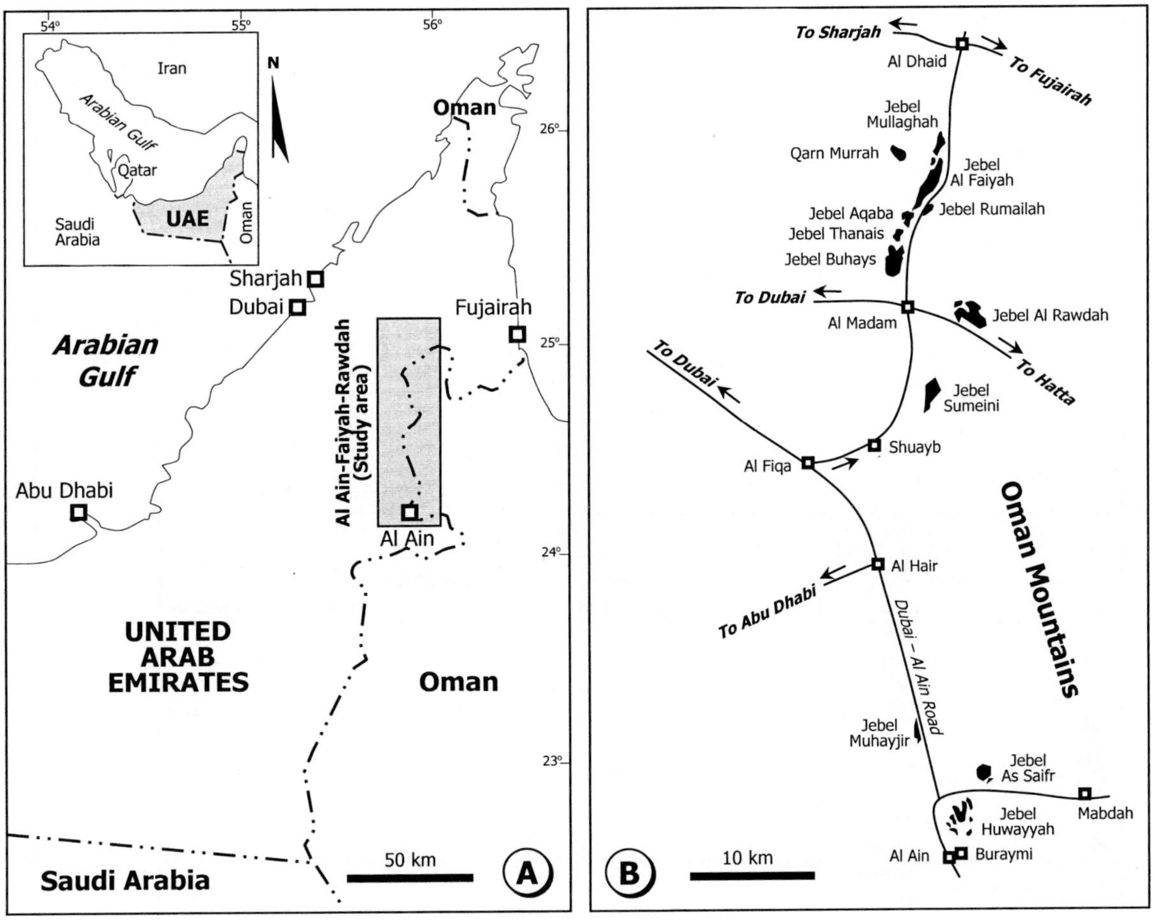


FIG. 2.—A) Location map of the study area in northwestern Oman Mountains in the United Arab Emirates. B) Detailed map of inset area showing locations of different Jebels mentioned in the text.

## *Lower Member*

*Type Reference:* Jebel Buhays (in Al Faiyah region), lat. 25°02'N, long. 55°47"E (Fig. 3), 4 km north of Al Madam village, U.A.E., Dhayed sheet 1:100,000, NG-40-107.

*Thickness:* About 34 m.

*Geographic Distribution:* All over the Jebel Al Faiyah anticlines from Jebel Mullaghah in its northern outcrops to southward through Jebel Al Faiyah, Thanais, Aqaba, Rumailah, and Buhays (the type locality). Further occurrences are recorded from Jebel Al Rawdah (Oman).

*Lithology:* Orange yellow, moderately hard algal, orbitoidal limestone, with a rich rudist fauna (Fig. 4A), either fragmented or disarticulated, or in life position as extended banks. It is very rich in macrofossils.

*Fossil Content:* This member is characterized by the presence of *Dictyoptychus morgani* and *Durania* sp. (Fig. 4B, C, D). It also includes other radiolitid rudists, but these are less abundant than the dictyoptychids. This member includes both the *Dictyoptychus* and the *Durania* facies of Skelton et al. (1990).

The following fauna is found in the type reference: calcareous worms encrusting the echinoids; bivalves; and gastropods. A very similar fauna is recorded at Jebel Al Rawdah and other areas reported by Ali (1989), Skelton et al. (1990), and Metwalli (1992, 1993).

1. Rudists: *Dictyoptychus morgani* (Douvillé), *Durania* sp., and *Hippurites* sp.
2. Other bivalves: *Spondylus* sp., *Mytilus* sp., *Neithea* sp., *Agerostrea* sp., *Pterotrigonia* sp., and many others.
3. Gastropods: *Actaeonella* sp. (Fig. 4F), *Trochus* sp., *Keilostomata* sp., *Architectonica* sp., *Aporrhais* sp., *Strombus* sp., among others (most gastropods are badly preserved internal molds).
4. Scleractinian corals: Few colonial (Fig. 4E, 4H) and solitary, e.g., *Cyclolites* sp.
5. Calcareous sponges.
6. Foraminifera: *Orbitoides media* (d'Archiac) (Fig. 4G), *Omphalocyclus macroporus* (Lamarck) (Fig. 5A), *Lepidorbitoides* sp., L. *minor* (Schlumberger) (Fig. 5B), *Siderolites* sp. (Fig. 5C), *Loftusia minor* (Cox).
7. Echinoids: many regular and irregular echinoids such as *Faujasia eccentripora* (Lees), *Hemipeneustes arabicus* (Ali), and *Noetlingaster emiratescus* (Ali).
8. Algae: abundant rhodolithic algae and green algae (Fig. 5D).

*Boundaries*: The Lower Member lies unconformably on the Qahlah Formation, and its boundary is sharp and planar (Fig. 5 F, G). The upper boundary in contrast is gradational and is based on the presence of a distinct local erosional surface and decrease in *Dictyoptychus.*

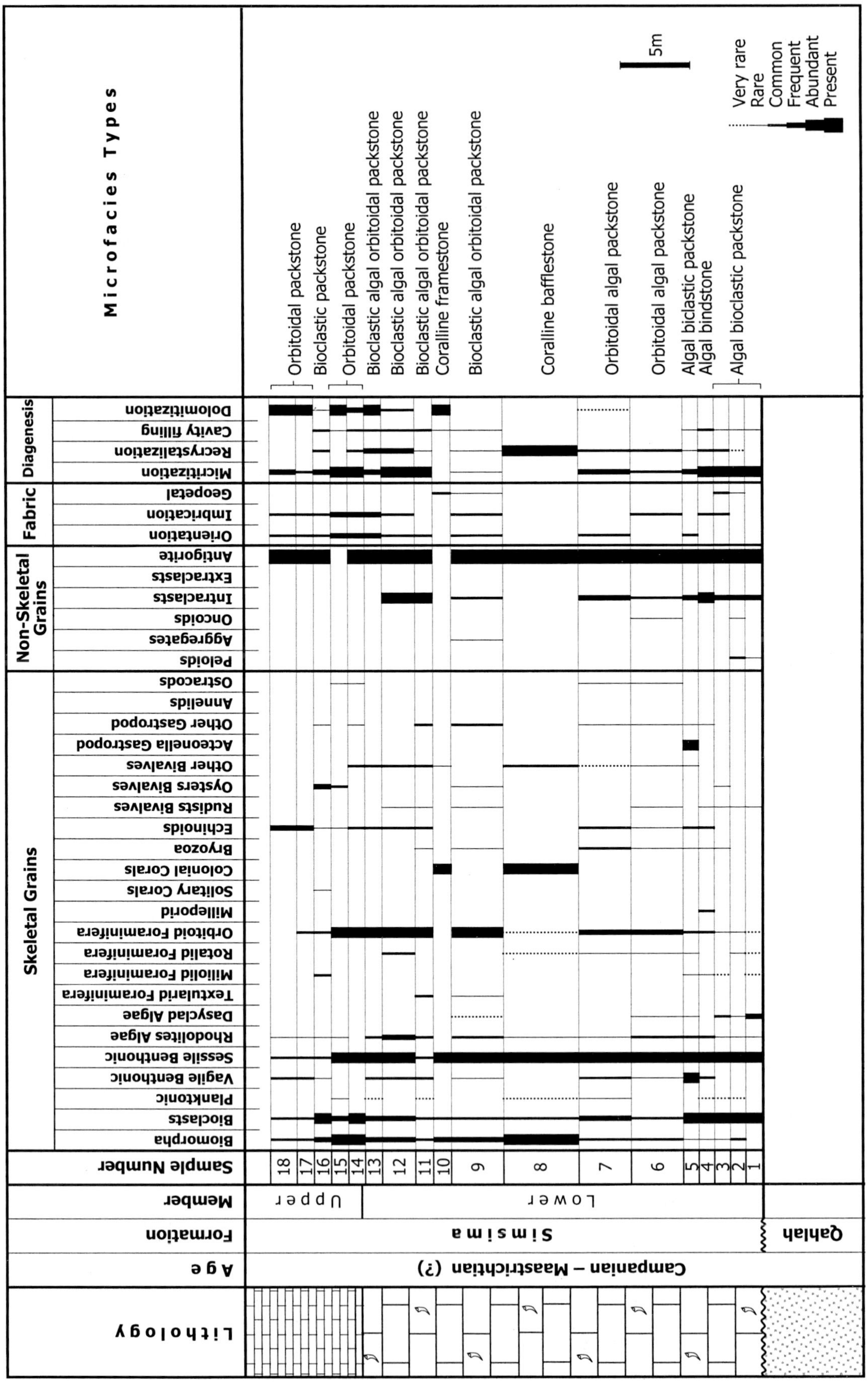


FIG. 3.—Lithostratigraphy and facies distribution of the Simsima Formation at Jebel Buhays al Faiyah in northwestern Oman Mountains.

*Sedimentary Environment:* The Lower Member was deposited in a shallow, open marine setting with normal marine salinities and moderate to high energy. This conclusion is based on the following criteria.

1. The dominant dictyoptychids and occasional radiolitid rudists *(Durania)* (Fig. 4B) grew in abundance in a relatively shallow marine environment. The radiolitid rudists were more common at shelf margins and in foreslopes, where they were associated with and were gradually replaced by corals, spongiomorphic hydrozoans, red algae, and sponges; all biota inhabited water of normal marine salinity. Comparable conditions are thought to have existed in the study area, where scleractinian colonial mounds have replaced the radiolitid rudists (in the Jebel Al Faiyah section).
The Lower member is comparable with facies of shelf-margin deposits. The position within this wide facies zone is controlled by the paleorelief, which varied irregularly and changed rapidly from one place to another. In areas of deeper water, *Dictyoptychus morgani, Plagioptychus* sp., and radiolitids accumulated. In shallower areas, bivalves and gastropods were the dominant fauna associated with large foraminifera (orbitoids) and red algae.
2. The lower member is composed of bioclastic packstones containing worn grains and rare to very rare planktonic foraminifera that confirm open-marine depositional conditions.

### *Upper Member*

*Reference Section:* It is chosen at Jebel Al Faiyah, lat. 25°04'N, long. 55°49'E (Fig. 6), 8 km N of Al Madam village, U.A.E., Dhayed sheet 1:100,000, NG-40-107.

*Thickness:* About 60 m.

*Geographic Distribution:* It is well exposed in the Jebel Al Faiyah area, at Jebel Al Rawdah, and southward of the Al Ain-Buraymi area (Fig. 2B).

*Lithology:* It is composed of thick, well-bedded, highly fossiliferous limestone. The limestone is nodular, bioturbated at the base, cross-bedded in the middle, and bioturbated, thin-bedded and dolomitic in its upper parts. *Orbitoides* are the most abundant microfossils. The macrofossils decrease rapidly upward.

*Fossil Content:* Except for the absence of rudistid banks, few scattered, disarticulated *Durania* sp. are recorded within the lowermost 5 m of the member. Macrofossils dominate the lower part of the member and include clusters of *Spondylus* sp. The well cemented character of the member prevents collection of well preserved macrofossils. However, these latter are present in some diversity, especially in the lower part of the member.

*Boundaries:* The lower boundary of the Upper Member is based on gradational contact with the Lower Member, while in this region the upper boundary can be easily separated from the Muthaymimah carbonate above by an angular unconformity.

*Sedimentary Environment:* The Upper Member with its gradational contact with the underlying Lower Member represents a gradual transgressive phase of the Simsima Sea in the study area. This phase reached its maximum extent in Al Rawdah area, where the orbitoidal packstones change gradually upward to planktonic foraminiferal packstones. The upper member is interpreted to have accumulated in an open-shallow-shelf environment, below wave base with moderate currents and normal salinity, and in the photic zone without (or with very rare) terrigenous influx. This conclusion is based on the following criteria.

1. The orbitoidal packstone (partly grainstone) is the most dominant type. It is associated with the following types: bioclastic packstone, algal bioclastic packstone, and peloidal bioclastic packstone. These microfacies types consist mainly of bioclasts (no micrite), and most *Orbitoides* sp. (or any other elongated skeletal grains) are oriented or even imbricated.
2. The gradual upward decrease of algae and the gradual increase of planktonic foraminifera.
3. The absence of interbedded terrigenous influx.
4. The presence of lamination and cross-bedding in the lower half of the member.

## SEDIMENTARY FACIES ANALYSIS

Eighty-five thin sections from the Simsima Formation were used for facies analysis. They are described systematically using the check list proposed by Flügel (1982) (see Figs. 3 and 6). The system of nomenclature and classification is after Dunham (1962) and Embry and Klovan (1971).

The lithology is consistent throughout the various areas sampled. Grainstone and packstone dominates, and wackestones are rare. Bioclastic grains predominate except in a few samples in which oolites and peloids are subordinate. The peloids may be highly micritized skeletal fragments (e.g., echinoid spines), but the lack of internal fabric leaves open the possibility of a nonskeletal origin. The matrix consists most commonly of fine to medium calcite spar; micrite mud makes up the matrix of some packstones.

The Simsima Formation contains a wide range of fauna. *Orbitoides* sp. and rudist and scleractinian corals are the most abundant skeletal grains. Echinoderm platelets and spines, dasycladacean green algae, and a number of foraminifera, including miliolids and textularids are all common. Most facies are poorly sorted; larger grains, such as coral and brachiopod fragments, *Discocyclina* sp. tests, large echinoderm platelets, bryozoans, and red algae were commonly mixed with the small planktonic foraminifera and bioclasts.

## MICROFABRIC TYPES

*Skeletal Grains*: These grains show the greatest variation and high diversity; both biomorphs and bioclasts are well represented. The skeletal grain assemblages are mainly polymictic, but monomictic groups also occur, including *Orbitoides* and rhodolites. The rhodoliths are the most abundant algal group; the dasycladacean algae are either secondary or, in a few cases, are primary rock constituents. The skeletal grains are dominated by the benthonic sessile groups; the vagile groups; such as regular echinoids and gastropods, are less common. Planktonic foraminifera are present in negligible percentages.

The roles of the skeletal grains can be arranged as follows, in order of decreasing importance: essential rock-building constituents (most systematic groups), sediment binders (algal boundstone), sediment bafflers (milleporid and/or coralline bafflestone), and framework builders (colonial corals framestone). In a few cases biogenic encrustation of bioclasts of bivalves is present. The main encrusting taxa are membraniporiform bryozoans, red algae, and encrusting forams. Most skeletal grains are enveloped by thin micritic rims. Aragonitic skeletal grains, such as molluscs, scleractinian corals, and millepores, are usually recrystallized. The original microstructure of the oysters is normally preserved, but partial recrystallization is common. The polymictic nature of the skeletal grains are in most cases the reason for their poorly sorted state (from annelids to rudist debris). Rounded bioclasts are very rare. The following is the systematic grouping of the skeletal grains, arranged according to their frequencies: *Orbitoides* sp., rhodoliths, rudists and other bivalves, gastropods, miliolids, rotalids, textularids, dasycladacean algae, corals, bryozoans, ostracodes, and annelids.

*Peloids*: These occur generally in small quantities. Both algal peloids and fecal pellets are present. They are mostly sorted, oval or subrounded, scattered, and composed mainly of homogeneous micrite. The peloids are usually associated with rhodoliths.

*Aggregate Grains*: These are entirely lumps, agglutinated particles mostly of bioclasts and rarely extraclasts. The agglutinated particles are either truncated or not, are enclosed by a sparitic cement, and are dominant only in the orbitoidal bioclastic microfacies. Where they are present, the aggregate grains constitute less than 10% of the total particles.

*Oncoids*: These occur as a tiny percentage in a very few of the samples. They are represented by algal oncoids (rhodoliths) embedded in sparite. They have a deformed, irregular shape. They are up to 10 mm in size where they have a regular structure. The laminae of the pisolites are crinkled. The nuclei correspond to particles of the associated materials.

*Cortoids*: Many bioclasts have micrite envelopes. In some cases the whole particle is micritized. These ovate grains, or cortoids, are always found in association with rhodoliths, with highly diverse faunal assemblage. They show no indication of transportation. Locally echinoid fragments have micritic envelopes.

*Lithoclasts*: Both intraclasts and extraclasts are present, and where this mixture occurs the former are dominant. In a few cases the clasts are truncated along their edges. The components that form the lithoclasts are of the same material as the associated particles and have the same color. Few features have been recognized. Both the lithoclasts and the other particles have similar cement. They are characterized by an irregular nodular, rounded shape, reminiscent of grapestones. No parallel orientation or cross-bedding can be recognized. These grains are common throughout the studied sections.

*Terrigenous Particles*: These particles are of ophiolitic origin and are concentrated in the lower parts of the Simsima, especially near the contacts with the underlying Qahlah Formation or Semail Ophiolite. These particles are mainly composed of antigorite and are ill-sorted, disoriented, subrounded to subangular, and disseminated. A few small, sorted, subangular, disoriented, and disseminated chromite particles have been recognized throughout the studied section and are common especially in the upper parts.

*Fabric*: The particles in almost all studied sections are grain-supported. The contact types between the particles include point and tangential. Orientation, imbrication, and geopetal structures are common in the studied sections.

*Carbonate Cement*: Cements include a blocky cement, of which several generations can be recognized. A micritic cement is common, and cements in optical continuity with the echinoid fragments are rare. The cement between the particles does not correspond to the cement within the particles. There is a sharp contact between the cements and the particles. The crystal size of the cements increases away from the walls of the original cavities being filled.

## MICROFACIES TYPES

In general the Simsima Formation in the studied sections is dominated by three groups of carbonate: packstone, boundstone, and dolostone. Each group can be subdivided on the basis of the frequency of the major carbonate particle components, except for the dolostone group.

### *A. Packstone Group*

This group occurs in about 90% of the studied samples and comprises predominantly *Orbitoides* sp., miliolids, planktonic foraminifera, coralline algae, and echinoderms. Also present are bryozoans, brachiopods, and nonskeletal components, such as fecal pellets and peloids.

1. *Orbitoidal Packstones Microfacies:* The orbitoidal packstone microfacies is the dominant microfacies in the Simsima Formation and occurs in Jebels Buhays, Thanais, Aqaba, Al Faiyah, and Al Rawdah. It constitutes 45% of the rocks studied in thin section. The most abundant skeletal grains and/or carbonate grains are used to divide the packstones into the following microfacies subtypes: bioclastic orbitoidal packstones (Fig. 7A), bioclastic algal orbitoidal packstones (Fig. 7B), orbitoidal packstones (Fig. 7C), algal orbitoidal packstones (Fig. 7D), and peloidal bioclastic orbitoidal packstones (Fig. 7E).

2. *Bioclastic Packstones Microfacies:* The bioclastic packstone microfacies is the second most abundant microfacies type, constituting 38% of the thin sections studied in Jebels Buhays, Thanais, Aqaba, Al Faiyah, and Al Rawdah. The following microfacies subtypes can be distinguished: algal bioclastic packstones (Fig. 7F), peloidal bioclastic packstones (Fig. 7G), algal orbitoidal bioclastic packstones (Fig. 7H), and algal peloidal bioclastic packstones (Fig.7I).

3. *Algal Packstones Microfacies:* Packstone with abundant algal remains constitutes 4% of the studied thin sections and occurs in Jebels Buhays and Aqaba only. It includes the following two subtypes: orbitoidal algal packstones (Fig. 8A) and bioclastic algal packstones (Fig. 8B).

4. *Microporous Planktonic Foraminiferal Bioclastic Packstone:* Four percent of the studied thin sections can be classified as planktonic foraminiferal bioclastic packstone microfacies (Fig. 8C–H). This facies type is recorded only from the uppermost part of the Simsima Formation at the Jebel Al Rawdah. *Orbitoides biomorpha*, foraminiferal bioclasts, and echinoid bioclasts are associated with it.

### *B. Boundstone Group*

The boundstone type constitutes 7% of the studied thin sections and is recorded in Jebels Buhays, Aqaba, and Al Faiyah. It is of limited occurrence and can be considered as a lateral and local replacement of the packstone group. Scleractinian corals, oncolites (rhodoliths), and rudists are the main constituents of

FIG. 4. (opposite page)—A) *Durania* sp. in the Lower Member at Jebel Aqaba section. B) *Durania* sp. in association with rudists in the Lower Member of the Simsima Formation at Jebel Aqaba Section. C) Bioclastic packstone with large rudistid fragment (width of field about 45 mm). D) Monopleurid rudists within *Dictyoptychus* sp. in the Lower Member at Jebel Aqaba section. E) Coralline colonies in the Lower Member at Jebel Aqaba. F) Nearly axial section of *Actaeonella* sp. (x 1.7) at Jebel Aqaba. G) *Orbitoides media* (D'Archaic) (x 20) at Jebel Aqaba. H) Coralline fragments (x 1.7) at Jebel Buhays.

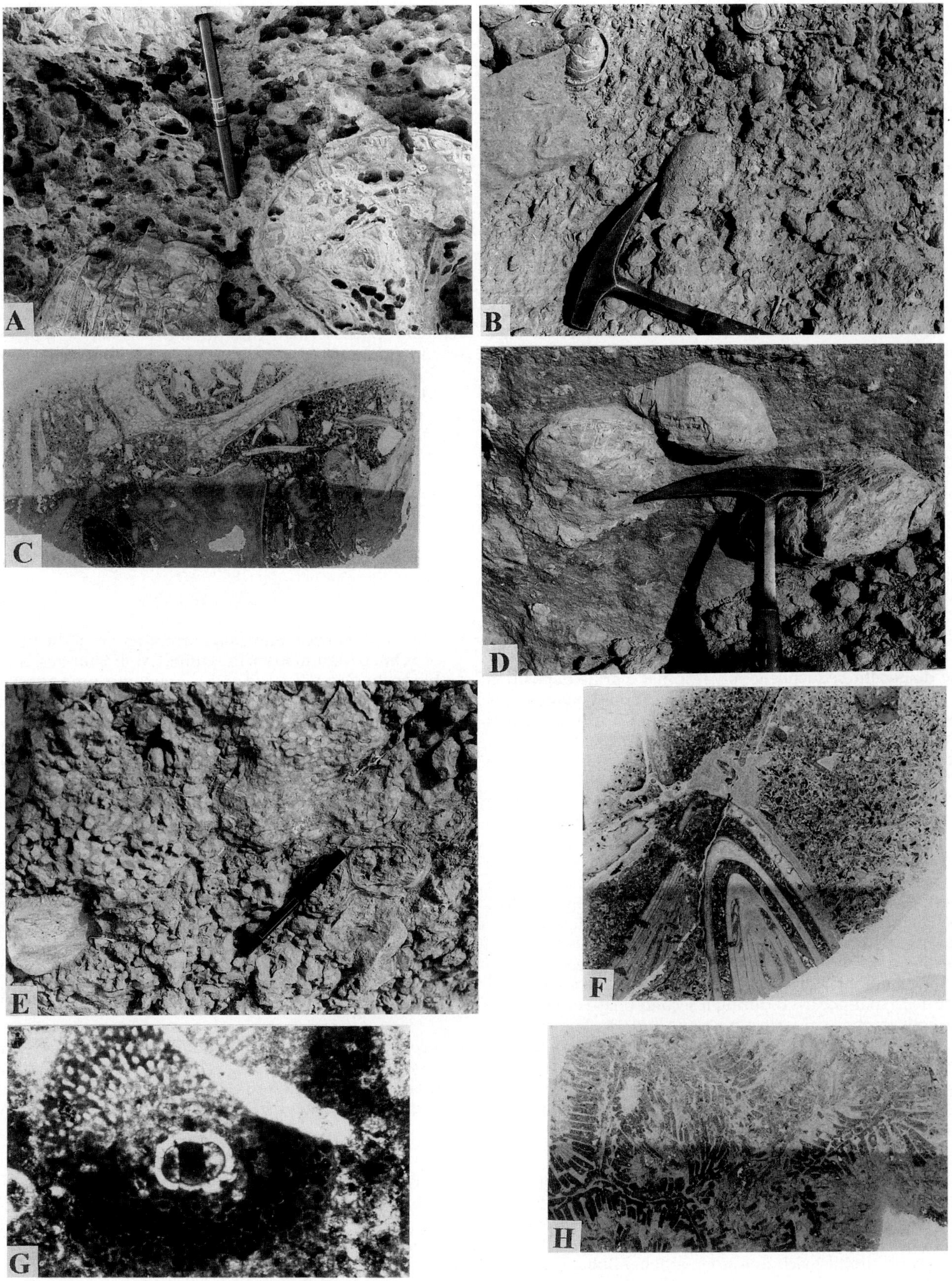
A
B
C
D
E
F
G
H

these boundstones. Milleporids are also present but are subordinate. The following microfacies types can be distinguished.

1. *Milleporid:* Milleporid hydrozoans are one of the important constituents of recent and old reefs. Few large milleporid bioclasts are recorded from the basal part of the Lower Member. Its reticulate structure as well as the presence of large and small pores (gastopores and dachtylopores, respectively) confirms its identification.

2. *Coralline:* Fragments of scleractinian corals are recognized in the field, as in Jebel Aqaba, and in thin sections, as in the lower part of Jebel Buhays section. Dendroid scleractinian corals also occur and act as sediments bafflers.

3. *Algae:* The frequent occurrence of rhodoliths in the basal part of the Lower Member presents a good example of the algal boundstone microfacies type.

### C. Dolostone (Dolosparite) Group

The dolostones have a very limited occurrence (1%) and are very patchy in their distribution. Some samples were completely dolomitized by subhedral to euhedral, zoned rhombic dolomite (Fig. 8I). There are no clues as to their precursor. Moreover, few packstones are partially dolomitized. Dedolomitization has occurred, and neomorphic calcite crystals may cut across the dolomite crystal boundaries. Within the dolomite crystals limonite may parallel growth lines, or in some places limonite infills secondary porosity associated with the dolomite.

## DIAGENETIC HISTORY

The diagenetic sequence involves initial marine deposition followed by marine cementation, evidenced by fibrous cements encrusting echinoid spines and the interiors of foraminiferal chambers. Micritization, which has obliterated much of the original fabric of many skeletal grains, and the formation of micrite envelopes also occurred at this time (Friedman, 1964; Bathurst, 1964, 1966; Friedman et al., 1971). Recrystallization of marine cements to fine- to medium-grained calcite spar is attributed to meteoric diagenesis (James and Choquette, 1990), which has occluded some of the original porosity. However, subsequent leaching of bioclasts and matrix spar has created good moldic and intraparticle porosity, which is now commonly cement-filled. Coarse, equant calcite spar precipitated within moldic and interparticle pores during early burial diagenesis.

Reexposure of the rocks at the surface has resulted in a suite of late diagenetic changes. Partial dissolution of the void-filling coarse calcite spar occurred during a second phase of leaching. The partially leached voids do not contain hydrocarbons and are thus inferred either to postdate hydrocarbon emplacement or to have been outside of migration paths.

Precipitation of coarse calcite spar was followed by local formation of rhombic, zoned dolomite within the limestone matrix and is embedded within coarse calcite spar. The rhombs occurred scattered throughout the matrix of muddy packstones and clustered within the matrix of sparry grainstones. Those embedded within the coarse calcite spar void-fill occurred as isolated single crystals. Rare unzoned rhombic dolomite was also observed in association with zoned rhombic dolomite.

Iron staining and hematite formation postdate emplacement of hydrocarbons and are probably related to a period at or near the groundwater table. Rare dedolomitization also occurred during a second phase of meteoric diagenesis, because few of the zoned rhombs were shown to be calcitic through staining of petrographic thin sections with Alizarin Red-S.

Finally, local chertification is in the form of megaquartz and chalcedonic laths in many samples, and in rare cases quartz silt fills matrix vugs. Chertification occurred mainly in compaction and fractures and preferentially within the intraparticle pores of echinoid spines. Evidence for the late diagenetic timing for chert was found where chert fills open seams and voids created by leaching of coarse calcite spar.

Moldic and interparticle pores are the most important porosity types in the Simsima. Subordinate intraparticle porosity also occurs where chambers of principally foraminiferal and bryozoan tests are not filled by calcite spar. Cementation by coarse calcite spar during early burial diagenesis reduced porosity in some samples. Leaching of coarse calcite spar from pores and from many fractures, and pressure solution seams have created additional porosity, although this occurred after hydrocarbon migration and entrapment.

## CONCLUSIONS

1. The Maastrichtian Simsima Formation, from the northwestern Oman Mountains of the United Arab Emirates, consists of bioclastic packtones and grainstones with minor wackestones exhibiting great species diversity and representing shallow, open marine depositional conditions. The limestones are rich in foraminifera (e.g., *Orbitoides* sp. and miliolids) as well as rudists, scleractinia, and echinoderm platelets and spines. A range of other planktonic foraminifera, brachiopods, bryozoans, red algae, dasycladacian green algae, and bivalve fragments also occur.

2. Initial deposition in a shoal-water, open marine setting was followed by marine cementation and micritization. Recrystallization of marine cements led to fine- to medium-grained calcite spar, which occluded much of the original porosity. However, leaching of bioclasts and some of the sparry matrix created good moldic and interparticle porosity. Cementation by coarse, equant calcite spar within moldic and interparticle was also found.

3. Reexposure of the rocks at the surface has resulted in numerous late diagenetic changes: a) a second phase of leaching that has removed a significant proportion of the void-filling spar; b) formation of zoned dolomite has occurred locally,

FIG. 5. (opposite page)—A) *Omphalocyclus macroporus* (Lamarck) (x 20). B) *Lepidorbitoides* sp. cf. *L. minor* (Schlumberger) and recrystallized cavities of *Acteonella* sp. at the Lower Member of the Simsima Formation (x 20) at Jebel Buhays. C) *Siderolites* sp. (x 20) and *Clymphalocylus macroporus* (Lamarck). D) Dasycladacean algae in bioclastic peloidal packstone at Jebel Buhays (x 20). E) The lowermost part of the Lower Member, with *Actaeonella* sp. with its longitudinal axis parallel to this bedding plane in algal (rhodolith)-rich limestone at Jebel Buhays. F) The Lower Member (1) and the basal part of the Upper Member (2), Simsima Formation at Jebel Buhays. Note the presence of two *Dictyoptychus* in the Lower Member. G) The contact between the Simsima (S) and Qahlah (Q) Formations at Jebel Buhays. Note the sharp linear contact and *Acteonella* in algal (rhodolith) limestone in the lower part of the Lower Member.

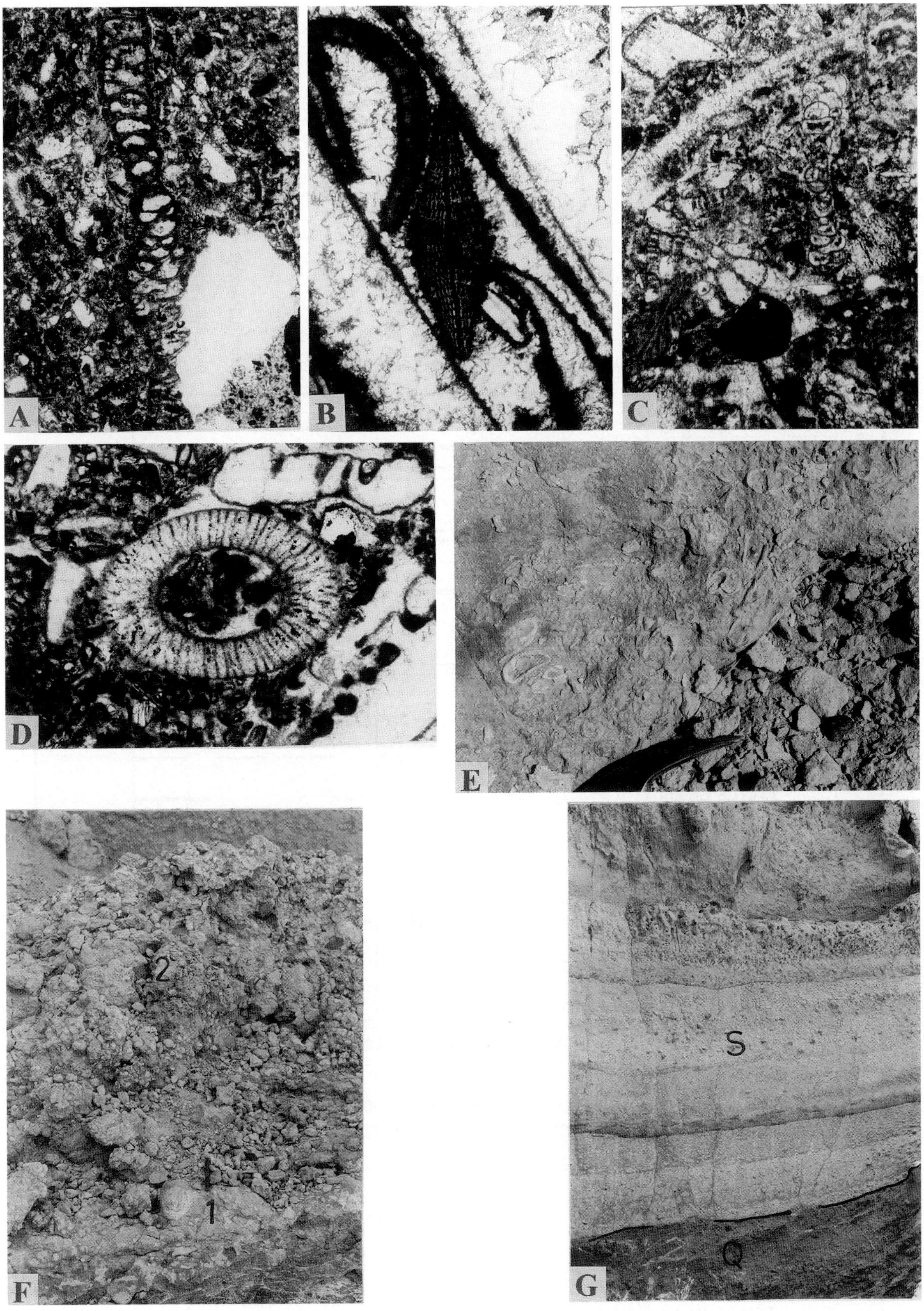
A
B
C
D
E
2
1
F
S
Q
G

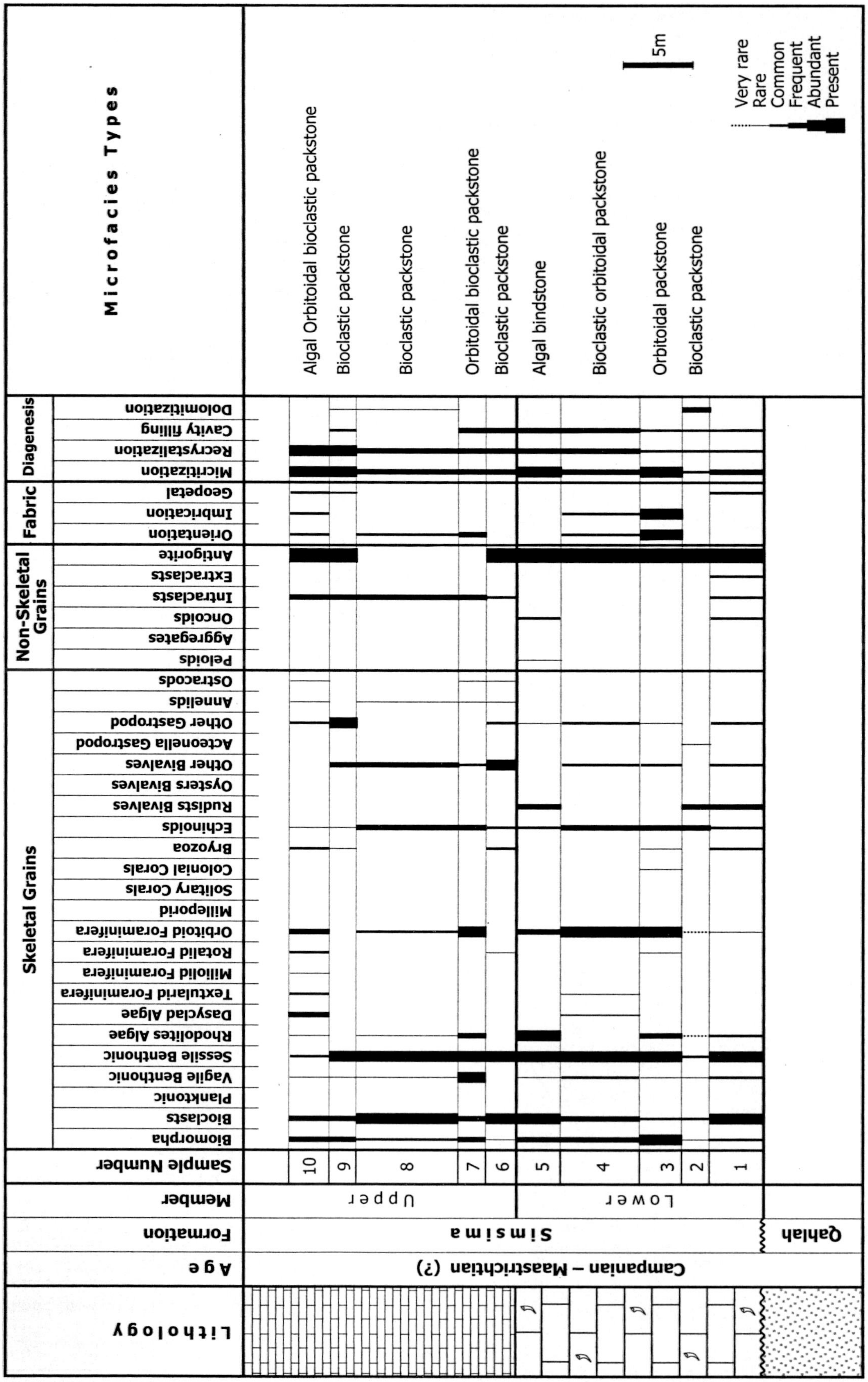


FIG. 6.—Lithostratigraphy and facies distribution of the Simsima Formation, northwestern Oman Mountains.

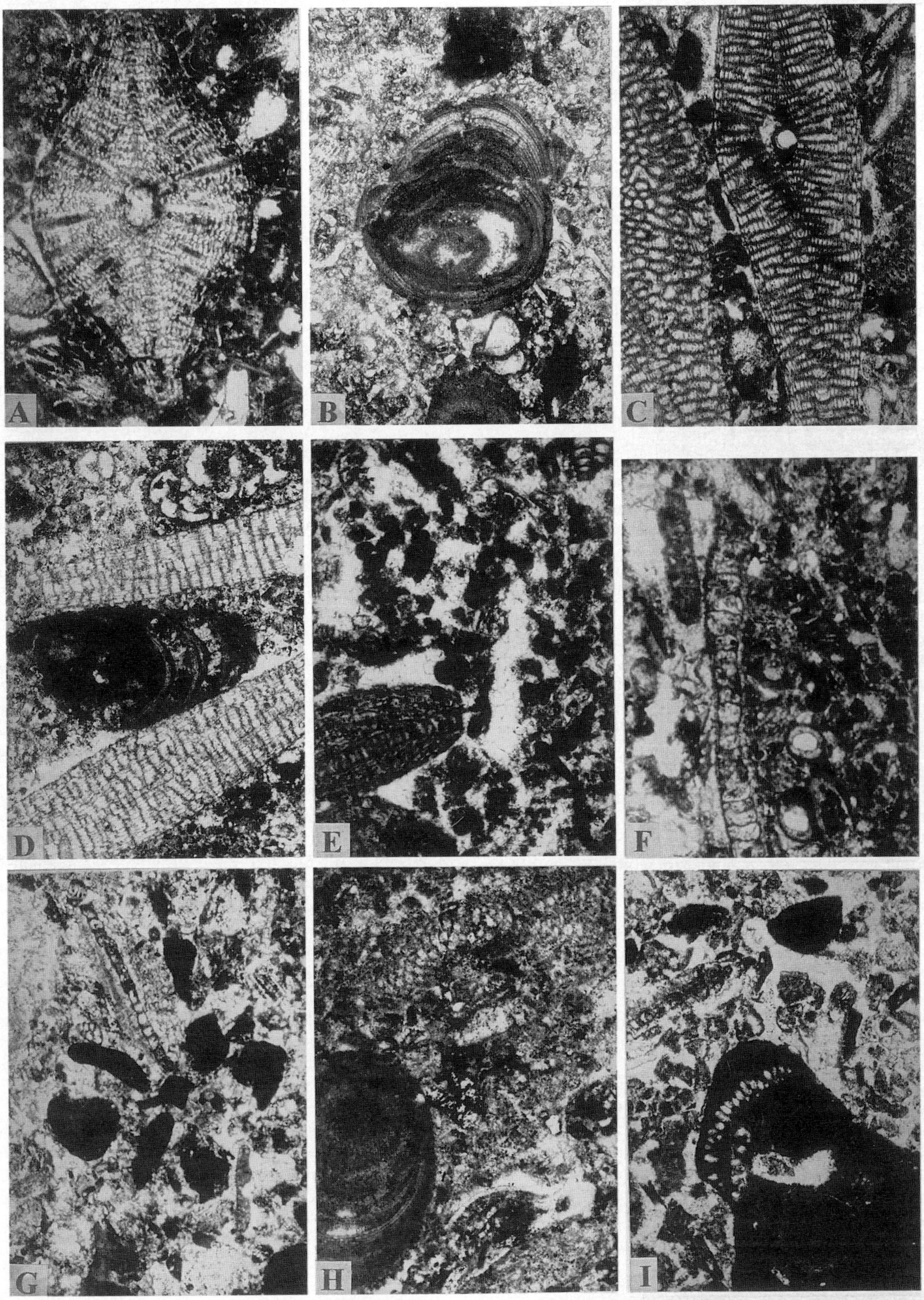


FIG. 7.—A) An axial section of *Orbitoides media* (D'Archaic) (7A to 7I x 20). B) Bioclastic algal orbitoidal packstone with rhodolith fragments. C) Axial sections in *Lepidorbitoides* sp.; notice the orientation and partial imbrication of the orbitoidal packstone. D) *Lepidorbitoides* sp. with rhodolith fragments of algae. E) Peloidal bioclastic orbitoidal packstone. F) Algal bioclastic packstone and *Omphalocycus macroporous* (Lamarck). G) Peloidal bioclastic packstone. H) Algal orbitoidal bioclastic packstone. I) Algal peloidal bioclastic packstone.

with some rhombs embedded in the void-filling calcite spar; c) iron staining and hematite formation; d) rare dedolomitization was observed where zoned rhombs are now calcitic; e) local chertification has occurred in the form of megaquartz and chalcedonic laths; and f) rare quartz silt fills matrix vugs.

## ACKNOWLEDGMENTS

We express our great appreciation to C. G. St. C. Kendall and M. A. Boukhary for their invaluable help in reading the manuscript and their many useful suggestions. Andrew Smith, Peter Skelton, and Robert Scott provided an excellect critical review and valuable comments which improved this paper. We thank Bruce A. Masters for identifying some of the Simsima forams.

## REFERENCES

ALI, M.S.M., 1989, Late Cretaceous echinoids from Gebel El Rawdah, Hatta area, U.A.E.: Neues Jahrbuch für Geologie und Paläontologie, Abhandlungen, v. 177, p. 393–412.

ALSHARHAN, A.S., 1995, Facies variation, diagenesis, and exploration potential of the Cretaceous rudist-bearing carbonates of the Arabian Gulf: American Association of Petroleum Geologists, Bulletin, v. 79, p. 531–550.

ALSHARHAN, A.S., AND NAIRN, A.E.M., 1986, A review of the Cretaceous formations in the Arabian Peninsula and Gulf, Part I. Lower Cretaceous (Thamama Group) stratigraphy and paleogeography: Journal of Petroleum Geology, v. 9, p. 365–392.

ALSHARHAN, A.S., AND NAIRN, A.E.M., 1988, A review of the Cretaceous Formations in the Arabian Peninsula and the Gulf: Part II. Mid-Cretaceous (Wasia Group) stratigraphy and paleogeography: Journal of Petroleum Geology, v. 11, p. 89–112.

ALSHARHAN, A.S., AND NAIRN, A.E.M., 1990, A review of the Cretaceous Formations in the Arabian Peninsula and Gulf: Part III. Upper Cretaceous (Aruma Group) stratigraphy and paleogeography: Journal of Petroleum Geology, v. 13, p. 247–266.

ALSHARHAN, A.S., AND NAIRN, A.E.M., 1993, Carbonate platform models of Arabian Cretaceous reservoir, *in* Simo, J.A., Scott, R.W., and Masse, J.P., eds., Cretaceous Carbonate Platforms: American Association of Petroleum Geologists, Memoir 56, p. 173–184.

ALSHARHAN, A.S., AND NAIRN, A.E.M., 1997, Sedimentary Basins and Petroleum Geology of the Middle East: Amsterdam, Elsevier Science, 940 p.

ALSHARHAN, A.S., AND NASR, S.J.Y., 1996, Sedimentological and geochemical interpretation of a transgressive sequence: the Late Cretaceous Qahlah Formation in Western Oman Mountains, United Arab Emirates: Sedimentary Geology, v. 101, p. 227–242.

BATHURST, R.G.C., 1964, The replacement of aragonite by calcite in the molluscan shell wall, *in* Imbrie, J., and Newell, N.D., eds., Approaches to Paleoecology: New York, Wiley, p. 357–376.

BATHURST, R.G.C., 1966, Boring algae, micrite envelopes, and lithification of molluscan biosparites: Journal of Geology, v. 5, p. 15–32.

DUNHAM, R.J., 1962, Classification of carbonate rocks according to depositional texture, *in* Ham, W.E., ed., Classification of Carbonate Rocks: American Association of Petroleum Geologists, Memoir 1, p. 108–122.

EMBRY, A.F., AND KLOVAN, J.E., 1971, A Late Devonian reef tract on northeastern Banks Island, Northwest Territories: Bulletin of Canadian Petroleum Geology, v. 19, p. 750–781.

FLÜGEL, E.F., 1982, Microfacies Analysis of Limestone: Berlin, Springer-Verlag, 633 p.

FRIEDMAN, G.M., 1964, Early diagenesis and lithification in carbonate sediments: Journal of Sedimentary Petrology, v. 34, p. 777–813.

FRIEDMAN, G.M., GEBELEIN, C.D., AND SANDERS, J.E., 1971, Micrite envelopes of carbonate grains are not exclusively of photosynthetic algal origin: Sedimentology, v. 16, p. 89–96.

GLENNIE, K.W., BOEUF, M.G.A., HUGHES CLARKE, M.W., MOODY-STAURT, M., PILAAR, W.H.F., AND REINHARDT, B.M., 1974, Geology of the Oman Mountain: Royal Geological and Mining Society [Netherlands], Transactions, v. 31, 423 p.

HARRIS, P.M., FROST, S., SEIGLIE, G.A., AND SCHNEIDERMANN, N., 1984, Regional unconformities and depositional cycles; Cretaceous of the Arabian Peninsula, *in* Schlee, J.S., ed., Interregional Unconformities and Hydrocarbon Accumulation: American Association of Petroleum Geologists, Memoir 36, p. 67–80.

JAMES, N.P., AND CHOQUETTE, P.W., 1990, Limestones—the sea floor diagenetic environment, *in* McIlreath, I.A., and Morrow, D.W., eds., Diagenesis: Geoscience Canada, Reprint series no. 4, p. 75–112.

KENNEDY, W.J., 1995, Maastrichtian ammonites from the United Arab Emirates–Oman Border region: Natural History Museum [London], Bulletin, v. 51, no. 2, p. 241–250.

METWALLY, M.H., 1992, Late Cretaceous (Maastrichtian) bivalves from the northwestern flank of the Oman Mountains, United Arab Emirates: Neues Jahrbuch für Geologie und Paläontologie, Abhandlungen, v. 184, p. 123–139.

METWALLY, M.H., 1993, Cretaceous gastropods from the northwestern flank of the Oman Mountains, United Arab Emirates: Zagazig University, Bulletin of the Faculty of Science, v. 15, p. 333–359.

MORRIS, N.J., 1995a, Maastrichtian nautiloids from the United Arab Emirates–Oman border region: Natural History Museum [London], Bulletin, v. 51, no. 2, p. 251–256.

MORRIS, N.J., 1995b, Maastrichtian Inoceramidae from the United Arab Emirates–Oman border region: Natural History Museum [London], Bulletin, v. 51, no. 2, p. 257–266.

MORRIS, N.J., AND SKELTON, P.W., 1995, Late Campanian–Maastrichtian rudists from the United Arab Emirates–Oman border region: Natural History Museum [London], Bulletin , v. 51, no. 2, p. 277–305.

MURRIS, N.J., 1980, Middle East: Stratigraphic evolution and oil habitat: American Association of Petroleum Geologists, Bulletin, v. 64, p. 597–618.

NOLAN, S.C., SKELTON, P.W., CLISSOLD, B.P., AND SMEWING, J.D., 1990, Maastrichtian to early Tertiary stratigraphy and Palaeogeography of the Central and northern Oman Mountains, *in* Robertson, A.H.F., Searle, M.P., and Ries, A.S., eds., The Geology and Tectonics of the Oman Region: Geological Society of London, Special Publication 49, p. 495–519.

NOWEIR, M.A., ALSHARHAN, A.S., AND BOUKHARY, M.A., 1998, Structural setting and stratigraphic evolution of the northwestern Oman Mountain Front, United Arab Emirates: GeoArabia, v. 3, p. 387–398.

SCOTT, R.W., 1990, Chronostratigraphy of the Cretaceous carbonate shelf, southeastern Arabia, *in* Robertson, A.H.F., Searle, M.P., and Ries, A.C., eds., The Geology and Tectonics of the Oman Region: Geological Society of London, Special Publication 49, p. 89–108.

FIG. 8. (opposite page)—A) Orbitoidal algal packstone (8A to 8I x 20). B) Bioclastic algal packstone. C) Planktonic foraminiferal bioclastic packstone with echinoid and pelecypod. D) *Pseudoguembelina palpebra*? (Bronnimann and Brown); note the micritic rim envelopes the specimen. E, F) *Rugoglobogerina* sp. cf. *R. hexacamerata* (Bronnimann); note the micritic filling of the planktonic foraminifera and the surrounding sparitic cement. G) *Globotruncana* sp. cf. *G. aegyptiaca* (Nakkady); note the micritized planktonic foraminifera and the loosely packed skeletal grains supported in sparitic cement. H) *Globotruncanita* sp. cf. *G. elevata* (Brotzen). I) Fine- to medium-crystalline rhombic dolomite crystals.

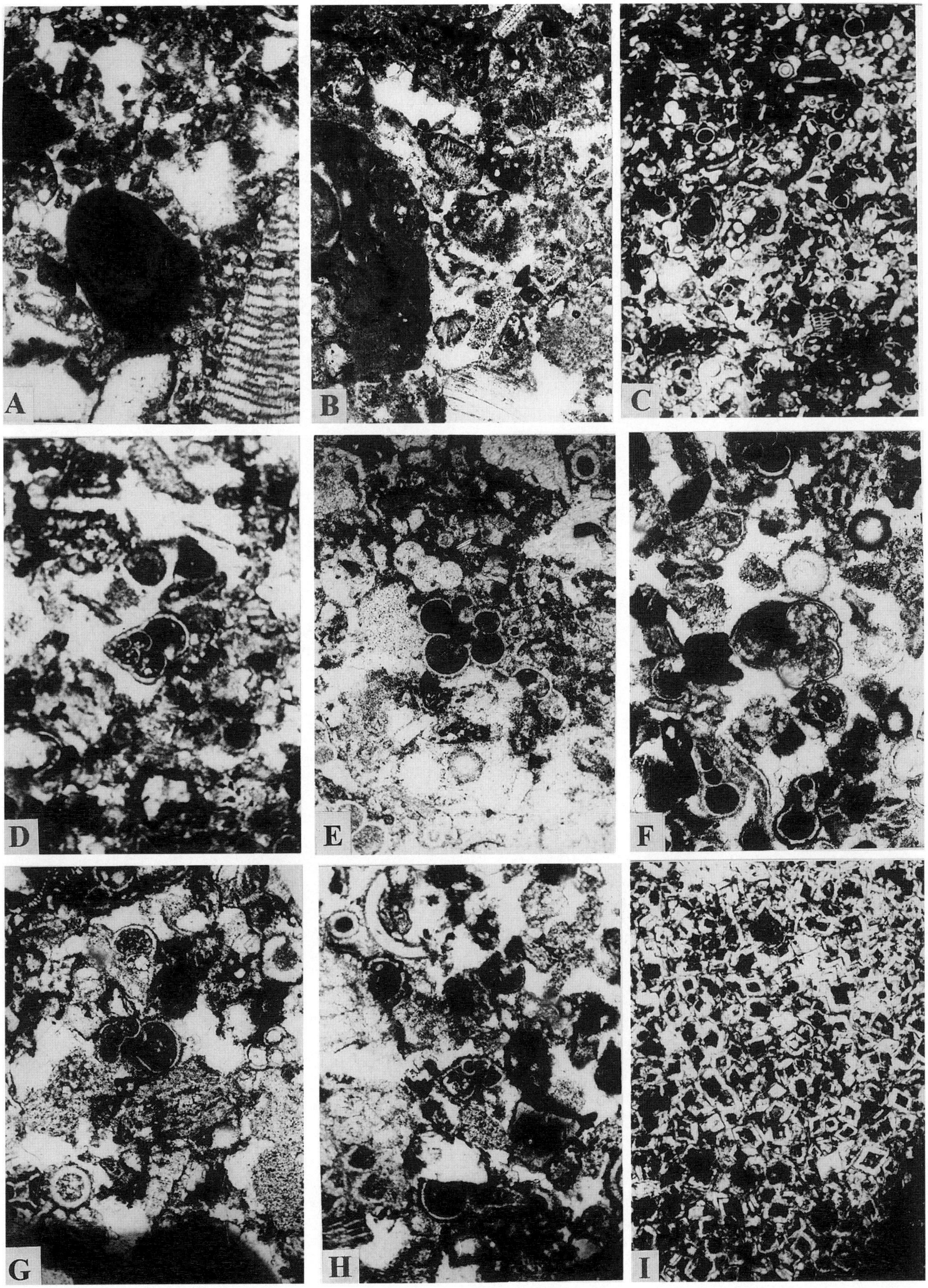
A
B
C
D
E
F
G
H
I

SKELTON, P.W., NOLAN, S.C., AND SCOTT, R.W., 1990, The Maastrichtian transgression onto the northwestern flank of the Proto–Oman Mountains: Sequences of rudist-bearing beach to open shelf facies, *in* Robertson, A.H.F., Searle, M.P., and Ries, A.S., eds., The Geology and Tectonics of the Oman Region: Geological Society of London, Special Publication 49, p. 521–547.

SMITH, A.B., 1995, Late Campanian–Maastrichtian echinoids from the United Arab Emirates–Oman border region: Natural History Museum [London], Bulletin, v. 51, no. 2, p. 121–240.

SMITH, A.B., MORRIS, N.J., KENNEDY, W.J., AND GALE, A.S., 1995a, Late Cretaceous carbonate platform faunas of the United Arab Emirates–Oman border region: Natural History Museum [London], Bulletin, v. 51, n. 2, p. 91–120.

SMITH, A.B., MORRIS, N.J., GALE, A.S., AND ROSEN, B.R., 1995b, Late Cretaceous (Maastrichtian) echinoid–mollusc–coral assemblages and palaeoenvironments from a Tethyan carbonate platform succession, northern Oman Mountains: Palaeogeography, Palaeoclimatology, Palaeoecology, v. 119, p. 155–168.

SUGDEN, W., AND STANDRING, A.J., 1975, Qatar Peninsula: Lexique Stratigraphique International, Centre Naturale Recherche Scientifique, Paris, III, Asie, fasc. 10b3, 120 p.

# PALEOECOLOGY OF LATE CRETACEOUS RUDIST SETTLEMENTS IN CENTRAL OMAN

DIETRICH SCHUMANN

*Geologisch-Paläontologisches Institut, Technische Universität Darmstadt, Schnittspahnstr. 9, D-64287 Darmstadt, Germany*

ABSTRACT: Rudist associations were normally destroyed during or shortly after their development. Consequently, exact paleobiological examinations are very difficult or impossible. The in-situ associations of Central Oman are often completely preserved and enable a precise paleobiological and paleoecological description to be made. They developed during a transgression onto the Arabian platform. Microfacies analysis based on extensive field observations reveal that vigorous turbulent conditions prevailed very frequently throughout the period. The development and preservation of in-situ rudist associations are rare. As a rule episodic turbulence prevented the development of such structures. In the same way, turbulence normally destroyed rudist associations. The model of constratal growth (Perkins 1974; Gili et al., 1995a; Skelton et al., 1995) is discussed, including the question whether thick, extensive rudist associations should be named reefs. The absence of vast rudist reefs is believed not to be a consequence of an inherent inability of vertically growing rudists to build such structures, but can be mainly attributed to exogenous, abiotic factors. The associations investigated here inhabited restrictive, shallow-marine environments. Rapid growth of a few rudist species is verified. Vertically growing rudist associations of 5–8 successive in-situ generations represent a period of only two to three hundred years.

## INTRODUCTION

The Mesozoic transgressions in the Central Oman region took place over the Paleozoic Haushi–Huqf paleohigh (Platel et al., 1994). Broad shelves developed with these transgressions, which are subdivided into open shelves, intrashelf basins, and inner shelves (Murris 1980; Harris et al., 1984). During the Late Cretaceous the investigated areas were situated at the Equator. Computer-aided modeling yielded a position in an area with a strong E–W ocean current (Barron 1990). Throughout the Cretaceous Period the Arabian platform was situated off the Paleo–Indian Ocean. The distribution of continent and ocean has remained largely similar in this region since Late Cretaceous times. In the Campanian, as today, the coast of SE Arabia bordered directly on a wide Paleo–Indian Ocean.

The sedimentology and stratigraphy of the rudist-bearing sections in the Saiwan district of Central Oman are not described in this article. The profiles include parts of the Campanian (Samhan Formation) and the Maastrichtian (Simsima Formation). Details of the stratigraphy and biozonation have been published by Dubreuilh et al. (1992), Hughes Clarke (1988), Platel et al. (1994), and Philip and Platel (1995). In relation to the widespread outcrops of the Cretaceous formations, complete in-situ rudist associations are not very common. The aim of this paper is to find an explanation for this relative rareness even in thick rudist limestone profiles.

## LITHOLOGICAL ASPECTS OF THE LOWER CAMPANIAN SAMHAN FORMATION

An important transgression took place in the Early Campanian over terrigenous sandstones. The profiles of the Samhan Formation of Saiwan represent the *Globotruncana elevata* zone, which, according to current knowledge, indicates a period of 4–5 million years.

Rudists are extremely rare within the lowermost few meters overlying the terrigenous sandstones. They are also missing as fragments in the microfacies. Throughout the next 30–40 m of the lower Campanian, rudists are found in all but a few horizons, usually as reworked rubble. This indicates that, as a rule, there was no primary development of dense rudist associations. The sediments of the Samhan Formation of Saiwan were interpreted in context with geodynamics and environments by Platel et al. (1994). Schumann (1995) investigated numerous sites in the vicinity of Saiwan. Six of these profiles have been published. Therefore, only a few basic comments on the facies are given here (Fig. 1A). The spectrum of the microfacies is not very broad. The very shallow-water sediments, consisting of sand and silt in the lower part of the profile, contain numerous thick-walled ostreids, cyclolitid corals, rare small cone-shaped solitary corals, some colonial corals, pectinids, and microfauna. Rudists are extremely rare in this facies.

Dense, occasionally very compact rudist associations (V1/V2) are found in marly limestones. As a rule these limestones are intensely weathered as a result of the modern climate aided by the salty air and a frequent stiff breeze from the east (slowly working salt wedging). The cryptic sediments of the rudist horizons (samples taken from the spaces between rudists) contain very sparse associated fauna. The horizons of the section between the rudist horizons V1/V2 and the striking stromatoporoid horizon (part 5) contain numerous colonial corals, stromatoporoids, and solitary rudists, which are only very rarely grouped to small lenticular structures. This facies contains numerous small and large foraminifera, and the bedding was influenced by strongly agitated water. Local accumulations of a large *Durania* sp. are found above the stromatoporoid horizon S. They are grouped into clusters of a few individuals or patches with dozens of individuals (Fig. 4), but also into associations of several hundred (Fig. 7) and possibly, according to a three-dimensional reconstruction of the exposure areas, even several thousand individuals. The primary deposition within this horizon was locally chaotic: even very heavy shells were toppled during temporary high-energy conditions. An extensive continuous increase in carbonate content can be found in the upper section of the profiles (parts 9 and 10). The hard and nearly 100% pure yellow–white micrites form characteristic escarpments in the landscape. These limestones show a lenticular bedding and locally have a character similar to nodular limestones (Schumann 1995, pl. 40, fig. 6). In most localities the uppermost part of the section consists of dense micrite, often with silification. According to Platel et al. (1994) the

Middle East Models of Jurassic/Cretaceous Carbonate Systems
SEPM Special Publication No. 69, Copyright © 2000
SEPM (Society for Sedimentary Geology), ISBN 1-56576-075-1, p. 143–153.

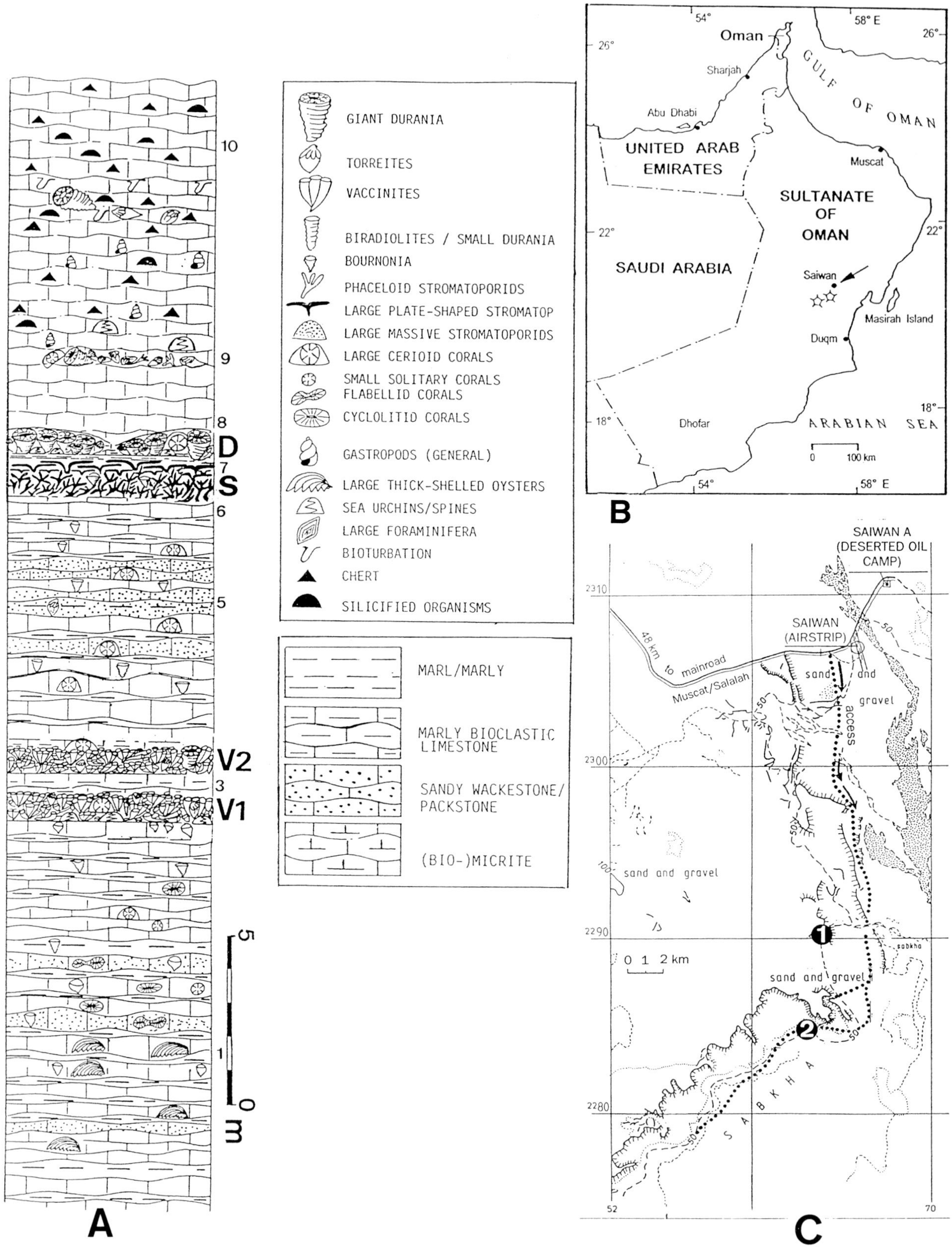


FIG. 1.—A) Typical lithologic section of the lower Campanian of the Saiwan area (section 2, Schumann 1995). V1/V2, *Vaccinites* associations; S, stromatoporoid association; D, *Durania* patch reefs. B) Location of Saiwan in Central Oman. C) Access from the Saiwan track to rudist localities 1 and 2.

top of this horizon represents a maximum flooding surface with hardground development. These hardgrounds were not observed in the vicinity of the rudist associations south of Saiwan. In summary, the depositional conditions were generally dominated by agitated water (see section on shell size, gregariousness, and water energy later in this paper).

## GEOMETRY OF THE RUDIST ASSOCIATIONS

Among the outstanding features of the Saiwan region are the unique rudist exposures. They enable vertical and horizontal observations of all details. In contrast to most rudist associations of other localities, those of Saiwan have remarkable lateral extent. The *Vaccinites vesiculosus*–dominated associations completely cover areas up to about 100 km$^2$. There is no known locality throughout the Saiwan region where these associations are more than 2 m thick. As a rule, they are no thicker than 80 cm.

The most successful species, although preserved only twice as massive in-situ associations in approximately 5 million years, is *Vaccinites vesiculosus* (Woodward). The slender shells grew up to 50 cm high. The associations may be bushy to very dense and can consist of several layers of in-situ generations (Fig. 2). This species is nearly 100% dominant. However, the following can occasionally be found growing within the *Vaccinites vesiculosus* associations: a *Torreites sanchezi milanovici* Grubic, a *Radiolites* sp., or a *Biradiolites* sp., a 50-cm-long and 35-cm-wide *Durania*, or a 70-cm-high, 30-cm-wide colonial coral. It must be emphasized that the *Vaccinites vesiculosus* outnumber the other species by 1000 to 1 or more! Therefore, large areas of the *Vaccinites* associations are monospecific.

A notable variation of the normal *Vaccinites* association is found in the same horizon 5 km to the north of the community shown in Figure 3. Here, *Torreites sanchezi milanovici* constitutes approximately 50% of the association. The shells of this species grew up to 40 cm high and are frequently found in originally compactly cemented groups of up to 10 individuals. The proportion of head-size stromatoporoids and colonial corals is remarkable. This locality is the most extensive *Torreites*-dominated area in Saiwan and actually in the world. In view of the intense weathering and modern detritus production, the character of this community can be revealed only by excavations.

A very large species of *Durania* was found in the upper parts of the profiles of Saiwan. In the area around the locality 2 (Fig. 1) this species forms dozens of patches of varying dimensions (Figs. 4, 7).

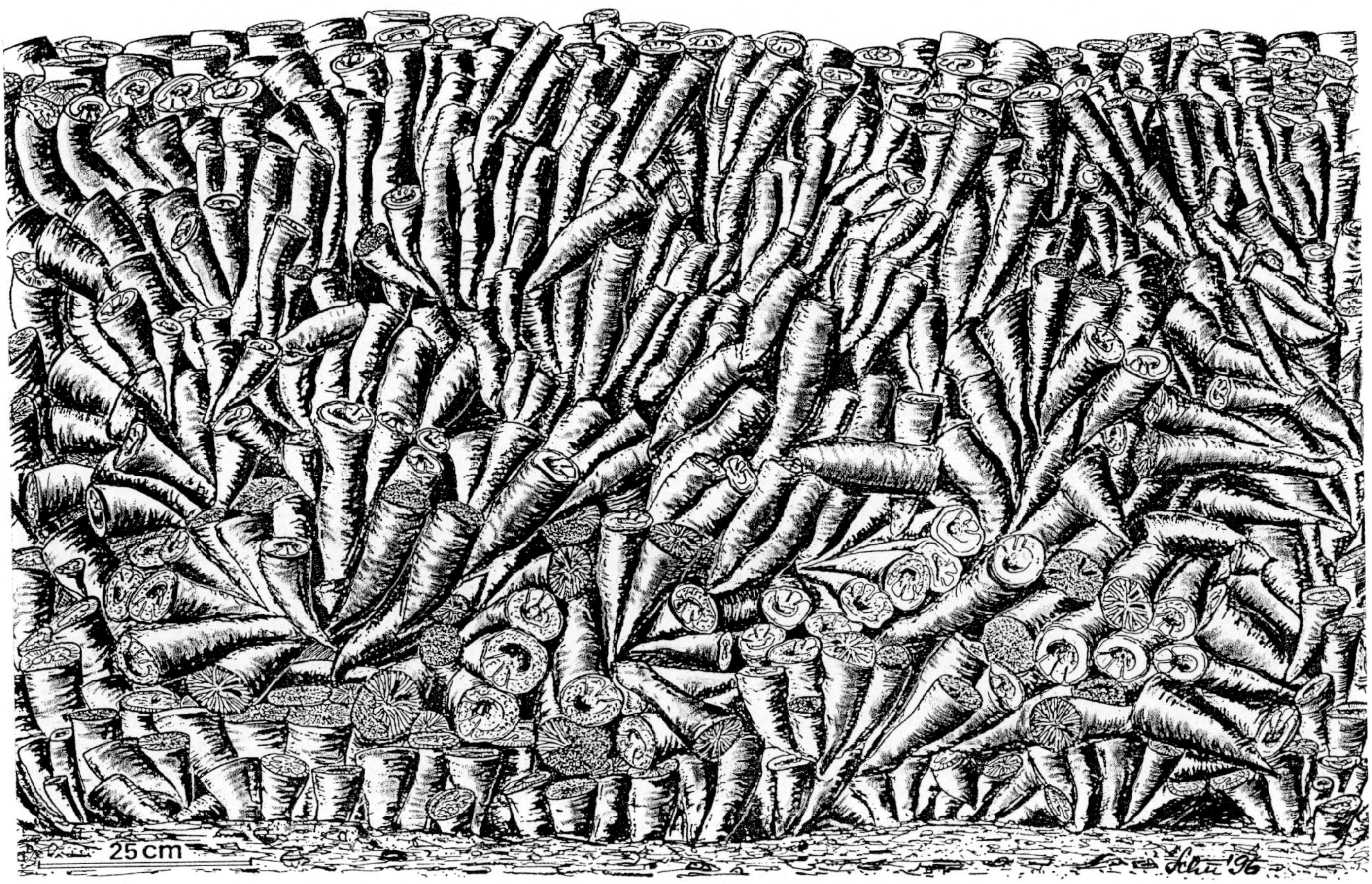


FIG. 2.—Most of the *Vaccinites vesiculosus* associations of Saiwan are monospecific. Up to 8 generations can settle one upon the other and form locally thick very compact constructions, nearly without sediment. The thickness shown here is developed only in a few localities in the Saiwan area (compare Fig. 6). Because of to the rapid growth of *Vaccinites* at least the last generation was above the sedimentary sea floor. Sediment support aiding the vertical growth of the individuals was not necessary, because of the dense, co-cemented gregariousness. At some localities (not documented here) tropical storms uprooted large parts of the communities, leaving distinct discontinuities. Around locality 2 (Fig. 1).

FIG. 3.—At very few localities only the normally *Vaccinites vesiculosus*–dominated associations show a differing composition. At locality 1 (Fig. 1) the ratio of *Vaccinites vesiculosus* to *Torreites sanchezi milanovici* is about 50%:50%. Approximately 5–10% of the community are represented by small and large stromatoporoids (dotted) and colonial corals. All *Vaccinites* and *Torreites* shells are completely cemented with each other forming very dense and robust constructions. This reconstruction is based on four excavations of about 1 $m^3$ each and a distance of 2–4 m between them. Piece by piece was recorded. In most of the pieces exposed to weathering the upper valve was lost. In contrast, excavated individuals have been recorded much more often with their upper valve in place. This points to a rapid covering with sediment during storm events. The reconstruction shows only the part of the community that as a result of successive rapid growth existed above the sea floor. A couple of analogous communities existed below. It is not believed that the primary upright life position of the individuals was supported by a constratal growth (Gili et al., 1995a). Not only neighboring *Vaccinites* shells but also *Vaccinites* and *Torreites* shells are co-cemented very strongly. Almost all of the pieces are in Darmstadt.

The individuals attain a maximum height of 80 cm and a maximum diameter of 60 cm. Large, massive stromatoporoids, cerioid corals, and massive *Torreites* shells are less common. No other rudists were found in these patches. A few sea urchins and many spines occur in the cryptic niches at the base. Single large individuals of *Durania* at the same stratigraphic level regularly appear in nearly all sites, but they do not form *Durania* patches anywhere. It seems that optimal ecological conditions for this very characteristic species existed in the region around locality 2. It is obvious, however, that it appears at some localities in completely different environments with very large specimens, e.g., among the in-situ *Vaccinites vesiculosus* communities or in the yellowish-white micrites of horizons 9 and 10 (Fig. 1A). In contrast, *Vaccinites vesiculosus* specimens have not been found in any of the numerous *Durania* patches. This indicates that this large *Durania* sp. had a higher ecological tolerance. Apart from the central rise, the upper valve of this *Durania* sp. is extremely thin, presumably translucent.

## A *DURANIA* ASSOCIATION

A *Durania*-dominated association of the Simsima Formation covering approximately 1 $km^2$ was found on a tectonically isolated block 2 km NE of locality 2. This *Durania* sp. is probably identical to the species classified as *Durania* Form A by Morris and Skelton (1995). The shells of this species have strong vertical ribs and grew up to 50 cm high. They formed densely interlocked and mutually cemented communities with new generations continuously growing on top of the last. The associations are up to 1.8 m thick. Solitary, notably massive *Torreites sanchezi milanovici*, *Vaccinites* sp., and, very rarely, *Pironaea* aff. *syriaca* shells occur in this associations. Apart from rudists, there are also isolated large colonial corals and stromatoporoids. The character of the basal sediment and the occurrence of small sand channels in a number of places indicate a very shallow-water environment. According to current knowledge this is the only outcrop of the Simsima Formation in the western part of the Haushi–Huqf paleohigh. Much better exposures of the Simsima Formation can be found in the Filim region, where remarkable associations of the same *Durania* species (Fig. 5) can be found in the Wadi Halfayn near the town of Hayy.

## SEDIMENTATION RATE, COLONIZATION, AND SPATIAL REQUIREMENTS

The lower Campanian profiles of Saiwan are 40 m thick and represent 4–5 million years of sediment accumulation. Consequently, the average sedimentation rate is estimated at 1 cm per

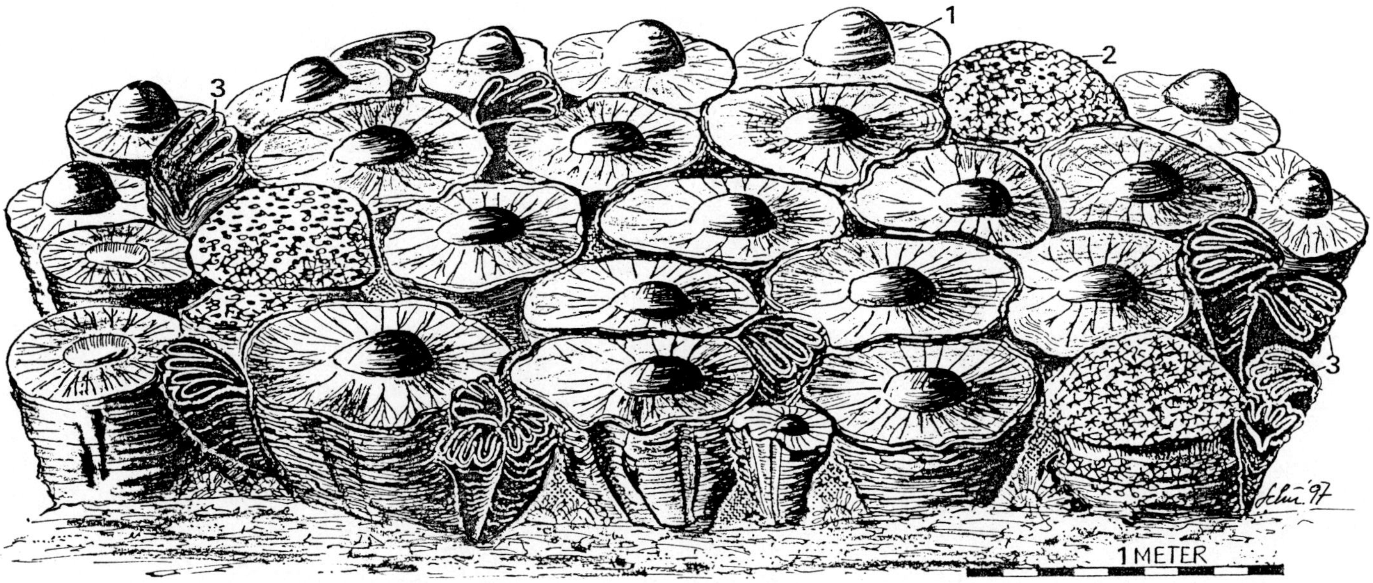


FIG. 4.—Patches dominated by a giant *Durania* sp. (1) are common in the Saiwan area. Besides this species large colonial corals (2) and robust *Torreites sanchezi milanovici* (3) occur, but never *Vaccinites*. The percentage of colonial corals differs from patch to patch but never exceeds 10%. The depicted community is one of the smaller ones. Hundreds and thousands of individuals could participate, and a couple of generations could grow successively on top of each other. Around locality 2 (Fig. 1); compare Figure 4.

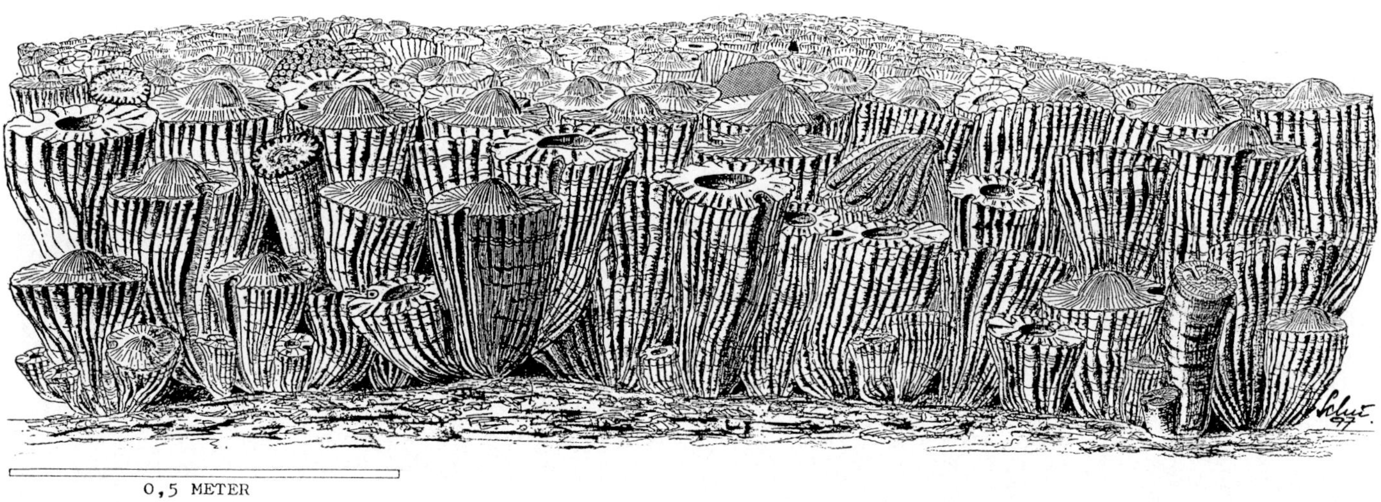


FIG. 5.—A Maastrichtian association of *Durania* sp. Very rare thick-shelled *Vaccinites vesiculosus*, *Vaccinites* sp., *Pironaea* aff. *syriaca*, extremely robust *Torreites sanchezi milanovici*, large colonial corals, and massive stromatoporoids occur in an outcrop of about one km$^2$ NE of locality 1 (Fig. 1). The dense settlement of larvae caused an extraordinary competition for space and in this way led to a compact, massive structure of the community that could resist high energies. A couple of generations settled on top of each other; of these only the uppermost is shown.

FIG. 6.—Association of *Vaccinites vesiculosus*, 1 km W of locality 2 (Fig. 1). Compare Figure 2. Length of hammer is 32 cm.

1000 years. This is not an unusual value. The character of the sediments indicates that only gradual changes occurred over a period of several million years. Furthermore, fragments in the microfacies indicate that the same rudist species of the in-situ associations existed virtually throughout the entire period represented by the profiles under investigation. In spite of this, dense or compact rudist associations were able to develop and to be preserved on only three occasions. The reason for this must be sought in the nearly permanent active redeposition of material. The characteristics of the bedding in the profiles is typified by a sequence of discontinuities. The question arises why in-situ associations are so rarely preserved. In similarity to recent pelecypods, rudists probably also developed via a free-swimming larval stage. These larvae settled onto hard particles. On the Saiwan shelf the chances for growth over a period of decades were very unfavorable, because of the nearly permanent redeposition. Evidently, dense in-situ associations could develop only during a rarely occurring phase of somewhat lower-energy conditions lasting at least some weeks or months. One can assume the existence of an abundant supply of larvae for colonization from the vicinity of the Paleo–Indian Ocean. Essential for the successful development of a dense association is a massive larval colonization of a given area. In this way, during temporarily lower-energy conditions, a biogenic pavement consisting of millions of joined juvenile rudist shell stages could have developed within a few weeks.

The spatial requirements of the growing individuals were very high (compare Skelton 1979). For example, in Saiwan, a juvenile colony of radiolitid individuals developed from 60–70 larvae growing on a few square millimeters. Once the colony grew to a height of 2.5 cm, only 45 individuals survived, covering an area of 30 cm$^2$ (Fig. 10). Thus, the spatial requirements increased by 400–500% within a few months.

FIG. 7.—Association of several hundreds of individuals of the large *Durania* sp., locality 2, horizon D of the section (Fig. 1). Average diameter of the individuals is 40 cm. Compare Figure 4.

FIG. 8.—*Vaccinites vesiculosus.* Cluster of 27 co-cemented individuals. 3 km E of locality 2. Collection Geological Institute Darmstadt. x 0.2.

FIG. 9.—Three co-cemented individuals of *Torreites sanchezi milanovici.* Locality 1 (Fig. 1). Compare Figure 3. Collection Geological Institute Darmstadt. x 0.3.

Another colony of four *Vaccinites* individuals also started growing on a few square millimeters of substrate, and increased to 18 $cm^2$ by the time they reached 2.5 cm in height and 195 $cm^2$ at 15 cm height. This means that the spatial requirements increased by a factor of 4000–5000 within a few years. These examples illustrate that massive larval colonization can actually produce a biogenic pavement within a short period of time. Once such a pavement exists, much stronger water movements are required for reworking. Somewhat calmer periods are presumably represented only by the preserved, successful and extensive colonies of dense rudist associations. Their brief existence, as well as the presence of rubble layers (caused by storm events) often found within the structures document a rapid recurrence of strong water movements.

## GROWTH AND DEATH OF THE ASSOCIATIONS

The growth of several rudist species has been the subject of a number of investigations (Amico 1978; Schumann 1995; Steuber 1995, 1996, 1997). These have shown that a growth rate of around 1–4 cm per year, depending on the ontogenetic stage, is possible. The adult individuals of *Vaccinites vesiculosus*, the main constructors of the extensive monospecific associations of Saiwan, are estimated to be about 30 years old. The thickest *Vaccinites* associations of Saiwan comprise 8 successive generations. Thus, such a rudist bed developed in 200–300 years.

There is no locality known throughout the whole Saiwan region where these associations are more than 2 m thick. However, in contrast to Gili et al. (1995a) this does not mean that rudists are not potentially capable of building thicker reefs. The structure allows unlimited colonial succession upon existing surfaces. This principal capability is demonstrated in many cases. The reason why such massive reef structures, as seen in coral reefs, did not develop on a worldwide scale is suggested to be solely due to exogenous factors. Colonization evidently took place during a transgression in very shallow waters with brief quieter conditions. The factors and levels of tolerance controlling growth differ widely from those influencing coral growth. Extensive flat bodies, such as the rudist associations of the Saiwan type, could develop only on low-angle open shelf margins, a relief that existed along the Arabian platform margin.

The disappearance of the rudist associations of Saiwan was due to an increase in sediment supply. The associations were premortally covered with sediment. In correspondence to this, excavations usually reveal shells with the upper valve preserved. This would not be possible if individuals remained in contact with the ocean after death.

Surprisingly, only three cases of constructive growth of extensive rudist associations are known during the entire 4–5 million year period. Therefore, the chances for the development and preservation of such associations were near zero. On the Cretaceous shelves of the Arabian platform rudists usually have been producers and not constructors.

## PALEOCOLOGICAL INTERPRETATIONS

Considering the large quantities of preserved organisms, the diversity in the *Vaccinites* associations is extremely low. Numerous cryptic niches existed. In the case of normal marine conditions such niches would be populated with highly diversified associations. The examined areas are located near the Paleo–Indian Ocean. Therefore, a permanent influx of planktonic larvae can be expected, which were unable to develop in the vicinity of the *Vaccinites* areas. Oysters cemented on *Vaccinites vesiculosus* are very rare. Serpulids are frequently found on the inner side of the upper valve. Many shells are heavily bored by clionids, and pre-mortal borings of mussels are occasionally found on rudists and corals. During the pioneer settling of *Vaccinites*, *Torreites*, and *Durania* sea urchins were obviously quite frequent, but they are not co-inhabitants of the rudist associations. The result remains, however, that bryozoans, brachiopods, normally growing pelecypods, gastropods, calcareous algae, etc., normally found in great numbers in Cretaceous shelf seas, are missing completely (cf. Skelton et al., 1995). Numerous washed samples also confirm this result for microorganisms. Foraminifera and ostracods are rare in the cryptic niches. Outside the rudist communities foraminifera are present in many layers, and miliolids are very frequently found in the yellowish–white, nearly nodular limestones (see Schumann 1995, p. 199, fig. 6). Other small foraminifera play a secondary role. Actaeonellids, known from similar facies in other regions, are very rare here.

These conditions are difficult to interpret. The low diversity and the often nearly 100% dominance of a single species indicate a very restrictive biotope. *Vaccinites vesiculosus* was a suspension feeder, with obvious capabilities for developing in such an extreme biotope.

The absence of nearly all other groups of organisms can be explained only hypothetically. I presume that the rudist associations of Saiwan developed in a very shallow inner-shelf zone. Diurnal heating of this zone resulted in hypersalinity and led to very low oxygen conditions. Low-amplitude tides enabled an exchange of water with the open ocean, especially in the night. Presumably this was also the main filter-feeding period of *Vaccinites vesiculosus.* All larvae of other organisms, settling under the more favorable night time conditions, were not able to survive the ecologically restrictive conditions of the next days. It follows that *Vaccinites vesiculosus* was a rudist able to adapt also to extreme ecological conditions.

The presence of a contemporary higher-diversity community only a few kilometers away (Fig. 3) indicates a local shelf zone with better water exchange with the open ocean. The hypothesis of an extreme biotope for the *Vaccinites vesiculosus* communities is somewhat limited by the very rare occurrence of a large colonial coral, as well as the very rare presence of a giant *Durania* and a massive stromatoporoid. There is no doubt that these were contemporary members of the community. This is similar to the data of Gili et al. (1995b). However, the proportion of rudists and corals is very clear and indicates that there was no competition between these two organisms in the Saiwan communities.

The part of the profiles preceding and following the *Vaccinites vesiculosus* communities show a higher organism diversity and contain many corals. However, there is no evidence of a constructive coral reef development.

### *Were Rudists Able to Build Reefs?*

In connection with rudist associations almost every author has used the term reef. In addition the terms bioherm, biostrome, buildup, bioconstruction, congregation, lithosome, and meadows have been used. It cannot be the object of this paper to deal with the overall controversy concerning this terminology. Nonetheless I would like to comment on some of the statements made in the publications of Gili et al. (1995a) and Skelton et al. (1995). On the basis of detailed explanations both sets of authors propose that the term reef is not to be used for rudist associations, as a matter of principle.

According to both publications true reefs are characterized by a robust biogenetic framework, topographical relief, clonal growth, special pattern of nutrition, occupation of substrata, greater supply of upwelling water, animal–plant symbiosis, indefinite potential for longevity and continuous growth, growth of primary framework, bioerosion, infilling sediments, burial, cementation, and some further subordinate characteristics.

According to this definition only the Holocene coral–algal reefs and their fossil analogues would be true reefs. In the sense of a comparison, rudist associations have been contrasted to coral–algal reefs. Thereafter some of the characteristic features of coral–algal reefs are not present in rudist communities. Indeed, I believe that hitherto no one working with rudists has treated rudist associations as structurally analogous to modern coral reefs. The more aspects that are included in the definition of a binding reef, the more seldom the term "reef" will be used, as a consequence. If such a definition would incorporate as an obligation e.g. that only clonal organisms are capable of building true reef structures, then from the start many organisms are not reef builders per definition. This is presumably the intention and proposal of the authors named. However, such a concept would have to gain general approval. As a part of the Priority Program "Global and Regional Controls on Biogenic Sedimentation and Reef Evolution" of the German Science Foundation (DFG), repeated and comprehensive discussions with over 150 participants took place debating the terminological definition of "reef" without achieving a generally accepted outcome. Briefly I would like to give my opinion on the following points.

### *Relief of Rudist Associations*

Most rudist structures have had little or no positive topographic relief. There are several exceptions. I would like to mention only some of these. Outcrop photographs by Perkins (1974) show a distinct relief and display the structural organization of a reef, also interpreted by the author. Scott (1990) gives an account of the "best rudist-dominated reefs ..." and "a rudist reef belt" (p. 43) and therefore reconstructed rudist reefs with relief (p. 19, 21, 55). Fagerstrom (1987) states that rudists were "important reefbuilders from the Late Aptian ... through the Maastrichtian." (p. 314). Kauffman and Johnson (1988) give detailed account of "reef-building rudistids". Of course none of these reported rudist associations had a relief equal to Holocene coral–algal reefs.

### *Growth Fabric of Rudists*

The lack of binding and limited mutual attachment of shells, which would implicate sediment support, are suggested to be important features of vertically growing rudists (Gili et al., 1995a, p. 257). But there are numerous examples proving the opposite. My collection alone yields dozens of examples, where the entire length of the shells of *Vaccinites*, *Biradiolites*, *Torreites*, and *Durania* are massively grown together. It is absolutely certain that the generative zone of the mantle tissue of adjoining individuals have secreted (and co-cemented) their shells. Figure 8 shows a massive cluster of 27 individuals firmly intergrown. This cluster (now in Darmstadt) is a minor part of hundreds of individuals in Saiwan that are grown together in this way. These associations are packed so densely that virtually no space remains for sediment.

### *Constratal Growth*

The term "constratal growth" was proposed by Gili et al. (1995a) and is suggested to be valid in general. Perkins (1974) was probably the first to suspect the vertically growing bouquets to remain in life position through sediment support. According to my observations vertically growing bouquets or associations are normally very densely packed. There is very little space remaining. Under the assumption of constratal growth a major problem arises if the sedimentation rate is to be calculated. Compare the growth of a vertically growing hippuritid, in our case *Vaccinites*, to the rate of sedimentation. If a 40–50 cm high shell of *Vaccinites* grew in 20–30 years it would equal an average sedimentation rate

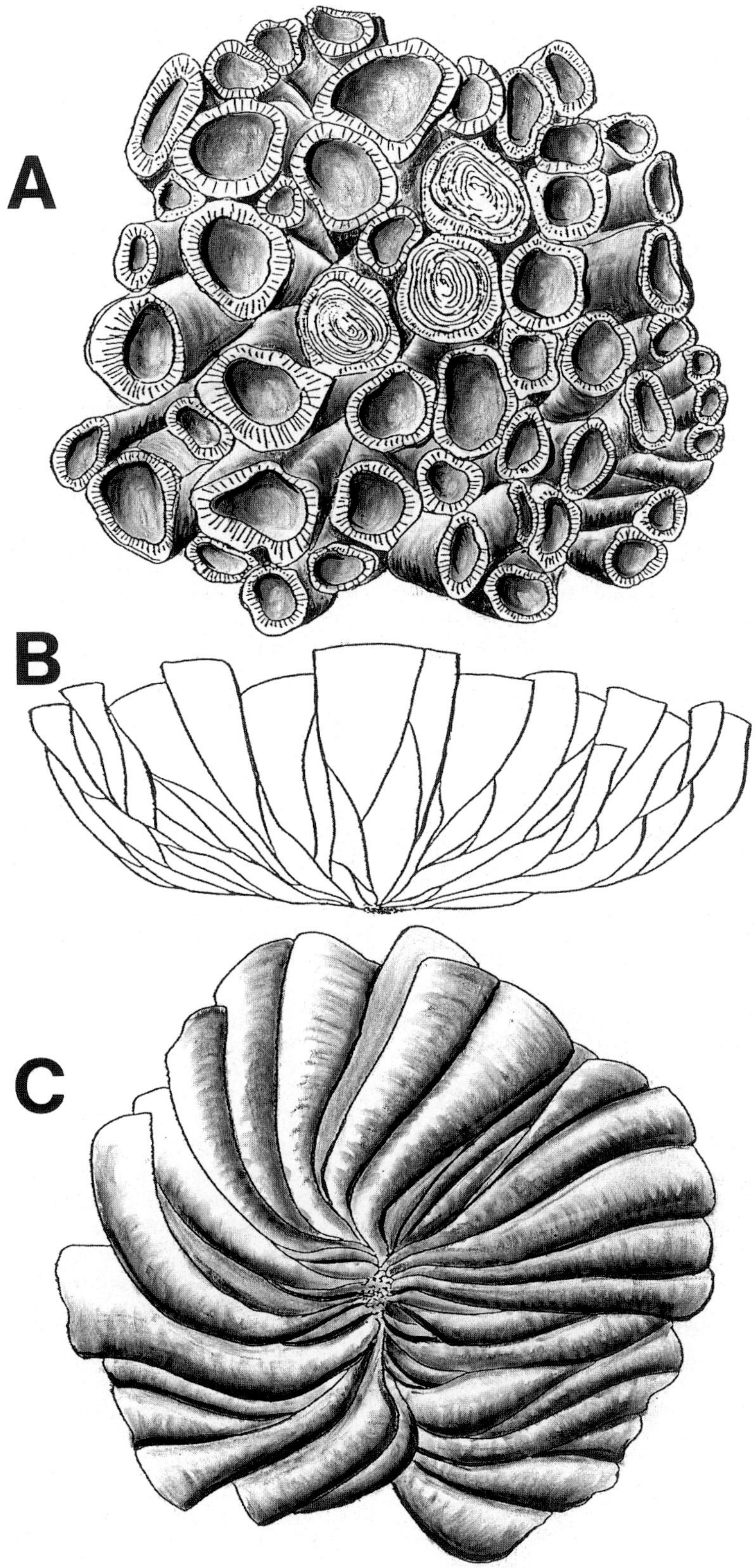


FIG. 10.—Space requirements of a juvenile radiolitid colony. A) Surface. B) Transverse section (reconstruction). C) Bottom view. Locality 2 (Fig.1).

of approximately 13 meters in about 1000 years under constratal growth. Even eight succeeding generations (several special places in the Saiwan area) including the undersized individuals building structures of 180 cm height in 200–300 years would still result in a sedimentation rate of 6 meters in 1000 years. Even lateral redeposition of sediment does not lead to an understandable solution. Constratal growth in the sense of Gili et al. (1995a) implies that the growth rate is equal to the rate of sedimentation. This, however, is highly improbable. It seems more likely to me that vertically growing bouquets and associations are characterized by superstratal growth. Only the basal cavities were filled by a mixture of bioclastic sediment, particles of biodegradation (if taking place), and mud of the background sedimentation. Because of their population density the associations are very resistant. For large areas of the Saiwan associations I cannot support the view of Skelton et al. (1995, p. 124) that "there is no evidence for any upstanding, self-supporting framework". However, in spite of superstratal growth the Saiwan associations did not develop a significant relief.

Local horizontal concentrations of borings on the shells could be one feature indicating steps of the presumed constratal growth. But among thousands of individuals not one horizontal mark could be observed. The shells of *Torreites* are frequently heavily destroyed by clionid borings without showing horizontal marks of concentration, also pointing to superstratal existence during the entire life span.

Furthermore it is questionable why a screen filtering species as *Vaccinites vesiculosus*, being the most prominent species in Saiwan, could be so successful in areas with high sedimentation rates as required by constratal growth.

### *Shell Size, Gregariousness, and Water Energy*

The *Torreites sanchezi milanovici* of Saiwan in the *Vaccinites*-dominated associations are conspicuously less robust than the ones in the *Durania* sp. associations (see Fig. 1A, horizon D, and Schumann, 1995). The absolutely thickest shells of *Torreites sanchezi milanovici* occur in the *Durania* association of the Simsima Formation (Fig. 5). According to my assessment all three associations belong to high-energy environments. In the *Vaccinites*-dominated associations the average energy was slightly lower. In the mid–Cretaceous and therefore probably also in the Late Cretaceous, the shelves and intracratonic lagoons of modern Central Oman were situated in an area of E–W prevailing winds. The same holds true for the mean annual surface currents (Barron 1990). Therefore one has to account for a average high energy in shallow environments of the Saiwan shelf, including the very large lagoons. Nonetheless, years or decades of lesser energy existed, and this statement involves no contradiction. In modern equatorial regions violent storms happen on average every 10 years and hurricanes every 20 years. Applied to the Late Cretaceous rudist environments of Saiwan this would give the *Vaccinites* associations 10 years time to grow relatively undisturbed before they were again subjected to a violent storm. This is consistent with many field observations (Platel et al., 1994; Schumann 1995, p. 195, Fig. 1). In contrast to Gili et al. (1995a, p. 262) the larger rudists in the Saiwan area with thick, strong shells were commonly not solitary forms or formed sparse or loosely structured communities. They formed very massive patches (Fig. 4, 7) or extensive square kilometer-size structures of great solidity (Fig. 5). Therefore I tend towards the opinion that the larger species with thicker shells were exposed to higher energy than the *Vaccinites*-dominated associations in which the large *Durania* occurs extremely rarely.

The densely packed rudist associations lack clingers. In comparison they are apparent in the coral-dominated horizons above and below V1 and V2 (Fig. 1A) in most cases solitary, rarely in small co-cemented clusters.

In conclusion it should be mentioned that on average 1 m of sediment in the Saiwan sections represents 100,000 years and therefore has been subjected to about 10,000 violent storms. Although this is just a statistical number it explains many of the sedimentological structures of the Saiwan sections.

## CONCLUSIONS

The Late Cretaceous rudist associations of Central Oman are among the best preserved in-situ associations in the world. They developed during cyclic transgression phases in very shallow and usually agitated water, on an exceptionally wide shelf. The very extensive monospecific *Vaccinites* associations probably developed in a very restrictive environment. Successive growth periods of the thickest associations continued for 200–300 years. An increased sediment supply stopped the growth period.

The general validity of "constratal growth" applied to successive, vertically growing rudist generations (Gili et al., 1995a) is doubted, because it would require improbably high sedimentation rates. In doubt as well is whether rudist associations in general were not able to build reefs. Hermatypic corals are extremely rare in rudist associations, although they can grow very successfully. There were no coral reefs in the Campanian sections of Saiwan. Colonial corals were very common below and above the *Vaccinites*-dominated associations, but they did not build coalescing structures. In the beds containing corals only flat-lying clingers occur, mostly radiolitids. Obviously a suitable substrate for vertically growing solitary corals and rudists was lacking. Sediment character and sediment structure point to water slightly deeper than in the rudist environments.

## ACKNOWLEDGMENTS

I would like to thank cordially K. and Dr. S. Engel for introducing me to the Saiwan area and for all their hospitality in Muscat, just as much W. Herget, his family, and his colleagues from Wintershall AG (Muscat) for their warm reception and important assistance. Just so I would like to thank Mr. K. Thomas and his colleagues from Conquest Oil Exploration Company (Muscat) for their extensive logistical help. For the most valuable discussions and instructions I thank Dr. J.-M. Pons (Barcelona) and Dr. S. Feist-Burkhardt (Darmstadt). Furthermore I would like to thank M. Walker and R. Borkhataria for their help in translation. Special thanks again to Dr. J.-M. Pons (Barcelona), Dr. P. W. Skelton (Milton Keynes) and Dr. J. Southard (Massachusetts Institute of Technology) for their time-consuming review and for the many valuable and critical comments on the first version of the manuscript. The responsibility for all remaining errors rests on myself. Last but not least I would like to thank my sons Bernhard and Ralf, R. and U. Bredner (Berlin), and J. Wittmann (Darmstadt) for their help and the most pleasant and cooperative ambiance during the field work. The investigation was generously supported by the Deutsche Forschungsgemeinschaft, Priority Program "Global and Regional Controls on Biogenic Sedimentation and Reef Evolution" (Grant Schu 410/10-1,2,3).

This publication is authorized by the Ministry of Petroleum and Minerals of the Sultanate of Oman.

## REFERENCES

AMICO, S., 1978, Récherches sur la structure du test des Radiolitidae: Université de Provence, Travaux Laboratoire de Géologie Historique et de Paléontologie, 131 p.

BARRON, E.J., 1990, Atlas of Cretaceous Climate Model Results: Pennsylvania State University, 32 p.

DUBREUILH, J., BÉCHENNEC, F., BERTHIAUX, A., LE MÉTOUR, J., PLATEL, J.P., ROGER, J., AND WYNS, R., 1992, Geological Map of Oman at 1/250 000, Al Khaluf, Sheet NF 40-15 and Explanatory Notes: Ministry of Petroleum and Minerals, Muscat, Oman.

FAGERSTROM, J.A., 1987, The Evolution of Reef Communities: New York, John Wiley & Sons, 600 p.

GILI, E., MASSE, J.-P., AND SKELTON, P.W., 1995a, Rudists as gregarious sediment-dwellers, not reef-builders, on Cretaceous carbonate platforms: Palaeogeography, Palaeoclimatology, Palaeoecology, v. 118, p. 245–267.

GILI, E., SKELTON, P.W., VICENS, E., AND OBRADOR, A., 1995b, Corals to rudists—an enviromentally induced assemblage succession: Palaeogeography, Palaeoclimatology, Palaeoecology, v. 119, p. 127–136.

HARRIS, P.M., FROST, S.H., SEIGLIE, G.A., AND SCHNEIDERMANN, N., 1984, Regional unconformities and depositional cycles, Cretaceous of the Arabian Peninsula, *in* Schlee, J.S., ed., Interregional Unconformities and Hydrocarbon Accumulation: American Association of Petroleum Geologists, Memoir 36, p. 67–80.

HUGHES CLARKE, M.W., 1988, Stratigraphy and rock unit nomenclature in the oil-producing area of Interior Oman: Journal of Petroleum Geology, v. 11, p. 5–60.

KAUFFMAN, E.G., AND JOHNSON, C.C., 1988, The morphological and ecological evolution of Middle and Upper Cretaceous reef-building rudistids: Palaios, v. 3, p. 194–216.

MORRIS, N.J., AND SKELTON, P.W., 1995, Late Campanian–Maastrichtian rudists from the United Arab Emirates–Oman border region: Natural History Museum [London], Geology Series, v. 51, no. 2, p. 277–305.

MURRIS, R.J., 1980, Middle East—stratigraphic evolution and oil habitat: American Association of Petroleum Geologists, Bulletin, v. 64, p. 597–618.

PERKINS, B.F., 1974, Paleoecology of a rudist reef complex in the Comanche Cretaceous Glen Rose limestone of Central Texas: Geoscience and Man, v. 8, p. 131–173.

PHILIP, J.M., AND PLATEL, J.-P., 1995, Stratigraphy and Rudist Biozonation of the Campanian and the Maastrichtian of Eastern Oman: Revista Mexicana de Ciencias Geológicas, v. 2, p. 257–266.

PLATEL, J.-P., PHILIP, J., BOURDILLON DE GRISSAC, C., BABINOT, J.-F., ROGER, J., AND MERCADIER, C., 1994, Modalités de la transgression campanienne sur le massif du Haushi–Huqf (Oman), Stratigraphie, contexte géodynamique et paléo-enviroments: Societé Géologique de France, Bulletin, v. 165, p. 147–161.

SCHUMANN, D., 1995, Upper Cretaceous rudist and stromatoporoid associations of Central Oman (Arabian Peninsula): Facies, v. 32, p. 189–202.

SCOTT, R.W., 1990, Models and stratigraphy of mid–Cretaceous reef communities, Gulf of Mexico: SEPM, Concepts in Sedimentology and Paleontology, v. 2, 102 p.

SKELTON, P.W., 1979, Gregariousness and proto-cooperation in rudists (Bivalvia), *in* Larwood, G., and Rosen, B.R., eds., Biology and Systematics of Colonial Organisms: London, Academic Press, p. 257–279.

SKELTON, P.W., GILI, E., VICENS, E., AND OBRADOR, A., 1995, The growth fabric of gregarious rudist elevators (hippuritids) in a Santonian carbonate platform in the southern Central Pyrenees: Palaeogeography, Palaeoclimatology, Palaeoecology, v. 119, p. 107–126.

STEUBER, T., 1995, Stable isotope sclerochronology of *Vaccinites cornuvaccinum* (Hippuritidae) from Beotia (Greece): Revista Mexicana de Ciencias Geológicas, v. 12, p. 307–314.

STEUBER, T., 1996, Stable isotope sclerochronology of rudist bivalves: Growth rates and Late Cretaceous seasonality: Geology, v. 40, p. 315–318.

STEUBER, T., 1997, Hippuritid rudist bivalves in siliciclastic settings—functional adaptions, growth rates and strategies: 8th International Coral Reef Symposium, Panama, Proceedings, p. 1761–1766.

# DIAGENESIS OF THE MIDDLE–UPPER JURASSIC CARBONATE PLATFORM (AMRAN GROUP), YEMEN

KHALID A. AL-THOUR
*Department of Earth and Environmental Sciences, Sana'a University, P.O. Box 2027, Sana'a, Yemen*
*email: hayssan@y.net.ye*

ABSTRACT: The Middle to Upper Jurassic shallow marine carbonate platform (Amran Group) is predominantly limestone to the west and northwest of Sana'a and limestone and dolomite to the east and northeast of Sana'a. Diagenesis of the Amran Group encompasses many processes with conspicuous effects, including cementation, dissolution, neomorphism, and compaction (both physical and chemical), producing secondary microporosity, micritization, and dolomitization. Dolomite cements are common and were precipitated mostly during later diagenesis in cavities and fractures. Replacive dolomitization occurred during shallow burial (small rhombic types) and during burial diagenesis with the formation of saddle dolomite. Integration of field, petrographic, and geochemical analysis (ICP) indicates that lithification of these carbonates occurred during synsedimentary and burial diagenesis, with much of the alteration controlled by eustatic sealevel change and regional tectonism.

Four major subenvironments, in which diagenesis of the Amran Group was operative, can be recognized. (1) Synsedimentary diagenesis is characterized by the formation of isopachous and syntaxial cements, hardgrounds (with associated borings and burrows, and shelter, fenestral, framework, interparticle, and intraparticle porosity), geopetal structures, and intraclasts, indicating deposition under marine conditions. (2) Shallow burial diagenesis shows other specific features such as leaching, recrystallization, and early dolomitization (both replacive and void-filling) and mold-filling cements. Moldic and vuggy porosity distribution, early compaction, collapse breccia, and silt deposition indicates that the Amran Group continued to receive meteoric water following sediment stabilization, enlarging some molds and vugs by solution. (3) Deep burial diagenesis is characterized by dissolution, blocky calcite cement, late compaction (fractures and sutured grains), and saddle dolomite. (4) Uplift diagenesis is characterized by reopening of stylolites along fractures and development of dolomitization under meteoric conditions. The occurrence of nonferroan calcite and ferric oxides in rhombohedral zones in dolomite indicates that dedolomitization was driven by oxidation and alteration of ferroan dolomite zones and probably reflects alteration related to recent weathering.

## INTRODUCTION

The Middle–Upper Jurassic carbonate sediments of the Amran Group extend from outcrops in the Sana'a region to the east into the subsurface in the central part of Yemen in the Wadi Al-Jawf-Marib–Shabwa'a Basin (Fig. 1). Full details of the sedimentology and stratigraphy of the measured sections, together with a complete review of previous works, are given in Al-Thour (1988), El-Anbaawy and Al-Thour (1989), Al-Ganad (1991), Al-Thour (1992b), Simmons and Al-Thour (1994), and Al-Thour (1997). These sediments either unconformably overlie the Precambrian basement rocks or disconformably overlie the Kohlan Formation and unconformably underlie the Tawilah Group and/or the Sabatayn Group. The thickness of the Amran Group is variable, ranging from 300 to 520 m. Through correlation between the different sections of each traverse, a composite lithostratigraphic log of the Amran Group was drawn up, and has a total thickness of 900 m (Fig. 2). Correlation between traverses is based on the specific sedimentary characteristics of the key horizons, as well as fossils (e.g., algae, foraminifera, corals, stromatoporoids, and palynomorphs), in different traverses (Al-Thour, 1988; Al-Ganad, 1991; Al-Thour, 1992b; Simmons and Al-Thour, 1994).

Studies of the diagenetic history of ancient carbonate platforms suggest a number of basic controls on the nature and degree of diagenesis, including the original facies and mineralogy, climatic conditions, sea-level changes, and burial history. Periods of emergence and flooding of a platform brought about by eustatic and/or tectonically induced changes in relative sea level produce significant variations in the patterns of diagenesis. Climate has been shown to exert a major control on early diagenesis and diagenetic history (Hird and Tucker, 1988; Wright, 1988; Moss and Tucker, 1995).

The carbonate sediments of the Amran Group exhibit evidence of great variety of diagenetic processes that caused porosity reduction through physical compaction, pressure solution, and mineral precipitation. These processes reflect the initial cycle of burial of the sediment, which then endured multiple episodes of burial and uplift. Despite the economic importance of these Jurassic carbonates, their diagenesis remains poorly understood. By detailing the overall diagenetic processes of the Middle–Upper Jurassic carbonate platform of the Amran Group, this paper aims to provide a diagenetic synthesis of this platform as a whole, through integration of field, petrographic, and geochemical analysis (ICP). It deals with the description and interpretation of the postdepositional processes and their sequences. Field observation and petrographic examination allow for more definitive statements about the extent of diagenesis.

## REGIONAL GEOLOGY

The sedimentary cover of the Arabian Peninsula is divided into three tectonic provinces (Powers et al., 1966). These are: (a) the Arabian Shield, where Precambrian (igneous and metamorphic) rocks are exposed, (b) the Arabian Shelf, including basins where the basement rocks are covered by a wedge of Phanerozoic sediments, and (c) the Mobile Belt, the unstable northern and eastern margins of the craton where the Taurus–Zagros–Oman Mountains are now emplaced (Fig. 1). The Phanerozoic sediments that cover the platform reflect a history of deposition during periods of marine transgression and regression of generally shallow shelf seas across the low-lying part of the Arabian Shield. The history of the carbonate platform during the Jurassic and Cretaceous of is overwhelming importance because of the extraordinary richness of hydrocarbons in these horizons (Shebl

Middle East Models of Jurassic/Cretaceous Carbonate Systems
SEPM Special Publication No. 69, Copyright © 2000
SEPM (Society for Sedimentary Geology), ISBN 1-56576-075-1, p. 155–171.

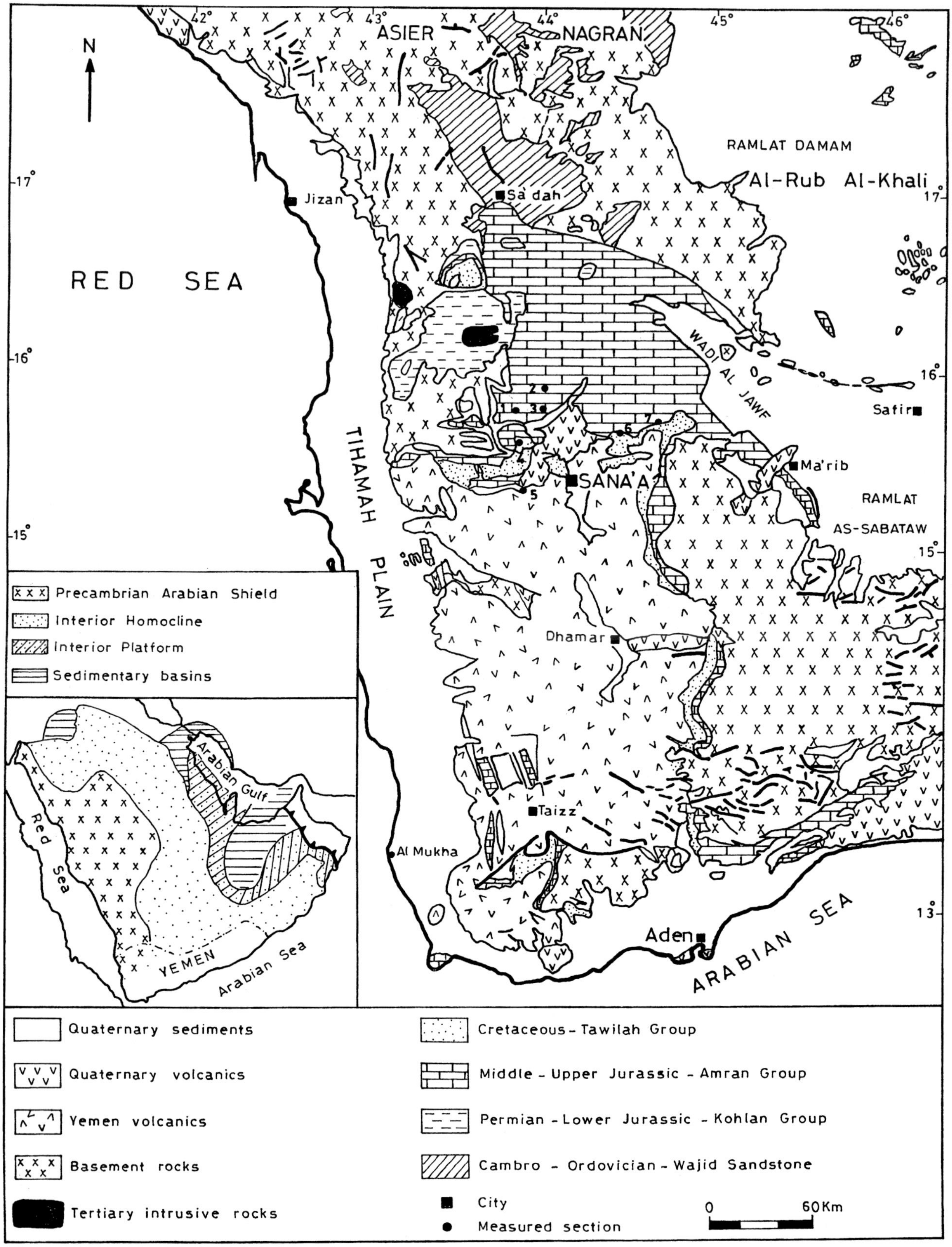


FIG. 1.—Simplified geological map of the central and western part of Yemen showing the location of the studied areas and measured traverses (1, Al Khothally; 2, Hamdah; 3, Wadi Nager; 4, Al Ayein; 5, Wadi Ahjur; 6, Al Gedáan; 7, Jabal Salab). Major tectono-sedimentary units in the Arabian Peninsula (after Alsharhan and Kendall, 1986) are shown on the left corner of the map.

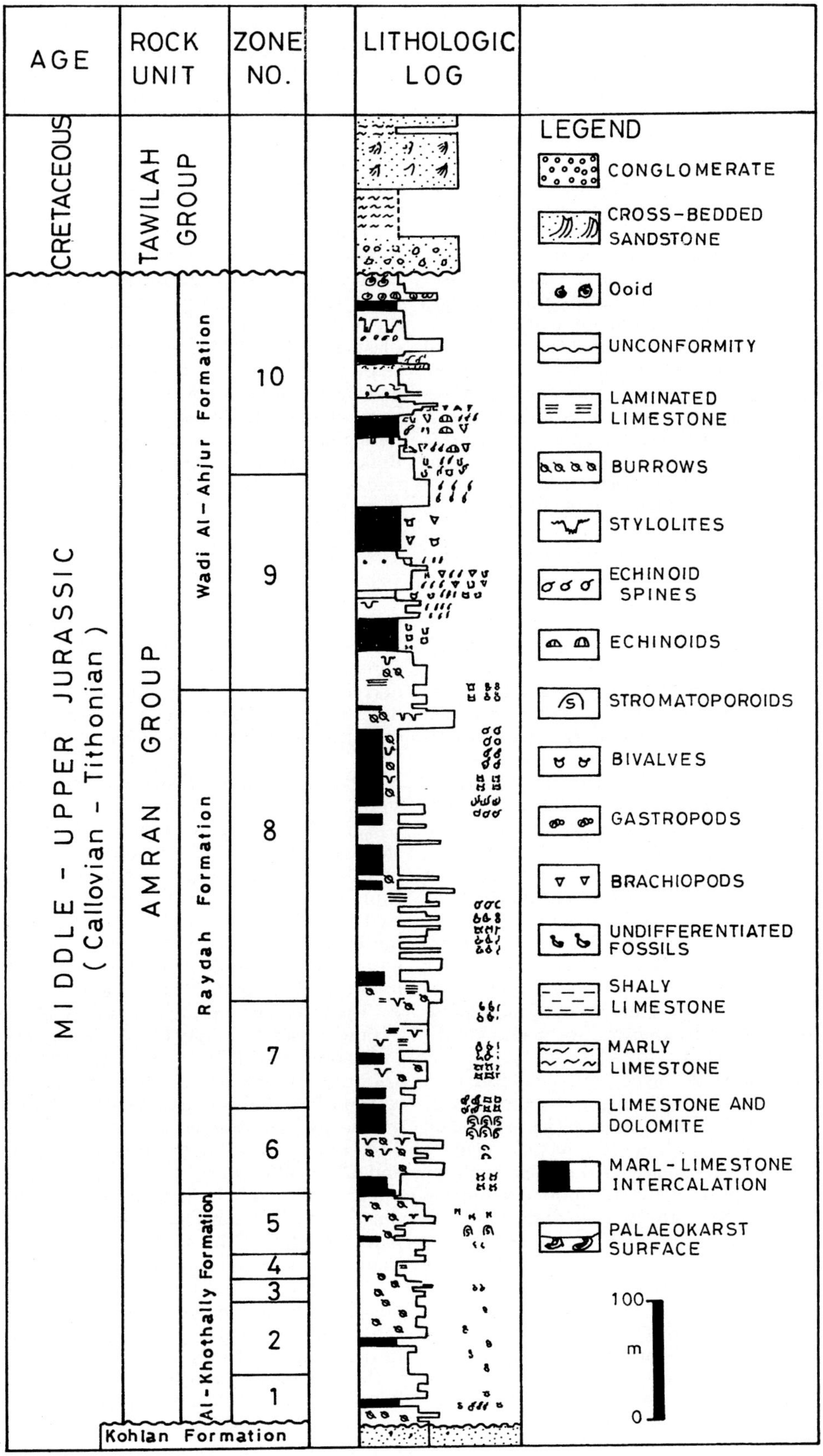


FIG. 2.—Composite lithostratigraphic log of the Jurassic Amran Group.

and Alsharhan, 1994; Alsharhan and Nairn, 1997). The oil and gas production from Mesozoic carbonates and siliciclastics in the central part of Yemen of the Wadi Al-Jawf-Marib–Shabwa'a Basin, regardless of age, is directly related to the occurrence of porous limestone, dolomite, and sandstone. In the Middle–Upper Jurassic sequence, the normally nonporous limestone became oil and gas reservoirs through fracturing or faulting and associated dolomitization (Al-Thour, in press).

The Middle to Late Jurassic in Yemen was a time of a gradual transgression of a shallow sea from the east and south (Al-Thour, 1997). At this time, the platform subsided steadily following continuous transgression. This subsidence occurred against a background of fluctuations in sea level. These could have been eustatic, but this is difficult to prove in the absence of refined biostratigraphic correlation. A quite stable pattern of sedimentation was established in the basin with abundant stromatoporoid and coral buildups. Mixed carbonate-siliciclastic sediments were deposited at a late stage, indicating tectonic differentiation of the platform. The first transgression phase led to deposition of the Al-Khothally Formation (Fig. 2). The second phase produced a deeper-water facies of the Raydah Formation. The relative sea-level changes together with increasing tectonic activity led to deposition of the mixed carbonate–siliciclastic sediments of the Wadi Al-Ahjur Formation, with the development of paleokarst surfaces at the top part of Al-Ayein traverse, indicating emergence.

Lithostratigraphic and paleogeographic studies by Beydoun (1988, 1991) and Alsharhan and Nairn (1997) show that the development of the carbonate platform, in the areas studied, began in Late Callovian time (Figs. 2, 3). The terrestrial clastic sandstone of the Kohlan Formation (Triassic–Middle Jurassic) is succeeded by shallow marine carbonate sediments and then mixed carbonate–siliciclastic deposits marking a general shallowing of the sea following an earlier phase of transgression. During Oxfordian–Kimmeridgian times some tectonic differentation was initiated. The carbonate platform was submerged in the Oxfordian–Kimmeridgian and the Raydah Formation was deposited, indicating a major rise of relative sea level. Gradual subsidence of the shelf could have been responsible for this submergence. However, the global picture suggests that this more likely reflects a eustatic rise of sea level (Hallam, 1978, 1988). The pattern of sedimentation changed sharply in the Tithonian, when a major sea-level fall occurred. Carbonate and siliciclastic deposits were established on the shelf, and the influx of terrigenous material increased sharply. As already indicated, however, a component of regional tectonism also increased (Al-Thour, 1997).

The overlying Upper Jurassic–Lower Cretaceous Sabatayn Group is composed of a very thick (6000 m) sedimentary succession of dominantly argillaceous and evaporitic facies overlain by detrital formations (Fig. 3). Synsedimentary tectonic activity from the Late Jurassic to the Early Cretaceous was responsible for the thick accumulation of argillaceous and evaporitic sediments in the subsiding rift basin. The unstable margin of the Wadi Al-Jawf-Marib–Shabwa'a rift basin was the site of rapid facies changes, local disconformities, and periods of emergence and dolomitization along the WNW- and NNW-striking boundary fault system (Al-Thour, 1988; El-Anba'awy and Al-Thour, 1989; Al-Ganad, 1991). The overlying Cretaceous Tawilah Group is composed of fluvial siliciclastic sediments (Al-Subbary et al., 1993; Alsharhan and Nairn, 1997).

## ANALYTICAL TECHNIQUES

The chemical analyses, by inductively coupled plasma emission spectrometry (ICP), were made at the Earth Sciences laboratory of the University of Birmingham, UK. ICP analysis of carbonates was performed after petrographic study and was used to determine the chemical composition of fifty-three samples from seven traverses. About 0.02–0.04 g of the finely powdered sample drilled from the pure micrites was homogenized and dried at 75°C for 2–3 hours. About 10 mg of the powdered samples was digested in 10 ml of 5% Aristar-grade HCl and the insoluble residue content determined. About 5 ml of the soluble part of the solution was separated and stored in polyethylene containers for ICP analysis.

The standards used were prepared from individual atomic absorption metal solution, and several standard reference sediment samples were analyzed to check the accuracy of the instrument. To run the analysis, the dissolved sample is nebulized into the ICP torch, where it is atomized in a high-temperature argon plasma (8000–10,000 K) and emits an atomic spectrum, which is resolved into the atomic lines of each element by an optical grating in an optical spectrometer. Each element-selective atomic spectral line is measured with a photomultiplier detector on the local plane of the spectrometer. The signal from each photomultiplier is amplified and printed through a microprocessor. All trace-element compositions are given in ppm of the carbonate fraction.

## FACIES SEQUENCES IN THE AMRAN GROUP

The Amran Group carbonate platform is divisible into three formations (Al-Thour, 1997) (Figs. 2, 4): (1) The lower Al-Khothally Formation is upper Callovian–Oxfordian in age (Zones 1–5 of Simmons and Al-Thour, 1994). This unit consists of sandy, oolitic, oncolitic, peloidal, partly dolomitic, massive to thick-bedded limestone about 200 m thick. (2) The middle Raydah Formation is early Kimmeridgian in age (Zones 6–8 of Simmons and Al-Thour, 1994). This unit shows several lithological facies beginning with massive, thick, cherty, fossiliferous, bituminous grainstone, packstone, wackestone, mudstone, and silty limestones intercalated with marly fossiliferous limestones, capped by massive mudstones with echinoids, stromatoporoids, and corals. It is about 400 m thick. (3) The upper Wadi Al-Ahjur Formation is late Kimmeridgian–Tithonian in age (Zones 9 and 10 of Simmons and Al-Thour, 1994) and consists of several carbonate intervals of mudstone, wackestone, packstone, and grainstone intercalated with sandy dolomitic marly fossiliferous shale and partly silty limestone. It is about 300 m thick.

The carbonate sediments of the Amran Group are composed of outcrop-scale sedimentary cycles of various types and scales (Al-Thour, 1992b). They comprise three facies associations (Al-Thour, 1997). Their organization is characterized by interfingering relationships, indicating that these sediments were deposited in a variety of depositional settings. The facies are grouped on the basis of shared sedimentological and paleontological features and named after the dominant environment in which they were interpreted to have been deposited. The three facies associations (Fig. 5) are distinguished in the Amran Group: (F1) carbonate platform facies, (F2) carbonate–marl alternations facies, and (F3) shallow-water coral and stromatoporoid buildup facies.

F1 dominates in all formations of the Amran Group. F2 is well developed in the Raydah and Wadi Al-Ahjur Formations, and F3 can be recognized in the upper part of the Raydah Formation but its distribution is limited.

The lateral and vertical changes of facies are interpreted to be due to relative sea-level changes responding to associated tectonism and eustasy. The repetition and interfingering of both fining-upward and shallowing-upward cycles in the study areas suggest that deposition occurred during a period fluctuating eustatic sea level.

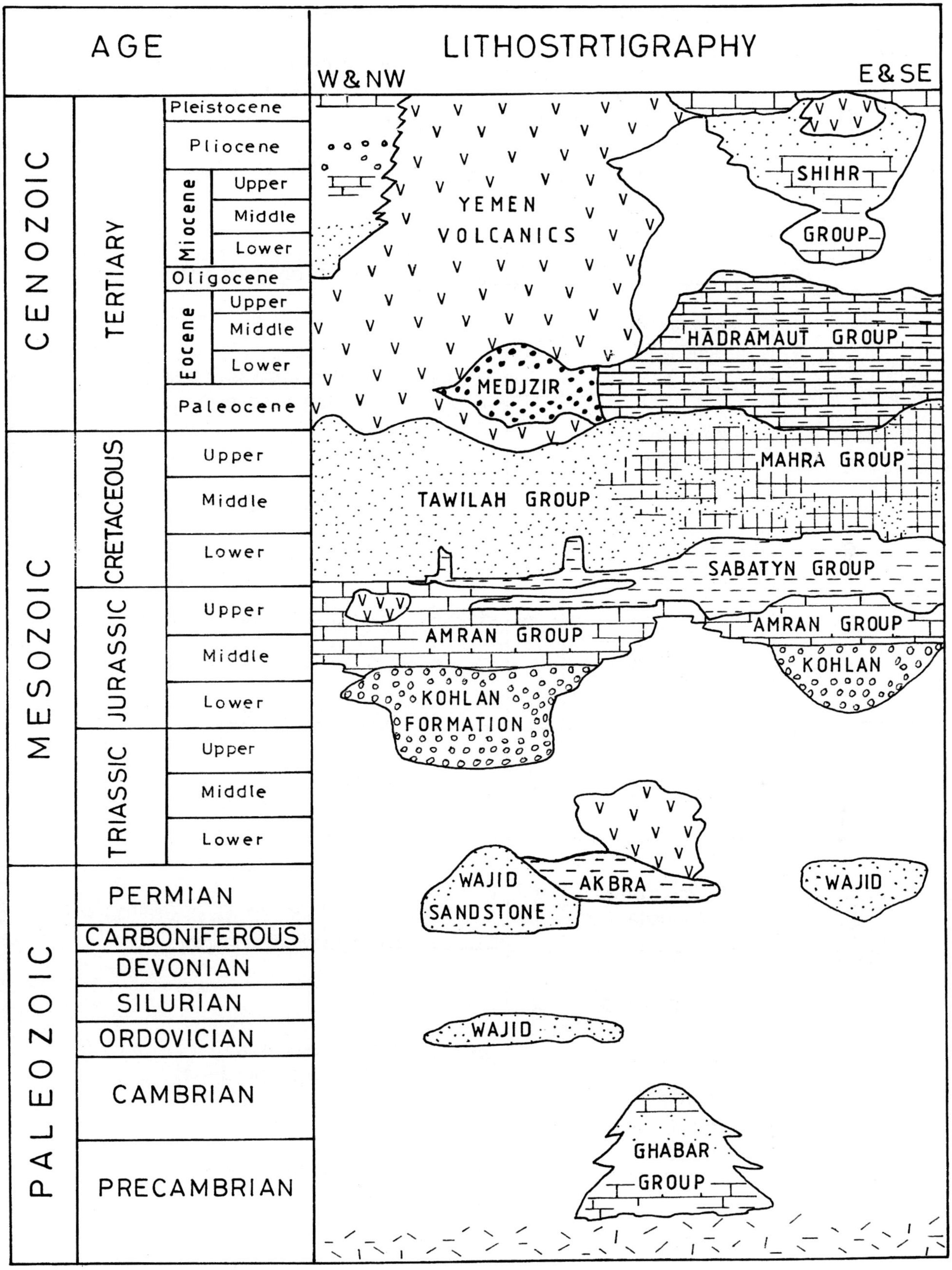


FIG. 3.—Generalized geologic column of various rock units in Yemen (modified from Beydoun, 1988, 1991; copyright material reporduced with permission of Scientific Press).

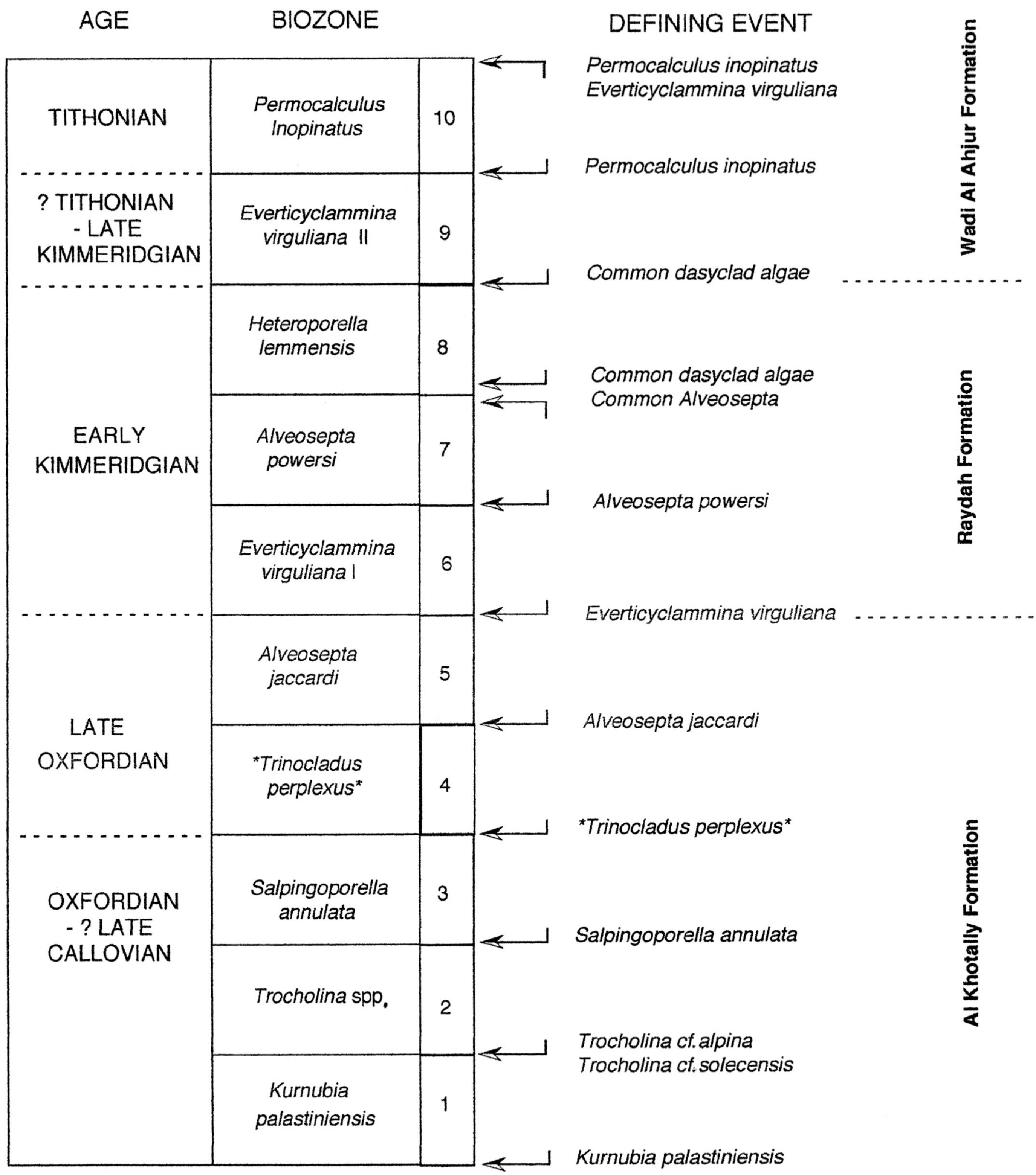


FIG. 4.—Micropaleontological biozonation of the Amran Group (after Simmons and Al-Thour, 1994).

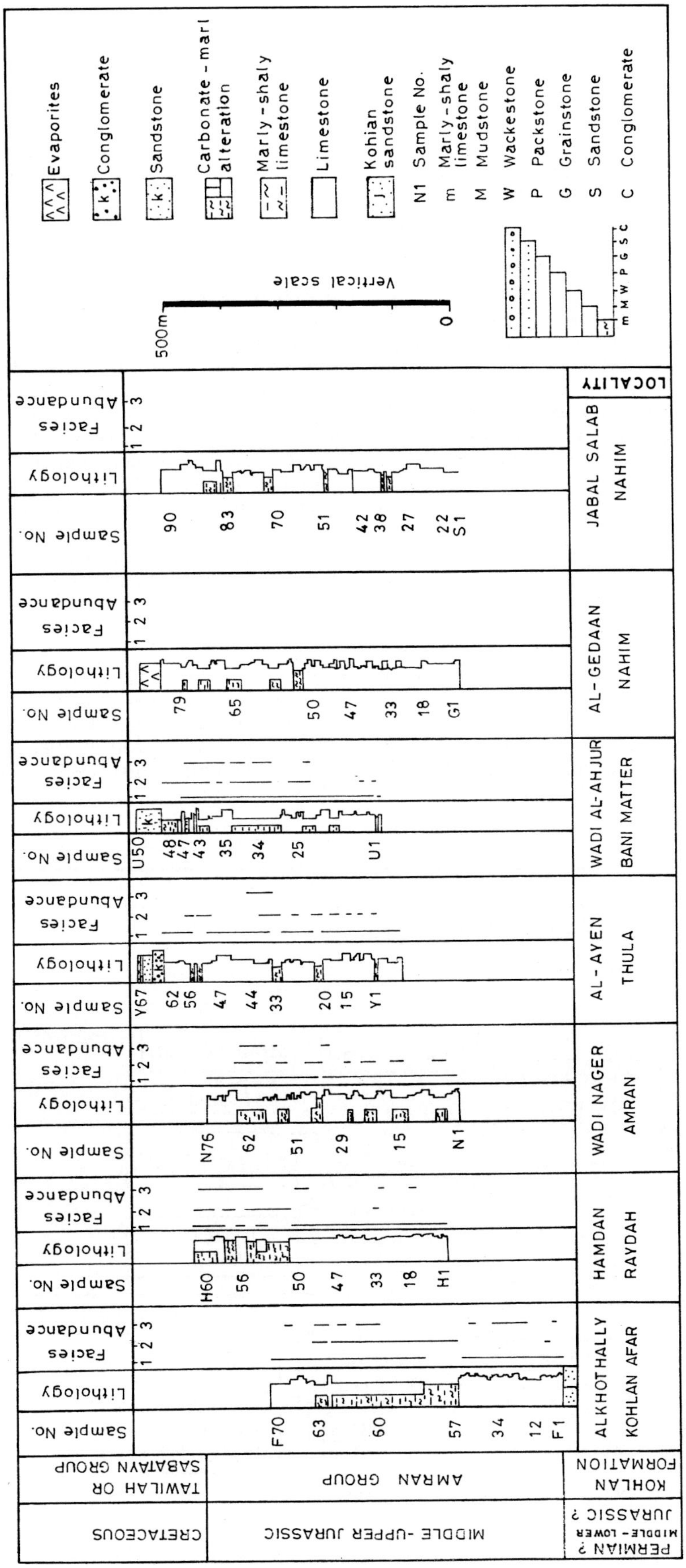


FIG. 5.—Stratigraphic sections showing facies associations and abundance within the different rock units of the Amran Group (modified after Al-Thour, 1997).

## DIAGENETIC SYNTHESIS OF THE AMRAN GROUP

Four major diagenetic sub-environments affected the Amran Group: (1) synsedimentary diagenesis, (2) shallow burial diagenesis, (3) deep burial diagenesis, and (4) uplift diagenesis. A generalized diagenetic scheme for the Amran Group (Fig. 6) showing the different types of cements and their relation to diagenetic sequences is based on study of over 450 thin sections using both standard petrography and cathodoluminescence. Diagenesis of the Amran Group encompasses many processes with conspicuous effects on carbonate sediments, such as cementation and dissolution to produce secondary porosity (Table 1), including the development of secondary microporosity (e.g., when unstable minerals such as aragonite were dissolved to leave molds) (Figs. 7, 8). Early cementation in the sediments is seen in the nuclei of ooids and oolitic coated grains and the internal framework of colonial organisms (Table 2). Cementation has replaced the nucleus of the ooids and/or skeletal grains prior to physical compaction. Compaction (both physical and chemical) becomes an increasingly significant process as a result of increasing overburden pressure. In the early stages of burial, physical compaction is more important and results in closer packing and fracturing of grains (Table 3). Eventually, chemical compaction leads to dissolution at grain contacts and the formation of planar and/or anastomosing stylolites by dissolution.

Late cements are coarser than early cements and are up to several millimeters in size. They can be seen in the field usually occluding secondary porosity developed by dissolution within the skeletal grains. The same cement occludes primary porosity (e.g., fenestrae) and secondary pore spaces developed by dissolution of colonial organisms. Petrographically, these cements are nonferroan and have a drusy fabric. They infill both interparticle and intraparticle porosity. Dolomite cements are common in these limestones, and most were precipitated during later diagenesis in cavities and fractures (Table 4). Replacive dolomitization (Sellwood et al., 1987) occurred during shallow burial (small rhombic crystals 36–280 μm in size) and during burial diagenesis with the formation of saddle dolomite. Saddle dolomite crystals are perpendicular to the walls of allochems, e.g., foraminifera and bivalves, and/or in fractures as a result of dissolution and/or tectonism. In Jabal Salab two major periods of dolomitization are recognized. The first period was widespread and affected large volumes of the carbonates. It occurred before emergence and deep karstic erosion. The second period was associated with fractures and microcavities filled during sulfide mineralization (Al-Ganad, 1991).

The sediments of the Amran Group also show some evidence of dedolomitization (Table 4). A most distinctive petrographic texture present within the Amran Group is the rhombohedral zones of calcite and iron oxide. The rhombohedral shape suggests that selected dolomite zones have been either replaced or that calcite and iron oxide grew on dolomite substrates during interruptions in dolomite development. Dedolomitization occurred when meteoric waters invaded rocks and washed the dolomite crystals, removing the magnesium while helping the precipitation of calcite. Dedolomitization occurred by oxidation and alteration of ferroan dolomite zones and probably reflects alteration related to recent weathering.

### *Synsedimentary Diagenesis*

Synsedimentary diagenesis is characterized by the formation of syntaxial and isopachous cements in developing hardgrounds (with associated borings and burrows, and primary shelter, fenestral, framework, interparticle, and intraparticle porosity), geopetal structure, intraclasts, and cementation. Columnar stubby cement is associated with high-energy facies.

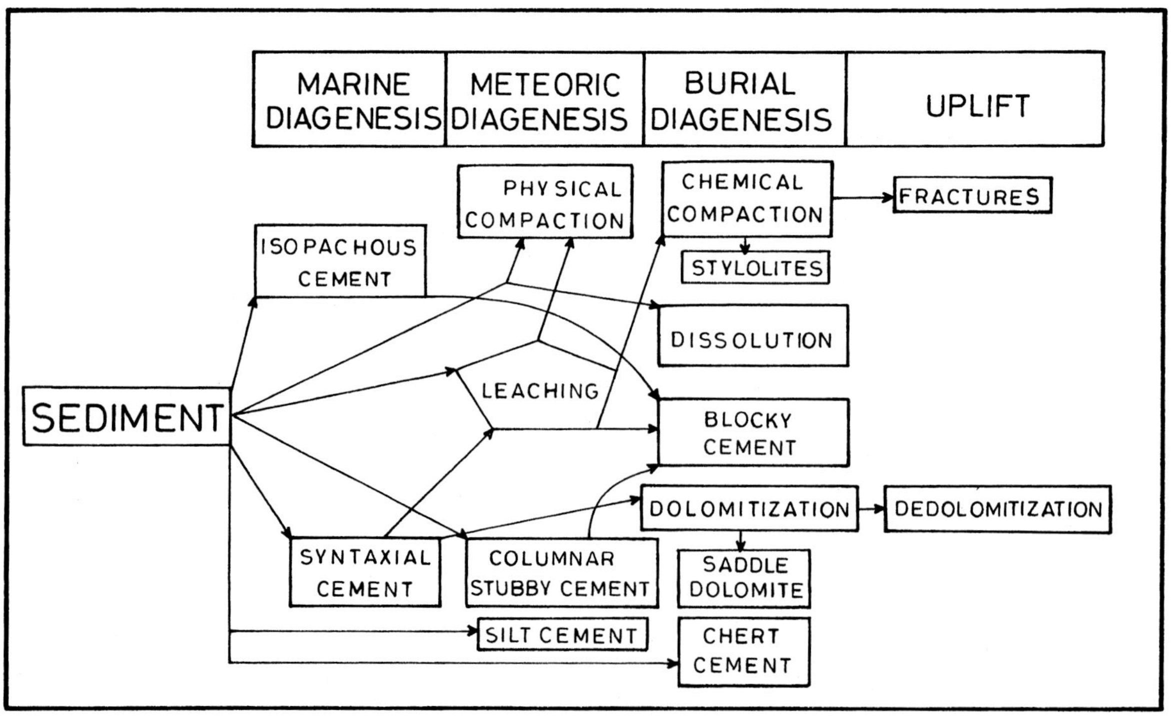


FIG. 6.—Diagram showing the different types of cements and their relation to diagenetic sequences.

TABLE 1.—Porosity in the Middle–Upper Jurassic Carbonates of the Amran Group.

| DIAGENETIC FEATURES | DESCRIPTION | INTERPRETATION | OCCURRENCES | FIGURES |
|---|---|---|---|---|
| Interparticle | Grains develop in mud-free carbonate sediments; their distribution is very limited. Have spherical shape, their size ranges between 2 mm and 1.5 cm. | The shape variation of grains is a function of their biological origin. The type of fossil grains indicate marine or nonmarine environment. | Common in all traverses. | 7A |
| Intraparticle | Developed within grains e.g. chambers of foraminifera, gastropods, bryozoa, corals and stromatoporoids. They show well-developed pore spaces. | Grains tend to become quickly filled during early reef development by both coarse and fine-grained internal sediments, leading to a complex depositional pore system. The development of this porosity within grains before, during, and shortly after deposition indicate a marine environment. | Common in Al-Khothally, middle-upper part of the Hamdah, middle part of Wadi Nager, and lower-middle part of Al-Ayein, Jabal Salab, and rare in Wadi Al-Ahjur and Al-Gedaán traverses. | 7B |
| Fenestral | Pores are small and vary from one microfacies to another. Rock surfaces have a distinctive mottled structure. | They are a fabric-selective depositional indicating a marine environment. The mottled structure indicates intense bioturbation. | Common in Al-Khothally Hamdah, Al-Ayein, Wadi Al-Ahjur, Al-Gedáan, and Jabal Salab traverses, but rare in Wadi Nager traverse. | 7C |
| Shelter | Occurs below shell fragments in a convex-up position. Fossil fragments have been affected by dissolution, cementation, and compaction. Compaction may take place prior to cementation. | Fossil grains provide an umbrella from the area beneath, protecting it from finer interstitial detritus. Sediments are unable to fill the cavities because of the "umbrella" effect of the shell. It is a fabric-selective depositional type. | Common in the middle and upper part of Al-Khothally, upper part of Hamdah, middle part of Al-Ayein, Wadi Al-Ahjur, Al-Gedáan, and Jabal Salab traverses, but is rare in Wadi Nager traverse. | 7D |
| Vuggy | Irregular secondary holes formed by dissolution that cut across grains and/or cement boundaries. Associated with unconformities. | Developed as a result of dissolution under meteoric conditions. | Common in Al-Gedáan and rare in Al-Khothally Hamdah, middle part of Wadi Ager. | 7E |
| Moldic | Formed by the dissolution of aragonite bioclasts. Have different shapes and sizes depending on their mode of origin. | Represents the first phase of dissolution and may have been related to exposure and meteoric dissolution of the sediments at the end of the Amran Group times. | Well developed in all traverses. | 7F |
| Intercrystalline | Associated with replacive dolomites formed by recrystallization of limestones and is the result of dissolution and dolomitization processes. A second phase of dissolution occurred after dolomitization, created minor amounts of intercrystalline porosity by the partial dissolution of zoned dolomite. The crystal shape is variable and the size ranges between 30 and 280 µm. | Form an important reservoir type in a number of settings ranging from supratidal/sabkha to normal marine sequences. | Developed in Jabal Salab, Al-Gedáan, lower part of Al-Khothally, upper part of Al-Ayein, Wadi Al-Ahjur, and Jabal Salab traverses. | 7G |

TABLE 1 (continued).—Porosity in the Middle–Upper Jurassic Carbonates of the Amran Group.

| DIAGENETIC FEATURES | DESCRIPTION | INTERPRETATION | OCCURRENCES | FIGURES |
|---|---|---|---|---|
| Fracture | Occurs in brittle, normally homogeneous carbonates. A tensional phase causes the opening of fractures or suture seams or if there are mineralogical changes in the rock such as dolomitization. | Indicate tectonic deformation or solution collapse associated with dissolution of limestone. Developed at any time during burial history of a carbonate sequence starting with shallow burial because of common early lithification. | Common in all traverses. | 7H |
| Porosity associated with breccia | Formed as a result of tectonic activity that led to the development of karst surfaces. Associated with limestone solution collapse, often resulting in enhanced porosity that may form either a reservoir for hydrocarbons or a host for mineralization. | Indicates deposition under shallow marine environment. Brecciation of carbonate rock sequences can occur in a number of situations, including limestone solution collapse, faulting, or soil formation. These breccia zones developed during episodic subaerial exposure. | Well developed in Al-Ayein, Thula, Al-Gedáan, and Jabal Salab traverses. | 7I |

Both isopachous and columnar stubby cements are nonluminescent under CL, and their better preservation suggests an original calcite mineralogy. Specific features can be related to the basic regional development. Hardgrounds were formed repeatedly at several horizons in all formations of the Amran Group, indicating that these carbonate deposits received early cement during deposition. Truncation of shells indicates that the sediment must have been cemented at the time of boring. Burrows are commonly filled with calcite. Leached burrows may show fenestral structure. The shelter porosity was developed under marine conditions very close to shoals. Physical compaction affected the sediment and was followed by dissolution and finally cementation. Micritization of grains was particularly common in the low-energy facies. Micritization and marine cementation were controlled by the facies and the depositional setting of the sediments on the platform. For example the packstone microfacies shows more micritization and less marine cementation than the skeletal oolitic grainstone microfacies. The relationship of fractured and collapsed micrite envelopes, fractured remnant grain material, infilling of dissolved grains, fractured marine cements, and subsequent spar generations and stylolitization provide petrographic evidence to constrain the diagenetic setting of the dissolution. Spar infilling voids produced by dissolution of grain material is postdated by stylolitization, showing that these cements predate significant burial. Thus, the processes responsible for creating the pore spaces must have occurred at shallow burial depths, suggesting that meteoric fluids caused dissolution (Moss and Tucker, 1995).

### *Shallow Burial Diagenesis*

Shallow burial diagenesis shows other specific features in all formations. Leaching, recrystallization, and early dolomitization (both replacive and void-filling) are the main features developed during shallow burial diagenesis of the Amran Group. The mold-filling cements are likely to have been precipitated during very shallow burial and before any compaction. The moldic and vuggy porosity distribution of early compaction, collapse breccia, and silt deposition indicate that the Amran Group continued to receive meteoric water after sediment stabilization, enlarging some molds and vugs.

The earliest of these late diagenetic phenomena is leaching (Bathurst, 1975; Choquette and Pray, 1970; Druckman and Moore, 1985; Moore, 1989). Evidence of dissolution is found in approximately sixty percent of the samples examined. It is important to mention that porosity, developed by leaching, has been obscured by the development of both columnar stubby and blocky calcite cement. Much of this cement was created in the shallow burial environment with the influence of meteoric water, but

FIG. 7 (opposite page).—A) Photomicrograph of skeletal packstone microfacies showing interparticle porosity and the occurrence of micritization around bivalve and echinoid fragments, scale is 1 mm; B) Photomicrograph of skeletal wackestone microfacies showing the development of framework porosity within algae (*Rivularia piae*), scale is 0.5 mm; C) Photomicrograph of fenestral mudstone microfacies showing fenestrae which forms by dissolution; scale is 4 mm; D) Photomicrograph of boundstone microfacies showing sheltering effects of compacted shallow marine gastropods; scale is 10 mm; E) Photomicrograph of dolomite microfacies showing the development of intercrystalline porosity; incipient dedolomitization has occurred; scale is 1 mm.; F) View of the Jurassic–Cretaceous paleokarst surface at the top of the Al-Ayein traverse, Thula area, Wadi Al-Ahjur Formation.

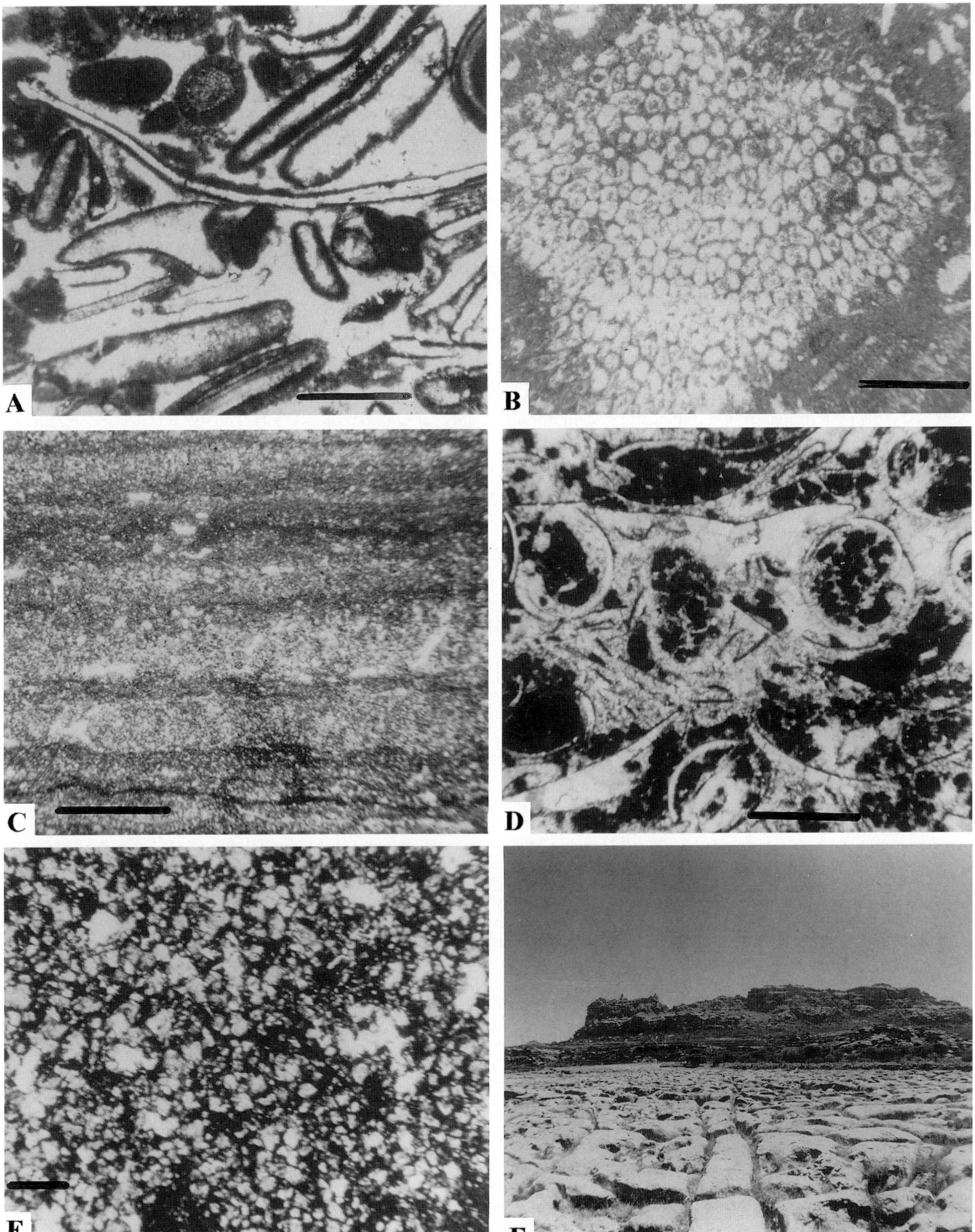
A
B
C
D
E
F

undoubtedly some solution of shallow burial cement has taken place. In some cases the secondary porosity thus created was partly destroyed by dolomite cements, but when dolomite content reached 50%, porosity is expected to increase with increasing dolomite percentage. The dolomite cement that fills leached porosity is the same as that filling nearby primary pores. The highest proportion of leaching effects is seen in coated grains and algal debris.

### *Deep Burial Diagenesis*

The deep burial diagenesis is characterized by some specific features such as dissolution, blocky calcite cement, compaction (fractures, sutured grains, and stylolites) and the formation of saddle dolomite (Scholle and Halley, 1985; Choquette and James, 1987). Chemical compaction was followed by further cementation of blocky calcite in void spaces. Burial pressure solution is the most likely explanation both for stylolites with organic material along grain boundaries and locally highly compacted nodular lithologies. Blocky calcite cements exhibit a consistent luminescent zonation, with an initial nonluminescent zone followed by a bright one. This pattern may reflect a meteoric system initially open to oxygenated waters. Fe and Mn cannot be incorporated into the calcite lattice under oxygenated conditions, so the initial calcite zones exhibit no luminescence. The bright luminescence of the second calcite zone indicates that precipitation fluids later became reducing, allowing incorporation of Fe and Mn into the crystal lattice (Moore, 1989). Finally, the development of dolomite either in veins, fractures, and/or molds or as a replacement of micrite have been recognized under burial conditions. Saddle dolomite and fracturing were developed indicating post-burial conditions.

### *Uplift Diagenesis*

Uplift diagenesis is characterized by reopening of stylolites through fractures and development of dolomitization under meteoric conditions. Frank (1981) suggested that the occurrence of nonferroan calcite and ferric oxides in rhombohedral zones in dolomite indicates that dedolomitization occurred by oxidation and alteration of ferroan dolomite zones and probably reflects alteration related to recent weathering. Subaerial exposure punctuated the development of the platform and led to the formation of a paleokarst surface at the top part of Al-Ayein Traverse, Thula area, Wadi Al-Ahjur Formation.

## CONCLUSIONS

Diagenesis of the Middle–Upper Jurassic carbonates of the Amran Group involved many processes and produced conspicuous effects, such as cementation and dissolution, resulting in secondary porosity, including the development of secondary microporosity. Compaction (both physical and chemical) became an increasingly significant process as a result of increasing overburden pressure. In the early stages of burial, physical compaction was more important and resulted in closer packing and fracturing of grains. Eventually, chemical compaction led to dissolution at grain contacts and the formation of planar and/or anastomosing stylolites by dissolution. Dolomite cements are common in these limestones and were mostly precipitated during later diagenesis in cavities and fractures. Replacive dolomitization occurred during shallow burial (small rhombic types) and during burial diagenesis with the formation of saddle dolomite. Dedolomitization occurred in response to oxidation and alteration of ferroan dolomite zones and probably reflects alteration related to recent weathering.

However, the diagenetic events in the Amran Group carbonate platform reveals the importance of several controls on carbonate diagenesis as a site for cementation and dolomitization of platform carbonates. Diagenesis in the Amran Group sediments was controlled by several factors. Original mineralogy and position of relative sea level controlled both early meteoric and later shallow-burial diagenesis by preferential dissolution of the less stable aragonitic components. The secondary porosity then provided sites for further calcite cementation. Climate was important in controlling the extent of meteoric diagenesis that the platform experienced during periods of exposure, so that karstification and meteoric diagenesis were limited. Sea-level changes were important for determining the timing, extent, and duration of exposure of the platform. Late burial was important for determining the possible sources of fluids for shallow-burial and deep-burial cementation as well as dolomitization. The major phase of calcite cementation, in most samples, postdates early physical compaction features, and this calcite occludes the porosity produced during meteoric exposures and synsedimentary deposition. Dolomitization can also be seen to postdate early compactional features.

Four major subenvironments where diagenesis of the Amran Group were operative are: (1) synsedimentary diagenesis, (2) shallow burial diagenesis, (3) deep burial diagenesis, and (4) uplift diagenesis.

## ACKNOWLEDGMENTS

The author thanks his numerous colleagues for a stimulating discussion. Drs. C.G.St. Kendall, Ihsan Al-Aasm, A.S. Alsharhan, and R.W. Scott provided excellent reviews and suggestions which improved the paper.

## REFERENCES

AL-GANAD, I., 1991, Etude geologique et metallogenique du Gesment ZnPbAg de Jabali (Bordure sue du basin de Wadi Al-Jawf, Republic de Yemen): Ph.D. thesis, Université d'Orleans, 173 p.

FIG. 8 (opposite page).—A) Photomicrograph of skeletal grainstone-wackestone microfacies showing columnar stubby cement that developed within leached shells and grew out from grain to meet another similar stubby spar growing from opposite grain, scale is 1 mm; B) Photomicrograph of packstone-wackestone microfacies showing small stubby crystals that form crudely isopachous rinds of cement, particularly within leached shells; scale is 1 mm; C) Photomicrograph of oolitic grainstone microfacies showing the effect of physical compaction and dissolution of aragonite; the compaction of the outer ooid layer suggests dissolution of the aragonite layer. Note that most of the ooid has Mg-calcite; scale is 1 mm; D) Scanning electron micrograph showing microporosity within mudstones. Note that the micrite is inhomogeneous, which indicates several different sources of the micrite, scale is 0.1 mm; E) Photomicrograph of skeletal wackestone microfacies showing the development of dolomite within micrite. Note that dolomite rhombs are partly dedolomitized; scale is 1 m; F) Jurassic calcrete paleosols from Al-Geda'an section, Naihm area to the east of Sana'a, Yemen. Note the presence of black pebbles in both the calcrete and the overlying carbonate; object in center is about 10 cm.

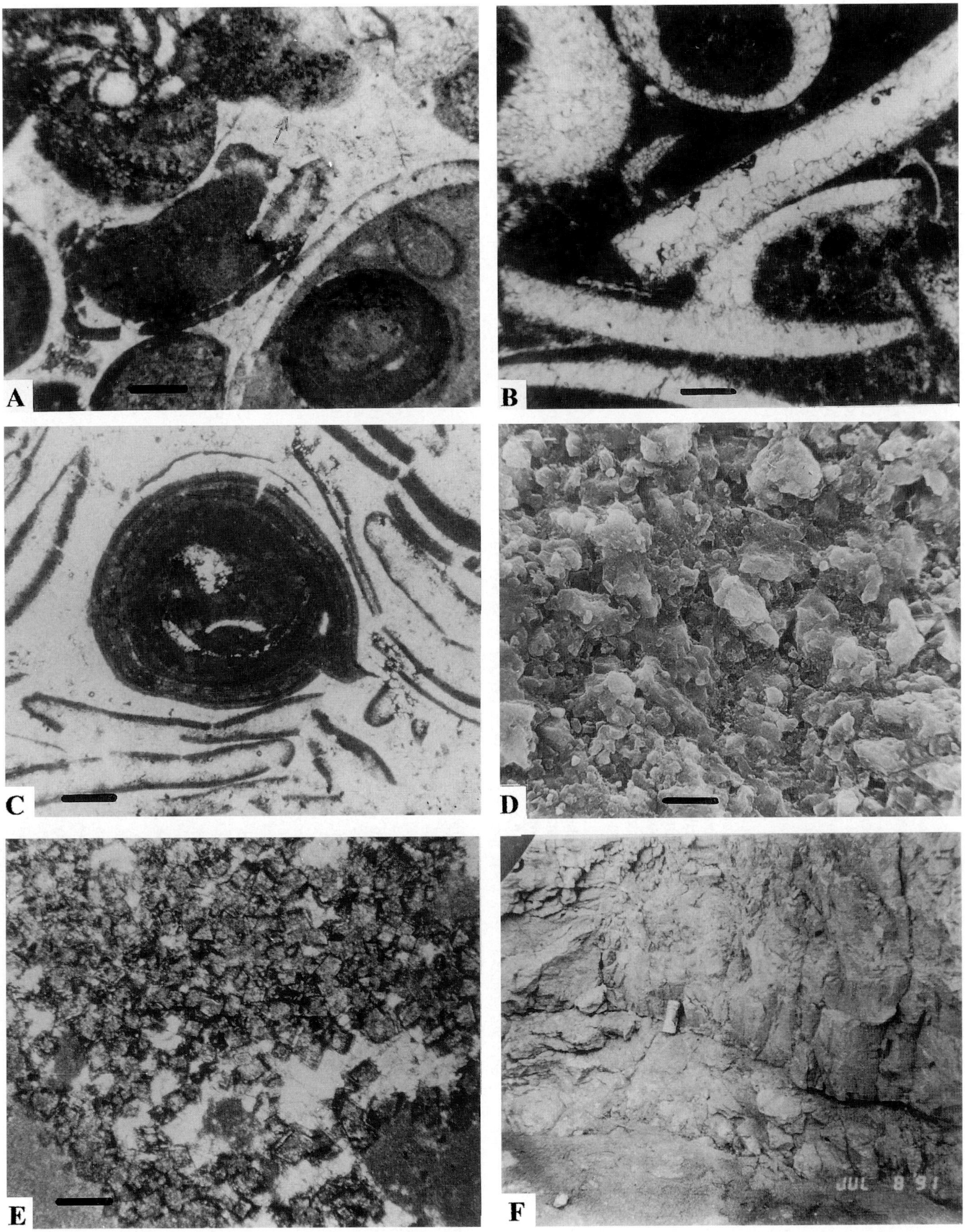
A
B
C
D
E
F
JUL 8 91

TABLE 2.—Cementation in the Middle–Upper Jurassic Carbonates of the Amran Group.

| DIAGENETIC FEATURES | DESCRIPTION | INTERPRETATION | OCCURRENCES | FIGURES |
|---|---|---|---|---|
| Isopachous | Consists of a uniform fringe of fibrous calcite cement around micritized grains. The size of the crystals is 1 mm. Grew perpendicular to the edges of the grains. | Represents the first generation of cement which can be interpreted as marine. | Common in the upper part of the Al-Khothally, middle part of the Hamdah, Al-Ayein, Al-Gedáan, Jabal Salab, and lower part of Wadi Al-Ahjur traverses. | 8A |
| Columnar stubby | Consists of small crystals (2–4 mm) that form crudely isopachous rims of cement, particularly in interparticle pores. The length-to-width ratio of these crystals is relatively small, ranging from 1:1 to 2:1. Developed within leached originally aragonitic shells and grew out from grain surfaces to meet similar stubby spar growing from the opposite grain. Sometimes stubby cement followed by blocky cement. | Represents the second phase of cementation after the development of isopachous cement, which can be interpreted as marine. It is also very common in meteoric environment. | Well developed in all studied traverses. | 8B |
| Blocky calcite | Forms a drusy texture in which the precipitation of calcite cement starts with the formation of small crystals on the edges of the particles, then these crystals ultimately grow away from the particle margin and their size increases, while they decrease in number. Fills the pore spaces inside and outside particles. | Represents the second and/or third phase of cementation, after the development of either isopachous cement, columnar stubby, cement or both, which can be interpreted as a shallow burial and/or deep burial environment. | Well developed in all studied traverses. | 8C |
| Syntaxial overgrowth | Found around the remains of echinoderms in optical continuity with the original single crystals. Have a porous structure progressively filled by the growth of the cement. The growth of rim cement is prevented by the micrite envelope, and it is not extended outside the grain. | Syntaxial calcite overgrowths on echinoderm debris associated with hardgrounds indicating seafloor cement. The overgrowth cement is regarded as synsedimentary in origin because it is cut by other grains or borings. Indication of synsedimentary cement overgrowths is shown by the growth of cement towards the boring walls. | Common in Al-Khothally, upper part of Hamdah, Wadi Nager, Al-Ayein, Wadi Al-Ahjur, Al-Gedáan and Jabal Salab traverses. | 8D |
| Micritization | Skeletal fragments appear to be sheathed in an envelope of micrite. The envelope is seen as patchy micrite and composed of micritic calcite. The outer surface coat has gentle smooth contours, while the inner surface is irregular. Grows in two directions, both inside and outside the grain. They are fine-grained carbonates, with crystal size less than 4 μm. This carbonate appears as a dense or translucent groundmass in thin sections. | Micritization occurs mainly by repeated boring of algae and fungi, infilling of the microborings with microcrystalline precipitates, or as a result of recrystallization. Micritization and diminution of skeletal grains in shallow seas provide a major source of lime mud. | Recognized in all traverses. | 8E |

TABLE 2 (continued).—Cementation in the Middle–Upper Jurassic Carbonates of the Amran Group.

| DIAGENETIC FEATURES | DESCRIPTION | INTERPRETATION | OCCURRENCES | FIGURES |
|---|---|---|---|---|
| Chert nodules | Characterized, in the field, by their nodular ellipsoidal shape; size is variable, but can reach 60 cm long and 6 cm wide. Have sharp boundaries which truncate the contained fossils and are enriched in iron and/or manganese minerals. | Their occurrence at the top of wackestone beds indicates shallow burial depth. The absence of this kind of cement in high-energy shallowing-upward sequences could indicate a submarine process. | Middle and top part of Al-Khothally, the lower and middle part of Hamdah, lower part of Wadi Nager, Jabal Salab, the upper part of Al-Ayein, and in the middle part of Wadi Al-Ahjur traverses. | 8F |
| Silt deposition | Represented by massive beds 20 cm thick. At certain horizons, the marly shaly fossiliferous limestone are intercalated with silty mudstones. | A storm-induced depositional mechanism is the reasonable and accepted interpretation for this sort of alternation. | Occurs at Wadi Al-Ahjur and in the middle part of Al-Ayein traverses. | 8G |
| Jurassic calcrete | The calcrete beds, 20-30 cm thick, are associated with ferruginous hardground surfaces, fenestrae, and geopetal structures. Lamination around spar-filled alveoli and grains can be recognized and may reflect the first stage of calcrete formation. These laminae may surround all or only part of the void and/or grain. Their thickness is about 2 mm. Irregularly shaped black pebbles consist of angular fragments of soilstone crusts. Jurassic calcrete is dense, dull, and dirty. | Common feature in semi-arid to arid climate regimes. Black pebbles occur above unconformities indicating meteoric conditions. | Well developed in the middle and top part of Wadi Al-Ahjur, middle part of Hamdah and middle-upper part of Al-Gedáan traverses. | 8H |
| Carbonate mud (micrite) | A fine-grained, usually dark matrix is common and susceptible to diagenetic alteration and may be replaced by coarser mosaics of microspar through aggrading neomorphism, which results in lower porosities. Micrite is not homogeneous, which means that it did not come from one source. Shows evidence of periods of subaerial exposure that led to a influx of meteoric water during and at the end of the Jurassic. | Micrites (carbonate mud) occur in the shallow marine, less agitated central parts of the platforms. The micropores in micritic limestone could be primary or secondary. Altered skeletal grains are the major source of the mud. The ICP results suggest that these micrites were originally composed of carbonate mud from differing sources (e.g. aragonite needles from algal disintegration, primary HMC precipitates, and/or possibly early microcrystalline cements formed in sediments during submarine lithification). The micrite was derived from many different sources for example (algae, bioerosion), and subsequently was recrystallized during shallow burial, or earlier during subaerial exposure and freshwater influx. | Occurs in most of the traverses. | 8I |

TABLE 3.—Compaction in the Middle–Upper Jurassic Carbonates of the Amran Group.

| DIAGENETIC FEATURES | DESCRIPTION | INTERPRETATION | OCCURRENCES | FIGURES |
|---|---|---|---|---|
| Physical | Represented by fractured and flattened grains. The timing of compaction events relative to cementation is both important and relatively easy to observe. The lack of porosity in mudstone and wackestone micro-facies indicates that the carbonate sediments of the Amran Group underwent substantial physical compaction. | Reflects the increase in pressure brought upon sediments as a result of deep burial. | Recognized in all traverses. | 9A |
| Chemical | Stylolites well developed and vary much in amplitude and in form from one occurrence to another. Marked by concentrations of organic matter, clays, pyrite, and detrital silicates and commonly dolomite residue of syn-stylolite origin. Formed along contacts between rigid grains or nodules or in clay-rich zones and clay partings. Both wispy seam stylolites and anastomosing wispy seams can be recognized. | Burial depth with attendant lithostatic load is one of the most critical factors determining the onset and ultimate efficiency of chemical compaction. The original mineralogy of the sediments undergoing compaction is also a very important control over the nature of its early chemical compactional history. | Recognized in all traverses. | 9B |


ALSHARHAN, A.S., AND KENDALL, C.G.ST.C., 1986, Precambrian to Jurassic rocks of the Arabian Gulf and Adjacent areas: their facies, depositional setting, and hydrocarbon habitat: American Association of Petroleum Geologists, Bulletin, v. 70, p. 977–1002.

ALSHARHAN, A.S., AND NAIRN, A.E.M., 1997, Sedimentary Basins and Petroleum Geology of the Middle East: Amsterdam, Elsevier, 945 p.

AL-SUBBARY, A., NICHOLS, G., AND BOSENCE, D., 1993, Cretaceous–Tertiary Pre-rift fluvial shallow marine sediments in Yemen, *in* Philobbos, A., and Purser, B., eds., Geodynamic and Sedimentation of the Red Sea–Gulf of Aden Rift System: Geological Society of Egypt, Special Publication, p. 383–408.

AL-THOUR, K.A., 1988, Geological and sedimentological studies on Al-Balaq area, Marib, Yemen Arab Republic: M.Sc. thesis, Sana'a University, 214 p.

AL-THOUR, K.A., 1992a, Sabatain Group: The product of main and post rift phase during Jurassic–Cretaceous time, Yemen Republic (abstract), *in* Marshall, J., convenor, 9th meeting of carbonate sedimentologists: The Bathurst Meeting, The University of Liverpool, U.K., Book of abstracts, 2 p.

AL-THOUR, K.A., 1992b, Stratigraphy, sedimentology and diagenesis of the Amran Group (Jurassic) of the region to the west and northwest of Sana'a, Yemen Republic: Ph.D. thesis, The University of Birmingham, 293 p.

AL-THOUR, K.A., 1997, Facies sequences of the Middle–Upper Jurassic carbonate platform (Amran Group) in the Sana'a region, Republic of Yemen: Marine and Petroleum Geology, v. 14, p. 643–660.

AL-THOUR, K.A., in press, Sedimentary facies and hydrocarbon potential of the Middle–Upper Jurassic carbonates (Amran Group) in the Wadi Al-Jawf-Mareb Shabwa'a Basin, Yemen: The Science Conference, Yemeni Scientific Research Foundation Special Publication series, Sana'a.

BATHURST, R.G., 1975, Carbonate Sediments and Their Diagenesis, 2nd Edition: Amsterdam, Elsevier, 658 p.

BEYDOUN, Z.R., 1988, The Middle East: Regional Geology and Petroleum Resources: London, Scientific Press Ltd., 292 p.

BEYDOUN, Z.R., 1991, Arabian plate hydrocarbon geology and potential—A plate tectonic approach: American Association of Petroleum Geologists, Studies in Geology, no. 33, 77 p.

CHOQUETTE, P.W., AND JAMES, N.P., 1987, Diagenesis # 12: Diagenesis in limestones—3, The deep burial environment: Geoscience Canada, v. 14, p. 335.

CHOQUETTE, P.W., AND PRAY, L.C., 1970, Geological nomenclature and classification of porosity in sedimentary carbonates: American Association of Petroleum Geologists, Bulletin, v. 54, p. 207–250.

DRUCKMAN, Y., AND MOORE, C.H., 1985, Late subsurface secondary porosity in a Jurassic grainstone reservoir, Smackover Formation, Mt. Vernon Field, Southern Arkansas, *in* Roehl, P.O., and Choquette, P.W., eds., Carbonate Petroleum Reservoirs: New York, Springer-Verlag, p. 369–383.

EL-ANBAAWY, M.I., AND AL-THOUR, K.A., 1989, Sedimentological evolution, diagenesis and hydrocarbon potentiality of Late Jurassic carbonates, Eastern region, Yemen Arab Republic: The Journal of the University of Kuwait (Science), v. 16, p. 401–423.

FRANK, J.R., 1981, Dedolomitization in the Taum Sauk limestone (Upper Cambrian), Southeast Missouri: Journal of Sedimentary Petrology, v. 51, p. 7–18.

HALLAM, A., 1978, Eustatic cycles in the Jurassic: Palaeogeography, Palaeoclimatology and Palaeoecology, v. 23, p. 1–32.

HALLAM, A., 1988, A reevaluation of Jurassic eustasy in the light of new data and the revised Exxon curve, *in* Wilgus, C.K., Hastings, B.S., Kendall, C.G.ST.C., Posamentier, H.W., Ross, C.A., and Van Wagoner, J.C., eds., Sea-Level Changes: An Integrated Approach: SEPM, Special Publication 42, p. 261–273.

TABLE 4.—Dolomitization and Dedolomitization in the Middle–Upper Jurassic Carbonates of the Amran Group.

| DIAGENETIC FEATURES | DESCRIPTION | INTERPRETATION | OCCURRENCES | FIGURES |
|---|---|---|---|---|
| Dolomitization | Several types have been recognized. The first type comprises a replacive dolomite; crystals range in size from 36 to 280 µm. The second type shows the development of saddle dolomite in voids and/or fractures, crystals occurr perpendicular to the walls of allochems, e.g., foraminifera, bivalves, and/or in fractures. Occurred before emergence and deep karstic erosion. Associated with fractures and microcavities withsulfide mineralization. | Dolomite distribution of various shapes indicates that the precipitation did not take place in one event, but over a much longer period of time, ranging from shallow burial to deeper burial. The coarse dolomite crystal size suggests a slow crystallization rate, and the weak sweeping extension of crystals seen in saddle dolomite indicates a formation temperature between 60°C and 150°C. These facts suggest a hydrothermal or deep burial environment of dolomitization. | Well developed in Wadi Al-Khothally, Wadi Al-Ahjur, and Al-Ayein traverses. | 10A |
| Dedolomitization | Rhombohedral zones of calcite and iron oxide. The rhombohedral shape suggests either that selected dolomite zones have been replaced or that calcite and iron oxide grew on dolomite substrates during interruption in dolomite development. | Dedolomitization process can be recognized when meteoric waters invade dolomite rocks and remove magnesium to help in the precipitation of calcite. Occurrence of nonferroan calcite and ferric oxides in rhombohedral zones in dolomite indicate that dedolomitization occurred by oxidation and alteration of ferroan dolomite zones and probably reflects alteration related to recent weathering. | Occurs in the upper part of Al-Khothally, Al-Ayein, Wadi Al-Ahjur, and Jabal Salab traverses. | 10B |

HIRD, K., AND TUCKER, M., 1988, Contrasting diagenesis of two Carboniferous oolites from South Wales: a tale of climatic influence: Sedimentology, v. 35, p. 587–602.

MOORE, C.H., 1989, Carbonate Diagenesis and Porosity: Amsterdam, Elsevier, Developments in Sedimentology 46, 388 p.

MOSS, S., AND TUCKER, M., 1995, Diagenesis of Barremian–Aptian platform carbonates (the Urgonian Limestone Formation of SE France): near-surface and shallow-burial diagenesis: Sedimentology, v. 42, p. 853–874.

POWERS, R.W., RAMIREZ, L.F., REDMOND, C.D., AND ELBERG, J.E., 1966, Geology of the Arabian Peninsula (Sedimentary Geology of Saudi Arabia): U.S. Geological Survey, Professional Paper 560D, 150 p.

SCHOLLE, P., AND HALLEY, R., 1985, Burial diagenesis: out of sight, out of mind, *in* Schneidermann, N., and Harris, P.M., eds., Carbonate Cements: SEPM, Special Publication 36, p. 309–334.

SELLWOOD, B.W., SCOTT, J., JAMES, B., EVANS, R., AND MARSHALL, J., 1987, Regional significance of dedolomitization in Great Oolite reservoir facies of Southern England, *in* Brooks, J., and Glennie, K.W., eds., Petroleum Geology of Northwest Europe: London, Graham & Trotman, p. 129–137.

SHEBL, H.T., AND ALSHARHAN, A.S., 1994, Sedimentary facies and hydrocarbon potential of Berriasian–Hauterivian carbonates in Central Arabia, *in* Simmons, M.D., ed., Micropalaeontology and Hydrocarbon Exploration in the Middle East: London, Chapman & Hall, p.159–174.

SIMMONS, M.D., AND AL-THOUR, K.,1994, Micropaleontological biozonation of the Amran Series (Jurassic) in the Sana'a Region, Yemen Republic, *in* Simmons, M.D., ed., Micropalaeontology and Hydrocarbon Exploration in the Middle East: London, Chapman & Hall, p. 43–60.

WRIGHT, V.P., 1988, Paleokarsts and paleosols as indicators of paleoclimate and porosity evolution: a case study from the Carboniferous of South Wales, *in* James, N.P., and Choquette, P.W., eds., Paleokarst: New York, Springer-Verlag, p. 329–341.

# DIAGENESIS OF THE APTIAN SHUAIBA FORMATION AT GHABA NORTH FIELD, OMAN

ABDULRAHMAN A. AL-AWAR

*Dubai Petroleum Company, P.O.Box: 2222, Dubai, United Arab Emirates*

*E-mail: Abdulrahman.Al-Awar@arc.conoco.com*

AND

JOHN D. HUMPHREY

*Department of Geology and Geological Engineering, Colorado School of Mines, Golden, CO 80401, U.S.A.*

ABSTRACT: The Aptian Shuaiba Formation is subdivided into two major reservoir zones in Ghaba North Field, Oman. The lower Shuaiba zone was deposited in a lagoonal setting with restricted to partially restricted marine circulation conditions. The upper Shuaiba zone was deposited under open circulation conditions in a shallow marine setting. Facies succession in the Shuaiba indicates an overall major shallowing-upward cycle, culminating in a regional unconformity at the top of the formation.

The Shuaiba reservoir is heterogeneous, and diagenesis has been responsible for both heterogeneity and enhancement of reservoir properties. Production is from secondary macroporosity, matrix microporosity, and fracture porosity. Reservoir storage capacity is dependent mainly upon matrix microporosity, which occurs as intercrystalline porosity within micro-rhombic calcite matrix. Reservoir quality is enhanced vertically in the proximity of two recognized subaerial exposure surfaces. Meteoric-water diagenesis at local exposure horizons and the regional unconformity led to matrix stabilization and development of both secondary macroporosity and microporosity within micro-rhombic calcite matrix. This interpretation is supported by the decrease in secondary macroporosity and matrix permeability downward from the exposure surfaces. Stable-isotope data, in conjunction with stratigraphic and sedimentologic observations, support the diagenetic influence of meteoric water in creation of favorable reservoir properties. The lower Shuaiba zone shows relatively invariant stable-isotope compositions and is interpreted to have been diagenetically stabilized in a confined aquifer under conditions of low water/rock ratio. The upper Shuaiba zone shows carbon-isotope depletions at the local subaerial exposure surface and at the regional unconformity. Attenuation of carbon- isotope shifts likely resulted from poor development of vegetation and soil in an arid setting. The potential for improvement of reservoir properties by local subaerial exposure should be considered in reservoir development plans.

## INTRODUCTION

The Shuaiba Formation is a prolific hydrocarbon reservoir in the Arabian Gulf, in general, and in northern and central Oman in particular. The formation is Aptian in age and is the uppermost formation of the Lower Cretaceous Thamama Group. It was deposited during an extensive marine inundation of the eastern Arabian Shield, and was marked by a series of thick rudist grainstone buildups (Alsharhan, 1985). Murris (1980), and Alsharhan (1995) suggested that during the Aptian, in the southern part of the Arabian Gulf and United Arab Emirates in particular, the intrashelf basin of Abu Dhabi had developed and fringing shelf-margin rudist buildups grew where conditions were favorable, creating the prolific Shuaiba reservoirs.

Hughes Clarke (1988) interpreted the Shuaiba Formation in Oman as a full cycle of carbonate sedimentation that can be subdivided into two parts. The lower part is described as an algal (*Lithocodium*) skeletal wackestone to boundstone. This passes up into argillaceous carbonates, which then grade up into cleaner foraminiferal, and in some places rudist-rich, wackestones to packstones. The Shuaiba in Oman and the United Arab Emirates overlies the Kharaib Formation, which consists of shoal deposits on a carbonate shelf.

The Shuaiba is terminated in the southern part of the Arabian Gulf by a shale seal called the Nahr Umr Formation, the lower formation in the Wasia Group of Middle Cretaceous age. The Nahr Umr Formation (Albian) records a shift to a mixed siliciclastic–carbonate system. Pratt and Smewing (1993) believe that it was deposited during drowning of the shelf in the Albian. Detrital sediment constituting the Nahr Umr Formation was generated by uplift on the west side of the platform. Large-scale evolution of the margin in the central Oman Mountains imposed major sedimentological changes on the platform in generating the siliciclastics of the Nahr Umr Formation (Pratt and Smewing, 1993). A regional unconformity, representing erosion or nondeposition of the middle Aptian section, marks the contact between the Shuaiba and the overlying Nahr Umr Formations in this area (Frost et al., 1983; Harris et al., 1984).

The present study focuses on Ghaba North Field of central Oman, discovered in 1972 (Fig. 1). The field lies within the Ghaba Salt Basin (Petroleum Development Oman report, 1992), where great thicknesses of Precambrian to early Paleozoic clastics, salts, and marine carbonates accumulated (Alsharhan and Kendall, 1986). Abundant structures formed by salt tectonism in this basin have developed prolific oil traps in the area.

The structural setting of the Ghaba North Field (Fig. 2) indicates an anticlinal structure trending NE–SW with a fractured and faulted crest at the Shuaiba Formation level (Al-Bastaki, 1993; Petroleum Development Oman report, 1992). The structure likely formed by deep-seated movement of the Ara Salt. Stresses caused by the salt movement, together with wrench faulting associated with the Maradi Fault Zone, created a NE–SW striking fault system consisting of normal faults (Al-Bastaki, 1993; Uitentuis, 1981).

## METHODS OF THE STUDY

In order to understand the controls on porosity evolution and the development of reservoir quality, this study focuses on the sedimentological history and diagenesis of the Shuaiba Formation. Much of the work for the present study came from detailed analysis of Ghaba North core No. 23 (GN-23). This core, taken in 1990, is 50 m long and 19.5 cm in diameter and is still in good condition. This large-diameter core affords an excellent opportu-

Middle East Models of Jurassic/Cretaceous Carbonate Systems
SEPM Special Publication No. 69, Copyright © 2000
SEPM (Society for Sedimentary Geology), ISBN 1-56576-075-1, p. 173–184.

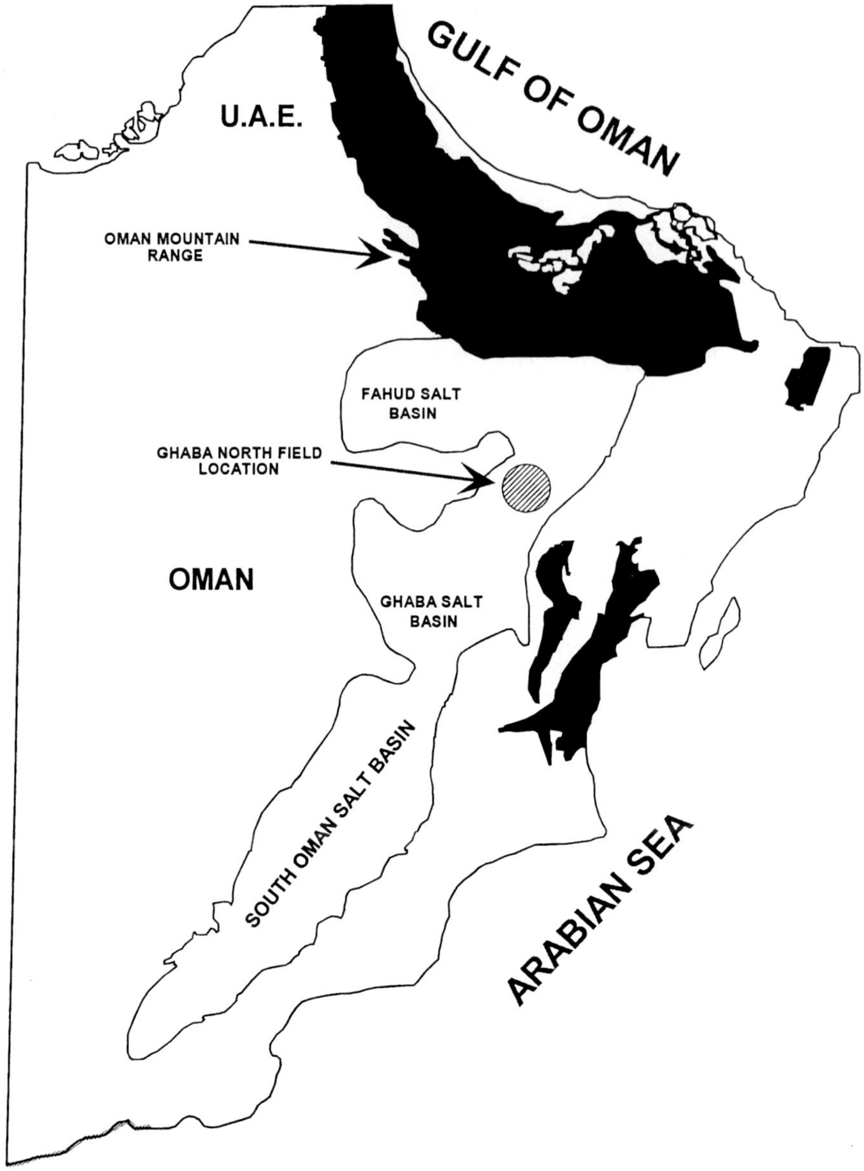


FIG. 1.—Simplified tectonic map of Oman showing configuration of Ghaba Salt Basin and location of Ghaba North Field (after Al-Bastaki, 1993; Hartmans, 1986).

nity for megascopic study. Moreover, three other cores from this field are used in this study; GN-1, GN-2, and GN-13 (see Fig. 2 for locations), all of which are 10.2 cm in diameter.

Eighty-five core samples were taken from core GN-23 at an approximate interval of 0.5 m. All representative facies were sampled in GN-23, with the emphasis being on recognition of sediment types, bounding features, and diagenetic phases. On the basis of the detailed sampling in GN-23, representative samples were also taken from the other three cores for comparison. Thirteen, four, and eight representative samples were taken from the cores GN-1, GN-2, and GN-13, respectively.

Thin sections were prepared from the core samples, some were stained, and all were examined using standard transmitted-light petrography. Broken surfaces of five samples from GN-23 were chosen for a scanning electron microscope (SEM) investigation. Three of these samples come from intervals with high porosities and two were from intervals with relatively low porosities. These high-porosity and low-porosity intervals alternate with each other from the bottom of the core to the top.

Powdered samples for stable carbon and oxygen isotopic analyses of matrix and cements were collected from slabbed core plugs with a dental drill. Samples of the matrix were repeatedly cleansed of hydrocarbons in a toluene bath. All samples were prepared for isotopic analysis by off-line extraction of $CO_2$ in 100% orthophosphoric acid at 90°C. Stable carbon and oxygen isotopes were analyzed on a VG Micromass 602E mass spectrometer according to standard procedures outlined in McCrea (1950). Isotopic ratios are expressed in the standard delta notation and are given relative to PDB standard in per mil (‰) notation. Precision on blind duplicates for 10% of the samples yields a mean half range of 0.20‰ for $\delta^{18}O$ and 0.07‰ for $\delta^{13}C$.

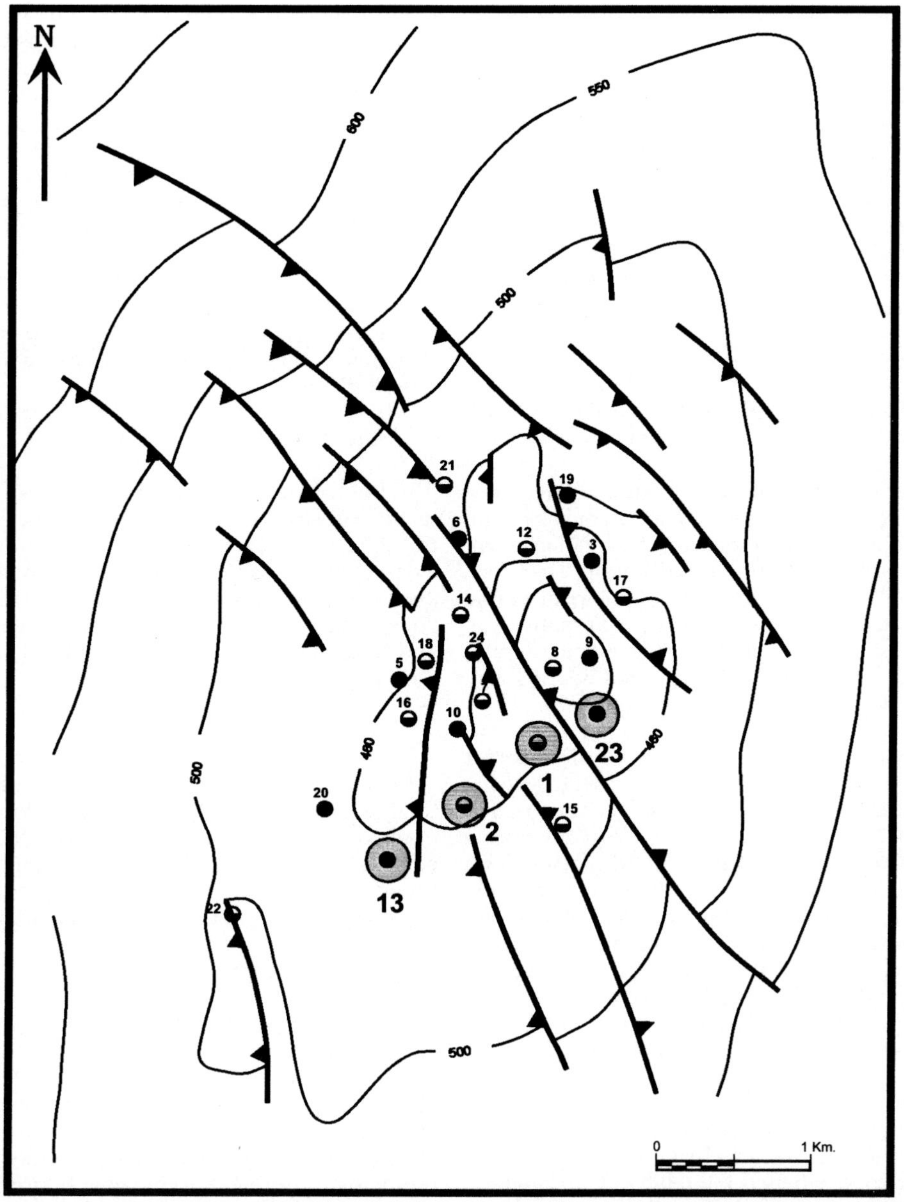


FIG. 2.—Structure contour map of top of Shuaiba Formation and well locations in Ghaba North Field (after Al-Bastaki, 1993; Petroleum Development Oman, 1992).

## RESULTS

### *Biofacies and Lithofacies*

*Biofacies.*—

Three biofacies were identified at Ghaba North Field. Specific types of allochems characterize each biofacies, but the boundaries between these biofacies are transitional and may display faunal mixing.

1. Foram Biofacies: This biofacies is dominated by *Orbitolina* and small benthic forams, including miliolids, uniserial and biserial forams, peneropolids, and agglutinated and spiral forams. The faunal assemblage is indicative of a restricted to partially restricted shallow marine setting (e.g., Wilson, 1975). This biofacies lacks any rudists or rudist fragments, which are indicative of higher-energy conditions.

2. Algae–Foram–Rudist Fragment Biofacies: This biofacies is characterized by abundant green algae (*Lithocodium*), *Orbitolina*, small forams, and small-sized rudist fragments. In addition, a few scattered whole rudists are present. Skeletal debris of echinoderms and mollusks is common as well. The algae are present as either nodules or layers. This faunal assemblage is indicative of a restricted to partially restricted, shallow marine setting (Wilson, 1975). However, in this case, the presence of rudist fragments indicates that water-circulation conditions were slightly better than in the previous biofacies. This indicates that the depositional site on the shelf/

ramp was closer to some rudist bioherms, in inter-biohermal sites, because of the presence of transported rudist fragments in wacke-packstone microfacies.

3. Rudist–*Orbitolina* Biofacies: Rudists and rudist fragments dominate this facies and in places are present in growth position. The benthic foram *Orbitolina* is abundant in this biofacies; however, other taxa of small benthic forams are rare or absent. The abundance of calcareous algae is low and, if present, occur as nodules only. This biofacies is indicative of a less restricted, much shallower setting. It represents deposition near, or within, rudist bioherms on the shelf.

*Lithofacies.—*

Two major lithofacies groups were identified in Ghaba North Field, with each group composed of a variety of subfacies.

1. Restricted Lagoon and/or Below Fair-Weather Wave Base Lithofacies: This lithofacies comprises the foram biofacies and the algae–foram–rudist fragment biofacies. This lithofacies was deposited on the landward side of the shelf/ramp. Therefore, it is indicative of relatively shallow waters but with partial restriction of circulation in a lagoonal setting. Alternatively, it might indicate deposition under slightly deeper water depths, below fair-weather wave base. It is represented by the lower Shuaiba zone, which occurs below 619.45 m in GN-23.

2. Relatively Open Shelf/Ramp and/or Above Fair-Weather Wave Base Lithofacies: This lithofacies is represented by the rudist–*Orbitolina* biofacies. It was deposited above fair-weather wave base, in shallower water and more open-circulation conditions than the restricted lagoonal lithofacies. This lithofacies is represented by the upper Shuaiba. In general, rudist boundstones, rudstones, and minor rudist floatstones dominate this zone. In addition, a grainier bioturbated packstone microfacies is present. The lithofacies formed by vertical stacking of rudist biohermal/biostromal facies, intercalated with inter-biohermal floatstone facies. Several high-frequency cycles of bioherm deposition occur in the upper Shuaiba zone.

Lithofacies succession at Ghaba North Field is indicative of a shallowing-upward trend. The restricted-lagoon lithofacies predominate in the lower zone, whereas the relatively open shelf/ramp lithofacies represents the upper Shuaiba zone. Stacked, high-frequency cycles constituting the upper zone culminated in grainstone deposition and inferred subaerial exposure. This overall shallowing-upward trend ended with a major drop in sea level and subaerial exposure, represented by the regional unconformity between the Shuaiba and Nahr Umr formations (Harris et al., 1984; Alsharhan, 1995).

## *Reservoir Properties*

Reservoir properties considered include total measured porosity, macroporosity, microporosity, and horizontal permeability. Total measured porosity and horizontal permeability were obtained from an unpublished report (Petroleum Development Oman, 1990). Grain volume, for plugs taken from the core, was determined using a helium porosimeter. Subsequently, the total measured porosity and grain density of each plug were calculated. Macroporosity represents the porosity observed and estimated for all intervals in the Shuaiba Formation from the core, plugs, and thin sections in this study. Microporosity percentages are calculated from the difference between the total measured porosity and the estimated macroporosity for each interval.

*Total Measured Porosity.—*

The Shuaiba lower and upper zones have similar characteristics regarding variation in total measured porosity (Fig. 3A). Both of these zones start with high total measured porosity at the base, which then decreases gradually through the middle sections of the zones. A gradual increase in the porosity then occurs as the top of each zone is approached. However, the lower zone has a lower average total porosity (27%) than the upper zone, which has an average of greater than 30%. Thus, the Shuaiba Formation, in general, shows a trend of increasing total measured porosity toward the top.

*Macroporosity.—*

The lower Shuaiba zone has relatively low average macroporosity (5%). This percentage holds relatively constant throughout the lower zone except for few intervals that show even lower values (Fig. 3B). Macroporosity increases gradually above the contact between the lower and upper zones. The upper zone is characterized by a higher average macroporosity (> 8%), and it reaches a maximum of greater than 15% at the top of the zone (top of the Shuaiba Formation).

*Microporosity.—*

Microporosity estimates represent the difference between total measured porosity and estimated macroporosity. The plot of microporosity values versus depth (Fig. 3C) is similar to the plot of the total measured porosity variation with depth (Fig. 3A). Trends are not obvious; however, average microporosity values are slightly lower in the lower Shuaiba zone.

*Horizontal Permeability.—*

Horizontal permeability is shown in Figure 3D. In general, the Shuaiba Formation is characterized by relatively low permeability values. The permeability is lower at the bottom and becomes slightly higher toward the top of the formation. In order to look at permeability variations in the microporous matrix, data in Figure 3D were replotted using a cutoff of 35 md, effectively removing the interconnected macroporosity (Fig. 3E). The lower zone shows that matrix horizontal permeability values are generally less than 10 md, averaging 6 md, and increasing to about 15 md at the top of the zone. The upper zone exhibits higher and more variable permeability values, with an average estimated to be about 15 md. The highest permeabilities, over 200 md, occur at or near the top of the upper Shuaiba zone (Fig. 3D).

## *Diagenesis of the Shuaiba Formation*

*Calcite Cements.—*

Macroscopic cements are volumetrically minor at Ghaba North Field. Original interparticle macropore spaces were sparse because of the predominance of wackestone fabric in the Shuaiba sediments of the area. The few cements observed are calcitic in composition but with variable morphologies. Cementation patterns vary with depth and facies of the Shuaiba Formation.

Two major types of calcite cements are observed. The first occurs as a euhedral to subhedral, fine-crystalline calcite, and the second occurs as a coarse-crystalline pore-filling calcite. The fine-

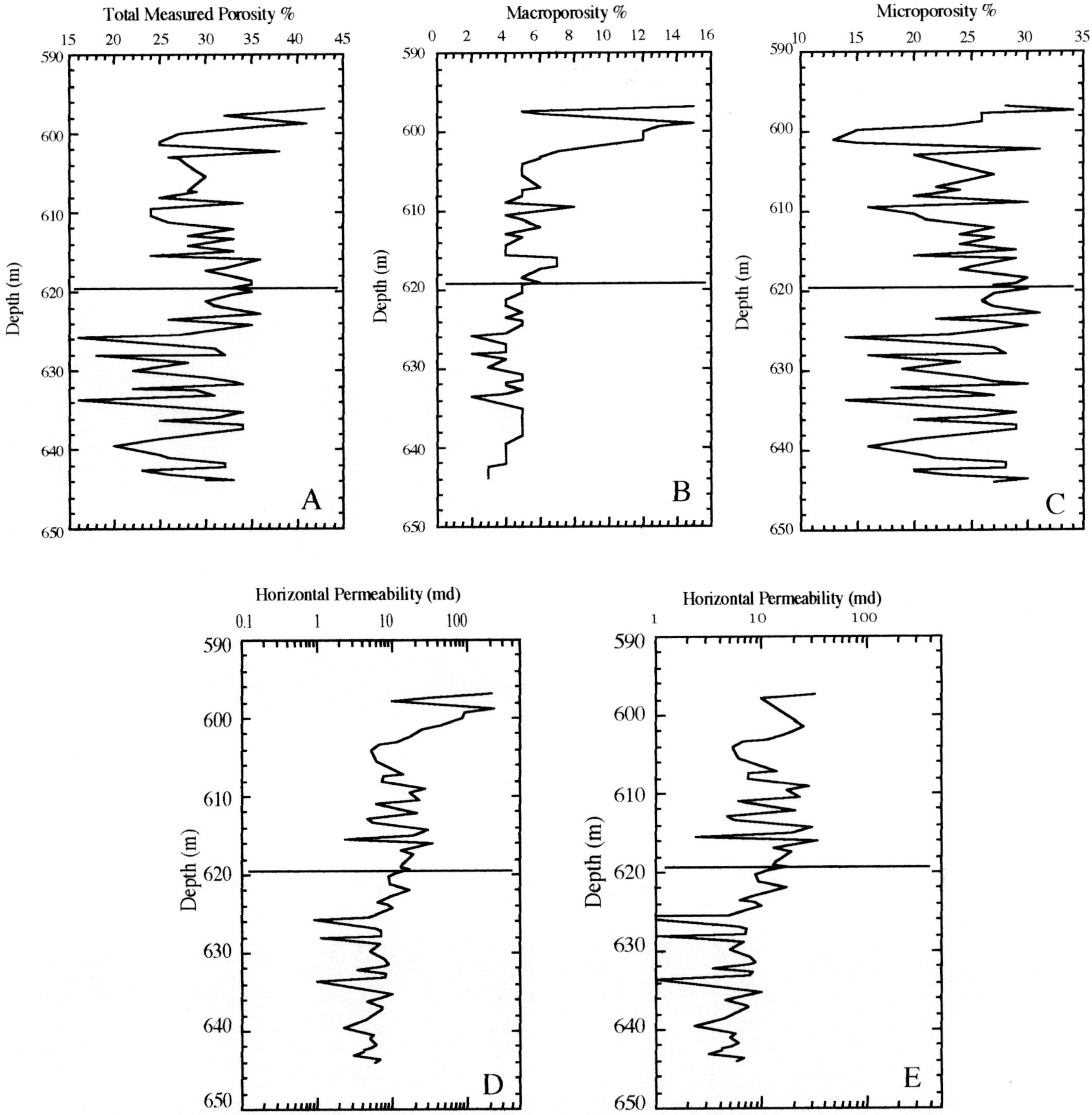


FIG. 3.—Reservoir properties of GN-23. (A) Total measured porosity versus depth. (B) Macroporosity versus depth. (C) Microporosity versus depth. (D) Horizontal permeability versus depth. (E) Horizontal permeability less than 35 md versus depth. Horizontal line represents the boundary between the lower and upper Shuaiba zones in each plot.

crystalline calcite cements occur in intraparticle pores of algae, *Orbitolina*, and other forams, as well as within microfractures, borings, shelter porosity, and primary interparticle pores. These cements partially to completely fill the pores. In partially filled pores, fine-crystalline calcite cement lines the pores and the rest is filled by hydrocarbons. This fine-crystalline calcite cement has a patchy appearance in the core and plugs. This is due mainly to localization of the cements within algae intraparticle pores, and therefore it corresponds to the distribution of algae. Microfractures filled with fine-crystalline calcite are later than these pore-filling cements, because they crosscut allochems that contain the earlier cement.

The coarse-crystalline calcite cements occur mainly in secondary moldic porosity or primary macroporosity, such as in rudist internal pore spaces. These cements are generally equant-blocky and subhedral to euhedral, and show a trend of increasing crystal size toward the centers of the pores. The presence of broken rudist fragments, of spalled micrite envelopes, and of brecciated fragments of matrix floating in these cements indicates that these cements occur later in the paragenesis. Higher abundance of these cements occurs in the upper Shuaiba zone and corresponds to rudist-rich intervals.

*Dissolution Effects.—*

Rudists and rudist fragments in the Shuaiba may either be preserved as replaced or recrystallized skeletal material, or partially to completely leached. Almost all of the rudists and rudist fragments in the lower zone are preserved with no signs of leaching. Here, some rudists retain their wall microstructures, while others lack such preservation because of pervasive recrystallization.

Data and observations from cores, plugs, and thin sections of the upper zone provide unequivocal evidence for dissolution and creation of moldic pores. This moldic porosity is present either as whole molds, now filled by hydrocarbons, or as partially cemented molds. The latter contain coarse-crystalline calcite cements lining the pores (Fig. 4). Development of this moldic porosity resulted in an increase in both macroporosity and horizontal permeability in the upper Shuaiba zone. Importantly, dissolution and leaching features are observed to be intense and abundant below distinct stratigraphic boundaries and decrease with depth away from them. These boundaries are the regional unconformity, which occurs at the top of the Shuaiba and separates it from the overlying Nahr Umr Shales, and a subaerial exposure surface identified at the top of a rudist rudstone facies at a depth of 609.45 m in GN-23.

*Micro-Rhombic Calcite Matrix.—*

Matrix microporosity accounts for most of the hydrocarbon storage capacity in the reservoir at Ghaba North. SEM investigation of matrix samples shows that this microporosity occurs because of abundant intercrystalline pores associated with micro-rhombic calcite (Fig. 5). Micro-rhombic calcite crystals are generally equant and may show partial rounding of crystal terminations, reflecting limited dissolution. Micro-rhombic calcite occurs in two fabrics, a blocky-crystal framework (Fig. 5) and a crystal mosaic (Fig. 6). Individual micropores average about 5 to 10 μm in diameter; however, some micropores approach 62 μm in diameter. The latter are secondary in nature and include micromolds and microvugs.

5

FIG. 5.—SEM photograph of micro-rhombic calcite matrix and intercrystalline micro-porosity. White bar is 1 micron long.

4

FIG. 4.—SEM photograph of a mold with coarse crystalline calcite crystals partially filling the void. Long bar in the photo is 10 microns.

6

FIG. 6.—SEM photograph of crystal mosaic texture in the Shuaiba matrix. Long bar is 10 microns.

### *Stable-Isotope Data*

Eighty-four samples of micro-rhombic calcite matrix, comprising all the plugs from GN-23, were analyzed for stable carbon and oxygen isotopes. In addition, three samples of late (post-leaching) coarse-crystalline calcite cements were analyzed. The $\delta^{13}C$ of the micro-rhombic calcite matrix ranges from +3.11 to +5.95‰ and $\delta^{18}O$ ranges from -7.31 to -4.10‰ (all relative to PDB) (Fig. 7). The three cement samples represent the most depleted carbon and oxygen isotope values.

## DISCUSSION

### *Reservoir Characteristics*

As discussed previously, the lower Shuaiba zone is dominated by the foram and the algae–foram–rudist fragment biofacies, and is represented by the restricted lagoonal lithofacies. The upper zone is dominated by the rudist–*Orbitolina* biofacies, and is represented by the relatively open shelf/ramp lithofacies. Analysis of the variability trends of total measured porosity, macroporosity, microporosity, and horizontal permeability in Figures 3A–E, indicates that the upper zone, in general, has better reservoir characteristics. The upper zone has higher averages of total measured porosity, macroporosity, microporosity, and horizontal permeability. It is important to note that the upper zone contains fewer algae, more rudists, and more *Orbitolina* and other allochems than the lower zone. This difference is attributed to the depositional conditions that dominated during deposition of each zone. Obviously, sedimentology, allochemical constituents, and biofacies forming the lithofacies of the Shuaiba Formation, which are dependent on the depositional conditions, play an important role in determining the quality of the reservoir. The Shuaiba, in general, has very high total measured porosity (Petroleum Development Oman, 1990). However, examination of cores, plugs, and thin sections shows that macroporosity is very low in the lower zone and becomes higher in the upper zone. Cores and samples examined in this study were saturated with hydrocarbons throughout the Shuaiba interval, with heavier hydrocarbon staining toward the top.

The Shuaiba Formation has been described as a chalky limestone (Moshier, 1989a, 1989b; Al-Bastaki, 1993; Gallagher et al., 1990) because of its micritic texture. Chalky-textured limestone petroleum reservoirs in the Mesozoic and Tertiary deposits of the Middle East are very common (Wilson, 1975). The chalky texture is due to abundant and disseminated microporosity existing within the micritic matrix of typical carbonate platform sediments (Moshier, 1989a). The Shuaiba in Ghaba North Field has high porosities, which are attributed to the microporosity of the micro-rhombic calcite matrix; however, there is an observed variability in this microporosity with depth in both zones of the Shuaiba (Fig. 3C). Development of secondary moldic porosity in the upper zone, due to the leaching of rudists and rudist fragments, enhanced the horizontal permeability of the matrix, thus augmenting the reservoir quality and capacity to hold hydrocarbons. Therefore, in addition to the role played by the depositional environment in determining reservoir properties, there is a diagenetic role that further enhances these properties.

#### *Controls on Microporosity.—*

Similarity between the vertical variability of microporosity and of total measured porosity is indicative of the control that microporosity has on hydrocarbon storage capacity in this reservoir. Microporosity in the lower and upper Shuaiba zones is highly variable (Fig. 3C), although the lower zone has a slightly lower overall average porosity in comparison with the upper zone.

Further analysis of microporosity variability in the Shuaiba Formation was done by selecting five samples for SEM investigation. These samples correspond to alternating intervals with high and low excursions in the microporosity plot (Fig. 3C, Table 1). SEM observations of these samples indicates that microporosity occurs as intercrystalline porosity within the micro-rhombic calcite matrix. Sample No. 19, which corresponds to an extensively cemented rudstone interval, has relatively low microporosity because of the presence of a neomorphic calcite microspar that heals matrix microporosity. Sample No. 77 shows very high macroporosity but relatively low microporosity. This is due to the presence of two types of fabrics in the matrix: a blocky-crystal framework fabric and a crystal mosaic fabric (Figs. 5 and 6, respectively). The former has high microporosity values, whereas the latter has low microporosity values.

#### *Controls on Macroporosity.—*

The fact that the upper zone exhibits a higher average macroporosity than the lower zone (Fig. 3B) is consistent with the higher rudist content in the upper zone. Nearly all macroporosity occurs as a secondary moldic porosity that developed because of leaching of rudists and rudist fragments by undersaturated diagenetic fluids. The upper zone also shows a trend toward increasing macroporosity toward the top of the upper zone. This reflects the presence of abundant caprinid and caprotinid rudists rather than monopleurid rudists. Caprinid and caprotinid rudist shells had thick aragonitic layers and thin calcitic layers (Peter Skelton, 1997, personal communication); thus, the rate of stabilization of these rudists to low-Mg calcite (LMC) is low during diagenesis (Al-Bastaki, 1993; Borgomano, 1991). Therefore, these rudists were leached more easily than rudists that stabilize to LMC at a faster rate (e.g., monopleurids).

The upper zone is capped by a regional unconformity, separating the Shuaiba Formation from the Nahr Umr Formation shales (Harris et al., 1984), which acted as a subaerial exposure

TABLE 1.—Reservoir properties of SEM samples.

| Sample | Depth (m) | Total Porosity % | Macroporosity % | Microporosity % | Horizontal Permeability md |
|---|---|---|---|---|---|
| 2 | 643.66 | 33 | 3 | 30 | 6.8 |
| 19 | 633.60 | 16 | 2 | 14 | 1 |
| 47 | 618.49 | 35 | 5 | 30 | 14 |
| 77 | 601.06 | 25 | 12 | 12 | 43 |
| 81 | 598.83 | 41 | 15 | 26 | 230 |

surface from which potential diagenetic fluids invaded the Shuaiba. Thus, the lower zone has lower macroporosity values because it either contained lesser amounts of rudists for leaching and/or it was far from the reach of undersaturated meteoric fluids that caused leaching of rudists in the upper zone.

*Controls on Horizontal Permeability.—*

Note the similarity between the plots for macroporosity (Fig. 3B) and for matrix permeability (Fig. 3E). Higher permeability values in the upper zone and the increase in these values toward the top of the zone correspond to the increase in macroporosity and effective porosity, defined as the interconnected pores. This results from leaching of rudists and rudist fragments, where leaching creates moldic porosity and enhances the connectivity of the pore system (Al-Bastaki, 1993; Borgomano, 1991). The upward increase in rudist abundance also means an increase in the amount of coarser-grained sediments at the expense of finer-grained materials, also leading to enhancement of matrix permeability. Anomalously high values of horizontal permeability are observed near the top of the formation in Figure 3D, and result from moldic macroporosity and the presence of open fractures.

## *Diagenesis at Ghaba North Field*

*Rudist Leaching and Macroporosity Development.—*

Rudists are a major skeletal constituent of the Shuaiba Formation and are represented by different taxa and morphologies and with different original mineralogies. Rudists and rudist fragments which have original shells that have thicker high-Mg calcite (HMC) layers (such as monopleurids) are more easily and rapidly stabilized to LMC than rudists originally with thicker aragonitic shell mineralogy (caprinids and caprotinids) (Al-Bastaki, 1993; Borgomano, 1991; P.W. Skelton, personal communication, 1997). The latter type are more commonly subjected to dissolution than to stabilization. Saturation with respect to LMC is attained much faster in sediments containing mixed stabilized and unstabilized constituents, because of higher solubilities of metastable mineralogies relative to LMC. This mechanism is fabric-selective and is termed "mineral-controlled alteration", which acts in low-activity phreatic realms (low water/rock ratio) (James and Choquette, 1984). However, if the phreatic realm was very active or all the constituents of the sediment were already stabilized, then alteration of the sediment occurs depending on the nature of the fluid (saturated or undersaturated). This process is non–fabric selective and is termed "water-controlled alteration". Leaching here might occur if large volumes of undersaturated fluids flowed through the sediment (high water/rock ratio) (James and Choquette, 1984). This process then proceeds in the sediment depending on saturation state, flow rates, porosity, permeability, and grain size. Rudists and rudist fragments, in intervals below the 619.45 m boundary separating the lower and upper Shuaiba zones in GN-23, are preserved and extensively cemented. Above 619.45 m rudists are only rarely preserved and the majority show development of moldic porosity. The degree of moldic porosity creation increases with proximity to the local subaerial exposure surface at 609.45 m and the regional unconformity at the top of the Shuaiba Formation. Different styles of rudist modification in the upper and lower Shuaiba zones suggest that the formation was affected by more than one diagenetic event in the Ghaba North area.

*Development of Micro-Rhombic Calcite Matrix and Microporosity.—*

Development of micro-rhombic calcite matrix in the Shuaiba is a very important event because intercrystalline microporosity of this matrix accounts for most of the hydrocarbon storage capacity in the reservoir. This micro-rhombic calcite has previously been identified in the Shuaiba formation in the United Arab Emirates by Moshier (1989b) and Budd (1989).

Most lime mud accumulating in modern shallow-platform settings is formed from post-mortem disintegration of calcareous green algae, precipitation from seawater, and bioerosion and abrasion of skeletal debris (e.g., Bathurst, 1975). Calcite microrhombs and associated microporosity in ancient rocks, on the other hand, most likely are the products of stabilization of this type of original metastable lime mud. However, the diagenetic environment in which stabilization occurred has been debated by Moshier (1989b) and Budd (1989). On the basis of geochemical data, Moshier (1989b) suggested that microporosity of the Shuaiba and other Lower Cretaceous formations in the United Arab Emirates formed by stabilization in a closed system of marine pore waters. On the other hand, Budd (1989) argued for the development of micro-rhombic calcite by meteoric fluids through a two-step process involving subaerial exposure and shallow burial diagenesis.

Matrix in the Shuaiba Formation at Ghaba North Field has both types of fabrics, blocky-crystal framework and crystal mosaic fabrics, identified and described by Moshier (1989a, 1989b). In general, blocky-crystal framework texture has higher microporosity and permeability, whereas crystal mosaic texture has lower values of both. Dissolution along crystal boundaries creates secondary intercrystalline microporosity and enhances permeability by enlargement of interconnected porethroats (Moshier, 1989a). Moshier suggested that a framework fabric developed by stabilization of wet, unconsolidated lime muds. Additional cementation and compaction of crystals of the framework resulted in formation of the mosaic texture with less microporosity. In addition, he suggested that framework texture forms in systems with low water/rock ratio, whereas crystal mosaic texture forms in (open) systems with high water/rock ratio. His argument was that open systems introduce excess carbonate for thorough cementation.

*Analysis of Stable Carbon and Oxygen Isotopes.—*

A cross-plot of $\delta^{13}C$ and $\delta^{18}O$ of both the matrix and late coarse-crystalline calcite cements (Fig. 7) shows ranges of +3.11 to +5.95‰ $\delta^{13}C$ and -8.25 to -4.10‰ $\delta^{18}O$ (PDB). The more enriched values for carbon and oxygen are generally taken to be closer to the original marine composition, while the more depleted values could be interpreted in different ways. For instance, Moshier (1989b) interpreted the $\delta^{13}C$ range of +3.4 to +4.0‰ and the $\delta^{18}O$ range of -4.0 to -5.0‰ of micro-rhombic calcite matrix of the Shuaiba in the Sajaa Field in the United Arab Emirates to be indicative of stabilization of matrix in a closed system with marine porewaters. Because a range of -2 to -2.5‰ PDB was reported for $\delta^{18}O$ of Aptian marine calcite (Scholle and Arthur, 1980; Moldovanyi and Lohmann, 1984), he suggested that his more depleted $\delta^{18}O$ values are the result of dissolution and reprecipitation through a process of replacement of original matrix materials during diagenesis. Thus, Moshier (1989b) suggested that the Shuaiba matrix was depleted by 1.5 to 2.0‰ relative to its precursor marine sediment.

On the other hand, Budd (1989) interpreted the $\delta^{13}C$ range of micro-rhombic calcite matrix of +3.1 to 4.5‰ PDB to be primary

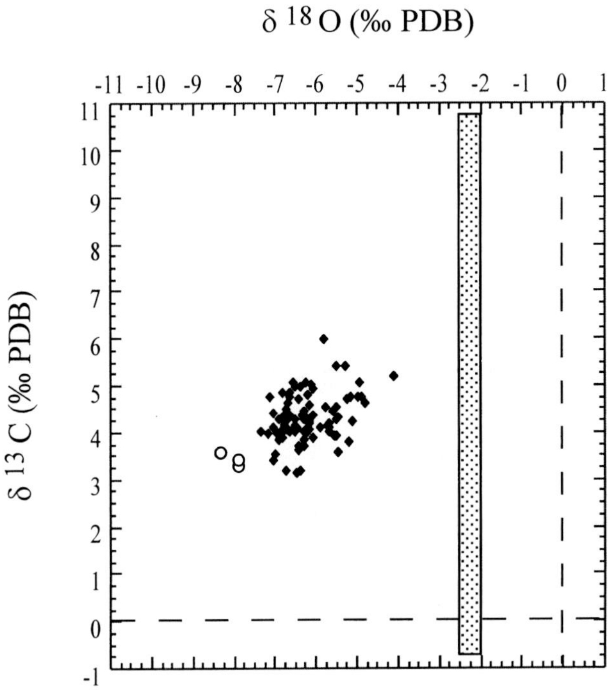


FIG. 7.—Cross-plot of $\delta^{13}C$ versus $\delta^{18}O$ of the Shuaiba matrix (solid diamonds) and late calcite cements (open circles) in GN-23 core. Solid vertical bar is the reported $\delta^{18}O$ composition of Aptian marine calcite (see text for further discussion).

and unaltered by diagenesis. However, the $\delta^{18}O$ range of -4.5 to -6.0‰ PDB was interpreted as a reflection of alteration by meteoric waters. Inasmuch as isotopic values similar to those in both the Budd (1989) and the Moshier (1989b) studies occur at Ghaba North Field, one could interpret these data as either indicative of stabilization in a system with marine pore-water composition or in meteoric water.

Evidence exists for subaerial exposure and communication with meteoric waters in Ghaba North Field. Two surfaces recognized in GN-23 have potential importance for this argument: a subaerial exposure surface recognized at 609.45 m and the regional unconformity at the top of the Shuaiba. Further evidence comes from observation of extensive leaching beneath these surfaces and from calcite cement fabrics. Other workers have recognized subaerial exposure both within and at the top of the Shuaiba Formation (Wagner, 1990; Vahrenkamp, 1996).

Isotopic data from micro-rhombic calcite (Figs. 7, 8B) can be separated into two clusters between those samples with $\delta^{18}O$ values more depleted or more enriched than a $\delta^{18}O$ value of -6‰. Although this is a mostly arbitrary demarcation, virtually all samples above 609.45 m in GN-23 are more enriched than -6‰ $\delta^{18}O$ (Fig. 8B). These data likely suggest that the calcite rhombohedra developed in more than one event. If this is the case, it explains the deviation of data from the pattern suggested by Allan and Matthews (1977, 1982) for the variation expected in carbon and oxygen isotopic compositions of limestones during fresh-water diagenesis. Typically, a narrower range of $\delta^{18}O$ values and a wider range of $\delta^{13}C$ values results from fresh-water diagenesis, in part because water has a large reservoir for oxygen relative to the rock (Lohmann, 1988). Thus, most of the oxygen incorporated in a diagenetic product is sourced from water, and the oxygen-isotope composition should be relatively invariant for a single diagenetic event. On the other hand, carbon reflects a combination of sources, such as soil-gas $CO_2$, precursor $CaCO_3$, and atmospheric $CO_2$. The carbon-isotope composition reflects water–rock interaction and proximity to an exposure surface and therefore varies with depth (Allan and Matthews, 1982; Lohmann, 1988). However, if more than one diagenetic event affected the sediments, then this pattern would not hold, and compositions of diagenetic carbonates would be dependent upon the isotopic composition of the different waters involved.

*Stratigraphic Geochemistry.—*

Variations of $\delta^{13}C$ and $\delta^{18}O$ with depth in the Shuaiba Formation (Fig. 8A and 8B) should be used together with petrographic and stratigraphic observations. The two plots can be divided into three zones (Zones I–III) with different isotopic characteristics. The first zone starts from the bottom upward to a depth of 619.45 m. The second zone begins at 619.45 m and ends at 609.45 m. The third zone starts at 609.45 m and ends at the top of the Shuaiba at 596 m. Each one of these diagenetic zones is discussed below.

Zone I is distinguished from the rest of the Shuaiba by its nearly invariant $\delta^{18}O$ and $\delta^{13}C$ values. The 619.45 m upper boundary was also recognized as the stratigraphic boundary, a marine hardground, between the lower Shuaiba zone (interpreted as restricted-lagoon lithofacies) and the upper Shuaiba zone (interpreted as relatively open-shelf lithofacies). Because -2 to -2.5‰ PDB was suggested for the $\delta^{18}O$ value of calcitic sediments deposited from Aptian seawater (Scholle and Arthur, 1980; Moldovanyi and Lohmann, 1984), then an average $\delta^{18}O$ value of -6.7‰ PDB of the first zone is depleted at least 4‰ PDB relative to marine values. If matrix stabilization had occurred in marine pore waters, then $\delta^{18}O$ of the matrix should be around -4‰ PDB, which indicates depletion by 1.5 to 2.0‰ relative to its marine precursor (Moshier, 1989b). We suggest that stabilization in meteoric waters was responsible for the additional 2.7‰ depletion beyond this value.

The average $\delta^{13}C$ of micro-rhombic matrix samples in Zone I is approximately +4‰ PDB, similar to values reported by other workers studying Shuaiba reservoirs (Moshier, 1989b; Budd, 1989; Wagner, 1990; Vahrenkamp, 1996). Carbon incorporated in the diagenetic stabilization process that produced the micro-rhombic calcite was dominated by carbon sourced from the rock. This suggests diagenesis under conditions of low water/rock ratio and stabilization far away from a source of depleted carbon, such as a subaerial exposure horizon. A hardground present at 619.45 m likely acted as a permeability barrier separating Zone I sediments from overlying subaerial conditions. Confined flow of meteoric water and stabilization below the hardground could have produced the relatively invariant isotopic compositions of the lower zone. In such a scenario, flow rates are low, water/rock ratio is low, oxygen-isotope values are depleted relative to marine conditions, and the carbon-isotope composition is buffered by the composition of the precursor limestone. A low water/rock ratio also would have the effect of producing slow stabilization of rudist material rather than leaching. Although timing of meteoric diagenesis of Zone I is unclear, it probably occurred prior to subaerial exposure within and atop the upper Shuaiba zone.

Zone II extends from 619.45 m to 609.45 m, where an erosional surface is recognized in core. The $\delta^{18}O$ of Zone II averages about -6.4‰ PDB. Apart from the $\delta^{18}O$ value at the base of this zone, other values are almost constant and then become enriched toward the top of the zone. Carbon-isotope values in this interval become progressively depleted toward the top. Carbon-isotope depletion and oxygen-isotope enrichment upward toward an erosional surface is a common pattern of isotope variability

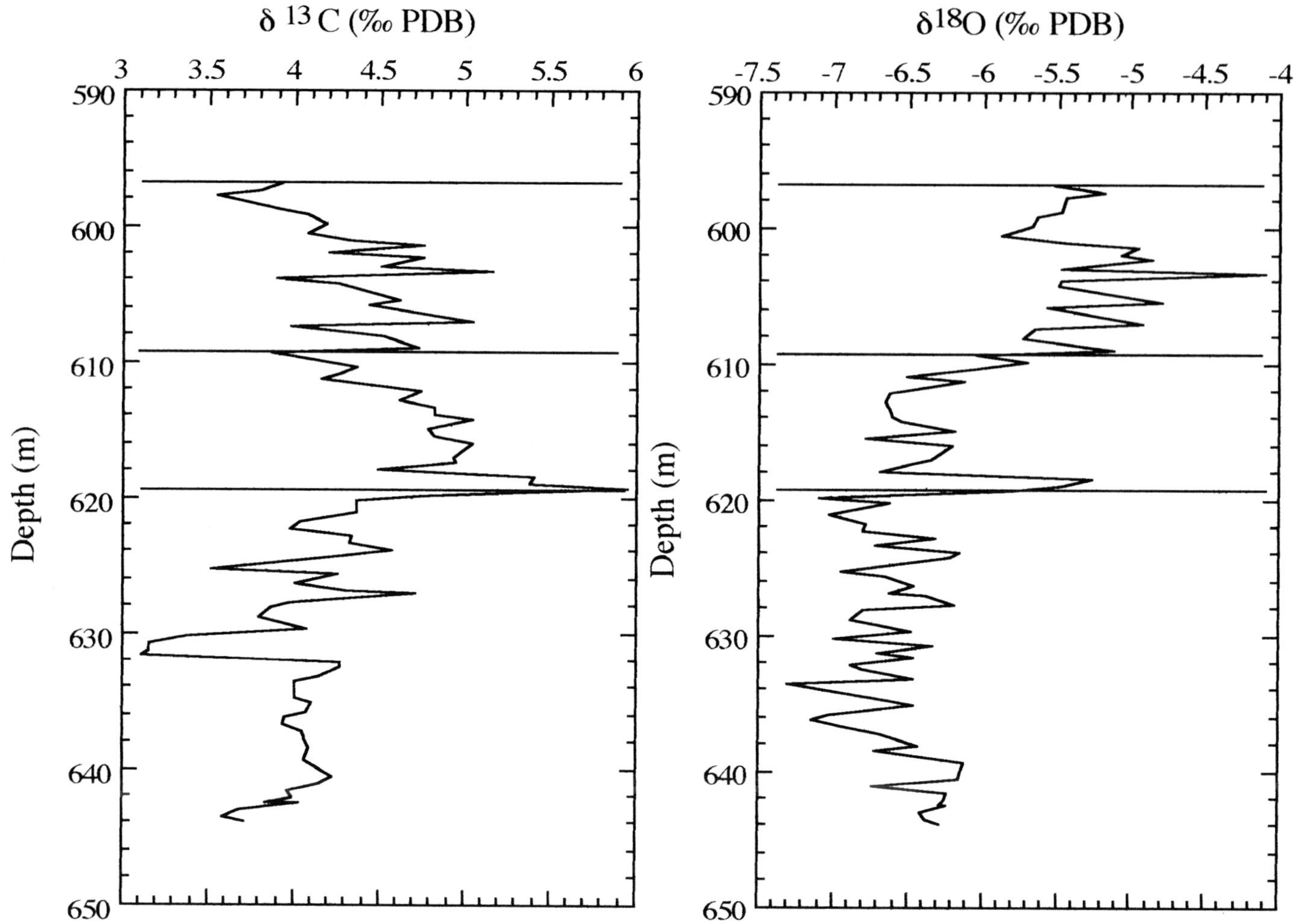


FIG. 8.—Matrix isotopic variation in GN-23. (A) $\delta^{13}C$ versus depth. (B) $\delta^{18}O$ versus depth. Lower horizontal line indicates boundary between lower and upper Shuaiba zones. Middle horizontal line marks local subaerial exposure surface. Upper horizontal line marks top of the Shuaiba Formation. See text for discussion.

imparted by meteoric diagenesis at a subaerial exposure surface (Allan and Matthews, 1978, 1982). In an arid climate with only limited plant and soil-zone activity, carbon-isotope depletion is attenuated at subaerial exposure surfaces (Budd, 1989). This may account for the limited depletion in carbon-isotope composition at the exposure surface. Depletion may also be attenuated because most arid vegetation would be expected to follow the C4 photosynthetic pathway, and soil-gas $CO_2$ in arid regions may be enriched in $\delta^{13}C$ by as much as 8‰ relative to soil gas in humid regions (Rightmire and Hanshaw, 1973). It is also likely that subsequent marine submergence of this surface would have led to erosion of whatever thin soil zone may have been present.

Stratigraphic evidence, represented by the presence of the erosional surface, and petrographic evidence, represented by extensive leaching of rudists and rudist fragments below the surface, corroborate these isotopic data. We envision localized subaerial exposure and development of a small freshwater lens in a rudist bioherm. This may have occurred at a number of places on the Shuaiba shelf where local seafloor topography was exposed during a minor relative sea-level fall.

Zone III extends from 609.45 m upward to the regional unconformity separating the Shuaiba from the overlying Nahr Umr Formation. Vertical $\delta^{18}O$ and $\delta^{13}C$ variations again show a pattern indicative of subaerial exposure and meteoric diagenesis. The presence of extensive leaching and moldic secondary porosity, especially below the unconformity, is further support for subaerial exposure at the end of Shuaiba time. The occurrence of a boundary at 600.4 m in this zone is obvious from $\delta^{13}C$ and $\delta^{18}O$ variation plots with depth (Fig. 8A, B). Sharp enrichment in $\delta^{13}C$ and $\delta^{18}O$ directly below 600.4 m and the depletion trend of these above it toward the top is a pattern indicative of a vadose–phreatic boundary (Allan and Matthews, 1982). The Shuaiba above 600.4 m has very high macroporosity and abundant unconsolidated rubbly intervals, indicative of leaching and a lack of extensive cementation, characteristics of subaerial diagenesis.

*Late Coarse-Crystalline Calcite Cements.—*

Coarse-crystalline calcite cements postdate compaction but predate hydrocarbon migration, and therefore precipitated at intermediate burial depths. These cements occupy secondary moldic pores that we interpret to have formed during early subaerial exposure and meteoric diagenesis. Samples of these cements have an average $\delta^{18}O$ of -8‰ PDB and an average $\delta^{13}C$

of +3.3‰ (Fig. 7). These cements have the most depleted oxygen values in this study, and are interpreted to have precipitated under slightly higher-temperature, moderate burial conditions.

It is likely that these cements were precipitated after deposition of the Nahr-Umr Formation. Migration of hydrocarbons into the Shuaiba Formation prevented these cements from completely occluding moldic porosity. Hydrocarbons would not have been trapped in the Shuaiba at Ghaba North Field unless Nahr Umr Shales had sealed the Shuaiba and the structure had already been developed. Thus, coarse-crystalline calcite cements were precipitated after the formation of the anticlinal trap at Ghaba North, which occurred during uplift associated with salt tectonics.

## SUMMARY AND CONCLUSIONS

The Shuaiba Formation (Aptian) in Ghaba North Field consists of two stratigraphically and sedimentologically distinct zones. The lower zone represents deposition below fair-weather wave base under restricted to partially restricted water circulation in a lagoonal setting. The upper zone represents deposition in a shallow relatively open shelf/ramp setting with more open (less restricted) circulation conditions, and above fair-weather wave base. Overall, this stratigraphic stacking represents a shallowing-upward trend.

Favorable reservoir characteristics are best developed in the upper Shuaiba zone. Development of secondary moldic porosity enhances reservoir properties; however, matrix microporosity is responsible for most of the hydrocarbon-holding capacity of the reservoir. The upper Shuaiba zone has higher macroporosity and horizontal permeability, and thus better reservoir quality. Microporosity is present as intercrystalline pores within micro-rhombic calcite matrix.

Three diagenetic zones are identified by stable-isotope data and supported by stratigraphic and petrographic evidence in the GN-23 core. These zones are indicative of different diagenetic events and the complexity of the Shuaiba diagenetic history. The lower zone was stabilized in a confined meteoric aquifer. The second zone experienced local subaerial exposure and meteoric diagenesis. The upper zone was affected by exposure at a regional unconformity marking the end of Shuaiba time. From isotopic evidence, presence of subaerial exposure surfaces, and the action of leaching, it can be concluded that meteoric waters were the main diagenetic fluids responsible for diagenesis of the Shuaiba Formation at Ghaba North Field. Matrix stabilization, development of micro-rhombic calcite, and the formation of intercrystalline microporosity occurred in the presence of meteoric water. Development of secondary moldic porosity in rudists and rudist fragments was controlled by the presence of abundant rudists, original mineralogy of the rudists, and proximity of rudist-rich facies to subaerial exposure surfaces.

Stabilization of the Shuaiba matrix, development of moldic macroporosity associated with the exposure surfaces, and establishment of the anticlinal trap in Ghaba North occurred relatively rapidly. The presence of partially cemented molds by coarse-crystalline calcite cement indicates that hydrocarbons migrated into the reservoir soon after precipitation of this late cement. Hydrocarbon migration thus prevented occlusion of porosity by cementation or compaction under heavy overburden pressures at deep burial depths.

This study of the Shuaiba Formation is fundamental for a development plan for the Ghaba North Field. This is a heterogeneous reservoir, with the upper Shuaiba zone providing the best reservoir properties. Most hydrocarbon storage is within matrix microporosity; however, reservoir quality is enhanced in the presence of secondary moldic porosity.

The presence of two subaerial exposure surfaces is significant for further exploration and exploitation. The subaerial exposure surface at 609.45 m in GN-23, although it likely records a localized exposure event, suggests that others may be present. Local growth of rudist bioherms or buildups above sea level, or a minor fall in relative sea level, could have occurred at other localities on the shelf. Such localized exposure, with subsequent meteoric diagenesis, would provide enhancement of the Shuaiba reservoir, similar to that developed in GN-23. Development of secondary moldic porosity should be denoted by an amplitude decrease in the top Shuaiba seismic reflectors, thus reducing the ordinary velocity contrast between the higher-velocity limestone of the Shuaiba and the overlying lower-velocity Nahr Umr shale (Frost et al., 1983). Variation of reservoir properties should be considered in any development or flooding program of the Shuaiba Formation in Ghaba North Field.

## ACKNOWLEDGMENTS

We wish to acknowledge Petroleum Development Oman for providing the data and support of the first author's Master's thesis project at Colorado School of Mines. J. Curtis and J. Warme provided constructive criticism during various stages of this work. Editors of the special publication, A. S. Alsharhan and R. W. Scott, are thanked for their encouragement and patience. Insightful reviews by P. M. Harris and V. C. Vahrenkamp significantly improved this manuscript.

## REFERENCES

Al-Bastaki, A.A., 1993, Sedimentology and diagenesis of the Shuáiba Formation (Lower Cretaceous; Aptian), Ghaba North Field, Oman: Unpublished M.Sc. thesis, Colorado School of Mines, Golden, Colorado, 161 p.

Allan, J.R., and Matthews, R.K., 1977, Carbon and oxygen isotopes as diagenetic and stratigraphic tools in surface and subsurface data, Barbados, West Indies: Geology, v. 5, p. 16–20.

Allan, J.R., and Matthews, R.K., 1982, Isotope signatures associated with early meteoric diagenesis: Sedimentology, v. 29, p. 797–817.

Alsharhan, A.S., 1985, Depositional environment, reservoir unit evolution, and hydrocarbon habitat of Shuaiba Formation, Lower Cretaceous, Abu Dhabi, United Arab Emirates: American Association of Petroleum Geologists, Bulletin, v. 69, p. 899–912.

Alsharhan, A.S., and Kendall, C.G.St.C., 1986, Precambrian to Jurassic rocks of the Arabian Gulf and adjacent areas: American Association of Petroleum Geologists, Bulletin, v. 70, p. 977–1002.

Alsharhan, A.S., 1995, Facies variation, diagenesis, and exploration potential of the Cretaceous rudist-bearing carbonates of the Arabian Gulf: American Association of Petroleum Geologists, Bulletin, v. 79, p. 531–550.

Bathurst, R.G.C., 1975, Carbonate Sediments and Their Diagenesis: Amsterdam, Elsevier Science Publishers, Developments in Sedimentology 12, 658 p.

Borgomano, J.R., 1991, Quick-look sedimentological study on selected Shuaiba cores: Petroleum Development Oman, Exploration Laboratory Note no. 24, 5 p.

Budd, D.A., 1989, Micro-rhombic calcite microporosity in limestones: a geochemical study of the Lower Cretaceous Thammama Group, United Arab Emirates: Sedimentary Geology, v. 63, p. 293–311.

Frost, S.H., Bliefnick, D.M., and Harris, P.M., 1983, Deposition and porosity evolution of a Lower Cretaceous rudist buildup, Shuaiba Formation of eastern Arabian Peninsula, *in* Harris, P.M., ed., Carbonate Buildups: Society of Economic Paleontologists and Mineralogists, Core Workshop 4, p. 381–410.

GALLAGHER, J., MERCADIER, C., MCFADYEN, K., AND MEIJ, R., 1990, Ghaba North Shuaiba incremental development: Petroleum Development Oman Report no. Pef/90/146, 16 p.

HARRIS, P.M., FROST, S.H., SEIGLIE, G.A., AND SCHNEIDERMANN, N., 1984, Regional unconformities and depositional cycles, Cretaceous of the Arabian Peninsula, *in* Schlee, J.S., ed., Interregional Unconformities and Hydrocarbon Accumulation: American Association of Petroleum Geologists, Memoir 36, p. 67–80.

HARTMANS, R.F., 1986, Fahud-223 core and Shuaiba reservoir geological model, Fahud oil field, Oman: Petroleum Development Oman, Production Geology Report no. 43, 18 p.

HUGHES CLARKE, M.H., 1988, Stratigraphy and rock unit nomenclature in the oil-producing area of interior Oman: Journal of Petroleum Geology, v. 11, p. 5–60.

JAMES, N.P., AND CHOQUETTE, P.M., 1984, Diagenesis 9: limestones—the meteoric diagenetic environment: Geoscience Canada, v. 11, p. 161–194.

LOHMANN, K.C, 1988, Geochemical patterns of meteoric diagenetic systems and their application to studies of paleokarst, *in* Choquette, P.W., and James, N.P., eds., Paleokarst: New York, Springer-Verlag, p. 55–80.

MCCREA, J.M., 1950, The isotope chemistry of carbonates and a paleotemperature scale: Journal of Chemical Physics, v. 18, p. 849–857.

MOLDOVANYI, E.P., AND LOHMANN, K.C, 1984, Isotopic and petrographic record of phreatic diagenesis: Lower Cretaceous Sligo and Cupido Formations: Journal of Sedimentary Petrology, v. 54, p. 972–985.

MOSHIER, S.O., 1989a, Microporosity in micritic limestones: a review: Sedimentary Geology, v. 63, p. 191–213.

MOSHIER, S.O., 1989b, Development of microporosity in a micritic limestone reservoir, Lower Cretaceous, Middle East: Sedimentary Geology, v. 63, p. 217–240.

MURRIS, R.J., 1980, Middle East: stratigraphic evolution and oil habitat: American Association of Petroleum Geologists, Bulletin, v. 64, p. 597–618.

PETROLEUM DEVELOPMENT OMAN, 1990, Core analysis for well Ghaba North-23, Oman: Core Laboratories Report, 31 p.

PETROLEUM DEVELOPMENT OMAN, 1992, Ghaba North Field: Production Reference Data Book, v. 1, 27 p.

PRATT, B.R., AND SMEWING, J.D., 1993, Early Cretaceous platform-margin configuration and evolution in the central Oman Mountains, Arabian Peninsula: American Association of Petroleum Geologists, Bulletin, v. 77, p. 225–244.

RIGHTMIRE, C.T., AND HANSHAW, B.B., 1973, Relationship between the carbon isotope composition of soil $CO_2$ and dissolved carbonate species in ground waters: Water Resources Research, v. 9, p. 958–967.

SCHOLLE, P.A., AND ARTHUR, M.A., 1980, Carbon isotope fluctuations in Cretaceous pelagic limestones: Potential stratigraphic and petroleum exploration tool: American Association of Petroleum Geologists, Bulletin, v. 64, p. 67–87.

UITENTUIS, E., 1981, Ghaba North Field production geological review: Petroleum Development Oman, Internal Report, 10 p.

VAHRENKAMP, V.C., 1996, Carbon isotope stratigraphy of the Upper Kharaib and Shuaiba Formations: Implications for the early Cretaceous evolution of the Arabian Gulf region: American Association of Petroleum Geologists, Bulletin, v. 80, p. 647–662.

WAGNER, P.D., 1990, Geochemical stratigraphy and porosity controls in Cretaceous carbonates near the Oman Mountains, *in* Robertson, A.H.F., Searle, M.P., and Ries, A.C., eds., The Geology and Tectonics of the Oman Region: Geological Society of London, Special Publication 49, p. 127–137.

WILSON, J.L., 1975, Carbonate Facies in Geologic History: New York, Springer-Verlag, 469 p.

# STYLOLITES IN LOWER CRETACEOUS CARBONATE RESERVOIRS, U.A.E.

ABDULRAHMAN S. ALSHARHAN
*Faculty of Science, United Arab Emirates University, P.O. Box 17551, Al Ain, U.A.E.*
AND
JAMES L. SADD
*Department of Geology, Occidental College, 1600 Campus Road, Los Angeles, CA 90041-3314, U.S.A.*

ABSTRACT: The major Lower Cretaceous reservoirs of the United Arab Emirates (U.A.E.) are characterized by stylolite-rich zones. Although their distribution and frequency are variable, stylolites tend to be most common and abundant toward the flanks of the fields. Three main types of stylolites are found in these reservoirs, each characterized by variations in amplitude, morphology, lateral continuity, and thickness of the accumulated insoluble residue. The recorded stylolites are classified as rectangular or high amplitude, solution seams or wave-like, and wispy seams or horsetail. The composition of the relatively insoluble seam material is variable, and depends mainly upon the composition of the nearby host rock. Most of the seams are composed of clay minerals, black bitumen, pyrite, and fine- to medium-grained calcite crystals or dolomite rhombs. Stylolites affect the petrophysical characteristics (porosity/permeability) and thickness reduction (compaction). The lower porosity and permeability values are found associated with the well-developed stylolites. Most stylolites observed in cores are parallel to subparallel to bedding (the horizontal type), indicating the predominance of vertical stress imposed by overburden pressure in stylolite formation. This also suggests a relative absence of tectonic or metamorphic activity in the area, which might produce inclined to vertical stylolites, although vertical to subvertical tension fractures are found associated with the well developed stylolites.

## INTRODUCTION

Stylolites are lithologic manifestations of pressure dissolution in carbonate reservoir rocks. Stylolitization usually results in reduction of unit thickness and also in porosity and permeability of the reservoir (Dunnington, 1967; Johnson and Budd, 1975; Koepnick, 1984; Alsharhan, 1985; Oswald et al., 1995). This paper focuses on stylolite-bearing zones in the Lower Cretaceous carbonate reservoirs in the U.A.E. The objectives are 1) to document the distribution and morphology of the stylolites, 2) to evaluate the importance of stylolites as barriers to fluid flow, 3) to evaluate the relationship between porosity, permeability, and stylolite distribution in the reservoir, and 4) to evaluate effects of stylolitization and fracturing on porosity and permeability of carbonate reservoirs in the Lower Cretaceous Thamama Group, one of the main producing carbonate formations of the U.A.E. Petroleum production is obtained from large anticlines in these limestone reservoirs (Fig. 1).

In the U.A.E., the Lower Cretaceous Thamama Group is characterized by interbedded porous and dense limestones (Fig. 2). Porous hydrocarbon-bearing limestones are divided into several zones. Each zone is subdivided into several reservoirs separated by dense impermeable to poorly permeable limestones. The latter act as barriers to vertical fluid migration and are of prime importance in oil production efficiency and reservoir development. The stylolites observed in the Thamama Group occur along lithologic transitions or partings between these two lithotypes.

The Thamama Group ranges in thickness from 800 to 1000 m, and includes beds of Berriasian to Aptian age, a time span of nearly 30 million years. It consists of four formations: the Habshan, Lekhwair, Kharaib, and Shuaiba formations, in ascending order (Fig. 2). This thick sequence mainly consists of shallow shelf carbonates, although some deep-water carbonates occur locally in association with an Aptian intrashelf basin (Alsharhan and Nairn, 1993).

## MATERIALS STUDIED AND METHODS USED

This study is based on Lower Cretaceous reservoir core samples from the U.A.E. fields shown in Figure 1. Data from these cores are presented to illustrate the field-wide distribution of stylolites in selected reservoirs. Also, an analysis of the wireline log response to stylolite-bearing intervals is presented to identify stylolite-bearing zones in uncored wells. Core porosity and permeability data from some wells are used to produce a composite log to show the relationship between the petrophysical characteristics (porosity and permeability) of the stylolites and log response. One hundred thin sections were studied using the polarizing characteristics of the stylolite zones in the studied cores.

The stylolite intensity (SI) approach of Johnson and Budd (1975) is also used to evaluate stylolite-bearing horizons. This relationship is expressed by the empirical formula

$$\mathrm{SI} = H \times \frac{100}{\phi}$$

where $H$ = thickness of stylolite zone on logs (feet), and $\phi$ = porosity within stylolite interval (%). This assumes that the petrophysical character of the rocks is related to stylolite development, but in many cases, this is not the case.

A stylolite intensity value was obtained for each stylolite-bearing horizon in each well. These wells were related to adjacent ones, including those with measured vertical permeability, and a map was prepared for each major stylolite zone. This provides a means to analyze the effectiveness of the zone as a barrier or impediment to fluid movement (Fig. 3). These stylolite intensity maps also can aid the development of reservoir simulation and field development of plans, especially where secondary-recovery water or water-injection programs are envisioned.

The percentage of compaction due to stylolitization (which also gives the minimum thinning) of a stylolitized limestone unit is calculated by comparing the average porosity before and after

Middle East Models of Jurassic/Cretaceous Carbonate Systems
SEPM Special Publication No. 69, Copyright © 2000
SEPM (Society for Sedimentary Geology), ISBN 1-56576-075-1, p. 185–207.

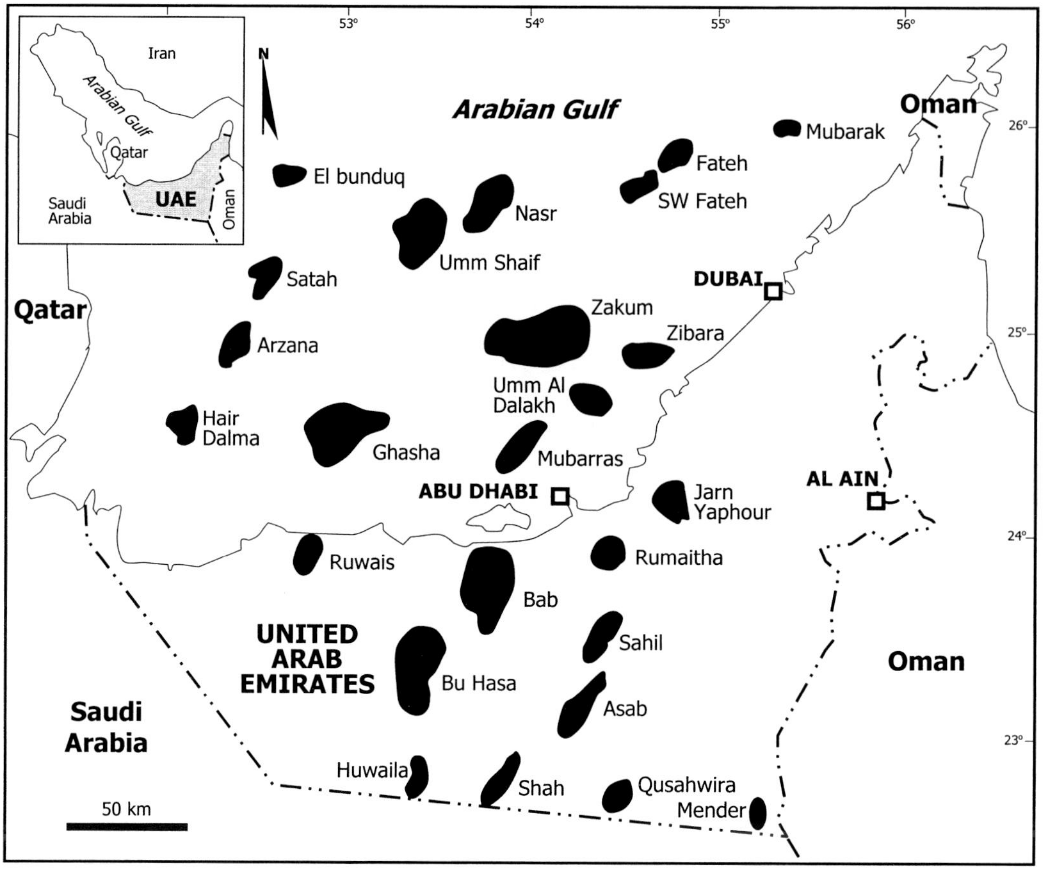


FIG. 1.—Location map of the United Arab Emirates and its major oil fields.

stylolitization by use of the formula developed by Dunnington (1967):

$$\phi = T + \bar{\phi} - \frac{T\ \bar{\phi}}{100}$$

where $\phi$ is the present porosity outside the stylolite zone, $\bar{\phi}$ is the present porosity within the stylolite zone, and $T$ is the percentage thinning of the unit due to stylolitization.

The porosity outside the stylolite zone ($\phi$) implies only short-distance transport of solute. In fact, stylolite-generated cementation may occur well away from an obviously cemented stylolite horizon. The formula implies that all porosity loss, relative to adjacent units, is stylolite related. In many cases, stylolites develop within or along beds that have been cemented at the sea floor. In this case, use of this formula requires caution.

The amplitude of stylolites was measured in the core samples, from the apex to the lowest point of the stylolites. It was found that the measured amplitude varied with the type of stylolite, e.g., the rectangular types had the largest amplitude and the wispy seams had the smallest.

In order to determine whether stylolite type is related to degree of pressure dissolution and loss of unit thickness, we examined many cores from the Shuaiba and Kharaib formations. Also, seven cores containing well-developed examples of the three stylolite types from the Bu Hasa, Bab, Asab, and Zakum fields (Fig. 1) were selected for analysis of mineralogy and trace-element geochemistry. Three cores from the Kharaib Formation in the Bab, Asab, and Zakum fields displayed rectangular stylolites and wave like stylolites, and one core from the Shuaiba Formation in the Bu Hasa Field displayed horsetail stylolites. Each core was first spot sampled using a 1 mm stainless steel drill to isolate portions of host micrite, sparry cement, and insoluble stylolitic residue within the core at various stratigraphic distances from a stylolite. Location of spot samples was aided using petrographic thin sections to locate relatively pure regions of each core. These spot samples were split analytically in preparation for mineralogic and trace-element analysis. One split was powdered by slow grinding in acetone using a diamond mortar, and analyzed on a General Electric XRD7 Xray diffractometer using copper $K\alpha$ radiation and scanning over a $2\theta$ range of 0 to 60 degrees. Refraction data were collected digitally and analyzed using dedicated software to correlate each peak with a specific mineral from the JCPDS files (American Society of Testing and Materials, 1967).

The second split of each subsample was analyzed semiquantitatively for a variety of trace metals. Powdered samples were extracted for geochemical analysis using analytical microwave extraction in $HNO_3$ following the standard United States Environmental Protection Agency analytical test methods 6010/200.8, 7471, and 8080 (http://www.epa.gov/Standards.html) using a CEM MD2000 microwave digestion apparatus. Solutions were analyzed in vacuum using an ARL 2410+ ICP, and all results are reported in parts per million dry weight or as ratios normalized to calcium content. Twelve trace metals (Al, Be, Cr, Cu, Fe, Mg, Mn, Na, Ni, Pb, V, Zn) were found

Fig. 2

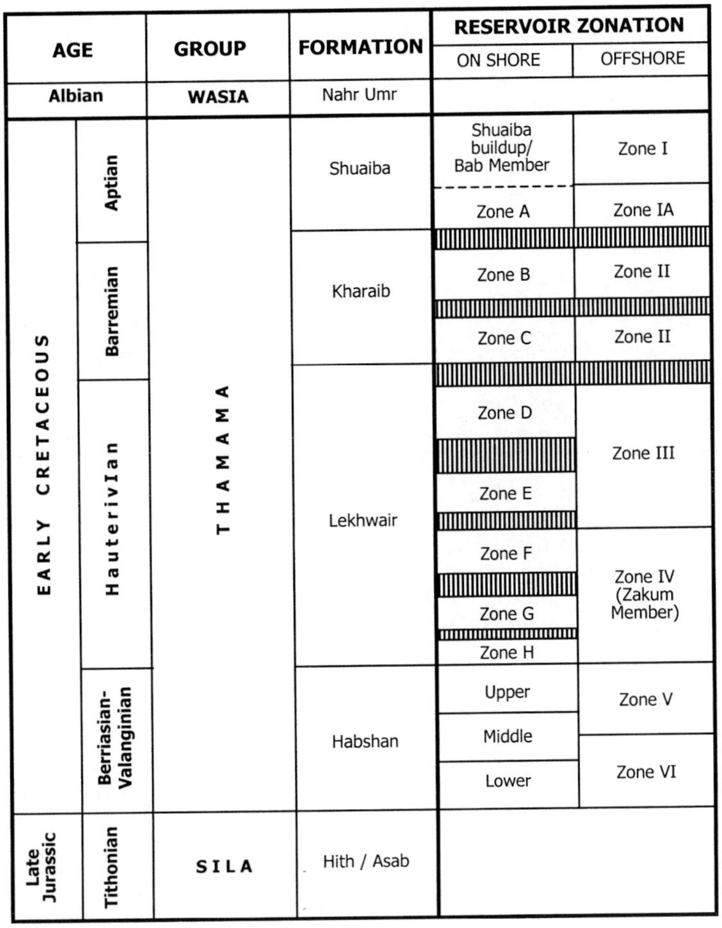


FIG. 2.—Simplified stratigraphy and reservoir zonation of the Lower Cretaceous (Thamama Group) in Abu Dhabi, U.A.E., (Alsharhan, 1985).

in analytically detectable concentrations (10x d.l.) in most spot samples, and were chosen for more detailed study of stratigraphic core subsamples. A 1-cm-wide longitudinal section was cut from each core oriented perpendicular to the stylolite, and this core section was then divided into 1 cm subsamples to allow analysis of portions of the core at various stratigraphic distances from the stylolite. These core subsamples were powdered and prepared for analysis of the concentration of the twelve selected trace metals. The mean concentration of each trace metal in the stylolitic zone was compared statistically to the mean concentration in the host rock above and below the stylolitic interval using the t–test (equal variance assumed). Results of these analyses, reported as ratios normalized to calcium content, are summarized in Table 1.

## ORIGIN AND CLASSIFICATION OF STYLOLITE TYPES

### *Historical Review*

The timing of stylolite formation is still controversial. Several authors have accepted a pre-induration origin (Shaub, 1939; Prokopovich, 1952; Park and Schot, 1968a, 1968b), but others have postulated a post-induration origin (Stockdale, 1943; Dunnington, 1954; Brown, 1959; Manten, 1968). A third group has claimed that although some stylolites are of pre-induration origin, the well-developed, sutured, rectangular types are of post-induration origin (Wanless, 1979; Buxton and Sibley, 1981), whereas Shinn and Robin (1983) suggested that those in fine-grained muds form simultaneously with the onset of lithification.

An early idea on the formation of stylolites was provided by Marsh (1867), who refuted the "wood" theory of Eaton and the "salt" theory of Vanuxem, which had prevailed since the 1830s, and instead concluded that stylolites could be ascribed to differential vertical compression before lithification. The argument for pressure dissolution was analyzed by Stockdale (1922), who concluded that stylolites form in lithified rocks and that clay caps result as a solution residue. He calculated the magnitude of vertical shortening in rocks in Indiana (USA) by measuring the thickness of the clay cap and clay concentrations in host rocks. Shaub (1939) contradicted Stockdale's (1922) solution theory and claimed that the clay cap is sedimentary and that stylolites form while sediments are still plastic and beginning to dewater. Prokopovich (1952) suggested that stylolites form in soft sediments on the sea bottom in local areas where pH is low. Deelman (1975) proposed that plastic deformation, due to crystal lattice dislocation rather than pressure dissolution, caused post-lithification stylolites. Garrison and Kennedy (1977) demonstrated that flaser structure in the Upper Cretaceous chalks of

TABLE 1.—Summary of results of geochemical analysis of the three stylolite types in the U.A.E.

| Trace Element | Rectangular Stylolites | Wave-like Stylolites | Horsetail Stylolites |
|---|---|---|---|
| Cr | 2.0X | 3.2X | very low |
| Fe | 5.9X | 3.0X | very low |
| Ni | 7.1X | 3.4X | very low |
| Al | 1.9X | 4.5X | very low |
| Cu | 2.0X | 3.2X | very low |
| Mg | 1.5X | 3.0X | very low |
| V | 1.8X | 1.8X | very low |
| Pb | 1.8X | 1.5X | very low |
| Be | very low | very low | very low |
| Na | very low | very low | very low |
| Mn | very low | 1.6X | very low |
| Zn | 1.5X | very low | very low |

For each stylolite type, the mean Me/Ca ratio for each element in the stylolite zone was compared to the mean ratio in the host rock above and below the stylolite. If this difference was statistically significant (95% confidence level using t-test; equal variance assumed), the mean concentration factor from all cores analyzed is shown. Both rectangular and wave-like stylolites show significant concentration in the stylolite zone of a similar suite of trace elements, although the degree of concentration varies substantially for some elements (Fe, Ni, Al, Mg). Horsetail stylolites show no significant concentration values for all elements.

Southern England probably predated stylolite formation and resulted from interaction between compaction, dissolution, and partial lithification.

It would appear that stylolites commonly, but not always, form along lithologic transitions or partings as a result of stress due to either overburden pressure or tectonism. Stress causes an increase in pressure, and when a critical level is reached, rock components start to dissolve in order to relieve the stress. Although the irregular shape of stylolites is initiated by differential solubility of the rock components, they are not completely irregular, and some have a wave-like form or a periodicity in them.

The conditions that determine the initial generation of stylolites are not well understood. Experiments carried out by Rutter (1983) showed that it is extraordinarily difficult to reproduce natural microstructures, whereas Shinn and Robin (1983) were able to produce wispy stylolite-like seams of organic matter in a series of sediments by compaction processes. Buxton and Sibley (1981) showed that stylolites may also abut porous zones, implying low gradients and transport over relatively long distances away from the sites of dissolution. The type of stylolites observed by these authors reflect the degree of cementation at the time of dissolution.

In this study all stylolite-rich zones observed in the Thamama Group are parallel or subparallel to bedding (termed horizontal or stratiform). The presence of horizontal stylolites indicates the absence of tectonic activity. Tectonism would give rise to inclined and/or vertical stylolites, and this fabric is documented in many parts of the world by Dunnington (1954, 1967), Park and Schot (1968a, 1968b), Bathurst (1975), and Wanless (1979). Purser (1984) attributes the difference in intensity of stylolites between the flanks and crests of the Abu Dhabi oilfields to the distribution of early diagenetic processes. He suggested that the reservoirs situated on paleohighs were more susceptible to early cementation (or leaching) than those on the flanks of the structures, which remained submerged and received less preburial cement.

## *Characteristics of Stylolites and Their Types*

Stylolites are undulose surfaces, often but not always highly irregular on a scale of centimeters. They occur at surfaces along which dissolution of calcium carbonate occurred as a direct result of stress exerted by the weight of overburden (Nelson, 1979; Burgess and Peter, 1985). Study of the development of stylolites in the Lower Cretaceous carbonate reservoirs in U.A.E. field areas is of importance to the understanding of the effective reservoir capacity. Clearly a better definition of the sedimentologic, petrographic, and diagenetic character of rocks would help in the evaluation of the total effective pore spaces in each reservoir so as to evaluate more accurately its hydrocarbon content.

Stylolite amplitude is generally considered to be a measure of the minimum thickness of the sedimentary section removed along a pressure-solution surface (Stockdale, 1922). As pressure solution proceeds, insoluble constituents gradually accumulate along the solution surface. As this residue increases, significant reduction in fluid cross-flow can occur. Measurement of stylolite amplitude generally provides a minimum estimate of section removal, i.e., an estimate of the minimum magnitude of section thinning caused by pressure solution. The amplitude of pressure-solution surfaces may not show a positive correlation with the degree of matrix cementation; high-amplitude stylolites may be weakly cemented, whereas low-amplitude stylolites may be strongly cemented (Koepnick, 1988). We agree with many workers such as Dunnington (1967), Koepnick (1984, 1987, 1988), Purser (1984), and Johnson and Budd (1975) that stylolite frequency and cumulative stylolite amplitude increase from the crest to the flanks of the anticlinal closure that forms the field. Koepnick (1988) concluded that the amplitudes of stylolites are controlled by the pattern of stress concentration developed during growth of the anticlinal structures, and by the inhibition of pressure solution by hydrocarbons along the anticlinal crest.

Dunnington (1967) recognized that at Bab Field stylolite frequency and stylolite-related reservoir thinning increase from the crest to the flanks of the structure. Stylolite distribution at this

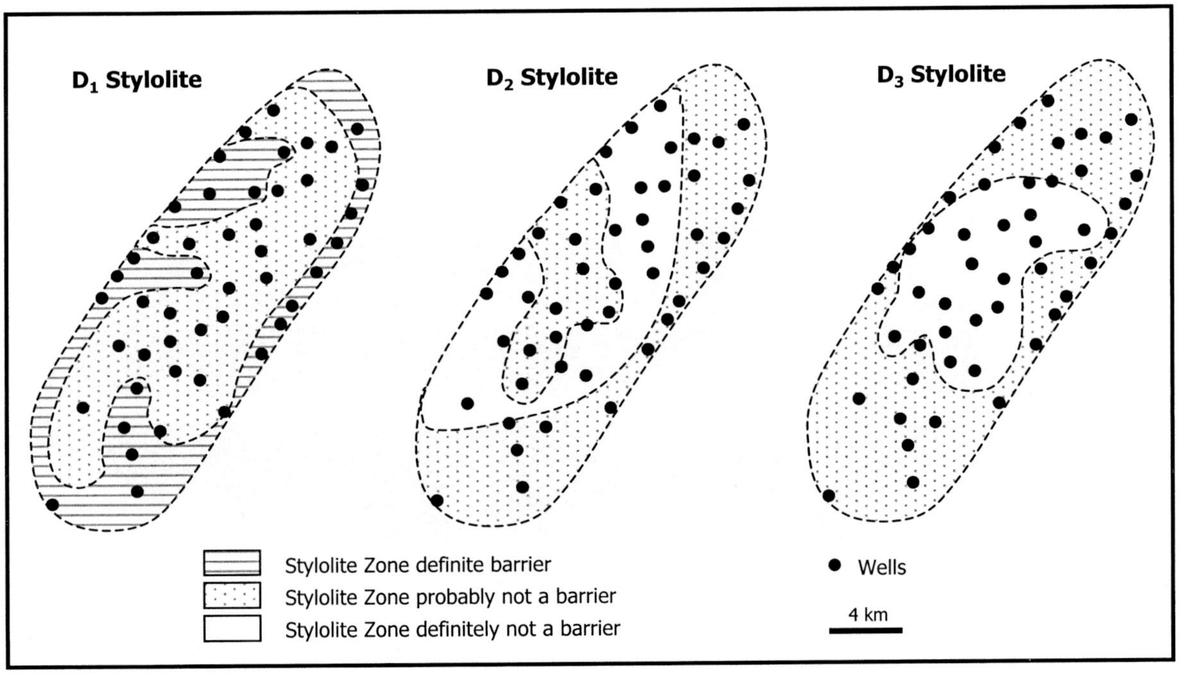


FIG. 3.—Stylolite intensity map of Asab Field for D1, D2, and D3 stylolite zones (after Johnson and Budd, 1975).

field is proposed to reflect a combination of stress concentration patterns developed during the growth of the reservoir structures, and of pressure-solution inhibition by hydrocarbon entrapment. The cumulative stylolite amplitude values cannot account for the total isopach variation, because the amplitude is only a measure of the minimum thickness of sedimentary section removed along the solution surface (Stockdale 1922, 1926) and because a certain thickness of the section was also removed during development of the wispy seams and solution seams.

In the Bab oilfield, high-amplitude stylolites of more than 35 cm amplitude can be observed in the most lithified facies, whereas the wispy stylolitic partings of less than 1 cm amplitude are present in the least lithified facies. The stylolite amplitude in this field is much greater in the water zone (up to 35 cm, with an average of 17.0 cm) than in the oil zones (maximum 5 cm). This probably reflects the inhibition of the wet chemical dissolution–reprecipitation reaction associated with stylolites developed during high hydrocarbon saturation.

The application of Johnson's and Budd's (1975) equation to the Bab Field revealed that stylolite intensity is inversely proportional to the porosity and permeability of the reported stylolites zone. Thus a high intensity of stylolites acts as a barrier to vertical fluid flow and reflects decreasing porosity and/or permeability, thus preventing the development of more continuous petroleum reservoir zones. In other petroleum fields, highly stylolitized carbonates can be excellent reservoirs.

The application of Dunnington's (1967) formula for assessing compaction, which uses porosity outside and within the stylolite zone on the studied core samples from Bab Field (Fig. 4), revealed that the stylolite zone that lies between subzones BI and BII in the Kharaib Formation showed a reduction in thickness. Thinning of the subcropping section reached about 23% and was due to active stylolitization. The total loss of porosity within the zone with more intense stylolites was mainly a response to chemical compaction. A small part of the dissolved material was redistributed within this zone, but most of this material must have passed out of the system to be precipitated elsewhere. This thinning caused by stylolitization, when separated from sedimentary compaction by porosity occlusion, reveals that the stylolites are preexisting sedimentary structures.

The petrographic features attributable to the pressure-dissolution effect are very common in the studied core samples, and the low-amplitude and wispy solution seams occur both above and below the oil/water contacts in the Kharaib Formation in Bab and Asab fields. The seams reflect pressure dissolution prior to complete lithification of the carbonate sediments. The stylolites are thought to have formed as a result of pressure solution among carbonate rock masses. Assuming that stylolites developed from low-relief seams, the greater the amplitude of these stylolites the greater the volume of limestone dissolved.

Published articles on the classification of stylolites, in general, are numerous and confusing. For example, Park and Schot (1968a) classified the stylolites on the basis of the two-dimensional geometry expressed by the seam and its relation to sedimentary bedding. Wanless (1979) classified stylolites on the basis of the presence and degree of suturing of seam material. In contrast Trurnit (1968) based his classification on the relative solubility and size/shape considerations of the adjacent rock or grain unit. This paper uses a modified version of the classification of Koepnick (1984). It is based on amplitude and morphology of the pressure-dissolution surface within stylolites, its lateral continuity, and the thickness of the accumulated insoluble residue in the stylolites.

The detailed sedimentologic amplitude analyses and petrographic study made for this paper resulted in the recognition of three kinds of stylolites in the Lower Cretaceous (Thamama Group). These are: 1) rectangular or high-amplitude stylolites, (2) solution seams or wave-like stylolites, and (3) wispy seam or horsetail stylolites (Fig. 5).

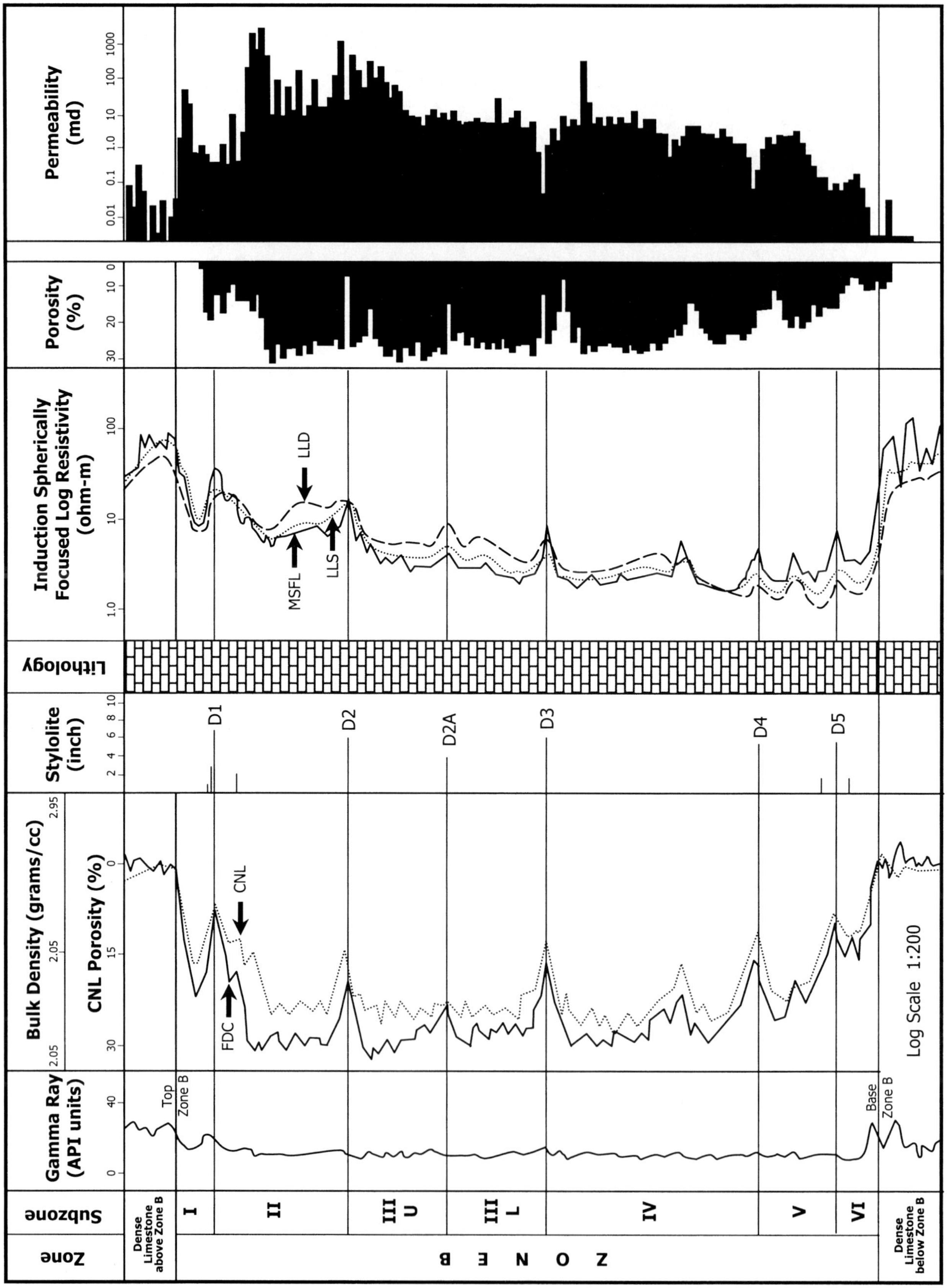


FIG. 4.—Logs and petrophysical characteristics of stylolites zones (D1 to D5) in the Thamama Zone B reservoir, Bab field, U.A.E., (after Alsharhan, 1985).

*Rectangular or High-Amplitude Stylolites.—*

These stylolites are characterized by relatively high amplitudes (Figs. 5A, 6). They are associated with the latest stage of stylolitization and are associated with fractures that are unrelated to structural uplift. Dissolution of calcium carbonate associated with the high-amplitude stylolite causes weakening of the adjacent section, which often becomes a site of subvertical fractures. Insoluble residue appears to become concentrated at the stylolite peaks (crests), whereas typically less or no residue occurs along columns. This is because the insoluble residue is a measure of local dissolution, and this type of stylolite is considered to be a poor barrier to vertical fluid flow. In contrast, stylolites with thick, laterally continuous insoluble residues constitute more effective barriers to fluid cross-flow than stylolites with either thin, laterally continuous insoluble residue or with discontinuous residues as described by Koepnick (1988).

*Solution Seams or Wave–Like Stylolites.—*

This type of stylolite is characterized by a smooth or undulatory pattern with a continuous coating of seam material along its surface (Figs. 5B, 6). The amplitudes of these stylolites are less than 1 cm and the thicknesses of the insoluble residue accumulations can reach 1 cm. These seams are similar to the composite seams of Garrison and Kennedy (1977). These stylolites appear to form during the early stage of chemical compaction. They occur at the boundary between porous and less porous intervals of limestone, either singly or in groups. They are observed to subdivide and meet again, forming an anastomosing interconnecting network. Park and Schot (1968a, 1968b), reported that such an association indicates several periods of stylolitization. Solution seams are laterally continuous on the core scale and are often associated with networks of wispy seams and small stylolites, and are commonly associated with impermeable stratigraphic intervals (e.g., Koepnick, 1988). These seams are important barriers to fluid cross-flow because of their lateral continuity and the thick buildup of insoluble residue.

*Wispy Seams or Horsetail Stylolites.—*

This low-amplitude (< 1 mm) stylolite most commonly occurs in the very thinly laminated, argillaceous limestones (Figs. 5C, 6) and is often disconnected (e.g., Bab Member of the Shuaiba Formation). This type of stylolite is similar to the "horse-tail stylolite" of Mossop (1972) and the simple seams of Garrison and Kennedy (1977). Wispy seams may form significant barriers to fluid cross-flow and are commonly associated with impermeable stratigraphic intervals in the reservoir (Koepnick, 1988).

### *Factors That Control the Development of Stylolites*

The variation in the style, intensity, and amplitude of stylolites in the Lower Cretaceous from one field to another is a response to many factors, the most common of which are pressure dissolution, fluid types, and lithology.

*Pressure Dissolution.—*

Pressure dissolution occurs in limestone under the stress of burial or tectonic movement. This begins with as little as 300 m of lithostatic overburden (Sellier, 1979). It is recognized that the overburden pressure is essential for the formation of stylolites (Dunnington, 1954, 1967), and thus the processes of stylolitization assumes greater importance in the moderate to deep burial setting (Choquette and James, 1987).

Bathurst (1980) suggested that pressure dissolution does not begin until a carbonate rock has undergone considerable burial. This idea is accepted (but not proven) by many workers. Factors such as water chemistry, presence of clay minerals, and presence of organic matter may also promote stylolitization regardless of burial depth. Dunnington (1967) placed the depth of pressure dissolution between 600 and 900 m for carbonate reservoirs in the Middle East. In contrast, Sellier (1979) suggested 300 m as the depth for the onset of pressure dissolution for the Aquitaine Basin of France. Schlanger (1964) presented evidence that in some cases burial need not to be greater than 90 m, particularly in limestones with a high clay content, i.e., primary lithology is as critical as depth of burial.

In south and southeast Abu Dhabi, where the top of the Lower Cretaceous is at shallow depths of burial (between 2000 and 3500 m), the stylolites are poorly developed or absent and have little influence on porosity or thickness reduction. They are best developed in fine-grained, porous and argillaceous limestones. Stylolite swarms are better developed in the central and western Abu Dhabi, with different stylolite intensities corresponding to different burial depths.

*Fluid Type.—*

The development of stylolites in the reservoirs under the same lithostatic pressure can vary in response to the water saturation. The presence of hydrocarbons prevents and localizes stylolite development. The stylolites are less well developed in gas-bearing zones of Bab and Asab fields than in oil-bearing zones. Probably differences in water saturation ratio and salinity are the controlling factors. Stylolites are better developed in the water-bearing formations at great depth, as in the Sahil oilfield.

When oil migrates into the oil-zone reservoir, it effectively limits or stops carbonate diagenesis, because cementation requires that a large number of pore volumes of water be moved through each pore to precipitate cement. In the water zone, in contrast, diagenesis and stylolite formation may continue freely, and it would be expected that carbonates below the oil/water contact are considerably reduced in porosity relative to those above (Burgess and Peter, 1985).

*Lithotypes.—*

Stylolites are relatively rare in the coarse-grained facies of the studied fields areas. In compacted oolitic limestones pressure dissolution is commonly manifested as micro-stylolites at grain boundaries, though stylolites occur mainly in the fine-grained or argillaceous intervals. Most of the stylolites are found in matrix-supported textures (mudstone or wackestone), but a few were observed in cemented grain-supported sediments (packstone and grainstone). The wave-like and wispy stylolites are concentrated in the laminated facies rich in clay. Stylolites commonly occur at bed boundaries or where mechanical layering created local stress concentrations.

## EFFECTS OF STYLOLITIZATION AND CEMENTATION ON POROSITY AND PERMEABILITY

The pore pattern in the studied pay zones resulted from a complex interplay of factors. The pore geometry can be characterized on the basis of pore size, pore shape, nature of the connec-

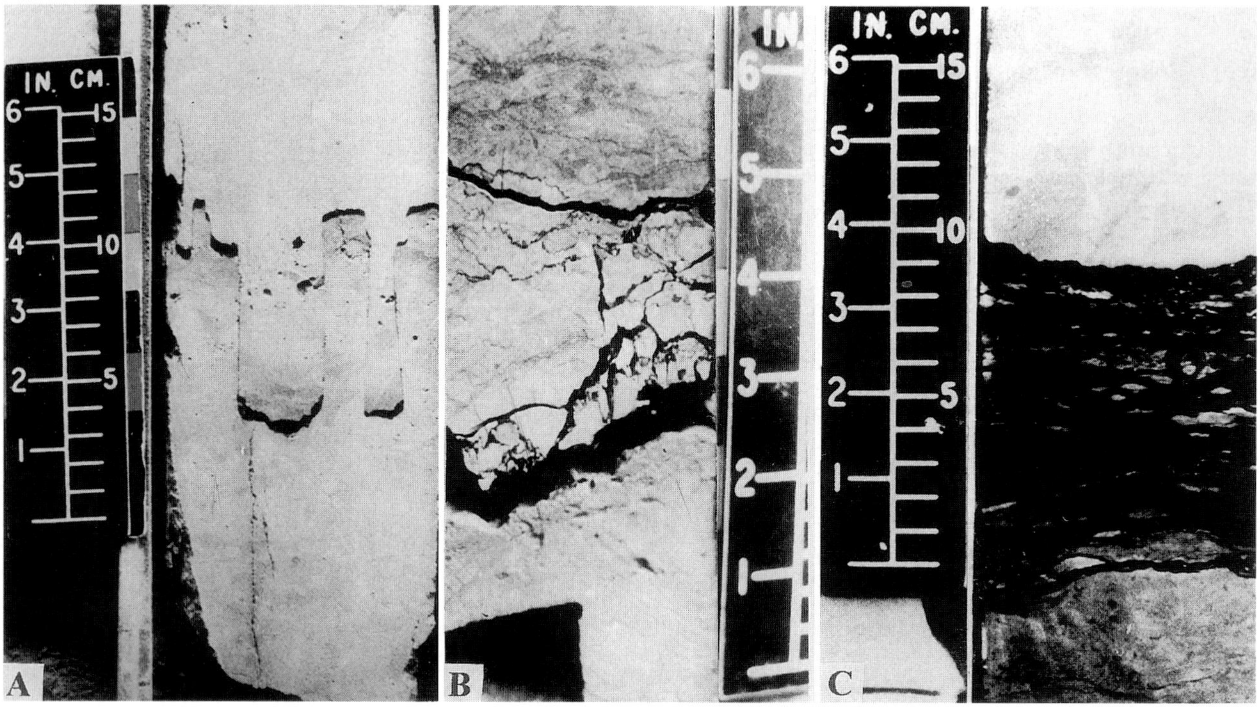


FIG. 5.—Core photographs showing the three types of stylolites. A) rectangular stylolite; B) solution seam (wave-like) stylolite; C) wispy-seam (horsetail) stylolite.

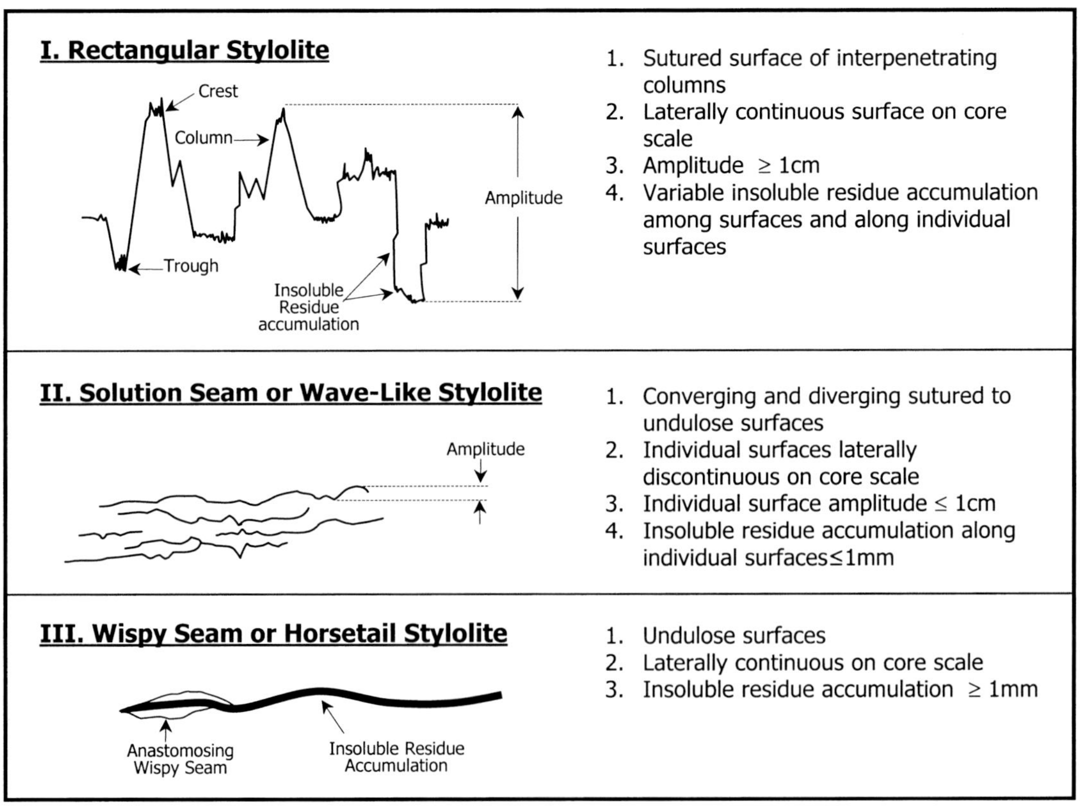


FIG. 6.—Types of stylolites in the Lower Cretaceous reservoirs of the U.A.E. (Modified from Koepnick, 1984, reproduced with permission from the author.)

tions between pores, character of the pore wall, and distribution and number of larger pores and their relationship to one another. The result of stylolitization is considerable loss of porosity, reduction of permeability, and overall lessening of bed thickness or rock volume. This relationship can diminish reservoir capacity (Kopenick, 1984; Oswald et al., 1995).

The importance of stylolites to reservoir geology is their association with a significant reduction in porosity and permeability and their effects on hydrocarbon fluid flow and migration. The limestone dissolved as the stylolite formed is reprecipitated as cement nearby, causing partial or total filling of the intergranular pore space. This phenomenon affects the reservoir capacity. Such reduction of pore space depends on the degree of stylolitization, as described by Dunnington (1967) and Bathurst (1975). Stylolites and the cements derived from dissolution have the potential to act as barriers to vertical fluid flow. Because the presence of barriers becomes important when a reservoir is exploited, it is necessary to consider whether the stylolites are well enough developed to constitute a barrier.

Numerous diagenetic phenomena affect the Lower Cretaceous reservoirs in U.A.E. (Fig. 7). In order to determine the relative importance of factors controlling porosity and permeability within the Thamama Group, all diagenetic phenomena, including stylolitization and fracturing, should be considered. In the Lower Cretaceous of the U.A.E. the initial void-filling cements were precipitated after mechanical compaction and grain breakage. These cements probably originated from solution transfer during development of wispy seams (the earliest pressure-dissolution features). Some cements are probably not related to pressure dissolution. It should be noted that petroleum residues coat the first-generation, post-compaction cements, suggesting that oil migration occurred after initiation of pressure dissolution. The formation of dense limestones probably led to progressive restriction of fluid movement away from the sites of stylolite generation and therefore to a decreased rate of cementation in surrounding lithologies. There was still a lot of fluid/cement movement in the vicinity of the stylolites even though it was decreased in surrounding lithologies. The dissolution rate along the stylolite increased with the amplitude, resulting in more calcium carbonate available to cement pores. The effect of dolomitization was limited because of the dispersed nature of dolomite rhombs in the studied intervals of some reservoirs, such as Zone II of the Kharaib Formation in Zakum Field. The effect of recrystallization and cementation can be observed only in mesopores and megapores, because of microscopic limitations.

Figure 8 documents the distribution of visually estimated thickness of cemented rock associated with the D1 and the D2 stylolite horizons in the Asab Zone B of the Kharaib Formation. Koepnick (1988) concluded that the D1 stylolite horizon constitutes a significant barrier to vertical fluid cross-flow by virtue of its lateral continuity and its associated zone of matrix cementation. The D2 and the D3 stylolite horizons appear to be less significant barriers to vertical fluid flow by virtue of their lateral discontinuity, weak cementation, and hydrocarbon saturation. Furthermore, Koepnick (1988) described the D1 stylolite horizon as largely devoid of hydrocarbon staining, which suggests that it underwent matrix cementation prior to substantial input of res-

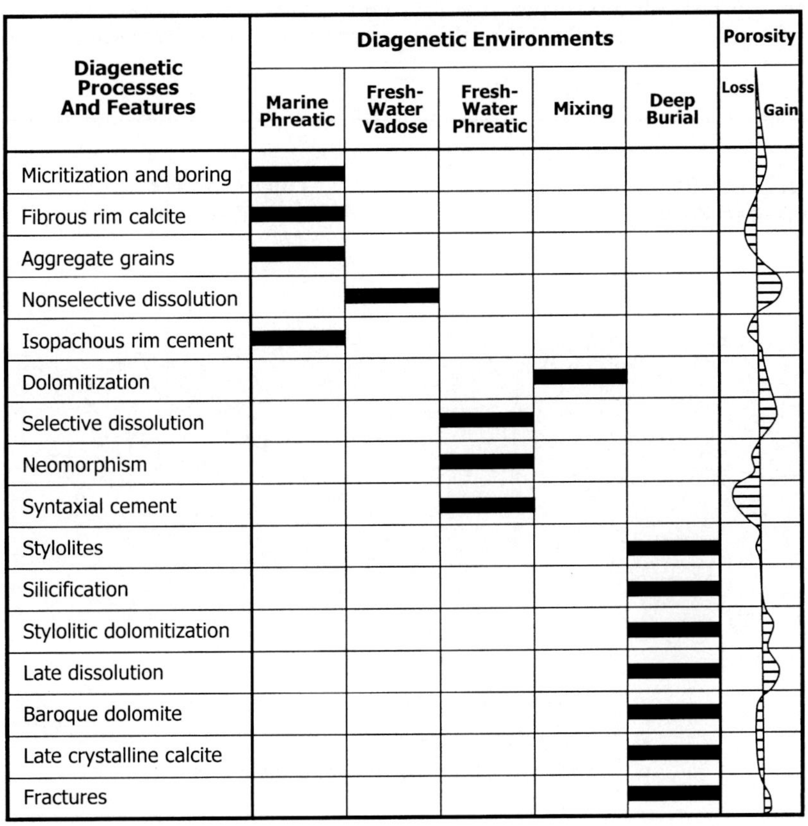


FIG. 7.—Relationship between diagenetic features, diagenetic environments, and porosity in the Lower Cretaceous reservoirs of U.A.E. fields. List of diagenetic sequence based on Carozzi and Von Bergen (1987).

ervoir hydrocarbons. Without hydrocarbons in the pore system, abundant nucleation sites are available nearby for precipitation of solute derived from pressure-solution surfaces. In contrast, the D2 stylolite horizon is uncemented and is saturated with hydrocarbons. These characteristics suggest that growth of the D2 stylolite horizon in Asab began during or after hydrocarbon entrapment. Hydrocarbons probably inhibited precipitation of solute derived from pressure-solution surfaces by coating many nucleation sites within the reservoir.

Carozzi and Von Bergen (1987) showed from petrographic observations that stylolitic porosity is discontinuous and corresponds to dissolution of carbonate material between the stylolite walls, which contained the least amount of insoluble residue. The effect of circulation of dissolving fluids along sutured stylolites controls the generation of the other types of burial secondary porosity. The hydrocarbons obviously migrated after the development of stylolitic porosity within a completely cemented and stabilized carbonate rock. Stylolitization in all instances results in liberation of calcium carbonate by pressure solution, and in concentration of insoluble residue along the stylolites (Carozzi and Von Bergen, 1987). Some stylolites are found in zones of relatively low porosity in the core material. This association could be initiated either early in diagenesis and localize subsequent stylolite formation, or could be late in diagenesis and due to reprecipitation of stylolite related pressure-dissolved material (Nelson, 1981, 1983). Dolomite commonly occurs as a replacement halo around individual pressure-solution features. Dolomitization associated with pressure-solution features drastically reduces matrix permeability and often is of greater significance as permeability barrier than adjacent undolomitized stylolite seams.

In the Zakum Field Zone II of the Kharaib Formation some stylolite swarms show that the greatest reduction in porosity and permeability is usually confined to a horizon 0.8–1.7 m thick (Alsharhan, 1990). These stylolite-rich horizons are generally thought to be too thin to constitute effective barriers, but much reduction in porosity and permeability results from stylolitization. This is because in such beds fractures, however uncommon, are apparently sufficient to enable fluid flow across the stylolitized intervals.

Cementation that preceded stylolitization also affects matrix porosity. Therefore, not all the dense intervals (from the porosity log) are caused by pressure dissolution. Permeability reduction is due to development of networks of the insoluble residues and to cementation of adjacent pore space resulting from solution transfer (Koepnick, 1984). The processes giving rise to solution transfer are: (1) pressure solution, which causes the minerals to dissolve along grain contacts or along more extensive surfaces under the influence of pressure or stress, and (2) solution transfer, which results in matrix cementation adjacent to pressure-solution surfaces.

Pressure or stress on deeply buried sediments dissolves the mineral along grain contacts or along more extensive surfaces. This pressure may exceed the average lithostatic pressure. Adjacent free areas of grain boundaries are subjected only to the pressure of the pore fluid, which is generally less than lithostatic pressure (Robin, 1978). Because mineral solubility increases with pressure, a decreasing concentration gradient between high-pressure grain contact areas and adjacent low-pressure areas develops across folds. If the insoluble residue continues to develop, significant permeability reduction results. Matrix cementation adjacent to pressure-solution seams reduces permeability. Extensive permeability reduction may result when matrix cementation occurs immediately adjacent to the pressure-solution surface. But if dissolved materials are transported beyond the immediate vicinity of the pressure solution surface, then adjacent matrix permeability remains.

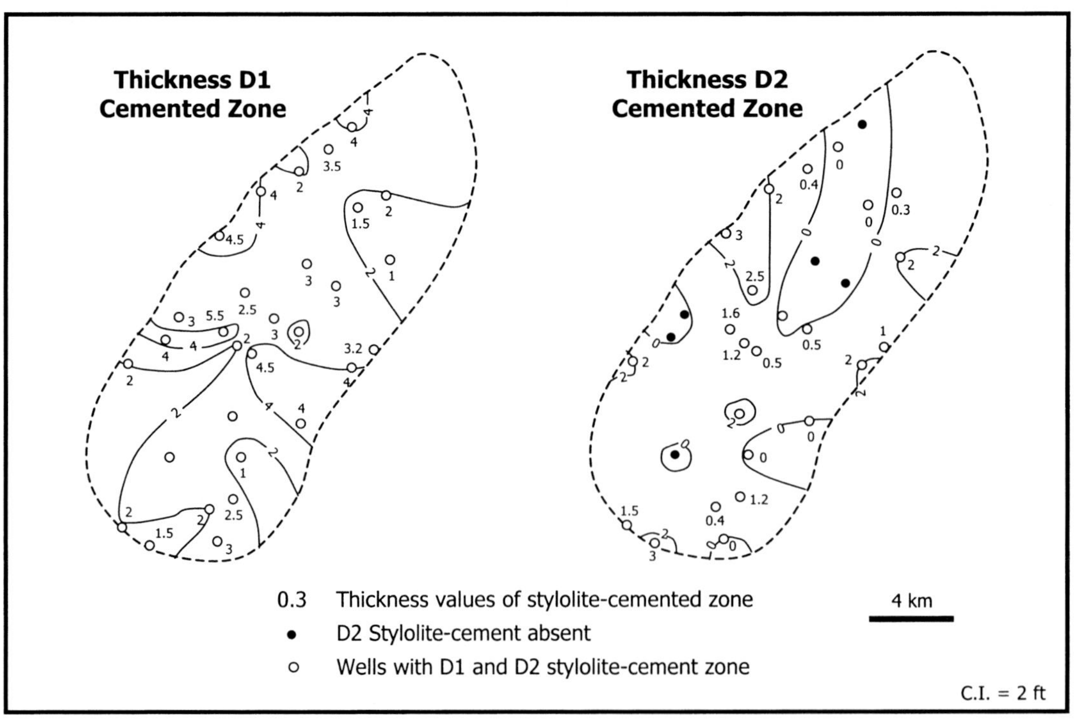


FIG. 8.—Contour map of thickness of D1 and D2 stylolite cemented zones. It shows display field outline, well locations and thickness values in feet of cemented zones for each well. (After Koepnick, 1988, reproduced with permission from the author.)

## SEAM MATERIAL AND DIAGENETIC PHENOMENA

The composition of the relatively insoluble seam material varies as a function of the composition of the nearby host rock and diagenetic processes. Most of the seams are composed largely of black bituminous hydrocarbon residue, and also contain insoluble rock residue (clay minerals, quartz silt) and diagenetic minerals (dolomite, calcite, pyrite, or other metallic sulfides; Fig. 9A). In the example presented here, the concentration of iron sulfide residue is proportional to the size of the stylolite.

The clay residue is largely a recrystallized kaolinite with minor quantities of illite, chlorite, and montmorillonite. Clays and organic matter are often associated together, and their proportions appear to vary within Lower Cretaceous swarm stylolites. The relative proportions of quartz and pyrite also tend to increase with the shale content of the stratigraphic horizon. Dolomite is common along stylolites (Fig. 9A) in the studied cores. Wanless (1979, 1982) believed that similar dolomite is formed during stylolite formation when pressure dissolution releases magnesium ions, but Pratt (1982) concluded that the dolomite forms either before or after stylolitization, but not during pressure solution. Calcite is occasionally observed as a fracture filling toward the tops of stylolites and probably formed from the reprecipitation of the carbonate dissolved at pressure-dissolution contacts.

The thicker and more laterally continuous the insoluble residue portions of the stylolites, the more effective a barrier they are to vertical fluid flow. In some reservoirs, as in Zone B of Asab Field, Johnson and Budd (1975) and Koepnick (1984) believed that swarms of wispy seams may form significant barriers to fluid cross flow because of the lateral continuity and thick buildup of insoluble residues.

Because stratigraphic intervals with large-amplitude stylolites are not always effective barriers to fluid flow, the amplitude may not be a good guide to loss of matrix porosity. It is more likely that the dense bituminous residue, with its incorporated noncarbonate minerals, which forms on the contact surfaces of the stylolitic seam, is the principal determinant of the effectiveness to fluid flow.

Examination of slabbed core samples revealed several examples of laminated bedding and undulatory wavy/nodular bedding. Thin sections studied show that the interlaminae consist of an anastomosing shale or organic-rich partings in lime mud matrix (Fig. 9B). Some dissolution seams appear to coincide with shaly organic-rich partings, which suggests that these sedimentary laminae might represent the first step towards stylolitization. The presence of organic matter promotes stylolite development.

Several studies have discussed the difficulty of distinguishing between these partings and stylolites or laminae. These have been variously termed clay seams, wispy or wavy laminae, pseudostylolites, microstylolites, hummocky structures, shaly or residual clay partings, stylolitized joints, and nonsutured to differentiated planes (Dunnington, 1967; Park and Schot, 1968a, 1968b; Trurnit, 1968; Coogan, 1970; Lucia, 1972; Shinn et al., 1977; Wanless, 1979; Buxton and Sibley, 1981; Pratt, 1982).

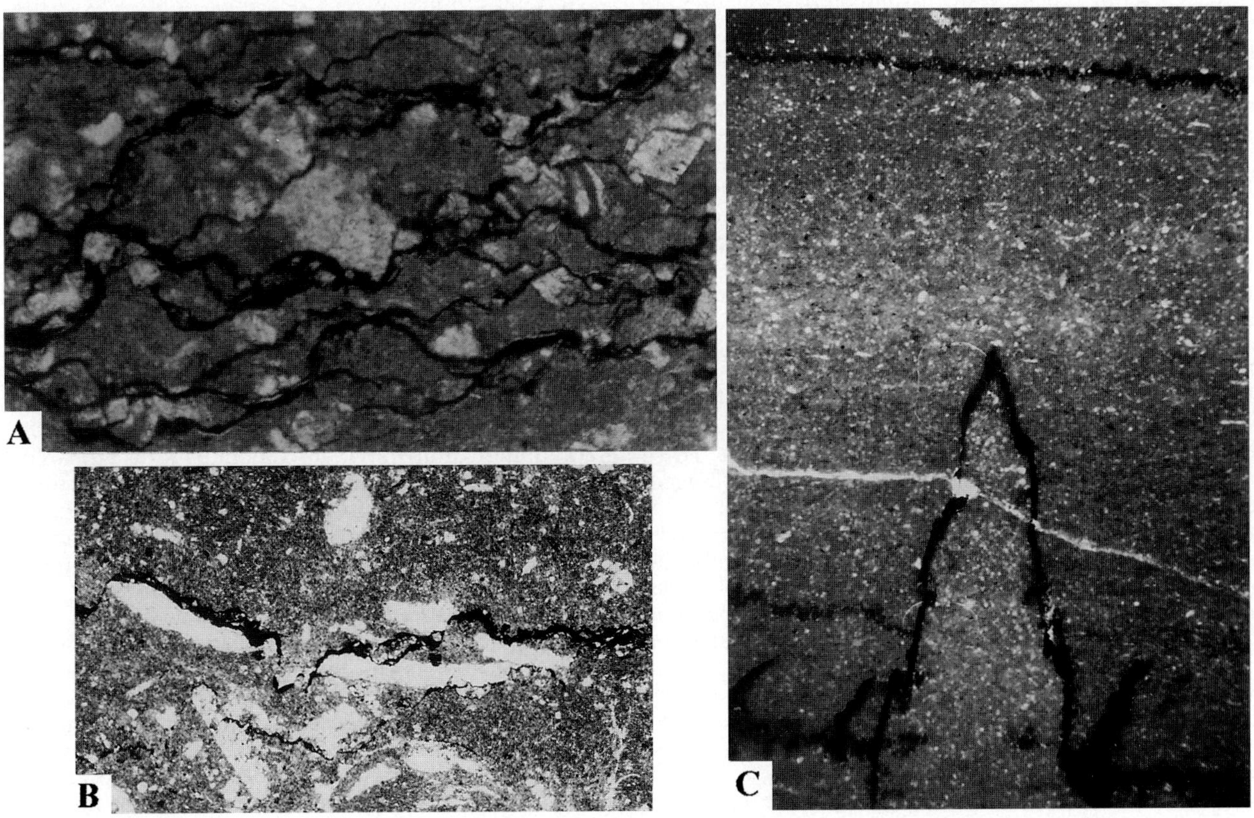


FIG. 9.—A) Photomicrograph showing microstylolites filled with calcite, dolomite, or residue in lime mudstone matrix (x28); B) Photomicrograph showing wispy-seam stylolites associated with scattered fossils in lime mudstones (x28); C) Photomicrograph showing two types of microfractures cutting each other open or filled by calcite and dolomite cements in lime mudstone (x28).

## FRACTURES ASSOCIATED WITH PRESSURE SOLUTION

Stylolite horizons commonly have a variety of associated natural and induced fractures, which originate from the same stress state that caused the stylolite. Nelson (1979, 1981) described two types of fractures associated with stylolites: tension gashes and unloading fractures (for more details see Nelson, 1981). The tension gashes are a set of parallel fractures associated with stylolites and also parallel to the maximum stress direction. These extension fractures are derived from the same overburden stress that controls stylolite development. This type of fracture is generally 5–10 cm long and parallel to columns of the stylolite. They are often wedge shaped and commonly filled with coarse, highly twinned calcite. Unloading fractures tend to be nearly parallel with the stylolite seam and are perpendicular to the maximum paleo–stress direction. They are either extension fractures or true tension fractures related to unloading or relaxation of the rock parallel with the maximum stress direction.

Numerous microfractures are commonly found associated with stylolites. Such an association might be important because of its effect on permeability. Uncemented fractures increase permeability but add relatively little to pore volume. Stylolite-related fractures are typically small (maximum length is around a few centimeters) and are parallel to stylolite peak or column orientation, i.e., perpendicular to bedding. The microfractures associated with stylolites may be open, partially cemented, or fully cemented (Fig. 9C). The cement is generally calcite, though in some intervals organic clay-rich material partially fills the fractures. Fractures are usually associated with the large well developed rectangular stylolites. Where a stylolite with a thick, continuous insoluble residue is crossed by fractures enlarged by dissolution (i.e., fractures are later than stylolite), the fracture counteracts the negative effect of the residue on vertical permeability. Opened fractures associated with stylolites therefore may increase permeability.

Cores from the Bab, Zakum, Asab, and Sahil fields in the Kharaib Formation show that the fractures are mainly subvertical, regularly spaced, and continuous, presumably because of tectonic extension. Another set of fractures, dipping about 60°, are more irregular but also more continuous, and usual extensions range from a few centimeters to a few decimeters. Fractures within oil zones are generally free of cement, except for a few isolated calcite crystals, and they are often enlarged by dissolution, thus improving reservoir quality.

Cores from the fields in the basinal Shuaiba Formation (Bab Member) show that most of the fractures are short, up to a maximum length of 15 cm, and have widths on the order of a few millimeters. They occur predominantly in the lime mudstone and wackestone intervals of the open and deep marine facies. These micro-fractures, which resemble calcite veins, are usually plugged, although some are open or only partially filled (Fig. 9C).

At Asab Field in Zone B of the Kharaib Formation, natural fractures are common, particularly in the muddy facies units, and in most cases closely associated with the D2 and D3 stylolite-rich horizons. The fractures are oriented at high angles to the bedding and are usually perpendicular to adjacent pressure-dissolution features. The orientation of the fractures, and their gradation into breccia horizons, suggest an extensional tectonic origin for these features. The extensional fractures and conjugate shear may develop within the same compressional stress regime, giving rise to pressure-solution features (Choukroune, 1969).

In some studied cores of the Kharaib Formation in the Zakum Field, the effect of fractures appears insignificant because no enhancement in vertical (compared to horizontal) permeability is observed (Alsharhan, 1990). This implies that fractures are important in other adjacent fields. The increase in vertical permeability in some wells is not due to fracturing because no major fractures were observed. Also, in Zakum Field, fractures are common throughout Zone V of the Habshan Formation. Horizontal, vertical, and inclined fractures were observed in cores. They vary from a few centimeters to a few decimeters. Microfractures were observed in thin sections. Some are filled with calcite or other material, but many often have good permeabilities.

The origin and timing of formation of the microfractures is controversial. Park and Schot (1968a) hypothesized a prelithification origin separate from that of stylolitization. Other studies have indicated that the microfractures occur simultaneously with stylolites and form after rock lithification (Groshanf, 1975; Nelson, 1981, 1983). These extensional microfractures are developed within the same pressure field as stylolites. The formation of the microfractures associated with stylolites is a mechanism for releasing stress caused by the overburden pressure, uneven chemical compaction, and uneven unbalanced stresses.

The microfractures observed in the Lower Cretaceous carbonates appear to be of post-lithification origin contemporaneous with stylolites, because all the fractures observed terminate at the stylolitic surface and no fractures were observed to cut across the stylolites. Also, the presence of organic, clay-rich material and few quartz grains in the fracture indicate the simultaneous formation of stylolites and fractures. These are not formed by pressure solution but by tension and were therefore filled with stylolite residues.

## STYLOLITE OCCURRENCES IN THE STUDY AREA

Stylolites in the Lower Cretaceous of the U.A.E. reservoirs are of post-lithification origin; these stylolites cut or penetrate micrite, fossils, and cemented pores indiscriminately. Some stylolites that pass through micrite and metasparite patches are of post-lithification origin.

The axial trends of increasing stylolite frequency and increasing cumulative stylolite amplitude from crest to the flank of Asab Field may partly explain by differential inhibition of pressure solution by reservoir hydrocarbon and emplacement. Reservoir hydrocarbons are an important component of stylolite insoluble residues, indicating post–oil migration growth of stylolites. Post–oil migration stylolite development very likely has been affected by increasing hydrocarbon saturation toward the crest of the structure (Koepnick, 1984). However, hydrocarbon inhibition of pressure solution does not explain the cross-axial trend, the preferential stylolite development on the southern structural nose, or the vertical sequencing of contour patterns. The stratigraphic alternation of stylolite contour patterns probably reflects differential response of stress of successive reservoir units developed during the structural growth of Asab Field.

In the oilfield the studied stylolite distribution is controlled by paleo–stress fields existing during development of the anticlinal closure, and by differential inhibition of stylolite growth by hydrocarbon saturation.

The hydrocarbons that were entrapped before stylolite formation prevented or at least localized the stylolite development, as is observed in "Zone B" of the Kharaib Formation of Asab and Sahil fields. This observation can be related to the counter-action of the reservoir pressure created by the accumulation of hydrocarbons. If the stylolite formation preceded hydrocarbon entrapment, however, it might have influenced the migration and accumulation of petroleum, as in the case of "Zone C" of the Kharaib Formation in Asab Field.

In the IA and II reservoirs zones (lower Shuaiba and upper Kharaib formations) of Zakum Field, all the stylolite-rich intervals exhibit stylolites that are oriented parallel to the bedding and their peaks are orthogonal to the seams. Crosscutting stylolites are also observed. Solution seams commonly converge into residual clay grooves that are slightly stylolitized. These are congruent to the bedding and commonly occur with horsetail structures.

### *Habshan Formation (Berriasian–Valanginian)*

The deposition of this formation started with a transgressive cycle, in a restricted or semi-restricted shallow-marine setting. The lower part is composed of argillaceous, dense lime mudstones and wackestones, which grade upward to dolomitic peloidal packstones and dolomitic peloidal grainstones. This unit is water-bearing, and stylolites are absent. The absence of stylolites in southeast Abu Dhabi fields (e.g., Mender) is due to insufficient burial. Here, the upper part of the Habshan Formation is composed of lagoonal to intertidal lime mudstones and wackestone, with subordinate dolomite. Habshan cores from Asab Field show a very poor stylolite development, but the stylolites are well developed in the Habshan of the Bab and Sahil fields.

To the north in offshore Abu Dhabi, stylolites are, in contrast, developed throughout the upper part of the Habshan reservoir (Zone V) of Zakum Field. Stylolites tend to be more common in the mud-supported limestones rather than in the grainstones and packstones. More than 80% of the stylolites in this reservoir are in mud-supported limestones, and their amplitudes are generally in the range of 1.5 cm. The majority of the stylolite in this reservoir are of small magnitude and do not constitute barriers to the vertical permeability.

### *Lekhwair Formation (Hauterivian–Lower Barremian)*

The Lekhwair Formation was deposited in an unrestricted marine shelf environment. It is composed of a number of cyclic sequences of alternating nonporous and porous limestones. In the onshore of Abu Dhabi, this formation splits into five reservoir zones (D–H), but in the Abu Dhabi offshore it is divided into two reservoir zones (III, IV). The onshore reservoir Zone H contains stylolites of wispy-seam type, which are associated with microfractures. Zone G has some stylolite development but less than the underlying Zone H reservoir in most fields. Zone F contains the most porous and permeable horizon, and the stylolites are more poorly developed than in Zones G and H.

Stylolite swarms occur in the lower part of the Lekhwair Formation in offshore (Zone IV or Zakum Member) of the Zakum Field (Fig. 10). Four stylolite-rich horizons (IVA–S, IVB 1–S, IVB 2–S, and IVC–S) are described by Hassan and Wada (1979). Although stylolites are present elsewhere in the reservoir of Zone IV, they are not laterally continuous and do not have a significant effect on porosity and permeability.

Within reservoir zone IVA–S, a stylolite-bearing interval some 1.0–2.5 m from the base of subzone IVA effectively subdivides this reservoir into two subzones of unequal thickness (IVA–1 and IV–2). Subzone IVA–S is well defined stratigraphically and is correlative throughout the Zakum Field, although its thickness and porosity are variable. The stylolite-rich horizon IVB 1–S is thin but can be correlated using porosity logs from a number of wells. This stylolite horizon causes only a small reduction in porosity and permeability and is characteristically present in mud-supported limestone. The stylolite-rich horizon IV B2–S has a constant thickness and can be correlated throughout the field. It occurs some 1.5–3 m above the base of subzone IVB and is too thin to form a permeability barrier within the reservoir. Although the stylolitic–rich horizon IVC–S occurs some 4.5 m below the top of subzone IVC, it is well developed along the flanks of the structure. Where it is well developed there is a significant reduction in porosity and permeability as a result of stylolitization across this interval. However, it is not sufficiently well developed that IVC–S constitutes a barrier to fluid communication between the upper and lower parts of subzone IV C.

### *Kharaib Formation (Barremian–Lower Aptian)*

This formation consists of a series of upward-shallowing cyclic units. The basal transgressive parts of the cycles are characterized by argillaceous lime mudstones/ wackestones with subordinate peloidal packstones, which represent normal open marine shelf conditions. The succeeding regressive parts of the cycles begin with a mud-supported limestone deposited as inner-shelf or deep-lagoon facies, which shallows upward into grain-supported limestones representing higher-energy conditions. Reservoir Zones B and C are the predominant hydrocarbon-bearing zones in the Bab and the Asab fields. These zones are divided into subzones by well-developed stylolitic intervals.

The Zone B reservoir is oil-bearing in most fields (e.g., Zakum, Bab, Asab, and Sahil). It is considered to be the principal reservoir in the onshore Abu Dhabi. This zone is characterized by a porous inner-shelf limestone capped by a regressive shoaling sequence of packstone and peloidal grainstone. It is divided into subzones by well-developed stylolitic intervals along muddy horizons. In the southeast area at the Mender, Qusahwira, and Shah fields, the stylolites in Zone B are poorly developed or absent, perhaps because of the shallow burial depths.

Stylolite development increases with depth of burial in the Bab and Asab fields, and the D1 stylolite swarms form a barrier to vertical fluid flow in most of the onshore producing fields. In general, compaction, cementation, and stylolitization are less in the Asab Field than in the Sahil and Bab fields because of the shallower depth of burial and finer grain size of the Asab reservoir sediments.

Offshore Abu Dhabi (in the central part of Zakum Field) stylolites in reservoir Zone II are poorly developed. The degree of stylolitization decreases from the flank toward the top of the structure, and stylolites are thought to be responsible for the overall reduction in thickness on the east and west flanks. Stylolite development in the northeast part of this field is diminished, because of either an anomaly in the oil-water contact, high hydrocarbon saturation, or probably post-oil-emplacement differential structural tilting. The highest-intensity and highest-amplitude stylolites occur in the mud-supported sediment intervals. The intensity and thickness of these stylolitic intervals are not uniform throughout the field. The five stylolite-rich horizons, S1–S5, separating the different reservoirs in Zone II of Zakum Field (Fig. 11) were defined by Alsharhan (1990) using both log data and core observations. The majority of stylolites in this zone exhibit low amplitude, and the few intervals that are stylolite-rich are not recognized on either electrical or porosity–permeability logs.

The Zone C reservoir is oil-bearing in the Asab and Sahil fields, gas-bearing in Bab Field, and water-bearing in the Bu Hasa and southeast Abu Dhabi fields. The zone is subdivided into two sub-reservoirs (CI and CII/CIII) by a well developed stylolite swarm (S1) that occurs between CI and CII (Fig. 12). In the fields where the zone is water saturated and burial depth is shallow, stylolites are not well developed. In Asab and Sahil fields and in the fields located in southeastern area, the lower

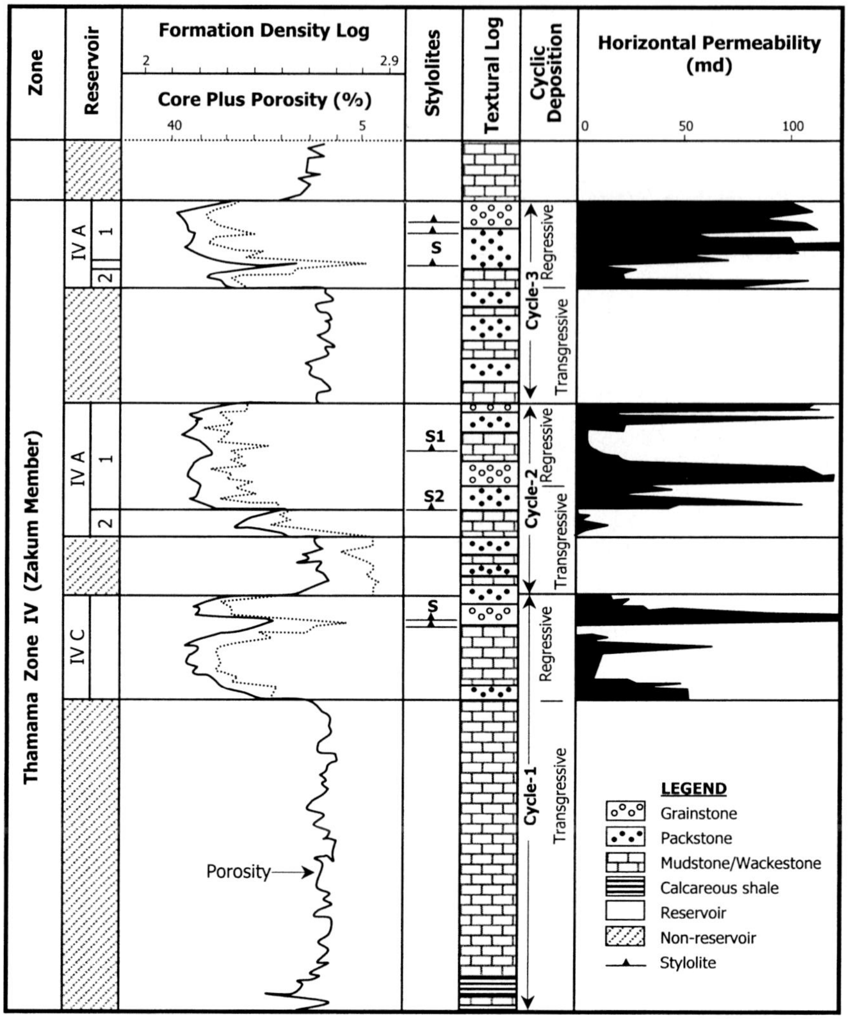


FIG. 10.—Thamama Zone IV (Zakum Member) reservoir characteristics in Zakum Field, U.A.E. Modified from Hassan and Wada (1979) (reproduced with permission from Society of Petroleum Engineers).

reservoir (CII/CIII) is further subdivided into subzones CII and CIII by the stylolite swarm CS2. At the Bab Field, the CS2 stylolite swarm is present in the northeastern part of the field and delineates subzone CII from CIII. At the crest and in the southern part of the structure, however, the CS2 stylolite swarm is poorly developed or absent and subzones CII/CIII are undivided. In Bu Hasa Field, the CS1 stylolite swarm is well developed, whereas the CS2 swarm is poorly developed or absent. This zone is water-bearing and deeply buried.

## *Shuaiba Formation (Early to Mid Aptian)*

The Shuaiba is the youngest formation of the Lower Cretaceous of the U.A.E. It is characterized by rudistid and algal carbonate platform sediments that have very good reservoir characteristics (e.g., Bu Hasa Field). Rudist/algal buildups are associated with the inner rims of the Abu Dhabi intrashelf basin, which has its center in the central part of Abu Dhabi and extends into western Oman and includes most of the offshore areas in Qatar. Sedimentary rocks in the intrashelf basin are known as the Bab Member and consist mainly of microporous lime mudstones and wackestones. Basinward and upward this Bab Member progressively increases in density and argillaceous content and exhibits calcareous shale intervals with deep, open marine fauna (Alsharhan, 1987).

The basal Shuaiba Formation is called the Zone A reservoir. It is composed of algal packstones and boundstones and is overlain by microporous lime mudstones and wackestones. The zone persists over most of the U.A.E. region. In the onshore area of Abu Dhabi, the Zone A reservoir is subdivided into three main subzones (AI–AIII) by two well-developed and regionally correlative stylolite swarms, AS1 and AS2 (Fig. 13).

In the Zone IA reservoir of Zakum Field, all the stylolites observed are parallel to the bedding, with peaks orthogonal to the seams (stratiform stylolites). Wave-like or undulatory, sutured and/or rectangular types are recognized but vary in intensity and amplitude. The wave-like forms are abundant and occur at pressure-dissolution contacts between grains. This type of stylolite commonly has converging solution seams, which themselves are slightly stylolitized. These are usually congruent to the bedding and occur with "horsetail" structures. It may be difficult, in some cases, to recognize whether or not certain clay grooves associated with bedding planes are actually dissolution residues or discrete depositonal shale–marl partings.

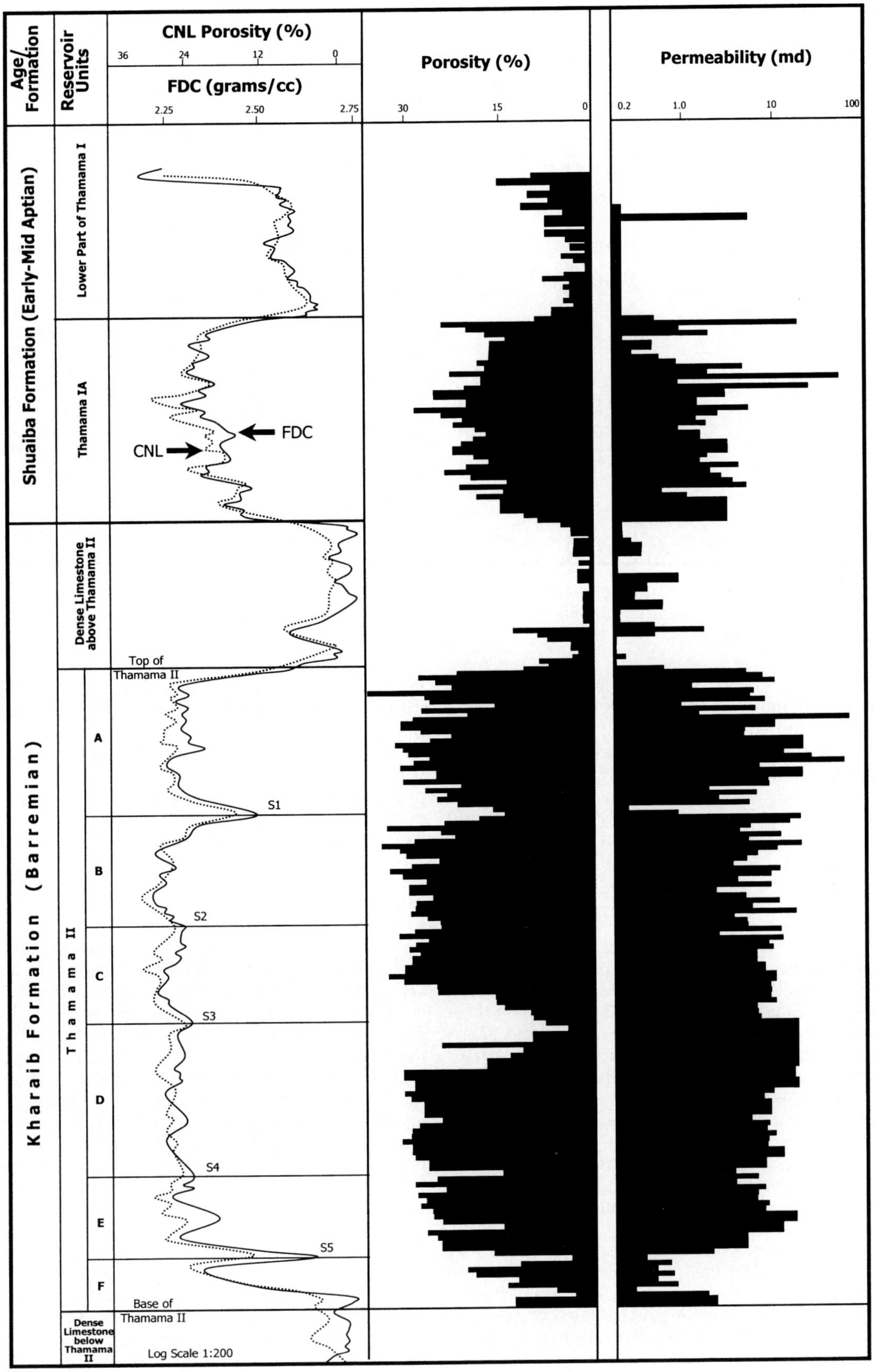


FIG. 11.—Logs and petrophysical characteristics and its relation with stylolites (S1–S5) in the Thamama Zone II reservoir, Zakum Field, U.A.E. (After Alsharhan, 1990, with permission of the publisher.)

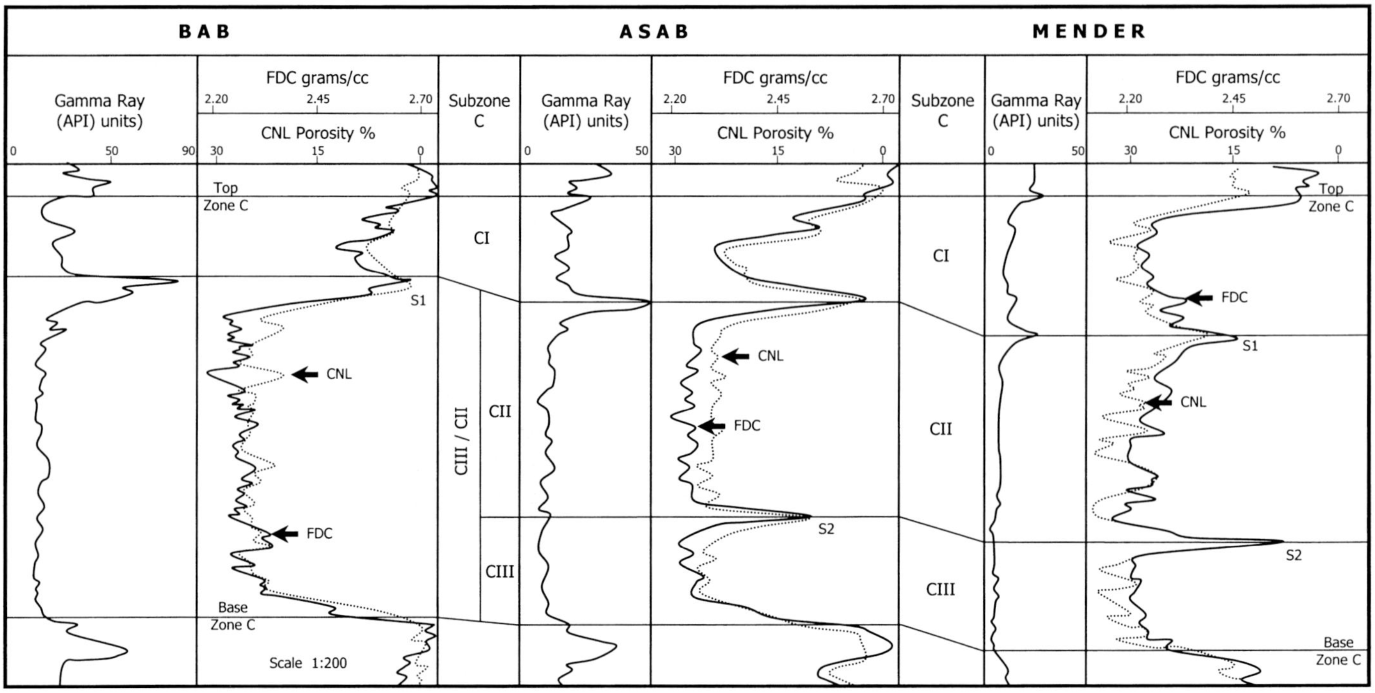


FIG. 12.—Generalized cross section through some fields, showing the subdivision of the Thamama Zone C reservoir by stylolites (S1 and S2). (Based on data from Alsharhan, 1985.)

In the Bab Member, stylolites occur mostly in the lime mudstone and wackestone intervals of the deep and open marine facies. Wave-like or sutured stylolite types, oriented parallel to bedding (stratiform stylolites), are the most common. The maximum amplitude observed in Zakum field in this member is about 17 cm (Fig. 14). The residue in the stylolite seam consists of organic matter, clay, pyrite, and quartz. Dolomite and microfractures associated with stylolitization are rare. The Bab Member is characterized by predominance of linear or undulatory laminae consisting mainly of organic-rich material.

### *Field Observations*

The Lower Cretaceous Thamama Group is divided into several reservoir zones (Fig. 2). The Zone A in Asab, Sahil, and Bab fields is divided into three subzones (A1, A2, and A3) by three intervening horizons of stylolites "swarms" (S0, S1, and S2). A poorly developed stylolite swarm (ASO) divides the zone to two subzones, AIU and AIL, except in SE Abu Dhabi at the Mender Field, where it is hard to separate the reservoir into these subzones (Fig. 13) because of the observed stylolite-related marker horizons. The AS1 swarm lies between subzones AI and AII, and the AS2 swarm lies between subzones AII and AIII.

The Zone B reservoir of the Asab and Mender fields has three important stylolite swarms (D1–D3), which subdivide the Zone B reservoir into four subzones (BI–BIV) (Fig. 15). At the Bab, Sahil, Bu Hasa, and Rumaitha fields the Zone B reservoir contains six main stylolite swarms (D1–D5); they subdivide the reservoir into seven porous subzones, (these are subzones BI to BVI) (Fig. 15). The D1 horizon is a stratigraphic interval containing stylolites. The typical low porosity is partly a product of sedimentology and partly pre-stylolite diagenesis.

Zone B reservoir of the Bab Field is cited as an example of how the stylolite intensity (SI) is distributed. Johnson and Budd's (1975) empirical formula is used. The D1 seam contains the most intense stylolite swarm of all major stylolites in Zone B (Fig. 16), and is present over the entire field. The D1 horizon is, in fact, a stratigraphic interval containing stylolites. The typical low porosity is the result partly of sedimentological effects and partly of pre-stylolite diagenesis. Stylolite intensity values vary from moderate to very strongly developed and increase generally from southwest to northeast. The highest intensity values, 220, occur on the northern flank; the lowest value, 4, was calculated for the southern flank. The amplitude of D1 increases toward all the flanks, except to the north and northeast. The highest amplitude, about 20 cm, occurs on the western flank, and shrinks to about 2 cm at the northeast. The amplitude of the D1 stylolite seam is low, whereas the stylolite intensity is high.

The D2 stylolite seam is present only locally and is poorly developed over most of the reservoir (Fig. 16). Stylolite development improves towards the northeastern and western flanks; the highest intensity, about 6, was calculated on the northeastern flank, and the lowest value is less than 1 on the crest. The stylolite amplitude of the stylolite seam D2 is much greater than of the D1 stylolite seam. Stylolite amplitude, in general, increases down–flank, except on the northeastern flank, where the lower value is 1 cm, and the highest amplitude of 35 cm is encountered on the southern flank.

The stylolite intensity value of the D2A stylolite swarm increases generally down flank except along the southern edge and part of the western flank (Fig. 16). However, the highest intensity calculated is about 15 for the northeastern flank and the lowest value is less than 1, and is near to the crest. The amplitude of the D2A stylolite swarm shows high amplitudes on the southern and northern flanks, whereas on the southwestern and northwestern flanks low-amplitudes have been observed and the amplitude varies from less than 1 cm to more than 30 cm.

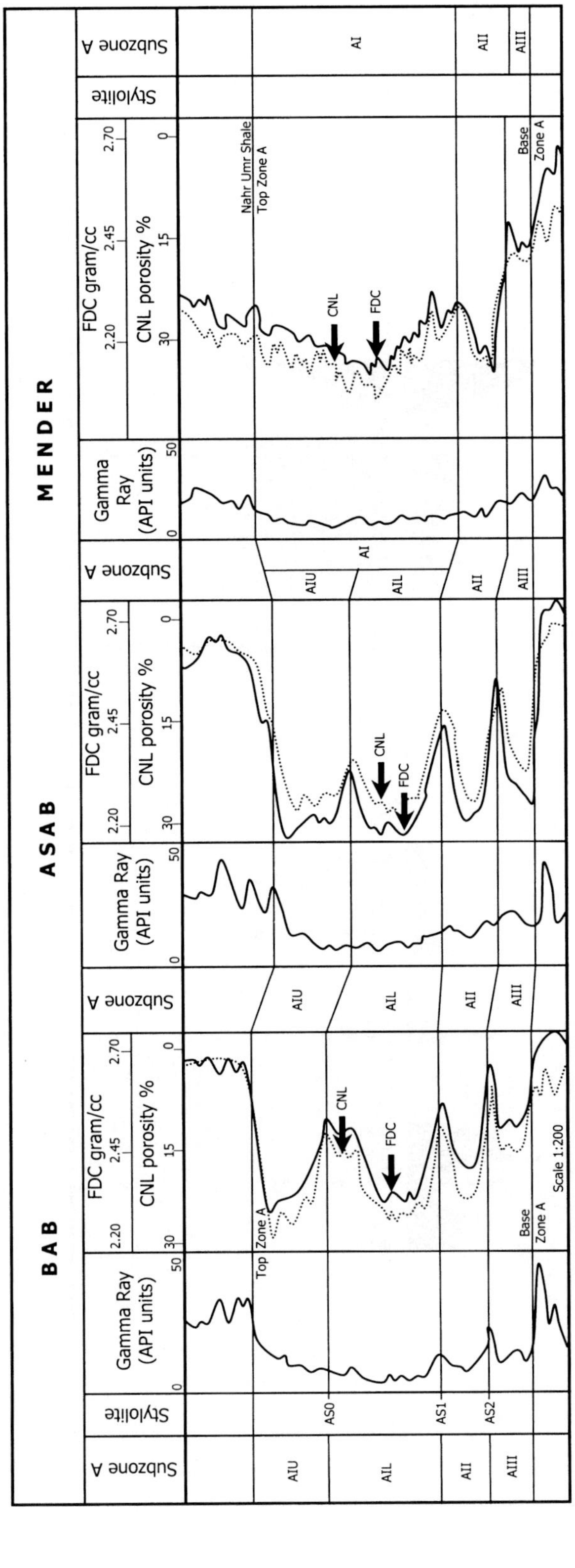


FIG. 13.—Generalized cross section through some fields shows the subdivision of the Thamama Zone A reservoir by stylolites (AS1 and AS2). (Based on data from Alsharhan, 1985.)

The D3 stylolite swarm is better developed than the D2, D2A, and D4 swarms and is locally absent on the crest. The stylolite intensity value ranges from moderate on the flanks to absent on the crest (Fig. 16). The highest stylolite intensity of about 8 is on the west flank, and the lowest is less than one at the crest. The stylolite intensity trend decreases toward the northeastern and southern flanks compared to the western and eastern flanks. The maximum amplitude of 35 cm is found on the eastern flank, and the smallest (2 cm) on the anticlinal crest and northeastern flank.

The D4 stylolite seam is somewhat better developed than the D2 and D2A swarms but not as well developed as the D3 swarms. It is not present over the whole field area. The highest intensity values, 13, were encountered on the northern flank, and the lowest (less than one) are at the crest (Fig. 16). The amplitude of this stylolite shows a trend similar to the D3 swarm, with an increase in amplitude towards the flank except on the northeastern and northwestern flanks. Again, the maximum amplitude (> 30 cm) is present on the eastern flank and the lowest (< 2 cm) is at the crest.

The D5 stylolite swarm is the second best developed stylolite in Zone B (after D1 swarm). The stylolite intensity values vary from weak to strong and generally increases from southwest to northeast (Fig.16). The highest calculated stylolite intensity is 36 at the northern flank, and the lowest is 1 on the southern flank. The greatest amplitude measured is about 23 cm at the eastern flank, and the smallest is about 5 cm at the crest.

Johnson and Budd (1975) and Koepnick (1984) concluded that the D1 stylolite horizon in Asab Field constitutes a significant barrier to vertical fluid migration because of its excellent lateral continuity and associated matrix cementation. The D2 and D3 stylolite swarms constitute less significant flow barriers because of their lateral discontinuity, cementation, and hydrocarbon saturation. The D1 stylolite swarm was developed largely before hydrocarbon entrapment, whereas the D2 and D3 stylolites grew mostly during and after hydrocarbon entrapment (Koepnick, 1984). Johnson and Budd (1975) reported that the D1 stylolite swarm in Zone B, over much of the Asab Field and particularly down flank, is a complete barrier to the movement of fluids but less effective toward the crest. The D2 and D3 stylolite swarms represent no real barrier over much of the field, even at their greatest intensity, where they probably impede vertical fluid movement.

Zone C in Asab, Sahil, and Mender fields is subdivided into three subzones (CI through CIII) by two stylolite swarms (Fig. 12). These are the SC1 swarm at the base of subzone CI and the SC2 swarm at the base of subzone CII. Stylolite swarm SCI is well developed over all the field areas and is considered to have a significant impact on inter-reservoir communication between subzones CI and CII. In Bab Field, Zone C is divided into two subzones, CI and CII/CIII (Fig. 12), by a stylolite swarm SC1 at the base of subzone CI. The stylolite swarm CS2 is also locally developed and consistently separates subzone CII from subzone CIII.

## *Mineralogy and Trace-Element Geochemistry*

Not all of the measured trace metals were found in stylolitic horizons in the studied wells at statistically significant concentrations when compared to the host rock, but where significant differences occur there is a relative concentration within the stylolitic interval (Fig. 17). In general, elements displaying the greatest degree of concentration include Al, Cr, Cu, Fe, Mg, and Ni. Less, but significant, relative concentration was detected in Mn, Pb, V, and Zn ratios; there is no significant concentration of Be and Na in any core. Trace-element concentrations associated

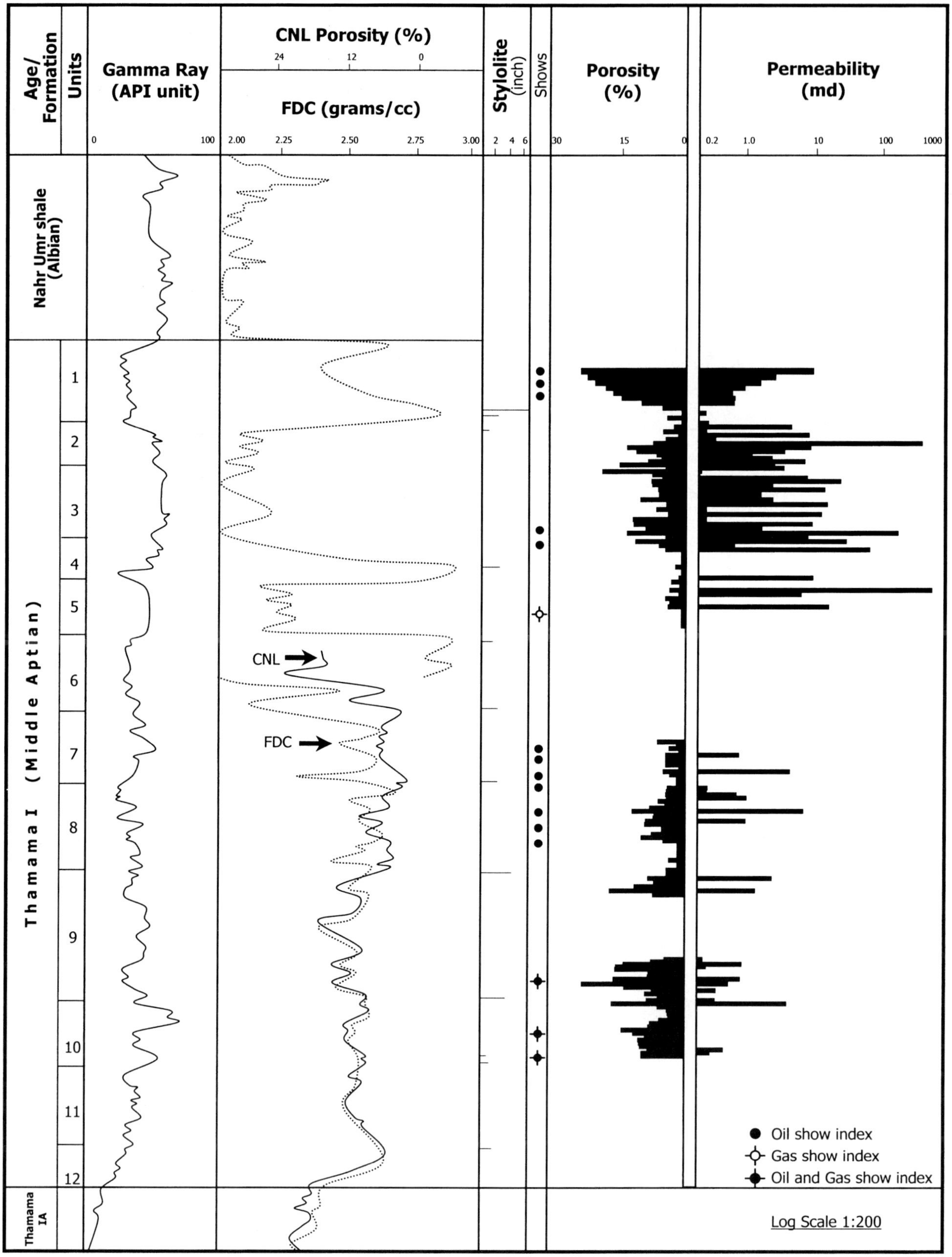


FIG. 14.—The relationship between logs, petrophysical characteristics, and stylolite swarms in the Thamama Zone I (Bab Member) reservoir in the Zakum Field (after Alsharhan, 1985).

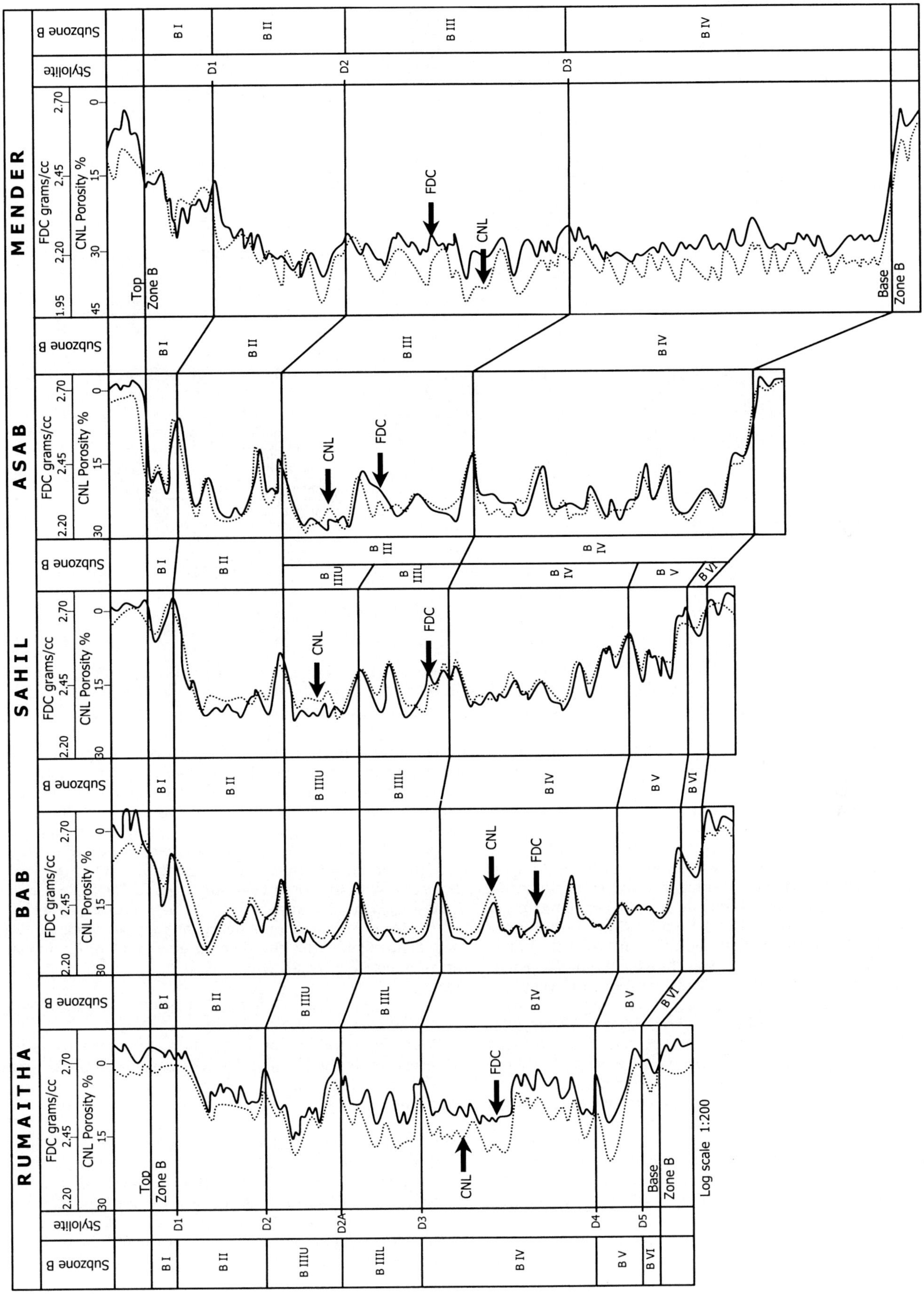


FIG. 15.—FDC/CNL Log correlation and stylolites occurrences in Thamama Zone B reservoir in some fields in Abu Dhabi. (Based on data from Alsharhan, 1985.)

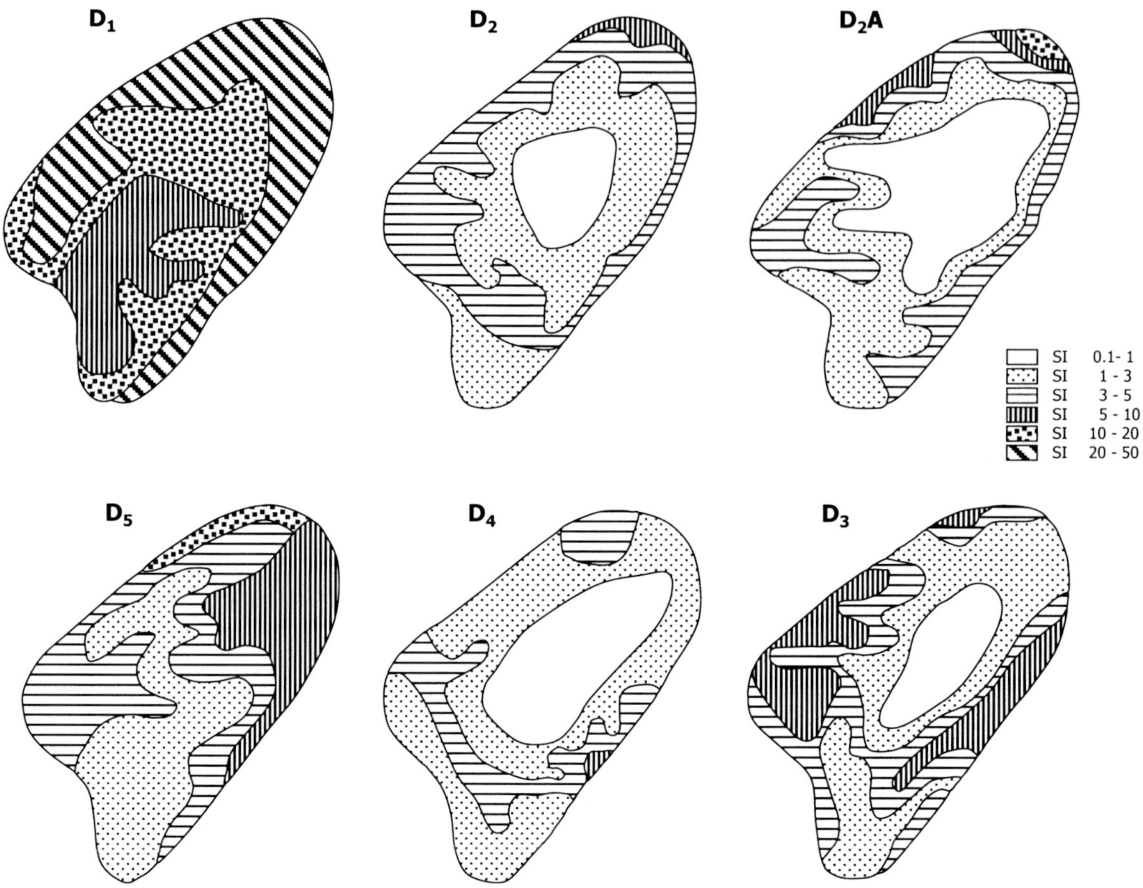


FIG. 16.—Distribution of stylolite intensity in the Thamama Zone B reservoir of Bab Field, U.A.E. (Based on log and core data from Alsharhan, 1985.) More data will definitely give more accurate interpretation and distribution of stylolytes.

with both rectangular and wave-like stylolites are common and roughly equal in magnitude. Horsetail stylolites do not show significant concentration for any element. This suggests that, considering a constant stylolite amplitude, the degree of dissolution and intraformational loss of carbonate minerals is much greater for rectangular and wave-like stylolites than for horsetail stylolites, as evidenced by the elevated ratios for trace elements (Mn, Fe, Zn, Pb) that tend to substitute for Ca in the calcite crystal lattice (Reeder, 1983). It suggests that the wave-like and rectangular stylolites may have formed under equilibrium conditions of diagenetic alteration different than conditions in which the horsetail stylolites formed. It is also possible, however, that diagenetic alteration took place in contact with a different subsurface water.

There also tend to be slightly greater mean concentrations in samples with wave-like stylolites than in those with rectangular stylolites (Table 1). This suggests that processes that produced wave-like forms in these fields have been somewhat more chemically aggressive than those that produced the rectangular stylolites. Further study of the trace-element chemistry at the crystal level using electron microprobe would be helpful in determining the degree to which this apparent difference is related in general to the processes that result in formation of these two different types of stylolites, because trace-element ratios of bulk subsamples could be strongly affected by variations in mineralogy and mineral composition.

The mineralogy of the subsamples provides some clues to the differences in some trace-element ratios, and the chemical extraction technique used in this study may have partially dissolved some noncarbonate minerals. For example, rectangular stylolite seams appear to be richer in metallic sulfides than seams in cores with wave-like stylolites, and this is consistent with the relatively higher Fe/Ca and Ni/Ca ratios in samples with rectangular stylolites (Table 1). Similarly, wave-like stylolitic intervals appear to be richer in clay minerals and euhedral dolomite, which is consistent with the somewhat higher Al/Ca and Mg/Ca values in these samples. No minerals were detected using either XRD or petrography which contain significant copper or chromium, so there is no clear explanation for the elevated Cu/Ca and Cr/Ca ratios detected in these cores.

## CONCLUSIONS

1. Study of stylolitization in the Lower Cretaceous carbonate reservoirs in the U.A.E. oil field areas is of importance in the evaluation of the capacities of the reservoir characteristics for petroleum.

2. The presence of hydrocarbons prevents and localizes the stylolite development. The stylolite-rich horizons in the studied field areas constitute the effective barriers for petroleum migration and accumulation, because of the pronounced reduction in permeability resulting from stylolite formation and development within the fields and/or the geologic formation stratigraphic levels.

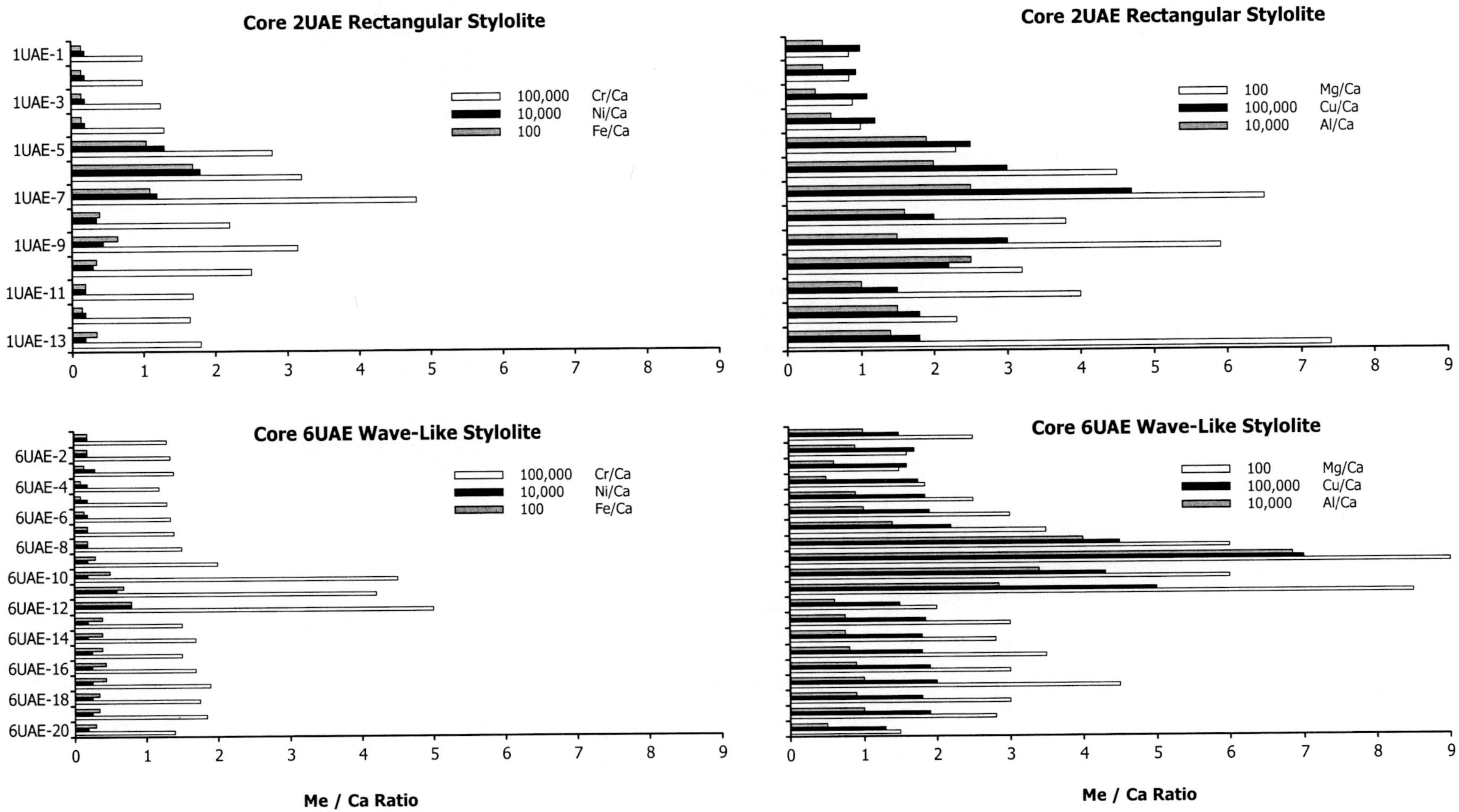


FIG. 17.—Results of geochemical analysis of stratigraphic core subsamples on two selected cores U.A.E.-1 (rectangular stylolites) from Kharaib Formation, Bab Field, and U.A.E.-6 (wave-like stylolites) from Kharaib Formation, Asab Field. Numbered subsample intervals show the Me/Ca ratio for various trace elements in each stratigraphic position in the core; the shaded part shows the subsamples in the stylolitic zone. Concentrations within the stylolite zone are highest for Al, Cr, Fe, and Ni; Cu and Mg also are concentrated in the stylolitic zone, but at lower levels and with no significant difference between stylolite types. There is no significant concentration of Be, Mn, Na, Pb, V, or Zn in either core, or in cores with horsetail stylolites (Table 1), so the results are not shown.

3. Matrix cementation and continuity of insoluble residue seams in the vicinity of stylolites is significant. It is true that some stylolite seams have not contributed to localized matrix cementation, but they are not the ones of concern. It is the cemented ones that could effect the reservoir.

4. Mineralogic and trace-element geochemical analysis of the three stylolite types from the fields studied suggests that reduction in unit thickness from diagenetic dissolution is significantly greater for wave-like and rectangular stylolites, with little detectable loss associated with horsetail stylolites.

## ACKNOWLEDGMENTS

The authors would like to thank Dr. C.G.St.C. Kendall for reading the paper and also to Drs. R. Koepnick and R.W. Scott for their detailed review and valuable comments which enhanced this manuscript.

## REFERENCES

ALSHARHAN, A.S., 1985, Petrography, sedimentology, diagenesis and reservoir characteristics of the Shuaiba and Kharaib Formations (Barremian–mid Aptian carbonate sediments of Abu Dhabi, United Arab Emirates: Unpublished Ph.D. thesis, University of South Carolina, Columbia, 175 p.

ALSHARHAN, A.S., 1987, Geology and reservoir characteristics of the carbonate buildup in the Giant Bu Hasa oilfield, Abu Dhabi, U.A.E.: American Association of Petroleum Geologists, Bulletin, v. 71, p. 1304–1318.

ALSHARHAN, A.S., 1990, Geology and reservoir characteristics of Lower Cretaceous Kharaib Formation in Zakum Field, Abu Dhabi, United Arab Emirates, *in* Brooks, J., ed., Classic Petroleum Provinces: Geological Society of London, Special Publication 50, p. 299–316.

ALSHARHAN, A.S., AND NAIRN, A.E.M., 1993, Carbonate platform models of Arabian Cretaceous reservoir, *in* Simo, J.A., Scott, R.W., and Masse, J.P., eds., Cretaceous Carbonate Platforms: American Association of Petroleum Geologists, Memoir 56, p. 173–184.

AMERICAN SOCIETY OF TESTING AND MATERIALS (ASTM), 1967, Fink Index to the Powder Diffraction File: ASTM Publication PD 1S–17F (with periodic updates), ASTM, Philadelphia, Pennsylvania.

BATHURST, R.G.C., 1975, Carbonate Sediments and Their Diagenesis, 2nd Edition: Amsterdam, Elsevier, Developments in Sedimentology 12, 658 p.

BATHURST, R.G.C., 1980, Deep crustal diagenesis in limestones: Universidad de Barcelona, Instituto de Investigaciones Geológicas, Revista, v. 34, p. 89–100.

BROWN, W.W.M., 1959, The origin of stylolites in the light of a Petrofabric study: Journal of Sedimentary Petrology, v. 29, p. 254–259.

BURGESS, C.J., AND PETER, C.K., 1985, Formation distribution and prediction of stylolites as permeability barriers in the Thamama Group, Abu Dhabi: 4th Middle East Oil Show, Bahrain, Society of Petroleum Engineers, SPE 13698, p. 165–174.

BUXTON, T.M., AND SIBLEY, D.F., 1981, Pressure solution features in a shallow buried limestone: Journal of Sedimentary Petrology, v. 51, p. 19–26.

CAROZZI, A.V., AND VON BERGEN, D., 1987, Stylolitic porosity in carbonates: a critical factor for deep hydrocarbon production: Journal of Petroleum Geology, v. 10, p. 267–282.

CHOQUETTE, P.W., AND JAMES, N.P., 1987, Diagenesis in limestones: The deep burial environment: Geoscience Canada, v. 12, p. 3–35.

CHOUKROUNE, P., 1969, An example of mesoscopic analysis of concentric folding in limestone series: Tectonophysics, v. 7, p. 57–70.

COOGAN, A.H., 1970, Measurements of compaction in oolitic grainstones: Journal of Sedimentary Petrology, v. 40, p. 721–929.

DEELMAN, J.C., 1975, Pressure solution or indentation?: Geology, v. 3, p. 23–34.

DUNNINGTON, H.V., 1954, Stylolite development postdates rock induration: Journal of Sedimentary Petrology, v. 24, p. 27–49.

DUNNINGTON, H.V., 1967, Aspects of diagenesis and shape change in stylolitic limestone reservoirs: 7th World Petroleum Congress, Mexico, Proceedings, v. 2, p. 339–352.

GARRISON, R.E., AND KENNEDY, W.J., 1977, Origin of solution seams and flaser structure in Upper Cretaceous chalks of southern England: Sedimentary Geology, v. 19, p. 107–138.

GORSHONG, R.H., JR., 1975, Strain, fractures and pressure solution in natural single-layer folds: Geological Society of America, Bulletin, v. 86, p. 1363–1376.

HASSAN, T.H., AND WADA, Y., 1979, The geology and development of the Thamama Zone IV of the Zakum field, Abu Dhabi: 1st Middle East Oil Show, Bahrain, Society of Petroleum Engineers, SPE 7779, p. 231–238.

JOHNSON, J.A.D., AND BUDD, S.R., 1975, The geology of the Zone B and Zone C Lower Cretaceous limestone reservoirs of Asab field, Abu Dhabi: 9th Arab Petroleum Congress, Dubai, U.A.E., 24 p.

KOEPNICK, R.B., 1984, Distribution and vertical permeability of stylolites within a Lower Cretaceous carbonate reservoir, Abu Dhabi, United Arab Emirates: Paper presented at the seminar on stylolite held in Abu Dhabi 6th–8th Nov., 1983, sponsored by Abu Dhabi National Reservoir Research Foundation, p. 261–278.

KOEPNICK, R.B., 1987, Distribution and permeability of stylolite-bearing horizons within a Lower Cretaceous carbonate reservoir in the Middle East: SPE Formation Evaluation, v. 2, p. 137–142.

KOEPNICK, R.B., 1988, Significance of stylolite development in hydrocarbon reservoirs with an emphasis on the Lower Cretaceous of the Middle East: Geological Society of Malaysia, Bulletin, v. 22, p. 23–43.

LUCIA, F.J., 1972, Recognition of evaporite–carbonate shoreline sedimentation, *in* Rigby, J.K., and Hamblin, W.K., eds., Recognition of Ancient Sedimentary Environments: Society of Economic Paleontologists and Mineralogists, Special Publication 16, p. 160–191.

MANTEN, A.A., 1968, Pre- or post-induration formation of stylolites seams—a reply: Geologie en Mijnbouw, v. 47, p. 114–115.

MARSH, D.C., 1867, On the origin of the so-called lignilitites or epsomites: American Association for the Advancement of Science, Proceedings, v. 16, p. 135–143.

MOSSOP, G.D., 1972, Origin of the peripheral rim, Redwater reef, Alberta: Bulletin of Canadian Petroleum Geology, v. 20, p. 238–280.

NELSON, R.A., 1979, Natural fracture systems, description and classification: American Association of Petroleum Geologists, Bulletin, v. 63, p. 2214–2221.

NELSON, R.A., 1981, Significance of fracture sets associated with stylolite zones: American Association of Petroleum Geologists, Bulletin, v. 65, p. 2417–2425.

NELSON, R.A., 1983, Localization of aggregate stylolites by rock properties: American Association of Petroleum Geologists, Bulletin, v. 67, p. 313–322.

OSWALD, E.J., MUELLER, H.W., III, GOFF, D.F., AL HABSHI, H., AND AL-MATROUSHI, S., 1995, Controls on porosity evolution in Thamama Group carbonate reservoirs in Abu Dhabi, U.A.E.: 9th Middle East Oil Show, Bahrain, Society of Petroleum Engineers, SPE 029797, p. 251–265.

PARK, W.C., AND SCHOT, E.H., 1968a, Stylolites: their nature and origin: Journal of Sedimentary Petrology, v. 38, p. 175–191.

PARK, W.C., AND SCHOT, E.H., 1968b, Stylolitization in carbonate rocks, *in* Müller, G., and Friedman, G.M., eds., Recent Developments in Carbonate Sedimentology in Central Europe: Berlin, Springer-Verlag, p. 66–74.

PRATT, B.R., 1982, Limestone response to stress: pressure solution and dolomitization—discussion and examples of compaction in carbonate sediments: Journal of Sedimentary Petrology, v. 52, p. 323–354.

PROKOPOVICH, N., 1952, The origin of stylolites: Journal of Sedimentary Petrology, v. 22, p. 212–220.

PURSER, B.H., 1984, Stratiform stylolites and the distribution of porosity: Examples from the middle Jurassic limestones of the Paris basin: Paper presented at the seminar on stylolite held in Abu Dhabi 6th–8th Nov., 1983, sponsored by Abu Dhabi National Reservoir Research Foundation, p. 203–220.

REEDER, R.J., ed., 1983, Carbonates: Mineralogy and Chemistry: Mineralogical Society of America, Reviews in Mineralogy, v. 11, 394 p.

ROBIN, P.Y.F., 1978, Pressure solution at grain-to-grain contacts: Geochimica et Cosmochimica Acta, v. 42, p. 1383–1389.

RUTTER, E.H., 1983, Pressure solution in Nature: theory and experiment: Geological Society of London, Journal, v. 140, p. 725–740.

SCHLANGER, S.O., 1964, Petrology of the limestone of Guam: U.S. Geological Survey, Professional paper 403-D, 52 p.

SELLIER, E., 1979, Contribution a l' étude Pétrologique expérimentale de la genèse des stylolites dans les formations calcaires du bassin d' Aquitaine: Unpublished Thesis, Université de Bordeaux III, 300 p.

SHAUB, B.M., 1939, The origin of stylolites: Journal of Sedimentary Petrology, v. 9, p. 47–61.

SHINN, E.A., HALLEY, R.B., HUDSON, J.H., AND LIDZ, B.H., 1977, Limestone compaction: an enigma: Geology, v. 5, p. 21–24.

SHINN, E.A., AND ROBIN, D.M., 1983, Mechanical and chemical compaction in fine-grained shallow-water limestones: Journal of Sedimentary Petrology, v. 53, p. 595–618.

STOCKDALE, P.B., 1922, Stylolites: their nature and origin: Indiana University, Studies, v. 9, 97 p.

STOCKDALE, P.B., 1926, The stratigraphic significance of solution in rocks: Journal of Geology, v. 34, p. 399–414.

STOCKDALE, P.B., 1943, Stylolites: primary or secondary: Journal of Sedimentary Petrology, v. 13, p. 3–12.

TRURNIT, P., 1968, Analysis of pressure solution contacts and classification of pressure solution phenomena, *in* Müller, G., and Friedman, G.M., eds., Recent Development in Carbonate Sedimentology in Central Europe: New York, Springer-Verlag, p. 75–84.

WANLESS, H.R., 1979, Limestone response to stress: pressure solution and dolomitization: Journal of Sedimentary Petrology, v. 49, p. 437–462.

WANLESS, H.R., 1982, Limestone response to stress: pressure solution and dolomitization—reply: Journal of Sedimentary Petrology, v. 52, p. 328–332.

# ORIGIN AND CORRELATION OF DISCONFORMITY SURFACES AND MARKER BEDS, NAHR UMR FORMATION, NORTHERN OMAN

ADRIAN IMMENHAUSER AND WOLFGANG SCHLAGER
*Vrije Universiteit, Earth Sciences, De Boelelaan 1085, 1081 HV Amsterdam, The Netherlands*
STEPHEN J. BURNS
*Geologisches Institut, Universität Bern, Baltzerstrasse 1, 3012 Bern, Switzerland*
ROBERT W. SCOTT
*Precision Stratigraphy Associates, RR 3 Box 103-3, Cleveland, OK 74020, U.S.A.*
TINI GEEL
*Vrije Universiteit, Earth Sciences, De Boelelaan 1085, 1081 HV Amsterdam, The Netherlands*
JENS LEHMANN
*Institut für Geologie and Paläontologie, Sigwartstrasse 10, 72076 Tübingen, Germany*
SJERRY VAN DER GAAST
*NIOZ, P.O. Box 591790. AB Den Burg, Texel, The Netherlands*
AND
LISELOTTE J. A. BOLDER-SCHRIJVER
*Vrije Universiteit, Earth Sciences, De Boelelaan 1085, 1081 HV Amsterdam, The Netherlands*

ABSTRACT: During Aptian to Albian times a major influx of terrigenous material from the emerged Arabian Shield led to the deposition of the Nahr Umr Formation within the intracratonic Bab Basin. This argillaceous facies interfingers in northern Oman with sediments shed from reefal complexes. We studied a 100 km transect across the zone of facies intercalation extending from the margin of the Neo-Tethys (northern Jabal Akhdar) into the fringes of the Bab Basin in northern Oman (Foothills). The age of formation boundaries and key beds and surfaces in a reference section (Wadi Bani Kharus; I) was dated by use of the graphic correlation method. Graphic correlation indicates that the Nahr Umr Formation is of uppermost Aptian to Upper Albian age. Two types of stratigraphic markers were recognized: limestone marker beds, successions, and disconformity surfaces. Calcareous marker beds are probably related to increased carbonate production during sea-level lowstands. Investigations of the geochemistry (stable isotopes, fluid inclusions) and sediment petrography of limestones beneath disconformities revealed that they are of combined submarine and subaerial origin. Evidence for a subaerial exposure stage is recognized along the basin margin (northern and southern Jabal Akhdar) but is found less frequently in the basinward settings of the Foothills. Marker-bed successions and disconformity surfaces are correlated throughout the sections and indicate coeval deposition of the Nahr Umr Formation in northern Oman.

## INTRODUCTION

Throughout the Gulf Region, mid-Cretaceous sedimentary rocks are of primary importance in terms of carbonate petroleum systems. Consequently, the mid-Cretaceous reservoir facies such as the Lower Aptian Shuaiba Formation in the U.A.E. or the Upper Albian to Lower Turonian Natih Formation in Oman have been the subject of numerous studies during the last few decades (cf. Christian, 1997, and references therein).

The Upper Aptian to Upper Albian Nahr Umr Formation of northern Oman, a regionally significant seal, has received comparably little attention, because it is not itself a source or reservoir rock. We have investigated the Nahr Umr Formation in Jabal Akhdar (northern Oman Mountains) and the Foothills to the south. Measured sections reach from the margin of the Neo-Tethys (Jabal Akhdar) into the fringes of the intracratonic Bab Basin to the south (Foothills; Figs. 1, 2).

Several aims governed this study: 1) to obtain detailed information on the paleoenvironmental setting of the Nahr Umr Formation; 2) to establish a chronostratigraphic framework by use of the graphic correlation method; 3) to correlate marker beds and surfaces throughout sections in Jabal Akhdar and into the poorly exposed sections in the Foothills. Marker surfaces were studied in detail in order to recognize their origin. The paleoceanographic and sea-level record of the Nahr Umr Formation is discussed in an associated paper (Immenhauser et al., 1999).

### *Regional Geotectonic and Paleoenvironmental Setting*

From the mid-Permian through the earliest Turonian (ca. 170 My) an extensive carbonate platform covered large parts of the Arabian craton (Fig. 2). During this period, some 3500 m of mainly carbonate strata accumulated (Hajar Supergroup; Fig. 1; Glennie et al., 1974; Murris 1980; Harris and Frost, 1984; Rabu, 1987; Scott et al., 1988; Hughes Clarke, 1988; Scott, 1990; Pratt and Smewing, 1993; Alsharhan, 1995).

The Arabian platform was bordered by basement highs along its southwestern, eastern, and northern margins (Fig. 2). Sedimentary units show a regional thinning and environmental shallowing towards the following structural axes: A) the uplifted and exposed Arabo–Nubian shield to the southwest; B) the uplifted Huqf–Haushi axis to the east (Hughes Clarke, 1988; Ries and Shackleton, 1990); and C) a shallow swell on the continental margin to the north.

During the mid-Cretaceous the Arabian Platform was subdivided into several paleodepositional belts (Fig. 2B). Alluvial to coastal deposits rimmed the Arabo–Nubian Shield in the southwest, grading northeastwards into the shallow marine carbonate shelf. Since the Middle Jurassic the Arabian Shelf was superimposed by the intracratonic Bab Basin, which was deepest to the north (present-day United Arab Emirates) and extended south to Oman (Vahrenkamp, 1996).

During Late Aptian to Late Albian time a major influx of terrigenous material from the emerged Arabo–Nubian Shield to

Middle East Models of Jurassic–Cretaceous Carbonate Systems
SEPM Special Publication No. 69, Copyright © 2000
SEPM (Society for Sedimentary Geology), ISBN 1-56576-075-1, p. 209–225.

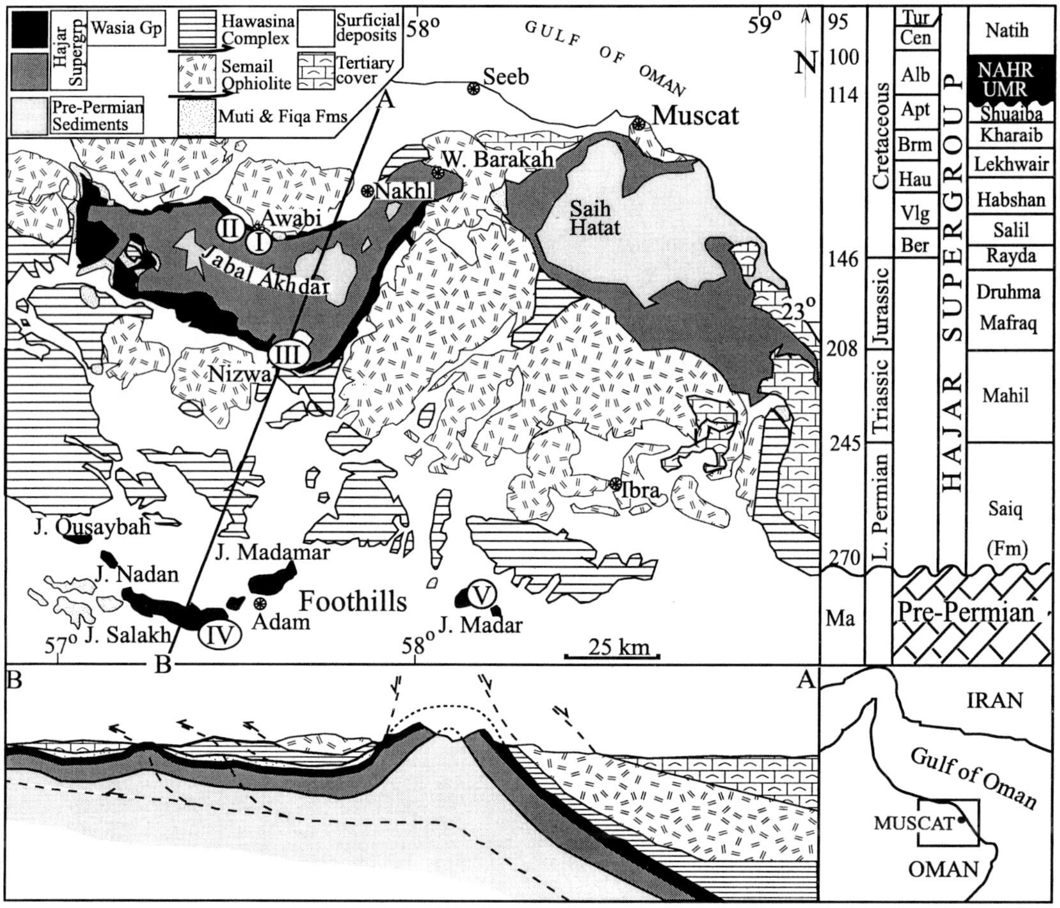


FIG. 1.—A) Geologic setting of the study area in northern Oman. The sedimentary cover of the Arabian Platform was buried by the Hawasina Nappes and the Semail Ophiolite in Campanian time and subsequently covered by lower Tertiary deposits. Mid-Tertiary compressional tectonism led to detachment and upfolding of autochthonous sediments, allochthonous nappes, and neo-autochthonous Tertiary deposits in the Jabal Akhdar anticline as well as in smaller anticlines (Foothills) to the south. The Wasia Group, of which the Nahr Umr Formation is a part, is shown in black. After Béchennec et al. (1992a) and Béchennec et al. (1992b), transect A–B after Mann et al. (1990) and Hanna (1990). Sections studied are indicated I = Wadi Bani Kharus; II = 2.5 km to the West of Wadi Bani Kharus; III = Wadi Mu'aydim; IV = Jabal Salakh; V = Jabal Madar. B) Stratigraphy of the Upper Permian to Lower Turonian Hajar Supergroup, based on Hughes Clarke (1988) and Simmons (1994).

the northwest onto the Arabian carbonate platform (Fig. 2; Murris, 1980; Hughes Clarke, 1988) led to the deposition of the argillaceous Nahr Umr Formation within the Bab Basin. Throughout the southern Gulf the Nahr Umr is principally an impermeable unit of clayey facies acting as a seal for the Lower Aptian Shuaiba–Upper Thamama carbonate reservoir rock (Murris, 1980; Hughes Clarke, 1988). In Oman the Nahr Umr is a unit of the parautochthonous Wasia Group of the Hajar Supergroup (Fig. 1; Glennie et al., 1974; Harris and Frost, 1984; Harris et al., 1984; Simmons and Hart, 1987; Rabu, 1987; Hughes Clarke, 1988; Pratt and Smewing, 1993).

Reefal buildups of Aptian to Albian age (Al Hassanat Formation; Masse et al., 1997) rimmed the Arabian Platform to the northeast and probably to the east (Fig. 2, black dots; Immenhauser et al., 1998). These rudist–algal reefs (Masse et al., 1997) grew on a shallow swell trending along the continental margin of Oman. The slope deposits of the Al Hassanat reefs interfingered basinwards (to the south) with Nahr Umr sediments.

The Hajar Supergroup (including the Nahr Umr Formation) was buried by the overthrusted Hawasina Complex and Semail nappes in the Campanian (Glennie et al., 1974). In the Early to Middle Miocene the Afro–Arabian plate collided with Eurasia in the Makran–Zagros area. This led to detachment of cratonic sediments from their basement and uplift of these units as part of the Jabal Akhdar dome and smaller anticlines to its south (Fig. 1; referred as the "Foothills" in this contribution; Rabu, 1987; Rabu et al., 1990; Loosveld et al., 1996).

### *Methodology*

X-ray powder diffraction was measured at 50% relative humidity with CuK$\alpha$ radiation using a custom-built goniometer equipped with automatic divergence and anti-scatter slits and a Si–Li energy-dispersive detector (Kevex). Doubly polished thick sections (150 μm) were analyzed for fluid-inclusion microthermometry (vapor and fluid phase) on a Linkam TP/91-THMS 600 gas-flow heating and freezing stage.

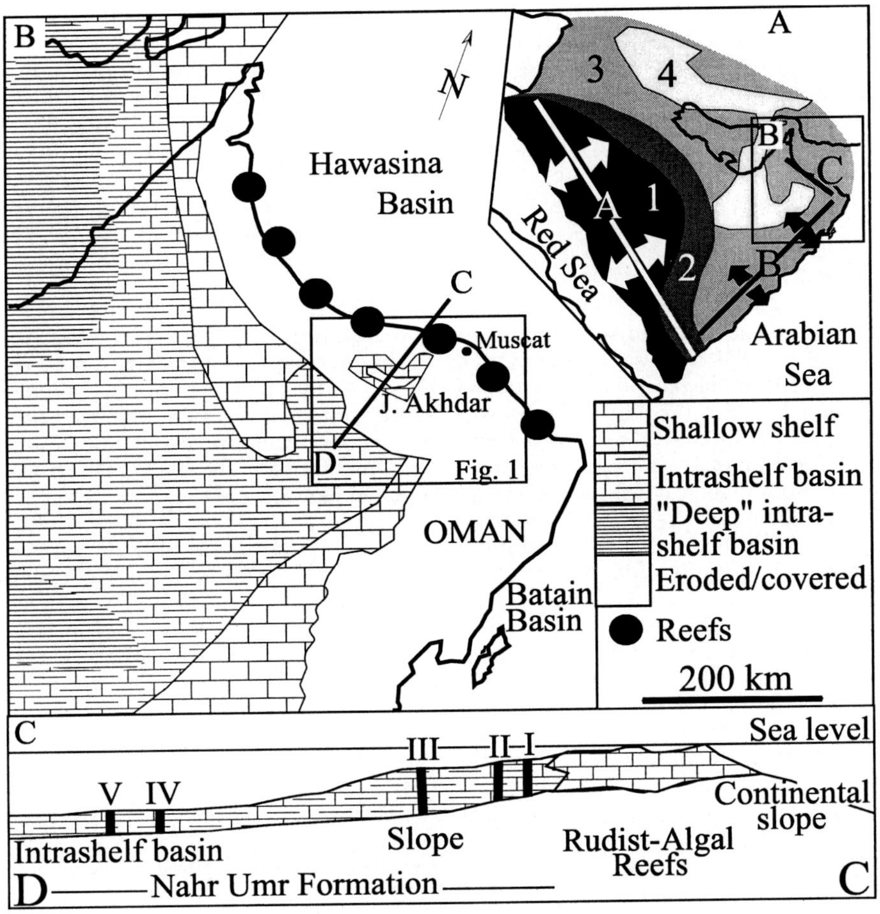


FIG. 2.—A) Reconstruction of the Albian carbonate platform in Arabia. Basement highs are marked A, B, and C. Facies belts: 1, Arabian shield; 2, coastal plain; 3, shallow marine inferred platform, 4, Lurestan Bab Basin. B) An enlargement of the northeastern part of the Arabian carbonate platform. Black dots mark the inferred location of Al Lurestan Formation reefs. Location of transect CD is shown. C) Transect from the continental margin into the intracratonic Bab Basin, with location of sections I to V (Fig. 1). After Murris (1980) and Harris et al. (1984); transect is based partly on Masse et al. (1997).

The samples were cooled until all of the liquid as well as the vapor phase in the individual inclusions was frozen (approximately -80 to -90°C) and then slowly heated at a rate of 0.5 to 5°C per minute until all of the ice was melted. 415 matrix micrite and 80 carbonate cement samples were analyzed on a VG Prism II ratio mass spectrometer for carbon and oxygen isotope ratios. Repeated analyses of standard material show a reproducibility of better than 0.1‰ for $\delta^{18}O$ and less than 0.05‰ for $\delta^{13}C$. All results are presented as parts per thousand relative to the VPDB standard. Abundance of Ca, Sr, Mg, Mn, Fe, and Ca in calcite and dolomite cement phases were measured at the Institute for Earth Sciences, Vrije Universiteit, Amsterdam by electron microprobe analysis operating at 20 kV, beam current 0.015 μA, and 13 μm beam diameter. Cathodoluminescence on carbonate cements was conducted on a cold cathode luminescence microscope operating under 10–14 kV accelerating voltage, 200–300 μA beam current, and a beam diameter of 4 mm.

## LITHOSTRATIGRAPHY AND SEDIMENTOLOGY

Along the northern slope of Jabal Akhdar, the Nahr Umr Formation is best exposed on the western bank of Wadi Bani Kharus about 1 km south of the village of Awabi (I; Fig. 1). This is the type locality of Rabu (1987) and our reference section.

Along the southern flank of Jabal Akhdar exposures are poor, as in Wadi Mu'aydim (section III), where only about 50% or less of the Nahr Umr is uncovered. Likewise, in the Foothills at Jabal Salakh (IV) only individual beds protrude from the gravel. However, at Jabal Madar (V) approximately 70% of a continuous section is exposed because of a landslide. Sections along the northern flank of Jabal Akhdar (105 m) and those measured in the Foothills (92 m) show a similar stratigraphic thickness, whereas sections in Wadi Mu'aydim reach almost 220 m in thickness. At this locality, an exceptional stratigraphic thickness is also found for the underlying Shuaiba Formation (Pratt and Smewing, 1993) and the overlying Natih Formation (Fig. 1; Philip et al., 1995).

### *Facies Associations and Related Paleo–Hydrodynamic Regimes*

On the basis of sedimentology and microfacies we distinguish three hydrodynamic regimes: a low-energy environment, an intermittently agitated (storm) environment, and a high-energy environment. Low-energy and high-energy environments are end members in terms of hydrodynamic energy and are characterized by their facies and, if present, sedimentary structures. Some of the facies associations were attributed to an intermittently storm-agitated regime because they indicate episodes of agitation in an otherwise calm regime. A main feature of the intermittently agitated environment is the presence of discrete layers, formed by locally coarse bivalve fragments (packstones), in an otherwise argillaceous facies with an in-situ fauna. These layers are interpreted as storm deposits obscured by bioturbation to various degrees.

*Low-Energy Environment.—*

Unbedded ochre-colored, greenish and reddish clays and, to a minor extent, also marls reflect the calmest hydrodynamic regime. This facies forms about 50% of the paleo-basinal sections measured in Jabal Madar (V) and Salakh (IV; Fig. 1), but is also present at several levels in the Jabal Akhdar sections (paleo–basin margin). The mineral content distinguishes two groups. Clays and marls from the northern slope of Jabal Akhdar contain large amounts of pyrophyllite and chlorite, not present in the Foothill sections. In contrast, samples from the Foothills contain kaolinite and goethite, which are not found at Jabal Akhdar, where illite–mica is mainly present.

A low-energy environment is also interpreted for two other facies types. The first is an ochre-colored, poorly consolidated argillaceous mudstones or fine-grained skeletal wackestone with an *in-situ* fauna of orbitolinids, few miliolids, ostracods, molluscs, oysters, and echinoderms. *Orbitolina texana* (Römer), *O. subconcava* (Lemeyre), and *O. aperta* (Erman) are most commonly found. For a more detailed list refer to Rabu (1987) and Simmons and Hart (1987). The echinoderm fauna consists of *Pseudodiadema carthusianum* (Gras) and *Coeholecyptus* aff. *simulis* (Desor; cf. Rabu et al., 1986). Locally, beds of this facies are of a condensed facies characterized by abundant iron oxides, phosphate pebbles, and rare ammonites.

The second facies type is ochre-colored and very shaly, and forms beds that are morphologically recessive. Orbitolinid foraminifera form up to 70 to 80% of the rock volume. *Orbitolina* packstones are the most characteristic facies of the Nahr Umr and correspond to the "*Orbitolina* facies" found elsewhere in mid-Cretaceous carbonate platforms (Vilas et al., 1995). *Orbitolina* packstones are interpreted as an *in-situ* facies rather than the result of winnowing (Immenhauser et al., 1999).

*Intermittently Agitated Environment.—*

Sedimentary rocks attributed to an intermittently agitated regime are a predominant facies of the Nahr Umr Formation in Jabal Akhdar, and to a minor degree in the Foothill sections. Beds of this facies type are very commonly nodular and between a few decimeters to several meters thick. Bases of individual beds are commonly shaly but become calcareous towards their upper parts. Bedding surfaces are characterized by a sharp contact between grayish limestones of the underlying bed and ochre-colored, recessive shales at the base of the overlying bed. Two main facies were recognized: a grayish to ochre wackestone to packstone with an in-situ fauna of echinoderms, oysters, serpulids, red algae, orbitolinids, and miliolids, and also fragmented bivalve debris (1–5 mm) in discrete layers of up 5 cm thickness. A second, similar facies type, but with few or no orbitolinids, is found towards the Nahr Umr–Natih formational boundary.

The packstone layers (fragmented bivalves), interpreted as storm deposits, are in many cases intermingled with the surrounding sediments because of subsequent bioturbation. In some exposures, well-defined storm layers could be traced laterally into heavily bioturbated zones, where they become blurred.

*High-Energy Environment.—*

The high-energy facies is found only at a few levels in the Jabal Akhdar sections and in the uppermost part of the underlying Shuaiba Formation (Fig. 4F). Bedding surfaces are commonly planar, and in a few cases also slightly nodular. The microfacies types are poorly washed peloidal and skeletal grainstones. In the case of marker succession C (Fig. 3), numerous rudists in growth position are found. They have been determined as *Prearadiolites* sp. by J.-P. Masse (Fig. 4D). Other beds with a high-energy facies are typified by their coarse (3 mm and larger) skeletal components such as fragments of gastropods, bivalves, oysters, and rudists and display meter-scale cross-bedding. Orbitolinids are sparse in the high-energy environments, and the few found are worn and encrusted by small serpulids. Centimeter-thick layers of bivalve fragments, present in the intermittently agitated environment, are a common feature of this facies. Bioturbation is generally less prominent than in the facies of the low-energy or intermittently agitated environments.

## MARKER-BED SUCCESSIONS AND DISCONFORMITY SURFACES

Three important marker-bed successions (A, B, and C, Fig. 3) were recognized in all sections measured except at Jabal Salakh (IV), where the top two marker-bed successions are not exposed. They are characterized by one or several meter-thick limestone beds. These marker beds are widely traceable on wireline logs throughout much of the interior Oman subsurface (Hughes Clarke 1988).

Marker successions A and B consist of massive peloidal packstone beds several meters thick. The diagnostic criteria are the massive beds forming conspicuous cliffs and the low clay content. Marker bed succession A reaches 9 m in thickness on the northern slope of Jabal Akhdar, and is 4.5–5 m thick in the Foothills. Marker succession B is about 4 m thick in the north and decreases to less than 3 m in the south. Marker succession C (2–3.5 m in thickness) is characterized by the rudist *Praeradiolites* sp. in growth position. This unique facies is recognized in all sections at Jabal Akhdar but is absent in the Foothills.

Numerous disconformity surfaces (hardbottoms) are found throughout all sections measured (Fig. 3). These surfaces are hummocky, often reddish, stained with goethite (FeO(OH)), covered locally with phosphate nodules (a few millimeters in size), and encrusted by black, bored oyster shells. Early lithification is demonstrated by numerous small borings (a few millimeters wide and a few centimeters deep) that truncate allochems, and by holes having the shape of the molluscs that etched them.

Ten surfaces were exposed in the Wadi Bani Kharus section (I), nine were found in section II to the west, six in Wadi Mu'aydim (III), four at Jabal Madar (V), and one at Jabal Salakh (IV; Fig. 3). Sedimentary rocks underneath these surfaces generally are intensely bioturbated, and some burrows are filled with poorly consolidated reddish and ochre micrite rich in secondary iron oxides and dolomite rhombs (Fig. 4G).

Despite careful investigations, no evidence for subaerial erosion was found in the field even though chemical evidence of the influx of light soil carbon and light oxygen from meteoric water is found in several sections. Only the first few centimeters below surfaces were affected by strong dissolution–precipitation processes and iron staining (Fig. 4F). However, dolomitization of micrite-filled burrows may reach several meters into the underlying rocks.

Firmgrounds and bioturbated horizons were recognized in most sections. Firmgrounds are relatively thin domains of well-

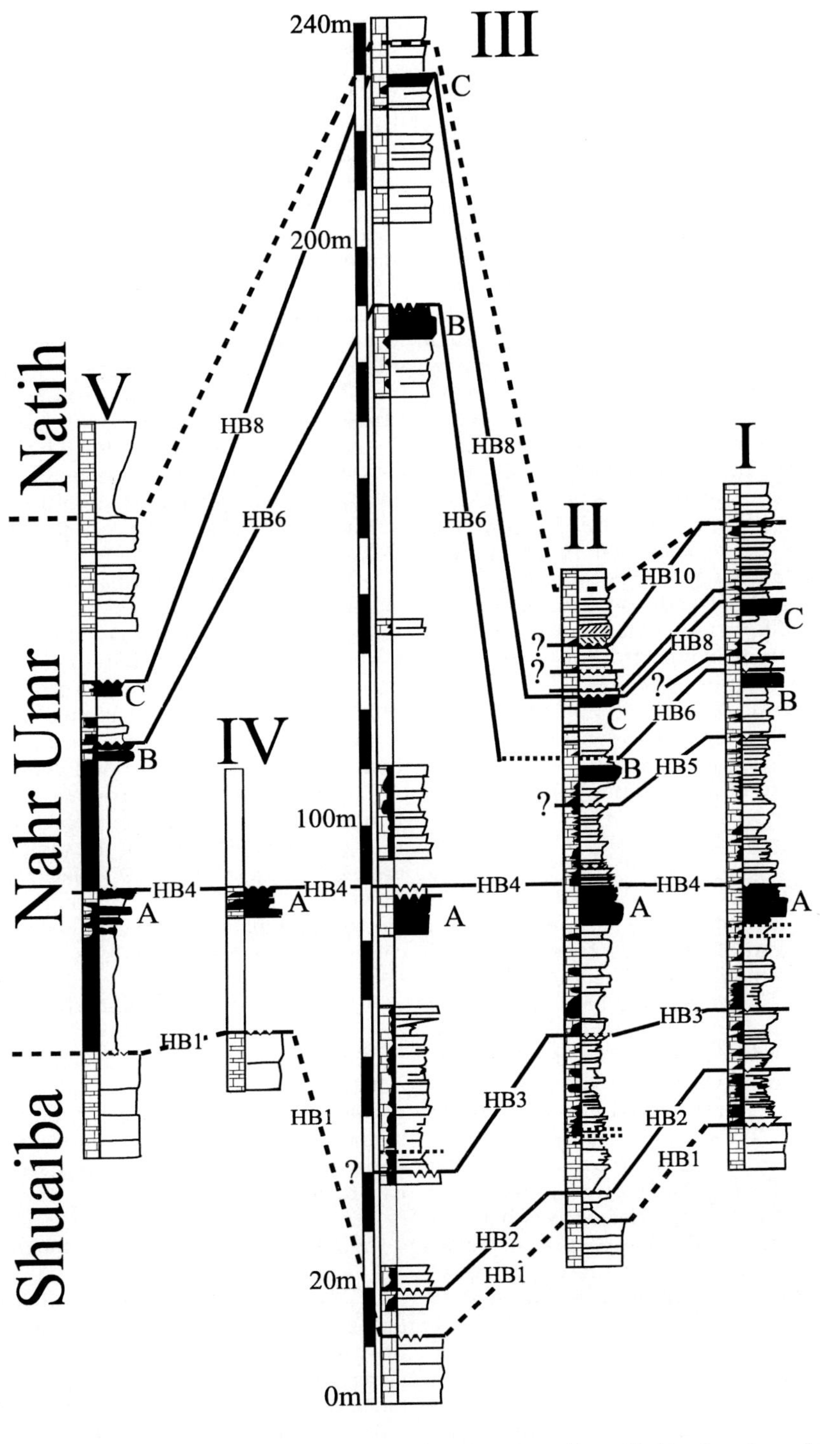


FIG. 3.—Correlation of hardbottoms (HB1–HB10) and marker-bed successions (A to C). For location of individual sections refer to Figures 1 and 2. A simplified lithostratigraphy is shown and the estimated shale–clay (black) vs. carbonate (limestone signature) composition is given at the left side of each lithostratigraphic column. Note the expanded stratigraphic thickness of the Wadi Mu'aydim Section (III).

preserved burrows at the top of an individual bed and are of a clayey, reddish and greenish appearance. Burrows are conspicuous because of their sharp outlines, probably caused by a "stiffness" of the sediment. This incipient submarine cementation confirms an interpretation as *firm*-grounds. The above characteristics are considered to be the result of slow (condensed) sedimentation, although evidence for early lithification (borings) such as in the case of hardgrounds is absent. Bioturbated horizons, on the other hand, may appear within individual beds. In both cases bioturbation is by far more prominent than in other beds.

## *Geochemistry*

### *Carbon Isotopes.—*

Light carbon-isotope ratios were obtained from matrix samples directly underneath hardbottoms (Fig. 5). Across these contacts $\delta^{13}C$ values fall from a range of +4 to +2.5‰ to values between +0.9 to -1.2‰. Besides a clear negative excursion associated with the disconformity at the Shuaiba–Nahr Umr Formation boundary, eight other excursions are present in Wadi Bani Kharus (I). A bundle of three minima appears between meters 0 and 20, respectively, and five of them are grouped into a second bundle between meters 65 and 104 (Fig. 5). The distance between individual hardbottoms within these bundles is remarkably uniform and on the order of 10 to 12 m. Similar but less pronounced negative excursions are observed in sections II and III. In Wadi Mu'aydim (III), however, the shift in $\delta^{13}C$ isotope ratios is much reduced to values of +1 to +2.5‰. Only the basal Shuaiba–Nahr Umr unconformity shows a negative excursion of over +5‰ (Fig. 5).

Microsamples drilled from matrix material display $\delta^{13}C$ values ranging from +2.5 to -1.2‰. These values are slightly heavier than background data from bulk rock samples (Fig. 6). $\delta^{13}C$ values measured in samples drilled from orbitolinids are well within the values of bulk-rock samples, but samples of dolomite-rich burrows underneath hardgrounds may show very light $\delta^{13}C$ values on the order of -3.8‰.

Heavy $\delta^{13}C$ values (+4 and +5‰) occur within the top ~ 20 m of the Shuaiba Formation in Wadi Bani Kharus (I). In the equivalent units of the Wadi Mu'aydim Section (III) $\delta^{13}C$ values of up to +5.5‰ were measured, whereas at Jabal Madar (V) $\delta^{13}C$ values range from +4 to +4.2‰ (Fig. 5).

### *Oxygen Isotopes.—*

In the lower 65 m of the Wadi Bani Kharus Section (I), oxygen isotopes largely follow the trend of carbon isotopes described above (Fig. 5). $\delta^{18}O$ values reach minima underneath hardbottom surfaces except at surface 3. The magnitude of drops in $\delta^{18}O$ values is on the order of 2.25 to 2.5‰ to values as low as -5 to -6.5‰. Oxygen isotope values become more homogeneous from meter 65 upward, and excursions associated with hardgrounds are on the order of 0.5‰ only. A pronounced negative shift to values of -7.6‰ is found in limestones in the top few meters of the section, which have been altered during initial burial beneath the hot metamorphic sole of the Semail Ophiolite.

Microsamples drilled from dolomite-rich burrows beneath hardgrounds show $\delta^{18}O$ ranging from -4.3 to -11.8‰. These values are in part more negative than the background values of -3.2 to -6.5‰ measured from bulk-rock samples of identical hand specimen (Fig. 6).

### *Fluid Inclusions.—*

The coarse blocky calcite that fills druses in the rudist *Praeradiolites* sp. (Fig. 4D) contained two-phase (liquid–vapor) fluid inclusions that were in part suitable for microthermometric measurements. Samples were taken beneath the hardbottom atop marker-bed succession C in the Wadi Bani Kharus (I) and the Wadi Mu'aydim (III) sections (Fig. 3). The vapor phase was most likely induced during Late Cretaceous burial beneath the hot metamorphic sole of the overriding Semail Ophiolite and indicates temperatures of about 70°C during formation. The size of most inclusions is small, between 1 and 10 µm. The shapes of inclusions are round to square, rectangular or irregular and rather flat. A preferred orientation of inclusions was not observed (Fig. 7).

Criteria summarized in Goldstein and Reynolds (1994) were used to recognize the origin of the inclusions that represent the fluid composition at the moment of trapping (under the assumption that no reequilibration has taken place). Fluid inclusions studied are very likely primary because they form clusters bound to specific growth zones within the blocky calcites, indicating that no recrystallization has occurred (Fouke and Reeder, 1992). The inclusion-rich cements are nonluminescent, typical of a near-surface environment of precipitation and thus are unlikely of burial origin.

Determination of the final melting of ice in small-sized inclusions, such as the case in this study, is difficult to recognize because the relief of the ice is almost identical to that of calcite. Zwart (1995) described a method for liquid–vapor inclusions using the tendency of the last remaining ice to stick to the walls of vapor bubbles. The sudden movement of the vapor bubbles associated with the disappearance of the last ice is commonly the only way to determine the melting-point depression.

The results of microthermometric analyses of final ice melting temperatures (freezing-point depression) of 31 inclusions is shown in Figure 8. Conversion of the temperature range between -1.4 and -1.6°C into salinity (wt. % NaCl) after Bodnar (1993) indicates brackish water (less than 2.8% salinity). Inclusions with a depression of the freezing point below values of -1.6°C comprise normal marine to hypersaline fluids.

## *Carbonate Cements Beneath Hardbottoms*

Six major cement phases were recognized. The geochemical data are presented in Table 1. These phases are, from old to young: (1) early marine isopachous cement; (2) coarse, iron-stained, blocky sparite; (3) zoned euhedral dolomite; (4) sparry equant calcite; (5) mosaic sparite in root molds; and (6) late calcite veins.

### *Early Marine Isopachous Cements.—*

Early marine cements are found in the high-energy grainstone facies of the topmost Shuaiba Formation and, less commonly, in the upper few meters of the Nahr Umr Formation. They form isopachous fringes, 50 µm thick, of short bladed crystals coating bioclasts. Early marine cements are nonluminescent, and individual crystals are too small to be sampled by conventional methods for stable-isotope analysis.

FIG. 4. (facing page)—A) Hole of boring bivalve (hardbottom 2; Wadi Mu'aydim, III) encircled with black marker pen. Coin is 2 cm across. B) Bored oyster (hardbottom 3; Wadi Bani Kharus); pen is 15 cm long. C) Thin section of borehole (hardbottom IV; Section II) sealed with a crust of secondary iron oxides (i); width of view is 8 mm. D) Rudist shell filled with blocky cement used for microthermometry studies (hardbottom 3; Wadi Bani Kharus, I); width of view is 7 cm. The approximate location of Figure 7 is indicated. E) *Microcodium* colony in root molds (hardbottom 4; Wadi Mu'aydim, III); width of view is 6 mm. F) High-energy facies of the uppermost Shuaiba Formation in Jabal Madar (V). Note the secondary intraparticle porosity (p) due to selective leaching during the subaerial exposure stage. Width of view is 6.5 mm. G) Phase 3 dolomites in matrix of burrows and replacing sediment (hardbottom 9; Wadi Bani Kharus, I); width of view is 6.5 mm. H) Phase 4 calcites (arrows) related to the subaerial exposure stage. This cement fills fissures in secondary iron oxides (i). Width of view is 5.5 mm (hardbottom 4, Wadi Mu'aydim, III).

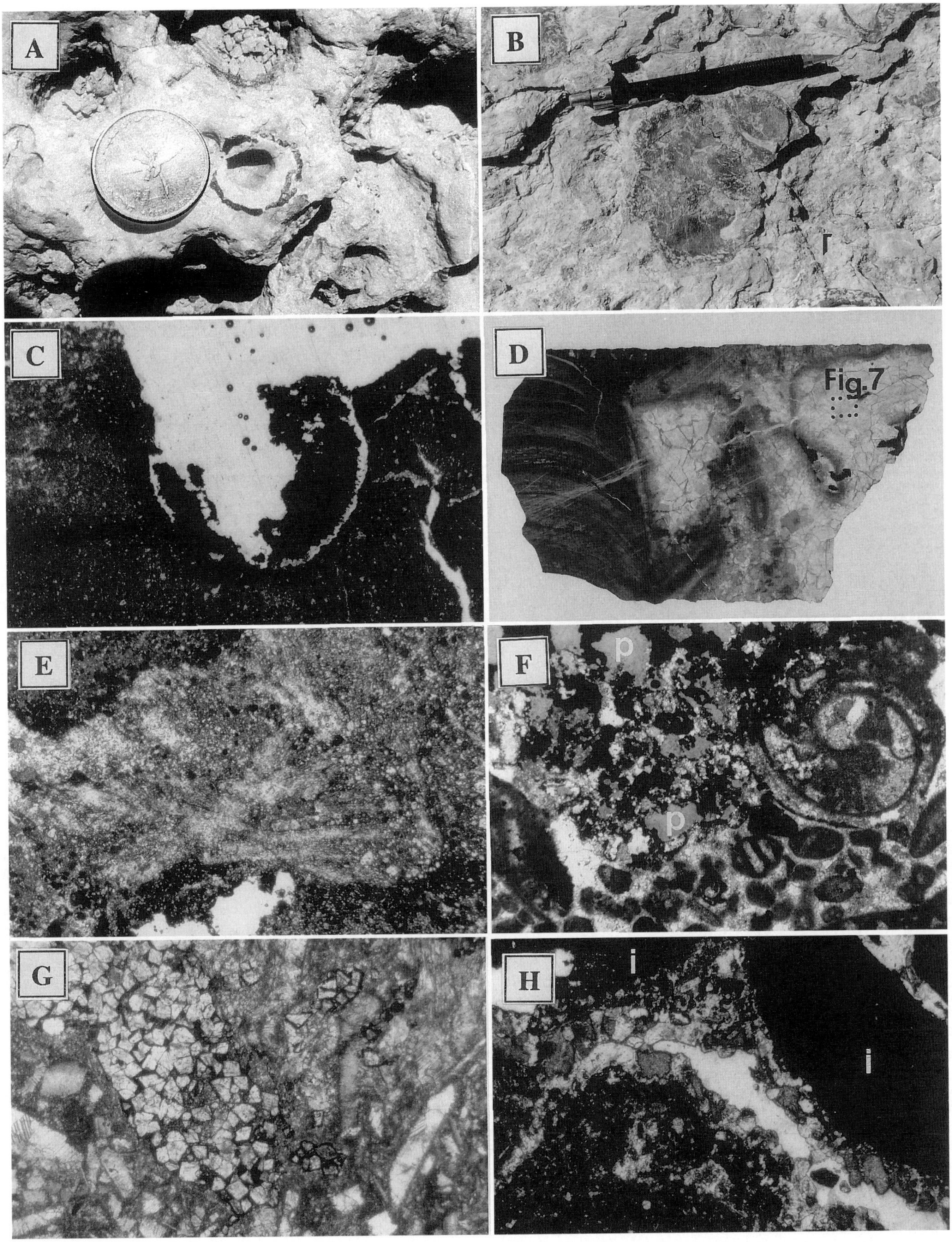
A
B
r
C
D
Fig.7
E
F
p
p
G
H
i
i

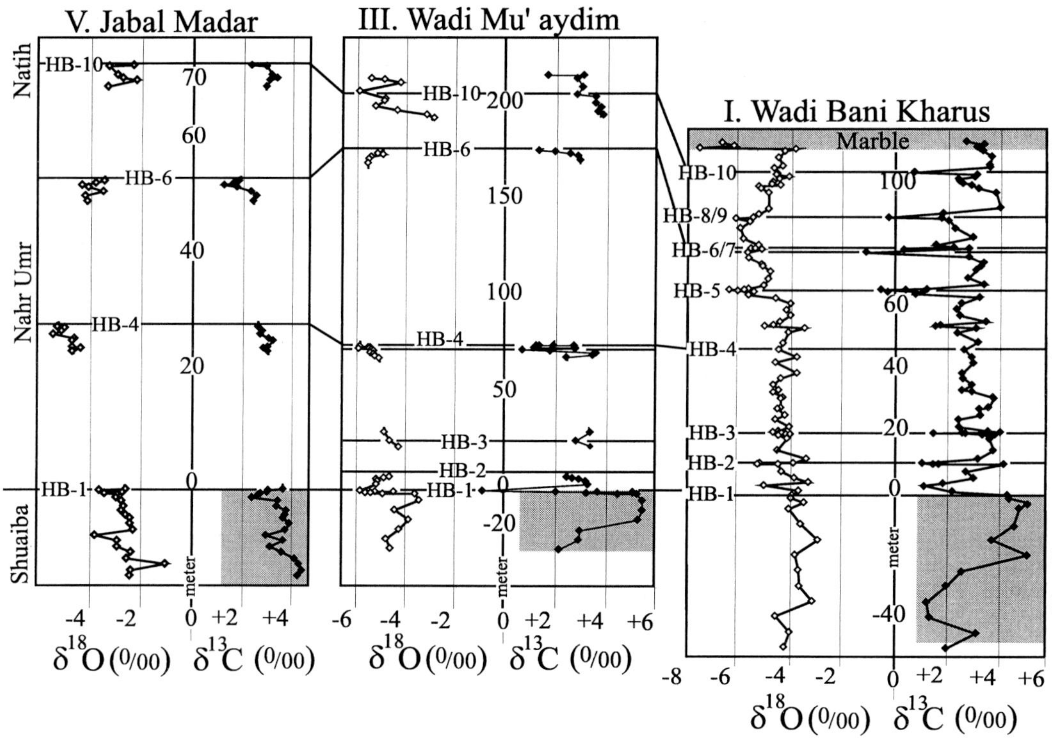


FIG. 5.—Chemostratigraphy of the Nahr Umr Formation and the top of the underlying Shuaiba Formation in wadis Bani Kharus (I), Mu'aydim (III), and Jabal Madar (V). Note the negative excursions in $\delta^{13}C$ and partly in $\delta^{18}O$ isotope values beneath hardgrounds in the Wadi Bani Kharus Section (I). The negative excursion in $\delta^{18}O$ values within the top few meters in Wadi Bani Kharus (I) is related to metamorphism beneath the overriding hot sole of the Semail Ophiolite.

*Coarse, Iron-Stained Blocky Sparites.—*

Ochre-colored to slightly reddish, ferroan and nonferroan, blocky calcite cements fill solution cavities and occlude part of the pore space in rudist shells. The characteristic reddish color is due to an Fe content up to ten times higher than in other cement phases. This phase is also enriched in Mn relative to other cements (Table 1). This cement phase is very dull to nonluminescent. Stable-isotope analyses reveal mean $\delta^{18}O$ values of -7.6‰ and $\delta^{13}C$ values of 1.0‰. Secondary iron oxide (goethite) replaces this cement in many thin sections.

*Euhedral Dolomites.—*

Euhedral dolomites (Fig. 4G), about 200 µm in length, replace matrix and bioclasts as well as cement phases one and two. They appear in great number in burrows underneath hardbottoms and fill voids of dissolved bioclasts in the grainstone facies towards the top of sections in the northern Jabal Akhdar. Dolomite crystals are nonluminescent with the exception of an outer, bright luminescent rim of goethite and dolomite, 10–25 µm thick. No stable-isotope data could be measured from individual dolomite crystals, because of their small size. Data obtained are from bulk rock microsamples drilled from dolomite-rich zones. The resulting isotope data are light relative to background values of the matrix. The range in $\delta^{18}O$ values is -8.3 to -4.4‰, and $\delta^{13}C$ values of 1.54 to -0.13‰ were found (Fig. 6).

*Sparry, Equant Calcite–Quartz Intergrowth.—*

Inclusion-free, equant sparite occludes voids, covers walls of small fissures in sediments or secondary iron oxides, and replaces blocky cement of phase two (Fig. 4H). Individual crystals in fissures are on average 0.5 mm long. This cement locally replaces phase three dolomites. Phase four calcites are commonly enriched in Mg relative to earlier cement phases and are nonluminescent (Table 1). Stable-isotope analyses reveal mean $\delta^{18}O$ values of -6.3‰ and $\delta^{13}C$ values of -1.8‰ (Fig. 6). As observed under the electron microprobe, calcites of phase 4 form a complex commingling with authigenic quartz. The average calcite:quartz ratio is about 7:3. The coexistence of calcite and quartz cements was not observed in other phases. A fine-grained phase of the calcite–quartz intergrowth is also present in the goethite crusts that seal the hardbottoms.

*Mosaic Sparite in Root Molds.—*

Root molds are characterized e.g., by the appearance of *Microcodium* (Fig. 4E) and are filled with a poorly consolidated reddish to ochre micritic sediment. Patches of mosaic, inclusion-rich sparites replace *Microcodium* and matrix. Stable-isotope analyses of this cement phase display mean $\delta^{18}O$ values between -6.2 and -7.3‰ and $\delta^{13}C$ values between +0.8 and +1.2‰. This cement phase is nonluminescent and differs geochemically from the preceding phase four by its lower Mg content.

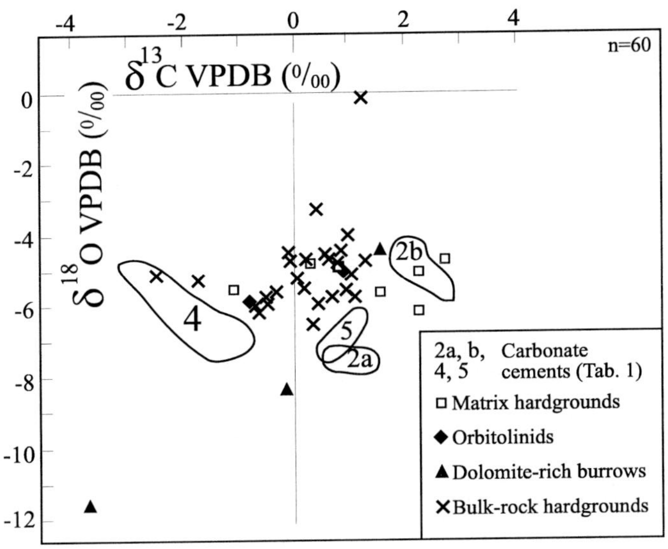


FIG. 6.—Comparison of carbon and oxygen isotope composition in selected microsamples obtained from sedimentary rocks directly beneath surfaces. Note the isotopically light composition of cement phase four, related to a subaerial exposure stage.

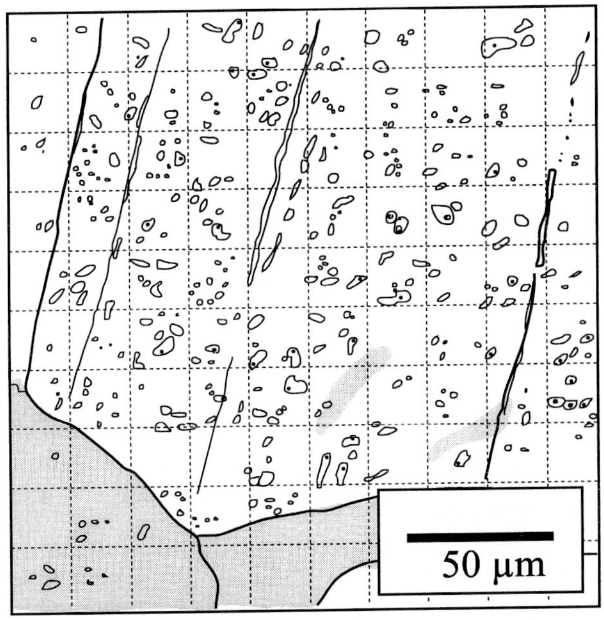


FIG. 7.—Detailed sketch of a clear calcite crystal rich in primary liquid–vapor inclusions. Inclusions show no preferred orientation and are randomly dispersed. Adjacent crystals in the lower left corner are cloudy and grayish and yield few inclusions. The position of Figure 7 is indicated in Figure 4D.

*Late Calcite Veins.—*

Late diagenetic features such as calcite veins and authigenic precipitates in the pressure shadows of bioclasts crosscut and thus postdate all previously mentioned cement phases. All veins show complex patterns of numerous subsequent phases of calcite, most of which display a bright luminescence.

## AGE OF THE NAHR UMR FORMATION IN NORTHERN OMAN

### *Graphic Correlation*

Graphic correlation is a quantitative stratigraphic technique that compares the fossil ranges in a section with the total ranges of these taxa composited in numerous sections. A result is that ranges of different groups are integrated. Graphic correlation of sections generates hypotheses of age relationships by testing which taxa in a section are near to their true appearance or extinction times. The fossil ranges in the section are compared with their ranges in a composite standard database of many sections (Shaw, 1964; Carney and Pierce, 1995). The mid-Cretaceous composite standard (MIDK CS) in this study is a database of 40 sections located in the Tethyan Realm in Europe, Africa, the Mid-East, and North America. Over one thousand taxa and event marker beds form this database. Stage boundaries are defined by European reference sections that have been graphed into the database and calibrated to the Harland et al. (1990) time scale. During the last few years new ages for the Aptian–Albian and Albian–Cenomanian boundaries have been proposed (cf. Obradovich, 1993; Gradstein et al., 1994), but until these proposals are fully evaluated, the Harland time scale is retained here.

Although the age-diagnostic biota in the Wadi Bani Kharus Section (I) are not diverse, they are typical of the Albian in Oman. The following biota were used to graphically correlate this section: *Orbitolina texana* (Roemer), the red algae *Permocalculus irenae* Elliott, *Orbitolina subconcava* Leymerie, *Orbitolina aperta* (Erman), *Eoradiolites lyratus* (Conrad), and *Hemicyclammina whitei* (Henson) (Fig. 9).

Our first data point is provided by *Orbitolina texana* found above the basal discontinuity at the Shuaiba–Nahr Umr Formation boundary, dated in the MIDK CS at 113.72 Ma (Fig. 9). This, however, does not date the onset of subaerial exposure but dates the initial sedimentation related to sea-level rise and drowning of this surface. The base of the Nahr Umr here is as old as latest Aptian, using a date of 112 Ma for the Aptian–Albian boundary. This is consistent with the Aptian age of the Shuaiba–Nahr Umr Formation boundary in Wadi Mu'aydim (III) and in good agreement with Late Aptian nannoplankton recovered from the basal Nahr Umr (Shaffer, *in* Scott 1990). On the basis of the sediment accumulation rate discussed below, the Aptian–Albian boundary is approximately located at meter 14 in the Wadi Bani Kharus Section (I; Fig. 9).

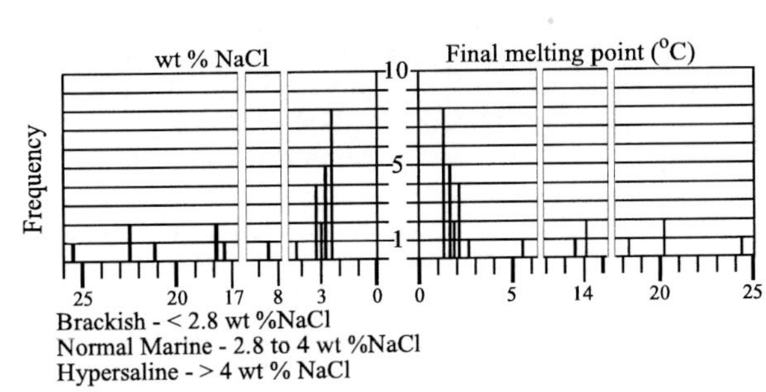


FIG. 8.—Histogram showing the distribution of final melting temperatures measured in fluid inclusions and the calculated weight percent of NaCl (Bodnar, 1993).

TABLE 1.—Trace-element and isotopic composition of calcite and dolomite cements in the sediments beneath hardbottoms.

| Cement phase | Mineralogy | $\delta^{18}$O ‰PDB | $\delta^{18}$C ‰PDB | MgO (ppm/%) | MnO (ppm) | FeO (ppm) | SrO (ppm) | CaO (wt %) |
|---|---|---|---|---|---|---|---|---|
| I. Isopachous, early marine cement fringes | calcite | n.d. | n.d. | n.d. | n.d. | n.d. | n.d. | n.d. |
| IIa. Coarse, iron-stained, blocky cement | calcite, goethite | -7.55 | 1.12 | 1120 (±80) | 810 (±200) | 8640 (±220) | b.d.l. | 56 |
| IIa. Coarse, iron-stained, blocky cement | calcite, goethite | -7.8 | 0.83 | 720 (±80) | 1900 (±200) | 7610 (±220) | 60 (±130) b.d.l. | 56 |
| IIa. Coarse, iron-stained, blocky cement | calcite, goethite | -7.59 | 1.09 | 1890 (±80) | 840 (±200) | 3890 (±220) | b.d.l. | 56 |
| IIb. Coarse, inclusion-rich, blocky sparite in rudists | calcite | -4.38 | 2.06 | 2400 (±80) | 970 (±210) | 840 (±210) | b.d.l. | 55.5 |
| IIb. Coarse, inclusion-rich, blocky sparite in rudists | calcite | -5.69 | 2.68 | 1670 (±80) | 560 (±210) | 7810 (±220) | b.d.l. | 55.1 |
| IIb. Coarse, inclusion-rich, blocky sparite in rudists | calcite | -4.33 | 2.03 | 1430 (±80) | 750 (±210) | 410 (±220) | b.d.l. | 56.2 |
| III. Zoned, euhedral dolomite rhombs | dolomite | n.d. | n.d. | 21.4 (%) | 540 (±190) | b.d.l. | b.d.l. | 29.9 |
| III. Zoned, euhedral dolomite rhombs | dolomite | n.d. | n.d. | 21.3 (%) | 710 (±190) | b.d.l. | b.d.l. | 30 |
| IV. Clear, equant sparite-quartz intergrowth | calcite, quartz | -7.3 | -1.4 | 5020 (±80) | 40 (±200) | 1630 (±220) | 110 (±130) b.d.l. | 56 |
| IV. Clear, equant sparite-quartz intergrowth | calcite, quartz | -7.19 | -1 | 3470 (±80) | b.d.l. | 1060 (±220) | b.d.l. | 56 |
| IV. Clear, equant sparite-quartz intergrowth | calcite, quartz | -4.89 | -2.92 | 2010 (±80) | b.d.l. | b.d.l. | 120 (±130) b.d.l. | 56 |
| V. Mosaic sparite filling root molds | calcite, goethite | -6.17 | 1.24 | 730 (±80) | 10 (±200) b.d.l. | 2130 (±220) | b.d.l. | 56 |
| V. Mosaic sparite filling root molds | calcite, goethite | -7.27 | 0.77 | 103 (±80) | 200 (±200) b.d.l. | 850 (±220) | b.d.l. | 56 |
| V. Mosaic sparite filling root molds | calcite, goethite | n.d. | n.d. | 81 (±80) | 130 (±200) b.d.l. | 2890 (±220) | 20 (±130) b.d.l. | 56 |
| Rudist shell, iron stained | calcite | -4.94 | 2.44 | 2440 (±80) | 750 (±200) | 11.730 (±220) | b.d.l. | 54.6 |

n.d. = no data; b.d.l. = below detection limit

The Nahr Umr–Natih Formation boundary plots at 99.9 Ma in the graphic correlation (Fig. 9). The implication of this is that the top of the Nahr Umr Formation in Wadi Bani Kharus (I) is Late Albian in age. This age is somewhat older than that found for the top of the Nahr Umr Formation in Wadi Mu'aydim (III; Scott, 1990). The Albian–Cenomanian boundary was placed at 97 Ma by Harland et al. (1990).

On the basis of this graphic correlation experiment, a simple calculation of the age vs. thickness in the Wadi Bani Kharus Section (I) provides a rough scale for the duration of sedimentary cycles. Each meter of section represents about 123 ky on average (8.13 m/million years = 8.13 bubnoffs; Fischer 1969). This value, however, includes the time represented by hiatuses, and also it assumes uniform depositional rates (which probably was not the case), and it assumes no erosion during deposition of the entire Nahr Umr Formation. The ages implied by graphic correlation of the other hardbottoms found in the Wadi Bani Kharus Section (I) are shown in Figure 9.

The above results are at variance with some previous work suggesting a Middle Albian to Cenomanian age for the Nahr Umr Formation (see Le Métour et al., 1995, and references therein). Hoogkamer (1979), Simmons and Hart (1987), Hughes Clarke (1988), and Scott (1990) attributed an Albian age to the Nahr Umr in Oman. Harris et al. (1984), Alsharhan and Nairn (1988), and Scott (1990) indicate a Late Aptian to Albian age for the Nahr Umr Formation in the United Arab Emirates, whereas Alsharhan (1991) suggested a Mid-Albian age for this unit.

### *Ammonite Biostratigraphy*

Few ammonites have been reported from the intracratonic basin of the Arabian Platform. We have found *Knemiceras uhligi*

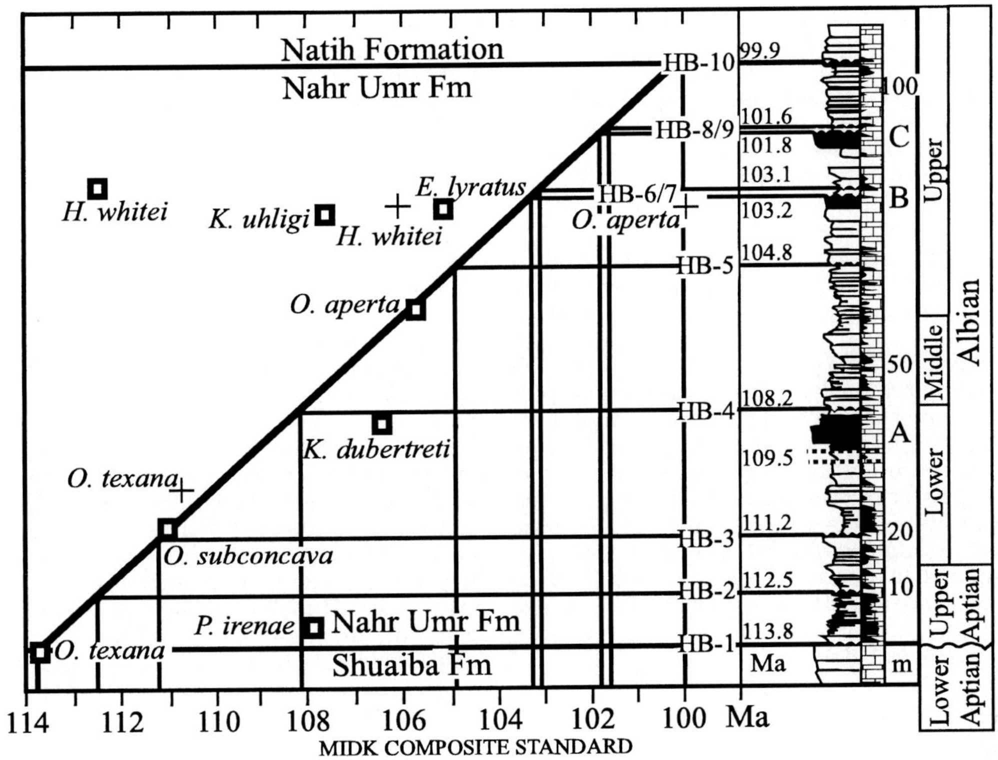


FIG. 9.—Graphic correlation of the Wadi Bani Kharus Section (I). The position of individual marker biota and the resulting line of correlation are shown. The vertical axis is the stratigraphic thickness of the section, and the horizontal axis is the time line. The chronostratigraphic result of this graph suggests a lower formational boundary at 113.8 Ma and an upper boundary at 99.9 Ma. This refers to a Late Aptian age for the base of the Nahr Umr Formation and a Late Albian age for its top in Wadi Bani Kharus (I); (Harland et al., 1990, time scale).

(Choffat) at meter 82 of our section in Wadi Bani Kharus (I) above HB 7 (Figs. 3, 10). Our specimens are described in the Appendix. This level is dated to be Upper Albian by graphic correlation. However, the group of *Knemiceras* is known generally to be of Lower to Mid-Albian age, although the substage boundaries are not accurately identified in the sections yielding specimens of *Knemiceras*. A range into the Upper Albian cannot be ruled out at this time.

In Lebanon and Israel Lewy and Raab (1976) reported two horizons with ammonites of this genus that are upper Lower Albian and probably Middle Albian in age (compare Kennedy and Simmons 1991). Kennedy and Simmons (1991) described *K. dubertreti* (Basse) from the lower part of the Nahr Umr Formation from Jabal Madar (V; Fig. 1; see fig. 3 of these authors) and dated the ammonite horizon following Lewy and Raab (1978) as upper Lower or lower Middle Albian, which is consistent with microfaunal data. According to our stratigraphy, the ammonite horizon of Kennedy and Simmons (1991) lies above of marker bed succession A in Wadi Bani Kharus (I) between meter 36–40 and directly below HB 4, which is also dated by graphic correlation to be upper Lower or lower Middle Albian at 108.2 Ma.

## INTERPRETATION AND DISCUSSION

### *Disconformity Surfaces? Marine versus Subaerial Origin*

On the basis of evidence presented in the previous chapters several stages in the evolution of disconformity surfaces in the Nahr Umr Formation of Northern Oman are distinguished. The marine stage(s) of most hardbottoms in the Nahr Umr Formation in Jabal Akhdar is easily recognizable and predominant in the field. Conspicuous features clearly relate to the marine environment. Petrographic evidence indicates early submarine lithification accompanied by phosphatization, penetration of these surfaces by numerous boring bivalves, and encrustation by well-preserved biota such as oysters (Fig. 4B) and serpulids. From the following lines of evidence, however, we propose that a subaerial exposure stage is superimposed on many of these surfaces in Jabal Akhdar.

1. The carbon and, in part, the oxygen chemostratigraphy of the Nahr Umr Formation is characterized by a shift towards lighter values beneath hardbottoms (Fig. 3). Following previous authors (Allan and Matthews, 1982; Humphrey et al., 1985; Beier, 1987; Marshall, 1992), we interpret the sharp decreases in carbon and oxygen isotope ratios associated with hardgrounds to be the result of diagenetic overprinting during subaerial exposure and associated paleosol formation (Immenhauser et al., 1999). The light $\delta^{13}C$ value is the opposite of the heavy $\delta^{13}C$ value commonly found in submarine hardgrounds (Marshall and Ashton 1980). The presence and magnitude of light excursions associated with subaerial exposure depends on the climate and the duration of exposure. Formation of these hardgrounds related to sea-level lowering is also indicated by a statistical study of the functional morphology of orbitolinids, which indicates a very shallow envi-

ronment *beneath* and a rapid deepening *above* hardground surfaces (Immenhauser et al., 1999).

2. Our study of fluid inclusions in blocky calcites beneath the marker-bed succession C in Wadi Bani Kharus (I) and Wadi Mu'aydim (III) shows the presence of a large number of inclusions with brackish water. This is considered to be the result of mixing of meteoric water with residual salt water during subaerial exposure (Fig. 8; Immenhauser et al., 1999).
3. Indirect evidence for paleosols above these hardgrounds is also provided by root molds, some of which still contain *Microcodium*, calcitized mycorrhizae which are interpreted to be the symbiotic association between soil fungi and cortical cells of roots of higher land plants (Fig. 4E; Klappa, 1978). The calcite–quartz intergrowth of cement phase four is possibly another indication for a pedogenic overprint of these surfaces. Authigenic quartz cements are precipitated during burial diagenesis because many deeper subsurface waters contain several tens of ppm of silica. This interpretation seems unlikely in this case, however, because the crystals are overlain by marine sediments (Fig. 4H). We therefore favor the interpretation that this calcite–quartz intergrowth is incipient silcrete, which is well known from many fossil soils forming mainly under semiarid to arid climatic conditions (Meyers, 1977; Summerfield, 1983).

All of the above lines of evidence are in contrast to the scarcity of macroscopic features, such as deep-cutting karst, commonly found at exposure surfaces (Wright, 1988; Mylroie and Carew, 1995). Locally, faint superficial karst penetrates a few centimeters into the limestones. It is also important to note that few upward-shoaling cycles were recognized beneath these surfaces, resting in many cases directly upon sediments indicating a low to intermittently agitated paleo-hydrodynamic regime (Immenhauser et al., 1999). For the time being, it is not understood why major karst features are absent. This applies especially to the Shuaiba–Nahr Umr Formation boundary, which was subjected to subaerial conditions for several millions of years but also lacks large karst features. Possible reasons for the absence of deep-cutting karst is discussed in Immenhauser et al. (1999).

Another key problem that merits discussion is whether subaerial exposure preceded a marine hardground stage or whether a marine hardground stage preceded subaerial exposure. Petrographic and field evidence strongly suggests that two marine hardground stages were separated by the subaerial exposure stage described above. Circumstantial evidence for a marine hardground stage following the subaerial exposure are the well preserved encrusting biota found in many locations (Fig. 4B). The preservation potential of such delicate features under conditions related to subaerial exposure is indeed negligible. It thus seems likely that these shells, as well as an unknown proportion of the borings, postdate subaerial exposure. But equally compelling evidence suggests a marine hardground stage *before* subaerial exposure. The following chain of arguments is based mainly on petrographic observations. As observed in thin sections, many of the borings were later sealed with crusts of goethite (Fig. 4C). The growth of large goethite crystals fractured the lithified sediments beneath hardground surfaces. These crusts fill and therefore postdate the borings. They are related either to a late phase of a marine hardground stage or, more likely, to a subsequent subaerial exposure stage. Late fault systems are locally also iron stained, but these thin veneers cannot be compared with the centimeter-thick crusts sealing hardgrounds. The assumption that crusts of secondary iron oxides are the product of subaerial exposure is supported

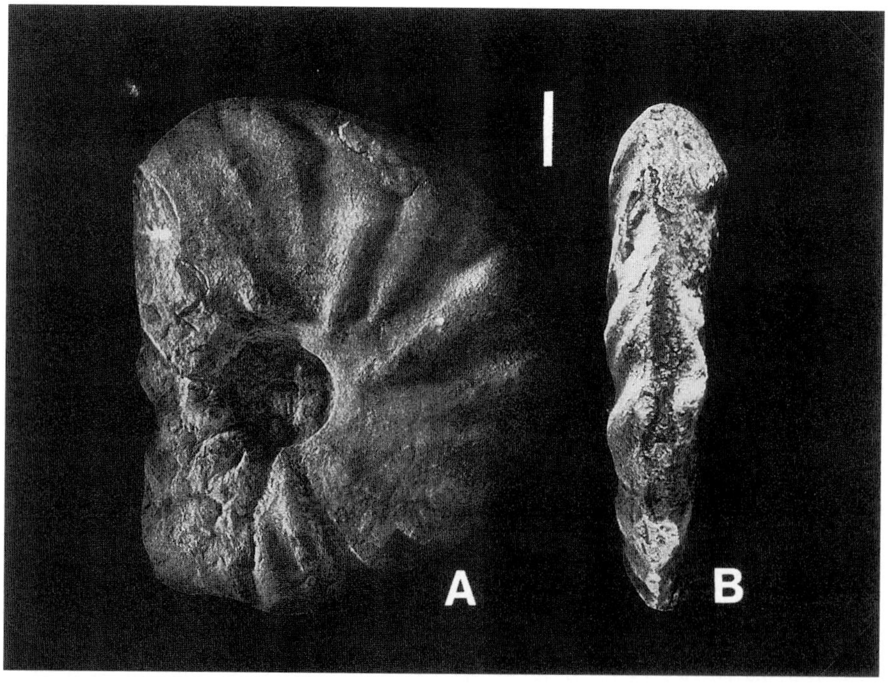


FIG. 10.—*Knemiceras uhligi* (CHOFFAT 1886) from the upper part of the Nahr Umr Formation (meter 82) in Wadi Bani Kharus (I). A) Ventral, B) lateral views. Scale bar is 20 mm.

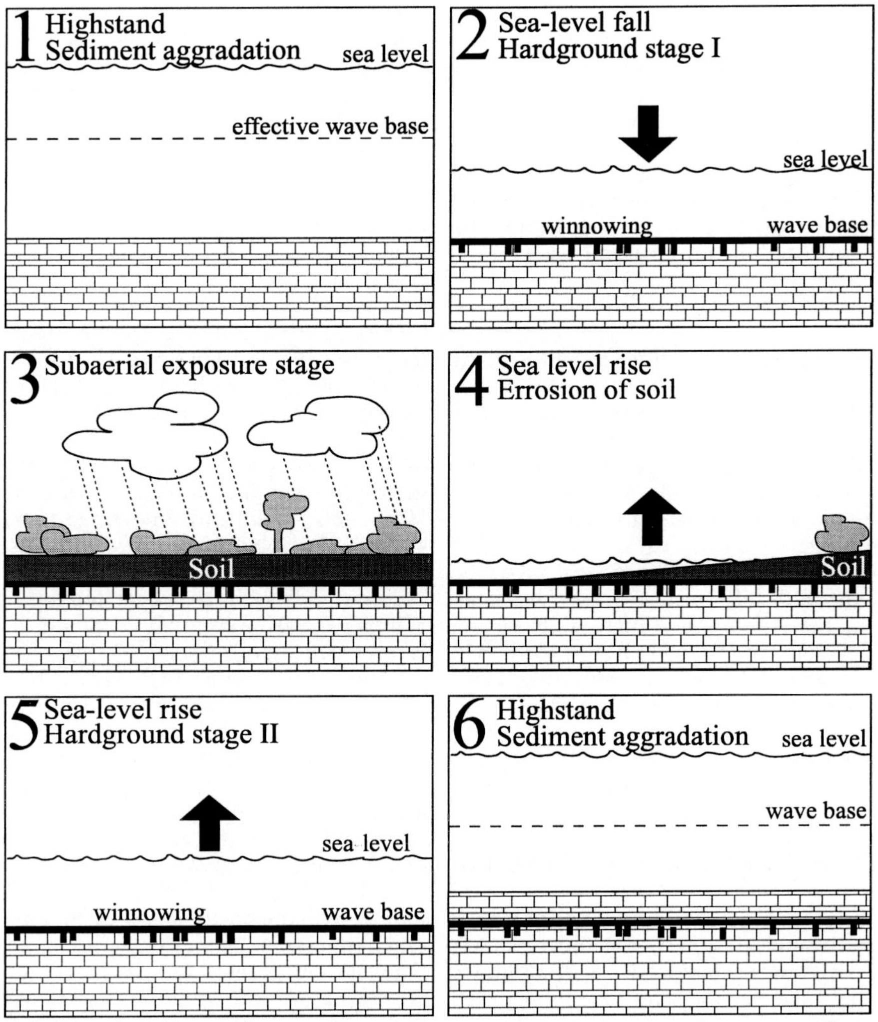


FIG. 11.—Origin of composite hardgrounds in the Nahr Umr Formation in Jabal Akhdar. 1) Sedimentation below effective wave base during highstand. 2) Sea-level fall lowers the effective wave base. Wave and currents winnow the seafloor and circulate water through pores, causing early marine cementation (hardground stage I). 3) Continuous sea-level drop leads to exposure of the hardground surface. Sediments beneath the surface are altered due to isotopically light soil $CO_2$ and meteoric water. 4) Sea level rises again and erodes the soil atop hardground surface. 5) Hardground stage II. Wave and currents initially inhibit sediment deposition. 6) Sea level and effective wave base rise above the sea floor; sedimentation continues.

by the following observation. In many cases, phase four calcite–quartz (Table 1) occludes pore space *within* the iron-oxide crusts that seal bore holes and also fills fissures created by the growth of large goethite crystals. The isotope composition of the cement phase four yields the light isotope signal measured underneath hardbottoms (Table 1, Fig. 6). The light isotope composition is related to subaerial exposure, soil formation, and meteoric water.

We conclude, in summary of the above discussion, that the composite surfaces of the Nahr Umr Formation initially formed as genuine marine hardgrounds, subsequently were turned into subaerial exposure surfaces during lowstand, and became marine hardgrounds again during subsequent sea-level rise. This accounts for the surfaces in Jabal Akhdar but probably not for those in the Foothills. Disconformity surfaces in the Foothills, with the exception of the Shuaiba–Nahr Umr Formation boundary, show only very faint or no signs of subaerial exposure and may thus be marine in origin.

### *Origin of Composite Surfaces*

The formation of marine hardgrounds is related to periods of nonsedimentation and early lithification (Purser, 1969; Kennedy and Garrison, 1975; Bromley, 1978; Brett and Brookfield, 1984). These environmental conditions may be related to either a rapid sea-level rise (maximum flooding surfaces), sea-level fall, and/or bottom currents (Tucker and Wright, 1990; Schlager et al., 1994).

In the case of the hardbottoms recognized in the Nahr Umr Formation of northern Oman it seems likely that the winnowing of these surfaces is related to a drop in sea level. The concomitant lowering of the effective wave base exposes the sea floor to wave action (Fig. 11). Waves and currents sweep the sea floor clean, pump water through pores of the sediment, and cause precipitation of early marine cements. The first result is a marine hardground stage. Under continuously falling sea level the hardground eventually emerges and becomes exposed to subaerial conditions, and soil forms and land plants leave root molds (Fig. 10). Eventually transgression takes place and removes the soil layer. Waves inhibit sediment deposition (second hardground stage) until the surface has subsided below wave base and sedimentation atop of the hardbottoms continues.

The above scenario (Fig. 11) applies to the Nahr Umr Formation in Jabal Akhdar only. Hardbottoms in the Foothills, situated in a more basinward and thus deeper setting, underwent only very short-lived subaerial exposure or remained submerged. The amplitude of sea-level drop was probably too small to expose the sea floor in the deeper domains of the intracratonic basin.

### *Origin of Marker-Bed Successions*

It seems likely that the marker-bed successions of the Nahr Umr Formation are the product of a change in environmental conditions related to a drop in sea level. Light intensity and organic production are, among other factors, related predominantly to water depth (Bosscher and Schlager 1992) and thus linked to sea-level fluctuations. A sea-level drop consequently leads to increased illumination and, linked to this, increased carbonate production by reefs.

Marker-bed succession C in Jabal Akhdar is recognized by the massive appearance of the rudist *Praeradiolites* sp. The same rudist is characteristic of the Aptian to Albian Al Hassanat Formation, which was deposited along the margins of the Bab Basin of northern Oman. A sea-level drop as a controlling factor for the formation of the marker-bed successions is consistent with an inferred progradation of the Al Hassanat reefs towards the intracratonic basin. An alternative explanation for the origin of the marker beds would be the temporary reduction of clay input into the basin.

### *Spatial and Temporal Correlation of the Nahr Umr Formation*

Some of the surfaces in sections of the Nahr Umr Formation occupy a characteristic stratigraphic position at the top of marker-bed successions (Fig. 3). A one-dimensional sequence stratigraphic interpretation suggests that many of the hardbottoms between marker-bed successions can be correlated throughout sections in Jabal Akhdar and in part also between Jabal Akhdar and the Foothills (Immenhauser et al., 1999). Marker-bed successions and hardbottoms are considered to be quasi-time-parallel event markers and allow the correlation of measured sections. They are the product of very specific paleoenvironmental conditions at a given time.

The correlation of two stratigraphic boundaries demands further attention: the Shuaiba–Nahr Umr boundary (hardbottom HB 1, Fig. 3) and the Nahr Umr–Natih boundary. The Shuaiba–Nahr Umr boundary is an important regional unconformity in northern Oman. This surface defines a major change in depositional style from thick bedded shallow-shelf limestones below (Shuaiba Formation) to a shaly succession of mixed clastic–carbonate deposits above (Nahr Umr Formation) and represents a hiatus of probably several million years duration. In northern Jabal Akhdar, where this surface can be traced over many kilometers, it remains parallel to bedding planes of the underlying Shuaiba and therefore is a disconformity or a very low-angle unconformity. The facies of sediments underlying the disconformity indicate an upward increase of wave energy and thus a shoaling of the environment. The Shuaiba–Nahr Umr Formation boundary may record both a pronounced global regression followed by rapid transgression (Murris, 1980) as well as regional tilting and uplift in the west (Burchette and Britton, 1985; Scott et al., 1988; Scott, 1990; Wagner, 1990).

The sedimentary rocks underneath the Shuaiba–Nahr Umr disconformity are characterized by relatively heavy carbon-isotope values (Fig. 5) and were dated as upper Lower Aptian (M. Simmons, personal communication, 1997; Scott et al., 1988, Scott, 1990). This is younger than the positive carbon-isotope excursion related to OAE 1a and older than OAE 1b (Erbacher et al., 1996). Stable-isotope curves from the Shuaiba in the Bab Basin (Vahrenkamp, 1996) and elsewhere (Grötsch et al., 1998) show similarly heavy isotope values for the Lower Aptian Shuaiba cycle 2, which is below a shift to "normal" carbon-isotope compositions and a drop in sea level in the Late Aptian. According to the Harland et al. (1990) time scale the boundary between Lower and Upper Aptian is interpolated at 118 Ma, and Gradstein et al. (1994) date this boundary at 117 Ma by assuming that the ammonite zones are of equal duration. Scott et al. (1988) dated this boundary at 116 Ma by graphic correlation and in the MIDK CS it is dated at about 114.6 Ma. The basal sedimentary rocks of the Nahr Umr above the basal unconformity were dated at 113.8 Ma by our graphic correlation model (Fig. 9). Thus, the duration of the hiatus in Wadi Bani Kharus (I) may range from 1 to 4 My. This upper range is a longer period of nonsedimentation than suggested by previous authors (Harris et al., 1984; Simmons and Hart, 1987; Pratt and Smewing, 1993), although Scott (1990) suggested that the hiatus ranged from 1 to 10 My. Scott (1990) suggested that the Shuaiba–Nahr Umr hiatus is diachronous and becomes increasingly longer towards the east and northeast. Within the 100 km scale of the transect, this assumed diachroneity is not visible in our data.

The second important marker is the Nahr Umr–Natih boundary (Fig. 3). In fact, it is not a surface but rather a zone, several meters thick, in which the Nahr Umr Formation grades into the Natih Formation. In the field, the Nahr Umr–Natih boundary was placed where the shaly, ochre-colored beds of the Nahr Umr grade into dark-gray, massive limestones of the basal Natih Formation. This change in facies is also related to a last appearance of important accumulations of orbitolinids. In Wadi Bani Kharus (I) and Wadi Mu'aydim (III) surface 10 (Fig. 3) coincides approximately with the position of the formation boundary. In Section II and in the Foothills the contact is a gradual change in facies and depositional pattern but no hardbottom was found. In both Sections I and II the top few meters of the Nahr Umr consist of high-energy, partly cross-bedded grainstones.

### *Paleobathymetry of the Bab Basin in Northern Oman*

Plate tectonic reconstructions (Scotese et al., 1988) indicate a subequatorial to equatorial paleo-latitude for Oman in the Albian. During the mid-Cretaceous, wind fields in Oman were probably very weak (Banner and Simmons, 1994), implying low-energy surface waves and maybe a low background water energy for the Bab Basin. Reefs of the Al Hassanat Formation (Masse et al., 1997) rimmed the outer margin of the Arabian Platform in Oman and protected the shelf setting from the impinging oceanic swell of the Neo-Tethys (Fig. 2). However, as discussed before much of the sediments of the Nahr Umr Formation in Jabal Akhdar were deposited below fair-weather wave base but within the reach of storm waves. This led to sediment deposition in an environment that we termed "intermittently agitated" (Immenhauser et al., 1999). The resulting wackestone facies is common in most sections. A very tentative estimate of the paleo–water depth for the Nahr Umr during highstand can be made by assuming a fair weather-wave base at 10 to 20 meters and a storm wave base at 30 to 50 meters. In northern Jabal Akhdar the paleo–water depth was therefore possibly in the range of 30 to 40 meters. Towards the south, the basin deepened. This is indicated by the argillaceous facies of the Foothill sections, which were deposited in water depths below the limit of even the largest storm waves.

The present-day Persian Gulf, 1000 km long and 200–300 km wide, is probably a modern analogue of the Bab Basin. The sea floor on the Arabian side of the Gulf averages 35 m deep and about 80 m in the basin center, which may well be comparable with the Bab Basin in Jabal Akhdar. Given a hypothetical paleo–water depth of 30 to 40 m in northern Jabal Akhdar and an average basinward gradient of 0.01° (which again is comparable to the Persian Gulf), 50 to 60 m water depth results for the Foothills (100 km vertical distance). This result is compatible with our assumption that wave base was lowered down to the basin floor during sea-level lowstands. The bathymetric changes recorded by each marker bed were in the same range of 30 to 50 m. None of the Upper Albian beds in Oman recorded a "dramatic" fall of sea level prior to a rise of even higher amplitude in the *Rotalipora appenninica* zone (Grötsch et al., 1993).

## CONCLUSIONS

The Nahr Umr Formation in northern Oman is of uppermost Aptian to Upper Albian age. In a 100 km transect measured from the margin of the Neo-Tethys (Jabal Akhdar) south into the fringes of the intracratonic Bab Basin (Foothills), formation boundaries are quasi-isochronous. Three marker-bed successions were recognized and correlated throughout all sections. The marker beds are pure, thick limestones deposited during periods of sea-level lowstand.

Disconformity surfaces were found in all sections and are of composite submarine–subaerial origin. In Jabal Akhdar, hardgrounds formed during sea-level drop and subsequently became exposed to subaerial conditions during the peak of regression, and were again overprinted as marine hardgrounds during transgression. Hardgrounds in the more basinward setting of the Foothills remained submerged or underwent only short-lived exposure. Several of the surfaces can be traced and correlated between Jabal Akhdar and the Foothills.

## ACKNOWLEDGMENTS

We gratefully acknowledge financial support from the Swiss National Foundation (Immenhauser; No. 83EU-051870) and the Vrije Universiteit Industrial Associates in sedimentology research grants held by Wolfgang Schlager. Fieldwork was possible due to the help of the Ministry of Commerce and Industry of Oman, especially Mohammed Bin Hussein Bin Kassim, Director General of Minerals, and Hilal Al Azri, Director of the Geological Survey, who provided logistical help and support. We would like to thank A. Matter, University of Bern, for support of stable-isotope measurements and Tj. Peters, University of Bern, for logistical help in the field. W. Gerber (GPI, Tübingen) took the ammonite photos. W. J. Lustenhouwer operated the electron microprobe. Elemental analyses were provided by the Vrije Universiteit Amsterdam and by NWO, the Netherlands Organization for Scientific Research. Development of the MIDK composite standard data base by R. Scott was initiated by support of W. Schlager's project at the Vrije Universiteit Amsterdam. This paper benefited greatly from discussions and inputs by J. Erbacher, L. Hottinger, H. Oterdoom, J.-P. Masse, J. Touret, and H. Weissert. Comments by SEPM reviewers R. Loucks and T. Dickson, and by technical editor J. Southard improved an earlier version of the manuscript.

## APPENDIX

Systematic paleontology and discussion of *Knemiceras uhligi* found in Wadi Bani Kharus (I)

Superfamily Hoplitaceae Douvillé, 1890
Family Engonoceratidae Hyatt, 1900
Genus *Knemiceras* Böhm, 1898
*Knemiceras uhligi* (Choffat 1886)

1995 *Knemiceras uhligi uhligi* (Choffat 1886); Geyer, p. 12, figs. 3 a & b, 4 a–d (for further synonymy).

1995 *Knemiceras uhligi subcompressum* (Hyatt) n. subsp.; Geyer, p. 12 (and synonymy).

1995 *Knemiceras uhligi choffati* n. subsp.; Geyer, p. 12, fig. 3 c & 5 a–f (and synonymy).

1997 *Knemiceras uhligi subcompressum* (Hyatt) Geyer 1995; Geyer et al. (1997), p. 226, fig. 2f–h, 3f–g.

*Material.—*

One incomplete specimen is figured and described here (inventory number GPIT 1823/1 at the University of Tübingen, Germany; Fig. 10). A badly preserved fragment of *Knemiceras* sp. without diagnostic species features is not considered (inventory number GPIT 1823/2; University of Tübingen, Germany).

*Description.—*

The new specimen is flattened and has about half a whorl of the body chamber preserved. Because of distortion it is difficult to estimate the original shape of the whorl section and its ventral tuberculation, but it appears to be highly compressed with a narrow venter. The greatest breadth is at mid flank in intercostal section, and has distinctly clavate ventrolateral tubercles. Coiling is evolute in comparison with other representatives of the genus, with the umbilicus constituting about 21% of the diameter. Umbilical tubercles are very prominent, especially on the phragmocone, where they are situated directly at the umbilical margin. On the body chamber the umbilical wall is smooth. Tubercles are situated more ventrally on the inner flank, up to one-third of the flank height. The distinct prosiradiate ribs arise from the umbilical inner-flank tubercles; there are two to three intercalated secondary ribs.

*Comparison and Discussion.—*

The prominent umbilical, inner-lateral tuberculation and fairly coarse ribs of the specimen described above resemble *Knemiceras compressum* var. *subcompressum* Hyatt (1903; Hyatt, pl. 16, fig. 14) which is, however, more involute and depressed. Geyer (1995) placed Hyatt's subspecies in the species, *K. uhligi subcompressum,* referring to figure 4 on pl. 16 of Hyatt (1903) as lectotype. This is obviously a misprint, fig. 4 is *K. syriacum* von Buch, and, as can be supposed from his synonymy, Geyer assigned fig. 14 of Hyatt (1903) as lectotype. We do not follow Geyer's (1995, 1997) subspecific divison of *K. uhligi,* however, because his subspecies are not stratigraphically or geographically separated. Although the stratigraphic ranges are not well known, this is at least true for the typical subspecies and *K. uhligi choffati* (*sensu* Geyer). Furthermore, it is more likely that these subspecies fall into the variation of a single morphotype, as indicated by figures 6–9 of Geyer (1995). This is confirmed by a reinvestigation of *Knemiceras* from the Albian of Iran by Reyment and Kennedy (1991), who showed that there is a great variability in inflation and ornamentation of specimens. These ammonites are probably referable to a single species, but have been referred to different taxa before. This great variability can be explained by the habitation of a shallow, labile (epicontinental) environment (Reyment and Kennedy, 1991).

## REFERENCES


ALLAN, J.R., AND MATTHEWS, R.K., 1982, Isotope signatures associated with early meteoric diagenesis: Sedimentology, v. 29, p. 797–817.

ALSHARHAN, A.S., 1991, Sedimentological interpretation of the Albian Nahr Umr Formation in the United Arab Emirates: Sedimentary Geology, v. 73, p. 317–327.

ALSHARHAN, A.S., 1995, Facies Variation, Diagenesis, and Exploration Potential of the Cretaceous Rudist-Bearing Carbonates of the Arabian Gulf: American Association of Petroleum Geologists, Bulletin, v. 79, p. 531–550.

ALSHARHAN, A.S., AND NAIRN, A.E.M., 1988, A review of the Cretaceous Formations in the Arabian Peninsula and Gulf: Part II. mid-Cretaceous (Wasia Group) Stratigraphy and Paleogeography: Journal of Petroleum Geology, v. 11, p. 89–112.

BANNER, F.T., AND SIMMONS, M.D., 1994, Calcareous algae and foraminifera as water-depth indicators: an example from the Early Cretaceous carbonates of northeast Arabia, *in* Simmons, M.D., ed., Micropalaeontology and Hydrocarbon Exploration in the Middle East: London, Chapman & Hall, p. 244–252.

BÉCHENNEC, F., ROGER, J., LE MÉTOUR, J., AND WYNS, R., 1992a, Geological Map of Seeb, Sheet NF 40-03, Scale 1:250,000: Oman Ministry of Petroleum and Minerals, Directorate General of Minerals.

BÉCHENNEC, F., WYNS, R., ROGER, J., LE MÉTOUR, J., AND CHEVREL, S., 1992b, Geological map of Nazwa, Sheet NF40-07, Scale: 1:250,000: Oman Ministry of Petroleum and Minerals, Directorate General of Minerals.

BEIER, J.A., 1987, Petrographic and geochemical analysis of caliche profiles in a Bahamian Pleistocene dune: Sedimentology, v. 34, p. 991–998.

BODNAR, R.J., 1993, Revised equation and table for determining the freezing point depression of $H_2O$-NaCl solutions: Geochimica and Cosmochimica Acta, v. 57, p. 1861–1873.

BOSSCHER, H., AND SCHLAGER, W., 1992, Computer simulation of reef growth: Sedimentology, v. 39, p. 503–512.

BRETT, C.E., AND BROOKFIELD, M.E., 1984, Morphology, faunas and genesis of Ordovician hardgrounds from southern Ontario, Canada: Palaeogeography, Palaeoclimatology, Palaeoecology, v. 46, p. 233–290.

BROMLEY, R.G., 1978, Hardground diagenesis, *in* Fairbridge, R.W., and Burgeois, J., eds., The Encyclopedia of Sedimentology: Stroudsburg, Pennsylvania, Dowen, Hutchinson & Ross, p. 397–400.

BURCHETTE, T.P., AND BRITTON, S.R., 1985, Carbonate facies analysis in the exploration for hydrocarbons: a case-study from the Cretaceous of the Middle East, *in* Brenchley, P.J., and Williams, B.P.J., eds., Sedimentology: Recent Developments and Applied Aspects: Geological Society of London, p. 311?338.

CARNEY, J.L., AND PIERCE, R.W., 1995, Graphic correlation and composite standard databases as tools for the exploration biostratigrapher, *in* Mann, K.O., and Lane, H.R., eds., Graphic correlation: SEPM, Special Publication 53, p. 23–43.

CHRISTIAN, L., 1997, Cretaceous subsurface geology of the Middle East Region: GeoArabia, v. 2, p. 239–256.

ERBACHER, J., THUROW, J., AND LITTKE, R., 1996, Evolution patterns of radiolaria and organic matter variations: A new approach to identify sea-level changes in mid-Cretaceous pelagic environments: Geology, v. 24, p. 499–502.

FISCHER, A.G., 1969, Geological time–distance rates: the Bubnoff unit: Geological Society of America Bulletin, v. 80, p. 549–552.

FOUKE, B.W., AND REEDER, R.J., 1992, Surface structural control on dolomite composition: Evidence from sectoral zoning: Geochimica et Cosmochimica Acta, v. 56, p. 4015–4024.

GEYER, O.F., 1995, *Knemiceras uhligi* (Choffat) dans la Chaîne Ibérique mérdionale (Albien supérieur, Espagne): Batalleria, v. 5, p. 5–17.

GEYER, O.F., KUSS, J., AND BACHMANN, M., 1997, On some Albian ammonites from Northeastern Sinai: Paläontologische Zeitschrift, v. 71, p. 221–229.

GLENNIE, K.W., BOEUF, M.G.A., HUGHES CLARKE, M.W., MOODY-STUART, M., PILAAR, W.F.H., AND REINHARDT, B.M., 1974, Geology of the Oman Mountains: Koninklijk Nederlands Geologisch Mijnbouwkundig Genootschap, Verhandelingen: The Hague, Shell Research BV, 423 p.

GOLDSTEIN, R.H., AND REYNOLDS, T.J., 1994, Systematics of fluid inclusions in diagenetic minerals: SEPM, Short Course 31, 199 p.

GRADSTEIN, F.M., AGTERBERG, F.P., OGG, J.G., HARDENBOL, J., VAN VEEN, P., THIERRY, J., AND HUANG, Z., 1994, A Mesozoic time scale: Journal of Geophysical Research, v. 99, p. 24,051–24,074.

GRÖTSCH, J., SCHRÖDER, R., NOE, S., AND FLÜGEL, E., 1993, Carbonate platforms as recorders of high-amplitude eustatic sea-level fluctuations: the Late Albian appenninica-event: Basin Research, v. 5, p. 191–212.

GRÖTSCH, J., BILLING, I., AND VAHRENKAMP, V., 1998, Carbon-isotope stratigraphy in shallow-water carbonates: implications for Cretaceous black-shale deposition: Sedimentology, v. 45, p. 623–634.

HANNA, S.S., 1990, The alpine deformation of the Central Oman Mountains, *in* Robertson, A.H.F., Searle, M.P., and Ries, A.C., eds., The Geology and Tectonics of the Oman Region: Geological Society of London, Special Publication 49, p. 341–360.

HARLAND, W.B., ARMSTRONG, R.L., COX, A.V., CRAIG, L.E., SMITH, A.G., AND SMITH, D.G., 1990, A Geologic Time Scale 1989: New York, Cambridge University Press, 265 p.

HARRIS, P.M., AND FROST, S.H., 1984, Middle Cretaceous carbonate reservoirs, Fahud Field and northwestern Oman: American Association of Petroleum Geologists, Bulletin, v. 68, p. 649–658.

HARRIS, P.M., FROST, S.H., SEIGLIE, G.A., AND SCHNEIDERMANN, N., 1984, Regional unconformities and depositional cycles, Cretaceous of the Arabian Peninsula, *in* Schlee, J.S., ed., Interegional Unconformities and Hydrocarbon Accumulation: American Association of Petroleum Geologists, Memoir 36, p. 67–80.

HOOGKAMER, P.J.C., 1979, Lithostratigraphic atlas of Oman, 2 volumes: Petroleum Development of Oman, internal report.

HUGHES CLARKE, M.W., 1988, Stratigraphy and rock unit nomenclature in the oil-producing area of interior Oman: Journal of Petroleum Geology, v. 11, p. 5–60.

HUMPHREY, J.D., RANSON, K.L., AND MATTHEWS, R.K., 1985, Early meteoric diagenetic control of Upper Smackover Production, Oaks Field, Louisiana: American Association of Petroleum Geologists, Bulletin, v. 70, p. 70–85.

HYATT, A., 1903, Pseudoceratites of the Cretaceous: U.S. Geological Survey, Monograph 44, 250 p.

IMMENHAUSER, A., SCHREURS, G., PETERS, T., MATTER, A., HAUSER, M., AND DUMITRICA, P., 1998, Stratigraphy, sedimentology and depositional environments of the Permian to uppermost Cretaceous Batain Group, eastern Oman: Eclogae Geologicae Helvetiae, v. 91, p. 217–236.

IMMENHAUSER, A., SCHLAGER, W., BURNS, S.J., SCOTT, R.W., GEEL, T., LEHMANN, J., VAN DER GAAST, S., AND BOLDER-SCHRIJVER, L.J.A., 1999, Late Aptian to Late Albian sea-level fluctuations constrained by geochemical and biological evidence (Nahr Umr Fm, Oman): Journal of Sedimentary Research, v. 69, p. 434–446.

KENNEDY, W.J., AND GARRISON, R.E., 1975, Morphology and genesis of nodular chalks and hardgrounds in the Upper Cretaceous of southern England: Sedimentology, v. 22, p. 311–386.

KENNEDY, W.J., AND SIMMONS, M.D., 1991, mid-Cretaceous ammonites and associated microfossils from the Central Oman Mountains: Newsletter on Stratigraphy, v. 25, p. 125–154.

KLAPPA, C.F., 1978, Biolithogenesis of *Microcodium*: elucidation: Sedimentology, v. 25, p. 489–522.

LE MÉTOUR, J., MICHEL, J.C., BÉCHENNEC, F., PLATEL, J.P., AND ROGER, J., 1995, Geology and Mineral Wealth of the Sultanate of Oman: Muscat, Ministry of Petroleum and Minerals, Directorate General of Minerals, Sultanate of Oman and Bureau de Recherches et Minières, France, 285 p.

LEWY, Z., AND RAAB, M., 1976, mid-Cretaceous stratigraphy of the Middle East: Muséum d'Histoire Naturelle de Nice, Annales, v. 4, p. 1–20.

LOOSVELD, R.J.H., BELL, A., AND TERKEN, J.J.M., 1996, The tectonic evolution of Interior Oman: GeoArabia, v. 1, p. 28–51.

MANN, A., HANNA, S.S., AND NOLAN, S.C., 1990, The post-Campanian tectonic evolution of the Central Oman Mountains, *in* Robertson, A.H.F., Searle, M.P., and Ries, A.C., eds., The Geology and Tectonics of the Oman Region: Geological Society of London, Special Publications, p. 549–564.

MARSHALL, J.D., 1992, Climatic and oceanographic isotopic signals from the carbonate rock record and their preservation: Geological Magazine, v. 129, p. 143–160.

MARSHALL, J.D., AND ASHTON, M., 1980, Isotopic and trace element evidence for submarine lithification of hardgrounds in the Jurassic of eastern England: Sedimentology, v. 27, p. 271–289.

MASSE, J.-P., BORGOMANO, J., AND AL-MASKIRY, S., 1997, Stratigraphy and tectonosedimentary evolution of an upper Aptian–Albian carbonate margin: the northeastern Jabal Akhdar (Sultanate of Oman): Sedimentary Geology, v. 113, p. 269–280.

MEYERS, W.J., 1977, Certification in the Mississippian Lake Valley Formation, Sacramento Mountains, New Mexico: Sedimentology, v. 24, p. 75–105.

MURRIS, R.J., 1980, Facts and principles of world petroleum occurrence: Hydrocarbon habitat of the Middle East, *in* Miall, A.D., ed., Facts and Principles of World Petroleum Occurrence: Canadian Society of Petroleum Geologists, Memoir 6, p. 765–800.

MYLROIE, J.E., AND CAREW, J.L., 1995, Karst development on carbonate islands, *in* Budd, D.A., Saller, A.H., and Harris, P.M., eds., Unconformities and Porosity in Carbonate Strata: American Association of Petroleum Geologists, Memoir 63 p. 55–76.

OBRADOVICH, J.D., 1993, A Cretaceous time scale, *in* Caldwell, W.G.E., and Kauffman, E.G., eds., Evolution of the Western Interior Basin: Geological Association of Canada, Special paper 39, p. 379–396.

PHILIP, J., BORGOMANO, J., AND AL-MASKIRY, S., 1995, Cenomanian–Early Turonian carbonate platform of Northern Oman: stratigraphy and paleoenvironments: Palaeogeography, Palaeoclimatology, Palaeoecology, v. 119, p. 77–92.

PRATT, B.R., AND SMEWING, J.D., 1993, Early Cretaceous platform-margin configuration and evolution in the central Oman Mountains, Arabian Peninsula: American Association of Petroleum Geologists, Bulletin, v. 77, p. 225–244.

PURSER, B.H., 1969, Syn-sedimentary marine lithification of Middle Jurassic limestones in the Paris Basin: Sedimentology, v. 12, p. 205–230.

RABU, D., 1987, Géologie de l'Autochthone des Montagnes d'Oman: la fenêtre de Jabal Akhdar. La semelle métamorphique de la Nappe ophiolitique de Samail dans les parties orientale et centrale des Montagnes d'Oman: une revue: Bureau des Recherches Géologiques et Minières, Document no. 130, 582 p.; unpublished Docteur es-Sciences thesis, Pierre et Marie Curie University, Paris.

RABU, D., BÉCHENNEC, F., BEURRIER, M., AND HUTIN, G., 1986, Geological map of Nakhl, Sheet NF 40-03E, Scale 1:100,000, with Explanatory Notes. Muscat, Oman, Ministry of Petroleum and Minerals, Directorate General of Minerals.

RABU, D., LE MÉTOUR, J., BÉCHENNEC, F., BEURRIER, M., VILLEY, M., AND BOURDILLON-JEUDI DE GRISSAC, C., 1990, Sedimentary aspects of the Eo-Alpine cycle on the northeast edge of the Arabian Platform (Oman Mountains), *in* Robertson, A.H.F., Searle, M.P., and Ries, A.C., eds., The Geology and Tectonics of the Oman Region: Geological Society of London, Special Publication 49, p. 49–68.

REYMENT, R.A., AND KENNEDY, W.Y., 1991, Phenotypic plasticity in a Cretaceous ammonite analyzed by multivariante statistical methods: A methodological study, *in* Hecht, M.K., Wallace, B., and MacIntyre, J., eds., Evolutionary Biology: New York, Plenum, p. 411–426.

RIES, A.C., AND SHACKLETON, R.M., 1990, Structures in the Huqf–Haushi Uplift, east Central Oman, *in* Robertson, A.H.F., Searle, M.P., and Ries, A.C., eds., The Geology and Tectonics of the Oman Region: Geological Society of London, Special Publication 49, p. 653–664.

SCHLAGER, W., REIJMER, J.J.G., AND DROXLER, A., 1994, Highstand shedding of carbonate platforms: Journal of Sedimentary Research, v. B64, p. 270–281.

SCOTESE, C.R., GAHAGAN, L.M., AND LARSON, R.L., 1988, Plate tectonic reconstructions of the Cretaceous and Cenozoic ocean basins: Tectonophysics, v. 155, p. 27–48.

SCOTT, R.W., 1990, Chronostratigraphy of the Cretaceous carbonate shelf, southeastern Arabia, *in* Robertson, A.H.F., Searle, M.P., and Ries, A.C., eds., The Geology and Tectonics of the Oman Region: Geological Society of London, Special Publication 49, p. 89–108.

SCOTT, R.W., FROST, S.H., AND SHAFFER, B.L., 1988, Early Cretaceous sea-level curves, Gulf Coast and Southeastern Arabia, *in* Wilgus, C.K., Hastings, B.S., Kendall, C.G.S.C., Posamentier, H.W., Ross, C.A., and Van Wagoner, J.C., eds., Sea-Level Changes: An Integrated Approach: SEPM, Special Publication 42, p. 275–284.

SHAW, A.B., 1964, Time in Stratigraphy: New York, McGraw-Hill, 365 p.

SIMMONS, M.D., AND HART, M.B., 1987, The biostratigraphy and microfacies of the Early to mid-Cretaceous carbonates of Wadi Mi'aidin, Central Oman Mountains, *in* Hart, M.B., ed., Micropalaeontology of Carbonate Environments: Chichester, U.K., Ellis Horwood, p. 176–207.

SIMMONS, M.D., 1994, Micropalaeontological biozonation of the Kahmah Group (Early Cretaceous), Central Oman Mountains, *in* Simmons, M.D., ed., Micropalaeontology and Hydrocarbon Exploration in the Middle East: London, Chapman & Hall, p. 177–206.

SUMMERFIELD, M.A., 1983, Petrography and diagenesis of silcrete from the Kalahari Basin and Cape Coastal Zone, Southern Africa: Journal of Sedimentary Petrography, v. 53, p. 895–909.

TUCKER, M.E., AND WRIGHT, V.P., 1990, Carbonate Sedimentology: Oxford, U.K., Blackwell, 482 p.

VAHRENKAMP, V.C., 1996, Carbon isotope stratigraphy of the Upper Kharaib and Shuaiba Formations: Implications for the Early Cretaceous evolution of the Arabian Gulf region: American Association of Petroleum Geologists, Bulletin, v. 80, p. 647–662.

VILAS, L., MASSE, J.P., AND ARIAS, C., 1995, *Orbitolina* episodes in carbonate platform evolution: the early Aptian model from SE Spain: Paleaogeography, Palaeoclimatology, Palaeoecology, v. 119, p. 35–45.

WAGNER, P.D., 1990, Geochemical stratigraphy and porosity controls in Cretaceous carbonates near the Oman Mountains, *in* Robertson, A.H.F., Searle, M.P., and Ries, A.C., eds., The Geology and Tectonics of the Oman Region: Geological Society of London, Special Publication 49, p. 127–137.

WRIGHT, V.P., 1988, Paleokarsts and paleosols as indicators of paleoclimate and porosity evolution: A case study from the Carboniferous of South Wales, *in* James, N.P., and Choquette, P.W., eds., Paleokarst: Berlin, Springer, p. 329–341.

ZWART, E.W., 1995, Fluid inclusions in carbonate rocks and calcite cements [unpublished Ph.D. thesis]: Vrije Universiteit, Amsterdam, 157 p.

# Part 3:
# Hydrocarbon Habitat and Exploration/Development Case Studies

# STRATIGRAPHIC FRAMEWORK AND RESERVOIR DEVELOPMENT OF THE UPPER JURASSIC IN ABU DHABI AREA, U.A.E.

MOHAMED R. AYOUB AND IBRAHIM M. EN NADI

*Abu Dhabi Company for Onshore Oil Operations (ADCO) ADCO, P.O. Box 270, Abu Dhabi, U.A.E.*

ABSTRACT: The Upper Jurassic succession exhibits lateral variations in both thickness and lithology over the Abu Dhabi area. These variations are related to eustatic changes in sea level, structural growth of the Qatar Arch and Mender–Lekhwair high, spreading of the Neo-Tethys, and local diapiric movements.

The Upper Jurassic section in the eastern part of Abu Dhabi is subdivided, from bottom to top, into the following formations and reservoirs: Tuwaiq Mountain Formation with Hadriya reservoir, Hanifa Formation with Hanifa reservoir in the southeast area, Jubaila dense section, and the Asab–Asab equivalent with the Asab Oolite reservoir. In the western area, the section contains the Tuwaiq Mountain, Hanifa, Jubaila with Arab D reservoir at the top overlain by the Arab Formation with Arab A, B, and C reservoirs and the Hith anhydrite, locally overlain by the Manifa reservoir at the top of the Hith Formation.

The Late Oxfordian structural growth of the Mender–Lekhwair high and the Qatar Arch led to the development of the main intrashelf basin in between, where the Hanifa Formation was deposited. The Hanifa is the main source rock for the Arab and the Thamama hydrocarbons.

Approximately 75% of the present structural configuration of Abu Dhabi onshore area resulted from the late Cretaceous structural compressional movements due to the Oman Nappes overthrust zone. The Late Cretaceous subsidence in the Falaha trough in central Abu Dhabi enhanced organic and inorganic gas generation.

## INTRODUCTION

Many of the major hydrocarbon-bearing reservoirs in the Arabian Peninsula are found in the Upper Jurassic rocks. In addition the Upper Jurassic Hanifa Formation is considered to be the main source rock for the Upper Jurassic and lower Cretaceous hydrocarbons (Alsharhan and Kendall, 1986). Hydrocarbon occurrence in the Upper Jurassic reservoirs in Abu Dhabi is controlled by the presence of the Hith Anhydrite (cap rock), which passes laterally across central Abu Dhabi along a NNW–SSE trend into the carbonates of the Asab Formation. This lateral facies change has directed and controlled the migration of the Hanifa hydrocarbons to the Thamama Group reservoirs (Alsharhan, 1989).

The Upper Jurassic sequence exhibits lateral variations in both thickness and lithology across onshore Abu Dhabi area (Fig. 1). These variations are related to eustatic changes in sea level, structural growth of the Qatar Arch and the Mender–Lekhwair high, spreading of the Neo-Tethys ocean, and local diapiric movements. The Upper Jurassic succession unconformably overlies the Middle Jurassic Araej Formation and underlies the Lower Cretaceous Habshan Formation (Fig. 2) (de Matos, 1994). The objectives of this paper are to review the lithostratigraphy and reservoirs development of the Upper Jurassic succession in the onshore Abu Dhabi area (ADCO Concession).

## TECTONIC SETTING

The Abu Dhabi onshore area is part of the Arabian platform. Gentle and simple anticlinal folds are the common features encountered in the area, though complexly faulted structures are also present. These structures gave birth to a significant number of hydrocarbon accumulations. The north/northeast–south/southwest direction is the main elongated trend of the onshore producing fields.

## STRATIGRAPHIC FRAMEWORK

### *Araej Formation*

During the Middle Jurassic (Bajocian to Callovian time), a shallow to very shallow warm-water carbonate ramp prevailed across the Abu Dhabi area and gave rise to the deposition of the relatively clean carbonate section of the Araej Formation (Fig. 3). This formation exhibits poor reservoir characteristics because of subsurface diagenesis, especially in the deeply buried area (central and northeastern part of Abu Dhabi).

The thickness of the Araej varies from 870 ft (265 m) in the southeast area to 680 ft (207m) towards the northwest, reflecting the effect of structural growth of the Qater Arch and possible diapiric movements in the offshore areas. In general, the Middle Jurassic succession shows good lateral continuity in terms of lithology and thickness, which reflects a passive tectonic period and a steady structural growth during deposition.

### *Tuwaiq Mountain Formation*

The Tuwaiq Mountain Formation is the lowest sequence of the Upper Jurassic section. It is Oxfordian in age, overlying the Middle Jurassic (Callovian) Araej Formation (Fig. 3). The contact is an unconformity surface placed at the first appearance of the clean, dense aphanitic lime mudstone interval directly above the slightly argillaceous packstone–grainstone section of the Araej Formation. The contact between the Araej and the Tuwaiq Mountain is marked by an irregular yellowish limonitic surface with some anhydrite-filled fractures and nodules marking the beginning of the first stratigraphic sequence within the Upper Jurassic section.

The structural growth of the Mender–Lekhwair high and Qatar Arch played an important role in controlling the stratigraphy (Fig. 4). Sea level rose rapidly during deposition of the Tuwaiq Mountain Formation. The carbonate production rate was

Middle East Models of Jurassic/Cretaceous Carbonate Systems
SEPM Special Publication No. 69, Copyright © 2000
SEPM (Society for Sedimentary Geology), ISBN 1-56576-075-1, p. 229–248.

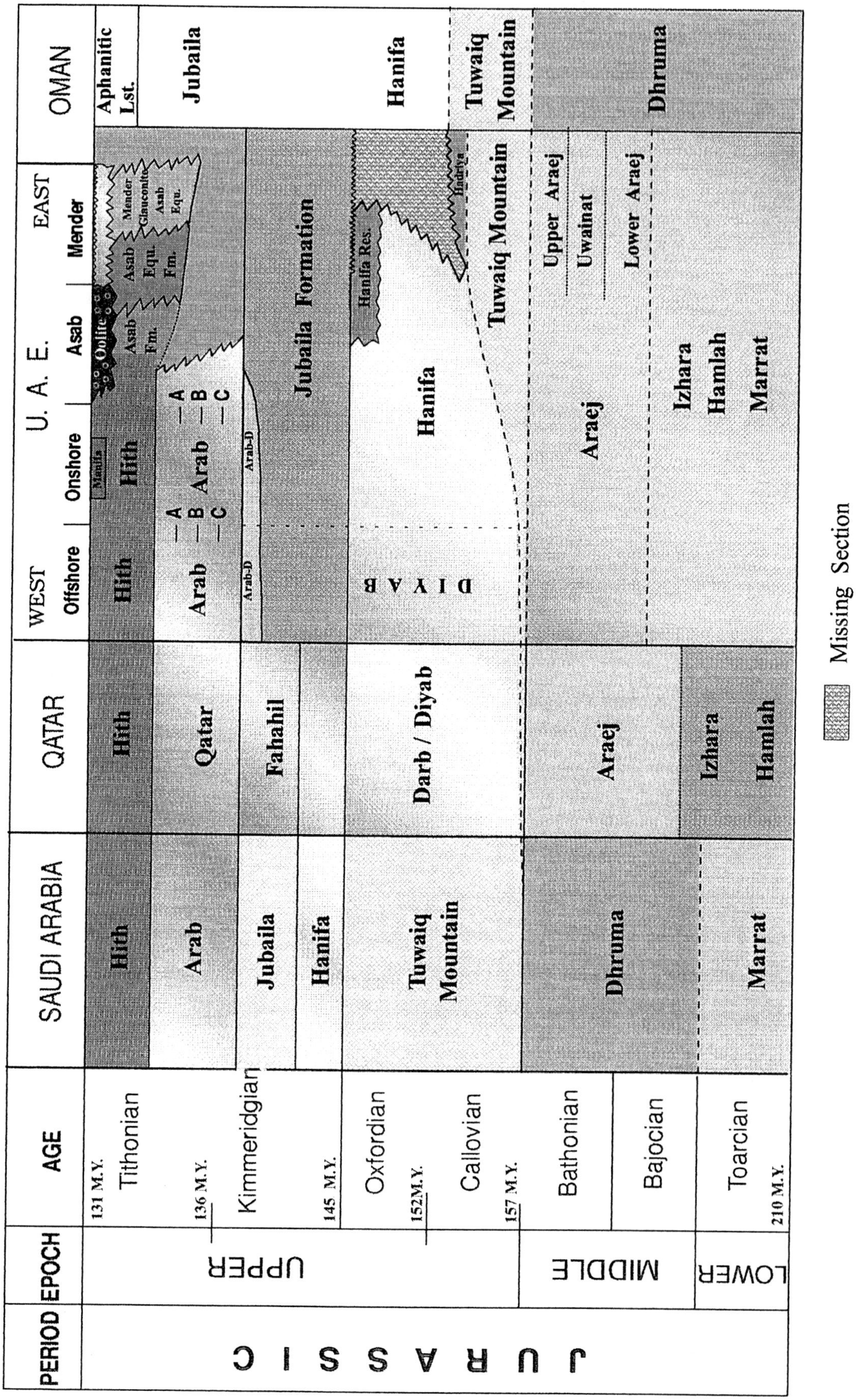


FIG. 1.—Schematic stratigraphic relations of Jurassic formations in eastern Arabia.

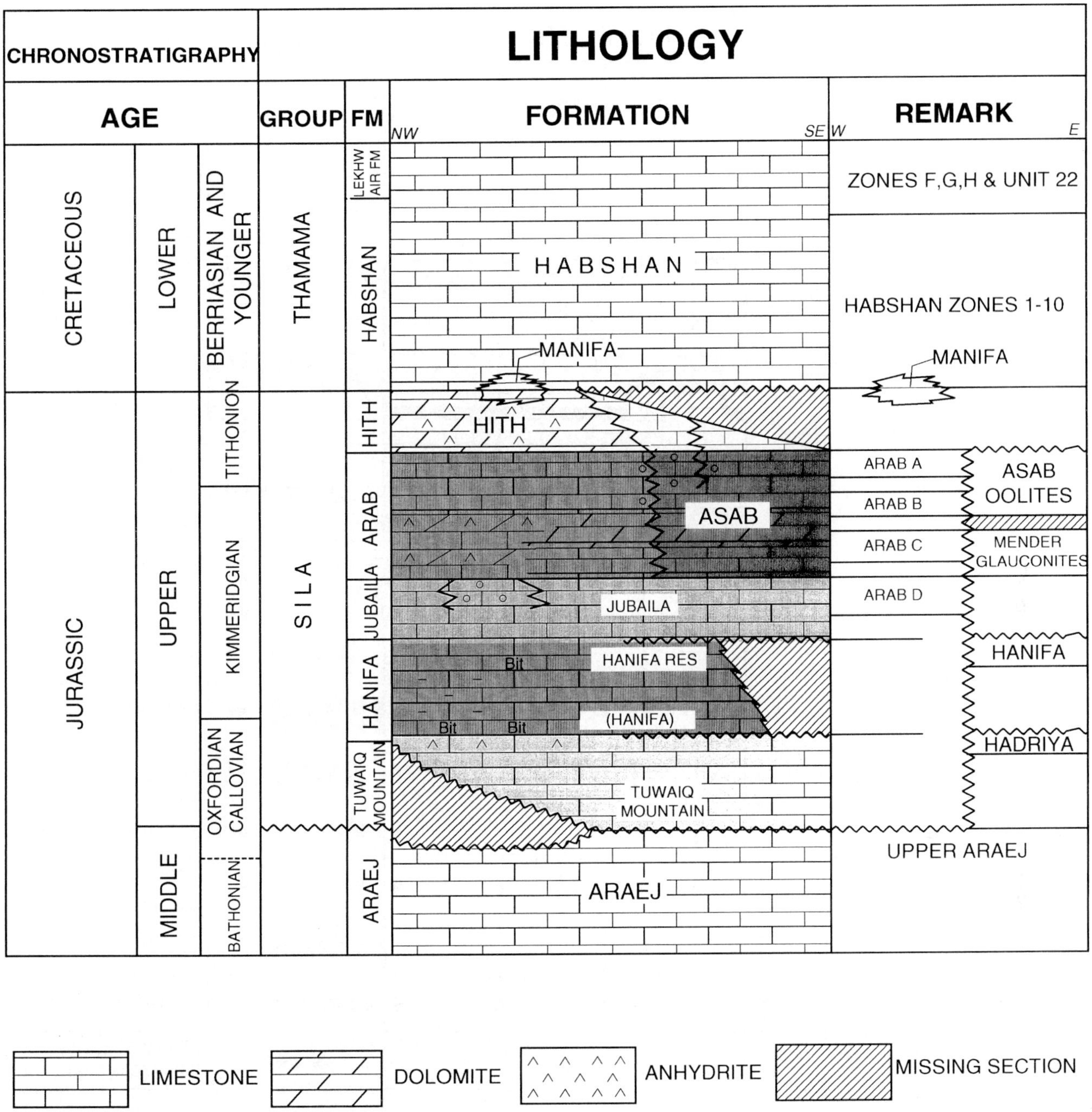


FIG. 2.—Lithostigraphic section of the Upper Jurassic in onshore Abu Dhabi.

high in the southeast section because of the structural growth in the Mender–Lekhwair high, whereas the rate of sedimentation was very low in the western and northern areas of Abu Dhabi, thus giving rise to a starved basin (Fig. 5). The thickness of the Tuwaiq Mountain section ranges from less than 50 ft (15 m) in the northern offshore area to over 700 ft (210 m) in the southeast area.

The structural growth in the Mender area resulted in deposition of a clean aphanitic dense lime mudstone interval (lower part), which gradually thickens towards the southeastern area, and it ended by the deposition of the Hydria reservoir at the top of the formation. This porous section pinches out basinward (Fig. 6).

The upper part of the Tuwaiq Mountain attests to a drop in sea level, which resulted in subaerial exposure (unconformity surface) in the southeast area and an arid hypersaline environment in the restricted intrashelf basin to the west. As a result, anhydrite streaks are the lateral equivalent of the unconformity surface in the western and northwestern area (Figs. 7, 8). However, locally, because of the increase in water depth (semi-restricted), a marly to highly argillaceous lime mudstone streak was developed as a lateral equivalent to the anhydrite streaks, thus marking the end of the first stratigraphic sequence in the Upper Jurassic succession.

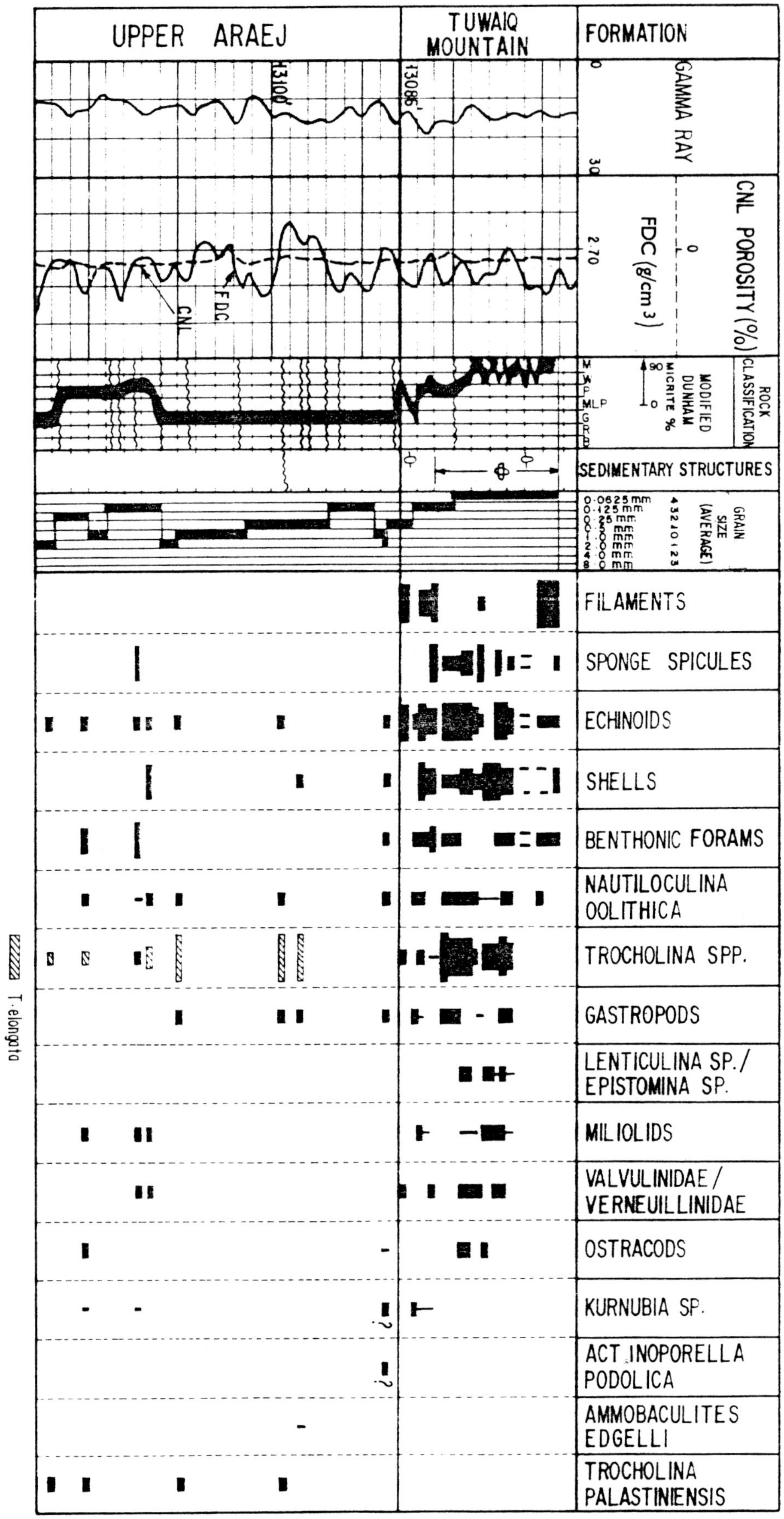


FIG. 3.—Log characteristics and fossil distribution in the Upper Araej Member and the Tuwaiq Mountain Formation in onshore Abu Dhabi.

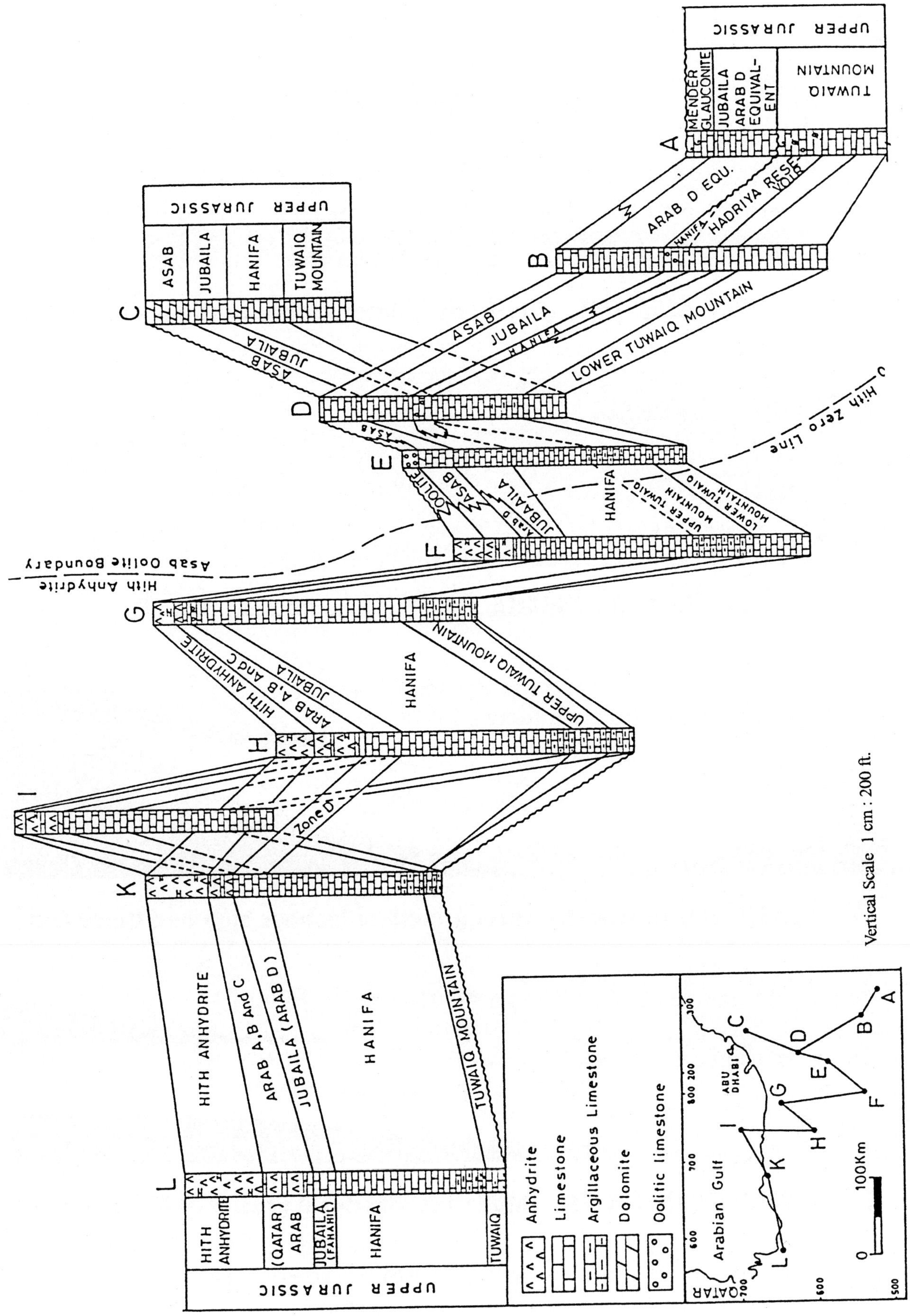


FIG. 4.—West–east lithostratigraphic section of the Upper Jurassic in onshore Abu Dhabi.

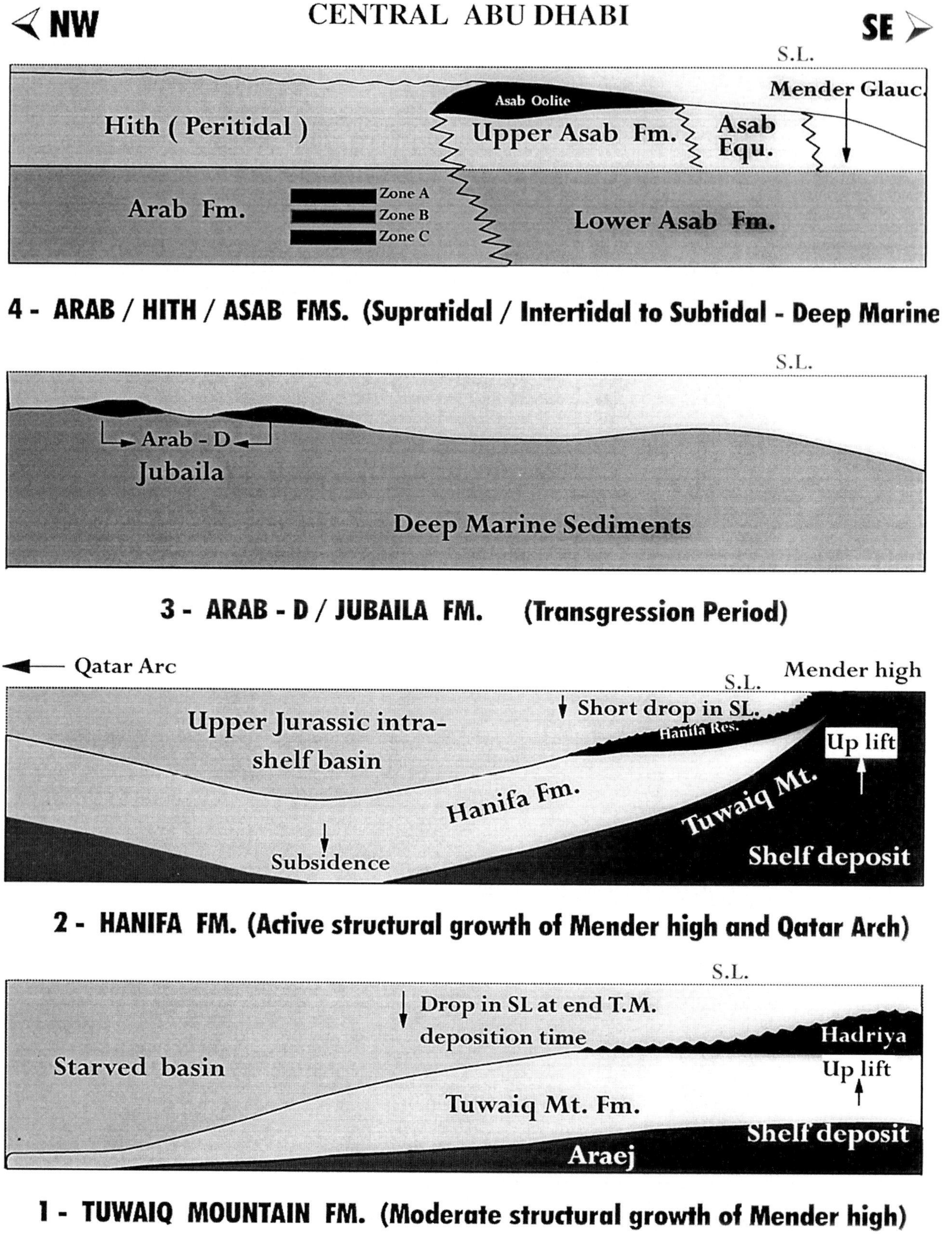


FIG. 5.—Northwest–southeast schematic sections showing depositional relationships of the Upper Jurassic formations in onshore Abu Dhabi.

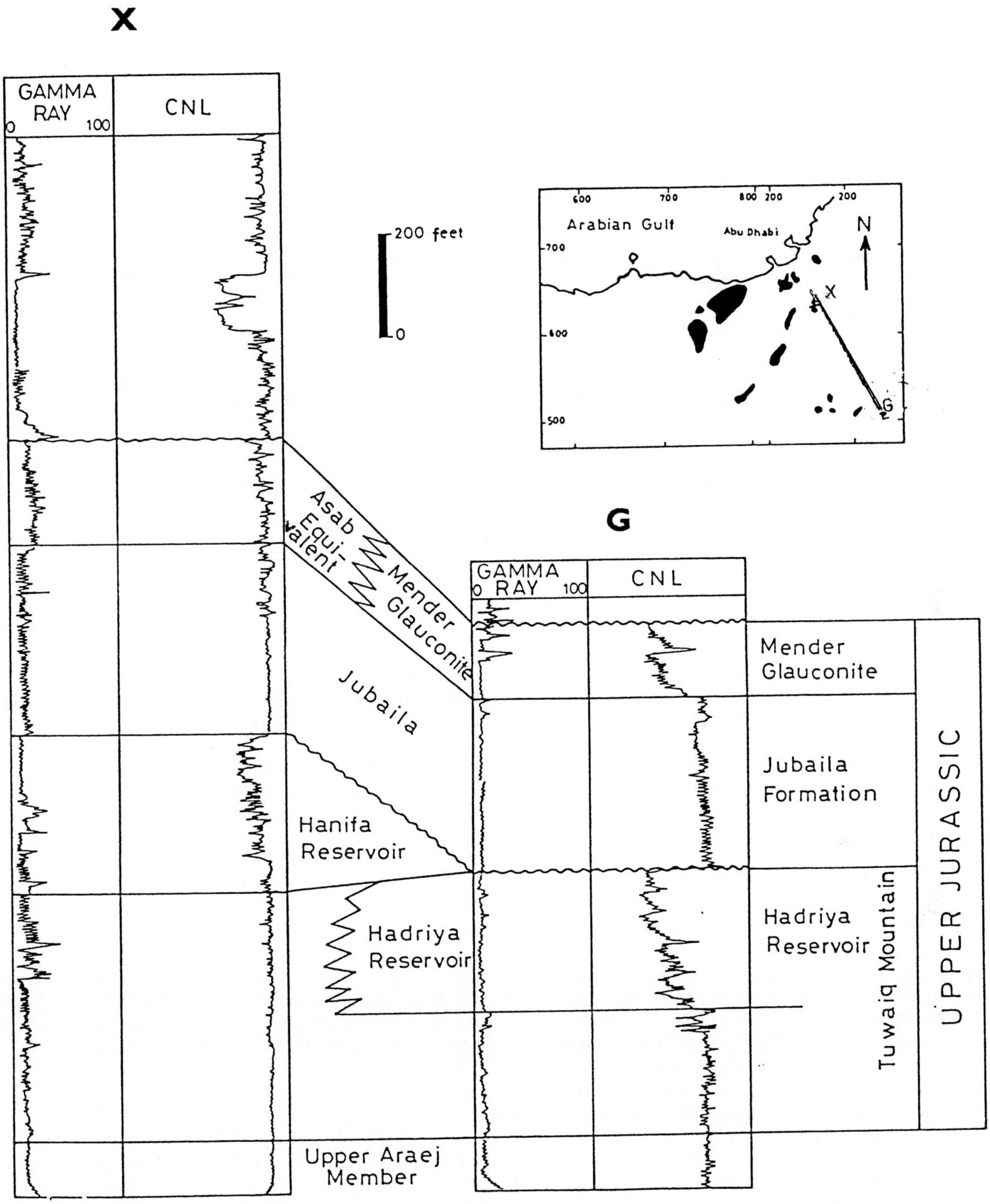


FIG. 6.—Log correlation section of Upper Jurassic formations along Mender and Agrab areas in southeastern Abu Dhabi.

### *Hanifa Formation*

The Late Jurassic intra-shelf basin was structurally active during the Late Oxfordian to Early Kimmeridgian time because of the structural growth of the Qatar Arch and the Mender–Lekhwair High. This intrashelf basin was short-lived as it was filled during the subsequent sea-level rise, post top Tuwaiq Mountain unconformity surface. During Oxfordian–Early Kimmeridgian time organic rich sediments accumulated in the intrashelf basin to form the main source-rock intervals in the Upper Jurassic section (Fig. 8). The depocenter is located in the western and southwestern areas of Abu Dhabi and east of the Qatar Arch. The thickness of the Hanifa Formation ranges from about 1000 ft (304 m) in the southwestern are to less than 700 ft (213 m) in the northern offshore, with gradual thinning towards the southeastern area.

The Hanifa reservoir was developed in the southeastern area (Fig. 6) as a result of upward shallowing associated with contin-

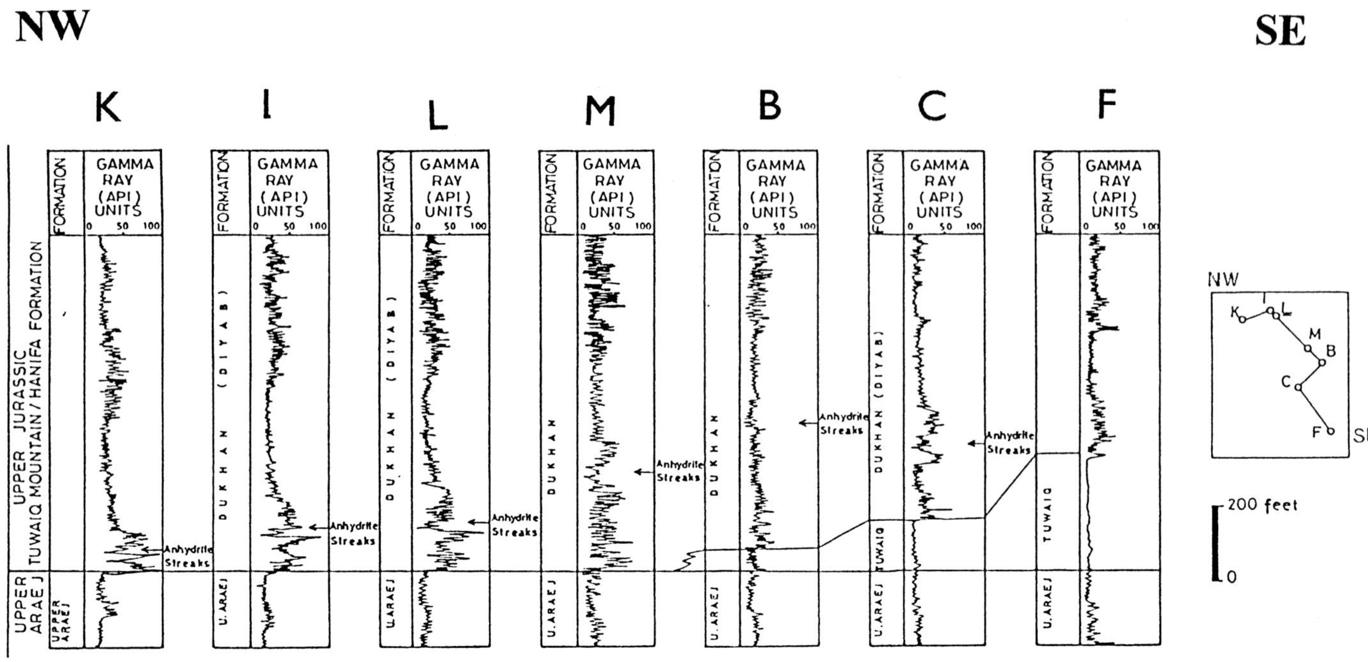


FIG. 7.—Northwest–southeast section of gamma-ray log correlation on top of the Tuwaiq Mountain anhydrite streaks in Abu Dhabi.

ued emergence of the Mender High. In the Mender area however, the Hanifa Formation is missing because of either nondeposition or erosion (Fig. 8). In the offshore area, because of contemporaneous salt doming, the section becomes less argillaceous and more porous and locally dolomitic toward the northeastern area. The top of the Hanifa Formation could be an unconformity surface associated with a drop in sea level and indicating a second stratigraphic sequence.

### *Jubaila Formation*

The Jubaila Formation is the upper part of what was previously called the Dukhan Formation. Its thickness ranges from 220 ft (67 m) in the southwest to 550 ft (168 m) in central Abu Dhabi (Fig. 9). After deposition of the Hanifa Formation a rise in sea level led to the deposition of slightly argillaceous and aphanitic deep-water lime mudstone facies throughout the area (Fig. 4) Because of shallowing to the northeast, toward the edge of the Hanifa intrashelf basin, however, the lithofacies passed into dolomitic limestone. The top of the Jubaila Formation has been picked at the base of the Arab Formation in the western area and at the base of the Asab Formation or Asab equivalent in the eastern area. The GR/CNL response shows some cyclicity and development of porosity in the offshore area which may be related to local structural growth (salt diapirs). The top of the Arab D reservoir might locally be considered to be an unconformity surface marking the end of the third sequence in the Upper Jurassic section.

### *Arab Formation*

In the central and western part of Abu Dhabi as well as most of Arabia, an arid climate prevailed during early Tithonian time and resulted in extensive evaporite deposits interbedded with carbonate beds (Fig. 5; Murris, 1980).

The Arab Formation ranges in thickness from zero beyond the pinchout of the Hith in central Abu Dhabi to more than 400 ft (122 m) in the western and northwestern area (Figs. 4, 10). It was deposited as a major regressive sequence that accumulated in a peritidal to supratidal environment. In general the Arab Formation consists of a series of depositional cycles each consisting of a lower bioclastic lime mudstone–wackestone and an upper thin tidal-flat dolomitic mudstone that grades upward into supratidal anhydrite units. It is dominated by porous dolomite and dense limestones interbedded with several streaks/beds of anhydrite. A relative drop in sea level in the western area resulted in deposition of the Arab-D oolitic packstone–grainstone section, and to the east (towards the deep marine area) a subtidal environment prevailed (Fig. 11). Three porous members of the Arab C, B, and A (in ascending order) are separated by anhydrite intervals (Fig. 12). The three zones show porosity development with a thinning trend towards the east (the Hith Anhydrite edge). It becomes difficult, therefore, to differentiate and correlate among the three Arab members toward the zero isopach.

### *Hith Formation*

The Hith Anhydrite is the top seal to the Arab hydrocarbon-bearing zones in the western part of Abu Dhabi. It is Late Tithonian in age and represents the final regressive supratidal stage of the major Late Jurassic depositional cycle. Toward Late Jurassic time, the climatic conditions became arid (restricted basin or sabkha) in western Abu Dhabi and shoal to deep marine conditions prevailed towards the eastern area and Oman.

The massive Hith Anhydrite thins gradually eastward across central Abu Dhabi and passes laterally into intertidal carbonate sediments of the Asab Formation (Fig. 13). The Hith seal plays an important role in the distribution of the Arab and Thamama group hydrocarbons. Most of the Thamama zones are hydrocarbon bearing in the absence of the anhydrite, and the Arab zones are hydrocarbon bearing where the anhydrite seal is present. Both the Arab and the Thamama are hydrocarbon bearing in the vicinity of the Hith Anhydrite edge (Alsharhan and Kendall, 1994).

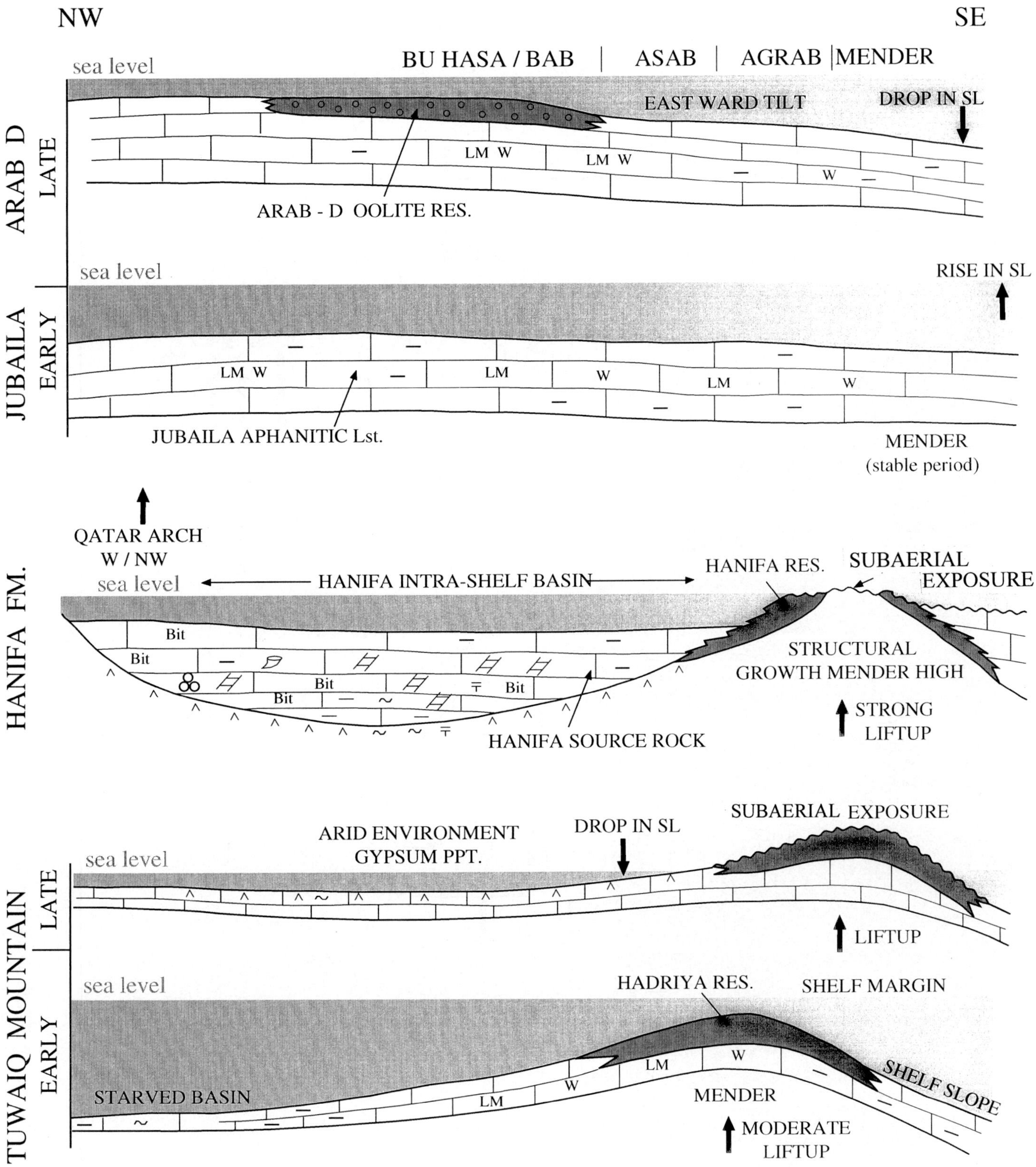


FIG. 8.—Northwest–southeast sketch depositional models of Upper Jurassic reservoirs in onshore Abu Dhabi.

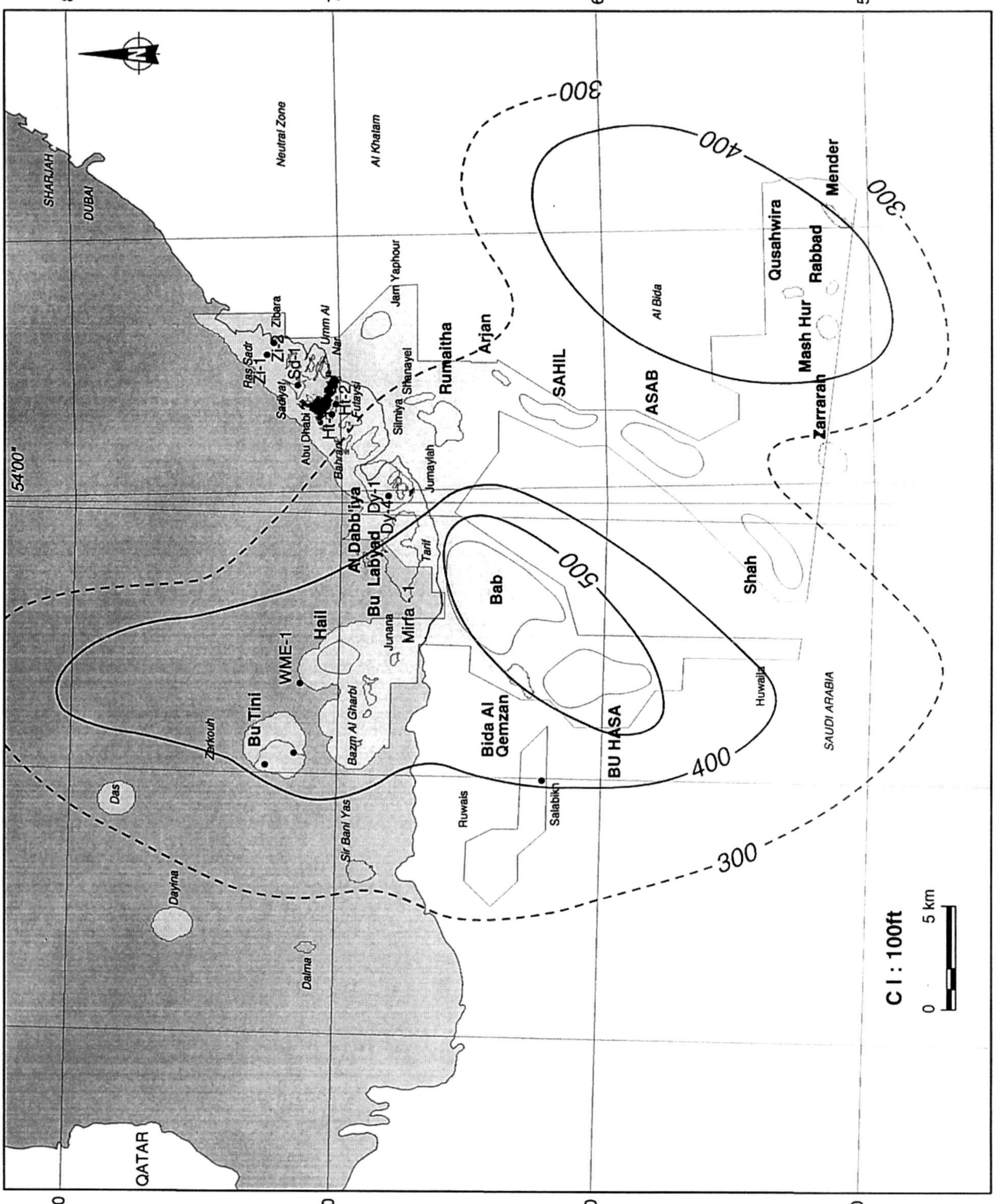


FIG. 9.—Isopach (ft) map of the Jubailah Formation in Abu Dhabi.

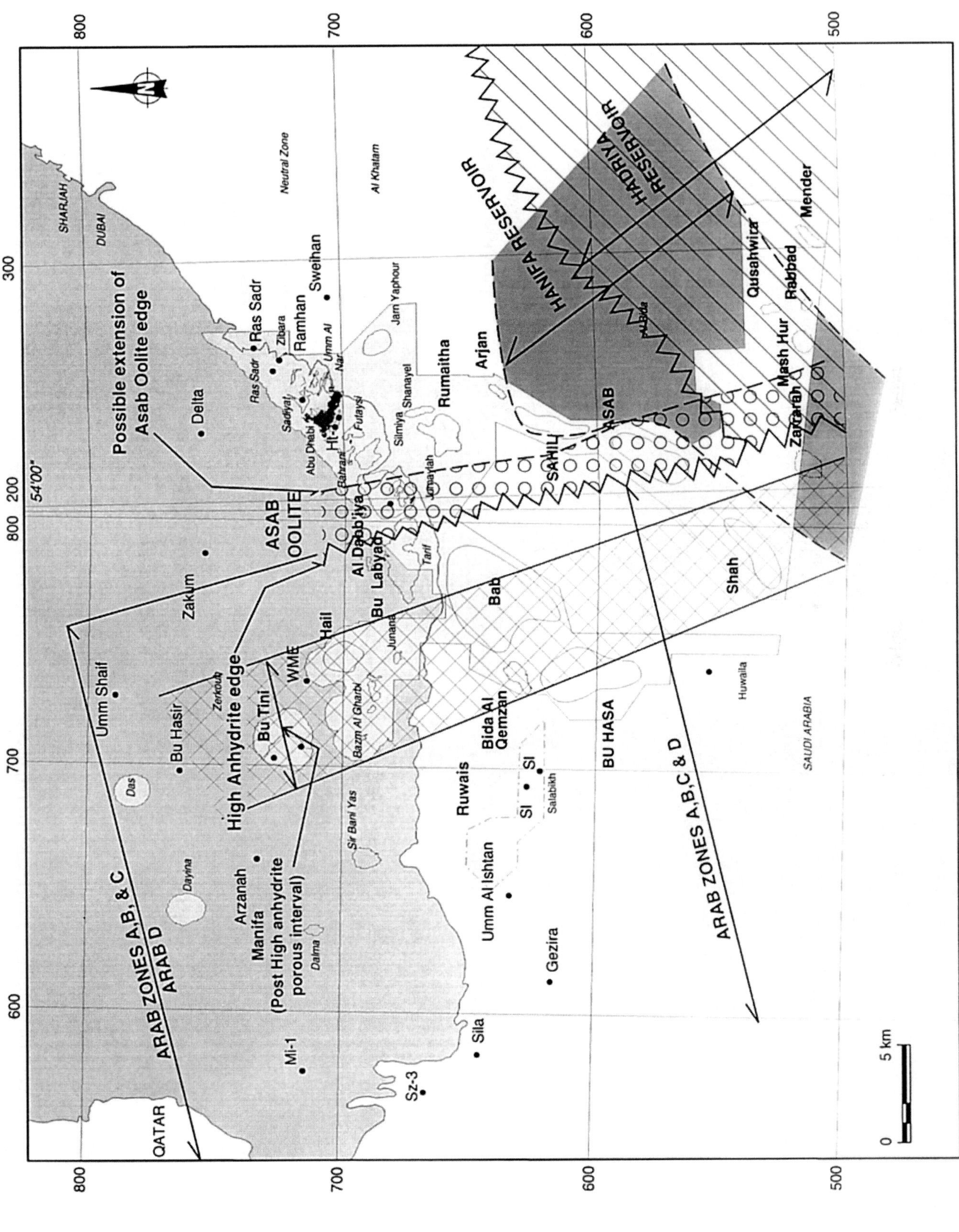


FIG. 10.—Areal distribution of Upper Jurassic porous zones and potential reservoir in onshore Abu Dhabi.

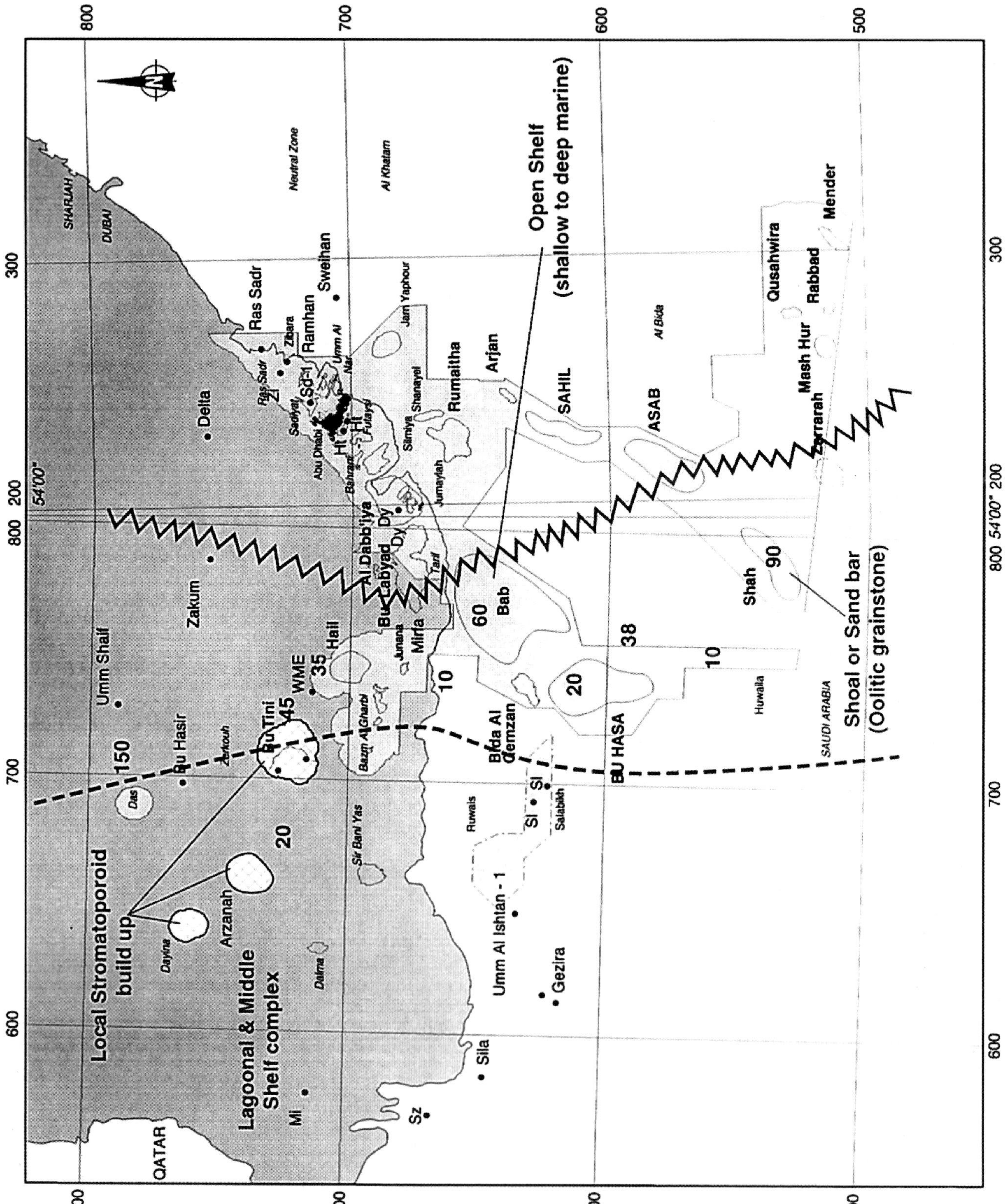


FIG. 11.—Depositional setting and thickness (ft) of Arab D reservoirs in onshore Abu Dhabi.

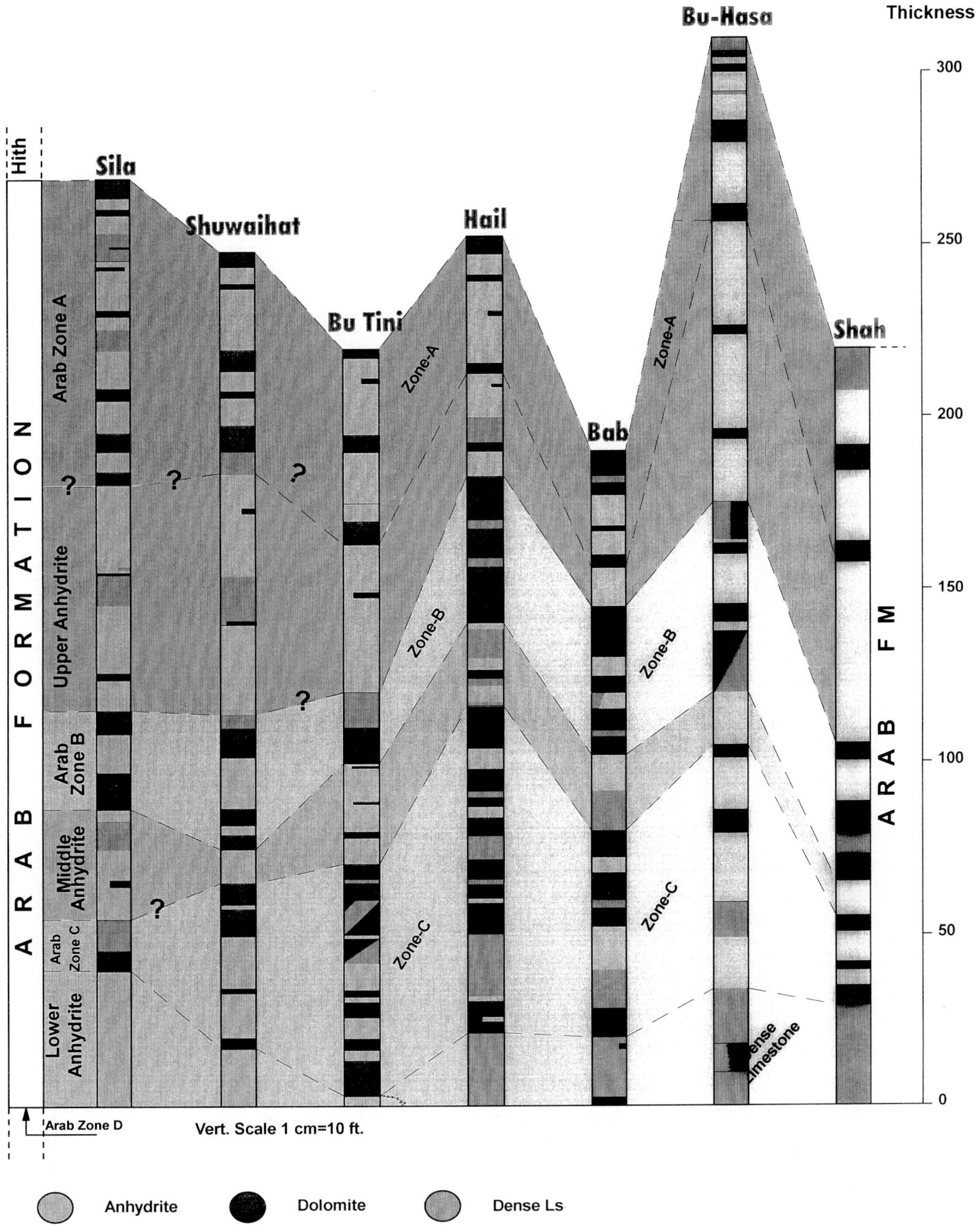


FIG. 12.—Lithostratigraphic correlation of Upper Jurassic Arab formations in Abu Dhabi.

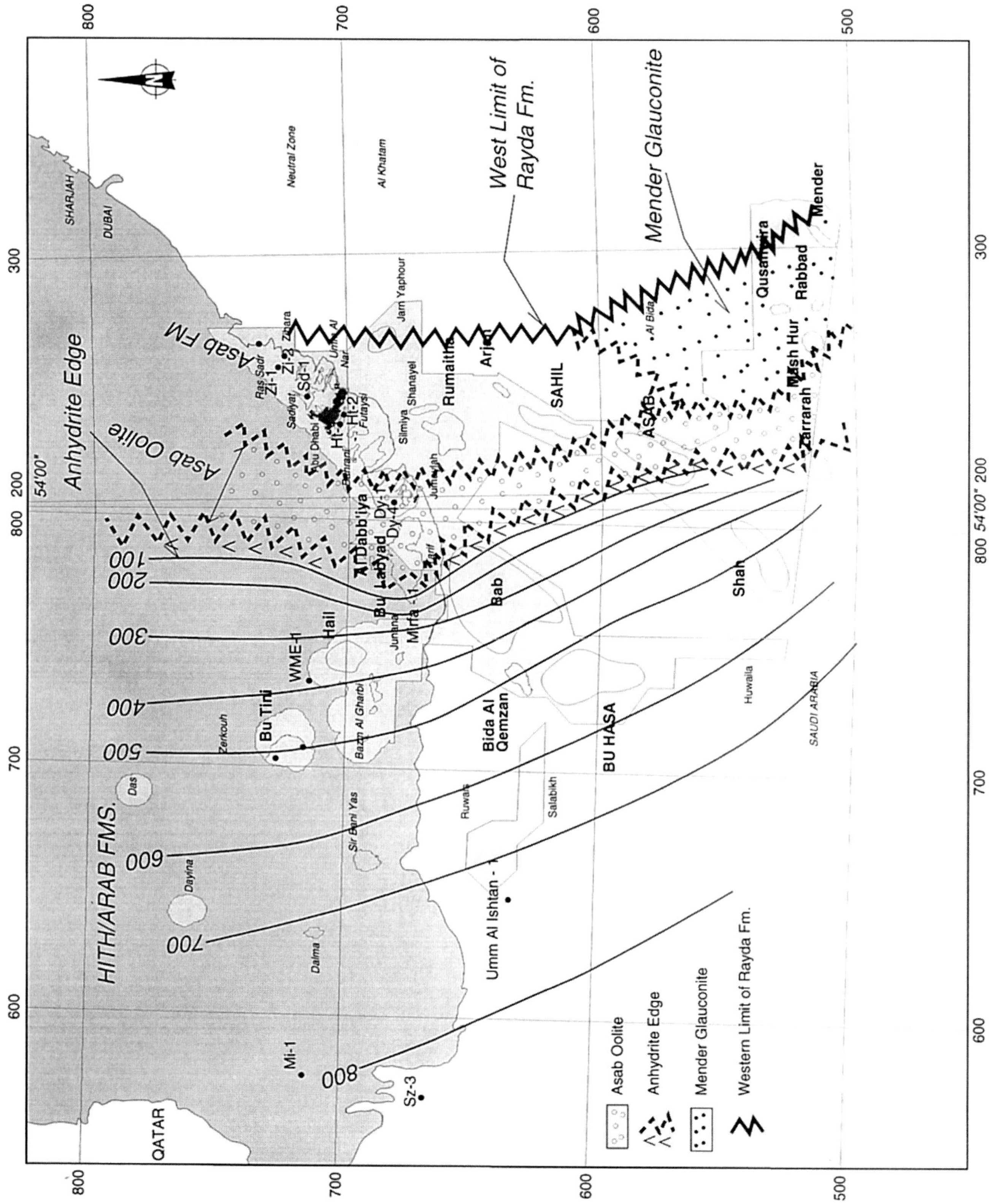


FIG. 13.—Isopach (ft) and facies map of Upper Jurassic Arab, Hith, and Asab formations in onshore Abu Dhabi.

The isopach map of the Hith Formation shows thickness variations from zero in central Abu Dhabi to 480 ft (146 m) in the western area. Also, differentiation between the Hith and the Arab Formations becomes less clear towards the Hith Zero line (Fig. 14). The top of the Hith (Jurassic) marks the fourth stratigraphic sequence in the Upper Jurassic section of western Abu Dhabi. East of the Hith anhydrite edge, the depositional environment ranged from shallow open marine to shoal, grading eastward to outer-shelf deeper open marine. The Asab Formation or Asab-equivalent strata were deposited in this area. The formation is encountered in a north–south trend updip of a carbonate ramp, and it is interpreted as the eastern lateral equivalent of the Hith and the Arab formations (Fig. 14).

The lower part of the Asab Formation is an argillaceous, hard and dense lime mudstone grading upward into dolomite with scattered anhydrite nodules, which in turn is overlain by the Asab Oolite near the shoreline margin. East of the Asab Formation, toward the relatively deeper marine environment, the Asab-equivalent section was deposited as lime mudstone facies. In the southeastern area, a deeper marine environment prevailed where glauconitic lime mudstone facies was deposited as a lateral equivalent of the Asab Formation (Fig. 13). The isopach map of the Asab Formation and its equivalent ranges from zero near the Hith anhydrite edge to 350 ft (107 m) in the onshore area (Fig. 15). The top of the Asab Formation (top Jurassic) could be an unconformity surface, which marks the fourth stratigraphic sequence in the eastern area of Abu Dhabi.

## RESERVOIR DEVELOPMENT

The Upper Jurassic section contains several reservoir intervals. The eustatic changes in sea level and the structural growth of the Mender–Lekhwair High together with local offshore salt diapirs played an important role in reservoir development (see de Matos, 1994; Alsharhan, 1989). Eight reservoirs are developed in the Upper Jurassic succession in onshore Abu Dhabi (Fig. 1): the Hadriya at the top of the Tuwaiq Mountain; the Hanifa at the top of the Hanifa Formation in the southeast area; the Asab Oolite in the central area (at the top of the Asab Formation); the Manifa reservoir at the top of the Hith Formation; the Arab Members A, B, and C in the Arab Formation; and Arab D at the top of the Jubaila Formation in the western area (west of the Hith Anhydrite edge) (Fig. 10).

### *Arab Reservoirs*

Ten to sixteen relatively porous limestone or dolomite streaks/beds can be recognized throughout the Arab Formation (Fig. 12). Three porous streaks are identified in the Arab Member A, and Members B and C are variable in cyclicity and reservoir development. Therefore the differentiation between the Arab members is less clear towards the Hith anhydrite edge. The Arab zones are hydrocarbon bearing in most of the Abu Dhabi developed anticlines.

The top of the Arab D is picked at the base of the lowest anhydrite streak or dense limestone bed of the Arab Formation. The lithology is mainly oolitic packstone–grainstone of variable reservoir development in the western area. The base, although not clear, is picked at the base of the oolite section (Fig. 12). The Arab D is the uppermost part of the Jubaila Formation.

In the eastern area, the facies equivalents of the Arab D were deposited in open-shelf, deep marine conditions. The thickness ranges from 10 to 90 ft (3 to 30 m) near the Hith anhydrite edge. The Arab D porosity is variable, from fair to good, and the permeability ranges from 10 to 30 md.

### *Asab Oolite and Manifa Reservoirs*

The Asab oolite reservoir, 15 to 170 ft (5 to 52 m) thick, constitutes the uppermost part of the Asab Formation. The section consists of well cemented oolitic packstone–grainstone underlain by dolomite, with scattered anhydrite nodules. The Asab oolite section exhibits good porosity (17–19%), and low permeability (0.9–3 md). The oolitic limestones were deposited in a north–south trend in a high-energy setting parallel to the Hith anhydrite edge (Fig. 14). The hydrocarbon generated from the Hanifa source rock in the presence of Hith anhydrite cap rock migrated vertically through the Asab oolite and laterally westward to the Lower Cretaceous Thamama Group (Fig. 16).

The Manifa reservoir is relatively porous wackestone–packstone interval, about 40 ft (12 m) thick, locally developed at the top of the Hith Formation, with average porosity of 6% and permeability of less than 4 md. The Manifa tested hydrocarbons in a few wells in the Abu Dhabi onshore area.

### *Hanifa Reservoir*

The Hanifa reservoir is the uppermost porous section at the top of the Hanifa Formation. It consists of a peloidal to oolitic packstone interval restricted to parts of the southeastern area of Abu Dhabi. It is predicted to form an updip stratigraphic pinchout along the western flank of the west Mender High. The thickness of the Hanifa reservoir reaches approximately 165 ft (50 m) in the vicinity of the Mender–Lekhwair High, and the average log porosity is 15%.

### *Hadriya Reservoir*

The Hadriya reservoir forms the uppermost porous section of the Tuwaiq Mountain in the southeast area. The interval is composed of medium- to coarse-grained, cemented pelletal packstones and grainstones. The Hadriya reservoir is up to 400 ft (122 m) thick; it has an average porosity of about 18%, and permeability ranges between 10 to 50 md.

## SOURCE-ROCK POTENTIAL

The Upper Jurassic section contains a triplet of source rocks, reservoirs, and seals in the western area of Abu Dhabi. Based on the available geochemical information, the Hanifa is the main source rock for the Upper Jurassic and Lower Cretaceous hydrocarbons in the area. The Hanifa source rock entered the oil-generation window some 100 million years ago during middle Cretaceous time (Fig. 17).

During the Late Cretaceous the depth of burial of the Hanifa source rock was about 9000 ft (2743 m) in the Falaha trough, with average reservoir temperature of 140°C (peak of oil generation). The gradual temperature increase in the Falaha and Ghurab troughs is considered the main factor for gas generation. The Late Cretaceous to Early Tertiary is the possible time for oil migration from the Hanifa source rock, and the gas-migration phase took place possibly during Tertiary time as a second phase of hydrocarbon generation. Souring occurred in a subsequent phase of geothermal processes when $H_2S$ was generated in the Falaha and Ghurab troughs, in the presence of the Hith Anhydrite and the Arab Formation. The Hanifa source rock [about 260 ft (79 m) thick] may have generated more than $150 \times 10^9$ barrels of oil from the Central Abu Dhabi area (40,000 $km^2$), assuming source rock potential (S1 + S2) equal to 10 kg per ton, 75% conversion factor, 2.55 $g/cm^3$ density, and 50% oil-migration efficiency (Taher, 1997; Hawas and Takezaki, 1995).

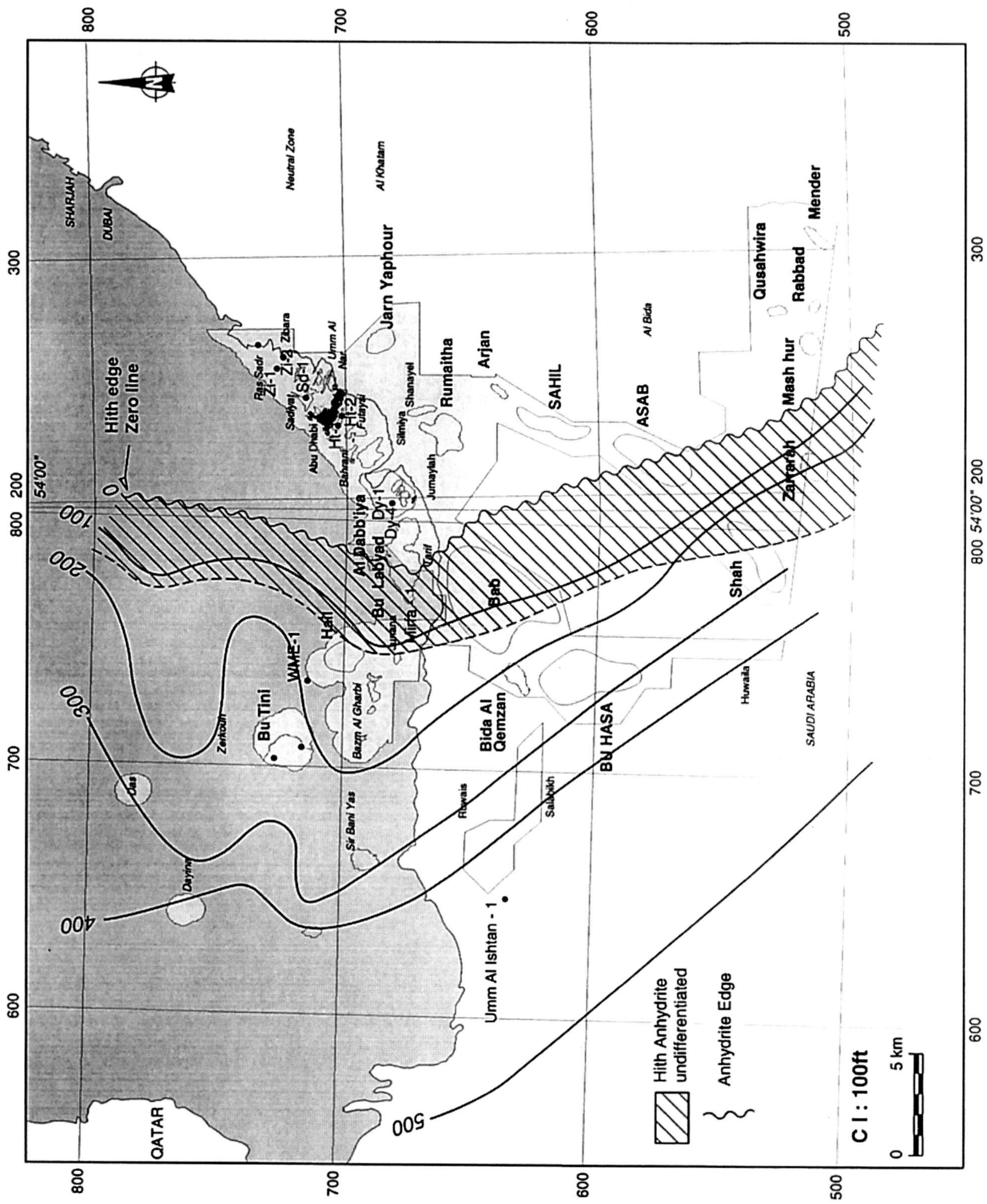


FIG. 14.—Isopach (ft) map of Hith Formation in onshore Abu Dhabi.

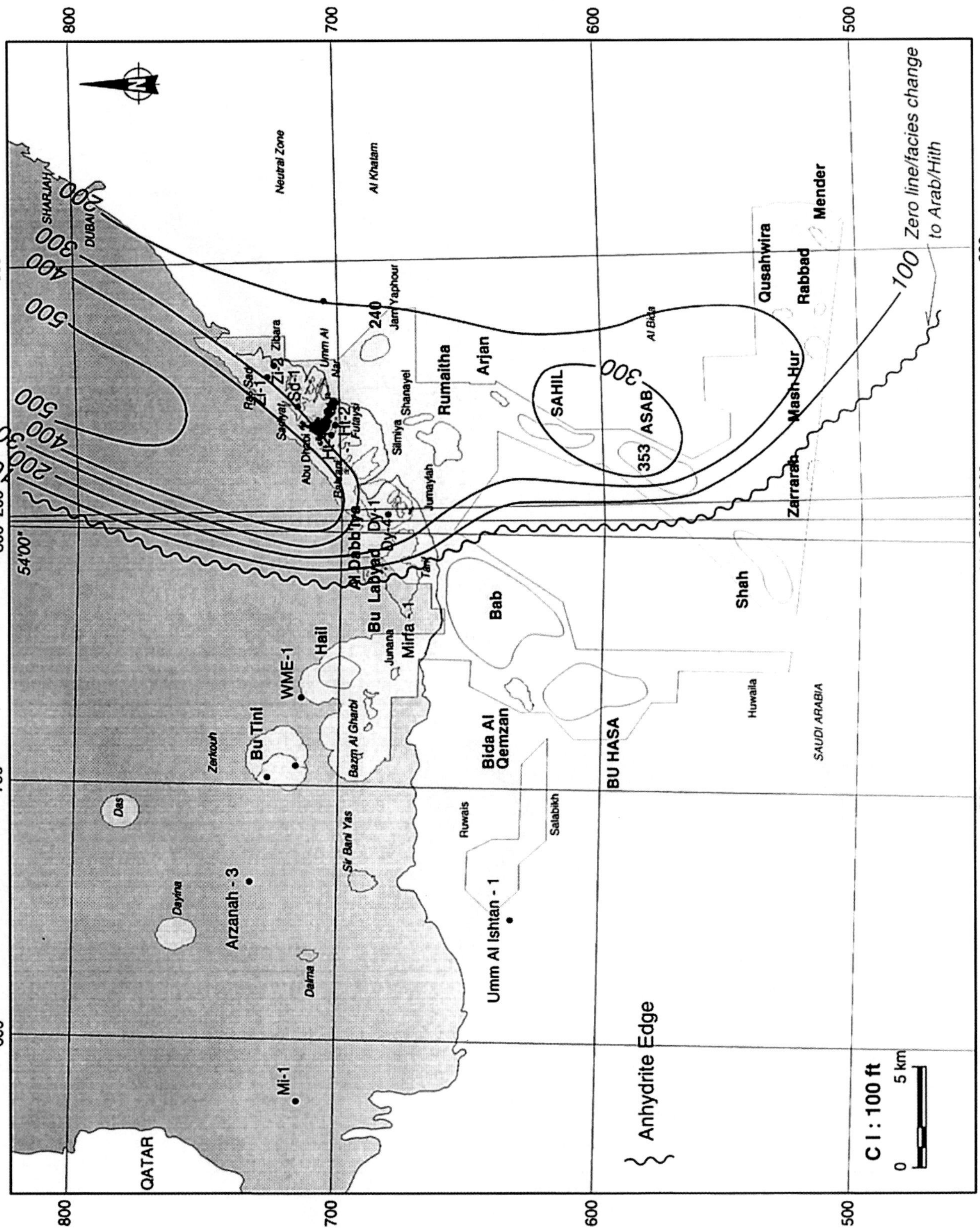


FIG. 15.—Isochore map of Asab Formation in onshore Abu Dhabi.

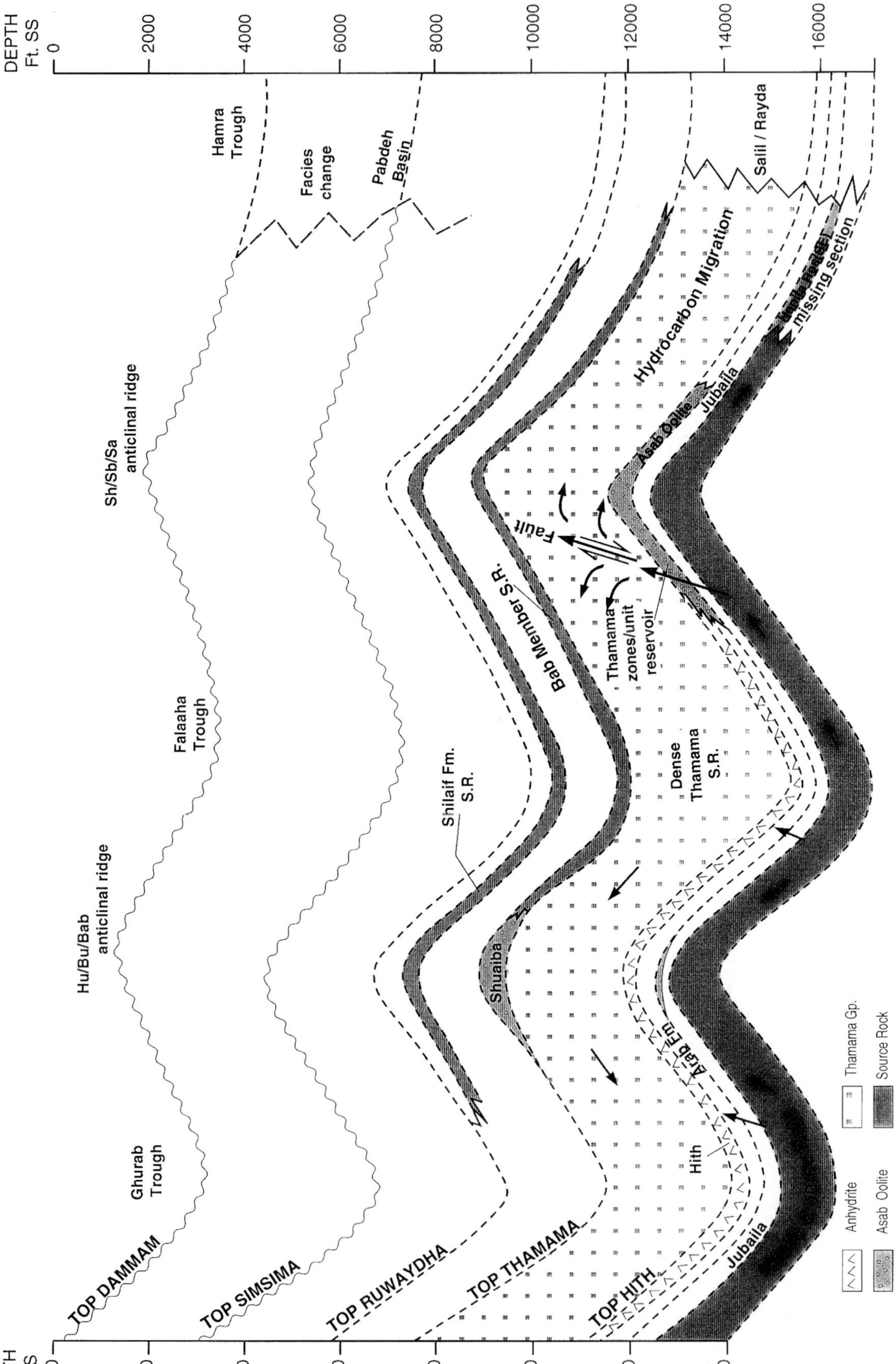


FIG. 16.—Northwest–southeast stratigraphic cross section of Eocene–Jurassic in onshore Abu Dhabi.

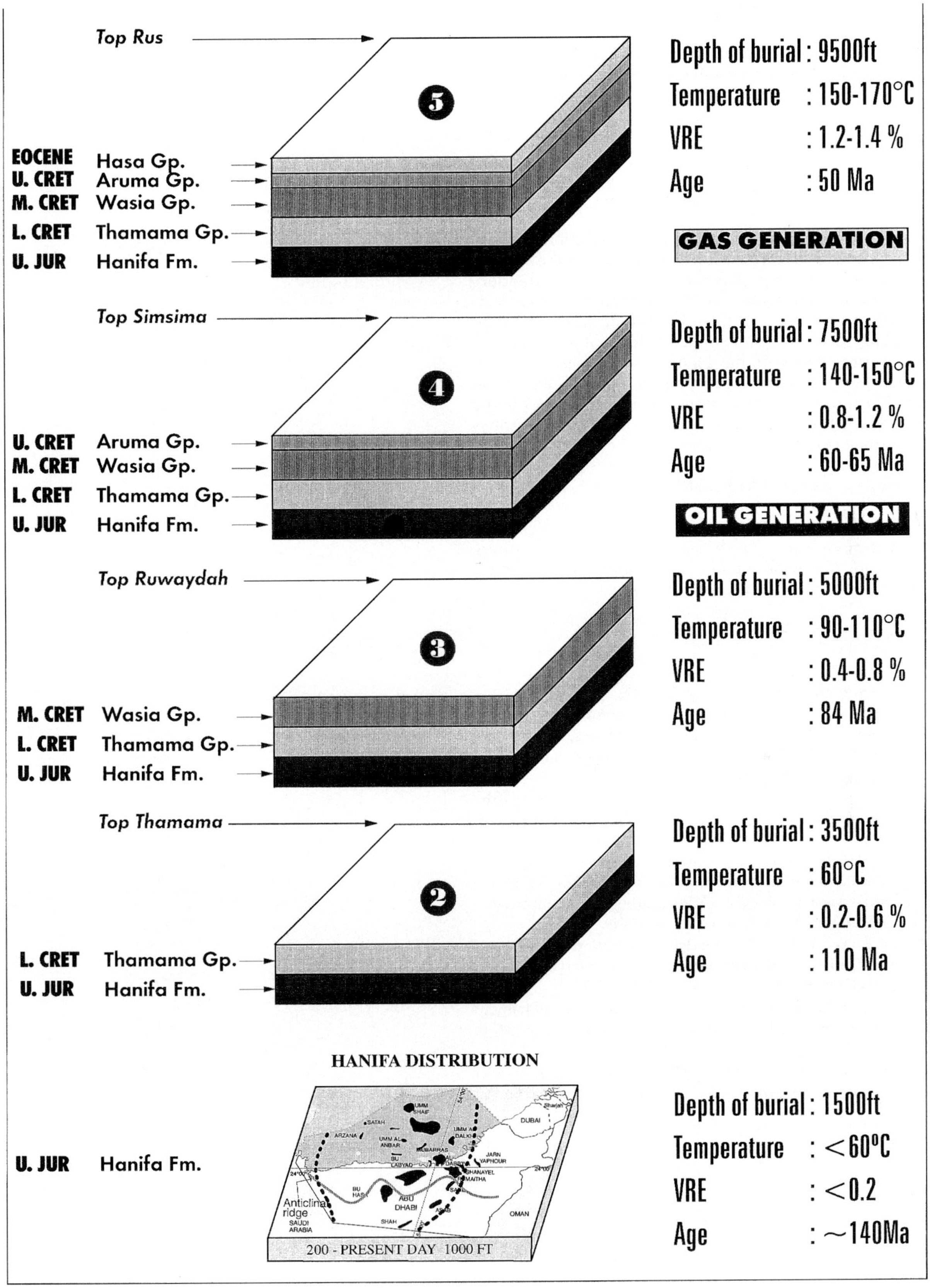


FIG. 17.—Burial models of Hanifa source rock and hydrocarbon generation in onshore Abu Dhabi.

## SEAL POTENTIAL

The Hith Anhydrite is a major hydrocarbon seal for the Arab Zones. It represents the final regressive supratidal stage of the last major Jurassic depositional cycle in the western area of Abu Dhabi. In the eastern area, east of the Hith edge, dense limestone of the Habshan Formation forms the seal for the Asab Oolite reservoir.

## CONCLUSIONS

The Upper Jurassic section exhibits lateral variation in thickness, lithology, and reservoir development because of the eustatic changes in sea level and the structural growth of the Mender–Lekhwair High and the Qatar Arch. The Upper Jurassic witnessed a relative subsidence of the intrashelf basin between the the Mender High and the Qatar Arch, which resulted in deposition of the source rocks of the Hanifa Formation. The Late Cretaceous to Early Tertiary is the main phase of hydrocarbon (oil) generation and migration from the Hanifa source rock to the Upper Jurassic and Lower Cretaceous reservoirs. The Hith Anhydrite forms a major seal for the Arab zones.

## ACKNOWLEDGMENTS

The authors wish to thank the Management of ADNOC and ADCO for permission to publish this paper. Also the authors thank R.W. Scott, A.S. Al sharahan, I.A. Al Jalal, and B. Granier for their fruitful comments on editing this paper.

## REFERENCES

ALSHARHAN, A.S., 1989, Petroleum Geology of the United Arab Emirates: Journal of Petroleum Geology, v. 12, p. 253–288.

ALSHARHAN, A.S., AND KENDALL, C.G.ST.C., 1986, Precambrian to Jurassic rocks of Arabian Gulf and adjacent area: Their facies, depositional setting and hydrocarbon habitat: American Association of Petroleum Geologists, Bulletin, v. 70, p. 977–1002.

ALSHARHAN, A.S., AND KENDALL, C.G.ST.C., 1994, Depositional setting of the Upper Jurassic Hith Anhydrite of the Arabian Gulf: American Association of Petroleum Geologists, Bulletin, v. 78, p. 1075–1096.

DE MATOS, J.E., 1994, Upper Jurassic–Lower Cretaceous stratigraphy: The Arab, Hith and Rayda Formations in Abu Dhabi, *in* Simmons, M.D., ed., Micropalaeontology and Hydrocarbon Exploration in the Middle East: London, Champan & Hall, p. 81–99.

HAWAS, M., AND TAKEZAKI, H., 1995, A model for migration and accumulation of hydrocarbons in the Thamama and Arab reservoirs in Abu Dhabi: Geo'94,Gulf PetroLink, Bahrain, v. 2, p. 483–495

MURRIS, R.J., 1980, Middle East: strtatigraphic evolution and oil habitat: American Association of Petroleum Geologists, Bulletin., v. 64, p. 597–618.

TAHER, A.K., 1997, Delineation of organic richness and thermal history of Thamama Group in east Abu Dhabi: A modelling approach for oil exploration: GeoArabia, v. 2, p. 65–88.

# STRATIGRAPHY AND GEOCHEMISTRY OF UPPER JURASSIC DIYAB FORMATION, ABU DHABI, U.A.E.

AHMED S. AL-SUWAIDI AND AHMED K. TAHER
*Abu Dhabi National Oil Company, Exploration Division, P.O. Box: 898, Abu Dhabi, U.A.E.*
AND
ABDULRAHMAN S. ALSHARHAN AND MOHAMED G. SALAH*
*Faculty of Science, United Arab Emirates University, P.O. Box: 17551, Al Ain, U.A.E.*
**Present address: Reservoir Characterization Research & Consulting (UK) Ltd., P.O. Box: 46800, Abu Dhabi, U.A.E.*

ABSTRACT: The Diyab Formation (Oxfordian to mid-Kimmeridgian) is widely distributed in Abu Dhabi. It consists of argillaceous lime mudstones/wackestones (rich in organic matter) in the west, which change laterally eastward into oolitic packstones and grainstones. Its thickness reaches up to 1300 ft (395 m) towards the southern onshore Abu Dhabi. Geochemical evaluation of the Diyab Formation was carried out using analytical data from five different laboratories. Source rock screening indicated the presence of a rich source-rock interval in the lower part of the formation in western Abu Dhabi, which becomes a lean source-rock in eastern Abu Dhabi. The Diyab source rock contains oil- and gas-prone kerogen, and was deposited in an intrashelf basin that was enclosed by the Diyab shelfal facies sediments, which restricted water circulation, causing anoxic conditions below wave base.

A combination of geological and geochemical information was used to reconstruct burial and thermal histories of the potentially petroliferous intervals in the Diyab source rock. This source rock was sufficiently mature in the southwestern onshore Abu Dhabi to generate hydrocarbons since Late Cretaceous time. Currently, this formation lies within the gas generation window for most of onshore and southern offshore Abu Dhabi.

Comparative analysis of oil and source-rock characteristics, using both bulk and molecular parameters, carbon-isotope analysis, gas chromatography (GC), and gas chromatography–mass spectrometry (GC–MS), was carried out. The oils from both the Thamama Group and the Arab Formation have the characteristic of oils derived from carbonate source rocks. However, Arab oils are less mature and have lower concentration of saturate, and are isotopically lighter than Thamama oils at equivalent reservoir depth.

The secondary migration scheme was based on a series of structure and paleostructure maps at the top of the Arab-D level. The migration direction of the Arab Formation was controlled mainly by the presence of western synclinal structures. East of these synclines, oil migration is predominantly northeastward. The early Tertiary (50–40 Ma) was the major expulsion phase, with more than 75% of the total oils being generated from Diyab source rock, approximately 95% of which comes from the Lower Diyab section.

Crude oil and extract analysis resulted in identification of four main oil families. The Simsima-reservoired oil appears to originate from the Shilaif source rock, whereas the Thamama oils are sourced from the Shuaiba Basinal Facies (Bab Member) and the Diyab Formation. The Arab oils were sourced mainly from the Diyab Formation, and the Araej oil was sourced mainly from the Diyab Formation and possibly from an unknown pre-Diyab source rock.

## INTRODUCTION

The main objectives of this study are (1) to evaluate the qualitative and quantitative source-rock potential of the Upper Jurassic Diyab Formation, (2) to conduct comprehensive maturation modeling to establish the timing of hydrocarbon expulsion from the Diyab source rock, (3) to investigate the lateral and vertical hydrocarbon migration and distribution, and (4) to detect the genetic oil families of Abu Dhabi and relate these oils to their probable sources. A source-rock evaluation of the Diyab Formation was carried out using integrated and averaged analytical data from five different laboratories. The data used in this part of the study include more than 70 oil samples and source-rock extracts, in addition to hundreds of core chips and thousands of cutting samples from more than 40 wells covering the study area (Fig. 1). Sequence stratigraphical analyses of well and core data and lithofacies description and depositional setting were used.

Very little published work exists on the Diyab Formation. Murris (1980), Ayres et al. (1982), Droste (1990), and de Matos and Hulstrand (1995) have addressed the origin and development of the Diyab intrashelf, and Hassan and Azer (1985) discussed the source-rock potential of the formation. The lithologic description of Diyab Formation on a microscopic scale and an environmental facies interpretation were carried out by Whittle and Alsharhan (1996).

The Upper Jurassic Diyab Formation in offshore Abu Dhabi is the equivalent of the Dukhan Formation of onshore areas. It is also equivalent to the Hanifa and Jubailah Formations of Qatar and Saudi Arabia, respectively (see Alsharhan and Nairn 1997).

The Lower Cretaceous and the Middle–Upper Jurassic contain some of the most important stratigraphic successions of Abu Dhabi. The porous carbonate sections constitute the main reservoirs, whereas the dense carbonate layers act as intraformational seals and, in part, potential source rocks. The Albian Nahr Umr Shale and the Tithonian Hith Anhydrite form the regional seals for the Thamama and Arab hydrocarbon accumulations (Fig. 2).

Late Jurassic tectonic differentiation in response to the combination of a eustatic sea-level rise and differential subsidence over the platform resulted in the development of the Diyab intrashelf basin in western Abu Dhabi (Alsharhan and Kendall, 1986). This Diyab basin was filled with bituminous lime mudstones, which pass laterally into non-source, shallow-water platform carbonates (de Matos and Hulstrand, 1995; Taher, 1997). As the intrashelf basin shallowed during the Late Jurassic, platform carbonates extended across the basin to form the major reservoir units of the

Middle East Models of Jurassic/Cretaceous Carbonate Systems
SEPM Special Publication No. 69, Copyright © 2000
SEPM (Society for Sedimentary Geology), ISBN 1-56576-075-1, p. 249–271.

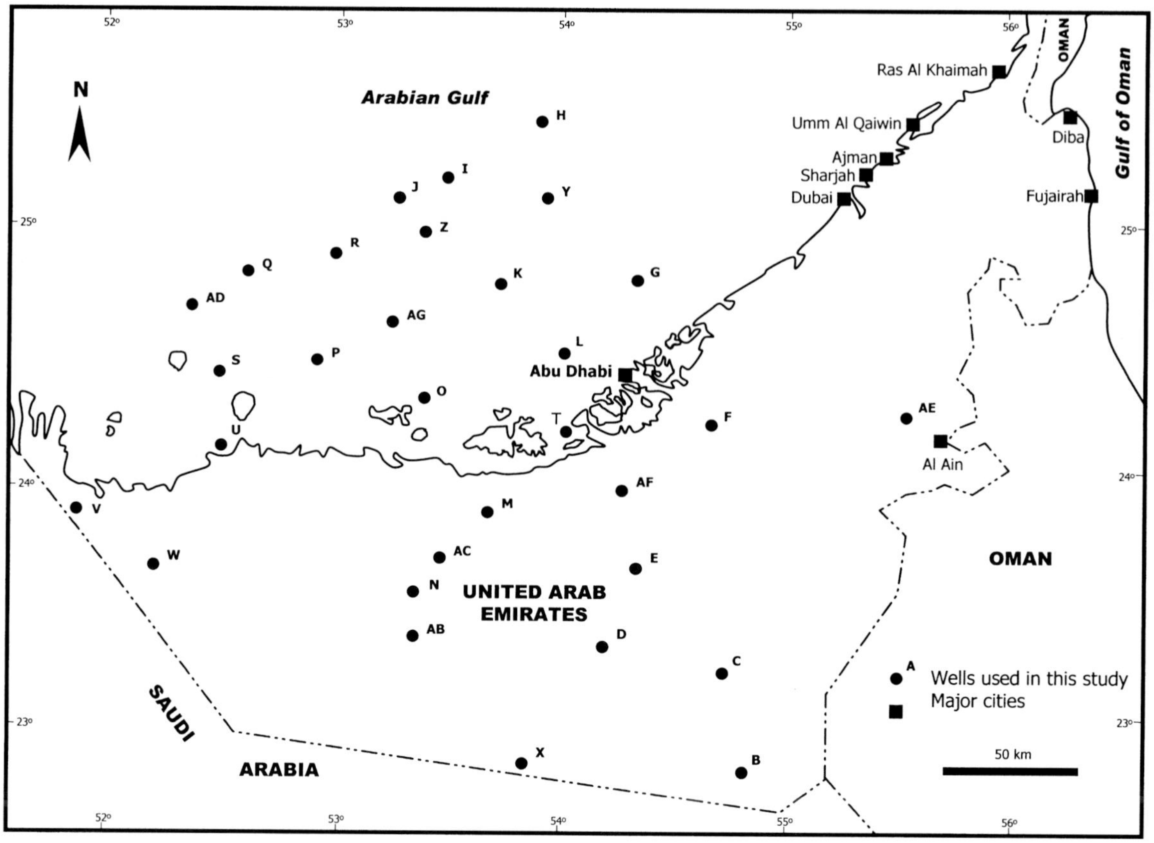


FIG. 1.—Map of the United Arab Emirates showing the area of study and the location of the wells.

Arab Formation. By Tithonian times, a sabkha environment had developed, and with increasing aridity the evaporites of the Hith Formation were widely deposited across western Abu Dhabi. A gradual return to a more humid climate in the Early Cretaceous coincided with the reestablishment of the Early and middle Cretaceous shelf carbonate depositional environment.

## STRATIGRAPHIC FRAMEWORK

The Diyab Formation is underlain by the Middle Jurassic Araej Formation and overlain by the upper Kimmeridgian Arab Formation. From paleontological and palynological evidence, the Diyab Formation has been dated as Oxfordian to mid-Kimmeridgian (Alsharhan, 1989). The Diyab Formation varies in thickness from 750 ft (229 m) at well I to 1350 ft (411 m) at well D among the studied wells (Figs. 3, 4) and is interpreted to have been deposited in an intrashelf basin, slope, shelf margin, and shelf interior (de Matos and Hulstrand, 1995; Whittle and Alsharhan, 1996; Taher, 1997).

The Diyab Formation consists of argillaceous lime mudstones wackestones (rich in organic matter) that change laterally eastward into peloidal packstones and grainstones. The formation thickens to over 1300 ft (411 m) southward (i.e., toward the south of present onshore Abu Dhabi), showing a gradational facies change, where it consists of lime mudstones. Grainstones and dolomitic limestones occur toward the east. In western and northwestern Abu Dhabi, the Diyab exhibits highly organic shales and carbonates.

Total organic carbon (TOC) values show a good correlation with the gamma-ray response (Figs. 3, 4). The regional distribution of the Diyab Formation argillaceous limestones indicates that it was deposited in an intrashelf basin that was enclosed by the Diyab shelf sediments (Figs. 5–7). However, part of the increased gamma ray can be attributed to other sources of radiation in the Diyab argillaceous limestones. In Abu Dhabi, the Diyab Formation was subdivided into three informal members: lower, middle, and upper (Figs. 3, 4) based on lithology and gamma-ray signatures (Azer, 1989; Whittle and Alsharhan, 1996; de Matos and Hulstrand, 1995).

The lower member thickens toward the east, ranging from 300 ft (91 m) in the western offshore to 700 ft (213 m) in the eastern onshore (Fig. 5A). The thickening found in the east is mainly related to the development of high-energy shelfal deposits of the Hadriya reservoir (Fig. 5B), and follows a nearly north–south belt that prograded laterally westward towards the Diyab intrashelf basin.

The lower member can be further subdivided into two lithological units: the lower and upper units. The lower unit, which is the lowermost part of the Diyab Formation, is overlain gradationally by the upper unit and is underlain by the Araej Formation (Figs. 3, 4). The lower unit in the eastern and parts of the central Abu Dhabi areas is characterized by an intermediate gamma-ray response compared with very high gamma-ray of the upper unit and low gamma-ray of the Upper Araej Member (Figs. 3, 4). The lower unit ranges in thickness from 8 ft at well S to approximately 300 ft at well X. It reaches its maximum thickness

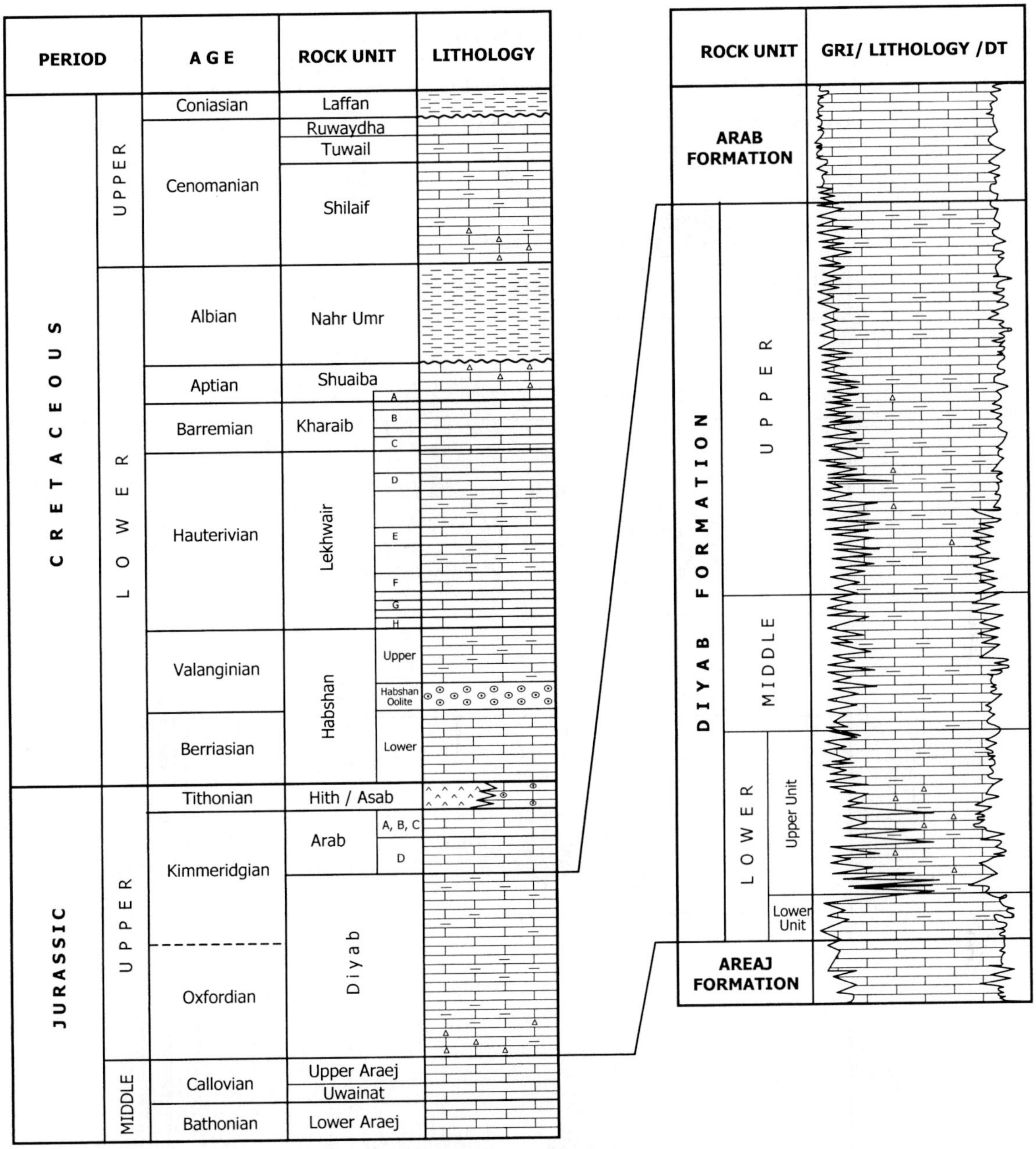


FIG. 2.—Lower Cretaceous and Upper Jurassic lithostratigraphy in Abu Dhabi, U.A.E.

in eastern Abu Dhabi (Figs. 3, 4), which suggests that the lower unit has a gradational relationship with the overlying upper unit.

The lower unit consists of peloidal bioclastic packstones and grainstones with minor wackestones over the study area. A facies change from an open-marine shelf during lower Diyab deposition to basinal slope condition is recorded in the western and central Abu Dhabi, whilst to the east the high-energy limestone section reflects deposition in an open marine shelf above wave base (Fig. 5B).

The upper unit is characterized by very high gamma-ray response in the western parts of Abu Dhabi. It is regarded as a very important potential source rock facies for the Arab and some Thamama oils in Abu Dhabi. The top of the upper unit over most of the western area and parts of the central area is easily picked at a contact of the beginning of increasing gamma-ray response with a relatively low one of the overlying cleaner middle member (Figs. 3, 4).

The upper unit has a lower gamma-ray response in the eastern area and parts of the central area than in the western area. However, the upper unit can be positively correlated in the western area on the basis of regional well-log correlation, and the interpreted paleoenvironmental model is consistent. The upper unit of the lower member in the western area (wells S, P, U, and W) and parts of the central area (well N) is composed of thinly laminated bituminous limestone, interbedded with distinct thin anhydrite layers, representing deposition in an intrashelf basin to basinal slope environment with euxinic conditions. In parts of the central and the eastern areas (wells X, D,

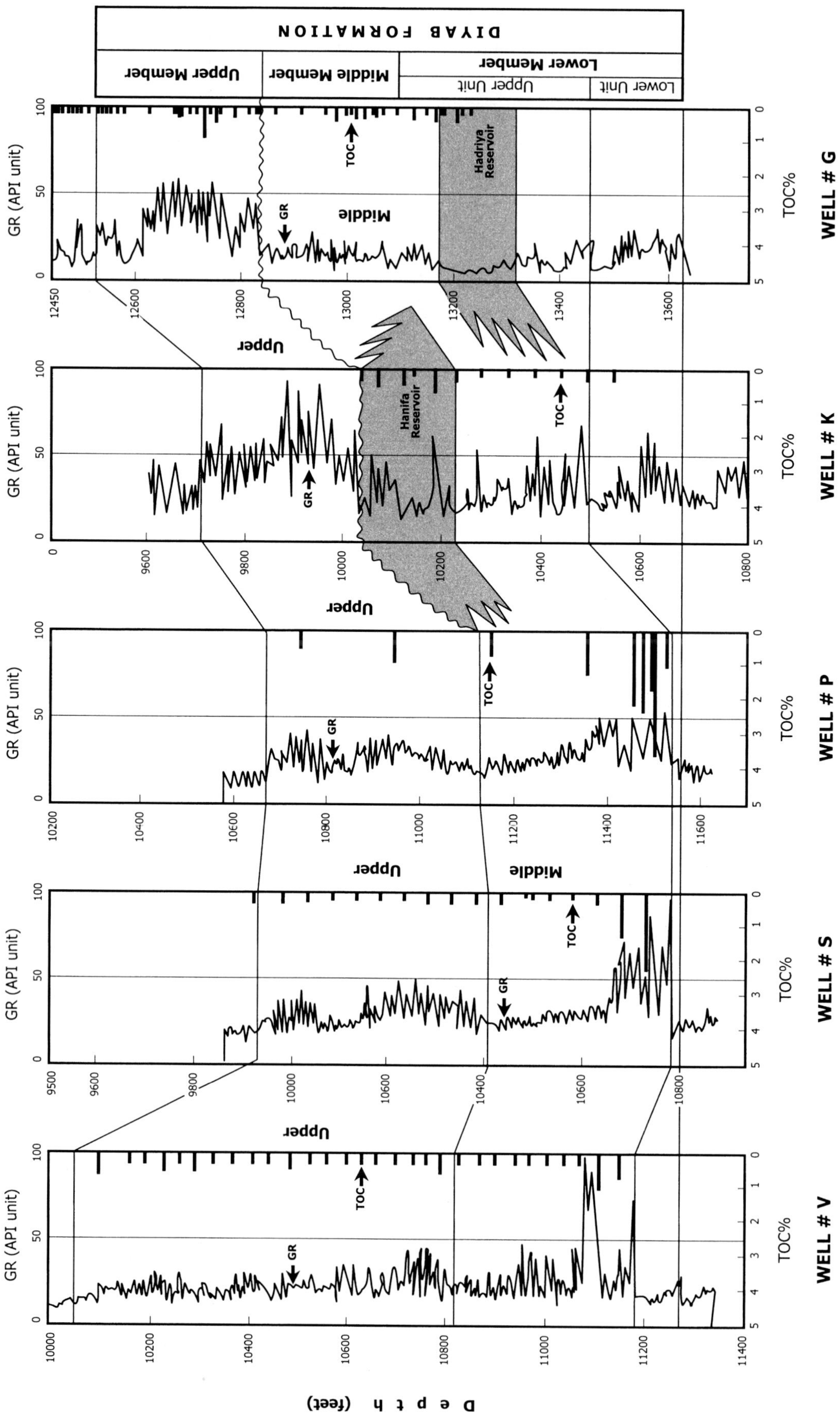


FIG. 3.—Gamma ray-total organic carbon (TOC%) correlation section, offshore Abu Dhabi.

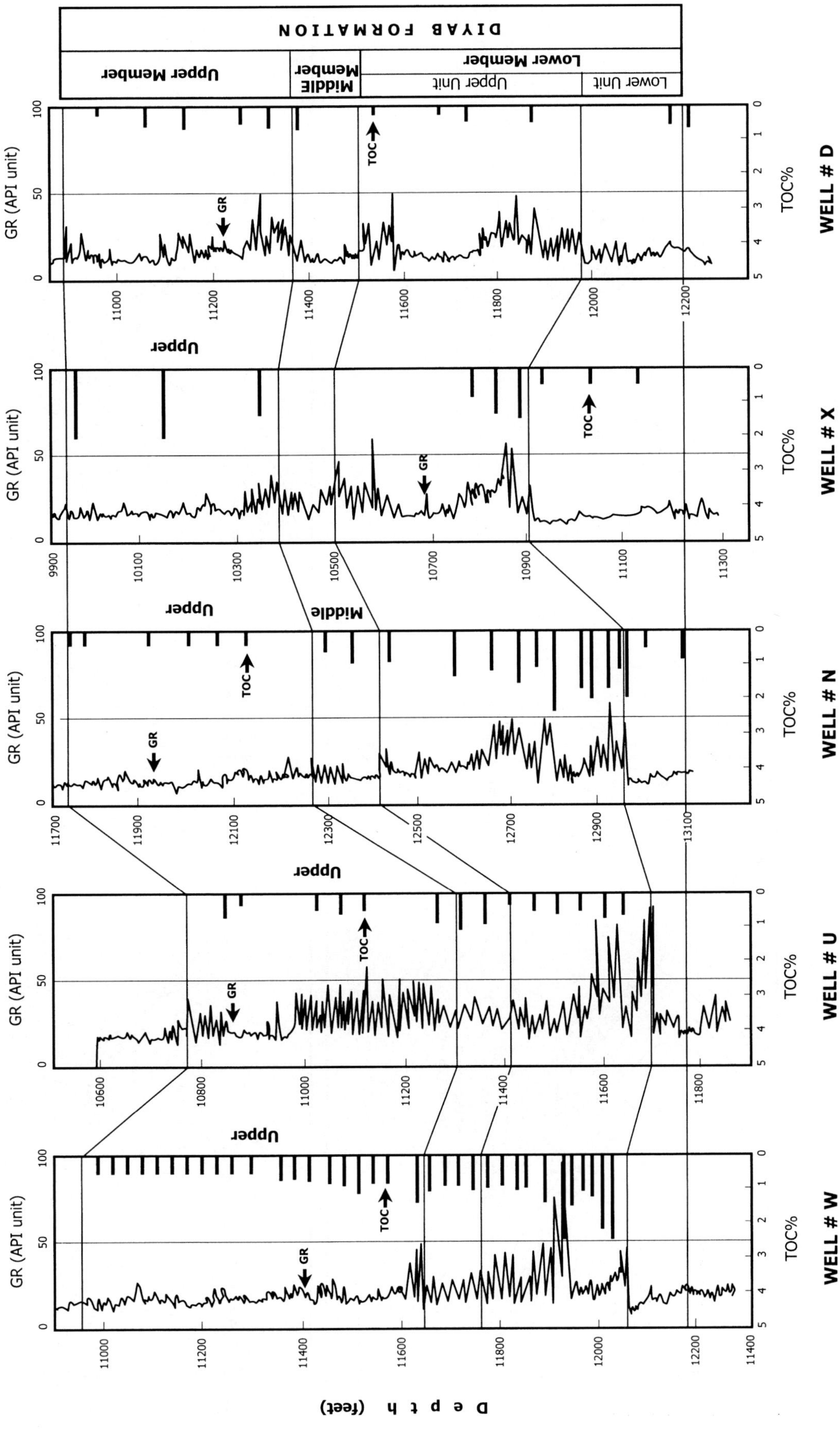


FIG. 4.—Gamma ray-total organic carbon (TOC%) correlation section, onshore Abu Dhabi.

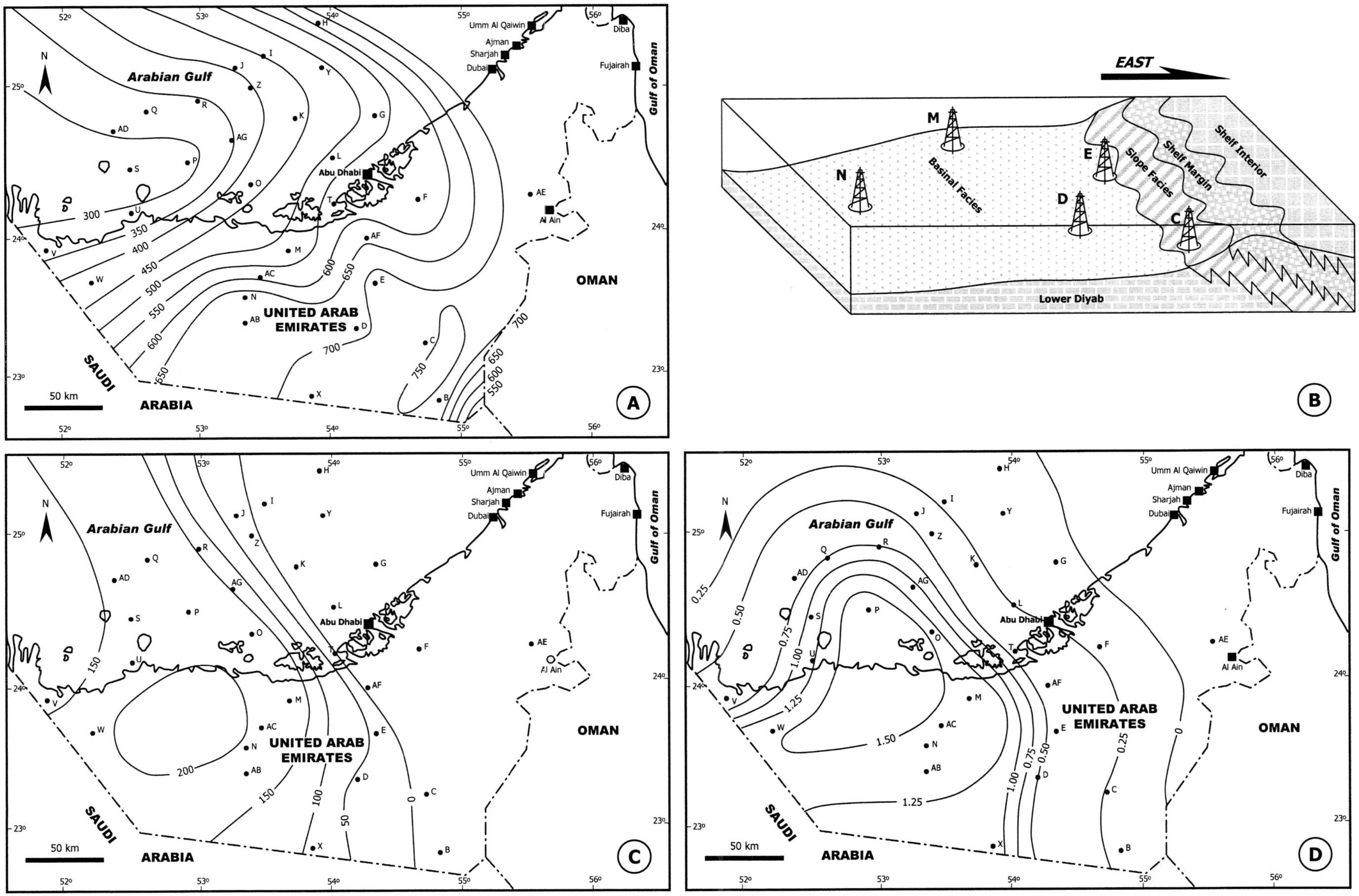


FIG. 5.—Isopach in feet (A), depositional model (B), and net source-rock thickness in feet (C), and average TOC % (D) of lower member, Diyab Formation in Abu Dhabi.

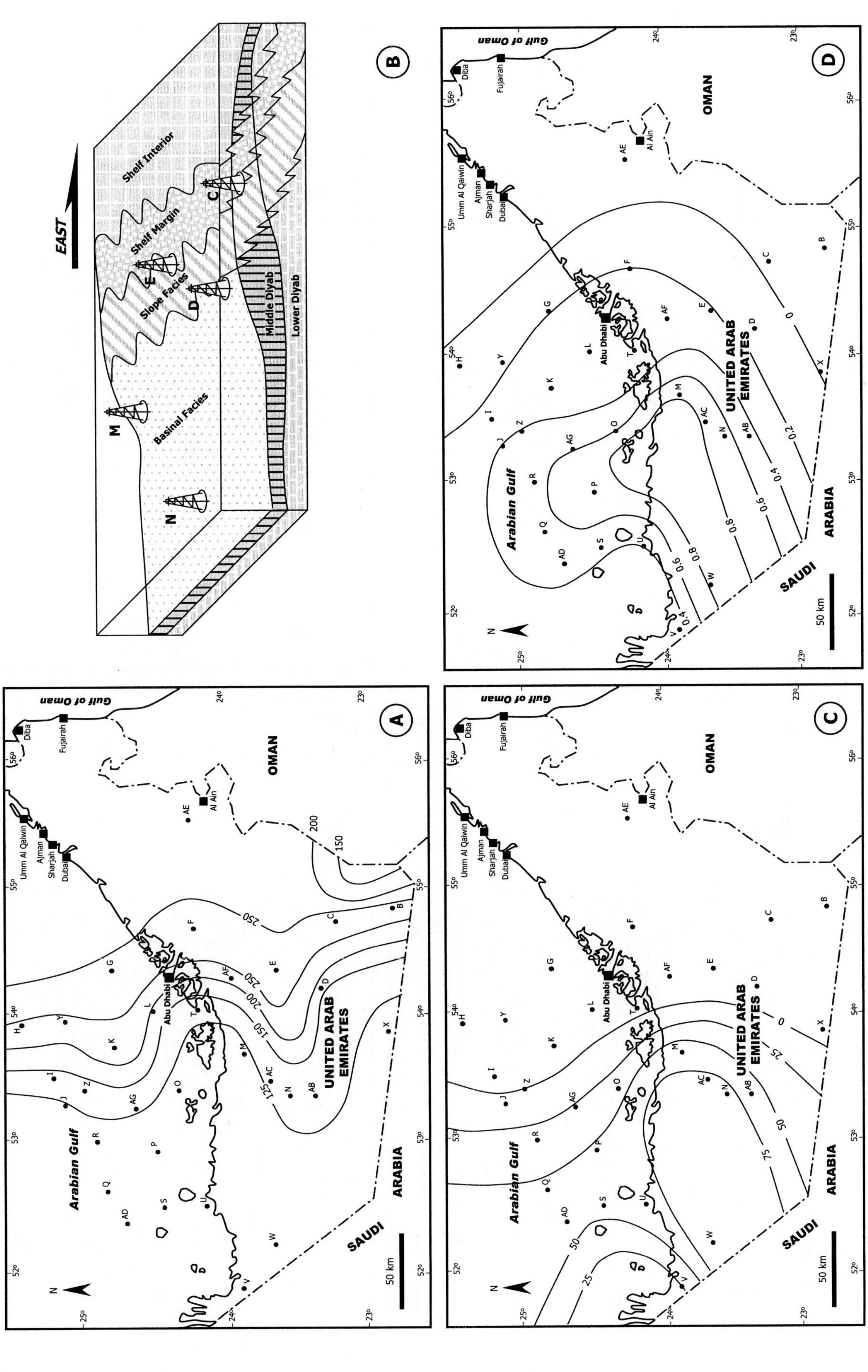


FIG. 6.—Isopach in feet (A), depositional model (B), net source-rock thickness in feet (C), and average TOC % (D) of middle member, Diyab Formation in Abu Dhabi.

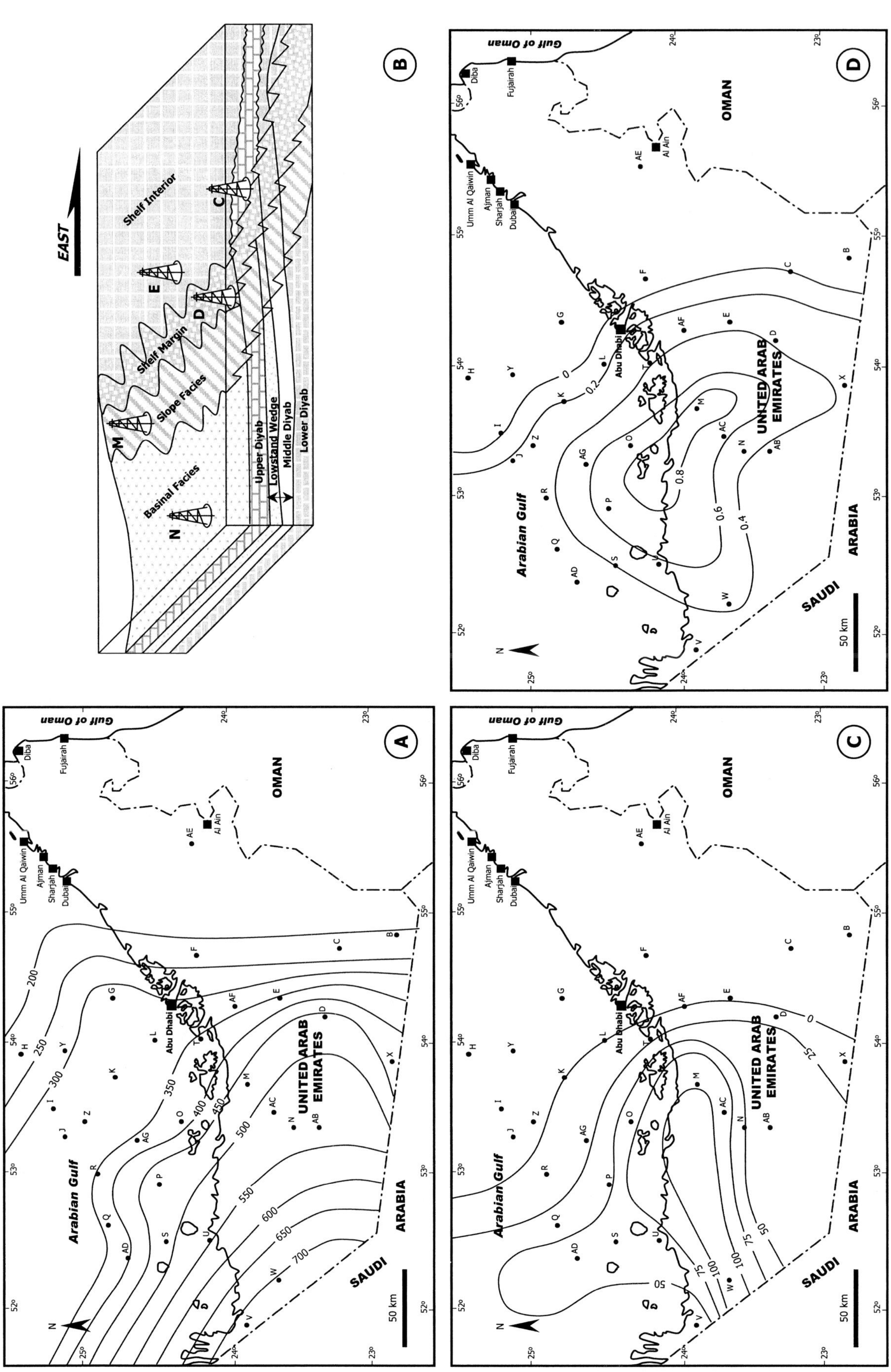


FIG. 7.—Isopach in feet (A), depositional model (B), net source-rock thickness in feet (C), and average TOC % (D), of upper member, Diyab Formation in Abu Dhabi.

and G; Figs. 3, 4), the upper unit consists predominantly of mudstone/wackestone deposited under an open-marine shelf condition, although it is completely dolomitized at wells Y and H.

The middle member is characterized by low gamma-ray response over the study area (Figs. 3, 4). The paleoenvironment can be interpreted as a series of regressive sequences. The member is composed of bioclastic, peloidal mudstone and wackestone in the western and parts of the central area, in wells V, S, and U, representing deposition under open marine conditions. A lagoonal condition, enabling development of oolite shoals, seems to have prevailed over the eastern Abu Dhabi, especially at wells D and K. In this area, the member consists mainly of oolitic, peloidal, bioclastic (corals, stromatoporoids, green algae, foraminifera) grainstones/packstones/wackestones, and considerable dolomitization (partly with oolite ghost texture) is present in wells Y and G. The member, which accumulated under in shoal conditions, has good reservoir characteristics (e.g., well I). The member thickens eastward to more than 250 ft (76 m), and thins westward to less than 125 ft (38 m) (Fig. 6A). The thickening in the east is related mainly to the development of high-energy shelf-margin deposits of Hanifa reservoir (Fig. 6B).

The upper member is characterized by relatively high gamma-ray response over the study area. In the western and eastern areas the gamma-ray response of the member is higher than that of the underlying middle member and the overlying lower Arab-D Member (Figs. 3, 4). In the central area (wells I, T and D), the gamma-ray response of the upper member is relatively low. On the basis of regional paleoenvironment analysis (Fig. 7B) the cleaner interval is interpreted to represent a shallow shelf with shoal environments. Lithologically, in the western area the member consists dominantly of bioclastic mudstone/wackestone, indicating deposition on an open marine shelf below wave base. In the central area, at wells J, Z, T and D, it is composed of bioclastic, peloidal, and oolitic grainstone/packstone/wackestone in the cleaner intervals, with subordinate lagoonal mudstone to wackestone. To the east, the upper member in wells Y and G is made up of dolomite, although its original texture is obscured by the considerable dolomitization; ghosts of pellets and ooids suggest deposition over banks with a lagoonal character.

The abnormal thickening of the upper member to the west, over the previous intrashelf basin (Fig. 7A) is most probably related to tectonic uplift and erosion in the east, while sedimentation continued in the western part of the study area, creating a lowstand wedge (Fig. 7B).

## SOURCE-ROCK POTENTIAL

The large geochemical database created by several contractors was reviewed and reinterpreted to evaluate the source-rock richness, distribution, and quality, which in turn allowed an assessment of the depositional setting and conditions favorable for the Diyab Formation source-rock potential.

### *Paleoenvironment*

The Diyab Formation was deposited during a marine transgression, resulting from regional subsidence. The sea-level rise allowed rapid up-building of a carbonate rim, while the basin floor lagged behind, because of lower rates of sedimentation (Read, 1985). The Diyab basinal facies of western Abu Dhabi consists of pelagic lime mudstone with abundant planktonic microfossils. The regional distribution of the Diyab basinal facies indicates that it was deposited in an intrashelf basin that was enclosed by the Diyab carbonate shelfal facies sediments. This resulted in restricted water circulation, anoxic conditions, and deposition below wave base (Murris, 1980; Whittle and Alsharhan, 1996). The shelfal facies are located mainly in the east and are represented by the present-day Hanifa and Hadriya reservoirs (Fig. 3). The lower and middle members of Diyab shelfal facies seem to have prograded laterally westward into their time-equivalent basinal facies deposits (Figs. 5B, 6B).

A drop of sea level, possibly associated with the tectonic uplift to the east, resulted in exposure of the middle member (Hanifa Reservoir) in the shelf interior and the shelf margin (Fig. 7B). The upper member (Jubaila) constitutes a low stand wedge, which is interpreted to have been deposited after the drop of sea level below the middle member (Hanifa) shelf break. This caused a lowstand wedge (Fig. 7B) infilling the bathymetric low of the middle member (de Matos and Hulstrand, 1995). Following deposition of the lowstand wedge, a marine transgression deposited open marine sediments of the Arab Formation with more oxygen-rich, less saline waters as carbonate production gradually increased.

### *Source-Rock Richness and Thickness Distribution*

#### *Lower Member.—*

This member shows significant organic richness in the western part of Abu Dhabi, with an average TOC of up to 1.2% by weight (Fig. 5D). However, the present-day high maturity of the analyzed rock samples suggests that the original source potential could have had higher TOC values than those recorded in the current analysis. Using estimates of kinetic maturity the thermal modeling method suggests that the original average may have been approximately 30–50% higher than recorded today.

Samples from the lower member in wells I, H, Y, E, D, and C of eastern Abu Dhabi have very low pyrolysis yields, ranging between 0.5 and 0.2 gm/kg. Also, average residual TOC contents are very low (< 0.3 wt %), indicating that the Diyab in eastern Abu Dhabi did not have any significant source potential.

The variation and distribution of net source thickness of the lower member reflect patterns similar to the distribution of organic richness. The lower-member source rock becomes thicker, about 150 ft (46 m) to the west, probably as a result of higher sedimentation rates, but the source rock thins rapidly to nothing towards the east (Fig. 5C).

The lower member has only poor source-rock productivity, with maximum S2 yields of only 1.1 and 2 kg/ton in wells X and O, respectively. However, the data suggest that the lower member may formerly have been rich source rock in wells N and M, with average source-rock potential yield up to 11.5 kg/ton (Table 1), and this might have been the original source rock potential of 5 kg/ton.

#### *Middle Member.—*

The member in well X is a moderately rich source rock in the west but becomes a lean source rock in the eastern part of the study area. The low source-rock potential (0.2–0.8% TOC by weight; Fig. 6D) as determined by previously analyzed rock samples (in 1982 and 1988) is confirmed by the latest analysis (in 1992). No significant organic richness was detected in any samples of the middle member analyzed from the H, Y, D, and C wells. It averages less that 0.2% TOC by weight (Fig. 6D).

The effective source thickness of the middle member varies from 25 to 50 ft (8 to 15 m) (Fig. 6C). The maximum source thickness of about 75 ft (23 m) is found in the west and adjacent

TABLE 1.—Volumetric calculations of oil migrated from Diyab source rocks.

**Table 1.1—Volumetric calculations of oil migrated from Diyab source rocks up to Maastrichtian (Simsima) time**

| DIYAB FORMATION | Drainage Areas | Mature Area ($km^2$) | Net Source Thick. (m) | Source Rock Potential (kg/ton) | Source Rock Conversion (%) | Oil Migration Efficiency (%) | Volumes of oil available to trap (Billion Barrels) |
|---|---|---|---|---|---|---|---|
| UPPER | Area-A | 6330 | 22.9 | 0.3 | 0.4 | 0.4 | 0.0 |
| | Area-B | 1265 | 24.4 | 0.5 | 0.45 | 0.4 | 0.0 |
| | Area-C | 4906 | 21.3 | 0.5 | 0.5 | 0.4 | 0.0 |
| | Area-D | 3797 | 7.6 | 0.0 | 0.2 | 0.4 | 0.0 |
| MIDDLE | Area-A | 6330 | 21.3 | 3.0 | 0.4 | 0.35 | 0.7 |
| | Area-B | 1265 | 22.9 | 5.0 | 0.45 | 0.35 | 0.6 |
| | Area-C | 4906 | 15.2 | 0.5 | 0.5 | 0.35 | 0.0 |
| | Area-D | 3797 | 4.0 | 0.0 | 0.2 | 0.35 | 0.0 |
| LOWER | Area-A | 6330 | 59.4 | 8.0 | 0.4 | 0.3 | 10.2 |
| | Area-B | 1265 | 61.0 | 11.0 | 0.45 | 0.3 | 3.6 |
| | Area-C | 4906 | 48.8 | 11.5 | 0.5 | 0.3 | 13.3 |
| | Area-D | 3797 | 21.3 | 3.0 | 0.2 | 0.3 | 0.0 |
| TOTAL | | | | | | | **28.3** |

**Table 1.2—Volumetric calculations of oil migrated from Diyab source rocks between Maastrichtian to Eocene (Dammam) time**

| DIYAB FORMATION | Drainage Areas | Mature Area ($km^2$) | Net Source Thick. (m) | Source Rock Potential (kg/ton) | Source Rock Conversion (%) | Oil Migration Efficiency (%) | Volumes of oil available to trap (Billion Barrels) |
|---|---|---|---|---|---|---|---|
| UPPER | Area-A | 16706 | 16.8 | 0.4 | 1.00 | 0.4 | 0.0 |
| | Area-B | 7340 | 25.9 | 1.5 | 0.80 | 0.4 | 1.1 |
| | Area-C | 9112 | 22.9 | 1.8 | 1.00 | 0.4 | 3.1 |
| | Area-D | 7594 | 7.6 | 0.4 | 0.85 | 0.4 | 0.0 |
| | Area-E | 3037 | 21.3 | 0.1 | 1.00 | 0.4 | 0.0 |
| MIDDLE | Area-A | 16706 | 18.3 | 1.0 | 1.00 | 0.35 | 0.8 |
| | Area-B | 7340 | 18.3 | 1.9 | 0.80 | 0.35 | 1.3 |
| | Area-C | 9112 | 15.2 | 1.8 | 1.00 | 0.35 | 1.8 |
| | Area-D | 7594 | 4.0 | 0.4 | 0.85 | 0.35 | 0.0 |
| | Area-E | 3037 | 15.2 | 0.0 | 1.00 | 0.35 | 0.0 |
| LOWER | Area-A | 16706 | 42.7 | 6.5 | 1.00 | 0.3 | 45.7 |
| | Area-B | 7340 | 53.3 | 8.0 | 0.80 | 0.3 | 24.7 |
| | Area-C | 9112 | 47.2 | 11.0 | 1.00 | 0.3 | 49.4 |
| | Area-D | 7594 | 25.9 | 2.0 | 0.85 | 0.3 | 2.0 |
| | Area-E | 3037 | 38.1 | 5.0 | 1.00 | 0.3 | 5.5 |
| TOTAL | | | | | | | **135.4** |

**Table 1.3—Volumetric calculations of oil migrated from Diyab source rocks between Dammam time until present day**

| DIYAB FORMATION | Drainage Areas | Mature Area ($km^2$) | Net Source Thick. (m) | Source Rock Potential (kg/ton) | Source Rock Conversion (%) | Oil Migration Efficiency (%) | Volumes of oil available to trap (Billion Barrels) |
|---|---|---|---|---|---|---|---|
| UPPER | Area-A | 17520 | 15.2 | 0.5 | 1.0 | 0.4 | 0.0 |
| | Area-B | 15232 | 19.8 | 1.0 | 0.8 | 0.4 | 0.0 |
| | Area-C | 10750 | 12.2 | 0.7 | 1.0 | 0.4 | 0.0 |
| | Area-D | 7040 | 6.1 | 0.5 | 0.85 | 0.4 | 0.0 |
| | Area-E | 4350 | 4.6 | 0.9 | 1.0 | 0.4 | 0.0 |
| MIDDLE | Area-A | 17520 | 15.2 | 1.0 | 1.0 | 0.35 | 0.7 |
| | Area-B | 15232 | 12.2 | 0.6 | 0.8 | 0.35 | 0.0 |
| | Area-C | 10750 | 10.7 | 0.0 | 1.0 | 0.35 | 0.0 |
| | Area-D | 7040 | 3.0 | 0.5 | 0.85 | 0.35 | 0.0 |
| | Area-E | 4350 | 3.0 | 0.9 | 1.0 | 0.35 | 0.0 |
| LOWER | Area-A | 17520 | 42.7 | 1.0 | 1.0 | 0.3 | 1.7 |
| | Area-B | 15232 | 39.6 | 2.6 | 0.8 | 0.3 | 8.7 |
| | Area-C | 10750 | 39.6 | 0.0 | 1.0 | 0.3 | 0.0 |
| | Area-D | 7040 | 18.3 | 1.0 | 0.85 | 0.3 | 1.0 |
| | Area-E | 4350 | 18.3 | 1.5 | 1.0 | 0.3 | 0.0 |
| TOTAL | | | | | | | **11.8** |

to the shoreline area, possibly representing a pinchout in the central part of Abu Dhabi.

### *Upper Member.—*

This member is generally organically lean and has a residual TOC of less than 0.8%, (Fig. 7D), with the average net source thickness of 25 to 100 ft (8 to 30 m) (Fig. 7C). It is not anticipated that this member has major source potential.

### *Source Quality and Kerogen Type*

In addition to organic richness, it is necessary to consider the nature and origin of the organic matter in the rocks of the Diyab Formation. The geochemical data show a general trend from high hydrogen indices (HI) at low maturity (around 200–250 HI) to low values at high maturity (around 50 HI). These values compare well with the mapped maturity (and depth) trends, and are consistent with the visual kerogen typing, which suggests that amorphous kerogen is the main type in most of the samples examined. In some cases, recognizable but very degraded marine sapropelic kerogen is present in much of the interval. A thick interval in wells V and W appears to contain spent oil-prone kerogen. From a study of the petroleum geochemistry of Abu Dhabi, Lijmbach et al. (1992) concluded that the Upper Jurassic Diyab (Dukhan) Formation is an excellent Type II, post-mature source rock for oil.

## MATURITY

Maturation modeling was used to investigate the time at which generation and expulsion of petroleum occurred in the Abu Dhabi area, in order to produce maps for a migration study. Geothermal modeling was undertaken for 27 wells throughout Abu Dhabi using an in-house (ADNOC) program based on the Lopatin–Waples algorithm (Waples, 1980) and the Arrhenius equation ( Wood, 1988; Hunt et al., 1991). The study was started by reconstructing the burial history of 27 key wells (Fig. 8), which are spaced such that they represent maturity variations. The input data for reconstruction of these burial-history profiles consist of present-day depth maps of the key horizons, formation thickness, and age and formation temperature. In the reconstruction of the burial history, unconformities were also accounted for. In southeast onshore Abu Dhabi, however, much of the Upper Cretaceous section is absent, resulting in a significant time gap of up to 45 million years (Taher, 1997). The paleo-thickness of the partially eroded Aruma and Wasia groups of southeastern onshore Abu Dhabi were estimated using a complete reference section in western Abu Dhabi.

The stratigraphic development of the area was taken as a passive infill, and thus a constant surface temperature of 86°F was assumed for offshore wells and 80°F for onshore wells. The input data for reconstruction of the thermal history consist of geothermal gradients, rock conductivity, fluid distribution, and porosity versus depth.

The data on geothermal gradient in the study area show a gradual increase in heat flow southwards that ranges between 1.5°F/100 ft in the north offshore and 2.3°F/100 ft in the southern onshore (Taher, 1997). The high geothermal gradient in the southern onshore is probably related to shallow depth of the basement, rock conductivity, and fluid distribution.

The thermal modeling was calibrated with available optical and statistical data to render it more useful for predicting maturation boundaries in the study area where no analytical data are available. The calibration includes the calculation of the thermal alteration index (TAI) and the vitrinite reflectance ($R_o$). Vitrinite-reflectance data for the Diyab Formation in ten key wells (Fig. 9) indicate that the offshore wells G, Q, P, K, and I are within the oil-generation window, and well H is still in the immature stage or at the onset of oil generation. Onshore wells X, M, and D are within the gas-generation window, and well AE is overmature.

Significant hydrocarbon generation and expulsion occured in southwestern onshore Abu Dhabi at the end of Maastrichtian time, about 65 Ma (Fig. 10A), with major expulsion of hydrocarbon at the end of Eocene time from most of the onshore and southern offshore area (Fig. 11A). The present-day maturity map of the Diyab Formation indicates that the study area is within the gas-generation window (Fig. 12A).

### *Volumetric Calculations*

Volumetric calculations were performed using Piggot's (1982) method and Goff's (1983) equation. The calculations require an estimate of source-rock richness, thickness, and productivity (kg/ton). In considering the evidence that the Diyab Formation is a potential oil source rock, the initial potential organic-matter content prior to maturation was estimated using present TOC % vs. percentage generation of hydrocarbon. It is concluded that the TOC values were doubling in the overmature areas, whereas in the immature areas, the present TOC values remain unchanged. The paleomaturity maps were created by projecting the maturity effect back to the Maastrichtian and Eocene (Figs. 10B, 11B). The resulting maturity kitchens were subdivided into five drainage polygons (A, B, C, D, and E; Fig. 11), which appear to be in different structural positions and isolated from each other. Volumetric calculations were made separately for the Maastrichtian, the Eocene, and the present day (Tables 1–3). Migration efficiency was assumed on the basis of similar cases studied previously in the area with consideration of the distance between the kitchens and the reservoirs. On the basis of these assumptions the total predicted oil-in-place that migrated from the Diyab source rock amounts to 175 billion barrels.

## SECONDARY MIGRATION MODEL

Migration models can be used as a predictive tool for exploration, enabling explanation of the known distribution of hydrocarbons in the study area. In addition, it should account for the absence of petroleum where valid structures are known to exist. The secondary-migration model relies on the paleostructure maps of the top Arab-D Member, because this is the carrier bed above the Diyab source rock and below the Hith Anhydrite regional seal.

The paleostructure maps were constructed from the present-day top of Arab-D Member structure map by subtracting isopach maps of shallower sections. The paleostructure maps produced were not corrected for compaction, but their overall form reflects the general features of the reservoir at any time and they are therefore a most useful tool for examining secondary migration directions.

The paleostructure maps of the Arab-D Member (Late Kimmeridgian) for the Maastrichtian, the Eocene, and the present day (Figs. 10–12) indicate that the study area can be subdivided into five drainage polygons (A, B, C, D, and E). These polygons are located in different structural positions and have been isolated from each other from Late Cretaceous time until the present.

### *Migration of Petroleum in Maastrichtian Time*

The paleo-maturity map of the Diyab Formation during Maastrichtian time indicates that the formation was mature for

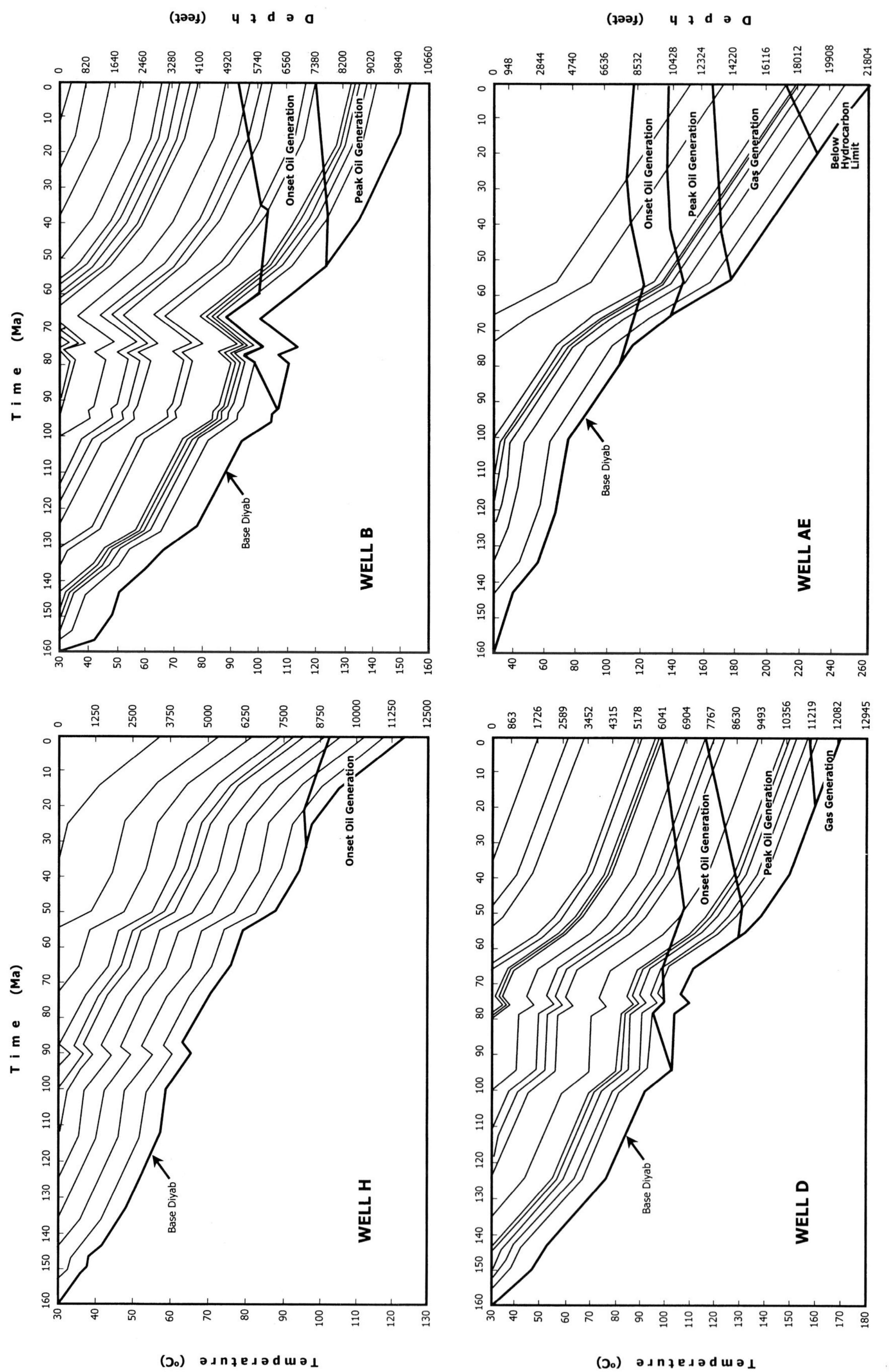


FIG. 8.—Burial history and hydrocarbon generation curves for four representative wells.

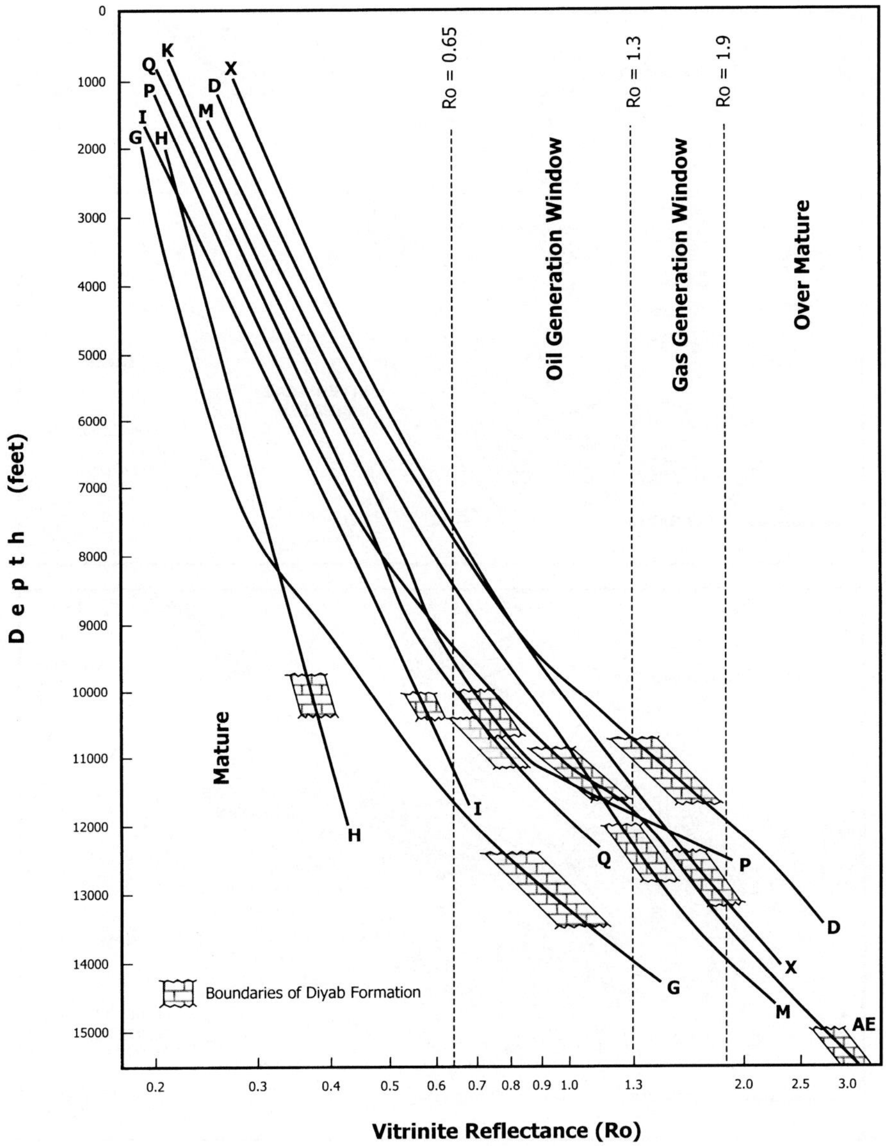


FIG. 9.—Maturity of the Diyab Formation in Abu Dhabi.

oil generation only in the southwestern onshore area (Fig. 10A). Oil generated in this area migrated into the Arab reservoir of the surrounding structures (e.g., wells X, M, N, and W; Fig. 10B).

From the assumptions for the volumetric calculations (Tables 1–3), approximately 95% of the predicted oil-in-place was generated and migrated from the lower Diyab member source rock of Abu Dhabi, of which 40% was generated and migrated within Area A, approximately 45% within Area C, and 15% within Area B (Tables 1–3).

### *Migration of Petroleum in Eocene Time*

The paleo-maturity map of the Diyab Formation during Eocene time shows that the Diyab source rock at that time was mature for oil generation over most of Abu Dhabi (Fig. 11A). The northern offshore was at the onset of oil generation, and the southwestern onshore synclines were within the gas-generation window.

The major expulsion phase of petroleum was in the early Tertiary (50–40 Ma) with more than 75% of the total oils generated from Diyab source rock having migrated during that time, with 95% having migrated from the lower member. Oil generated in Area A represents 35% of the total oil generated at that time, and this migrated mainly northwards towards the western offshore structures (Fig. 11B, Table 3). Migration direction in Area B (20%) was towards the northern offshore structures. Migration direction in Area C at that time was towards the onshore central highs into the N and M structures. The updip migration direction in the central high at that time was towards northeast Abu Dhabi. Approximately 15–20% of the oil accumulated in Area C possibly re-migrated out of this area towards the Hith edge, then towards the central offshore area into the K structure. Oil generated in Area D migrated into the southeast onshore structures, whereas oil generated in Area E migrated mainly into the K structure (Fig. 11B).

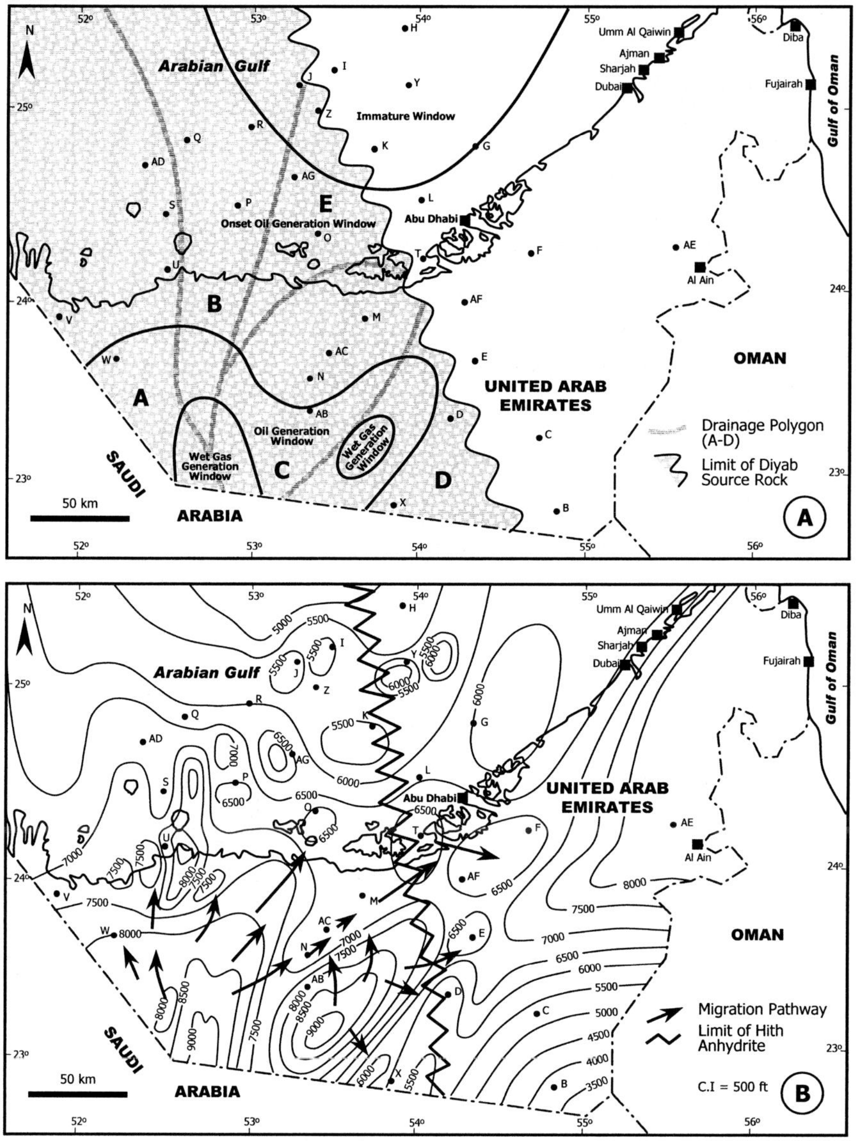


FIG. 10.—Diyab oil generation and migration. A) Diyab paleo-maturity map; B) top Arab-D paleo-structure map in Maastrichtian time.

## *Migration of Petroleum at Present*

The present-day maturity map of Diyab Formation (Fig. 12A) indicates that the formation is mature for gas generation in most onshore and southern offshore areas. However, the Diyab Formation is still actively generating oil in the northern offshore area.

The migration pathways were affected by late Tertiary tectonics. The late Tertiary seems to be characterized by a fill-and-spill migration mechanism between the structures rather than direct hydrocarbon migration from the Diyab source rock. During the late Tertiary to present day, a gas-generation phase occurred that possibly emplaced the previously trapped oil, especially in the onshore and southern offshore structures. Also, we have noticed that the late Tertiary tilting completely reversed the migration direction in the central onshore high (M and N structures) from the northeast (Fig. 11B) towards the south (Fig. 12B). Consequently, the main phase of oil spillage in Area C was southward rather than towards the Hith edge.

## ANALYSES OF OILS AND EXTRACTS

A geochemical investigation was carried out on the crude-oil samples and source-rock extracts from Abu Dhabi. The data include gas chromatography mass spectrometry (GSC-MS), and bulk and molecular parameters analyzed by five different laboratories. The data were integrated and reinterpreted in order to increase our understanding of the genetic oil families in Abu Dhabi area.

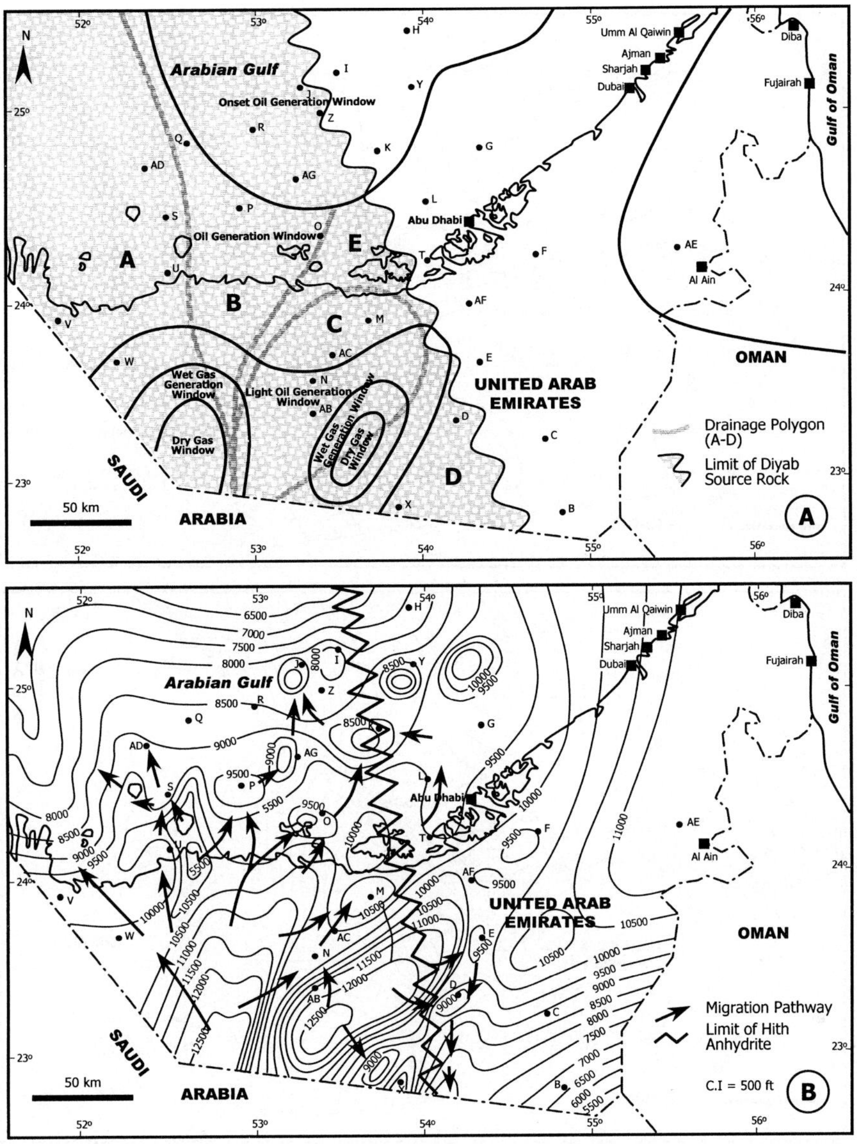


FIG. 11.—Diyab oil generation and migration. A) Diyab paleo-maturity map; B) top Arab-D paleo-structure map in Eocene time.

## *Oil–Oil Correlation*

### *Bulk Parameters.—*

The bulk parameters used in the oil–oil correlation include sulfur content, API gravity, and the $\delta^{13}C$ value of the saturate fractions (Fig. 13). The API gravity vs. depth shows considerable variations. These differences are presumably related to the maturity of the oil when API gravity increases with depth. These oils show also a decreasing sulfur content with increasing API gravity (Fig. 13). In addition the carbon-isotope saturates of these oils become isotopically heavier with increasing reservoir temperature. However, the reservoir temperatures are not sufficiently high to have caused significant cracking of the oils in the reservoirs. These differences may therefore be ascribed mainly to different sources, which were categorized into four groups (A, B, C, and D).

### *Molecular Parameters.—*

The molecular parameters, which include sterane distribution, saturates, total aromatics, and asphaltenes, indicate that the oil accumulations in the Lower Cretaceous reservoirs are different from those in the Upper Jurassic reservoirs (Fig. 14). The Arab oils (Group C) are generally less mature (Fig. 15), have a lower saturate and higher aromatic content, and are isotopically lighter than Thamama oils (Group B) at equivalent reservoir depth (Fig. 14).

The highly mature Thamama oil in structure F is isotopically similar to that found in the Middle Jurassic Araej oil (Group D) and the Upper Jurassic, Diyab/Hanifa oil in structure C (Fig. 15). These oils are isotopically similar to each other and most probably were sourced from pre-Diyab, Mesozoic sections. The Upper Cretaceous Simsima oils (Group A), which were sourced from the Shilaif Formation, are isotopically similar to the low-maturity Thamama oil of structure A.

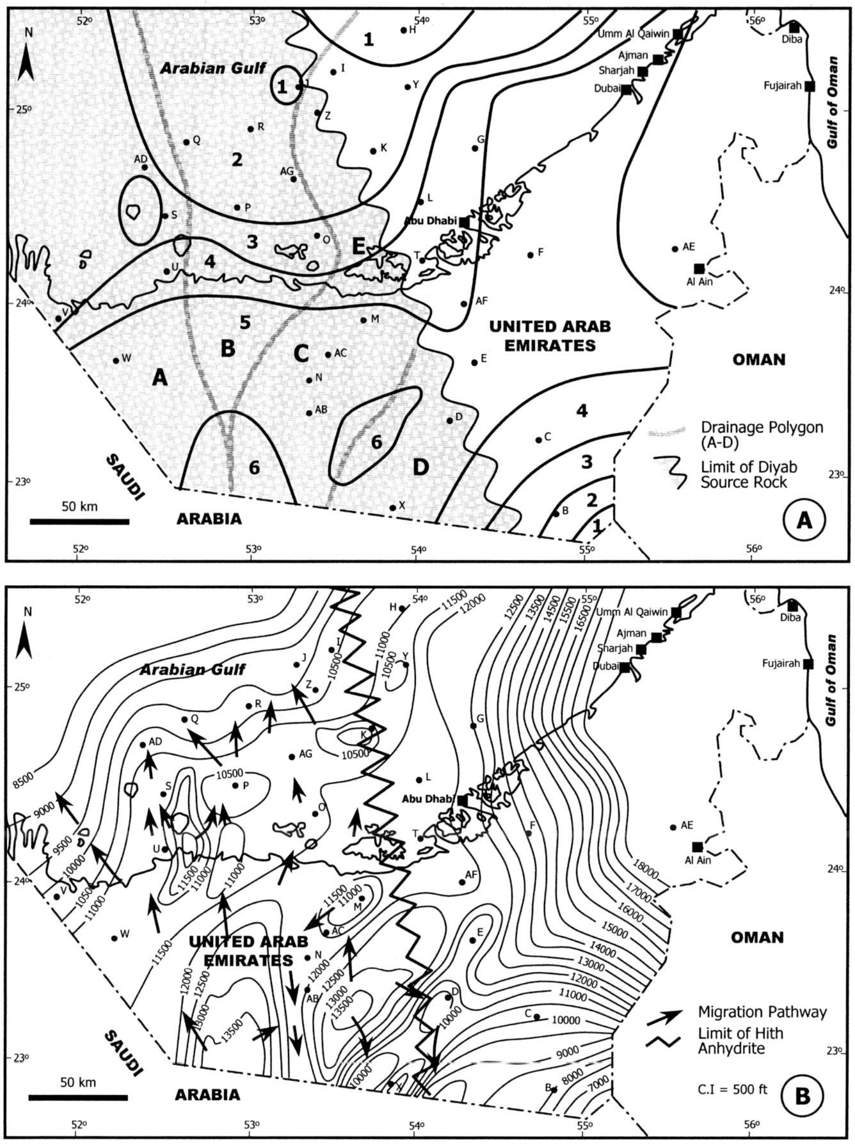


FIG. 12.—Diyab oil generation and migration. A) Diyab maturity map; B) top Arab-D structure map at present day.

## *Oil–Source Rock Correlation*

### *Carbon-Isotope Ratio.—*

The carbon-isotope compositions of the saturated and aromatic fractions (Fig. 15) display considerable variation, with $\delta^{13}C$ ranging from -28 to -23‰ PDB. These values reflect the genetic differences that existed between the samples analyzed. However, the distribution patterns show that Group B (Thamama oils) represents the major group, which is characterized by oil gravity between 25 and 45° API and $\delta^{13}C$ values from -27 to -24‰ PDB. Group A comprises the low-maturity Thamama oils in the southeast onshore Abu Dhabi (structures A, B, and C) and Simsima (Maastrichtian) oils in south onshore (structure A, B, and C) and are characterized by relatively low oil gravity (30–35°API) and high sulfur content (1.5%). Group C appears to be related mainly to Arab oils, which have moderate oil gravity (35°API) and moderate sulfur content (0.5–1%) and are isotopically heavier than Thamama oils. Group D comprises the Araej Formation, oils in the north offshore Abu Dhabi with highly mature Thamama oils in structure F and this group characterized by high oil gravity (40–50°API) with low sulfur content (< 0.5%).

The isotopic data of the different oils indicate that Group A oils are sourced mainly from the Cenomanian (Shilaif Formation). Group B comprises mainly the Lower Cretaceous, Thamama-reservoired oils, and seems to be sourced from the Oxfordian Diyab and Aptian basinal facies of the Bab Member, Shuaiba Formation (Fig. 15). Groups C and D oils are sourced mainly from the Diyab source rocks. It is also noted that Diyab and Shuaiba basinal facies source rocks are compositionally similar, which creates difficulties in differentiating between the oils from these two sources.

### *Gas Chromatography–Mass Spectrometry of Alkanes.—*

Sterane and triterpane distribution patterns obtained by GC-MS show many differences between the Thamama and Arab oils

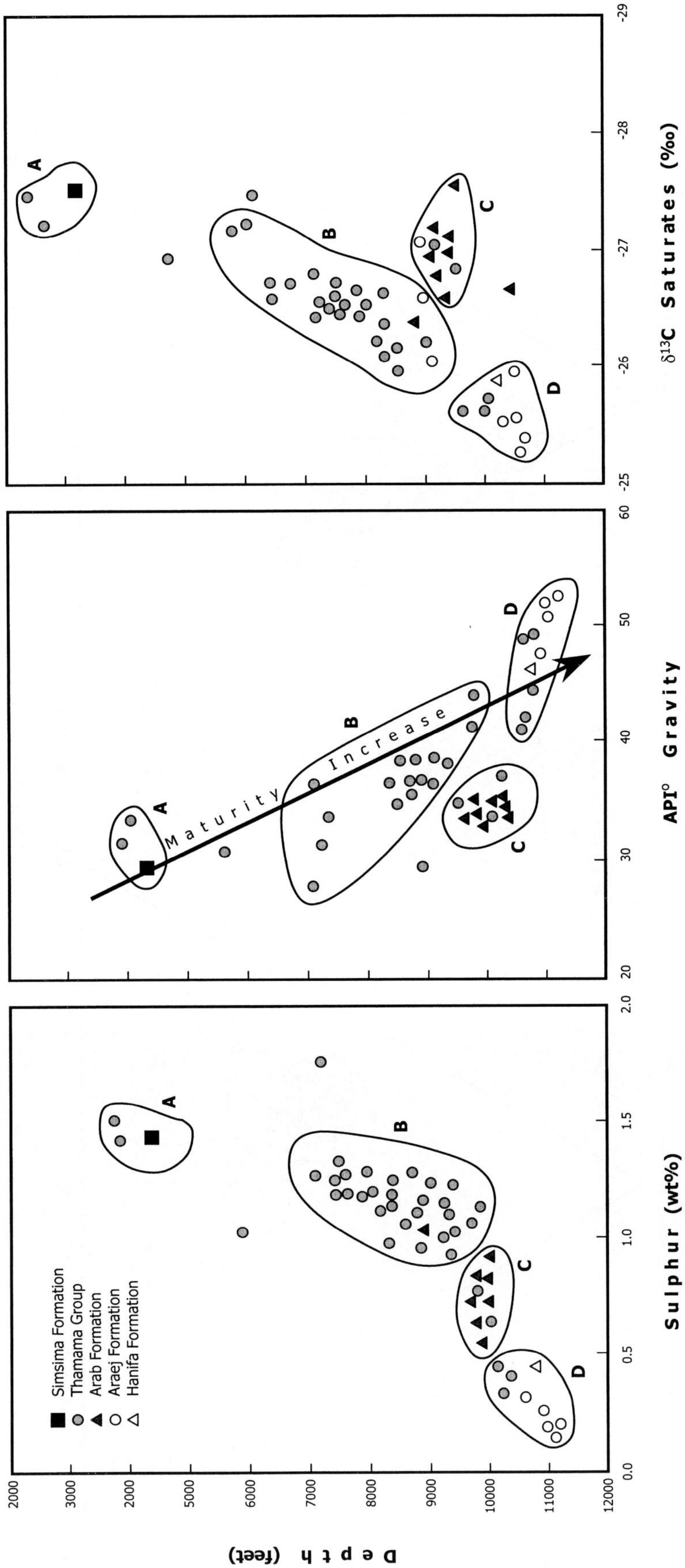


FIG. 13.—Relationship between sulfur (wt %) API gravity, and carbon saturates (‰ PDB) vs. depth for different oils in Abu Dhabi.

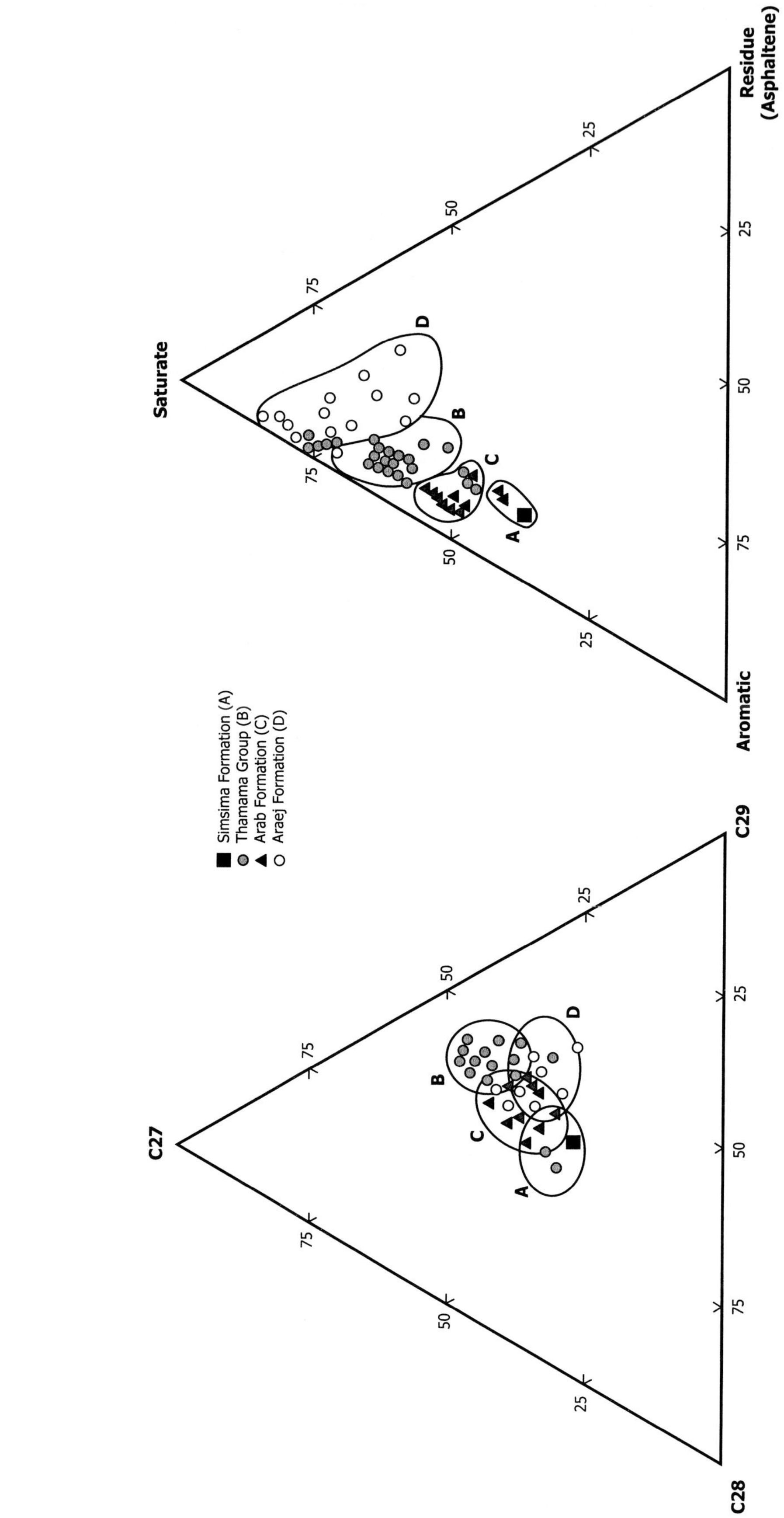


FIG. 14.—Ternary diagram showing the gross composition of Abu Dhabi oils.

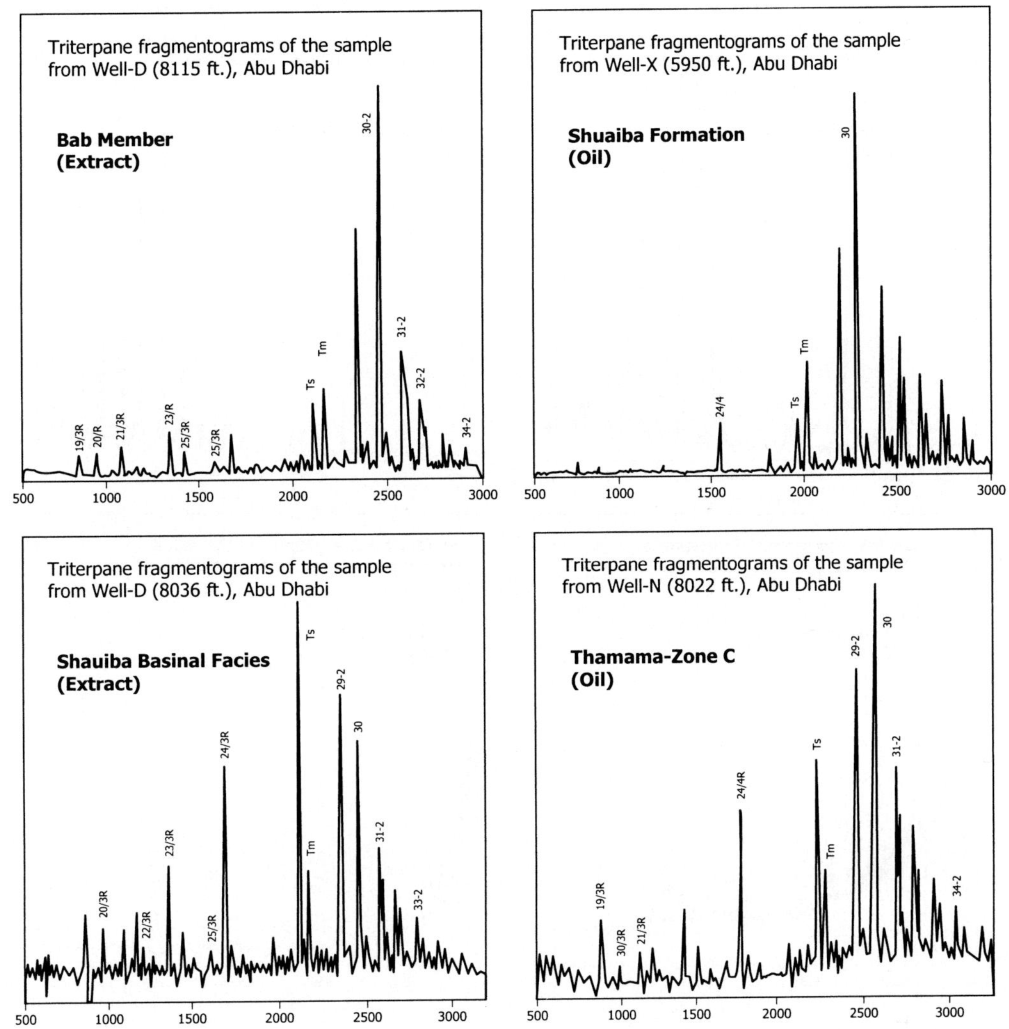


FIG. 15a.—Triterpane fragmentograms of the oil sample from the Shuaiba Formation and Thamama-Zone C, which correlate with the Bab Member and Shuaiba basinal facies, respectively.

and among Thamama oils themselves (Shuaiba and Thamama Zone C, Fig. 16). These differences cannot be simply related to the facies variation and different maturity levels.

The correlation between these oils and the different sources found in the study area indicates that the Simsima, Shuaiba, and Kharaib formations oils show a good correlation with Shilaif Formation and Bab Member (Shuaiba basinal facies) source rocks, respectively, and the Arab oils seems to be correlate with the Diyab source rock. The Diyab source rock correlation is influenced by the high maturity level of the formation. This is indicated by the fact that only post-mature Diyab samples are available.

## HYDROCARBON DISTRIBUTION

The north–south cross section of Abu Dhabi (Fig. 17) indicates that the mid- and Upper Cretaceous section seems to be completely absent in the southeast onshore Abu Dhabi. The Shilaif Formation source rock, which is part of the mid-Cretaceous Wasia Group, onlaps onto the Thamama Group. Therefore, a lateral hydrocarbon charge from the younger Shilaif into the Thamama reservoir is possible. The tectonic setting of southeast onshore Abu Dhabi is consistent with geochemical results suggesting that the Thamama oils of structure A are sourced from the Shilaif Formation.

Hydrocarbon distributions in the central onshore highs are difficult to account for solely by vertical hydrocarbon migration from the Diyab source rock, because at least six cycles of intercalated gas and oil can be recognized (Fig. 18), which reflect the presence of intraformational seals in this area. This intercalation between oil and gas reflects hydrocarbon migration from different mature kitchens that were located at different stratigraphic levels.

## CONCLUSIONS

1. The major expulsion phase of petroleum from Diyab Formation source rock was during the Early Tertiary.

2. The migration pathways of the oil generated from the Diyab source are interpreted to have been mainly towards the offshore and central onshore structures.

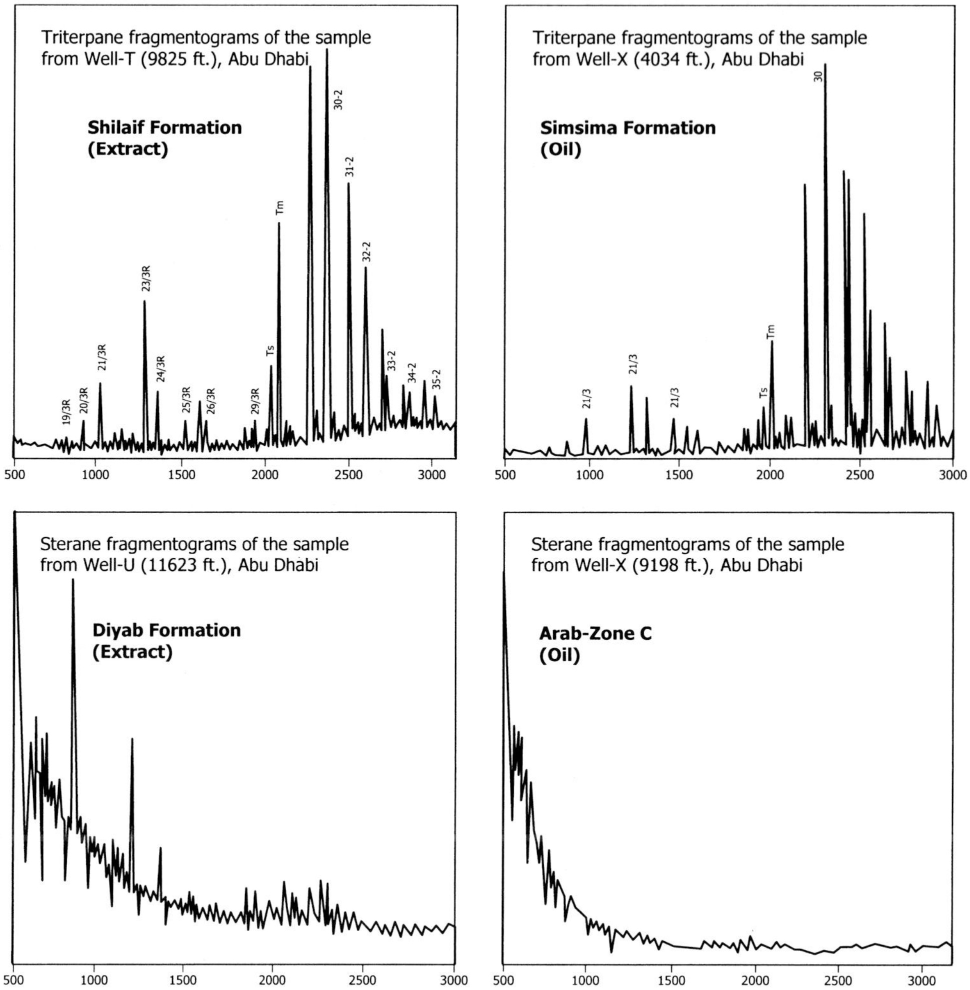


FIG. 15b.—Triterpane and sterane fragmentograms of the oil sample from Simsima Formation and Arab-Zone C, which correlate with Shilaif and Diyab Formations source rocks, respectively.

3. The amount of Diyab oil that migrated into the lower Thamama through the Hith edge represents less than 25% of the total oil migrated from Diyab source rock.

4. The upper unit of the lower Diyab member in the western part of Abu Dhabi is a prolific oil-prone source rock, which has generated more than 90% of the total oil expelled and migrated from the Diyab Formation.

5. Four major oil families (A, B, C, and D) are defined and correlated with the Diyab, Shuaiba basinal facies, Bab Member, and Shilaif source rocks.

## ACKNOWLEDGMENTS

The authors express sincere appreciation to the management of Abu Dhabi National Oil Company (ADNOC) for allowing us to publish this paper. Thanks also to many colleagues in ADNOC Exploration Division and U.A.E. University who provided significant stimulating discussions during the preparation stages of this paper. Our sincere thanks to Dr. C. G. St. C. Kendall, Dr. J. Connan, Dr. Peter Nederlof, and Dr. Robert W. Scott for reading and correcting the early drafts of this manuscript.

## REFERENCES


ALSHARHAN, A.S., 1989, Petroleum geology of the U.A.E.: Journal of Petroleum Geology, v. 12, p. 253–288.

ALSHARHAN, A.S., AND KENDALL, C.G.ST.C., 1986, Precambrian to Jurassic rocks of Arabian Gulf and adjacent areas: Their facies, depositional setting, and hydrocarbon habitat: American Association of Petroleum Geologists, Bulletin, v. 70, p. 977–1002.

ALSHARHAN, A.S., AND NAIRN, A.E.M., 1997, Sedimentary Basins and Petroleum Geology of the Middle East: Amsterdam, Elsevier Science, 940 p.

AYRES, M.G., BILAL, M., JONES, R.W., SLENTZ, L.W., TARTIR, M., AND WILSON, A.O., 1982, Hydrocarbon habitat in main producing areas, Saudi Arabia: American Association of Petroleum Geologists, Bulletin, v. 66, p. 1–9.

AZER, S., 1989, Preliminary investigation into possible stratigraphic traps, offshore Abu- Dhabi: 5th Middle East Oil Show, Bahrain, SPE 17999, p. 739–753.

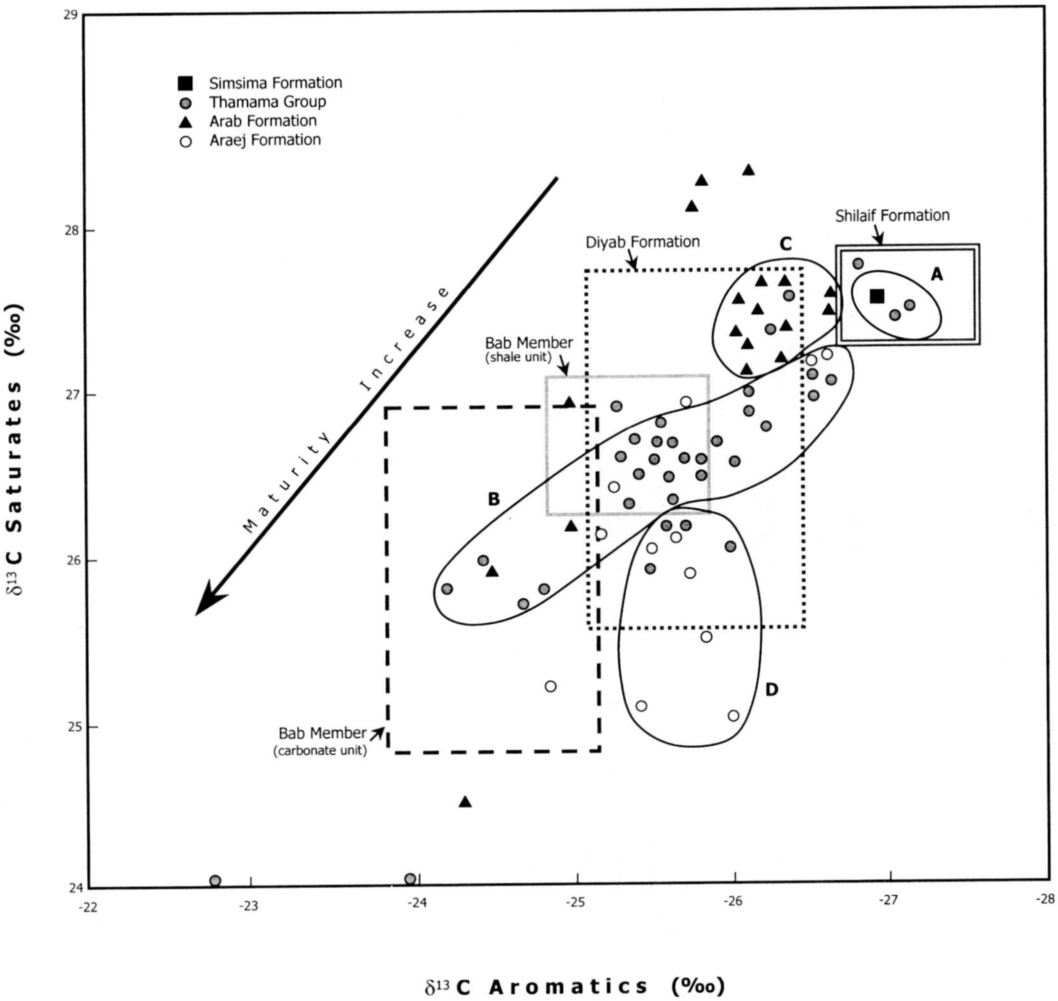


FIG. 16.—Carbon-isotope ratios of aromatics vs. saturates showing four major oil families in the Abu Dhabi area.

DE MATOS, J.E., AND HULSTRAND, R.F., 1995, Regional characteristics and depositional sequences of the Oxfordian and Kimmeridgian—Abu Dhabi, *in* Husseini, M.I., ed., Middle East Petroleum Geosciences (Geo 94): Gulf PetroLink, Bahrain, v. 1, p. 346–356.

DROSTE, H., 1990, Depositional cycles and source rock development in an epeiric intra-platform basin—The Hanifa Formation of the Arabian Peninsula: Sedimentary Geology, v. 69, p. 281–296.

GOFF, J.C., 1983, Hydrocarbon generation and migration from Jurassic source rock in the east Sheltand Basin and Viking Graben of the northern Sea, *in* Demaison, G., and Murris, R.J., eds., Petroleum Geochemistry and Basin Evaluation: American Association of Petroleum Geologists, Memoir 35, p. 273–302.

HASSAN, T.H., AND AZER, S., 1985, The occurrence and origin of oil in offshore Abu Dhabi: Proceedings of 4th SPE Middle East Oil Show, Bahrain, p. 143–155.

HUNT, J.M., LEWAN, M.D., AND HENNET, R.J.C., 1991, Modeling oil generation with time–temperature index graphs based on the Arrhenius equation: American Association of Petroleum Geologists, Bulletin, v. 75, p. 795–807.

LIJMBACH, G.W.M., BUISKOOL TOXOPEUS, J.M.A., RODENBURG, T., AND HERMANS, L.J.P.C.M., 1992, Geochemical study of crude oils and source rocks in onshore Abu Dhabi: 5th Abu Dhabi Petroleum Conference, Proceedings, SPE 24513 (ADSPE 305), p. 295–422.

MURRIS, R.J., 1980, Middle East: Stratigraphic evolution and oil habitat: American Association of Petroleum Geologists, Bulletin, v. 64, p. 597–618.

PIGGOT, N., 1982, Volumetric Assessment of Petroleum Generation in Basins: British Petroleum Exploration School, GCB/22/82.

READ, J.F., 1985, Carbonate platform facies models: American Association of Petroleum Geologists, Bulletin, v. 69, p. 1–21.

TAHER, A.K., 1997, Delineation of organic richness and thermal history of the Lower Cretaceous Thamama Group, east Abu Dhabi: GeoArabia, v. 2, no. 1, p. 65–88.

WAPLES, D.W., 1980, Time and temperature in petroleum formation: Application of Lopatin's method to petroleum exploration: American Association of Petroleum Geologists, Bulletin, v. 64, p. 916–926.

WHITTLE, G.L., AND ALSHARHAN, A.S., 1996, Diagenetic history and source rock potential of the Upper Jurassic Diyab Formation, offshore Abu Dhabi, United Arab Emirates: Carbonates and Evaporites, v. 11, p. 145–154.

WOOD, D.A., 1988, Relationships between thermal maturity indices calculated using Arrhenius equation and Lopatin's method: implications for petroleum exploration: American Association of Petroleum Geologists, Bulletin, v. 72, p. 115–134.

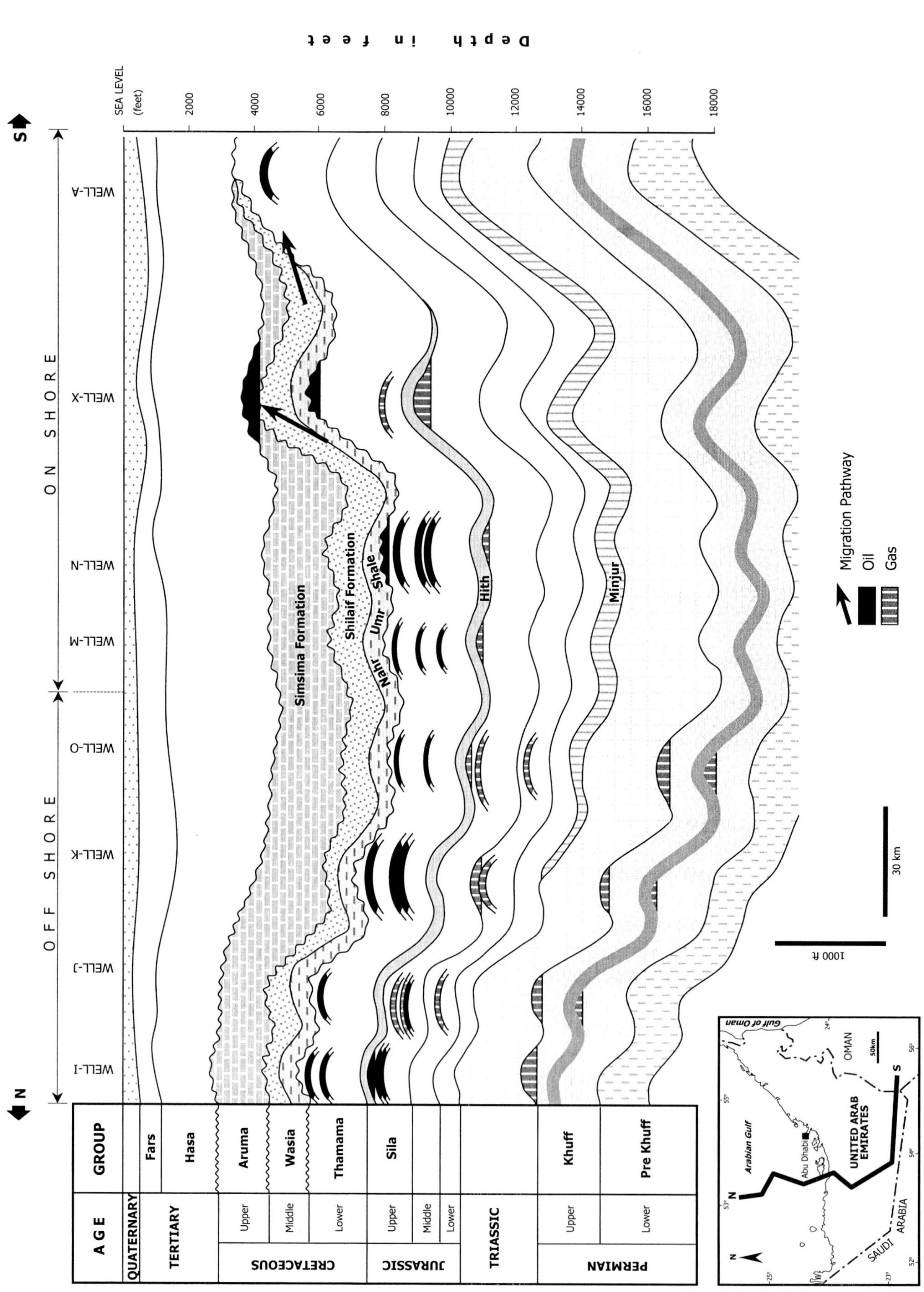


FIG. 17.—Lithostratigraphic–structural cross section in Abu Dhabi showing oil and gas accumulation in the Jurassic–Cretaceous and hydrocarbon migration pathway.

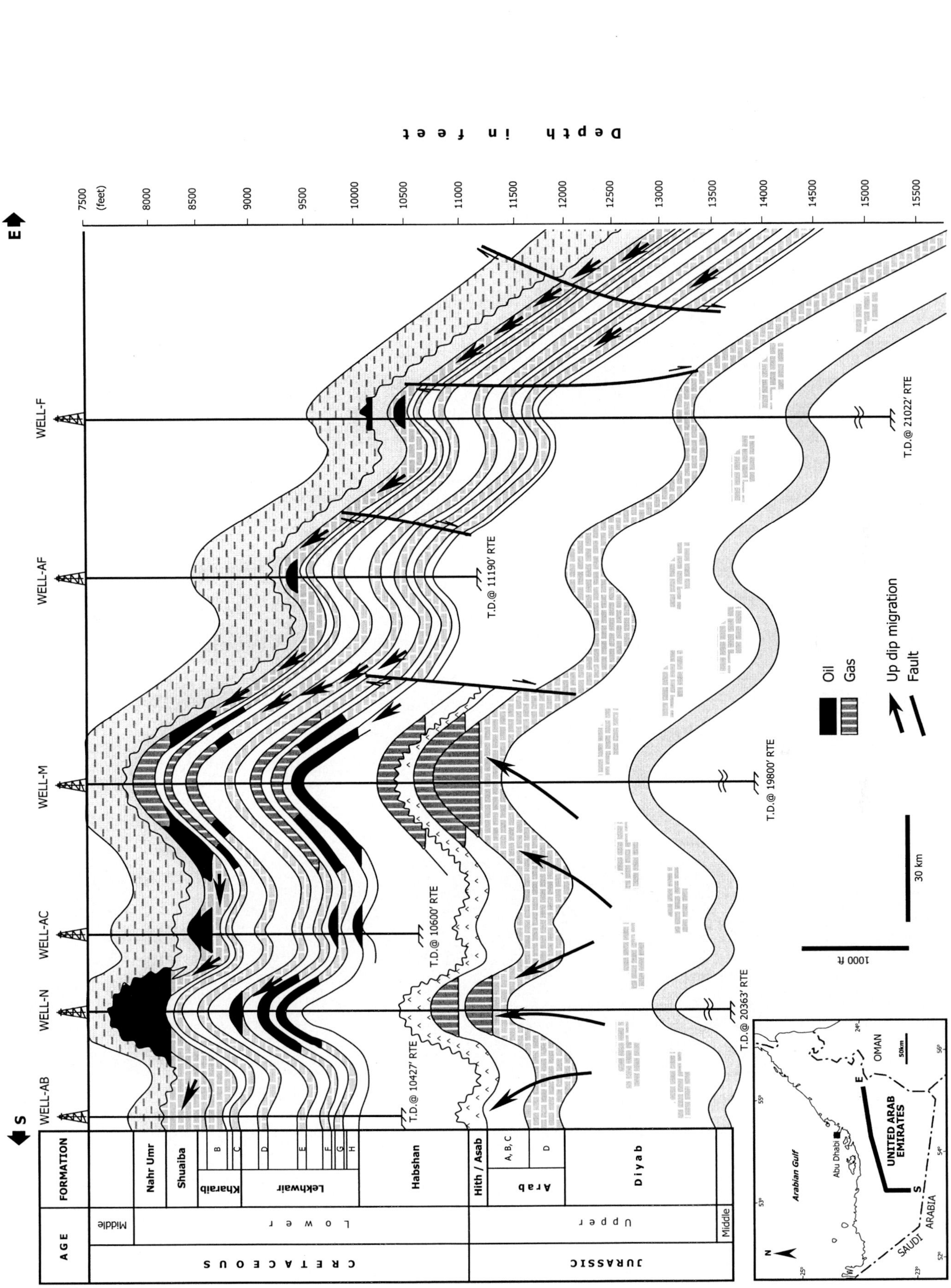


FIG. 18.—Lithostratigraphic–structural cross section in onshore Abu Dhabi showing updip hydrocarbon migration and distribution.

# RESERVOIR GEOLOGY OF THE MIDDLE MINAGISH MEMBER (MINAGISH OOLITE), UMM GUDAIR FIELD, KUWAIT

ROGER DAVIES AND CATHY HOLLIS*

*Badley Ashton and Associates, Ltd., Winceby House, Winceby, Horncastle, Lincolnshire, LN9 6PB, U.K.*

CLIVE BISHOP

*BP Kuwait, P.O.Box: 29335, Safat, Kuwait*

AND

RAM GAUR AND ABDUL AZIZ HAIDER

*Kuwait Oil Company, P.O.Box: 9758, Ahmadi, Kuwait*

**Name and address for correspondence*

ABSTRACT: The Minagish Oolite occurs in the Middle Minagish Member of the Minagish Formation (Berriasian–Valanginian) in Kuwait. Ten distinct lithofacies are recognized, which suggest sedimentation on a homoclinal carbonate ramp. A relatively small proportion of the Minagish Oolite (< 15%) consists of oolitic grainstone (Lithofacies 2), and this is confined to the lower part of the oil column. The dominant lithofacies comprises peloidal packstones to grainstones (Lithofacies 3). Sedimentation was highly storm-influenced, with significant reworking of shallow-water, inner-ramp skeletal allochems into the mid-ramp. The high level of reworking is believed to account for the relatively high proportion of grainstone and poorly washed packstones in the inner mid-ramp setting. The reservoir is interpreted as the product of sedimentation within late highstand, lowstand, and transgressive systems tracts, which together represent a low- (third?) order relative sea-level change. Within each systems tract, laterally correlatable flooding surfaces at the tops of parasequences are directly overlain by thin units of bioturbated wackestones to packstones (Lithofacies 7). These wackestones to packstones are interpreted as deeper-water, outer-ramp environments, and indicative of higher-frequency, fourth- or fifth-order, cyclicity. There is strong evidence of a southwestward lateral facies change into more argillaceous limestones ("marls") in the upper part of the Minagish Oolite. The geometry of the transition suggests that it marks the extreme fringe of a shallow-water clastic system. It represents the earliest evidence of delta progradation in the early Cretaceous of the Kuwait area. Evidence of associated shallowing is absent, and it seems that tectonic uplift in the hinterland was more influential than relative sea-level change.

Intense micritization has generated high proportions of microporosity, and it is the distribution of these micropores which mostly influences permeability. The best reservoir facies are grainstones of Lithofacies 2 and 3, where the pore network is macropore-dominated and microporosity is concentrated within micritized allochems. More heterogeneous packstones of Lithofacies 3 and 5 have mixed pore systems, whilst wackestones and packstones of Lithofacies 7 and 8 have micropore-dominated pore networks. In these samples, the pore network is dominated by interparticle micropores, and macroporosity is rare and isolated. These microporous facies typically form laterally correlatable beds above flooding surfaces and are capable of forming baffles and barriers to vertical transmissibility. Overall, the proportion of facies exhibiting mixed and microporous pore systems increases upwards through the reservoir, and hence there is a corresponding decrease in reservoir quality. During the later stages of production, as the oil–water contact rises, increasingly detailed understanding of the reservoir architecture wil be required to maintain production levels. The lateral facies change at the top of the reservoir allied to increased compartmentalization indicates that a more comprehensive secondary recovery scheme will be required in this part of the reservoir.

## INTRODUCTION

The Umm Gudair Field, Kuwait (Fig. 1) comprises two separate culminations, East and West Umm Gudair. Where East Umm Gudair straddles the boundary with the Partitioned Neutral Zone, a region jointly administered with Saudi Arabia, it is designated South Umm Gudair Field (Fig. 1). The discovery well was drilled in 1962, and there has been a major expansion of activity since 1993. Hydrocarbons are found at several levels in the Jurassic and Cretaceous sequence, but development has been focused largely on the Lower Cretaceous Middle Minagish Oolite Member reservoir (Fig. 2), which is the object of this study. This work builds upon previous reservoir description of the field (Sungur, 1996), and this paper aims to provide a detailed description of the Minagish Oolite reservoir, outlining lithofacies associations, diagenetic modification, and pore types in a sequence stratigraphic framework. These permit an evaluation of reservoir architecture and the controls on reservoir potential. The reservoir description will be a fundamental tool for long-term management of the Minagish Oolite reservoir.

## STRATIGRAPHY AND PALEOGEOGRAPHY

The Umm Gudair Field, Kuwait, is located on the southwestern margin of the Kuwait arch and is structurally simple, comprising the two separate anticlinal culminations of East and West Umm Gudair (Carman, 1996). The Minagish Formation was deposited on an eastward-prograding carbonate ramp system on the north–south trending Arabian Shelf (Alsharhan and Nairn, 1986; Longacre and Ginger, 1988). Sedimentation was initiated in the Cretaceous, following transgression of the late Jurassic intrashelf basin, within which interbedded salt and anhydrite–limestone facies of the Gotnia Formation and the overlying Hith Formation were deposited (Ali, 1994). The Minagish Oolite is the Middle Minagish Member of the Lower Cretaceous (Berriasian–Valanginian) Minagish Formation (Sungur, 1996) (Fig. 2). The Lower Minagish Member is a low-permeability, dolomitized section, and dense limestone and marls of the Upper Minagish Member form the seal to the productive Minagish Oolite (Sungur, 1996). The Minagish Oolite is equivalent to the Ratawi Oolite reservoir of the Partitioned Neutral Zone (Longacre and Ginger,

Middle East Models of Jurassic/Cretaceous Carbonate Systems
SEPM Special Publication No. 69, Copyright © 2000
SEPM (Society for Sedimentary Geology), ISBN 1-56576-075-1, p. 273–286.

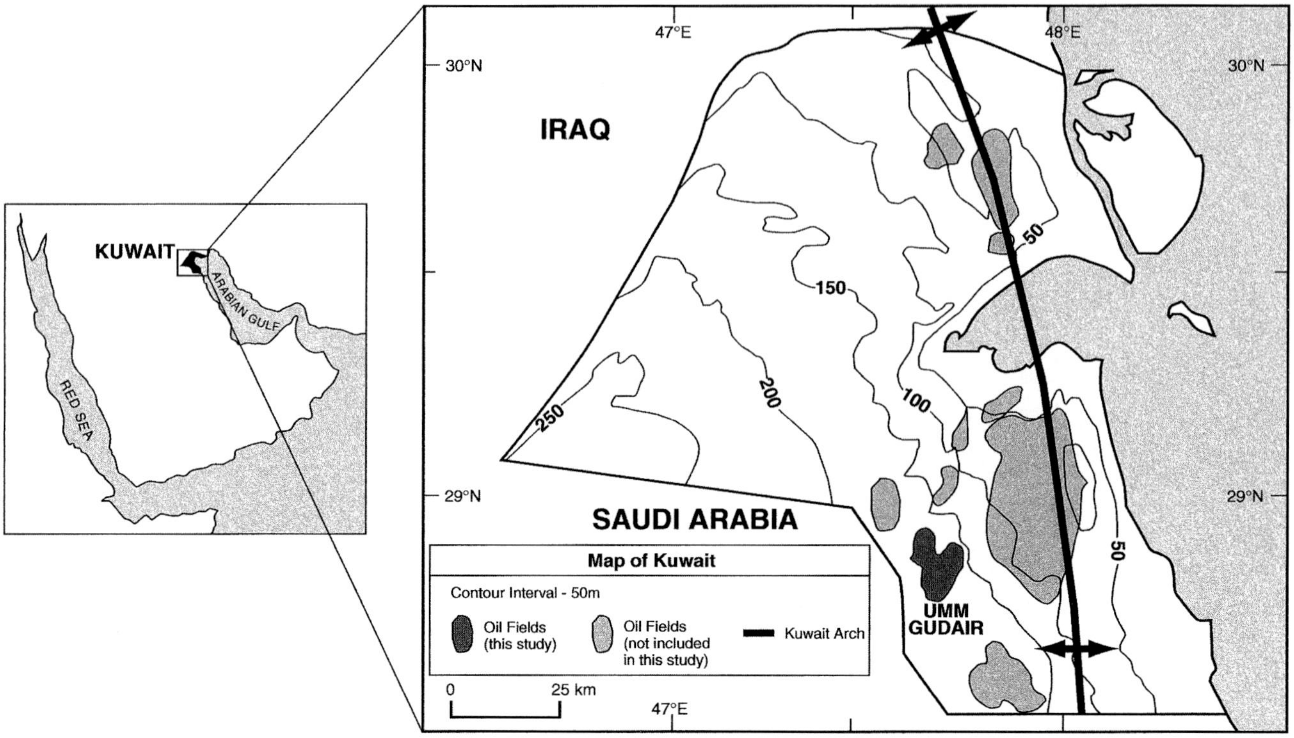


FIG. 1.—Location of the Umm Gudair Field, West Kuwait, showing topographic contours.

1988), the Ratawi reservoirs of northern Saudi Arabia (Ayres et al., 1982), and the Yamama Formation reservoirs of southern Iraq (Sadooni, 1993).

## DATABASE AND ANALYTICAL TECHNIQUES

Detailed core description was undertaken from the reservoir intervals in seven wells. One hundred and twelve thin sections, prepared from core plug trims, were examined from wells A, C, D, E, and F and a further thirty-eight thin sections from four wells. Thin sections were impregnated with blue-dyed resin and stained for carbonate with combined alizarin red-S and potassium ferricyanide (Dickson, 1965). Modal analysis was conducted on an all samples, by counting 200 points for mineral identification. Macroporosity was recorded over the same interval using isolator channels. A stage interval of 0.33 mm and track spacing of 2 mm was used to ensure a minimum of three traverses of the thin section. Core analysis data was available from all wells except well B.

Pore systems were examined in detail by scanning electron microscopy of thirty-five samples. Pieces of core that had been cleaned and gold coated prior to analysis were examined using a JEOL JSM-35CF microscope with a PGT System 4 EDS attachment. Operating conditions were set at 15 kV. Examination of ten samples under cathodoluminescence employed a Technosyn cold-cathode luminoscope 8200 MkII, fitted to a Nikon Optiphot-pol microscope. The vacuum was held at 0.2 atm and the beam current at approximately 650 μm. An accelerating voltage of 15–20 kV was used.

## LITHOFACIES ASSOCIATIONS AND SEQUENCE STRATIGRAPHIC FRAMEWORK

Ten lithofacies can be identified in the Minagish Oolite Member, based upon core description and supported by microfacies analysis (Table 1). These lithofacies effectively form a continuous spectrum, with gradational boundaries between different types, broadly relating to deposition in increasing water depths. Individual lithofacies are described below and their depositional environment assessed with respect to the nomenclature of Burchette and Wright (1992).

### *Lithofacies 1*

This lithofacies forms a minor component of the cored dataset (Table 1). It is markedly heterolithic and comprises a hybrid association of finely interlaminated peloidal packstones and grainstones and sparsely bioclastic mudstone beds (Fig. 3A). A relatively diverse, though not abundant, range of skeletal allochems within the packstone and grainstone components includes foraminifera (miliolids, uniserial and biserial), algal debris (*Lithocodium/Bacinella*, dasycladaceans, and *Permocalculus*), and echinoid fragments. Beds are mostly < 1 m thick, and this lithofacies is intimately associated with Lithofacies 2. Core analysis plugs are biased to coarser cleaner laminae, and are not representative of the mudstone facies. The packstones and grainstones typically exhibit very high helium porosities (23.4–24.7%; Fig. 6) and permeabilities (97–250 md; Fig. 4A). It is anticipated that the mudstone beds have lower porosities and insignificant permeabilities.

Lithofacies 1 caps upward-shoaling parasequences, directly overlying oolitic grainstones of Lithofacies 2 (Fig. 5A), implying a shallow-water origin. The sparse skeletal component to the mudstone beds suggests that deposition was within a low-energy, probably lagoonal, environment with cleaner, more skeletal packstone and grainstone laminae deposited by washover into the back-shoal, behind the oolitic sand sheet, during storms.

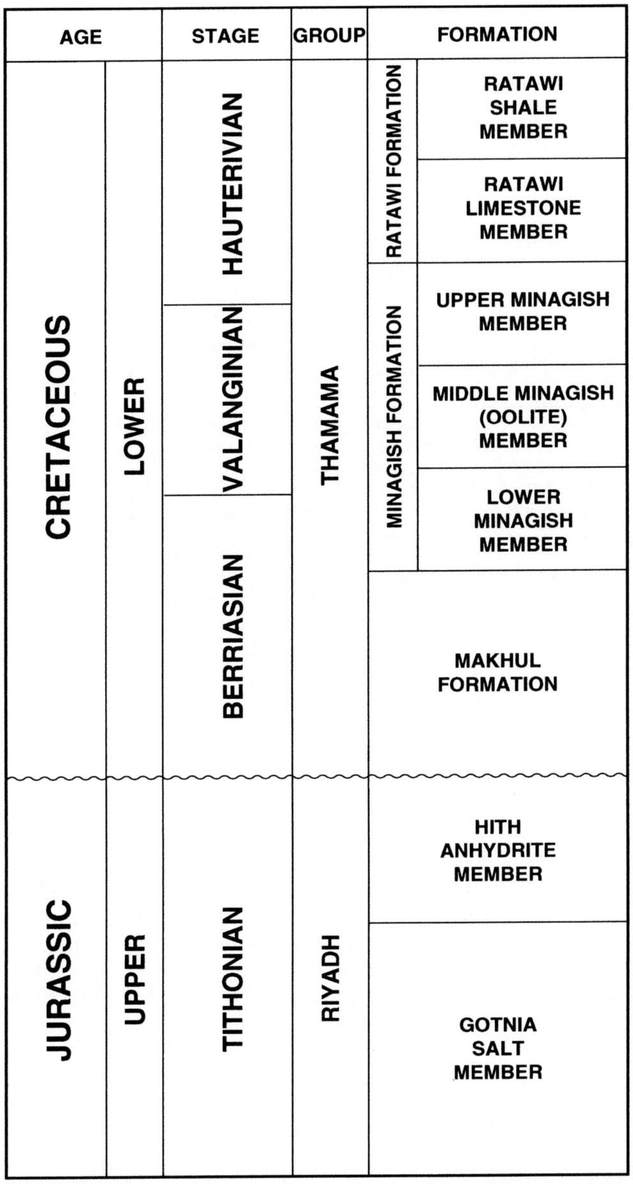


FIG. 2.—Stratigraphy of the lower part of the Thamama Group, Kuwait.

### *Lithofacies 2*

Lithofacies 2 accounts for a significant component of the cored dataset (Table 1). It comprises oolitic grainstones and peloidal oolitic grainstones (Fig. 3B), which are dominated by ooids, and rarely pisoids, with subordinate peloids. Although ooids usually maintain remnant concentric and radial microstructure, they are heavily micritized (Fig. 6A, B). Intraclasts and composite grains are also present, along with a low-abundance skeletal assemblage including foraminifera (dominantly miliolids) and echinoid and bivalve debris, all commonly with an oolitic coating. The dominant form of this lithofacies is as upward-fining beds, approximately 1 m thick (Fig. 5A). A second type, comprising trough cross-bedded oolites (locally bidirectional), is also recognized. Cross-bed sets are typically less than 0.5 m thick, and some set boundaries host simple burrows (Fig. 3B). Interlaminated medium- to coarse-grained ooids and very fine- to fine-grained peloids commonly define lamination (Fig. 6B).

Helium porosities (17.4–28.0%; Fig. 4A) and horizontal permeabilities (51.0–553 md; Fig. 4A) are mostly very good in Lithofacies 2. On open-hole logs, this facies expresses a distinctive bow-shaped sonic response and a very low gamma-ray signature.

The cross-bedded oolitic grainstones are interpreted as representative of *in situ* deposition in very shallow water, with sedimentation influenced by wave and tidal currents. Some units may represent tidal shoals and/or channels in flood/ebb deltas, suggestive of a nearshore inner-ramp setting, and possibly characteristic of a shallow barrier to a broad lagoon (cf. Wright, 1986). The more massive, upward-fining units are thought to be storm-reworked ooids, deposited offshore of the main area of ooid production, probably within the shallow mid-ramp.

### *Lithofacies 3*

Lithofacies 3 is the dominant lithofacies in the Minagish Oolite (Table 1). It comprises peloidal grainstone and skeletal peloidal packstone/grainstone (Figs. 3C, 6C, D), intraclastic peloidal packstone/grainstone, and peloidal packstone. A high-abundance, high-diversity, intensely micritized skeletal assemblage is recognized, which is dominated by benthic foraminifera (miliolids, textularids, rotalids, *Nautiloculina*, and *Trocholina*) and algal debris (principally *Lithocodium/Bacinella*, less commonly, *Permocalculus*, and dasycladaceans), with slightly less abundant bivalve, echinoid, gastropod, coral, and stromatoporoid debris (Fig. 6C, D). Partial micritization and allochem morphology suggests many peloids are micritized skeletal allochems (especially forams). Shallow water skeletal allochems, in particular miliolids and *Lithocodium/Bacinella*, are significantly more abundant in Lithofacies 3 in the tops of upward-coarsening parasequences than in the bases, within most of the reservoir. Towards the top of the reservoir, however, composition is not significantly different between the tops and the bases of individual parasequences. Locally, bimodal and trimodal beds ≤ 1 m thick are recognized which are characterized by very coarse-sand to pebble-grade, *Lithocodium*-encrusted, skeletal allochems and/or intraclasts, mostly of peloidal packstone and grainstone.

Beds are typically 1–2 m thick and upward fining, building up into larger-scale upward-coarsening sequences (Fig. 5B, C). Individual beds may be cross-bedded and are occasionally dissected on their upper surfaces by cylindrical and branching *Thalassinoides*-type burrows, implying firmground development. Very similar facies dominate the Ratawi Oolite reservoir of Wafra Field (Longacre and Ginger, 1988) and the Yamama Formation reservoirs of southern Iraq (Sadooni, 1993). Lithofacies 3 has heterogeneous reservoir properties, with poor to excellent helium porosities (2.5–35.3%; Fig. 4B) and horizontal permeabilities (0.01–4652 md; Fig. 4B). It is recognized by a consistently low gamma-ray curve, and somewhat variable sonic and density log porosity profiles.

The high degree of sediment reworking in these upward-fining sequences, with localized cross-bedding, suggests that these beds are amalgamated storm beds, forming a broad belt of carbonate sand in the mid-ramp (cf. Burchette and Wright, 1992). Very coarse-grained, trimodal beds probably represent individual storm events, and can be correlated to adjacent wells. Bathymetric ranges for both miliolids and *Lithocodium/Bacinella* suggest production of these allochems largely in water depths of up to 10 m, possibly up to 30 m, with production decreasing at increasing water depths as the depth of light penetration was

TABLE 1.—Summary of lithofacies characteristics for the Minagish Oolite in the Umm Gudair Field.

| Lithofacies | Component lithotypes | Biogenic/ sedimentary structures | Volume of core dataset | Depositional environment |
|---|---|---|---|---|
| 1 | Thinly interlaminated peloidal packstones, grainstones, and thin mudstone beds | thinly laminated | 0.1% | Peritidal/lagoonal |
| 2 | Oolitic grainstones | some trough cross-bedded | 13.9% | Inner-ramp sand shoal/ storm reworked into mid-ramp |
| 3 | Peloidal packstones–grainstones | bioturbated bed tops | 45.6% | Storm-dominated, mid-ramp sand sheet |
| 4 | Peloidal packstones–grainstones | cross-laminated–cross-bedded | 2.7% | Wave- and storm-dominated sedimentation in the inner to mid-ramp |
| 5 | Shell rich packstones and grainstones | — | 2.9% | Reworked rudist bioherms developed within a relatively low-energy, deep mid-ramp setting |
| 6 | Bioclastic peloidal packstones–grainstones | — | 6.6% | Mid- to outer-ramp |
| 7 | Bioclastic peloidal wackestones–packstones | bioturbated | 17.2% | Low energy, outer-ramp. Representative of deepening events — immediately overlie flooding surfaces |
| 8 | Bioclastic wackestones, peloidal packstones and grainstones | dissected by coarse-grained sediment-filled burrows | 9.4% | Usually associated with deepening and outer-ramp sedimentation. Firmgrounds |
| 9 | Nodular wackestones | burrowed | 1.2% | Deep-water–outer-ramp |
| 10 | Green-gray argillaceous mudstones | — | 0.5% | Distal fringes of fluviodeltaic system |

exceeded (Banner and Simmons, 1994). Deposition of Lithofacies 3 between fair-weather and storm-weather wave base at water depths of potentially up to 50 m (e.g., Tucker and Wright, 1990) implies significant offshore transport of shallower-water allochems, from above fair-weather wave base, into the mid ramp. The more diverse skeletal assemblage described from facies towards the tops of parasequences suggests a progressive loss of these allochems with depth, potentially a function not only of decreasing light penetration, but also of the waning effects of storm currents at depth. The decrease in skeletal diversity towards the top of the reservoir might appear to imply a subtle overall upward deepening. A lateral transition into argillaceous limestones of Lithofacies 10, however, indicates that the lower skeletal diversity is associated with an increased clay component, and that siliciclastic poisoning is also involved.

### *Lithofacies 4*

This lithofacies is volumetrically minor (Table 1) and comprises thinly laminated, very well sorted, and very fine- to fine-grained peloidal grainstones and packstones. The lamination is typically low angle to horizontal, but hummocky and swaly cross- stratification is also commonly recognized. These intervals have a very low gamma signature but very high porosities and high resistivities in the oil zone. Hummocky and swaly cross-stratification is consistent with wave- and storm-dominated deposition of this facies in the mid-ramp (cf., Burchette and Wright, 1992).

### *Lithofacies 5*

This is a distinctive lithofacies, which forms a minor component of the total cored interval (Table 1). Lithofacies 5 comprises almost entirely skeletal peloidal packstone (Fig. 3D). Skeletal allochems are dominated by rudist and "brain" coral fragments (Fig. 3D) and also include fragmented scleractinian corals, bivalves (including large fragments of *Inoceramus*), echinoid and crinoid fragments, and minor foraminifera (rotalids, rare miliolids) (Fig. 6E). No unequivocal examples of rudists and corals in life position were observed. Beds are typically < 1 m thick, commonly resting directly above flooding surfaces (Fig. 6C), and are structureless except for a weak background bioturbation mottling. Reservoir properties are variable for this lithofacies, but helium porosities are typically very good (12.8–24.2%; Fig. 4C) and horizontal permeabilities poor to moderate (1.6–395 md; Fig. 4C). Lithofacies 5 does not carry a distinctive open-hole log signature.

Longacre and Ginger (1988) interpreted a lithology similar to Lithofacies 5 as providing evidence of a reef a short distance beyond the eastern margin of the Wafra Field. A somewhat different interpretation is proposed here. The observed skeletal assemblage is suggestive of localized buildups, possibly bioherms, of rudists and corals. Fragmentation implies breakup of these bioherms potentially during storms, although the very coarse size of much of the skeletal debris suggests transportation was minor. Given that rudists largely require relatively low rates of sediment accumulation in order to remain stable on or within the sediment (Skelton, 1997), they often favor quite low-energy environments, and are easily fragmented. Elsewhere in the Middle East, rudist bioherms are found towards the top of upward-shallowing cycles (e.g., Burchette and Britton, 1985). The relative absence of shallow-water biota and the characteristic position of this facies towards the bases of upward-shallowing parasequences are more consistent, however, with quieter-water accumulation of these buildups in the deep mid-ramp. A similar association is seen in the Cretaceous of southwest Texas, where rudist bioherms are found closely overlying maximum flooding surfaces

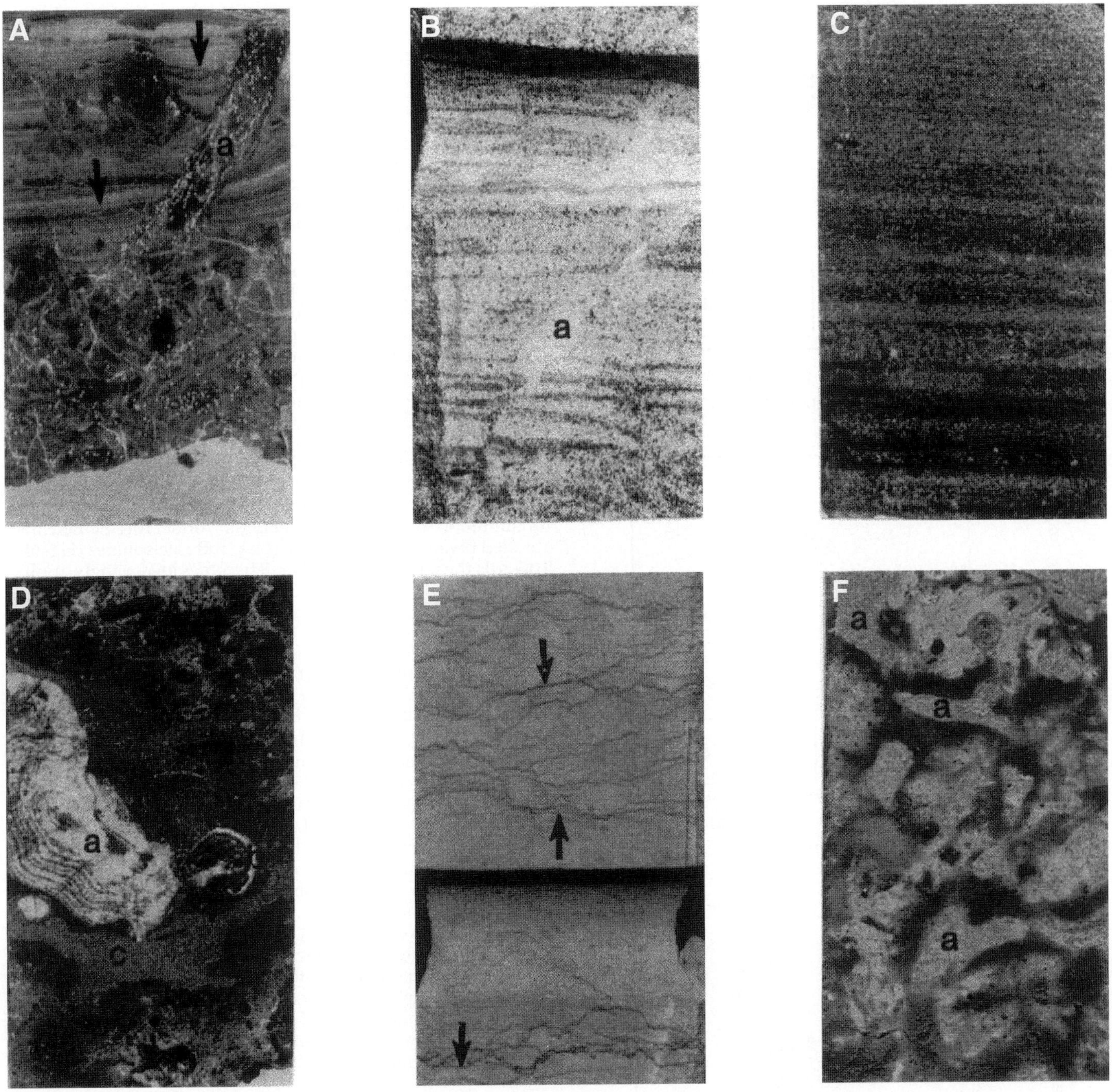


FIG. 3.—Selected core photos of lithofacies within the Minagish Oolite. A) Lithofacies 1. Mudstone with thin, coarse-grained peloidal and oolitic laminae (arrows). Dissected by burrow (a). B) Lithofacies 2. Laminated oolitic grainstone dissected by subtle burrow (a). C) Lithofacies 3. Laminated peloidal grainstone. D) Lithofacies 5. Skeletal packstone with large fragment of "brain" coral (a) and rudist shell (b). Matrix contains well defined burrows (c). E) Lithofacies 7. Skeletal wackestone with well developed stylolites (arrows). Note the intense cementation that surrounds these stylolitized zones. F. Lithofacies 8. Skeletal wackestone with well preserved *Thallasinoides*-type burrow network (a), which has focused differential cementation.

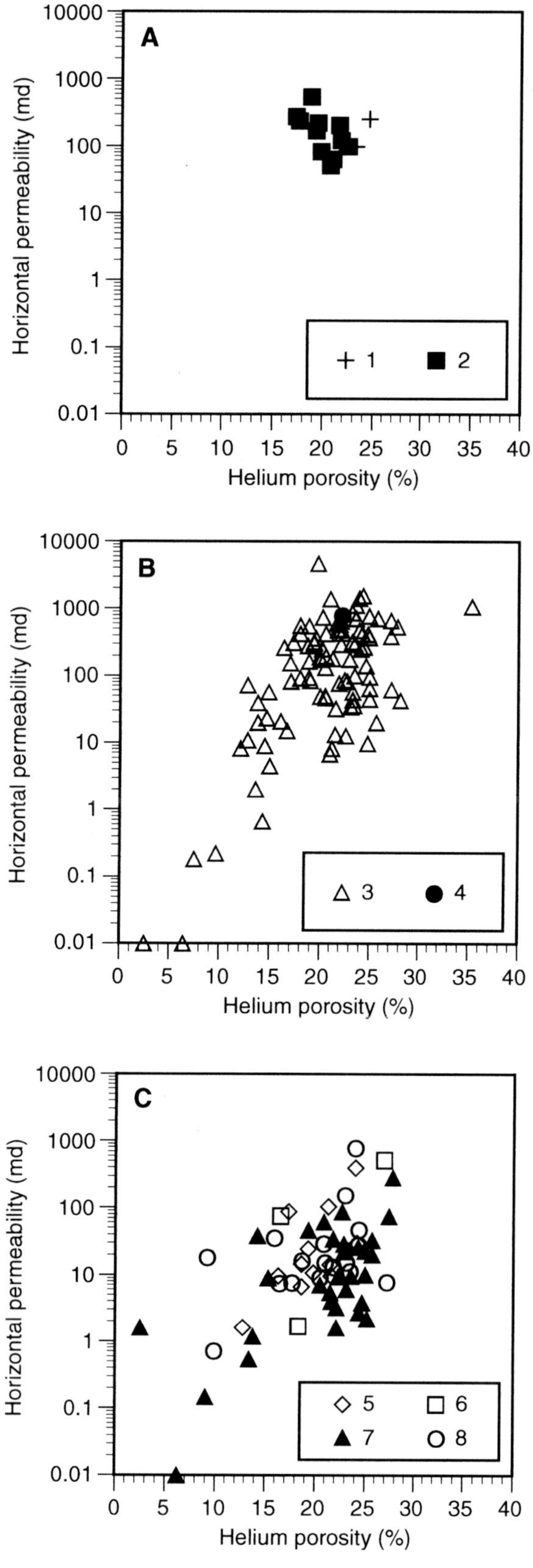


FIG. 4.—Reservoir properties of selected samples from the Minagish Oolite according to lithofacies. Symbols and numbers represent lithofacies.

(Trevor Burchette, personal communication). This would have permitted growth under relatively quiet conditions, with reworking only during major storms.

### *Lithofacies 6*

Lithofacies 6 is a minor component of the total cored interval of the Minagish Oolite (Table 1). It is dominated by skeletal peloidal packstones and typically contains abundant foraminifera (miliolids, textularids, and rotalids) as well as echinoid debris and *Lithocodium/Bacinella*. Helium porosities are mostly very good (16.5–26.9%; Fig. 4C), whilst horizontal permeabilities are more variable (1.70–507 md; Fig. 4C). Lithofacies 6 does not have a distinctive open-hole log signature for identification in uncored wells. The muddy texture and relatively diverse skeletal assemblage within this facies is suggestive of mid to outer ramp sedimentation with a continuous influence on deposition by storms.

### *Lithofacies 7*

Lithofacies 7 is a volumetrically significant lithofacies (Table 1). It comprises bioturbated peloidal packstone and wackestone (Fig. 3E), with a low-abundance skeletal assemblage including foraminifera (rotalids, rare miliolids), often fragmented and abraded bivalve and echinoid debris, and calcispheres (Fig. 6F). Towards the top of the reservoir, it becomes increasingly argillaceous, with a corresponding increase in stylolitization. Lithofacies 7 typically forms thin (< 1 m) beds that have a characteristic, pervasive bioturbation mottling. It abruptly overlies oolitic grainstones of Lithofacies 2 and skeletal peloidal packstones and grainstones of Lithofacies 3 (Fig. 5A–D). Lithofacies 7 has good measured helium porosities (2.5–27.8%; Fig. 4C) but poor horizontal permeabilities (0.01–284 md; Fig. 4C). Within the oil leg, this facies has lower resistivity than shallower, cleaner, grain-supported lithofacies, reflecting higher water saturations. Where present, the increase in argillaceous debris results in a gamma-ray response relative greater than other, cleaner facies.

The low abundance of skeletal allochems and the muddy, bioturbated fabric suggests low-energy sedimentation under slow rates of sediment accumulation. Such conditions may occur within a lagoonal environment in the inner ramp, or farther offshore in the outer ramp. Given the absence of a diagnostic shallow-water skeletal assemblage, the presence of calcispheres, and the often highly fragmented and abraded nature of the bivalve and echinoid debris, this facies is interpreted as a deep-water outer ramp packstone–wackestone. Because many of the basal contacts of Lithofacies 7 abruptly overlie Lithofacies 2 and 3 (Fig. 5), they mark the bases of parasequences and are interpreted as deepening events above flooding surfaces.

### *Lithofacies 8*

Lithofacies 8 makes up a significant component of the cored dataset (Table 1). It comprises peloidal and skeletal packstones and wackestones, with a low-abundance and low-diversity skeletal assemblage that includes foraminifera (principally rotalids) and echinoderm and bivalve debris. This lithofacies is distinguished by prominent coarser-grained burrow fills (including *Thalassinoides*), which dissect the fine-grained background lithologies (Fig. 3F). Towards the top of the reservoir section, many examples are thin (decimeter scale) and preferentially calcite cemented (Fig. 3F), with the degree of cementation increasing upwards. In several examples, preserved burrow diameters likewise increases upward, which is consistent with early cementation retarding compaction. Furthermore, some burrows contain

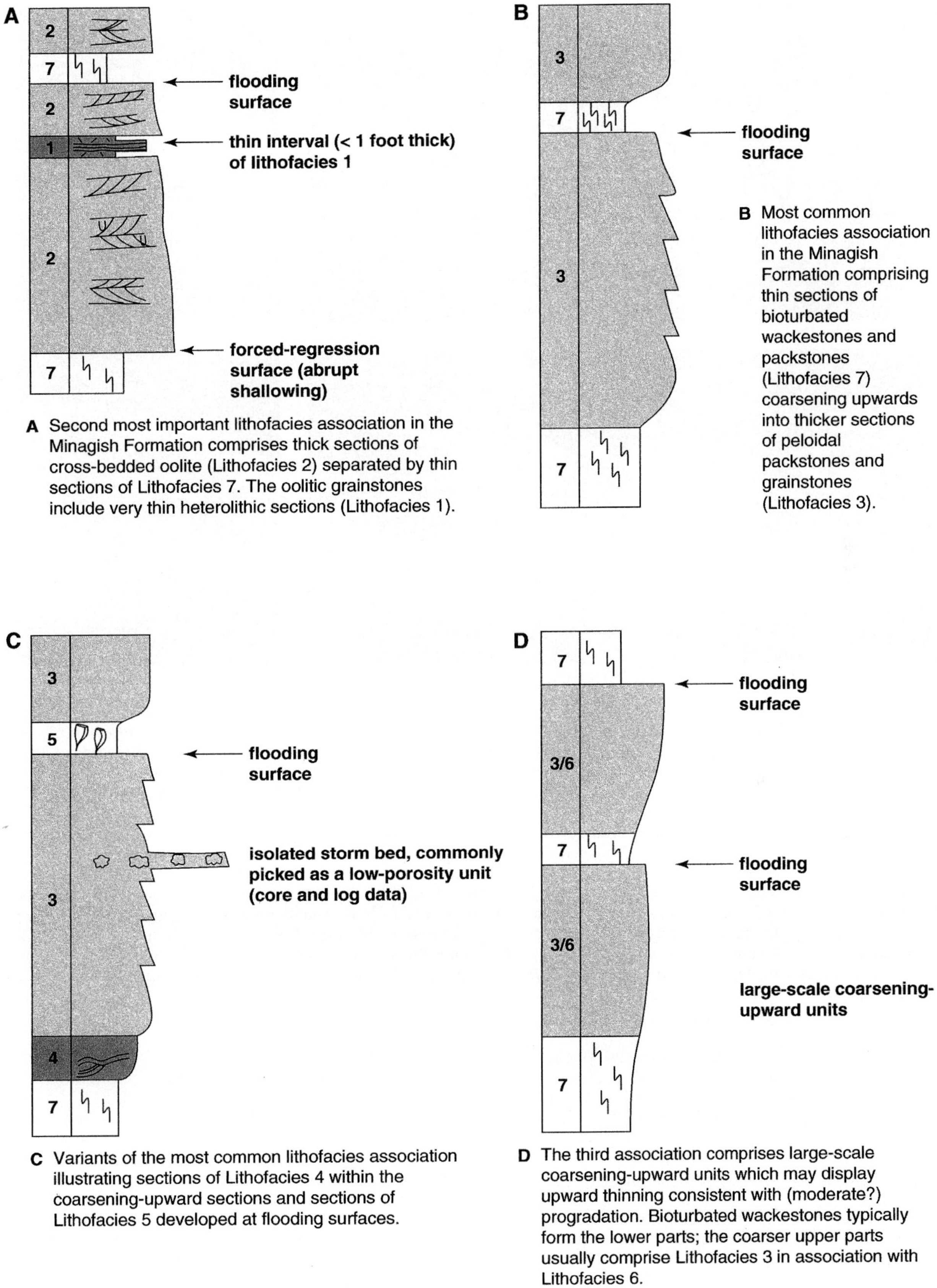


**A** Second most important lithofacies association in the Minagish Formation comprises thick sections of cross-bedded oolite (Lithofacies 2) separated by thin sections of Lithofacies 7. The oolitic grainstones include very thin heterolithic sections (Lithofacies 1).

**B** Most common lithofacies association in the Minagish Formation comprising thin sections of bioturbated wackestones and packstones (Lithofacies 7) coarsening upwards into thicker sections of peloidal packstones and grainstones (Lithofacies 3).

**C** Variants of the most common lithofacies association illustrating sections of Lithofacies 4 within the coarsening-upward sections and sections of Lithofacies 5 developed at flooding surfaces.

**D** The third association comprises large-scale coarsening-upward units which may display upward thinning consistent with (moderate?) progradation. Bioturbated wackestones typically form the lower parts; the coarser upper parts usually comprise Lithofacies 3 in association with Lithofacies 6.

FIG. 5.—Common lithofacies associations in the Minagish Oolite.

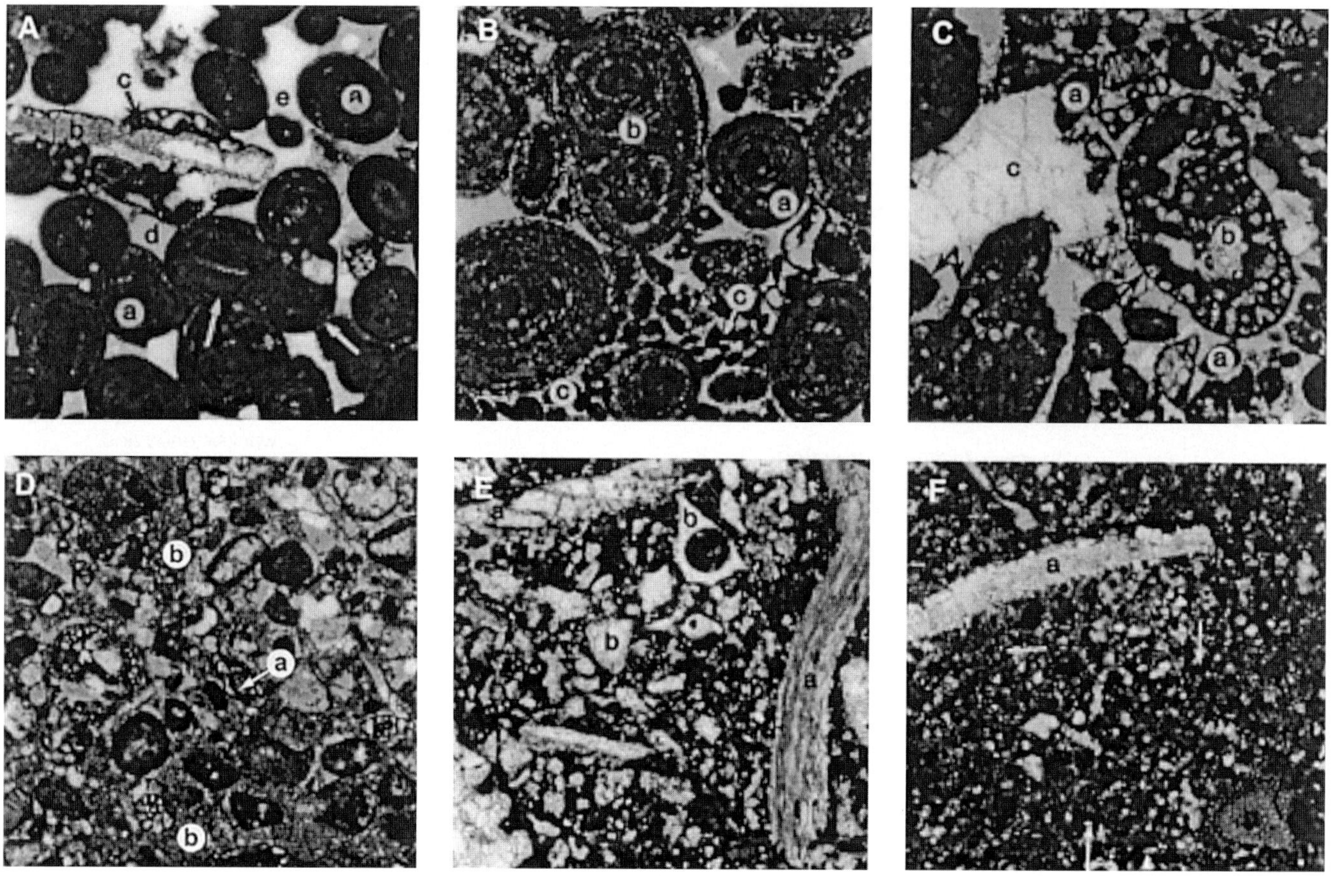

FIG. 6.—Selected photomicrographs of microfacies and diagenetic features. A) Oolitic grainstone (Lithofacies 2), dominated by heavily micritized ooids (a), which display some remnant texture. Bored bivalve shell is encrusted by *Lithocodium/Bacinella* (b, c). Interparticle porosity is partially occluded by nonferroan dolomite often following compaction (d), which has destroyed pore throats (e). Field of view, 4 mm. B) Peloidal oolitic grainstone (Lithofacies 2) comprising very coarse-grained ooids (a) and composite grains (b), and fine-grained peloids (c), which are principally distributed along laminae. Field of view, 4 mm. C) Skeletal peloidal packstone (Lithofacies 3), dominated by peloids(a) and benthic foraminifera (b, miliolids and *Trocholina*) with intraclast encrusted by *Lithocodium/Bacinella* (arrows). Allochems are coated by thin fringing nonferroan dolomite cement and primary and secondary interparticle porosity occluded by subsequent sparry nonferroan dolomite (c). Field of view, 4 mm. D) Skeletal peloidal packstone (Lithofacies 3) with diverse skeletal assemblage including foraminifera (a, dominantly miliolids), echinoid fragments, micritized bivalve fragments, and *Lithocodium/Bacinella*. Micrite matrix has been intensely neomorphosed (b). Field of view, 4 mm. E) Skeletal packstone (Lithofacies 5), with a less diverse skeletal assemblage than in lithofacies 3 and dominated by fragmented rudist debris (a). Note the complete occlusion of biomolds by nonferroan dolomite (b). Field of view, 4 mm. F) Skeletal peloidal packstone–wackestone, with a low-abundance, low-diversity skeletal assemblage comprising bivalve debris (a) and, in this sample, a reworked fragment of *Permocalculus* (b). Peloids are potentially intensely micritized forams. Field of view, 4 mm.

distinctive dark green glauconitic grains and phosphatized skeletal allochems. Beds are generally < 1 m thick. Helium porosities are mostly very good (9.2–27.2%; Fig. 4C) and horizontal permeabilities poor-moderate (0.70–771 md; Fig. 4C). Except where Lithofacies 8 is dissected by numerous stylolites, resulting in an increase in the gamma-ray response and a reduction in log porosity, it is difficult to differentiate this facies on open-hole logs.

The low-diversity skeletal assemblage and absence of shallow-water indicators in this lithofacies is consistent with sedimentation in the outer ramp. The burrows record decreases in sedimentation rate, either under decreasing flow following storms or, more significantly, during deepening. In these cases, the burrowed surface is a flooding surface, and the association of cementation, burrowing, upward increases in burrow diameter, and glauconite implies firmground development.

### *Lithofacies 9*

Lithofacies 9 makes up a minor proportion of the total cored interval (Table 1). It comprises characteristically dense, nodular wackestones and mudstones, which have a low-abundance, low-diversity skeletal assemblage comprising calcispheres and rare red algae and benthic forams. Common, incipient clay-rich stylolitic seams are common and result in low porosity zones and erratic gamma-ray profiles on open-hole logs. The presence of calcispheres and the muddy texture of this lithofacies implies relatively deep-water sedimentation, possibly at > 60 m water depth (Banner and Simmons, 1994). The association of this facies with better constrained deeper-water Lithofacies 7 and 8 is consistent with this and suggests sedimentation in the outer-ramp or basinal environment.

### *Lithofacies 10*

Lithofacies 10 forms a negligible component of the total cored interval (Table 1). It comprises green-gray argillaceous mudstones to wackestones and is largely featureless. A monospecific skeletal assemblage consists of localized concentrations of *in situ Exogyra*. A high but erratic gamma-ray signature and higher neutron than density porosities distinguish this lithofacies.

The muddy texture of this facies indicates a quiet-water environment, and the argillaceous content suggests close proximity to a contemporaneous clastic environment. *Exogyra* has been described from more basinal settings (e.g., Kimmeridge Clay of western Europe; N. Cross, personal communication), which seems at odds with an increase in clastic sediment supply. Conversely Hughes (1997), in a comparison between the Shuaiba Formation and the modern Great Pearl Bank Barrier of the Arabian Gulf, depicts *Exogyra* as a component of lagoonal faunas behind an Albian rudist-dominated barrier. It is difficult, therefore, to make a definitive interpretation on the basis of depositional character alone. A more helpful observation is that Lithofacies 10 forms the basal beds of the overlying Upper Middle Minagish Member, and that it grades laterally into mid- to outer-ramp sediments of Lithofacies 3, 7, and 8. The nature of the transition is such that similar water depths must be invoked for Lithofacies 10. The depositional geometries in particular render it unlikely that Lithofacies 10 represents a deeper-water (necessarily intrashelf-basin) setting. Instead it is best interpreted as representative of the extreme fringes of a shallow-water clastic system.

A similar lateral facies change has been recognized in the uppermost beds of the Ratawi Oolite reservoir in Wafra Field (Longacre and Ginger 1988), suggesting that the change is of regional significance. For these reasons, the argillaceous limestones can be considered to represent the earliest evidence of delta progradation in the Early Cretaceous in Kuwait.

### *Lithofacies Organization and Trends*

Lithofacies similar to those described in this study have been described from the Ratawi Oolite reservoir of Wafra Field (Longacre and Ginger, 1988) and the Yamama Formation reservoirs of southern Iraq (Sadooni, 1993). The Minagish Oolite can be broken down into four principal units, each comprising characteristic lithofacies associations.

The uppermost unit is separated from the underlying unit on the basis of the lateral facies change described for Lithofacies 10. Relatively thin interbeds of Lithofacies 3, 7, and 8 characterize the central to northeastern part of the field, whereas the southwestern part is composed exclusively of nonreservoir argillaceous limestones. The base of this unit is picked at an easily correlated, strongly cemented bed that equates to the base of the argillaceous limestones in the southwestern part of the field.

The second unit, which is also the main reservoir section, is marked by a basal flooding surface in which outer-ramp packstones of Lithofacies 7 abruptly overlie oolitic grainstones (Lithofacies 2). The most common lithofacies association in this second unit comprises meter-scale, stacked upward-fining beds of Lithofacies 3 overlying Lithofacies 7 (Fig. 6B). Less commonly Lithofacies 4 separates beds of Lithofacies 3 and 7, or Lithofacies 5 abruptly overlies Lithofacies 3 (Fig. 5C). Overall, this results in a common organization of broadly upward-coarsening parasequences.

A third unit (the so-called "main oolite"), beneath the flooding surface at the top of the upper unit, is separated from the basal unit by an abrupt facies change from outer-ramp packstones and wackestones (Lithofacies 7) to inner-ramp oolitic grainstones (Lithofacies 2). Within this main oolite, the principal lithofacies association is between Lithofacies 2 and Lithofacies 7, which forms the basal part (up to 1.5 m thick) of thick (up to 12 m) homogeneous sections of oolitic grainstone (Fig. 5A). Thin beds of Lithofacies 1 are recognized, particularly towards the top of this oolitic section (Fig. 5A). Finally, the lower unit, beneath the main oolite, comprises large-scale coarsening-upward units with Lithofacies 7 at the base, grading up to coarser-grained Lithofacies 3 and 6 (Fig. 5D) with no ooids recognized beneath the sharp base of the oolitic sandsheet.

## DIAGENETIC MODIFICATION

The Minagish Oolite Member has undergone significant diagenetic modification, and a number of events have exerted an important control on the reservoir potential of individual facies.

### *Marine Diagenesis*

The most striking feature of the Minagish Oolite is the intense micritization that both skeletal and nonskeletal allochems have undergone as a result of boring in the submarine environment (Fig. 5A–C). In many cases, this has rendered skeletal allochems unrecognizable, such that true peloids, for example, cannot be readily differentiated from completely micritized foraminifera (Fig. 6C). Isopachous, acicular, or prismatic nonferroan calcite and dolomite locally coat micritized skeletal and nonskeletal allochems (Fig. 6C), indicating that boring preceded cementation (Fig. 7). Isopachous cements are most common within grainstones of Lithofacies 2, 3, and 4, which were deposited in the relatively high-energy inner and shallow mid ramp. They are therefore interpreted as marine cements precipitated under agitated, wave- and storm-dominated conditions.

### *Burial Diagenesis*

Syntaxial overgrowths and equant sparry nonferroan dolomite and calcite (Fig. 6A, C, E) occlude primary and secondary macropores. Syntaxial overgrowths preserve interparticle volumes and inhibit compaction, whilst equant sparry cements occlude compactionally modified pores (Fig. 6A). Syntaxial overgrowths envelop isopachous marine cements and are themselves enveloped by equant sparry calcite. Under cathodoluminescence, syntaxial overgrowths are typically nonluminescent, suggesting precipitation from oxidizing fluids; potentially meteoric porewaters driven downdip from the west or southwest, which is perceived to be more landward (Sungur, 1996). In contrast, equant sparry cements exhibit a finely zoned, dull-bright luminescence more indicative of precipitation from evolved, reducing burial fluids (Machel and Burton, 1991).

Within finer-grained, muddier samples (e.g., Lithofacies 7 and 8), small, poorly connected primary interparticle macropores did not permit a ready throughput of diagenetic fluids. Significant pore-occluding cementation was, therefore, inhibited and compactional modification was more intense. Dense concentrations of stylolites are commonly observed in Lithofacies 7 and 8, particularly towards the top of the reservoir, and are surrounded by calcite cemented zones up to 10 cm thick (Fig. 3E). Micrite

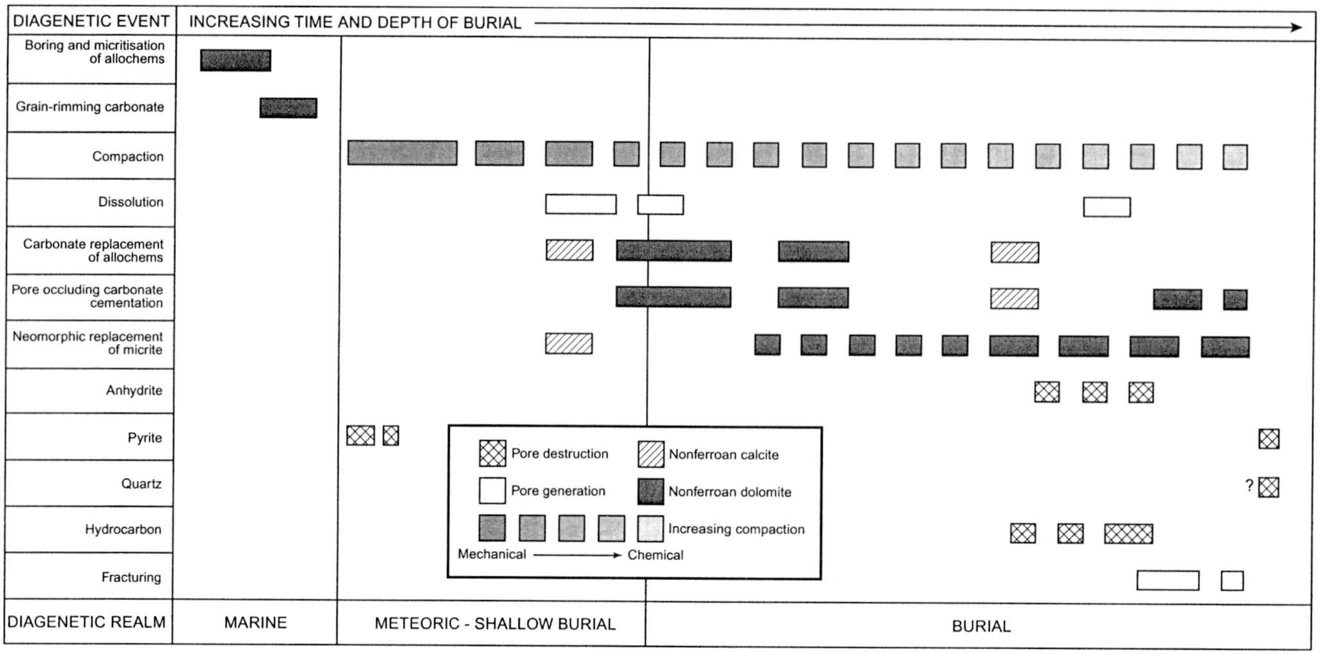


FIG.. 7.—Paragenetic sequence for the Minagish Oolite.

adjacent to stylolites has undergone intense neomorphism from the carbonate-saturated fluids generated by pressure dissolution. SEM analysis reveals that the size and crystallinity of micrite is increased from a few microns to up to 10 μm, such that interparticle microporosity has been destroyed.

Hydrocarbon is commonly concentrated along stylolites, where it is often diffusely replaced by pyrite (Fig. 7), suggesting that hydrocarbon emplacement at least partly preceded significant pressure dissolution. Few diagenetic events postdate compaction, but rare ferroan dolomite, with a saddle morphology, is precipitated along stylolites. Fracturing in the Minagish Oolite is not significant, but impersistent fractures are seen locally terminating against stylolites, implying a syncompactional or postcompactional origin (Fig. 7). These fractures are often partially cemented by nonferroan calcite, which also envelops saddle dolomite. A postcompactional hydrocarbon charge cannot be ruled out, and the corresponding circulation of organic acids may have generated some late-phase secondary macroporosity.

## PORE SYSTEMS

The Minagish Oolite Member is highly microporous, and therefore similar to many Lower Cretaceous carbonates in the Middle East (e.g., Longacre and Ginger, 1988; Moshier, 1989). Helium porosities differ little among the different lithofacies (Fig. 5) except where porosity has been occluded by carbonate cements. This has two important implications: the permeability of individual lithofacies cannot be implied by its porosity, and it is very difficult to differentiate individual lithofacies from porosity logs. It is therefore vital to understand the nature and distribution of pore types in order to understand the controls on reservoir potential.

The best reservoir rocks are the grainstones and clean packstones of Lithofacies 2 and 3, and less commonly Lithofacies 1, 4, and 6 (Fig. 4). These rocks have a well preserved primary interparticle macropore network, which may be enhanced by dissolution of skeletal allochems that were formerly composed of aragonite and high-magnesium calcite (Fig. 8A). Microporosity in these samples is concentrated largely in micritized allochems. Where there is a greater volume of micrite, mostly within Lithofacies 3, 5, and 6, the pore system is more heterogeneous. Primary interparticle macropores are smaller than in cleaner samples, and less well connected. Secondary macropores, in contrast, are typically well developed where large skeletal allochems have been dissolved. These biomolds have commonly been enhanced by matrix dissolution, creating channelized and vuggy pores, resulting in a strong contrast in pore sizes (Fig. 8B). Microporosity in these samples is both intraparticle, within micritized allochems, and interparticle, within the micrite, and permeabilities are therefore slightly lower than the more macroporous grainstones (Fig. 4). The poorest-quality rocks are the muddiest packstones and wackestones of Lithofacies 7 and 8 (Fig. 4C), which are dominated by interparticle micropores (Fig. 8C). These rocks contain only rare, isolated macropores, which are generally intraparticle, and therefore support only minimal flow. Sadooni (1993) described comparable pore systems in the equivalent Yamama Formation of Iraq, although he reported only microporosity (his "matrix porosity") in mud-grade sediments.

## DISCUSSION

### *Reservoir Architecture*

Previous studies have suggested that deposition of the Minagish Oolite Member was on a broad, eastward-prograding carbonate ramp (Alsharhan and Nairn, 1986; Longacre and Ginger, 1988; Sungur, 1996), and the results of this study are consistent with this. Understanding the depositional evolution of the Minagish Oolite is based upon interpretation of vertical successions and fieldwide correlations, themselves based upon recognition of key surfaces. These are principally flooding surfaces (Van Wagoner et al., 1990), which are most commonly recognized where outer-ramp packstones and wackestones of Lithofacies 7

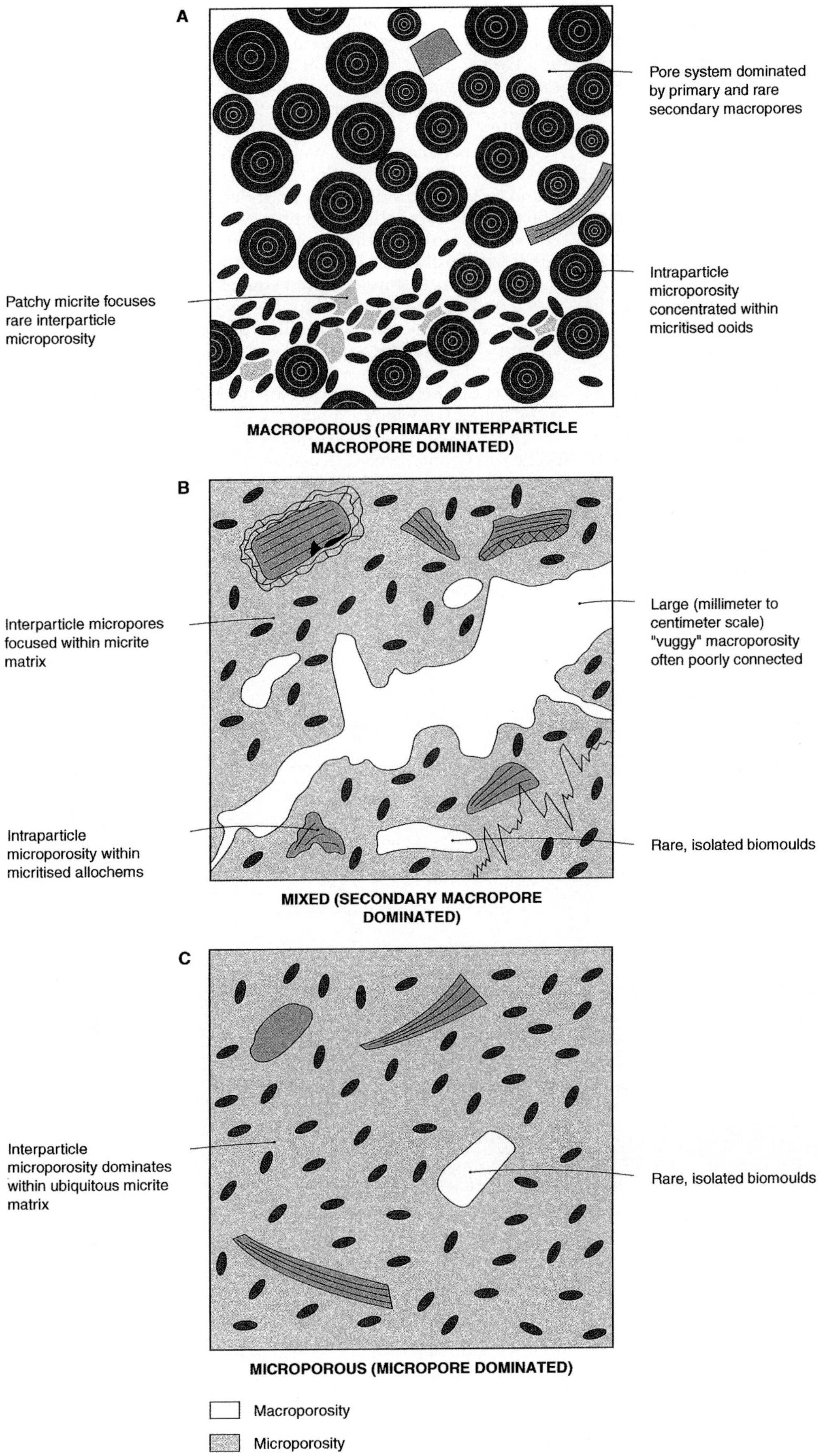


FIG. 8.—Generalized pore types in Minagish Oolite. A) Oolitic grainstone with well preserved primary interparticle macroporosity. Microporosity is dominantly intraparticle, within micritized allochems. B) Skeletal peloidal packstone dominated by large (millimeter-scale) macropores created by dissolution of skeletal allochems and enlarged by subsequent dissolution of the surrounding matrix. Isolated biomolds are also recognized. Microporosity is intraparticle, within micritized allochems, and interparticle, within micrite matrix. C) Peloidal packstone–wackestone, with a pore system dominated by interparticle micropores within the micrite matrix.

abruptly overlie shallow mid-ramp to inner-ramp packstones and grainstones of Lithofacies 2 and 3. Less commonly, forced-regression surfaces (Posamentier et al., 1992) are recognized, typically where inner-ramp coarse-grained grainstones of Lithofacies 2 or 3 directly overlie outer-ramp bioturbated packstones and wackestones of Lithofacies 7.

An approximate Berriasian–Valanginian age is assigned to the Minagish Oolite (Sungur, 1996). These two stages have a combined duration of approximately 10–12 million years (Haq et al., 1987; Harland et al., 1990). This duration suggests that up to three third-order cycles could be represented in the Minagish Formation as a whole. Haq et al. (1987) interpreted two major falls in global sea level at c. 126 and c. 128.5 Ma, respectively, sandwiching the Berriasian–Valanginian boundary. These two falls can be identified in the Cretaceous of Oman and U.A.E. (Aziz and El-Sattar, 1997), although absolute ages of c. 139 and c. 141.5 Ma are proposed there (Vahrenkamp, 1996, using time scale of Harland et al., 1990). Irrespective of their absolute age, these dates suggest that major falls in sea level could have influenced deposition and stratigraphic architecture of the Minagish Oolite. As described, detailed correlation suggests that the Umm Gudair Field can be divided into four distinct units: a basal unit, the main oolite, the main reservoir unit above the oolite, and a fourth unit characterized by a lateral facies change (Fig. 9). These units are interpreted as systems tracts within a single, low- (?third) order cycle. The component lithofacies associations form stacked, broadly upward-coarsening parasequences, interpreted to be representative of fourth- or fifth-order sea-level fluctuations.

Insufficient core data makes correlation of the unit below the oolite difficult. By comparison with the Minagish Field (Davies and Hollis, 1996), however, it appears to be progradational and may correspond to the upper part of a highstand systems tract (HST). The main oolite comprises three upward-shallowing and -thinning units, separated by deeper-water facies. Thickness variations between the two wells with the best core coverage over this interval are suggestive of shingled geometries between wells A and C (Fig. 9), though 3D seismic data to confirm this observation were not available at the time of the study. Shingled geometries, if proven, would imply progradation of the oolitic sandsheet, either within the late HST or within a lowstand systems tract (LST). The former model would suggest increased progradation of the oolitic sandsheet in response to a progressive decrease in accommodation space as the rate of sea-level rise decreases. The absence of a sequence boundary, either as an exposure surface or manifested by an increase in the abundance of meteoric cements, which could be representative of the subsequent lowstand, is inconsistent with such a model.

The marked basinward shift in facies at the base of the main oolite can be interpreted as a surface of forced regression at the top of the late HST, placing the main oolite within a lowstand systems tract. Progradation of the oolitic sandsheet would again have taken place in response to the limited accommodation space, and evidence of exposure might be anticipated to the southwest, in a more proximal position on the ramp. This model of deposition within the LST is consistent with the basinward migration of facies belts within the LST discussed by Tucker et al.

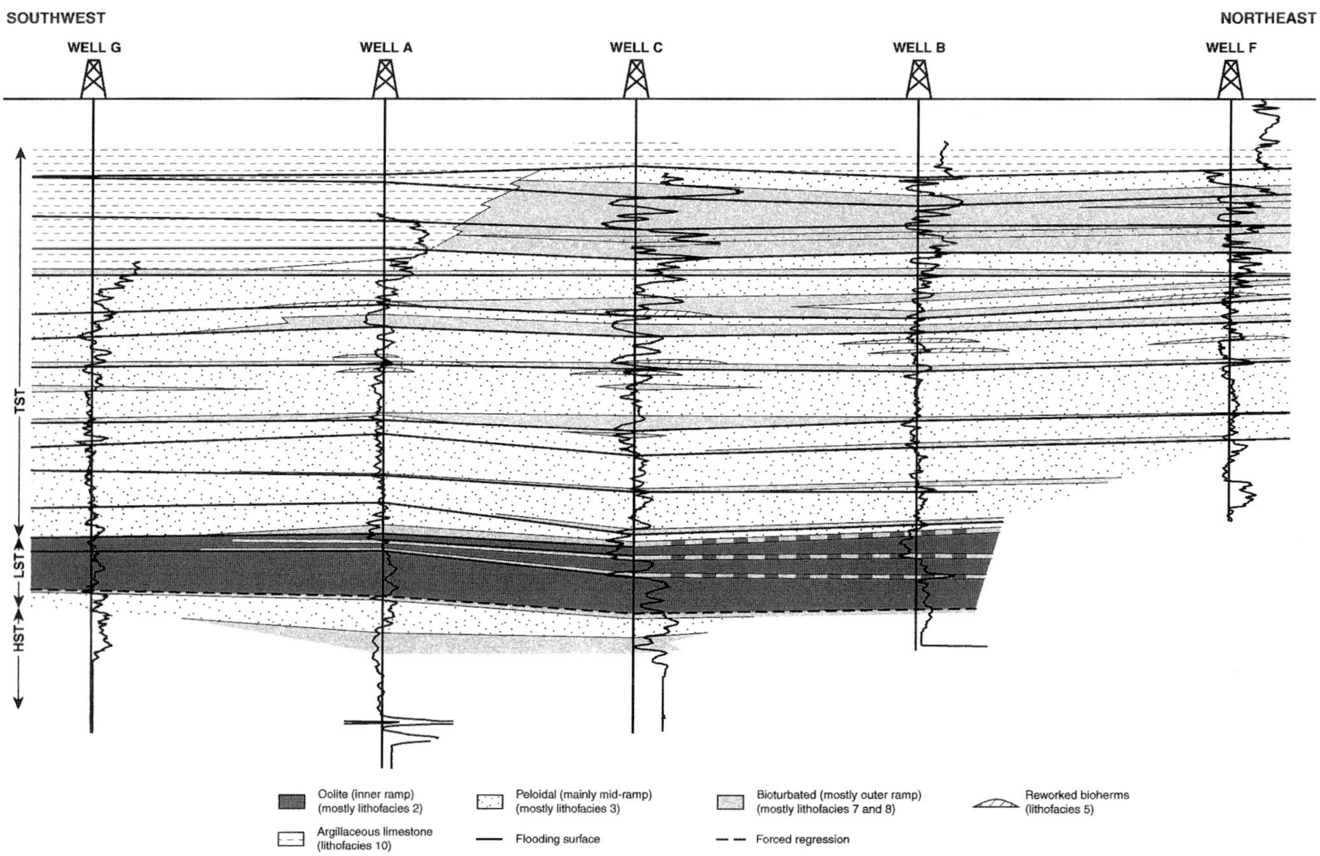


FIG. 9.—Schematic diagram illustrating the reservoir architecture for Minagish Oolite in the Umm Gudair Field.

(1993) and Burchette and Wright (1992). It is comparable to the shelf-margin systems tract described in the Upper Jurassic Hadriya Formation of northern Saudi Arabia by Koepnick et al. (1994). The latter terminology is not followed herein because there is no evidence of a shelf margin either in the Umm Gudair itself or on a more regional scale, in the time-equivalent Yamama Formation (Sadooni, 1993).

The unit above the main oolite can easily be correlated across the Umm Gudair Field and is dominated by stacked, broadly upward-coarsening parasequences (Fig. 9). These are composed largely of outer-ramp peloidal packstones and wackestones (Lithofacies 7), which shallow up into shallow mid-ramp skeletal peloidal packstones and grainstones (Lithofacies 3). At the base of this unit, however, parasequences are often capped by upward-fining beds of structureless oolitic grainstone (Lithofacies 2), which were reworked into the shallow mid-ramp from the inner-ramp oolitic sand sheet. This main reservoir unit, therefore, has a subtle retrogradational signature, with a trend towards thinner parasequences in its upper parts (Fig. 9). The entire unit is therefore interpreted as a transgressive systems tract (TST), and the gradual landward shift in facies upwards through the section suggests that carbonate production did not keep pace with sea-level rise (cf. Tucker et al., 1993).

In the central to northeastern parts of Umm Gudair, the uppermost unit continues the trend towards even thinner parasequences (Fig. 9). The parasequences largely comprise mid-ramp skeletal packstones (Lithofacies 3) interbedded with bioturbated wackestones that suggest outer-ramp settings. A regressive signature in the overlying Upper Minagish Member implies that the maximum flooding surface marking the top of the TST lies within the thinly interbedded, heterogeneous outer-ramp packstones and wackestones of Lithofacies 7 and 8, which dominate the uppermost unit of the Minagish Oolite. In this respect, it is significant that the continued upward-thinning trend is associated with the marked lateral facies change to more argillaceous limestone of Lithofacies 10 described previously (Fig. 9). The increased argillaceous content of these facies suggests that it is the distal expression of a deltaic system that prograded eastwards across the ramp from the southwest. This implies that siliciclastic poisoning rather than simple deepening alone gradually curbed carbonate sedimentation on the ramp. Increased suspended material in the water column would have suppressed the growth of algae, which were the dominant grain-forming organisms in the Minagish Oolite. A marked increase in the frequency of firmground development at parasequence tops suggests that significant periods of nonsedimentation or slowed sedimentation were associated with this poisoning. Given that the change from carbonate to deltaic sedimentation is located close to the maximum flooding surface, the progradation of this delta is likely to have been tectonically rather than eustatically controlled.

### *Controls on Fluid Movement and Implications for Reservoir Development*

The responses to relative sea-level change and the stratigraphic organization described imparts strong controls on permeability. Third-order and higher-frequency cyclicity imposes distinct reservoir properties. The highstand systems tract, lowermost in the reservoir, is a distinctive high porosity/low permeability zone, reflecting the dominance of microporous, muddy, mid- to outer-ramp facies. In contrast, the lowstand systems tract, characterized by oolitic grainstones, has lower porosities but higher permeabilities. The regressive signature to the transgressive systems tract, influenced most strongly by fourth-order sea-level fluctuations, imparts significant heterogeneity. In particular, there is an upward decrease in permeability towards the top of the reservoir, in response to deepening and a more intense diagenetic overprint. This upward degradation in permeability is exaggerated in the uppermost layers of the reservoir, but this seems to be related to siliciclastic poisoning and its associated diagenetic signature rather than to continued deepening.

Fourth-order and fifth-order cyclicity offers an important control on permeability distribution. Most importantly, finer-grained, muddy outer-ramp packstones and wackestones (Lithofacies 7–8) above flooding surfaces may have permeabilities of up to three orders of magnitude less than the clean mid-inner ramp packstones and grainstones (Lithofacies 2 and 3) they overlie (Fig. 4). Because flooding surfaces and the outer-ramp facies above them are correlatable field-wide, they are capable of forming important baffles to vertical transmissibility. Field data suggest that many are capable of limiting pressure communication, with common pressure shifts of tens (and some of hundreds) of psi occurring across correlatable flooding surfaces. This clearly limits vertical fluid movement. In the upper parts of the reservoir, an intense diagenetic overprint leads to consistent pressure differentials of several hundred psi and hence results in increasing reservoir compartmentalization. Increased production is likely to limit further communication across low-permeability beds, particularly in the uppermost part of the reservoir, where the lateral facies change to impermeable argillaceous mudstones limits the potential for natural aquifer support. It is anticipated, therefore, that a more comprehensive secondary recovery scheme will be required in this part of the reservoir.

The greatest postdepositional modifier of reservoir properties is compaction. Where diagenetic fluids have circulated easily within clean packstones and grainstones, marine-burial cements have provided a rigidity, which has inhibited compaction. Larger original pore sizes within these samples also means that the effects of compaction are less extreme, such that pore throats are narrowed or occluded rather than completely destroyed. The increasingly muddy textures of deeper-water lithofacies means that cementation is less likely to have imparted rigidity, and therefore the effects of compaction are more extreme. Small primary macropores and micropores are destroyed and originally poor reservoir properties significantly degraded further, both by compaction and by corresponding neomorphism of the micrite matrix. Where secondary interparticle macropores are recorded in packstones they may locally channelize flow, but they are often not large enough to focus fluids over more than a few centimeters.

The microporous nature of the Minagish Oolite requires a detailed understanding of the pore network in order to maximize production. Clearly the most efficient production will be from clean, macroporous skeletal peloidal and oolitic grainstones. These facies are focused towards the base of the reservoir, however, and the oil–water contact is now above the oolitic grainstones of the LST. The upward degradation in permeability within the reservoir corresponds to an overall decrease in pore sizes as facies become increasingly muddy. This means that sweep efficiencies will be reduced and enhanced recovery mechanisms may be necessary for effective hydrocarbon recovery later in field life. Much of the hydrocarbon may be reservoired within micropores, and therefore recovery mechanisms must aim to achieve an efficient sweep from these pores, even where they contrast markedly in size to surrounding millimeter- to centimeter-scale vuggy porosity.

## CONCLUSIONS

1. The Minagish Oolite reservoir of the Umm Gudair Field, Kuwait, represents storm-dominated sedimentation on a homoclinal carbonate ramp.

2. Ten lithofacies are identified, which typically form upward-shallowing parasequences separated by flooding surfaces.

3. Late highstand, lowstand, and transgressive systems tracts are recognized, which together represent a low-order, potentially third-order, stratigraphical cycle. Each systems tract comprises stacked, upward-shallowing parasequences, interpreted to be the products of fourth- or fifth-order sea-level fluctuations.

4. Intense micritization has generated highly microporous limestones, which significantly impact understanding of reservoir properties.

5. An upward degradation in permeability and increased heterogeneity through the reservoir means that a good understanding of pore-scale to reservoir-scale heterogeneities must be achieved later in field life to maintain recovery.

## ACKNOWLEDGMENTS

The authors gratefully acknowledge the Kuwait Oil Company for permission to publish this work. We would also like to thank Nigel Cross (Badley Ashton and Associates Ltd.), Christian Strohmenger (EXXON Exploration Company), and Harry Mueller (EXXON Production Research Company) whose comments greatly improved previous versions of this manuscript. We are also grateful to Arnout Everts and John Hopkins for thorough and detailed reviews of this paper. Gerry Hyde, Heather Nickson, and Dave Kemp (Badley Ashton and Associates) are thanked for preparing the diagrams.

## REFERENCES

Ali, M.A., 1994, Gotnia Salt and its structural implications in Kuwait, *in* Al-Husseini, M.I., ed., Geo'94 Middle East Geoscience: Gulf Petrolink, Manama, Bahrain, v. 1, p. 133–142.

Alsharhan, A.S., and Nairn, A.E.M., 1986, A review of the Cretaceous formations in the Arabian peninsula and Gulf: Part I. Lower Cretaceous (Thamama Group) stratigraphy and palaeogeography: Journal of Petroleum Geology, v. 9, p. 365–392.

Ayres, M.G., Bilal, M., Jones, R.W., Slentz, L.W., Tartir, M., and Wilson, A.O., 1982, Hydrocarbon habitat in the main producing areas, Saudi Arabia: American Association of Petroleum Geologists, Bulletin, v. 66, p. 1–9.

Aziz, S.K., and El-Sattar, M.M.A., 1997, Sequence stratigraphic modeling of the Lower Thamama Group, East Onshore Abu Dhabi, United Arab Emirates: GeoArabia, v. 2, p. 179–202.

Banner, F.T., and Simmons, M.D., 1994, Calcareous algae and foraminifera as water-depth indicators: an example from the Early Cretaceous carbonates of north-east Arabia, *in* Simmons, M.D., ed., Micropalaeontology and Hydrocarbon Exploration in the Middle East: British Micropalaeontology Society Publication Series, London, Chapman & Hall, p. 243–253.

Burchette, T.P., and Britton, S.R., 1985, Carbonate facies analysis in the exploration for hydrocarbons: a case study from the Cretaceous of the Middle East, *in* Brenchley, P., and Williams, B., eds., Sedimentology, Recent Developments and Applied Aspects: Blackwell Scientific, p. 311–338.

Burchette, T.P., and Wright, V.P., 1992, Carbonate ramp depositional systems: Sedimentary Geology, v. 79, p. 3–57.

Carman, G., 1996, Structural elements of Kuwait: GeoArabia, v. 1, p. 239–266.

Davies, R.B., and Hollis, C.E., 1996, Reservoir description of the Minagish Oolite Formation, based upon thirteen wells from the Minagish and Umm Gudair Fields, West Kuwait: Badley Ashton proprietary report for Kuwait Oil Company.

Dickson, J.A.D., 1965, A modified staining technique for carbonates in thin section: Nature, v. 205, p. 587.

Harland, W.B., Armstrong, R.L., Cox, A.V., Craig, L.E., Smith, A.G., and Smith, D.G., 1990, A Geologic Time Scale 1989: Cambridge, U.K., Cambridge University Press, 263 pp.

Haq, B.U., Hardenbol, J., and Vail, P.R., 1987, Chronology of fluctuating sea levels since the Triassic: Science, v. 235, p. 1156–1167.

Hughes, G.W., 1997, The Great Pearl Bank Barrier of the Arabian Gulf as a possible Shu'aiba Analogue: GeoArabia, v. 2, p. 279–304.

Koepnick, R.B., Waite, L.E., Kompanik, G.S ., Al-Shammery, M.J., and Al-Amoudi, M.O., 1994, Sequence stratal geometries and burial-related microporosity development: controls on performance of the Hadriya Reservoir, (Upper Jurassic), Berri Field, Saudi Arabia, *in* Al-Husseini, M.I., ed., Geo'94 Middle East Geoscience Conference: Gulf Petrolink, Manama, Bahrain, v.2, p. 655–677.

Longacre, S.A., and Ginger, E.P., 1988, Evolution of the Lower Cretaceous Ratawi Oolite reservoir, Wafra Field, Kuwait–Saudi Arabia Partitioned Neutral Zone, *in* Lomando, A., and Harris, P.M., eds., Giant Oil and Gas Fields: SEPM, Core Workshop 12, p. 273-331.

Machel, H.G., and Burton, E.A., 1991, Factors governing cathodoluminescence in calcite and dolomite and their implications for studies of carbonate diagenesis, *in* Barker C., and Kopp, O., eds., Luminescence Microscopy: Quantifications and Qualitative Aspects: SEPM, Short Course 25, p. 39–57.

Moshier, S.O., 1989, Development of microporosity in a micritic limestone reservoir, Lower Cretaceous, Middle East: Sedimentary Geology, v. 63, p. 217–240.

Posamentier, H.W., Allen, G.P., James, D.P., and Tesson, M., 1992, Forced regressions in a sequence stratigraphic framework: concepts, examples, and exploration significance: American Association of Petroleum Geologists, Bulletin, v. 76, p. 1687–1709.

Sadooni, F.N., 1993, Stratigraphic sequence, microfacies and petroleum prospects of the Yamama Formation, Lower Cretaceous, Southern Iraq: American Association of Petroleum Geologists, Bulletin, v. 77, p. 1971–1988.

Skelton, P., 1997, Lower Aptian rudists of Arabia: GeoArabia, v. 2, p. 494.

Sungur, D., 1996, Stratigraphic and facies definition of the Lower Cretaceous Minagish Oolite, Umm Gudair Field, West Kuwait (abstract): Geo'96, Middle East Geosciences Conference, April 1996, Manama, Bahrain.

Tucker, M., and Wright, V.P., 1990, Carbonate Sedimentology: Blackwell Scientfic Publications, 482 p.

Tucker, M.E., Calvet, F., and Hunt, D., 1993, Sequence stratigraphy of carbonate ramps: systems tracts, models and application to the Muschelkalk carbonate platforms of eastern Spain, *in* Posamentier, H., Summerhayes, C., Haq, B., and Allen, G., eds., Sequence Stratigraphy and Facies Associations: International Association of Sedimentologists, Special Publication 18, p. 397–418.

Vahrenkamp, V.C., 1996, Carbon isotope stratigraphy of the Upper Kharaib and Shuaiba Formations: implications for the early Cretaceous evolution of the Arabian Gulf region: American Association of Petroleum Geologists, Bulletin, v. 80, p. 647–442.

Van Wagoner, J.C., Mitchum, R.M., Campion, K.M., and Rahmanian, V.D., 1990, Siliciclastic sequence stratigraphy in well logs, cores, and outcrops: concepts for high resolution correlation of time and facies: American Association of Petroleum Geologists, Methods in Exploration Series, No. 7, 55 p.

Wright, V.P., 1986, Facies sequences on a carbonate ramp: the Carboniferous of South Wales: Sedimentology, v. 33, p. 221–241.

# GEOLOGICAL CONTROLS ON PORE GEOMETRY OF UPPER KHARAIB RESERVOIR, OFFSHORE ABU DHABI, U.A.E.

JUN-ICHI SAOTOME

*Bunduq Oil Company Ltd., P. O. Box 46015, Abu Dhabi, United Arab Emirates*

*Present address: Japan Oil Development Company Ltd., 1-7, Uchisaiwai-cho 1 Chome, Chiyoda-ku, Tokyo 100-0011, Japan*

TOSHIAKI SHIBASAKI

*Zakum Oil Development Company, P. O. Box 46808, Abu Dhabi, United Arab Emirates*

TSUYOSHI SHIKAMA

*Japan Oil Development Company Ltd., 1-7, Uchisaiwai-cho 1 Chome, Chiyoda-ku, Tokyo 100-0011, Japan*

AND

KEN-ICHI KAWAGUCHI

*Japan Oil Development Company Ltd., Abu Dhabi Branch, P. O. Box 2659, Abu Dhabi, United Arab Emirates*

ABSTRACT: The upper Kharaib reservoir of the Lower Cretaceous Thamama Group is one of the major oil-producing horizons offshore Abu Dhabi in the United Arab Emirates, and is traced over the area as a well-defined rock unit. This reservoir is composed of three lithofacies, rudistid–peloidal grainstone/packstone, algal–peloidal grainstone/packstone, and bioclastic wackestone/mudstone. Sediments were deposited on a broad carbonate shelf and show an upward-shallowing succession. Reservoir quality is determined by both original depositional fabric and subsequent diagenetic processes. In general, permeability is variable and porosity is constant throughout the sequence, except for stylolitized intervals.

In the "A" field, six high-permeability facies are identified in the grain-supported interval. These facies are characterized by unique pore-throat size distributions resulting from depositional textures and diagenetic modification. Grain sorting and preferential dissolution, as well as preservation of primary pores, are significant factors in controlling pore geometry. Dolomitization is also locally important. The highest permeability is present in well-sorted, coarse-sized grainstone at the basal part of upward-fining bed sets where well connected intergranular pores are preserved.

Thus, identification of high-permeability facies is crucial if one is to understand the performance of the reservoir under water flood. On the basis of the vertical and lateral distribution of high-permeability facies and pore-throat size distribution, the reservoir was divided into twenty-two sublayers as a basis for reservoir simulation study.

## INTRODUCTION

The "A" field is located offshore approximately 80 kilometers northwest of Abu Dhabi (Fig. 1). A structural contour on top of the upper Kharaib reservoir, defined as the Thamama II reservoir, shows a broad and gentle asymmetric anticlinal domal feature with dips of a few degrees. The relief in the western area of the field is almost flat, but slightly steeper dips (approximately 3–5°) are recognized on the north and northeast flanks. No major faults have been reported.

The Thamama II reservoir in this field is a heterogeneous reservoir with permeability as high as a few thousand millidarcies (Nishi and Shibasaki, 1996). Premature water breakthrough in the reservoir has been observed from some wells under the five-spot water injection pattern. Resistivity anomalies in open-hole logs and time-lapse pulsed neutron logs of some wells revealed that a preferential water movement has occurred along relatively thin layers. This evidence suggested the existence of thin, high-permeability streaks acting as conduits of injected water to producing wells.

This paper identifies and classifies high-permeability facies with data from pore-throat size distribution curves, lithological controls, diagenetic influences, and mercury-injection capillary pressure data analysis. This lithological and petrophysical information is integrated to develop a new fine-scale layering scheme for the reservoir.

## GEOLOGICAL SETTING

The Thamama II reservoir belongs to the upper part of the Kharaib Formation of the Thamama Group (Fig. 2) (Alsharhan, 1990). The Thamama Group of the Lower Cretaceous offshore Abu Dhabi is divided into four formations, the Habshan, Lekhwair, Kharaib, and Shuaiba formations, in ascending order. Sedimentation of these formations is mainly categorized into three depositional phases. The Habshan Formation is marked by a widespread marine transgression and is composed of sediments deposited under various environments from supratidal to shelf, in largely restricted conditions (Hassan et al., 1975; Alsharahan and Nairn, 1986). It is followed by cyclic sedimentation of the Lekhwair and Kharaib Formations resulting from eustatic sea-level changes in shelf environments of a carbonate ramp (Murris, 1980). Reservoir facies consist of regressive sediments having upward-coarsening sequences and are readily traced over the area. The Lekhwair Formation represents normal marine shelf sediments and is developed as a number of cyclic sequences. In general, the transgressive phase is dominant in the cycles. The Kharaib Formation was deposited in a vast, shallow, epicontinental shelf commencing with a regressive phase of cycles and characterized by predominantly regressive sediments. The Shuaiba Formation was deposited in various environments, influenced by paleogeography and sea-level changes, represented by deep-water shelf sediments and shallow-water rudist carbonates in some localized areas.

The Thamama II reservoir is characterized by an upward-shallowing system composed of three major lithofacies deposited on a shallow carbonate ramp during a major regression. The lower part of this reservoir consists mainly of bioclastic wackestone and mudstone deposited below wave base. The middle portion is dominated by algal–peloidal grainstone and packstone deposited on an undulating sea floor. Algal clasts were derived from algal bioherms developed as mounds. Distribution of intraclasts

Middle East Models of Jurassic/Cretaceous Carbonate Systems
SEPM Special Publication No. 69, Copyright © 2000
SEPM (Society for Sedimentary Geology), ISBN 1-56576-075-1, p. 287–297.

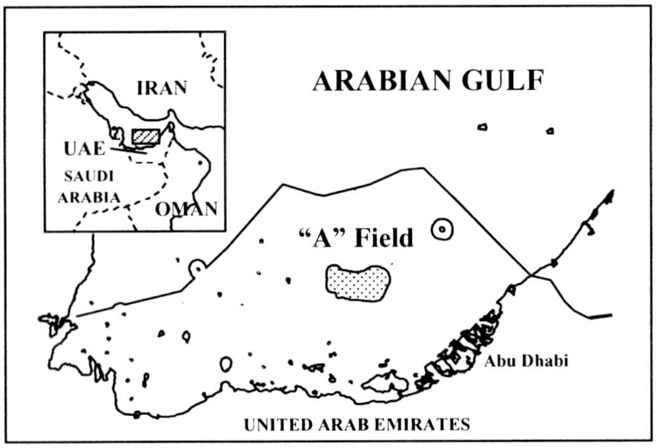


FIG. 1.—Location map of "A" Field, offshore U.A.E.

corresponds to topographical high relief just below sea level. The uppermost part is composed of rudist–peloidal grainstone/packstone interbedded with peloidal–foraminiferal grainstone/packstone, which is associated with rudist bioherms developed as small banks/mounds.

## RESERVOIR FACIES

### Facies Description

The Thamama II reservoir in the "A" field ranges in thickness approximately from 43 to 46 m (140 to 150 ft). Average porosity and permeability range from 25 to 30%, and 10 to 50 millidarcies, respectively. Stylolitization has occurred at certain horizons within the reservoir, forming five relatively tight layers that subdivide the reservoir into six units, i.e., IIA, IIB, IIC, IID, IIE, and IIF (Fig. 2). The tight layers are defined as "S" Zones: S1, S2, S3, S4, and S5, in descending order. They are correlative on porosity logs over the field, although no stylolites are recognized in S2 and S3 of some crestal wells. A typical seam is recognized in S5 of a thin and very tight layer, whereas low-amplitude stylolites are commonly developed in S1 to S4 in the grain-supported facies (units IIA to IID). Decrease of both porosity and permeability within "S" Zones is associated with stylolitic seams and cementation. Dolomite rhombs are variably present in the reservoir. Distinct dolomite bands are common within the middle of IIE. In some areas, dolomites have completely replaced the original sediments. Reservoir lithofacies are summarized as follows.

*Rudist–Peloidal Grainstone/Packstone (Units IIA and IIB in Part).—*

This is the most distinctive facies in the reservoir, occurring commonly in unit IIA and the upper part of unit IIB. It is characterized by a large amount of fragmented and intact rudists, interbedded with fine-grained, peloidal–foraminiferal grainstone/packstone. Rudist fragments are concentrated at certain horizons. Rolled fragments of rudists are observed sporadically throughout the interval. This accumulation of rudist bivalves could have resulted from reworking, although locally intact rudists are also present. Most of the rudists are partially to extensively bored.

*Algal–Peloidal Grainstone/Packstone (Units IIB, IIC, IID, and IIE in Part).—*

This facies is characterized by diverse grain constituents with a wide variation in size and sorting. Most of the grains are bioclasts and peloids. The former is composed of whole or fragmented rudists, benthonic foraminifera, and algae (*Lithocodium*). Echinoderms, dacyclad algae, and gastropods are also present. Relatively large grains ranging from granule to medium pebble size, represented mainly by bioclasts, algal clasts, and intraclasts, are incorporated into a porous matrix composed of silt- to fine-grained sized peloids. Generally, bioclastic grains show wide variation in size, ranging from very fine sand to pebble size. Granule- to small pebble-size bioclasts are commonly micritized. Algal patches and algal clasts are locally abundant. Intraclasts are the most distinct grains among these grain components because

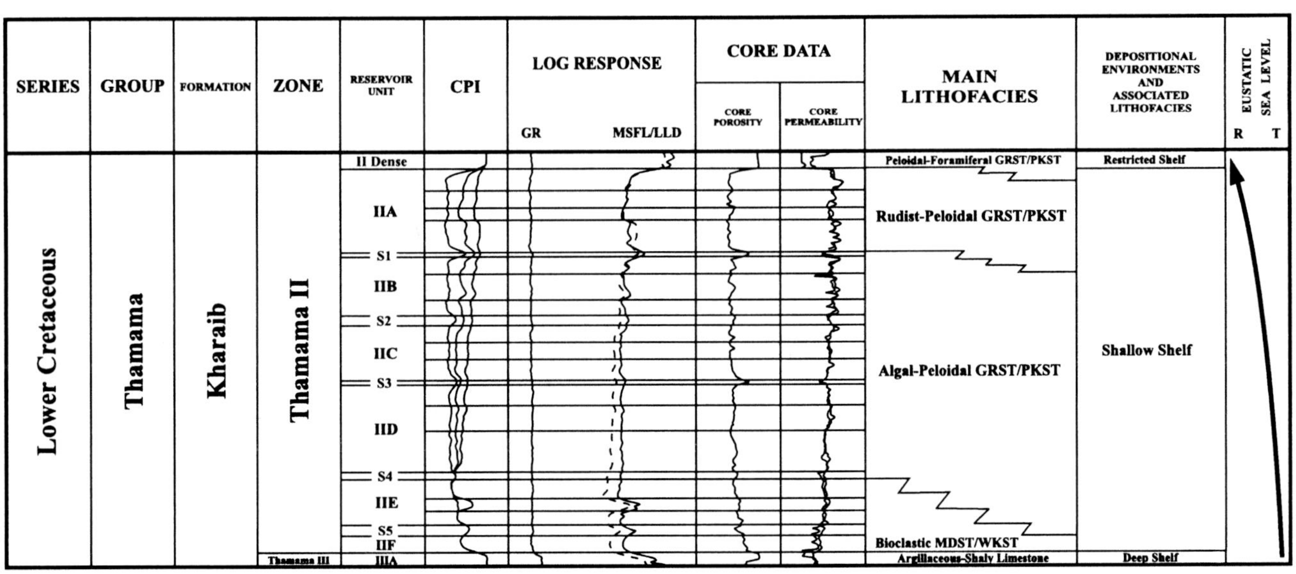


FIG. 2.—Generalized lithofacies and well-log signatures of the Thamama II reservoir.

of their size and shape. Intraclasts are present in "matrix" composed of very fine- to medium-grained, well sorted peloids as loosely packed muddy fragments. The origin of these intraclastic grains is indeterminate because almost all have been extensively micritized. In addition to the above-described components, subrounded, very fine- to medium-sand-size and well-sorted, coarse- to very coarse-sand-size grains (peloidal/algal coated) are subordinate grain constituents. Most peloids may be micritized foraminfera and bioclasts.

*Bioclastic Wackestone/Mudstone (Units IIE and IIF).—*

This lithofacies is distributed in the lowest part of the reservoir and is composed of wackestone and/or mudstone with scattered bioclasts and *Orbitolina*. Fine-grained bioclastic packstone is locally present. Porosity is present within the matrix as micropores and locally biomolds. Slight color mottling in texture caused by bioturbation, which is locally intense, is also recognized.

### *Significance of Sedimentological Features*

Small-scale, upward-fining bed sets are mainly recognized within the algal–peloidal grainstone/packstone, representing the facies of units IIB, IIC, IID, and IIE in part. These bed sets are characterized by the presence of very coarse sand- to granule-size grains at the base and by the appearance of cyclic features with individual thickness ranging from a half meter to a few meters (Fig. 3). Each cycle fines upward to a very fine- to fine-grained grainstone/packstone. The upper portion of each bed set is bioturbated, and burrows are commonly observed. The top and base of each bed set are commonly bounded by an irregular surface, which could have resulted from submarine erosion (disconformity?). Tidal currents and/or wave agitation may have controlled sorting of coarser grainstone and packstone observed in the basal parts of the bed sets

## HIGH-PERMEABILITY FACIES

The original sediments and subsequent diagenetic modification determine reservoir quality in the Thamama II reservoir. This section describes both lithological and diagenetic aspects controlling high-permeability facies in the reservoir.

### *Lithological Controls on High-Permeability Facies*

*Upward-Fining Sediments (Units IIB, IIC, IID, and IIE in Part).—*

The cyclic, upward-fining bed sets are a significant sedimentary feature in the algal–peloidal grainstone/packstone lithofacies. Higher permeability tends to occur at the bases of these coarse-grained, algal-rich bed sets. This facies is subdivided based upon grain sorting and size. Type A is moderately to well sorted algal–peloidal grainstone. Type B is poorly to moderately sorted algal–peloidal grainstone. Type C is poorly sorted algal–peloidal grainstone/packstone. A schematic textural profile of high-permeability facies shows the relationship between these three facies types and porosity/permeability types (Fig. 3). The highest permeability values correspond to the better sorted facies, typically Type A, and also to the coarser facies at the bases of the upward-fining bed sets. Relatively high permeability in the bed sets is determined

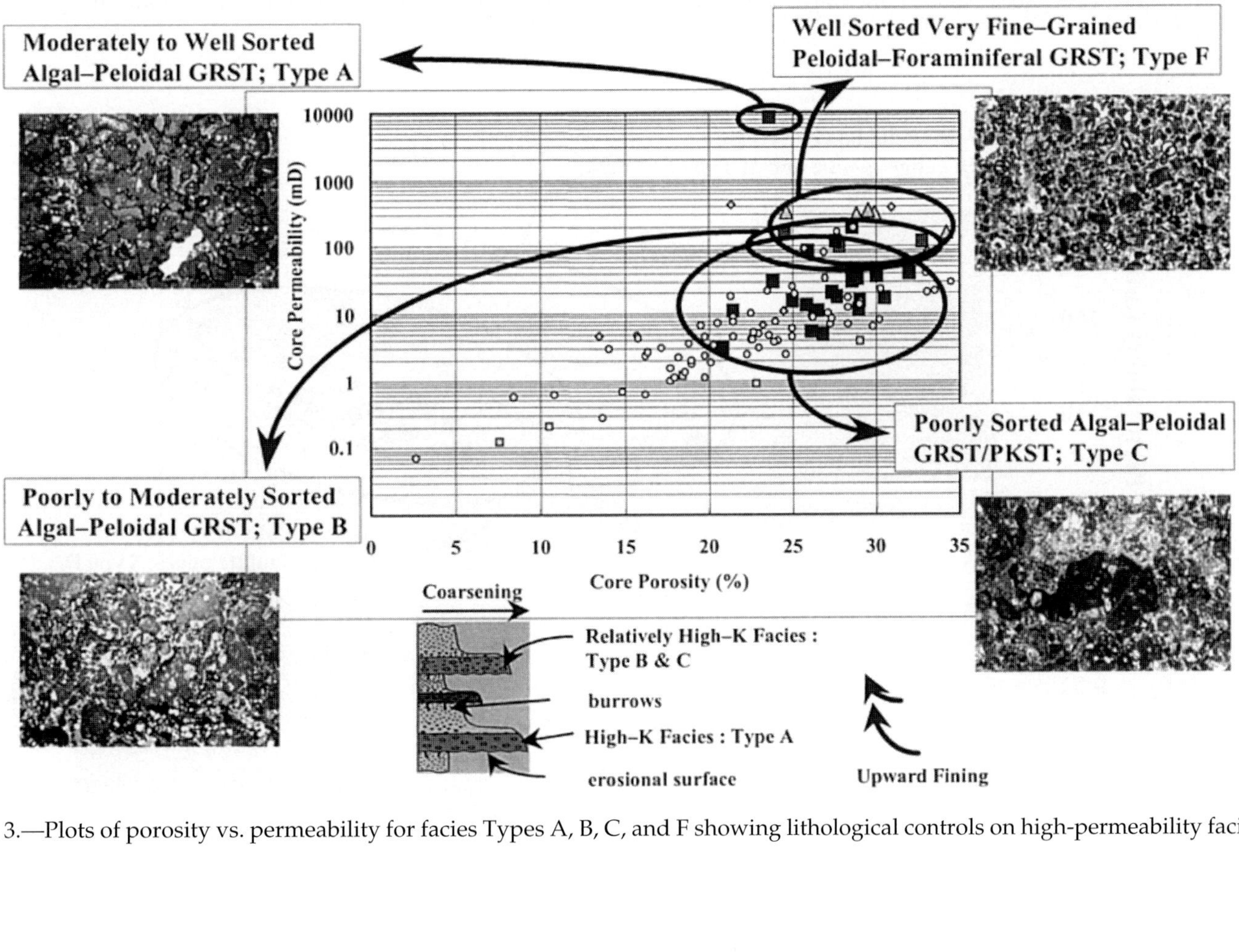


FIG. 3.—Plots of porosity vs. permeability for facies Types A, B, C, and F showing lithological controls on high-permeability facies.

principally by grain sorting and grain size rather than by diagenetic modification.

### *Well Sorted, Very Fine-Grained Peloidal–Foraminiferal Grainstone (Unit IIA).—*

This facies type is intercalated within the rudist–peloidal grainstone/packstone in unit IIA. This facies is defined as Facies Type F, and is characterized by well-sorted, very fine- to fine-sand-size peloids and micritized miliolids. Primary intergranular pores are well preserved with good pore connectivity resulting in enhancement of permeability (Fig. 3).

## *Diagenetic Modification of High-Permeability Facies*

Sediments in the reservoir may have undergone many diagenetic processes since deposition, therefore their fabrics may have significant influence on reservoir quality. In general, deterioration of reservoir quality in the western area of the "A" field might be related to intense cementation by calcite, but porosity and permeability have been preserved in the eastern area of the field. This section summarizes several diagenetic events that have improved both porosity and permeability in association with main facies types.

### *Very Fine to Fine-Grained Algal–Peloidal Grainstone/Packstone (Units IIB, IIC, IID, and IIE in Part).—*

Partial dissolution creating biomoldic and vuggy pores enhance permeability, although small equant calcite cements and large blocky calcite cements reduce permeability by occluding pores (Fig. 4). In addition, the upper parts of the upward-fining bed sets consist of very fine- to fine-grained algal–peloidal grainstone/packstone facies Type D, which generally has poor permeability. However, the facies affected by dissolution show permeability improvement up to approximately a few hundred millidarcies.

### *Rudist–Peloidal Grainstone/Packstone (Unit IIA).—*

Higher permeability corresponds to the presence and abundance of a secondary pore network. Preferential dissolution of rudists in facies Type E has created a secondary vuggy/moldic pore network, showing moderate to high permeability (Fig. 4). These secondary pores have enhanced pore connectivity, whereas the subsequent calcite cements, i.e., large blocky and small equant cements, have reduced the reservoir quality, although micropores and biomolds are still present. Finally, this facies is characterized by alternating low-permeability and high-permeability intervals as a result of these diagenetic processes.

## *Classification of High-Permeability Facies*

Facies distribution and subsequent diagenesis control reservoir quality. The porosity profiles of the lower part of the reservoir (units IIE and IIF), composed of mud-supported facies, tends to decrease towards the base of the Thamama II reservoir (Fig. 2). The upper and middle parts (units IIA to IID), in contrast, display a consistent porosity profile resulting from the dominance of grain-supported facies except for stylolitized layers (S1 to S5). On the other hand, permeability profile in the grain-supported facies is variable. The grain-supported interval has higher porosity and permeability than does in the mud-supported part. The reservoir is classified into nine main facies types in terms of permeability based on thin sections taken from 110 core samples, in which the following six types have been defined as significant high-permeability facies (Figs. 3 and 4).

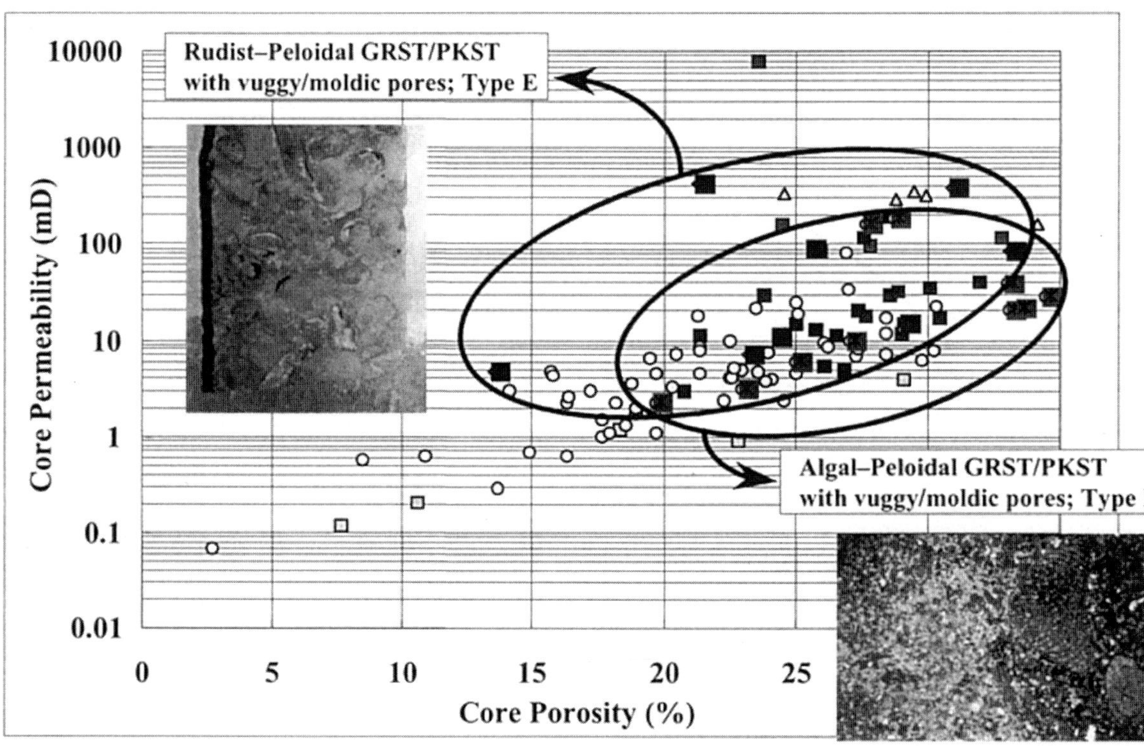


FIG. 4.—Plots of porosity vs. permeability for facies Types D and E showing diagenetic modifications on high-permeability facies.

**Facies Type A** (High-Permeability Facies)
Lithology: moderately to well sorted algal–peloidal grainstone
Pore type: intergranular pores
Permeability range: more than 1000 md

**Facies Type B** (Relatively High-Permeability Facies)
Lithology: poorly to moderately sorted algal–peloidal grainstone
Pore type: intergranular and vuggy pores
Permeability range: approximately from 50 md to 300 md

**Facies Type C** (Moderately High-Permeability Facies)
Lithology: poorly sorted algal–peloidal grainstone/packstone
Pore type: intergranular and vuggy pores
Permeability range: approximately from 1 md to 200 md

**Facies Type D** (Relatively High-Permeability Facies)
Lithology: very fine- to fine-grained algal–peloidal grainstone/packstone
Pore type: intergranular and vuggy pores
Permeability range: approximately from 1 md to 300 md

**Facies Type E** (Relatively High-Permeability Facies)
Lithology: rudist–peloidal grainstone/packstone
Pore type: vuggy and/or moldic pores
Permeability range: approximately from 1 md to 1000 md

**Facies Type F** (Relatively High-Permeability Facies)
Lithology: well sorted very fine-grained peloidal–foraminiferal grainstone
Pore type: intergranular pores
Permeability range: approximately from 50 md to 800 md

In addition to the above high-permeability facies, the following three facies have lower permeability values. However, both porosity and permeability vary in dolomite Type G, which locally has high permeability resulting from high connectivity in the intercrystalline pore network.

**Facies Type G**
Lithology: dolomite and/or dolomitic limestone
Pore type: intercrystalline pores
Permeability range: approximately from 1 md to 100 md

**Facies Type H**
Lithology: very fine- to fine-grained peloidal packstone (including stylolitized layers: S1 to S5)
Pore type: intergranular pores
Permeability range: approximately from 0.001 md to 10 md

**Facies Type I**
Lithology: bioclastic wackestone/mudstone
Pore type: matrix pores
Permeability range: approximately from 0.01 md to 10 md

### *Continuity of High-Permeability Facies*

*Units IIB, IIC, and IID.*—

The highest permeability in reservoir units IIB, IIC, and IID has a tendency to develop at the bases of the upward-fining cycles characterized by the relatively well sorted, coarse-sand-size grainstone with a dominant intergranular pore network. Such coarse grainstones vary in thickness from a meter ± to less than 30 cm, and tend to be more common in the eastern area of the field. However, they are difficult to correlate between wells. Judging from sedimentary features, this grainstone may locally have been deposited under high-energy conditions, as reworked sediments. These coarse grainstones consist mainly of algal grains that were possibly derived from relatively small-scale algal mounds. On the other hand, well sorted coarse grainstones are not generally recognized in the western area of the field, even though the upward-fining bed sets are present. This may indicate that subtle variation in depositional setting existed over the field. Diverse energy conditions of sedimentation can be inferred, compared with that of the eastern area, although the Thamama II reservoir was uniformly deposited under shallow-water conditions on a carbonate ramp (Alsharhan and Nairn, 1986). This interpretation is also based on evaluation of the available data and supports the idea that development of the better-permeability grainstone streaks is not continuous over the field and may be limited to the crestal area of the eastern part of the field.

*Unit IIA.*—

Rudist–peloidal grainstone/packstone in unit IIA is also an important high-permeability facies. Selective dissolution of the rudists occurred in locally discrete bands, and resulted in enhanced permeability, although subsequent coarse calcite cement partly fills these biomolds. The well sorted very fine-grained peloidal–foraminiferal grainstone having primary intergranular pores is a minor element. In addition, the intergranular pores in this facies are commonly filled by calcite cement. Thus, the reservoir characteristics of unit IIA facies are represented by heterogeneous rock with alternating low-permeability and high-permeability intervals. Considering all sedimentological and diagenetic features from both core observations and thin-section examination, it can be inferred that high-permeability facies in unit IIA are not laterally continuous over the field.

## PORE GEOMETRY IN HIGH-PERMEABILITY FACIES

### *MICP Data Conversion to Pore-Throat Size Distribution*

Carbonate reservoirs are characterized by complex pore systems, which are mainly categorized as primary and secondary pores. Pore geometry and its connectivity in a reservoir determine permeability. Although pore sizes, shapes, and connectivity are very complicated, mercury-injection capillary pressure data (MICP) provide useful information on the size of pores and throats. Capillary pressure curves indicate the percentage of total pores in the rock by plotting the mercury-injected volume against pressure. A pore-throat size distribution curve can be obtained from mercury-injection capillary pressure data. These pore-throat size distribution curves were also applied to categorize rock characteristics in this study. A total of 239 mercury-injection capillary pressure measurements from 23 wells represent all reservoir zones. Pore-throat size for each capillary pressure data set was calculated using the following formula, which was developed by Core Laboratories Inc. in 1982:

$$Ri = 2T\cosh(C/Pc)$$

where $Ri$ = pore-throat radius (µm); $T$ = air–mercury interfacial tension (= 480 dynes/cm); h = air-mercury constant angle of 140°; $C$ = unit conversion constant (= 0.145); $Pc$ = mercury-injection capillary pressure (psia).

### *Classification of Pore-Throat Size Distribution Curves*

The mercury-injection capillary pressure data were converted into pore-throat size distribution using the equation above. Pore-throat size distribution curves are classified into seven types based on curve shapes and peak location of pore-throat size. The pore-throat size distribution classes (Figs. 5 and 6) and are described as follows:

**Curve Type 1** (unimodal; 58 samples): Maximum probability density is higher than 10%, and the peak pore-throat size is located at less than 1 µm. Permeability ranges from 0.02 to 17 md.

**Curve Type 2** (unimodal; 91 samples): Maximum probability density is lower than 10%, and the peak pore-throat size is located at less than 1 µm. Permeability ranges from 0.05 to 25 md.

**Curve Type 3** (unimodal; 35 samples): The peak pore-throat size located at a point higher than 1 µm. Permeability ranges from 2.6 to 127 md.

**Curve Type 4** (unimodal; 18 samples): The distribution is symmetrical around 1 µm and has a broad curve shape. Permeability ranges from 4.4 to 44.7 md.

**Curve Type 5** (bimodal; 22 samples): Two peaks are recognized. One peak is at less than 1 µm and the second is greater than 1 µm. Permeability ranges from 4.7 to 1544 md.

**Curve Type 6** (trimodal; 8 samples): Three peaks are observed and the curve shape is very broad. Permeability ranges from 19.9 to 245 md.

**Curve Type 7** (flat; 6 samples): The curve shape is almost flat. Permeability ranges from 0.003 to 17.2 md.

In general for the Thamama II reservoir, as permeability increases, the curves become wider and the location of the distribution peak moves towards larger pore-throat size. Almost of all of the low-permeability samples show similar single peak curves with the peak at less than 1 µm. Each pore-throat size curve type has range of permeability (Fig. 6). Curve Types 1 and 2 show lower permeability ranges. Generally, low-permeability samples with less than 10 md are classified into either Curve Types 1 or 2. Curve Types 3, 4, 5, and 6 indicate relatively high permeability, and they can be roughly distinguished as two groups, a unimodal system (Curve Types 3 and 4) and a bimodal or trimodal system (Curve Types 5 and 6). On the other hand, many high-permeability samples have different curve types. The difference between curve types results from characteristics of the pore network of each facies.

## INTERRELATION OF PORE GEOMETRY AND FACIES/LAYERS

### *Relationship between Pore-Throat Size Distribution Curve Types and Facies*

An attempt was made to describe the pore geometry systems in the reservoir, especially in the high-permeability facies, after pore systems were classified comprehensively with emphasis on their genetic aspects. Based upon availability of thin sections, 99 mercury injection samples from 8 wells were used to clarify the relationship between facies type and pore geometry (Table 1). The relationships between facies and pore-throat size distribution curves are summarized as follows (Fig. 7).

**Facies Type A** (moderately to well sorted algal–peloidal grainstone): The mercury-injection data for this type are not available. Intergranular pores are the main visible pore type.

**Facies Type B** (poorly to moderately sorted algal–peloidal grainstone): Curve Types 5 (33.3%) and 6 (66.7%) are mainly found for this facies, in which both intergranular and vuggy pores are preserved as the main visible pore types. In particular, a large-sized intergranular pore network is well developed. In addition, micropores are probably present in the grains because the pore-throat size distribution curves with bimodal and/or trimodal shapes may indicate existence of more than two types of pores in the facies.

**Facies Type C** (poorly sorted algal–peloidal grainstone/packstone): Curve Types 1 (18.8%), 2 (50%), and 4 (18.8%) are the representative types for this facies. The grains of the facies are poorly sorted, and most pore spaces among large grains are filled with more fine grains and/or micrite as matrix. Most of the grains are affected by micritization, as in facies Type B. Therefore, pore-throat size distribution curves indicate a small pore throat. Inter-

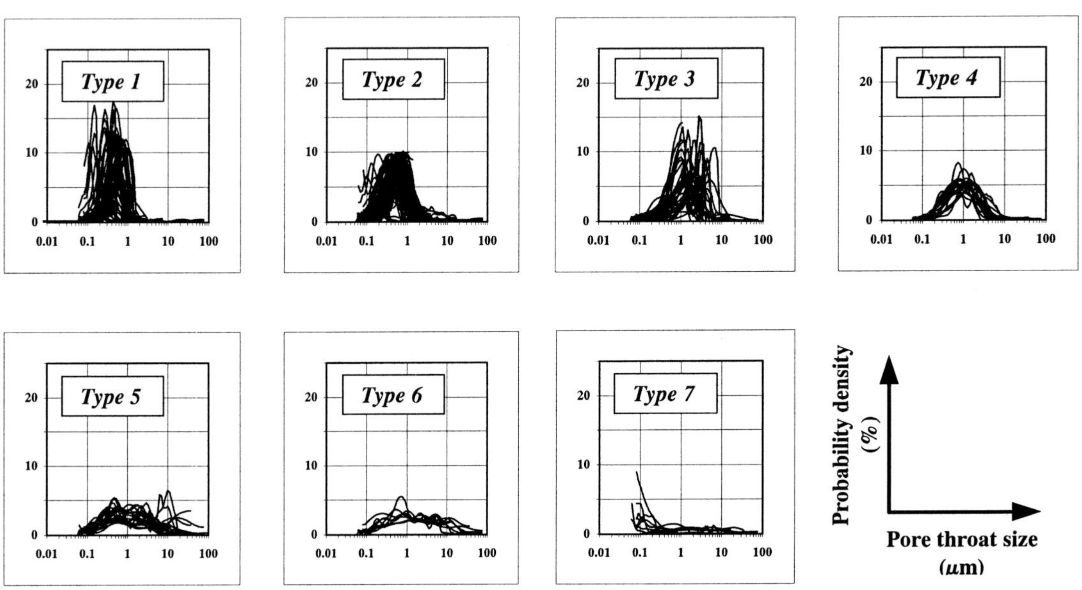


FIG. 5.—Classification of pore-throat size-distribution curves by plots of percent probability density to pore-throat size.

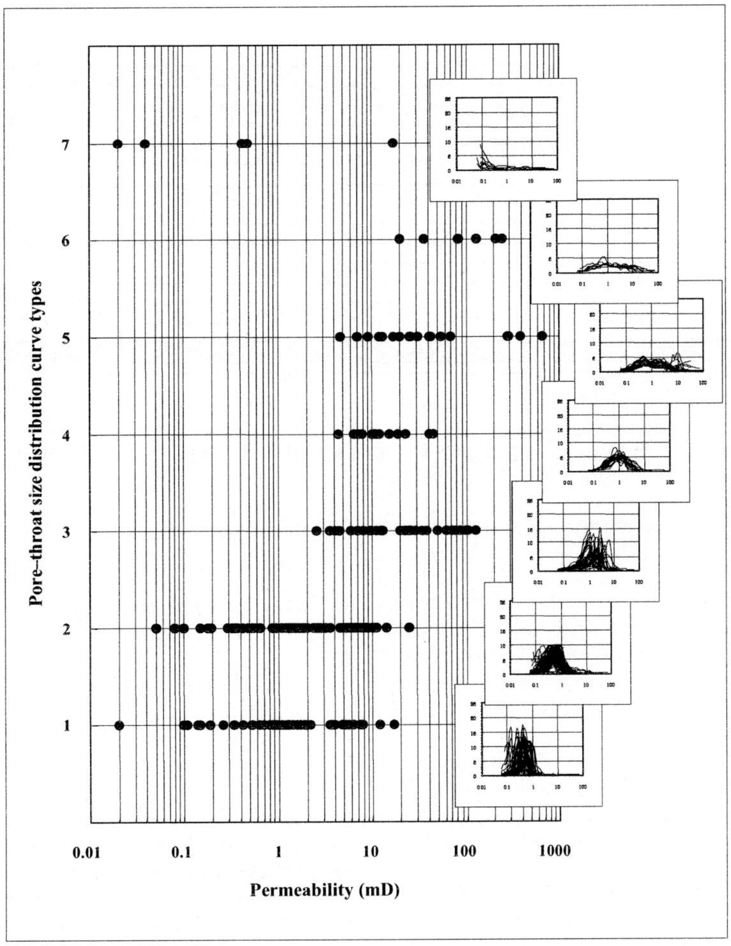


FIG. 6.—Permeability range of each pore-throat size-distribution curve type.

granular and vuggy pore networks are partially preserved and contribute to better permeability.

**Facies Type D** (very fine- to fine-grained algal–peloidal grainstone/packstone with vuggy pores): Curve Type 5 (62.5%) is the main type for this facies, but, Curve Types 2 (12.5%), 4 (12.5%), and 6 (12.5%) are also observed. This facies is affected by dissolution, and the large-sized pore network is formed partly by vuggy pores. Furthermore, micropores in the micritized grains and micritic matrix are possibly enlarged by leaching.

**Facies Type E** (rudist–peloidal grainstone/packstone): Curve Types 3 (42.9%) and 6 (28.6%) are the main types for this facies. Here, Curve Type 3 indicates the existence of small intergranular

TABLE 1.—Numbers of sample of each curve type in each facies type.

| | | Pore-throat size distribution curve types | | | | | | | |
|---|---|---|---|---|---|---|---|---|---|
| | | Type 1 | Type 2 | Type 3 | Type 4 | Type 5 | Type 6 | Type 7 | **Subtotal** |
| **Facies types** | Type A | 0 | 0 | 0 | 0 | 0 | 0 | 0 | 0 |
| | Type B | 0 | 0 | 0 | 0 | 1 | 2 | 0 | 3 |
| | Type C | 3 | 8 | 0 | 3 | 2 | 0 | 0 | 16 |
| | Type D | 0 | 1 | 0 | 1 | 5 | 1 | 0 | 8 |
| | Type E | 0 | 0 | 3 | 1 | 1 | 2 | 0 | 7 |
| | Type F | 0 | 0 | 4 | 0 | 1 | 0 | 0 | 5 |
| | Type G | 0 | 0 | 3 | 0 | 0 | 0 | 0 | 3 |
| | Type H | 19 | 23 | 0 | 4 | 0 | 0 | 1 | 47 |
| | Type I | 2 | 4 | 0 | 0 | 0 | 0 | 4 | 10 |
| **Subtotal** | | 24 | 36 | 10 | 9 | 10 | 5 | 5 | **Total 99** |

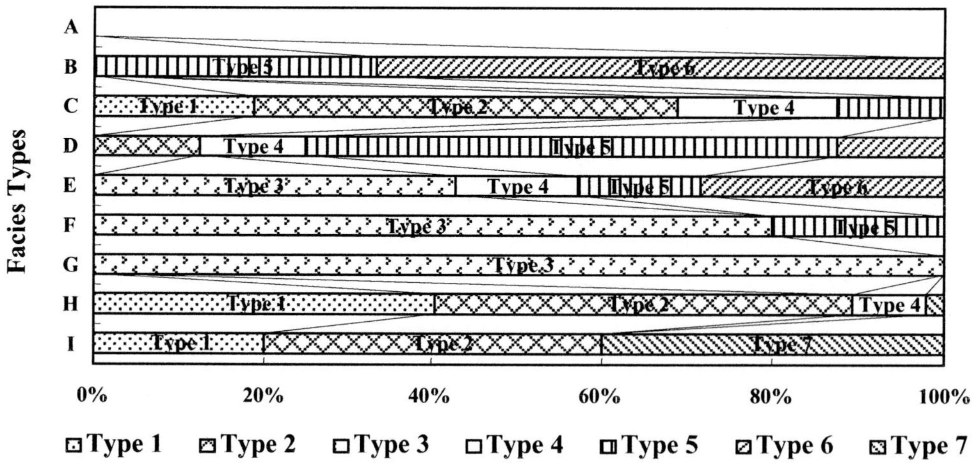


FIG. 7.—Relationship between facies types and pore-throat size-distribution types.

pores. However, according to observations on the slabbed cores, this facies is characterized by a relatively large secondary pore network caused by preferential dissolution of rudist bivalves. It should be noted that the individual plug samples used for measurement might not reflect the nature of the entire pore system.

**Facies Type F** (well sorted very fine-grained peloidal–foraminiferal grainstone): Curve Type 3 (80%) is the main type for this facies. The grains of this facies are well sorted, and small intergranular pores are well preserved as the main visible pore type.

**Facies Type G** (dolomite and/or dolomitic limestone): Pore-throat size distribution Curve Type 3 (100%) is the only type in this facies, and suggests the existence of intercrystalline pores.

**Facies Type H** (very fine- to fine-grained peloidal packstone, including stylolitized layers: S1 to S5): The main pore type is microporous, and most pore-throat size distribution curves are Curve Type 1 (40.4%) or Curve Type 2 (48.9%).

**Facies Type I** (bioclastic wackestone/mudstone): Curve Types 2 (40%) and 7 (40%) are the main types for this facies. Most of the pores are micropores.

As described above, each facies is characterized by a unique pore-throat size distribution(s) resulting from its own pore network. Some facies, however, have the same pore-throat distribution. This indicates the same petrophysical characteristics, even though the facies type is different. The main pore types for each facies and their relationship to permeability and pore-throat size distribution are illustrated in Figure 8.

### *Relationship between Pore-Throat Size Distribution Curve Types and Layers*

Each reservoir sublayer is characterized by a specific proportion of pore-throat size distribution curve types (Fig. 9 and Table 2). Curve Type 2 dominates units IIA to IID. Below IID, Type 2 is also common, but Curve Type 1 is the most prevalent. Curve Type 3 appears mainly in units IIA and IIE. Although the frequency of Curve Types 4, 5, and 6 is low, these types commonly appear in association with high-permeability facies in units IIA to IID. S1 and S5 include Curve Type 7. In general, Curve Types 1 and 2 increase for "S" zones compared with sublayers immediately above and below. The features of each layer are summarized as follows:

**Unit IIA:** This is characterized by a high proportion of Curve Types 2 and 3. Curve Types 1, 5, and 6 are a minor constituent. The main high-permeability facies in unit IIA are Facies Types E and F.

**Unit IIB:** This is characterized by Curve Types 2 and 5. Especially, the lower part (unit IIB3) is characterized by a high proportion of Curve Type 5. Facies Types B, C, and D are the main high-permeability facies in unit IIB, and pore geometry is characterized by bimodal or trimodal distributions.

**Unit IIC:** This is also characterized by Curve Types 2 and 5, and the high-permeability facies of unit IIC consists mainly of Facies Types B, C, and D. In particular, the upper part (unit IIC1) and lower part (unit IIC3) are characterized by a high proportion of Curve Type 5, with a bimodal system. On the other hand, the middle part (unit IIC2) consists mainly of Curve Types 3 and 4.

**Unit IID:** The upper part of unit IID is composed of various curve types, whereas Curve Types 1 and 2 are the main constituents of the middle part of unit IID. No data are available for the lower part of unit IID.

**Unit IIE:** The middle part of unit IIE is characterized by Curve Type 3, whereas the upper (unit IIE1) and lower (unit IIE3) parts are dominated by Curve Types 1 and/or 2.

**Unit IIF:** This is composed mainly of Curve Types 1 and 2.

As mentioned above, a reservoir sublayer is not represented by a specific pore-throat size distribution curve type but a combination of curve types. This may indicate that the pore-throat size distribution of a specific sublayer is characterized by specific, high-permeability facies types. In addition, the pore-throat size distribution curve is determined by the permeability range, which shows lateral variation in the sublayer. It is important to establish a pore-throat size distribution curve for a specific sublayer, because the seven curve types of pore-throat size distributions have different pore geometries and exhibit distinctive and characteristic capillary pressure curve shapes.

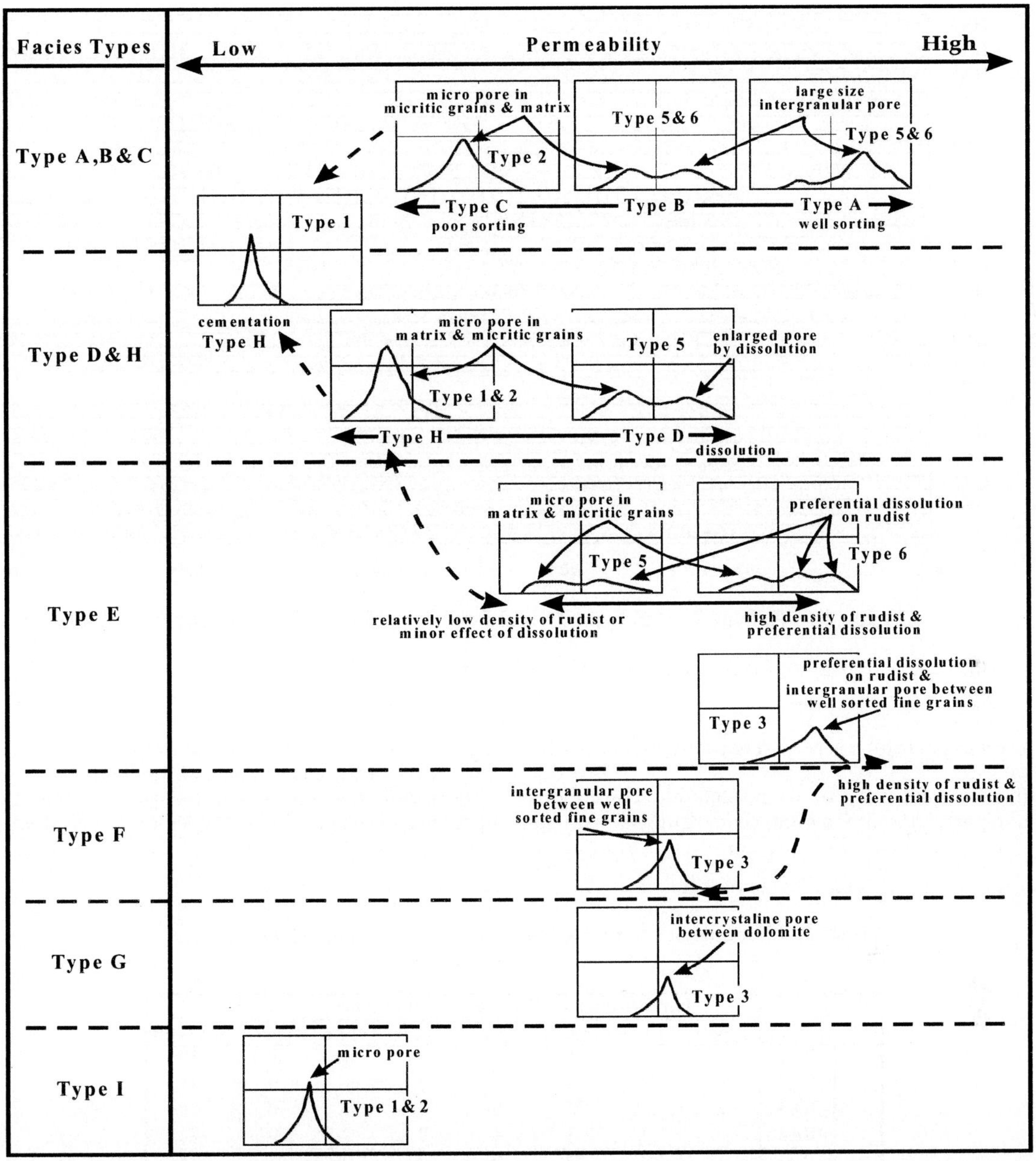


FIG. 8.—Relationship between facies types and pore geometry/permeability.

## RESERVOIR SIMULATION

Some porous zones in the eleven-layer scheme, which have been defined on the basis of porosity profiles, are further subdivided on the basis of log curve profiles and lithology, especially vertical and lateral distribution of high-permeability facies characterized by a pore-throat size distribution(s). A total of twenty-two sublayers are identified, e.g., unit IIA subdivided into four sublayers and units IIB, IIC, IID, and IIE each subdivided into three sublayers.

Each layer of the 22 sublayer scheme is categorized by a specific relationship between porosity and permeability (Fig. 10). This relationship can be applicable to permeability prediction of non-cored intervals and wells. Furthermore, a single pore-throat size distribution curve is selected on the basis of the permeability range in the specific sublayer of this 22 sublayer scheme to assign a specific capillary pressure to a reservoir simulation model. Finally, rock types are assigned using lithological, petrographical, and diagenetic and mercury-injection curve character.

## CONCLUSIONS

The Thamama II reservoir is classified into nine facies by lithology and permeability, in which six high-permeability facies are identified. Both original sediments and diagenetic modification control high permeability in these facies. High permeability in the algal–peloidal grainstone/packstone, which is characterized by fine-scaled, cyclic upward-fining bed sets, tends to be developed in the basal part of the bed sets and determined by grain size, shape, and sorting. High permeability in the rudist–

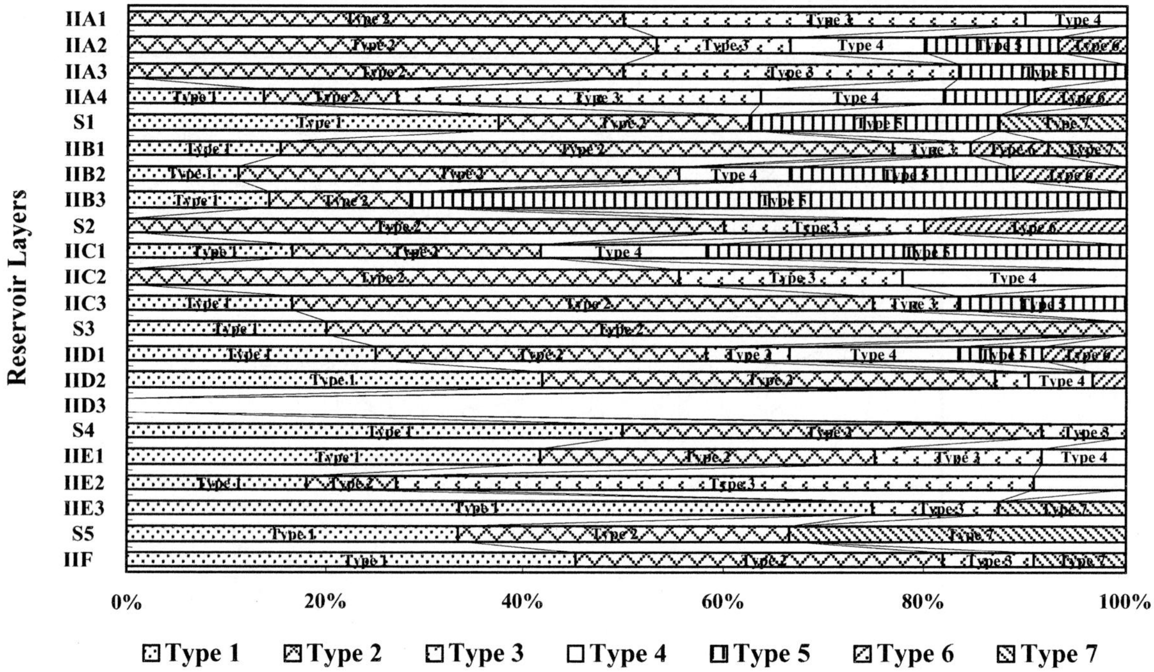


FIG. 9.—Relationship between reservoir layers and pore-throat size-distribution types.

peloidal grainstone/packstone is related to preferential dissolution of rudists.

These high-permeability facies are not laterally continuous over the field. In particular, well sorted, coarse grainstone in the basal parts of the bed sets, which are considered to have an important role on water breakthrough, may be limited to the crestal area of the eastern part of the field.

Pore-throat size distribution curves are classified into seven representative types, Curve Types 1 to 7, based on curve shapes. Curve Types 3, 4, 5, and 6 indicate relatively high permeability,

TABLE 2.—Numbers of sample of each curve type in each reservoir layer.

| | | Pore-throat size distribution curve types | | | | | | | |
|---|---|---|---|---|---|---|---|---|---|
| | | Type 1 | Type 2 | Type 3 | Type 4 | Type 5 | Type 6 | Type 7 | Subtotal |
| Reservoir layers | IIA1 | 0 | 5 | 4 | 1 | 0 | 0 | 0 | 10 |
| | IIA2 | 0 | 8 | 2 | 2 | 2 | 1 | 0 | 15 |
| | IIA3 | 0 | 3 | 2 | 0 | 1 | 0 | 0 | 6 |
| | IIA4 | 3 | 3 | 8 | 4 | 2 | 2 | 0 | 22 |
| | S1 | 3 | 2 | 0 | 0 | 2 | 0 | 1 | 8 |
| | IIB1 | 2 | 8 | 1 | 0 | 0 | 1 | 1 | 13 |
| | IIB2 | 1 | 4 | 0 | 1 | 2 | 1 | 0 | 9 |
| | IIB3 | 1 | 1 | 0 | 0 | 5 | 0 | 0 | 7 |
| | S2 | 0 | 3 | 1 | 0 | 0 | 1 | 0 | 5 |
| | IIC1 | 2 | 3 | 0 | 2 | 5 | 0 | 0 | 12 |
| | IIC2 | 0 | 5 | 2 | 2 | 0 | 0 | 0 | 9 |
| | IIC3 | 2 | 7 | 1 | 0 | 2 | 0 | 0 | 12 |
| | S3 | 1 | 4 | 0 | 0 | 0 | 0 | 0 | 5 |
| | IID1 | 3 | 4 | 1 | 2 | 1 | 1 | 0 | 12 |
| | IID2 | 13 | 14 | 1 | 2 | 0 | 1 | 0 | 31 |
| | IID3 | 0 | 0 | 0 | 0 | 0 | 0 | 0 | 0 |
| | S4 | 6 | 5 | 1 | 0 | 0 | 0 | 0 | 12 |
| | IIE1 | 5 | 4 | 2 | 1 | 0 | 0 | 0 | 12 |
| | IIE2 | 2 | 1 | 7 | 1 | 0 | 0 | 0 | 11 |
| | IIE3 | 6 | 0 | 1 | 0 | 0 | 0 | 1 | 8 |
| | S5 | 3 | 3 | 0 | 0 | 0 | 0 | 3 | 9 |
| | IIF | 5 | 4 | 1 | 0 | 0 | 0 | 1 | 11 |
| Subtotal | | 58 | 91 | 35 | 18 | 22 | 8 | 7 | Total 239 |

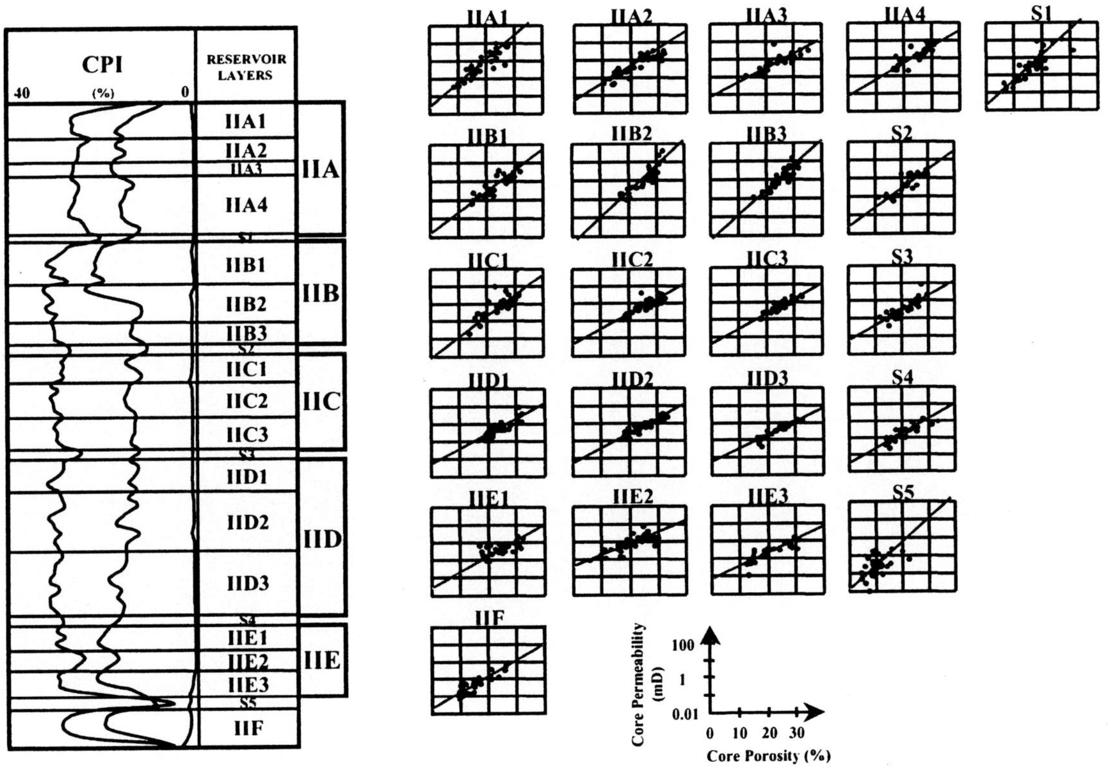


FIG. 10.—Plots of porosity vs. permeability for each reservoir layer.

and they can be roughly distinguished as a unimodal system (Curve Types 3 and 4) or a bimodal/trimodal system (Curve Types 5 and 6). Bimodal/trimodal curve types show a higher permeability range than the unimodal system. Almost all low-permeability samples (< 2 md) show similar single peak curves, with the peak lower than 1 μm.

Each facies type is characterized by a unique pore-throat distribution. High-permeability samples (> 100 md) in units IIB, IIC, and IID are characterized by either bimodal or trimodal pore systems. In units IIA and IIE, some high-permeability samples show unimodal and bimodal pore systems. Curve Type 2 is the most prevalent of the seven curve types in units IIA to IID. On the other hand, Curve Types 1 and 2 are common in units IIE and IIF.

The eleven-layer scheme, previously defined on the basis of porosity profiles, was further divided into sublayers based on log curve profiles, vertical and lateral distribution of high-permeability facies, and pore-throat size distribution curve types. A total of 22 sublayers were defined. A single pore-throat size distribution curve can be assigned for each sublayer according to facies distribution and permeability range. As a result, the relationship between porosity and permeability for each of the 22 sublayers is well characterized and will be applied to appropriate simulation modeling.

## ACKNOWLEDGMENTS

The authors are grateful to the management of the Abu Dhabi National Oil Company and the Zakum Development Company for permission to publish this paper. Thanks are also expressed to Japan Oil Development Co., Ltd. (JODCO) for support in preparing this manuscript. The authors are grateful to Mr. K. Nishi for comments and advice during the reservoir characterization study. We also very much appreciate the assistance and discussions from the Technical Department staff of JODCO during the study, and to the editors of this publication for final editing and suggestions to improve our paper.

## REFERENCES

ALSHARHAN, A.S., 1990, Geology and reservoir characteristics of Lower Cretaceous Kharaib Formation in Zakum Field, Abu Dhabi, United Arab Emirates, *in* Brooks, J., ed., Classic Petroleum Provinces: Geological Society of London, Special Publication 50, p. 299–316.

ALSHARHAN, A.S., AND NAIRN, A.E.M., 1986, A Review of the Cretaceous Formations in the Arabian Peninsula and Gulf: Part 1. Lower Cretaceous (Thamama Group): Stratigraphy and Paleogeography: Journal of Petroleum Geology , v. 9, p. 365–392.

HASSAN, T.H., MUDD, G.C., AND TWOMBLEY, B.N., 1975, The stratigraphy and sedimentation of the Thamama Group (Lower Cretaceous) of Abu Dhabi: 9th Arab Petroleum Congress, Dubai, United Arab Emirates, p. 1–11.

MURRIS, R.J., 1980, Middle East Stratigraphic evolution and oil habitat: American Association of Petroleum Geologists, Bulletin, v. 64, p. 597–618.

NISHI, N., AND SHIBASAKI, T., 1996, Capillary pressure effect on injected water movement and upscales relative permeability in a heterogeneous carboante reservoir: 7th Abu Dhabi International Petroleum Exhibition and Converence, Proceedings, SPE 36177, p. 26–36.

# STRATIGRAPHY, STABLE ISOTOPES, AND HYDROCARBON POTENTIAL OF THE APTIAN SHUAIBA FORMATION, U.A.E.

ABDULRAHMAN S. ALSHARHAN
*Faculty of Science, United Arab Emirates University, P.O. Box 17551, Al Ain, U.A.E.*
IHSAN S. AL-AASM
*College of Engineering and Sciences, Department of Earth Science, University of Windsor, Windsor, Ontario, N9B 3P4, Canada*
AND
MOHAMED G. SALAH
*Faculty of Science, United Arab Emirates University, P.O. Box 17551, Al Ain, U.A.E.*
*Present address: Reservoir Characterization Research & Consulting (UK) Ltd., P.O. Box: 46800, Abu Dhabi, U.A.E.*

ABSTRACT: The Shuaiba Formation of the United Arab Emirates (U.A.E.), forms one of the most important petroleum reservoirs in the Arabian Gulf. Its reservoir quality is controlled by diagenetic processes that were active during early shallow burial to late deep burial. Detailed well log evaluation, petrographic and geochemical studies of carbonate diagenesis, X-ray diffraction, cathodoluminescence microscopy, and oxygen and carbon isotopic determinations from cores and an outcrop section allowed evaluation of the stratigraphic and depositional framework of the Shuaiba Formation, its diagenetic history, and the prediction of its local reservoir potential. The Shuaiba Formation consists of two informal and one formal members: the lower Shuaiba and upper Shuaiba members and the Bab Member, in ascending order. The Shuaiba ranges in thickness from 45 to 145 m, having accumulated in a wide range of depositional settings from shallow to deep shelf. The lithofacies identified within the Shuaiba Formation include peloidal skeletal algal packstone/grainstone; ooidal-peloidal grainstone/packstone; skeletal algal (Lithocodioidea) floatstone; intraclastic and coated packstone/grainstone and skeletal wackestone/ packstone. These facies were deposited during third-order depositional sequences, including two transgression system tracts (TST) separated by a highstand system tracts (HST). Diagenetic alteration of the original carbonate components proceeded through marine, shallow burial, and deeper burial settings related to stabilization of the carbonate matrix, cements, and rudist shells. Oxygen and carbon isotopes of calcitic matrix have the least altered components of these rocks (av. $\delta^{18}O$ = 5.7‰ PDB; $\delta^{13}C$ = +2.5‰ PDB), whereas the calcite cements occluding shell porosity and veins have more depleted isotopic values (av. $\delta^{18}O$ = 8.8‰ PDB; $\delta^{13}C$ = +0.5‰ PDB). The variations of oxygen and carbon isotopes reflect changes in the water–rock interactions and increasing burial.

The carbonates of the lower and upper Shuaiba members have porosities between 12% and 32% and permeabilities between 1.0 and 160.0 md. The reservoir quality is highly affected by the diagenetic processes which include stabilization of metastable carbonate phases, cementation, dolomitization, stylolitization, and dissolution. The Bab Member, which was deposited in a basinal setting, is organically rich and forms substantial source rock in the eastern and northeastern parts of central UAE and is mature enough in deep troughs to generate and expel hydrocarbons to the reservoirs.

## INTRODUCTION

The Cretaceous rocks of the U.A.E. represent a time of carbonate platforms and offshore banks and buildups. Rudist bivalves colonized shallow marine settings with strong to moderate currents and turbulence (Alsharhan, 1995, Hughes, 1997) and assisted in creating localized bank complexes. The Shuaiba Formation represents the uppermost section of the Lower Cretaceous Thamama Group and locally forms carbonate buildups. These carbonates contain important reservoirs and produce oil and gas from a number of fields in U.A.E., Qatar, Oman, and Saudi Arabia.

This paper deals with the evaluation of diagenetic modifications using petrology and geochemistry of the Lower Cretaceous Shuaiba carbonates. It also delineates the sequence stratigraphic framework of these carbonates, and documents their depositional settings and hydrocarbon reservoir and source characteristics in the U.A.E.

## MATERIALS AND METHODS

The data and interpretations presented in this study were based on detailed examination of 120 samples from selected cores, sidewall cores, and ditch cuttings of the Shuaiba Formation. These samples came from both onshore and offshore fields in the U.A.E., including producing and nonproducing fields from Bu Hasa, Shah, Asab, Sahil, Jarn Yaphour and Fateh (Fig. 1). In addition, the best exposures of the Lower Cretaceous Shuaiba carbonates in the Wadi Dhayah, in the Northern Emirates (Fig. 1), were sampled, studied, and correlated with the subsurface data. Core samples were studied under a binocular microscope and in 88 thin sections cut from selected intervals. Polished slabs and uncovered thin sections were stained for Fe-calcite using a mixture of potassium ferricyanide and alizarin red-S according to the method of Lindholm and Finkelman (1972). Selective thin sections were examined by cathodoluminescence microscopy using a Technosyn 8200 MKII Model cold cathodoluminescence stage at a voltage of 15 kV and current intensity of about 450 μA. In addition, 47 thin sections were impregnated with an epoxy mixed with a blue fluorescent dye for porosity recognition. Various generations of calcite cements, fossil components, dolomite, and matrix were selected for carbon and oxygen isotopic analysis employing a microscope-mounted drill assembly to extract 0.5 to 4.0 mg of powdered carbonate from polished slabs. Fifty-one samples were treated in vacuo with 100% phosphoric acid at 25°C for calcite and at 50°C for dolomite. All gas extractions were made

Middle East Models of Jurassic/Cretaceous Carbonate Systems
SEPM Special Publication No. 69, Copyright © 2000
SEPM (Society for Sedimentary Geology), ISBN 1-56576-075-1, p. 299–314.

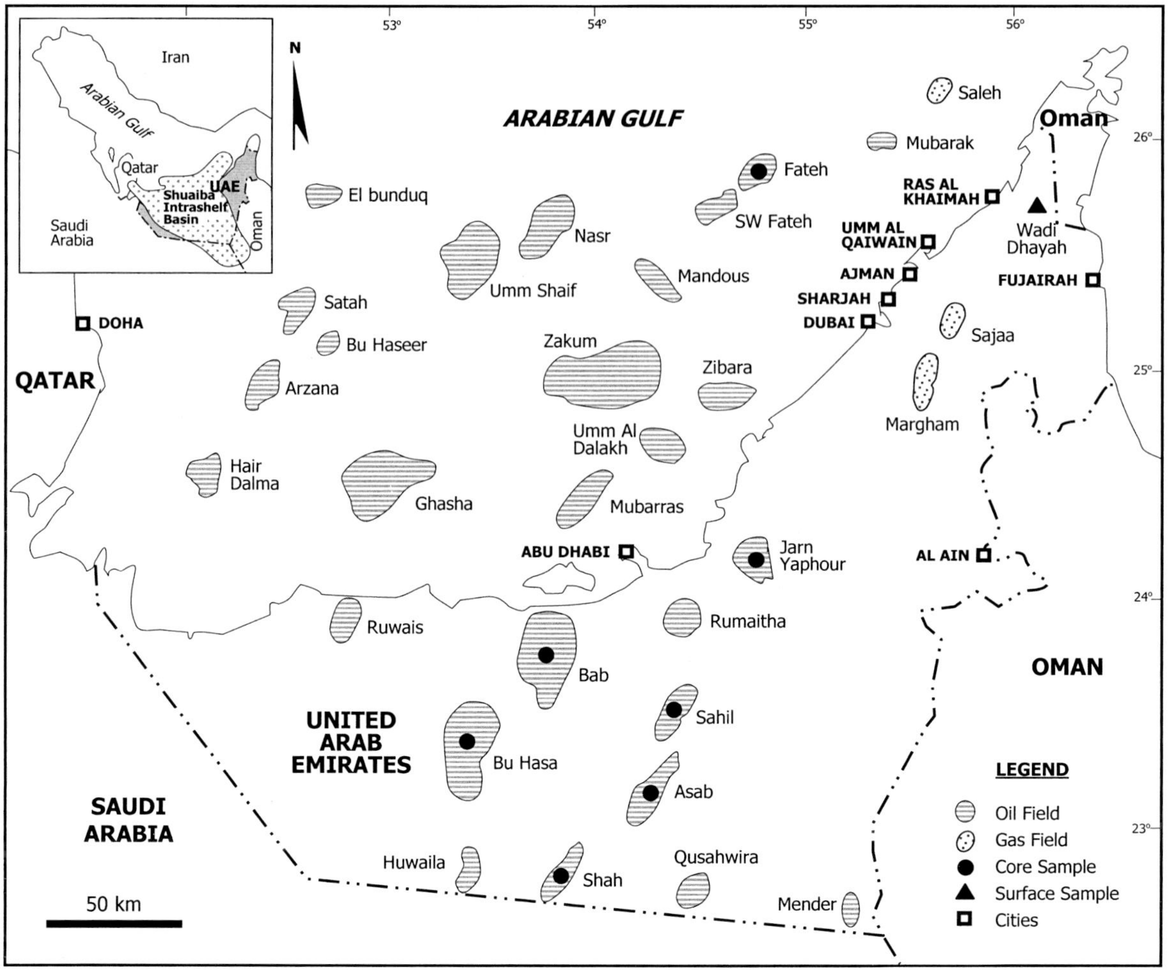


FIG. 1.—Location map of the United Arab Emirates showing the oil and gas fields and the surface samples used in this study.

in the Stable Isotope Laboratory of the University of Windsor, Canada, and the evolved $CO_2$ gas was analyzed for isotopic ratios on a SIRA-12 mass spectometer values of O and C isotopes are reported in per mil (‰) relative to the PDB standard. Precision was better than 0.1 ‰ for both isotopes.

Thirty-three surface and subsurface samples were analyzed by X-ray diffraction (Philips, model PW1840), using a Ni filter, Cu radiation (l = 1.542 Å) at 40 kV and 30 µA and scanning speed of 0.05‰. The reflection peaks between 2° and 70° $2\theta$ and the corresponding $d$ spacing (Å) and relative intensity (I/I°) were detected and compared with standard data of American Standards for Testing and Materials (A.S.T.M.).

Sonic, gamma ray, neutron, density, and resistivity logs for several wells covering the offshore and onshore parts of the U.A.E. were used, together with the available core and ditch samples, to subdivide the Shuaiba Formation, to create a depositional model and to define sequence boundaries.

## STRATIGRAPHY OF THE SHUAIBA FORMATION

The Shuaiba Formation is composed of thick, porous shelf carbonates (Fig. 2) which show considerable subsurface lateral and vertical lithofacies change over much of the Arabian Gulf. The Shuaiba Formation ranges from 45 to 145 m thick, increasing toward the west and the south. During Early Cretaceous time, the central part of the U.A.E. was a wide trough (intrashelf basin) oriented roughly northeast–southwest and deepening toward the northeast in the onshore area (Murris, 1980; Alsharhan, 1995). This trough was mainly a response to the influence of the deep structural features on the deposition of the Cretaceous sediments. The Shuaiba was deposited during an extensive Tethyan transgression on the northeastern part of the Arabian Shield during the early to mid Aptian. The Shuaiba intrashelf basin was affected by a long-term, second-order eustatic sea-level fall, formed during a third-order transgression in the early Aptian and was filled during the early to mid Aptian.

This formation rests conformably over the carbonates of the Kharaib Formation upper dense member and is unconformably overlain by the shale and argillaceous, silty limestone of the Albian Nahr Umr Formation. This upper boundary is marked upward by an abrupt increase in the gamma-ray readings and decrease in density readings (Fig. 3).

The Shuaiba Formation in the U.A.E. has been assigned to the Aptian by most of the researchers in the U.A.E. The upper Bab Member contains upper Aptian ammonites (Hassan et al., 1975), but has been dated as early Albian, based on palynological and paleontological analysis and also by strontium and potassium–argon dating (Azer and Toland, 1993). However, the Aptian age is

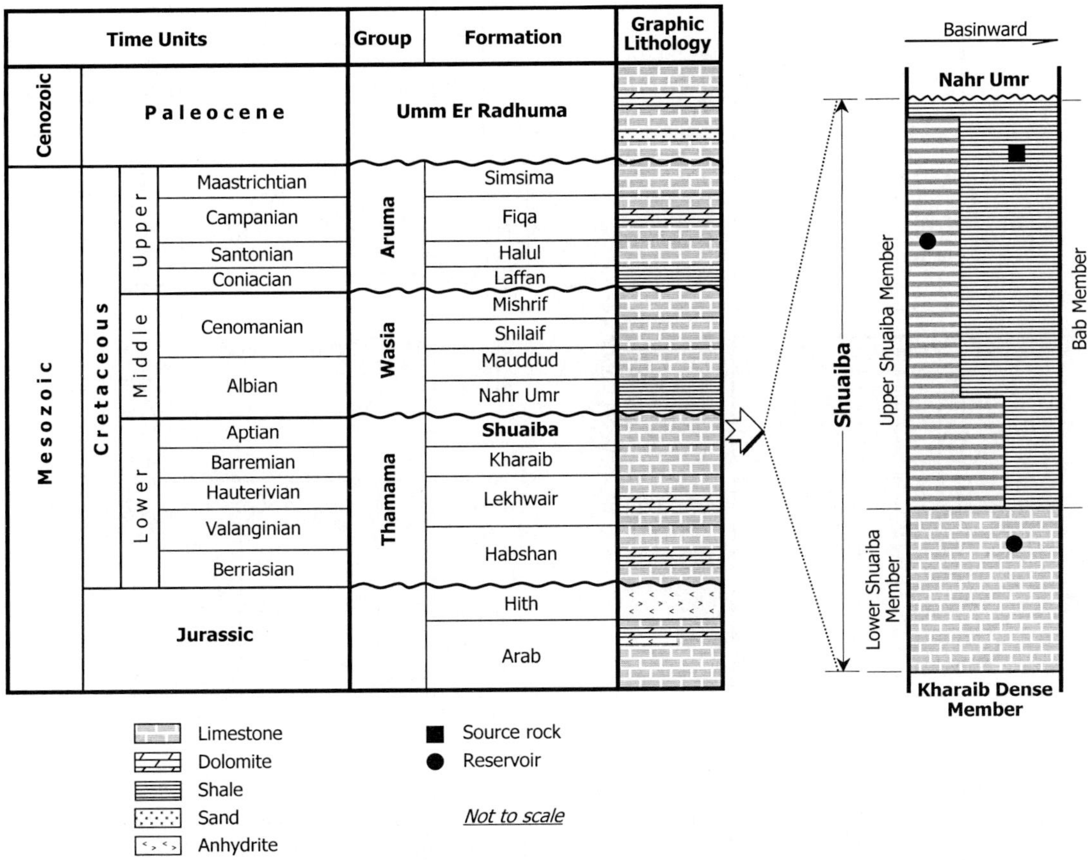


FIG. 2.—Generalized stratigraphic column of the U.A.E. showing the Shuaiba stratigraphy.

based on microfossils from the Shuaiba Formation in U.A.E. and Oman (Alsharhan and Nairn, 1986; Scott, 1990; Simmons, 1994; Witt and Gokdag, 1994), rudists (Skelton and Masse, this volume) and by isotopes (Vahrenkamp, 1996; Grötsch et al., 1998).

Although the Shuaiba Formation has been subdivided differently by some workers (e.g., Witt and Gokdag, 1994; Alsharhan, 1985) in the Oman and U.A.E., most divide the Shuaiba Formation into three members, two informal (lower and upper) and one formal (Bab Member). This subdivision is based on the lithostratigraphic and biostratigraphic data and the electric logs and summarized below.

### *Lower Member*

This informal member is known in the oil companies as Thamama I, Thamama A, or Thamama subzone IA, and forms the lowest reservoir facies (Units A and B in ascending order). It is 9 to 24 m thick and consists of blanket-type deposits that are generally present over the whole area. The lower part of this member is composed of wackestone and packstone with local grainstone and colonies of codiacian algae, overlain by microporous lime mudstone and wackestone that are slightly argillaceous, burrowed, and bioturbated. The abundance of algal colonies suggests that deposition occurred in shallow water in the photic zone. The occurrence of calcispheres, miliolids, and other arenaceous foraminifera suggests a lack of open marine circulation and a slight restriction. The upper part of this member is composed of medium-grained wackestone and boundstone with abundant codiacian algae. Occasional fine-grained microporous lime mudstone occurs with abundant stylolites and scattered pyrite. These sediments were deposited in offshore shelf-basin facies with open circulation.

Azer and Toland (1993) suggested a low rate of sedimentation accompanied by a fairly slow rise of the sea level during the deposition of this member, which is reflected as the sheet-like crusts of the *Lithocodium* algae.

The lower member conformably overlies the dense carbonates of the Kharaib Formation. The boundary is marked up section by a sharp increase in gamma ray, neutron porosity, and density logs (Fig. 3). This member grades conformably upward into the upper member in the west, south, and northwest, and it is overlain unconformably by the Bab Member in the east and northeast.

### *Upper Member*

The informal upper member contains several lithofacies and includes what are known as reservoir units C, D, E, F, G, H, $H_o$, and I. It is composed of medium- to coarse-grained packstone and grainstone with abundant rudistid fragments; *Orbitolina* packstone, dense lime mudstone with scattered fine bioclastic debris, bioturbated argillaceous microporous lime mudstone and wackestone, and fine- to medium-grained wackestone containing abundant and commonly large algal colonies. These different carbonates are recorded in most of the study area but are absent in the eastern and the northeastern parts of the U.A.E. This member represents a wide range of depositional settings from back-biostrome through biostrome to fore-biostrome.

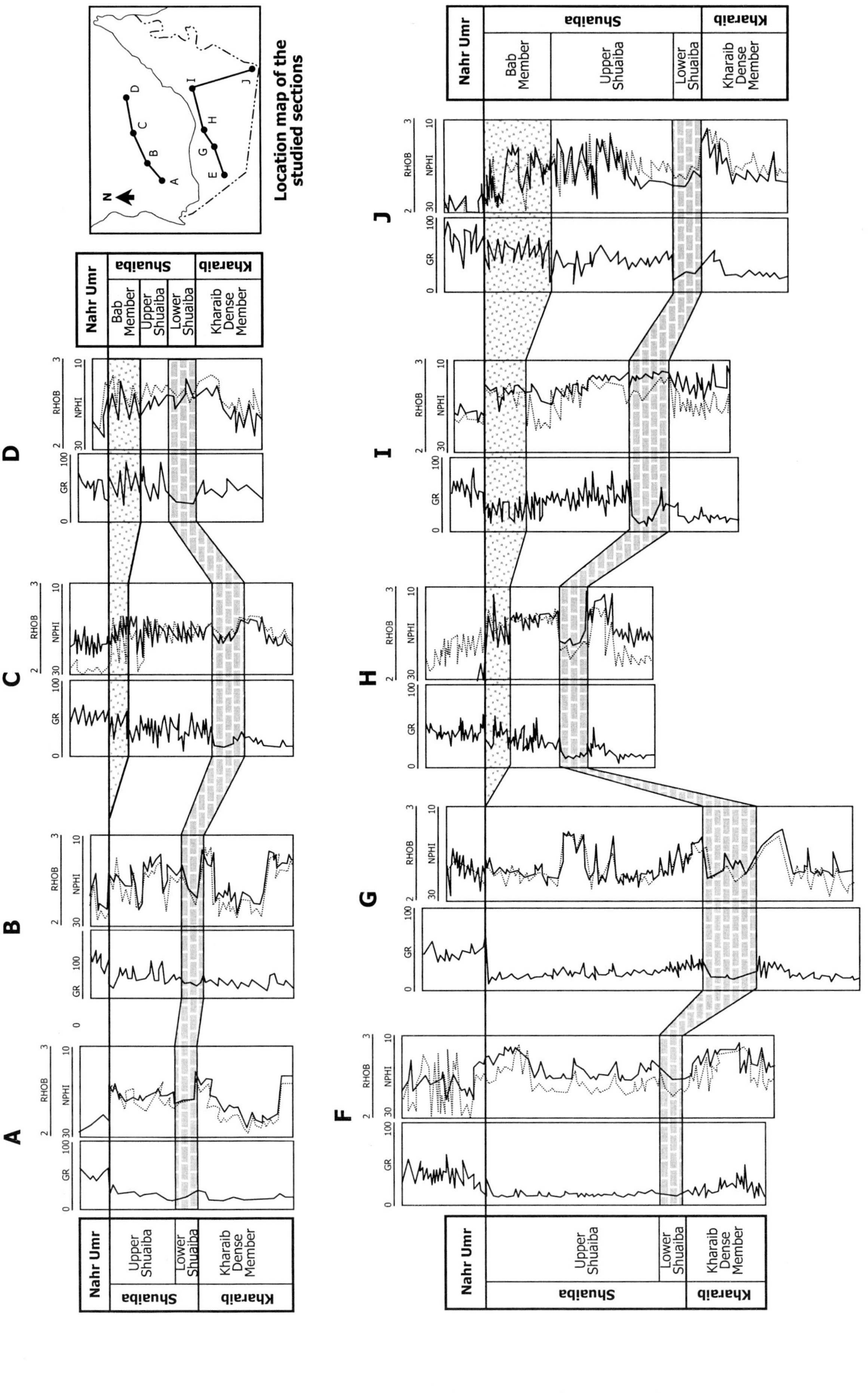


FIG. 3.—Electric log correlation diagram showing the lateral variations within the Shuaiba Formation. Vertical scale: 1 cm = 30 m (100 ft).

The upper member conformably overlies the lower Shuaiba member; it grades laterally and interfingers with the Bab Member in the east and northeast; and it is overlain unconformably by the Nahr Umr Formation, the contact of which is picked at a sharp uphole increase in gamma-ray log values. The upper boundary is picked at the sharp transition from the clean carbonates of this member to the shales of the Nahr Umr Formation in the west, north, and south, and the highly argillaceous limestone of the Bab Member in the northeast and east.

### Bab Member

The Bab Member, the youngest unit in the Shuaiba Formation, averages 70 m thick and is unconformably overlain by the Nahr Umr Formation. The Bab Member consists of a basinal argillaceous lime mudstone, with abundant planktonic foraminifera and moderate–high organic content (Hassan et al., 1975; Alsharhan, 1995; Taher, 1997). The Bab Member is restricted in the eastern and northeastern parts of the U.A.E. and was deposited in mid-ramp settings (intrashelf basin). The lower part of the Bab Member is a progradational carbonate cycle of deep-water argillaceous globigerinid limestone. This is overlain by Gymnocodiaceae, peloidal argillaceous packstone–wackestone, and skeletal *Everticyclammina* and *Pseudocyclammina* foraminiferal, argillaceous wackestone–packstone. The Shuaiba basinal facies is well developed in central U.A.E. and is known in the area as the Abu Dhabi Intrashelf basin (see Alsharhan and Nairn, 1986).

## SHUAIBA SURFACE SECTION

At Wadi Dhayah (surface sample A, Fig. 1), the upper member of the Shuaiba Formation crops out in the floor of the wadi and is overlain unconformably by the Albian Nahr Umr shales and carbonates (Fig. 4). This section has been discussed on two field trips (Russell, 1996; Alsharhan, 1997). The contact is a scree-covered surface and is picked at the transition from the light gray massive limestone of the Shuaiba Formation to thin-bedded, light to dark brown shales, marls, and limestones of the Nahr Umr Formation (Fig. 4). The Shuaiba Formation displays an average dip of 5° toward the west-southwest. Only the lower and upper members are exposed.

Only two meters of the lower member crop out in the floor of the wadi, and here it consists of light gray, hard, algal-rich (*Lithocodium* sp.) packstone to wackestone (Fig. 4). The upper member contains three different lithofacies, representing three reservoir facies units in ascending order based on core and outcrop descriptions.

a. *Lithocodium* packstone/mudstone, 12 m thick, is known informally in the oil companies as reservoir unit C. This consists mainly of nodular, stylolitic packstone/mudstone with abundant algae, foraminifera, and shell debris (Fig. 4).
b. *Chondrodonta* sp. Oyster floatstone/rudstone, 6 m thick is known as reservoir unit H, and consists predominantly of numerous caprinid and caprotinid rudists. It also contains *Chondrodonta* sp. and oysters in floatstone and rudstone (Fig. 4). The x-ray analysis of two samples from this unit showed a remarkable increase in the quartz content in the lower part of this unit (Fig. 4).
c. Dasyclad algal packstone/wackstone, 13 m thick, is known as reservoir unit I, and consists of fine-grained dasyclad algae, such as *Salpingoporella* (?) spp. and *Clypeina* sp., and foraminiferal packstone/ wackestone and is unconformably overlain by the Nahr Umr Formation.

## SEQUENCE STRATIGRAPHY

From our examination of cores and logs we conclude that the entire Shuaiba Formation is a part of a second-order super cycle that we correlate with Zuni B4 (LZB-4) (Haq et al., 1988). It has two third-order sequences: the LZB-4.1 and LZB-4.2 (Fig. 5). The lower part of the older cycle (LZB-4.1) is represented by the upper member of the underlying Kharaib Formation based on facies interpretation, areal extent, and depositional models. Also, correlation of upper dense Kharaib with basal Aptian is supported by Scott (1990), Simmons (1994), and Witt and Gokdag (1994). The Shuaiba Formation is represented by a highstand systems tract (HST) and a transgressive systems tract (TST), with one sequence boundary and two downlap surfaces (Fig. 5). A lower sequence boundary at 112 Ma separates the upper dense unit of the Kharaib Formation from the underlying Kharaib porous zones and we correlate it with LZB4.1 (Haq et al., 1988). The base of the Shuaiba Formation is the lower downlap surface calibrated with the Haq et al. (1988) 111 Ma event. The top of the lower Shuaiba member is a sequence boundary that we correlate with the 109.5 Ma of Haq et al. (1988). This sequence of the uppermost Kharaib and lower Shuaiba records a simultaneous subsidence accompanied by a slow rate of deposition.

The second sequence boundary separates the lower Shuaiba member from the upper Shuaiba member (Fig. 5), and we correlate it with the boundary between LZB-4.1 and LZB-4.2, dated at 109.5 Ma by Haq et al. (1988). We place the second downlap surface separating the HST from the TST at the unconformity between the Shuaiba and Nahr Umr formations and calibrated with the 108 Ma event of Haq et al. (1988). Because of the lack of seismic data the authors could not trace the Bab Member into the eastern Abu Dhabi subsurface. The Bab is absent in the Musandam outcrop because it onlaps and pinches out or because it was eroded (Vahrenkamp, 1996). An alternative sequence stratigraphy shows three sequence cycles in the Shuaiba Formation (Vahrenkamp, 1996) with lowstand facies in the uppermost Kharaib Formation, just below the Bab Member, and in the upper Bab Member. Kendall et al. (this volume) have conducted sedimentary simulations of the Shuaiba utilizing a similar sequence stratigraphy.

## LITHOFACIES

Examination of the available core and ditch samples revealed seven main lithofacies with bed thickness ranging from 10 cm to 10 m. This wide spectrum of lithofacies reflects the complex depositional settings and sub-environments that prevailed during the time of deposition. They were mainly deposited in inner, inner to middle, and middle ramp settings (Fig. 6). Similar depositional models for the Shuaiba were proposed by Hassan et al. (1975) and Hamdan and Alsharhan (1991).

1. Peloidal skeletal algal packstone/grainstone and algal floatstone/packstone. This facies represents the back shoal setting within the inner ramp (Fig. 6) and suggests a low-energy, semirestricted or restricted lagoonal environment. The skeletal grains include rounded, sand-size shell fragments (Fig. 7A). The average thickness of this lithofacies in the Bu Hasa oil field is 25 m.
2. Ooidal–peloidal grainstone/packstone. This facies accumulated as a prograding shoal (mobile) forming locally extensive sand sheets, and it constitutes limited parts of moderate-quality reservoirs. Common ooids, peloids, and molluscan fragments suggest shoal development within the inner ramp. It is best developed in the upper member, which is the best reservoir in the giant Bu Hasa oil field. The average thickness of this lithofacies in the Bu Hasa oil field is 67 m.

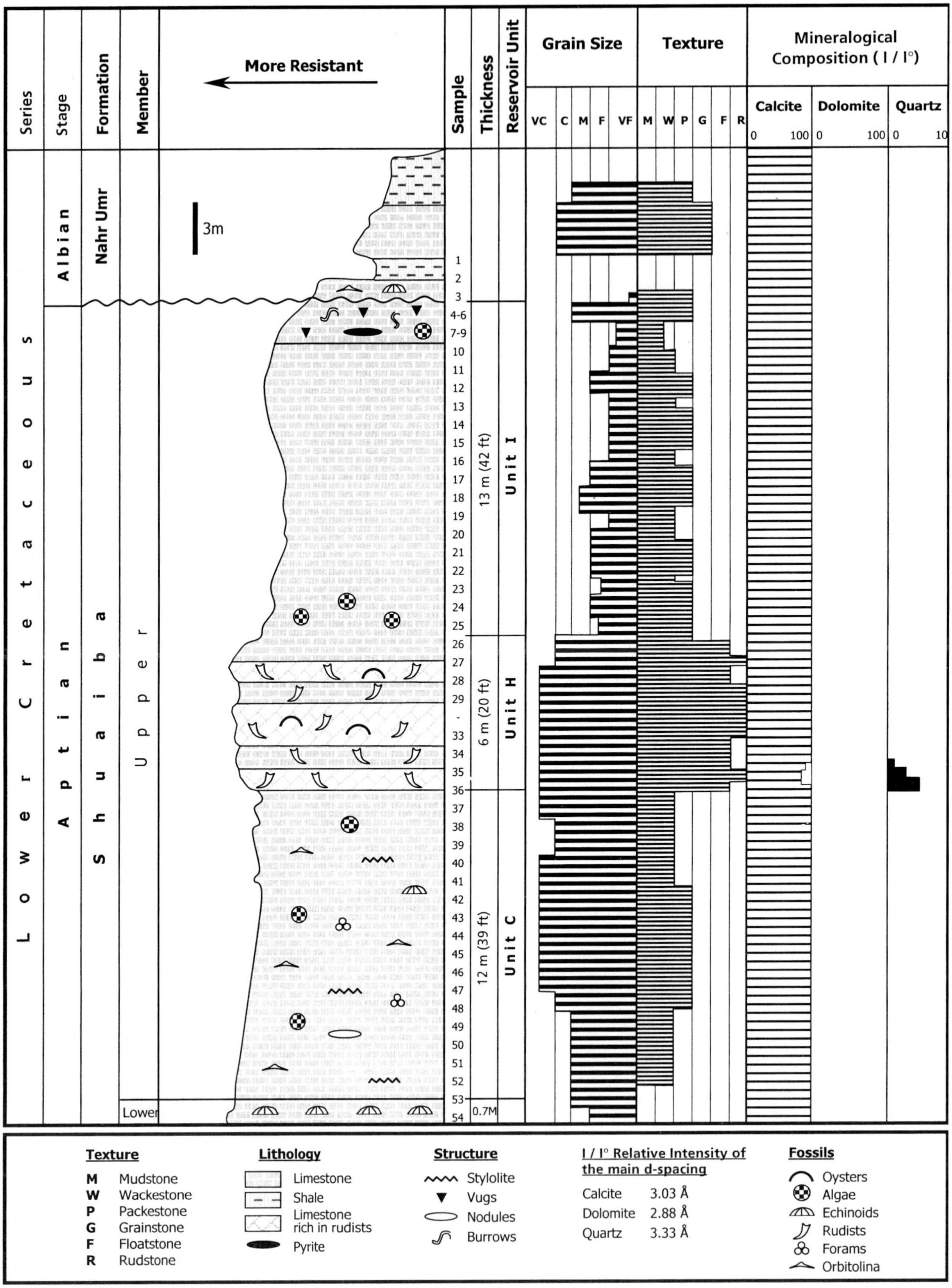


FIG. 4.—Stratigraphic column of the surface section in Wadi Dhayah (compiled with modification from Russell, 1996; Alsharhan, 1997; and our new field data).

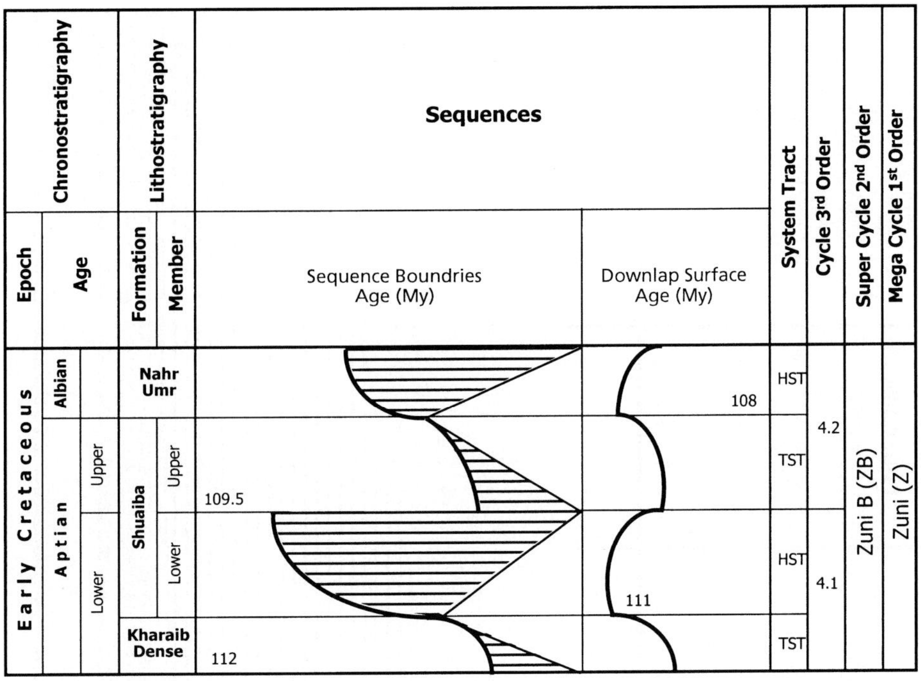


FIG. 5.—Correlation of sequence stratigraphic boundaries within the Shuaiba Formation cycles with the Haq et al. (1988) sea-level curve.

3. Skeletal algal (Lithocodioidea) floatstone/packstone/grainstone and skeletal peloidal algal (dasyclads) packstone/grainstone (Fig. 7B, D). These facies constitute considerable parts of the reservoir sediments. This lithofacies association suggests deposition in a foreshoal area in an inner-ramp environment (Fig. 6). It ranges in thickness from 3 to 78 m.
4. Intraclastic packstone/grainstone and cortoid packstone–grainstone (Fig. 7C, E). These facies comprise a minor portion of the reservoirs and occur mostly in dense limestone intervals. They were deposited in a backshoal that may have been developed between relatively high-relief areas of the sea floor in a backshoal lagoon environment. The average thickness of this lithofacies is 35 m.
5. Skeletal wackestone–packstone, algal (Lithocodioidea) floatstone wackestone, and algal wackestone. This lithofacies and the dense limestone intervals were deposited in a horizontal area extending over the inner fore-shoal to mid-ramp where local accumulations of algal *Lithocodium* mounds and complex ramp sediment were interbedded. This lithofacies ranges in thickness from 8 to 60 m.
6. Skeletal foraminiferal (*Pseudocyclammina*) argillaceous wackestone–packstone, skeletal peloidal (blackened grains) packstone/grainstone, and algal (Gymnocodiaceae) wackestone/packstone. These facies are most common in dense limestone intervals separating reservoirs. They are thought to have been deposited on a very gentle slope within the mid-ramp (Fig. 6). The range in thickness from 3 to 35 m.
7. Algal and skeletal packstone/*Lithocodium* boundstone and argillaceous wackestone/ mudstone. These facies represent the relatively deeper slope of the mid-ramp environment. They are typical of the lower member, informally known as reservoir units A and B or Thamama IA. These lithofacies have more stylolites than any other lithofacies in the Shuaiba Formation, which have an average amplitude of 10 cm. The lithofacies are about 15 m thick and overlie the dense limestone of the Kharaib Formation and contain abundant echinoderms, *Orbitolina* sp., and colonies of *Lithocodium* algae.

## PARAGENESIS AND STABLE ISOTOPES

### *Paragenetic Sequences*

The diagenetic history of the Shuaiba Formation in the study area includes multiple episodes of calcite cementation, mechanical and chemical compaction, stabilization of shell mineralogy, and dissolution. These diagenetic processes occurred during marine to shallow burial stages and culminated during intermediate to deep burial stages. The paragenetic sequence (Fig. 8) is based on petrographic observations, spatial distribution, cross-cutting relationships between depositional fabrics, cements, compactional features, and geochemical evidence.

The main skeletal framework in the studied limestones is composed of rudist bivalves, algae and algal aggregates, echinoderm parts, stromatoporoids, and diverse mollusks. These fossil components show variable degrees of preservation in response to their original shell mineralogy, shell structures, and the chemistry of diagenetic fluids (Al-Aasm and Veizer, 1986 a, 1986b; Al-Aasm and Azmy, 1996; Hendry et al., 1995). With the accretion of rudist buildups, marine borings in shells commenced contemporaneously accompanied by grain micritization, marine cementation, represented by nonluminescent fibrous calcite cement followed by inclusion-rich equant calcite, syntaxial-rim cement, and internal sedimentation.

Burial of Shuaiba carbonates at shallow depths was accompanied by the formation of minor compactional features such as microstylolites, in-situ broken shells, and close packing of grains. The precipitation of the early generation of dull, prismatic and drusy equant calcite cements (Fig. 7F) and partial replacement of

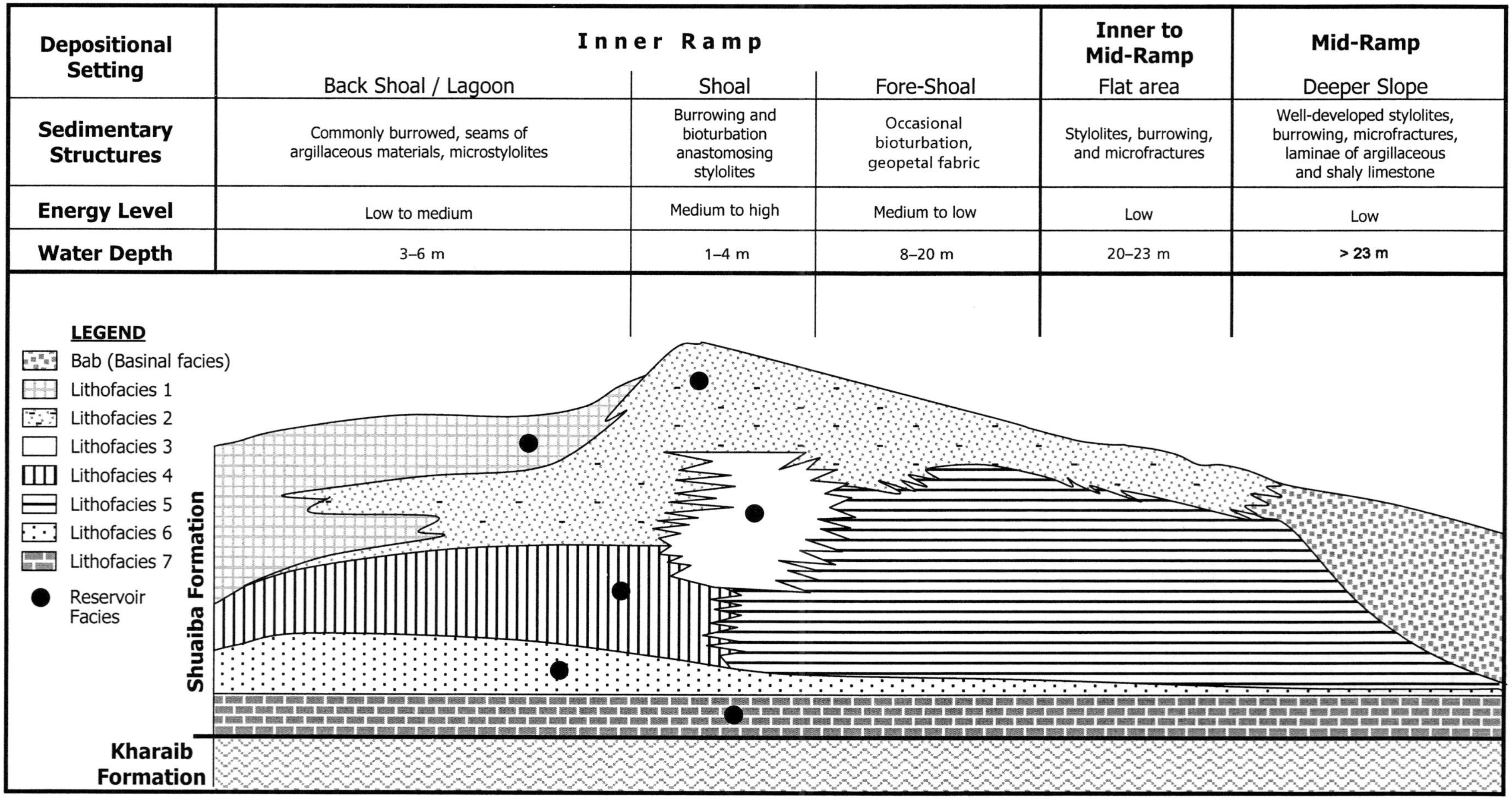


FIG. 6.—Depositional model of the Shuaiba Formation in the U.A.E.

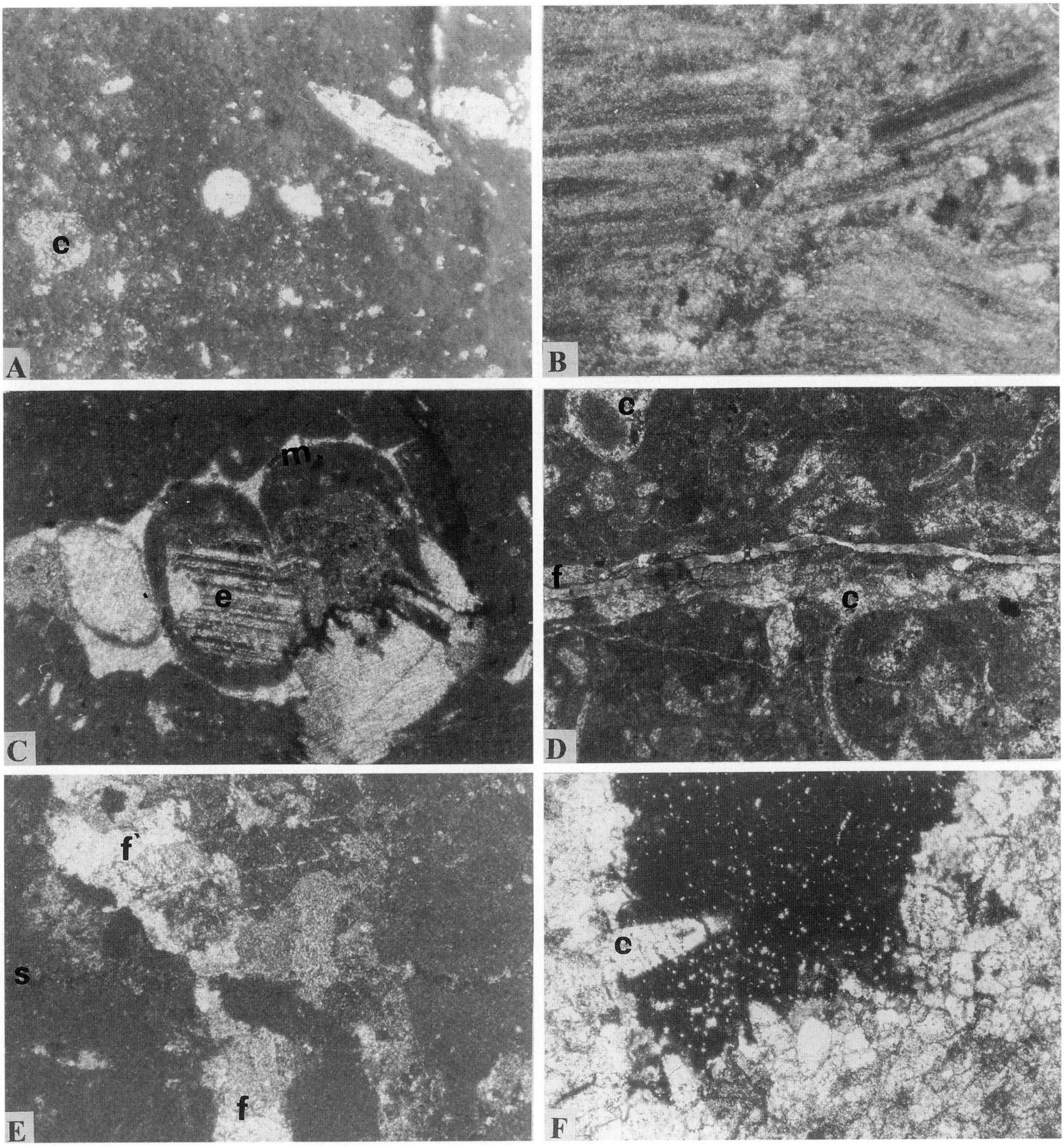


FIG. 7.—Photomicrographs from the Shuaiba representing different facies; scale: 1 cm = 0.5 mm. A) Algal packstone showing dolomitized matrix. Some micritic fossils molds are occupied by calcite cement, c. B) Floatstone showing well preserved rudist shell. C) Intraclastic packstone showing partially micritized gastropod shell that is leached and filled with the calcite crystals. D) Skeletal packstone showing calcite recrystallization, c, of fossils molds. Note the fracture, f, cutting across the photo. E) Intraclastic packstone showing two sets of fractures intersecting, f and f'. Both fractures are filled by recrystallized calcite. Note the stylolite(s), cutting across the fractures. F) Ooid grainstone showing calcite crystals, c, destroying some of the porosity.

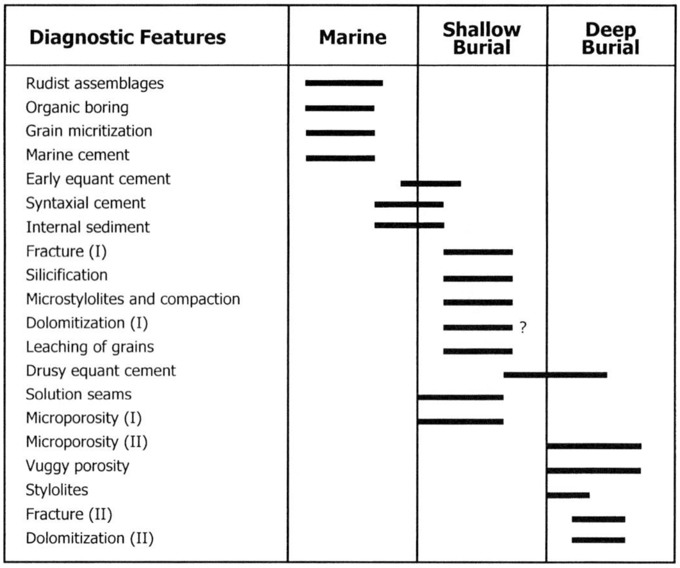


FIG. 8.—Paragenetic sequence of the diagenetic processes affecting the Shuaiba Formation.

the matrix with first-stage dolomite followed this. As a result of increasing compaction upon burial, solution seams and smooth stylolites developed, and diagenetic stabilization of the shell components was initiated with the neomorphic transformation of aragonite to low-Mg calcite, leaving behind relics of the original microstructure. At this stage also the first generation of microporosity developed (Moshier, 1989), in addition to leaching of some grains.

With increasing burial, a later phase of silicification commenced and microfractures developed in most lithofacies, cutting all cements and enhancing the conduit system where further active solutions and second generation of microporosity developed. At this stage low- and high-amplitude stylolites developed. The formation of the late wide fractures also occurred at this time, and were later infilled with red to dull drusy equant calcite cement (see also Burriss et al., 1983).

### *Stable Oxygen and Carbon Isotopes*

The stable carbon and oxygen isotope compositions of the skeletal and nonskeletal components of the Shuaiba Formation are plotted in Figure 9 (see also Table 1). Shown also on this plot is the distribution of these samples according to their location. Rudist shell layers of presumably aragonitic and calcitic original mineralogy (Skelton, 1979; Al-Aasm and Veizer, 1986a) have slightly depleted $\delta^{18}O$ values relative to the postulated oxygen isotopic composition of Lower Cretaceous carbonates (Moldovanyi and Lohmann, 1984; Morrison and Veizer, 1990), which may reflect partial reequilibration and mineralogical stabilization by diagenetic fluids (Brand and Veizer, 1980; Al-Aasm and Veizer, 1986b; Al-Aasm and Azmy, 1996). Carbon isotopes seem, however, to have been buffered by the original carbonates. The calcitic matrix of the studied carbonates show wider ranges in their $\delta^{18}O$ values and slightly depleted $\delta^{13}C$ values relative to the Lower Cretaceous carbonates by at least 1‰. The isotope values of matrix samples analyzed in this study overlap with values of micro-rhombic calcite reported by Moshier (1989) and Budd (1989) from Sajaa Field and of Shuaiba matrix in Oman (Wagner, 1990). However, our results show more depleted carbon and oxygen isotopes in this matrix (Fig. 9). Early generations of calcite cement have $\delta^{18}O$ and $\delta^{13}C$ values very comparable to the calcitic matrix, which may suggest similar fluids and were responsible for the stabilization of the matrix and precipitation of calcite cement. Later generations of calcite cements that occlude fractures, however, have more negative $\delta^{18}O$ and $\delta^{13}C$ values (Fig. 9). Dolomite partially replaces the calcite matrix and occurs as scattered euhedral rhombs. Its $\delta^{18}O$ values are similar to the calcitic matrix but its $\delta^{13}C$ signatures show wider and negative values for two samples. These negative $\delta^{13}C$ values may reflect some organic contributions during the formation of this dolomite (Durocher and Al-Aasm, 1997). There is a slight co-variation trend between oxygen and carbon isotopes in matrix and cement samples.

When isotopic values of the studied carbonates are plotted according to their locations an apparent trend appears. The isotopic composition of surface samples overlaps those from the subsurface. However, the deviation from this general trend arises in two areas. In one, the Bu Hasa samples show very narrow ranges in both oxygen and carbon isotopes (Fig. 9), but samples taken from the Fateh field show wider ranges in both isotopes. This may reflect variations in diagenetic pathways within the studied area, perhaps in response to variations in porosity, permeability, and accessibility of diagenetic fluids, assuming similar lithofacies. Otherwise, it may be reflect variable degrees of recrystallization of these carbonates with increasing depth of burial (Moshier, 1989).

Petrographic, burial-history, and isotope data can be used to infer changes in the oxygen isotope composition of diagenetic fluids in the Shuaiba carbonates during progressive diagenesis. The relationships among mineral $\delta^{18}O$ values, pore-water $\delta^{18}O$ values, and temperature for diagenetic phases from the Shuaiba carbonates are shown in Figure 10. Three possible pathways for pore-water evolution have been suggested and include (a) changes in temperature (increasing burial), (b) changes in fluid chemistry, or (c) a combination of increasing temperature and fluid–rock interaction. In the first scenario, temperature increase alone due to burial is the key factor for stabilization of the carbonate components and the formation of later diagenetic events. The diagenetic fluids of marine parentage did not change significantly during the burial history of these rocks. Moshier (1989) showed that the maximum burial of the Shuaiba Formation was over 3000 m. Accepting a geothermal gradient of 25°C/km and a surface temperature of about 20°C, this translates into ca. 90°C as a maximum burial temperature for the formation of diagenetic phases in the Shuaiba Formation. However, this model cannot explain the covariation of oxygen and carbon isotopes (Fig. 10). Furthermore, it assumes a closed diagenetic system, which may pose difficulty explaining the diagenetic repartitioning of stable isotopes and the formation of early and late calcite cement.

The other scenario assumes that most of the diagenetic changes occurred at lower temperatures due to changes in pore-water isotope chemistry, such as incursion of meteoric waters during exposure between deposition of the Shuaiba and the Nahr Umr Formations (Budd, 1989; Scott, 1990; Wagner, 1990; Alsharhan, 1995). This model may explain the stabilization of rudists and matrix mineralogy and development of microporosity, and may also help explain the covariant trend of stable isotopes in some components. Budd (1989) suggested that early diagenesis (formation of micro-rhombic calcite and early cement) of the Shuaiba carbonates in the U.A.E. occurred in an unconfined meteoric aquifer during subaerial exposure immediately after Shuaiba deposition. This suggestion was based on petrographic and isotopic studies of these carbonates. However, this model cannot explain adequately the slight variations in $\delta^{13}C$ values, nor does

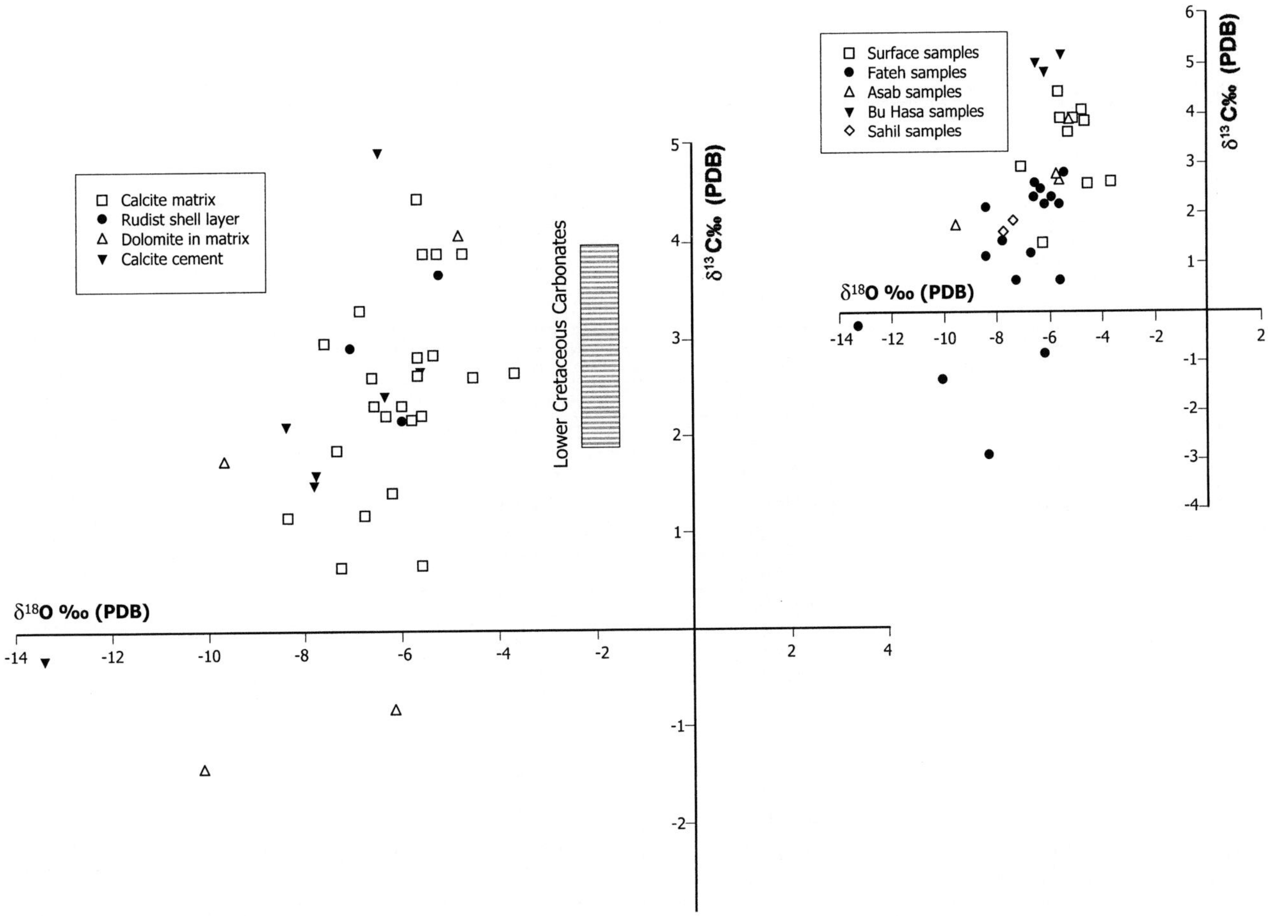


FIG. 9.—Carbon and oxygen isotopic compositions of carbonate components. The box represents the postulated ranges for Lower Cretaceous carbonates (Moldovanyi and Lohmann, 1984; Morrison and Veizer, 1990). Inset is carbon and oxygen isotope compositions of samples according to their locations. See text for details.

it explain the changes in oxygen isotopes in the later-precipitated diagenetic phases. It may have contributed earlier in the diagenetic history. However, the samples investigated in this study show no clear evidence of paleo-exposure, although their absence does not completely exclude this hypothetical exposure. Vahrenkamp (1996) discounted the possibility of meteoric exposure of the Upper Kharaib and Shuaiba Formations to explain carbon isotope variations. The stratigraphic variation in $\delta^{13}C$ of the Shuaiba has been proposed as a correlation tool (Wagner, 1990; Vahrenkamp, 1996; Grötsch et al., 1998).

Increasing temperature accompanied by progressive changes in pore-water isotopic chemistry due to water–rock interaction and/or mixing of fluids described by Frank and Lohmann (1995) should explain both the covariations in stable isotopes (Fig. 10) and the formation of later diagenetic events during burial. Modification of the original marine pore waters, whether due to incursion of meteoric waters at shallow burial or due to increasing water–rock interactions and temperature, will explain both the negative shift in oxygen isotopes and the buffering of carbon isotopes with the original values.

## HYDROCARBON POTENTIAL

The Shuaiba Formation carbonates are important reservoirs in the petroleum system of the U.A.E. (see Alsharhan and Nairn, 1997; Alsharhan and Scott, this volume). The future reservoir potential of the Shuaiba is a function of its reservoir characteristics associated with its source rocks and seals.

### *Reservoir Characteristics*

The carbonates of the lower and upper members of the Shuaiba Formation form one of the most important reservoir lithologies in the U.A.E., and produce oil from a number of giant fields, such as the Bu Hasa oil field and the Sajaa gas condensate field (Fig. 1). Their net pay thicknesses range between about 10 and 80 m; porosities range from 13% to 33%, and permeabilities from 1.0 to 160 md. Overall reservoir quality depends on the importance of diagenetic processes such as calcite recrystallization, dolomitization, fracturing, leaching, and secondary pore filling.

The Shuaiba Formation carbonates are well developed in the central sector of the U.A.E., where they are thick and extensive (e.g., the Bu Hasa field) (Fig. 1). In this giant oil field, the carbonates have a gross pay thickness of 68 m at a depth of 2346 m. Their porosity and permeability averages 23% and 100 md, respectively, and they contain almost 90% of the hydrocarbons present in this giant field (Alsharhan, 1993). The reservoir characteristics of the Shuaiba carbonates varies widely and are generated by the development of the rudist buildup complex over an algal platform.

TABLE 1.—Stable-isotope analysis of samples from the Shuaiba Formation in the U.A.E.

| Sample no. | Lithology | Mineralogy | $\delta^{18}O$, ‰ (SMOW) | $\delta^{18}O$, ‰ (PDB) | $\delta^{13}C$, ‰ (PDB) |
|---|---|---|---|---|---|
| Surf 0 | matrix | calcite | 27.19 | -3.61 | 2.66 |
| Surf 1 | algal mud–matrix | calcite | 24.54 | -6.18 | 1.41 |
| Surf 2 | algal mud–matrix | calcite | 25.67 | -5.08 | 3.89 |
| Surf 10 | algal matrix | calcite | 25.45 | -5.3 | 3.72 |
| Surf 19-1 | matrix | calcite | 25.32 | -5.42 | 3.87 |
| Surf 19-2 | Rudist shell | calcite | 23.67 | -7.02 | 2.93 |
| Surf 21-1 | matrix | calcite | 26.06 | -4.7 | 3.89 |
| Surf 21-2 | Rudist shell | calcite | 25.49 | -5.26 | 3.65 |
| Surf 24 | matrix | calcite | 25.12 | -5.62 | 4.45 |
| Surf 27 | matrix | calcite | 26.28 | -4.49 | 2.61 |
| Surf 26 | matrix | dolomite | 26.00 | -4.77 | 4.07 |
| FT 9081-1 | matrix | calcite | 25.16 | -5.58 | 2.19 |
| FT 9081-2 | large vein | calcite | 22.38 | -8.27 | -2.9 |
| FT 9081-1 | matrix | dolomite | 20.44 | -10.15 | -1.34 |
| FT 9083-1 | matrix | calcite | 24.77 | -5.95 | 2.26 |
| FT 9083-2 | vein | calcite | 22.87 | -7.8 | 1.5 |
| FT 9089-1 | matrix (fossil) | calcite | 24.16 | -6.55 | 2.62 |
| FT 9096 | matrix | calcite | 23.46 | -7.23 | 0.66 |
| FT 9099-1 | matrix | calcite | 24.92 | -5.81 | 2.17 |
| FT 9111 | matrix | calcite | 25.35 | -5.39 | 2.82 |
| FT 9119 | matrix | calcite | 25.07 | -5.66 | 2.66 |
| FT 9137-1 | matrix | dolomite | 24.56 | -6.16 | -0.82 |
| FT 9137-2 | fossil | calcite | 24.67 | -6.05 | 2.19 |
| FT 9151 | matrix | calcite | 22.32 | -8.34 | 1.17 |
| FT 9161-1 | matrix | calcite | 24.16 | -6.55 | 2.32 |
| FT 9161-2 | cement in fossil | calcite | 24.41 | -6.31 | 2.43 |
| FT 9166 | matrix–grainstone | calcite | 23.97 | -6.73 | 1.2 |
| FT 9172-1 | matrix | calcite | 24.53 | -6.19 | 2.22 |
| FT 9172-2 | cement in fossil | calcite | 17.05 | -13.44 | -0.28 |
| FT 9198-1 | matrix | calcite | 25.18 | -5.56 | 0.67 |
| FT 9198-2 | vein | calcite | 22.28 | -8.38 | 2.14 |
| Sa-65 (23) | matrix | calcite | 23.27 | -7.31 | 1.86 |
| Sa-65 (23) | cement in fossil | calcite | 22.92 | -7.75 | 1.61 |
| Sb-228 (18) | matrix | calcite | 25.56 | -5.19 | 3.89 |
| Sb-229 (17) | matrix | calcite | 25.07 | -5.67 | 2.80 |
| Sb-229 (17) | cement | dolomite | 20.96 | -9.65 | 1.74 |
| Sb-229 (17) | cement | calcite | 25.09 | -5.64 | 2.68 |
| Sy-16 (16) | matrix | calcite | 23.87 | -6.83 | 3.31 |
| Bb-133 (21) | matrix | calcite | 23.13 | -7.55 | 2.95 |
| Bu-410 (19) | fossil | calcite | 24.56 | -6.16 | 4.89 |
| Bu-410 (19) | cement in fossil | calcite | 25.25 | -5.49 | 5.21 |
| Bu-410 (20) | fossil | calcite | 24.03 | -6.67 | 4.99 |
| Bu-410 (20) | cement in fossil | calcite | 24.24 | -6.47 | 5.02 |

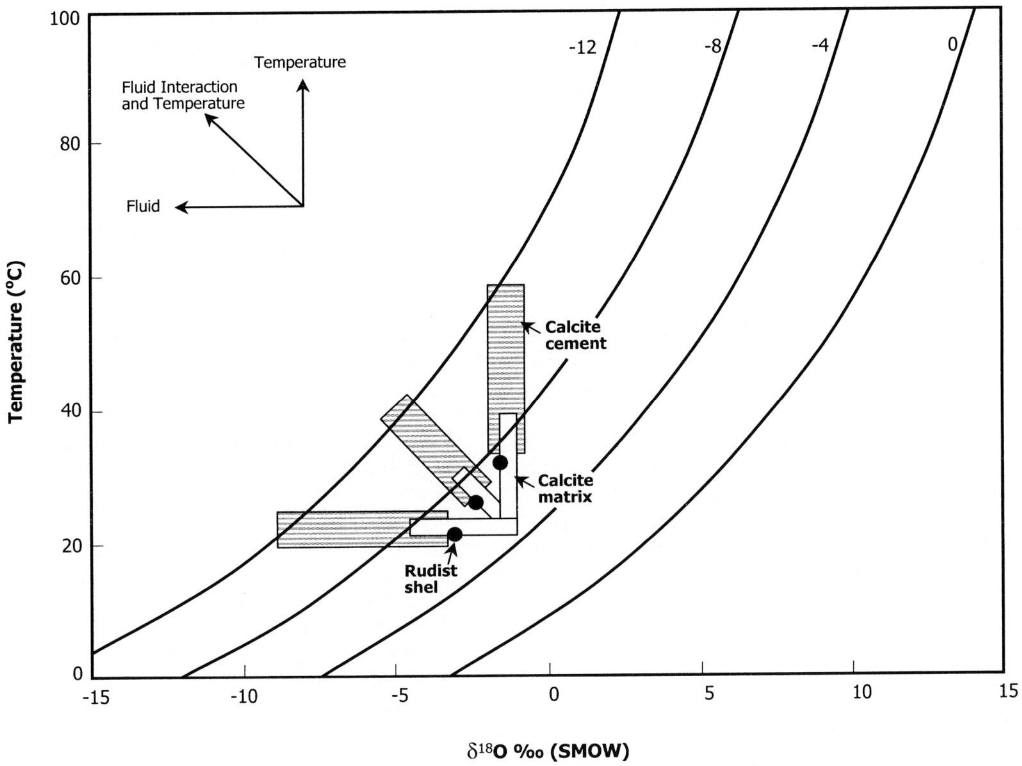


FIG. 10.—Oxygen isotope composition of pore water ($\delta^{18}O$, ‰ SMOW) vs. crystallization temperature (°C) for Shuaiba carbonates. Isotopic composition of calcite is shown as contours ($\delta^{18}O$, ‰ PDB). The equation relating temperature, $\delta^{18}O_{water}$, αvδ $\delta^{18}O_{mineral}$ is ($10^3 \ln \alpha_{cal\text{-}water} = 2.78 \times 10^6 T^{-2} - 2.89$) (Friedman and O'Neil 1977). Shown also the isotopic domains for micro-rhombic calcite from other studies (filled oval is from Budd, 1989; and the rectangle is from samples taken from Sajaa Field from Moshier, 1989). See text for details.

*Lower Member.* —

The pore types in this member include vuggy, moldic, and interparticle porosity. The average porosity is 18%. The abundance of stylolites in this member considerably decreases the porosity and permeability. Some of the pore spaces are filled with calcite, baroque dolomite (Alsharhan and Williams, 1987), and pyrite. The permeability is particularly affected by the abundance of stylolites, which decrease the permeability to 2 md. The permeability reaches its maximum 55 md within the algal-rich layers.

*Upper Member .*—

The porosity and permeability of this member are distributed erratically, corresponding to the heterogeneity of the lithofacies within this unit. The pore types are dominantly vuggy, moldic, and secondary interparticle porosity such as that formed by the dolomitization. The vuggy porosity and moldic porosity are formed by leaching of rudists, oysters, and foraminifers; however, some of these pores have been occluded by calcite cement. Most of the primary interparticle porosity that had been present throughout the sequence was filled by drusy calcite and microcrystalline cement. Dolomitization has produced a small amount of intercrystalline porosity, which in part has been reduced by later blocky and drusy calcite cementation and microcrystalline silica and pyrite. Some pores are fine to medium, formed by rhombic and mosaic dolomite with relic leached fossils. The porosity ranges from less than 3% in the dense limestones to 27% in the buildup facies of the upper member.

Cementation and sylolitization decrease to a great extent the permeability of the Shuaiba limestones to less than 1 md, forming the dense intervals of the member. This can be seen in an increase in the responses of the density and gamma-ray logs and decreases in the neutron-porosity log. The best permeability coincides with greater quantities of rudists, calcareous algae, and corals, and consequently reaches up to 150 md in the buildup facies.

## *Source Potential*

The major source rocks defined in the U.A.E. are the Diyab–Dukhan formations (Oxfordian–Lower Kimmeridgian) and the Shilaif–Khatiyah formations (Upper Albian to Cenomanian) (Alsharhan, 1989).

The source potential of the Aptian Shuaiba Formation has been studied in detail by Azzam and Taher (1995) and Taher (1997). They described the organic-rich carbonate intervals within the basinal facies Bab Member of the Shuaiba Formation with excellent to poor source potential, with total organic carbon (TOC) and thicknesses ranging from 4.3% and 30 m in the central part of the Shuaiba basin to 0.7% and 100 m toward the shelf margins, respectively. Some of the offshore wells such as in Umm Shaif, the Bab Member shows a pyrolysis yield up to 10 kg/ton (Alsharhan, 1989). In other areas, the Bab carbonates were not analyzed but are characterized by light brown to black, fine-grained, laminated limestones, and lower formation density and higher gamma ray in comparison with the underlying carbonates. These are the typical characters for carbonate source rocks described Meyer and Nederlof (1984) and Ohler (1984). The Bab Member, deposited in an intrashelf setting, was exposed to restricted water circulation, anoxic conditions, and deposition below wave base. The Bab carbonate extracts have a sulfur content of 4.5%, a pristane/phytane ratio of 1.4%, and an average $\delta^{13}C$ value of -25‰ (Mohamed and Ayoub, 1992), indicating shallow to deep marine environments of deposition.

Generally, the average geothermal gradient in the U.A.E. increases southward (Alsharhan, 1989). In the northern offshore part of the U.A.E., it is 1.5°F/100 ft and reaches up to 2.3°F/100 ft in the southern part. However, in the southeastern corner of the U.A.E., all source rocks are immature. Based on the Time–Temperature Index (TTI) method, Taher (1997) concluded that the basinal facies of the Shuaiba is immature only in the most northwestern and southeastern part of the U.A.E. and is mature everywhere else.

The geochemical characteristics of the Bab Member extracts were compared by Taher (1997) with those of the Jurassic Diyab source rock (the main source rock for the oils in the U.A.E.), using analytical gas chromatography, and showed significant differences. He concluded that extracts from the Bab basinal carbonates could be geochemically correlated with oil samples from the Shuaiba oils in some oil fields in the onshore part of the U.A.E. (Fig. 11).

### *Sealing mechanisms*

Two potential sealing rocks for hydrocarbon accumulations are associated with the Shuaiba Formation carbonates in the U.A.E., including (1) the fine clastics of the Albian Nahr Umr Formation and (2) the dense lime mudstone within the lower and upper members. The shales of the Nahr Umr Formation are considered to be the ultimate seal for hydrocarbon accumulations in the Shuaiba reservoirs. They form potential vertical seals above the underlying lower and upper members, and lateral seals where adjacent to the Shuaiba reservoirs across folds and/or faults. The dense lime mudstone, wackestones, and packstones in the lower and upper members act as seals for the reservoirs in the Shuaiba (see also Alsharhan and Scott, this volume).

## CONCLUSIONS

1. The Aptian Shuaiba Formation in the U.A.E. can be divided into the informal lower and upper members, and the Bab Member. The formation represents a wide range of depositional settings from shallow shelf to intrashelf starved basinal facies.

2. The carbonates within the formation comprise seven different lithofacies: peloidal skeletal algal packstone–grainstone; ooidal–peloidal grainstone/packstone; skeletal algal (Lithocodioideae) floatstone; intraclastic packstone–grainstone and coated packstone/grainstone; skeletal wackestone/packstone, algal (Lithocodioidea) floatstone; skeletal foraminiferal (*Pseudocyclammina*) argillaceous wackestone/packstone; and skeletal foraminifera/*Lithocodium* argillaceous wackestone/packstone.

3. The Shuaiba carbonates underwent various diagenetic modifications during shallow to deep burial stages. These modified their original altered textural and geochemical attributes and affected reservoir porosity. Diagenetic modification of the Shuaiba carbonates occurred at variable burial depths and water–rock interactions.

4 Oxygen and carbon isotopic data demonstrate that the calcitic matrix has the least altered values in both isotopes relative to the postulated Cretaceous marine carbonates (average $\delta^{18}O$ = -5.7‰ PDB; $\delta^{13}C$ = +2.5‰ PDB), whereas the calcite cements occluding shell porosity and infilling veins have more depleted isotopic values (average $\delta^{18}O$ = -8.8‰ PDB; $\delta^{13}C$ = +0.5‰ PDB). The isotopic data also show a covariant trend between these isotopes. The variations of oxygen and carbon isotopes reflect increasing water–rock interaction and/or fluid mixing during progressive burial.

5 The Shuaiba carbonates are one of the most prolific reservoir lithologies in the U.A.E. oil province, and produce oil and/or gas from several giant fields in the basin. The average porosity is 23%, and permeabilities range from 1.0 to 150 md. Their net pay thicknesses reach up to 78 m, as in the Bu Hasa field. Basinal facies within the formation (Bab Member) have poor to good source potential.

## ACKNOWLEDGMENTS

The authors are especially grateful to Abu Dhabi Company for Onshore Operation (ADCO) and Dubai Petroleum Company (DPC) for supplying some core data. They express their appreciation to Drs. Christopher Kendall, David Budd, G.W. Hughes, and Robert W. Scott for reviewing this manuscript; and to S.D. Russell (ADCO) who introduced us to the Shuaiba outcrop area. I.S. Al-Aasm is thankful to the Canadian Natural Sciences and Engineering Research Council (NSERC) for its continuous support.

## REFERENCES

Al-Aasm, I.S. and Azmy, K., 1996, Diagenesis and evolution of porosity of Middle–Upper Devonian Kee Scarp reefs, Norman Wells, Northwest Territories, Canada: petrographic and chemical evidence: American Association of Petroleum Geologists, Bulletin, v. 80, p. 82–100.

Al-Aasm, I.S., and Veizer, J., 1986a, Diagenetic stabilization of aragonite and low-Mg calcite, 1. trace elements in rudists: Journal of Sedimentary Petrology, v. 56, p. 138–152.

Al-Aasm, I.S., and Veizer, J., 1986b, Diagenetic stabilization of aragonite and low-Mg calcite, II: Journal of Sedimentary Petrology, v. 56, p. 763–770.

Alsharhan, A.S., 1985, Depositional environments, reservoir units evolution and hydrocarbon habitat of Shuaiba Formation, Lower Cretaceous, Abu Dhabi, U.A.E.: American Association of Petroleum Geologists, Bulletin, v. 69, p. 899–912.

Alsharhan, A.S., 1989, Petroleum geology of the United Arab Emirates: Journal of Petroleum Geology, v. 12, p. 253–288.

Alsharhan, A.S., 1993, Bu Hasa field, United Arab Emirates, Rub al Khali Basin, Abu Dhabi, *in* Foster, N.H., and Beaumont, E.A., compilers, Treatise of Petroleum Geology, Atlas of Oil and Gas Fields, Structural Traps VIII, American Association of Petroleum Geologists, p. 99–124.

Alsharhan, A.S., 1995, Facies variation, diagenesis and exploration potential of the Cretaceous rudist-bearing carbonates of the Arabian Gulf: American Association of Petroleum Geologists, Bulletin, v. 79, p. 531–550.

Alsharhan, A.S., 1997, Mesozoic carbonate platform-basin systems, United Arab Emirates: (Field Trips): SEPM Research Conference on Jurassic/Cretaceous Carbonate Platform–Basin Systems-Middle East Models, December 4–12, 75 p.

Alsharhan, A.S., and Nairn, A.E.M., 1986, A review to the Cretaceous formations in the Arabian Peninsula and the Gulf: Part I. Lower Cretaceous (Thamama Group), Stratigraphy and Paleogeography: Journal of Petroleum Geology, v. 9, p. 365–391.

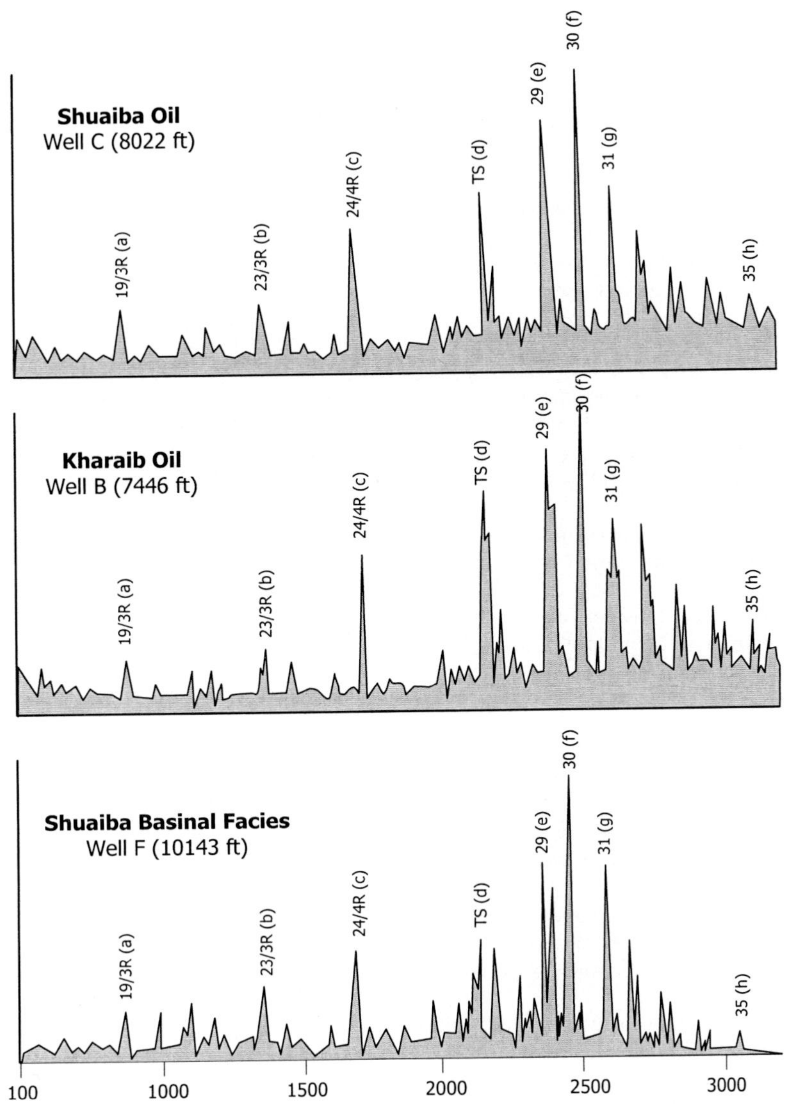


FIG. 11.—Gas chromatograph plots of oil/source correlation showing the genetic relationship between the Bab Member extracts and the Thamama oils.


ALSHARHAN, A.S. AND NAIRN, A.E.M., 1997, Sedimentary basins and petroleum geology of the Middle East: Amsterdam, Elsevier Science Publishers, 940 p.

ALSHARHAN, A.S., AND WILLIAMS, D.F., 1987, Petrography and stable isotopic composition of baroque dolomite from the Shuaiba Formation (Lower Cretaceous), Abu Dhabi, United Arab Emirates: Journal of African Earth Sciences, v. 6, p. 881–890.

AZER, S.R., AND TOLAND, C., 1993, Sea level changes in the Aptian and Barremian (Upper Thamama) of offshore Abu Dhabi, U.A.E.: Proceedings of SPE 8th Middle East Conference, Bahrain, v. 2, p. 141–153.

AZZAM, I., AND TAHER, A.K., 1995, Sequence stratigraphy and source rock potential of the Aptian in East onshore Abu Dhabi, A model approach to oil exploration: Proceedings of the 9th Middle East Oil Show, Bahrain, p. 305–317.

BRAND, U., AND VEIZER, J., 1980, Chemical diagenesis of multicomponent carbonate system, 1, trace elements: Journal of Sedimentary Petrology, v. 50, p. 1219–1236.

BUDD, D.A., 1989, Micro-rhombic calcite and microporosity in limestones: a geochemical study of the Lower Cretaceous Thamama Group U.A.E.: Journal of Sedimentary Geology, v. 63, p. 293–311.

BURRISS, R.C., CERCONE, K.R., AND HARRIS, P.M., 1983, Fluid inclusion petrography and tectonic–burial history of the Alali-2 well: Evidence for the timing of diagenesis and oil migration, northern Oman Foredeep: Geology, v. 11, p. 567–570.

DUROCHER, S., AND AL-AASM, I.S., 1997, Dolomitization and neomorphism of Mississippian (Visean) upper Debot Formation, Blueberry Field, NE British Columbia, Canada: Geologic, petrologic and chemical evidence: American Association of Petroleum Geologists, Bulletin, v. 81, p. 954–977.

FRANK, T.D., AND LOHMANN, K.C., 1995, Early cementation during marine–meteoric fluid mixing: Mississippian Lake Valley Formation, New Mexico: Journal of Sedimentary Petrology, v. 65, p. 263–273.

FRIEDMAN, I., AND O'NEIL, J.R., 1977, Compilation of stable isotope fractionation factors of geochemical interest, *in* Fleischer, M., ed., Data of Geochemistry, 6th Edition: U.S. Geological Survey, Professional Paper 440-KK.

GRÖTSCH, J., BILLING, I., AND VAHRENKAMP, V., 1998, Carbon-isotope stratigraphy in shallow-water carbonates—implications for Cretaceous black-shale deposition: Sedimentology, v. 45, p. 623–634.

HAMDAN, A.R.A., AND ALSHARHAN, A.S., 1991, Palaeoenvironments and palaeoecology of the rudists in the Shuaiba Formation (Aptian) United Arab Emirates: Journal of African Earth Sciences, v. 12, p. 569–581.

HAQ, B.U., HARDENBOL, J., AND VAIL, P.R., 1988, Mesozoic and Cenozoic chronostratigraphy and eustatic cycles, *in* Wilgus, C.K., Hastings, B.S., Kendall, C.G.St.C., Posamentier, H.W., Ross, C.A., and Van Wagoner, J.C., eds., Sea-Level Changes: An Integrated Approach: SEPM, Special Publication 42, p. 71–108.

HASSAN, T.H., MUDD, G.C., AND TWOMBLEY, B.N., 1975, The stratigraphy and sedimentation of the Thamama Group (Lower Cretaceous) of Abu Dhabi: 9th Arab Petroleum Congress, Dubai, United Arab Emirates, p. 1–11.

HENDRY, J.P., DITCHFIELD, P.W., AND MARSHALL, J.D., 1995, Two-stage neomorphism of Jurassic aragonitic bivalves: implication for early diagenesis: Journal of Sedimentary Research, v. A65, p. 214–224.

HUGHES, G.W., 1997, The Great Pearl Bank Barrier of the Arabian Gulf as a possible Shu'aiba analogue: GeoArabia Middle East Petroleum Geosicences, v. 2, p. 279–304.

LINDHOLM, R.C., AND FINKELMAN, R.B., 1972, Calcite staining: semiquantitative determination of ferreous iron: Journal of Sedimentary Petrology, v. 42, p. 614–617.

MEYER, B.L., AND NEDERLOF, M.H., 1984, Identification of source rocks on wireline logs, by density/resistivity and sonic transit time/resistivity crossplots: American Association of Petroleum Geologists, Bulletin, v. 64, p. 916–926.

MOHAMED, A.A., AND AYOUB, M.R., 1992, Onshore Abu Dhabi oil and source rocks geochemistry: selected samples: Proceedings of the 11th Exploration Conference sponsored by the Egyptian General Petroleum Corporation, p. 293–320.

MOLDOVANYI, E.P., AND LOHMANN, K.C., 1984, Isotopic and petrographic record of phreatic diagenesis within the Lower Cretaceous Sligo and Cupido Formation: Journal of Sedimentary Petrology, v. 54, p. 972–985.

MORRISON, J.O., AND VEIZER, J., 1990, Belemnite geochemistry: an indicator of secular and environmental variation of Cretaceous seawater (abstract): Geological Society of America, Abstracts with Program, p. A116.

MOSHIER, S.O., 1989, Development of microporosity in a micritic limestone reservoir, Lower Cretaceous, Middle East: Sedimentary Geology, 63, p. 217–240.

MURRIS, R.J., 1980, Middle East: Stratigraphic evolution and oil habitat: American Association of Petroleum Geologists, Bulletin, v. 64, p. 597–618.

OHLER, J.H., 1984, Carbonate source rocks in the Jurassic Smackover trend of Mississippi, Alabama, and Florida, *in* Palacas, J.G., ed., Petroleum Geochemistry and Source Rock Potential of Carbonate Rocks: American Association of Petroleum Geologists, Studies in Geology 18, p. 63–71.

RUSSELL, S.D., 1996, Wadi Dhayah/Wadi Rahabah Shuaiba and Kharaib Formations (Field Trip): Society of Explorationists in the Emirates, May 23–24, 8 p.

SCOTT, R.W., 1990, Chronostratigraphy of the Cretaceous carbonate shelf, southeastern Arabia, *in* Robertson, A.H.F., Searle, M.P., and Ries, A.C., eds., The Geology and Tectonics of the Oman Region: Geological Society of London, Special Publication 49, p. 89–108.

SIMMONS, M.D., 1994, Micropalaeontological biozonation of the Kahmah Group (Early Cretaceous), Central Oman Mountains, *in* Simmons, M.D., ed., Micropalaeontology and Hydrocarbon Exploration in the Middle East: London, Chapman & Hall, p. 177–220.

SKELTON, P.W., 1979, Preserved ligament in a radiolitid rudist bivalve and its implication of mantle marginal feeding in the group: Paleobiology, v. 5, p. 90–106.

TAHER, A.A., 1997, Delineation of organic richness and thermal history of the Lower Cretaceous Thamama Group., East Abu Dhabi: A modeling approach for oil exploration: GeoArabia Bulletin, v. 2, no. 1, p. 65–88.

VAHRENKAMP, V.C., 1996, Carbon isotope stratigraphy of the Upper Kharaib and Shuaiba Formations: Implications for the Early Cretaceous evolution of the Arabian Gulf region: American Association of Petroleum Geologists, Bulletin, v. 80, p. 647–662.

WAGNER, P.D., 1990, Geochemical stratigraphy and porosity controls in Cretaceous carbonates near the Oman Mountains, *in* Robertson, A.H.F., Searle, M.P., and Ries, A.C., eds., The Geology and Tectonics of the Oman Region: Geological Society of London, Special Publication 49, p. 127–137.

WITT, W., AND GOKDAG, H., 1994, Orbitolinid biostratigraphy of the Shuaiba Formation (Aptian), Oman—Implications for Reservoir Development, *in* Simmons, M.D., (ed.), Micropalaeontology and Hydrocarbon Exploration in the Middle East: Chichester, U.K., Chapman & Hall, p. 221–242.

# CRETACEOUS SEQUENCE STRATIGRAPHY AND PETROLEUM POTENTIAL OF THE MESOPOTAMIAN BASIN, IRAQ

FADHIL N. SADOONI

*Eden Petroleum Services Ltd., P.O. Box 11-559, Ellerslie, Auckland, New Zealand*

*Present address: Geology Department, Faculty of Science, Qatar University, P.O. Box 2713, Doha, Qatar*

AND

ADNAN A.M. AQRAWI

*Roxar ASA, Gamle Forusvei 17, P.O. Box 112, 4065 Stavanger, Norway*

ABSTRACT: Two major factors, in addition to the global eustatic sea-level changes, controlled the deposition of the Cretaceous stratigraphic sequences in the Mesopotamian Basin in central and southern Iraq: (1) the culmination of the Neo-Tethys rifting resulting from the movement of the Arabian Plate into the tropical and subtropical climatic zones during the Cretaceous, and (2) the contemporaneous growth of giant structures due to salt diapirism or relative basement-block displacement, particularly during middle and Late Cretaceous.

The Early Cretaceous carbonate ramp of the Yamama Formation underwent a gradual clastic invasion that started during the late Early Cretaceous. This was followed by tectonic activity beginning in the late middle Cretaceous that converted a ramp system into a block-faulted shelf/platform system by the end of Cretaceous.

Within the three groups of the Lower, middle, and Upper Cretaceous, referred to as the Thamama, Wasia, and Aruma Groups in the Arabian Gulf region, seven major sedimentary cycles are recognized. Three of these cycles or major sequences are attributed to the Lower Cretaceous Thamama Group, and each of the Wasia and Aruma Groups of the middle and Upper Cretaceous, respectively, are divided into two cycles. These major sequences are defined within their depositional time intervals as early Berriasian–Valanginian, Valanginian–Barremian, Barremian–middle Aptian, middle Aptian–Albian, Cenomainian–early Turonian, Turonian–Campanian, and Campanian–Maastrichtian. Each of the seven cycles can be subdivided into large- and medium-sized accommodation cycles on the basis of detailed sedimentological and paleontological data.

The Cretaceous sediments of the Mesopotamian Basin host one of the richest petroleum systems in the world. Each major sequence contains well-recognized stacked carbonate and sandstone reservoirs. These are overlain by multiple, tight carbonate and shale seals. The late Barremian–middle Aptian dolomitic sequence (i.e., the Shuaiba Formation) may be the only exception. Potential source rocks charging these reservoirs are either associated within the same major sequence or group, particularly in the Lower and middle Cretaceous, or are older, such as the Upper Jurassic Naokelekan and Sargelu Formations.

The Cretaceous might contain more plays either as new fields or at deeper levels in the present fields. More reserves are likely in the stratigraphic traps of the Mishrif, Zubair, and Nahr Umr formations. Formations like the Hartha, Ahmadi, Khasib, and Mauddud may prove to be productive outside the presently prospective areas.

## INTRODUCTION

The Cretaceous of Iraq, particularly in the Mesopotamian Basin of central and southern Iraq, has been subjected to intensive study since the middle of this century because it hosts around 80% of Iraq's proven oil reserves together with more than 70% of its probable unproven reserves (Fig. 1A). However, only a limited amount of this work has been published (e.g., Dunnington, 1967; Gaddo, 1971; Al-Siddiki, 1978; Ibrahim, 1981, 1983, 1984; Ali and Aziz, 1993; Sadooni, 1993, 1995b, 1996; Aqrawi, 1996; Aqrawi et al., 1998). These rocks form the main carbonate and sandstone reservoirs in most of the super-giant fields of the Mesopotamian Basin.

After more than a half-century of exploration in Iraq, the Cretaceous succession still produces more pleasant surprises than any other geological system. The Cretaceous comprises more than 3000 m of sediments distributed among 44 formations and includes by all standards the thickest, the most varied, and the most widespread sedimentary successions in the Middle East.

The Cretaceous Period encompassed the maximum rifting of the Neo-Tethys. The limits of the Cretaceous flooding extended beyond most of the previous Jurassic shorelines, and it evolved later into small isolated basins and playas during the Tertiary. Cretaceous deposits in the Mesopotamian Basin show an extensive horizontal and vertical facies differentiation, including the development of different reef types (sponge–stromatoporoid, algal, and rudist) as well as regional gradational spatial changes from carbonates to evaporites, or even to clastics. Carbonates are the major Cretaceous rocks in the Mesopotamian Basin, but some thick clastic units are also present. This was due in part to the low stand of sea level and the prevalence of humid conditions during certain periods of the Cretaceous (Murris, 1980; Beydoun, 1991). The stratigraphic relationships in the Cretaceous of Iraq, correlated to the Cretaceous of some other Arabian Gulf countries, are shown in Figure 2. In addition, a lithostratigraphic correlation of the Cretaceous succession in three selected wells, Majnoon-1, Ratawi-1, and Ghlaisan-1 (wells Mj-1, Rt-1, and Gh-1 in Fig.1B), is shown in Figure 3.

This study is an attempt to develop a preliminary sequence stratigraphic framework for the Cretaceous by distinguishing the major stratigraphic sequences in each of the three Cretaceous groups (i.e., the Thamama, Wasia, and Aruma) in the Mesopotamian Basin. This approach is based on integration of the available sedimentologic and chronostratigraphic data, and integrating these with established global eustatic sea-level changes (Haq et al., 1987), taking into consideration the basic concepts of sequence stratigraphy (e.g., Vail et al., 1977; Van Wagoner et al., 1990), particularly those applying to carbonate rocks (Loucks and Sarg, 1993). Some of these major sequences are also subdivided into large- and medium-sized accommodation cycles whenever more detailed sedimentologic and paleontologic data are available. We also discuss the petroleum prospects of each major

Middle East Models of Jurassic/Cretaceous Carbonate Systems
SEPM Special Publication No. 69, Copyright © 2000
SEPM (Society for Sedimentary Geology), ISBN 1-56576-075-1, p. 315–334.

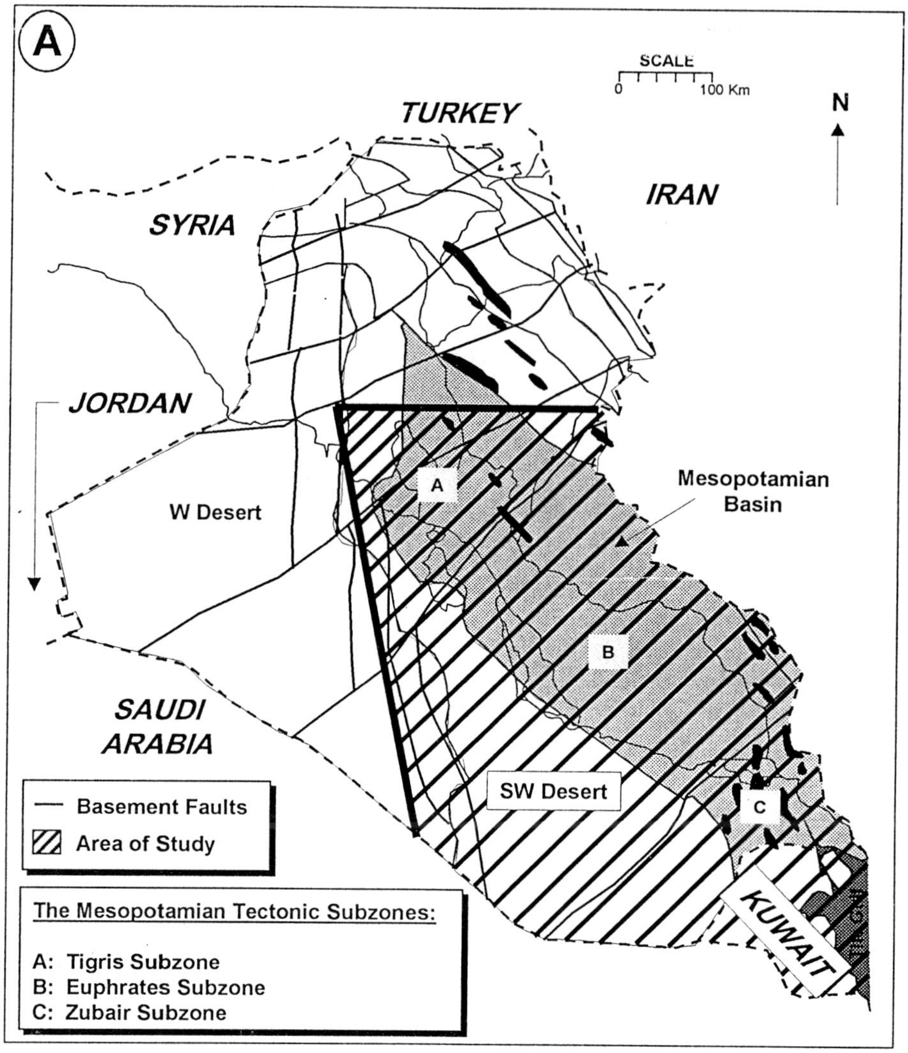


FIG. 1A.—A simplified basement tectonic map of Iraq (modified after Buday and Jassim, 1987) showing area of study and location of the Mesopotamian Basin among the longitudinal and transverse blocks separated by faults (dashed lines). Note the main fields as black areas.

sequence together with the future hydrocarbon potential of Cretaceous petroleum plays.

## FORMATION NOMENCLATURE: A HISTORICAL BACKGROUND

The stratigraphic framework of the Cretaceous in southern Iraq was originally formulated by Owen and Nasr (1958) (Fig. 2). This work was based on the extensive early fieldwork by geologists of the Iraq Petroleum Company (IPC) and the Basra Petroleum Company (BPC) during the Forties and Fifties. There were two stratigraphic nomenclatures for the northern (including central parts) and southern Iraq. This double nomenclature was the result of concession boundaries rather than real geological variations. The IPC was working in northern Iraq and the BPC was operating in southern Iraq, and each company relied on its own stratigraphic nomenclature.

Chatton and Hart (1961) attempted a regional correlation between the two provinces, but their findings remained unpublished. Since that time more data have been accumulated from two main sources: the exploration program conducted by the Iraqi National Oil Company (INOC) during the Seventies and the Eighties, and from the Regional Geological Survey (Geosurv), carried out by the State Establishment for Geological Survey and Minerals Investigation.

Some earlier results of Geosurv and INOC work appear in Buday (1980) and Buday and Jassim (1987), who also compiled much earlier data dating back to Bellen et al. (1959). The main findings of Geosurv survey were summarized in Jassim et al. (1984). The modifications proposed by Hart and Chatton (1960, 1961, 1962a, 1962b, 1962c), and the new findings of the INOC exploration program and the Geosurv survey, together with many individual studies, particularly those in the form of M.Sc. and Ph.D. theses in various universities across Iraq and elsewhere, brought some profound changes to the understanding of the Cretaceous stratigraphy of Iraq, particularly in the Mesopotamian Basin. Because most of these nomenclatural modifications have never been published before, a short summary of

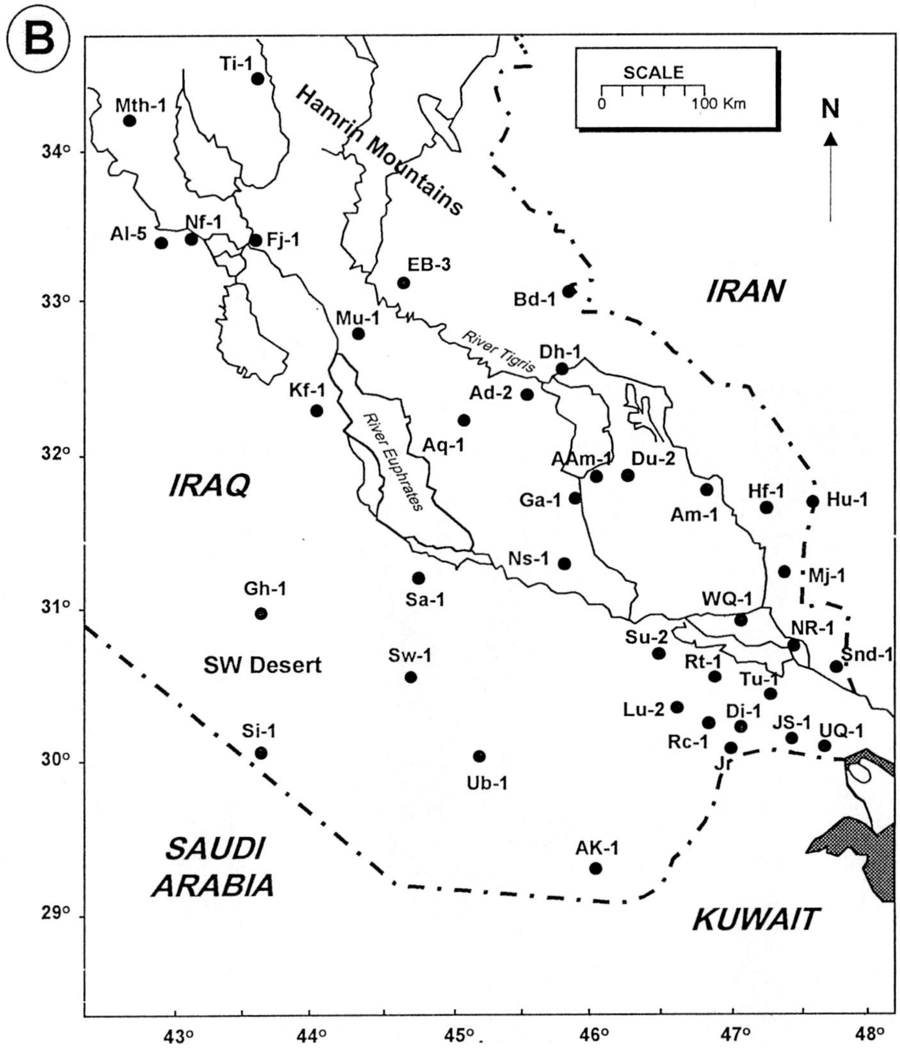


FIG. 1B.—Map of the study area showing the location of the studied wells mentioned in the text.

them is given in Table 1. These modifications (as in the current study) will reduce the confusion arising from the large number of formation names and will facilitate regional correlation for sequence stratigraphic purposes.

## CRETACEOUS SEQUENCE STRATIGRAPHY: THE CONTROLLING FACTORS

Cretaceous sedimentation in the Mesopotamian Basin was controlled by a combination of both regional and local factors in addition to global eustatic sea-level changes. Some regional factors such as the culmination of the Tethys rifting is evident from the widespread distribution of the Cretaceous sediments in most parts of the Arabian Plate. Also, the movement of the Arabian Plate to the tropical or the subtropical zone occurred at this time (Murris, 1980; Beydoun, 1991, 1993). Local factors are the effect of either salt diapirism or movement of basement blocks. As a result, three main categories of factors contributed to the final nature and distribution of the Cretaceous sediments in the Mesopotamian basin: tectonic, climatic, and eustatic.

### *Tectonic Influence*

A major local factor affecting sediment type in the Mesopotamian Basin of Iraq may have been the continuous growth of some structures since Early Jurassic time. By comparing the Mishrif Formation's thicknesses in two wells (one on the crest and the other on one flank) of the Rumaila Field, it is evident that the structure has grown since Early Cretaceous to the present, although the growth rate has declined with time (Fig. 4). This growth has created a difference of around 128 m between the two wells (Al-Sakini, 1992). Ala (1974) and Sadooni (1993) suggested that movement of the Infracambrian Hormuz Salt might have influenced sedimentation during most of the Cretaceous. Fuloria (1976) noticed a correspondence between reservoir-quality Mishrif facies and Yamama facies, which may reflect such tectonic control on both the Yamama and Mishrif Formations, of Early and middle Cretaceous ages, respectively. Patch reefs and oolite shoals are the two main reservoir facies that developed on the crestal parts of the field structures, whereas the flanks and low areas between these were filled with muddier facies (Sadooni, 1993; Aqrawi et al., 1998).

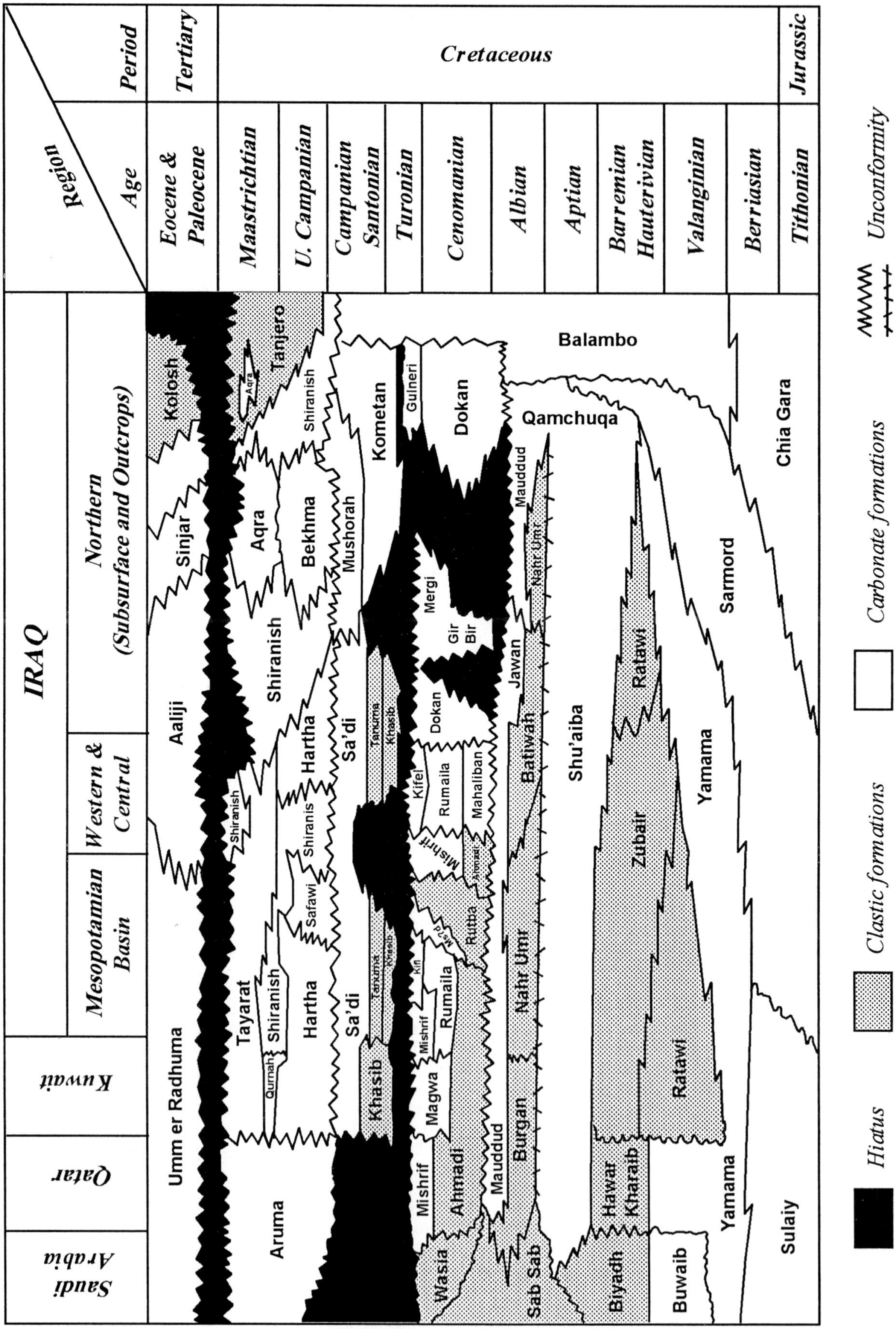


FIG. 2.—Stratigraphic relationships of the Cretaceous formations of Iraq and their equivalents in some Arabian Gulf countries, updated and amalgamated from various sources.

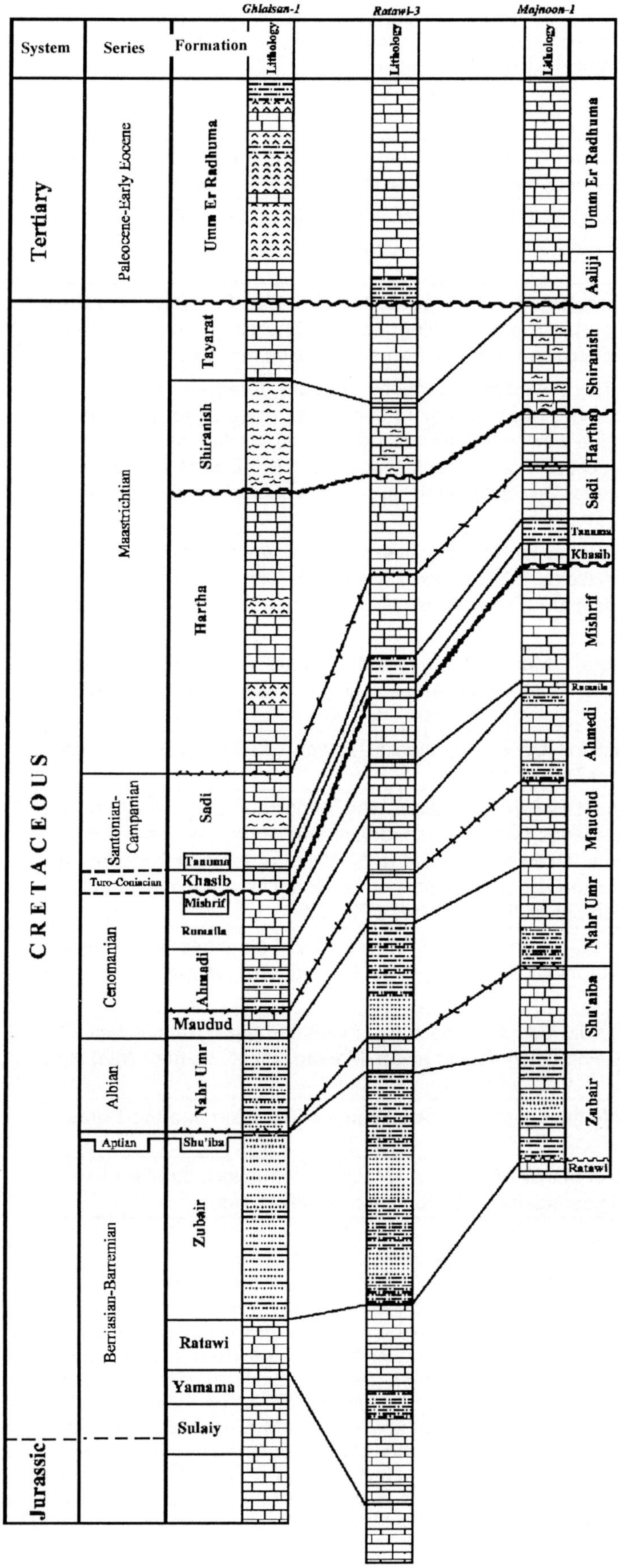


FIG. 3.—Stratigraphic correlation of the Cretaceous succession in three selected wells (Ghlaisan-1, Ratawi-3, and Majnoon-1, referred to as Mj-1, Rt-1, and Gh-1 in Fig. 1). For horizontal scale see Fig. 1. (Modified after Al-Siddiki, 1978a).

TABLE 1.—Some later unpublished modifications to the Cretaceous stratigraphic framework of the Mesopotamian Basin, after being formulated principally by Owen and Nasr (1958), and revised in more detail in the Iraq stratigraphic lexicon (Bellen et al., 1959).

| Author(s) | Modifications |
|---|---|
| Chatton and Hart (1961) | The Gir Bir (Cenomanian–?Turonian), Mergi (Turonian) and M'sad (?upper Cenomanian) are unnecessary synonyms of the Mishrif Formation. |
| Chatton and Hart (1962b), Aqrawi (1983), Sherwani (1983) | Fahad (Turonian), Maotsi (Turonian), and Mahliban (Upper Cenomanian) Formations are local variants of the intertonguing between the Rumaila and Mishrif Formations. |
| Chatton and Hart (1960), Sadooni (1978) | Upper Qamchuqa Formation is equivalent to the Mauddud Formation and the Lower Qamchuqa Formation is equivalent to the Shu'aiba Formation. The terms Mauddud and Shu'aiba Formations should be used in northern Iraq also. |
| Chatton and Hart (1960) | Upper Sarmord and Rim (?Albian) Formations are considered as parts of the Nahr Umr Formation, while Middle Sarmord should be attached to the Ratawi Formation and only the Lower Sarmord is considered as the real Sarmord facies as originally defined. The Karimia (?Tithonian–?Berriasian) and Makhul (?Tithonian) Formations are considered as local variants of the Sarmord Formation |
| Chatton and Hart (1960) | Garagu Formation (Valanginian–?Hauterivian) is considered as an equivalent to the Yamama Formation. Jib'ab (Upper Senonian), Digma (Maastrichtian), and Qurna (Maastrichtian) Formations are synonyms of the Shiranish Formation. |
| Chatton and Hart (1961) | Unification of the Aqra (?Maastrichtian), and Bekhme (Upper Campanian–?Lower Maastrichtian) Formations into a single Formation (i.e., Aqra Formation). |
| Chatton and Hart (1961) | The Pilsener Formation (Upper Campanian–Lower Maastrichtian), to be replaced by the Hartha Formation |
| Gaddo (1961) | Announcement of the Batiwah Formation, which consists of greenish gray calcareous shale interbedded with grayish, dense limestone. The formation is proposed for the neritic shale and limestone of Albian age, and represents a transition between the sandstone and shale of the Nahr Umr Formation and the limestone of the Albian part of the Qamchuqa Formation. |
| Chatton and Hart (1961) | Announcement of the Kifel Formation, which consists of anhydrite and pelletal limestone of lower Turonian, and represents the lagoonal equivalent of the Mishrif Formation. |
| Jassim et al. (1984) | Announcement of the Jid Formation as the phosphate-bearing equivalent of the Hartha–Tayarat Formation (Campanian–Maastrichtian) in the Western Desert |
| Al-Naqib (1967) | Upgrading the Dibs Member of the obsolete Pilsener Formation into the Safawi Formation. |
| Aqrawi (1983) and Sherwani (1983) | Replaced the names of Maotsi, Fahad, and Mahliban formations by Mishrif and Rumaila formations in central Iraq, as they are in southern Iraq. |

This halokinetic tectonism played a minor role in controlling sedimentation during most of the Early Cretaceous, but during the late Albian (i.e., at the beginning of the middle Cretaceous), tectonism became more conspicuous, with possible activation of the basement faults and relative displacement of blocks (Fig.1A) in addition to salt diapirism (e.g., Aqrawi, 1996; Sadooni, 1993, 1995b, 1996; Aqrawi et al., 1998). These local tectonic effects led to minor variations in facies, particularly in central Iraq, at the western edges of the Mesopotamian Basin, and some areas adjacent to the Iranian borders near Amarah. These contrasts resulted in the establishment of different formation names.

### *Climatic Factors*

It is generally agreed now that during most of the Cretaceous many parts of the Arabian Plate were in the tropical to subtropical zones (Beydoun, 1991). Murris (1980) notes the association of clastic materials with humid tropical periods. Aqrawi (1996) suggests the dominance of a humid tropical climate during the Late Cretaceous (particularly Turonian–Campanian) by study of the clay mineralogy of the Khasib, Tanuma, and Sa'di formations in the Mesopotamian Basin. Detrital illite and kaolinite are the main clay minerals detected in the clay fractions of these three

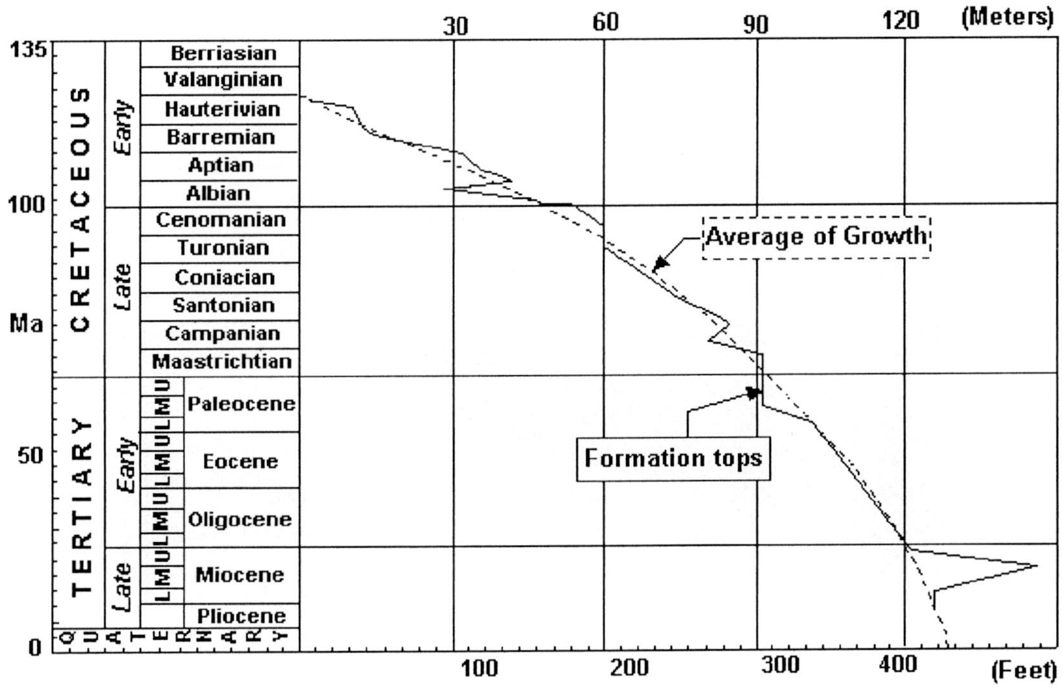


FIG. 4.—Growth of the Rumaila North oilfield during geologic time. (Modified after Al-Sakini, 1992.)

formations. However, periods of arid climate might have occurred occasionally, as during the late Cenomanian, as reflected by the deposition of Kifel Formation evaporites ending the regression of the Cenomanian–Early Turonian cycle.

Wind exerted a considerable influence on the nature, location, and distribution of sediments in the Mesopotamian Basin. With the present level of available information, however, it is not possible to comment on the wind direction and its intensity during the Cretaceous. What is evident from the careful examination of the nature of the sediments, however, is that major lateral differences in sediment nature take place across the different flanks of the same structure. As a result, the vertical differentiation in sediment types is usually accompanied by differences between the two flanks of the structures, as in the West Qurna and Ratawi fields, of southern Iraq (Figure 5A, B). A consistent pattern of facies distribution may indicate a windward/leeward effect. Alternatively this differentiation may be partially due to the difference in the gradient of the slope of the growing structures.

Furthermore, it seems that a gradual increase in the intensity of the energy level with time is reflected by the change from the low-energy patchy stromatoporoid–algal reef of the Yamama Formation (Lower Cretaceous) to the rudist reef of the Mishrif Formation (middle Cretaceous). It is difficult, however, to attribute this shift to a change in the wind intensity only, because it may be the result of several other factors, such as a profound shift in the whole climate (Sadooni, 1998), gradual evolution of the major structures, evolution of the carbonate platform to a more exposed setting, or even a change to shallow-water-dominated parasequences.

### *Eustatic Sea-Level Changes*

The stratigraphy of the Mesopotamian Basin is still far from being fully studied, and so this proposed sequence stratigraphic framework (Fig. 6) is a baseline step towards establishing detailed future studies on the high-resolution sequence stratigraphy.

Although regional and local factors together played an important role in shaping the Cretaceous depositional successions in the Mesopotamian Basin, it is possible to recognize the eustatic control on the major sequences or cycles. In some few cases, it is possible also to relate smaller, higher-frequency cycles to these events (e.g., Aqrawi et al., 1998). A full analysis of the role of eustatic factors requires a detailed chronostratigraphic analysis of the rock units, which is not yet possible.

Most previous paleontological work in Iraq has been concerned with the determination of broad formation age rather than the recognition of specific biozones. As a result, many single, homogeneous lithologic units may contain significant time breaks. Although the major sequences, which usually emcompass more than one formation and up to four formations, can be correlated with Haq et al. (1987) curve with reasonable confidence, other higher-frequency cycles might be the result of tectonism, particularly during middle and Late Cretaceous.

However, some more detailed analysis was carried out in order to subdivide some of the major sequences into large-scale and medium-scale shallowing-up cycles on the basis of the available lithological and wireline log information. The major cycles (or sequences) can be correlated regionally across the Middle East, particularly into neighboring countries (Alsharhan and Kendall, 1991). Lack of accurate time lines and identified boundary surfaces makes it more difficult to correlate the other smaller sequences, even within the Mesopotamian Basin itself (see Aqrawi et al., 1998, for a detailed case of the middle Cretaceous).

## MAJOR STRATIGRAPHIC SEQUENCES AND THEIR PETROLEUM PROSPECTS

The Cretaceous succession in the Mesopotamian Basin can be subdivided into three groups following the terminology used in the Arabian Gulf region (e.g., Powers et al., 1966; Harris et al., 1984; Kendall et al., 1991; Alsharhan and Nairn, 1986, 1988, 1990). These are the Thamama, Wasia, and Aruma Groups, which are

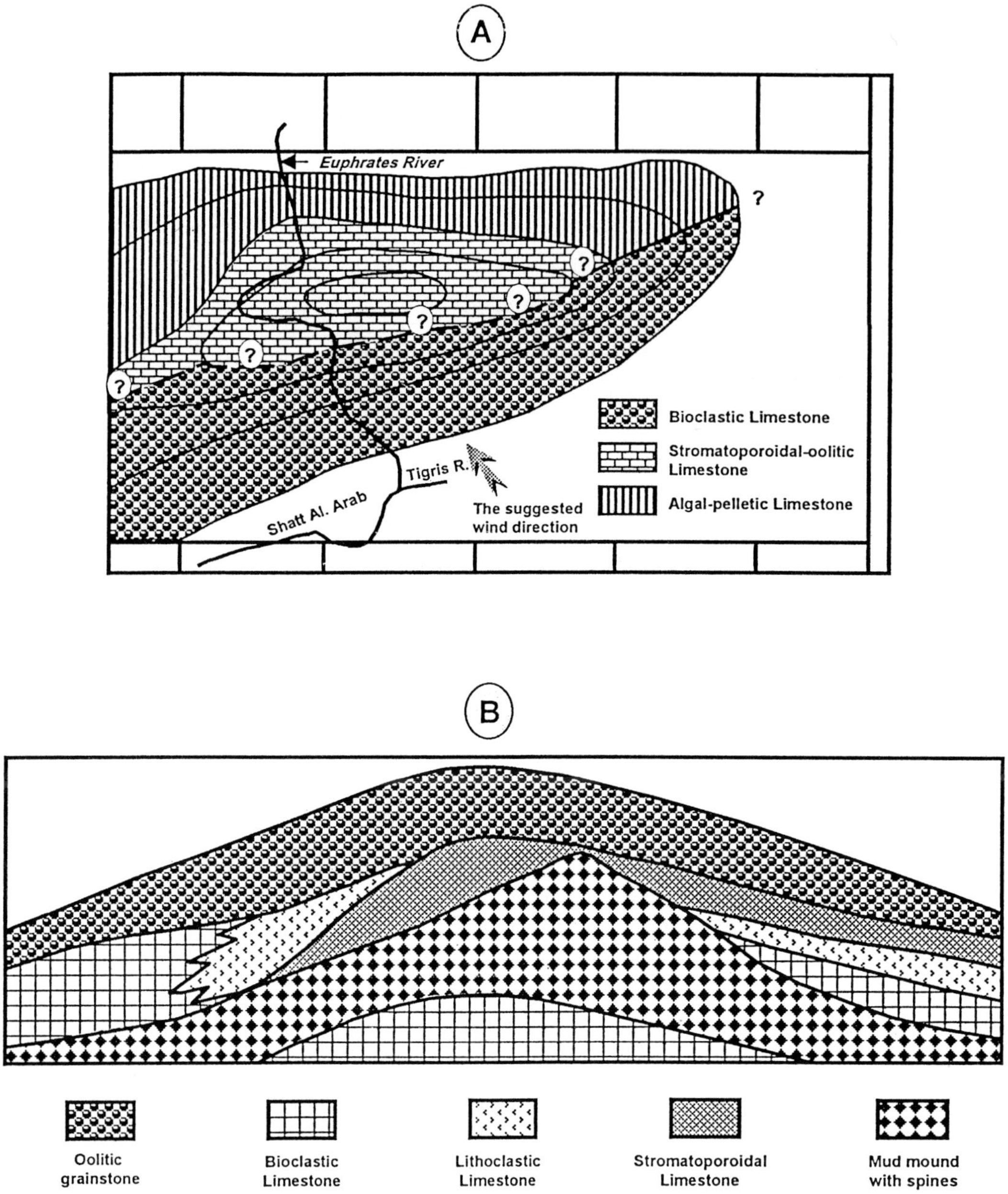


FIG. 5.—A) Vertical and horizontal facies differentiation of the Yamama Formation in the West Qurna Field, Mesopotamian Basin. These variations are probably the result of a combination of structural and climatic conditions. B) Vertical stacking of the Yamama Formation facies in the West Qurna Field may indicate a low-energy reef.

Lower, middle, and Upper Cretaceous, respectively (Fig. 2). These Groups are easily recognized in the Cretaceous successions of the Mesopotamian Basin, and are introduced to the Iraqi geological column in order to unify the general stratigraphic terminology of the Middle East. The bases for such division are the major unconformity surfaces, which are initially used as separating time lines between the main Cretaceous sedimentary cycles for regional studies (Bellen et al., 1959; Buday, 1980). However, such unconformities are usually prominent in the shallower stable shelf areas but are not easily traced into the deeper parts of the basin, such as the eastern Mesopotamian Foredeep region near the Iranian border. The recognizable sequences are named according to their time interval of deposition, as follows (Fig. 6).

### *Lower Berriasian–Valanginian Cycle*

The lower contact of this sequence is marked by the sharp facies change from the evaporites of the Upper Jurassic Gotnia Formation to the argillaceous limestones of the Lower Cretaceous Sulaiy Formation (Sadooni, 1997). However, this contact between the Jurassic and Cretaceous is gradational in the southern parts of the basin, and Sulaiy deposition might have began in the Late Jurassic. Redmond (1964) and Al-Siddiki (1978a) place the Juras-

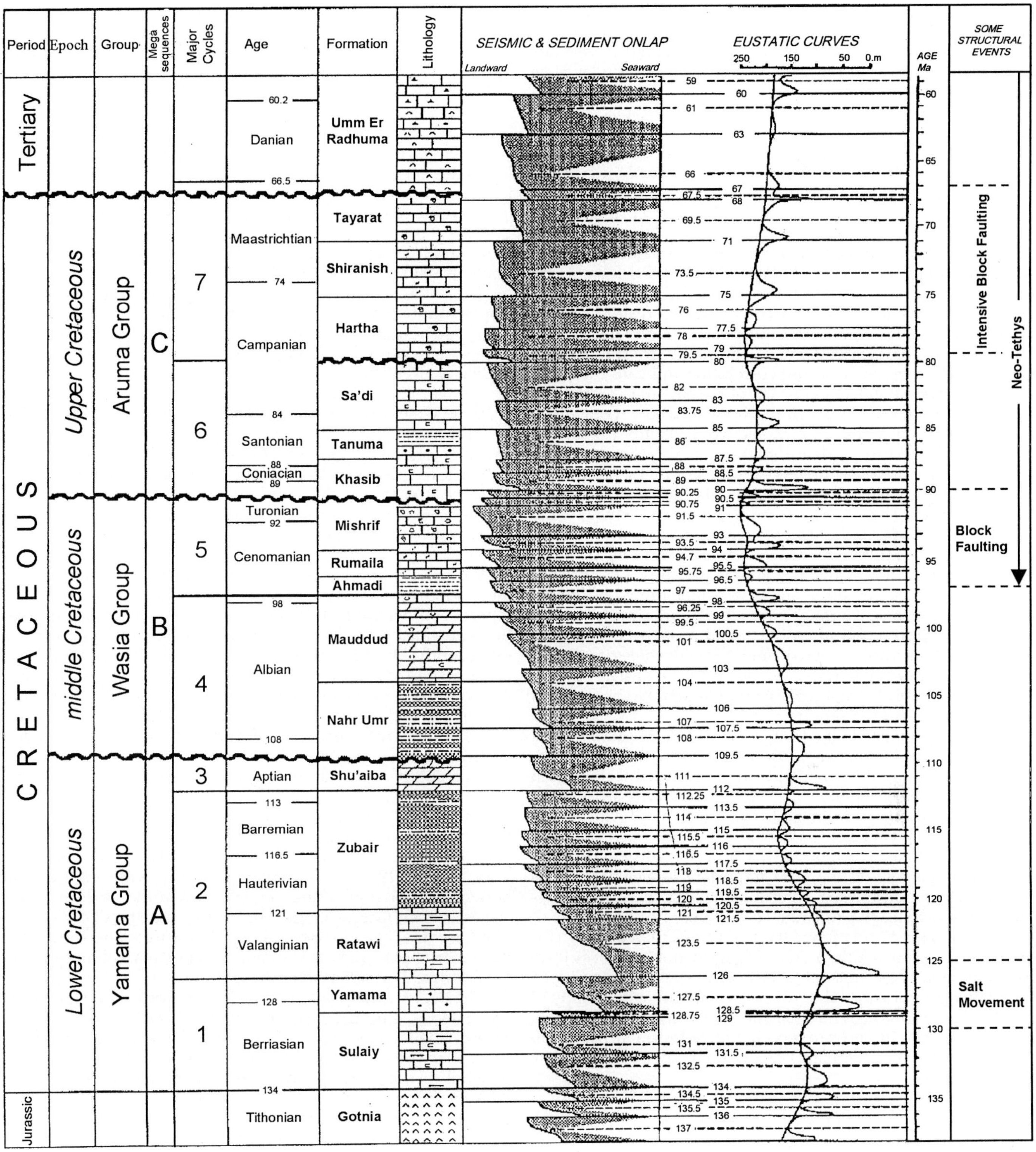


FIG. 6.—The seven major sequences distinguished throughout the Cretaceous sediments of the Mesopotamian Basin as compared to Haq et al. (1987) cycles.

sic–Cretaceous boundary in the lower part of the Sulaiy Formation. Accordingly, the *Bramkampella* sp. zone is characteristic of the lowermost parts of the Sulaiy, which is either lowermost Cretaceous or uppermost Jurassic, whereas the *Everticyclammina* sp. zone is characteristic of the lower parts of the Yamama Formation. The upper boundary of this sequence is placed at the base of the Ratawi Formation.

In addition to the Sulaiy and Yamama Formations of the Mesopotamian Basin, the sequence also includes a bathyal facies of organic-rich shales and marly limestones of the Chia Gara Formation, the lower part of the Balambo Formation, an outer-ramp, marly limestone of the Sarmord Formation, and inner-ramp, highstand oolitic limestone and patchy sponge–stromatoporoid reefs of the Sulaiy and Yamama formations (Fig. 5A, B). Although it was reported (Al-Siddiki, 1978a) that the Sulaiy,Yamama, and Ratawi formations form a single, homogeneous succession that is difficult to differentiate, detailed petrographic examination of materials from these formations and regional well-log correlation indicate the presence of numerous short breaks. These breaks (up to 14 of them just in the Yamama Formation) are characterized by concentrations of petrified plant remains and are probably associated with lowstand settings. Major breaks were used to subdivide the Yamama Formation into three reservoir units separated by two permeability barrier units formed of organic-rich, argillaceous materials. These breaks were correlated over most of the studied area and may be due to eustatic factors and, hence, could be correlated with the Haq et al. (1987) cycles. The vertical distribution of the Yamama facies indicates that the formation consists of three stacks of inner-ramp sediments separated by two outer-ramp facies. Each inner-ramp setting shallows up into either an oolitic shoal or a patch reef. The outer-ramp facies are formed of mudstones or shales (Sadooni, 1993). In the deeper parts of the basin to the east, it is very difficult to establish any break in what appears to be a record of continuous sedimentation.

According to Al-Sakini (1992), the Sulaiy Formation is a possible source rock in the major Cretaceous reservoirs of the northern Arabian region and extends northward to Hemreen Mountains, which separate the Mesopotamian tectonic zone from the Foothill zone in Iraq. He indicated that the formation enters the thermal maturation zone at a depth below 3000 m. Al-Haba and Abdulla (1989) reported that the Chia Gara and Balambo Formations are among the most important Lower Cretaceous source rocks of Iraq and suggested that the Balambo Formation is the main source for the Yamama Formation.

The Yamama Formation is the main Lower Cretaceous carbonate reservoir in southern Iraq. The formation produces from a combination of leached oolitic facies and slightly dolomitized horizons with porosity values of up to 22% and permeability of about 300 md. Although the formation is still under exploration, major finds of light oil were reported from West Qurna, North Rumaila, Rafedain, and many other oilfields (Al-Siddiki, 1978b; Sadooni, 1993).

### *Valanginian–Barremian Cycle*

This cycle starts with the first clastic pulse that invaded all the Cretaceous basins of Iraq, including the Mesopotamian Basin. This clastic front resulted in both the Ratawi (mainly argillaceous limestone and shale) and the Zubair (sandstone and shale) formations. It extended to Jambur Field in northern Iraq, where it was sandwiched between the carbonates of the Garagu Formation. In the deeper eastern parts of the basin, the Balambo Formation was still being deposited.

The Ratawi Formation was first introduced by Nasr (1950, unpublished report). Owen and Nasr (1958) described the formation from the Ratawi-1 well in southern Iraq, where it consists of massive, pyritic, greenish to dark shales interbedded with pseudo-oolitic, detrital, pyritic limestones. The main fossils are pseudochrysalinids, cyclamminids, trocholinids, and oysters. The age of the formation was considered to be Neocomian, but Bellen et al. (1959) and Buday (1980) suggested a Hauterivian–Valanginian age. Chatton and Hart (1962a) amended the definition of the formation by emphasizing the lithological differences between the Ratawi and Yamama formations. They defined the Yamama Formation as a calcarenite, and the Ratawi included a more argillaceous suite. Sadooni (1993) documented these distinctions with petrographic evidence.

The Ratawi Formation represents the beginning of the clastic front that culminated during deposition of the Zubair Formation. In its type locality at the Zubair-3 well, the Zubair Formation has been divided into five distinctive divisions, from top to bottom (Bellen et al., 1959):

1. shale with two zones of sandstones and a minor amount of siltstones;
2. sandstones with subordinate shales and siltstones;
3. black or greenish black, fissile, hard shale with local sandstone streaks;
4. sandstones with subsidiary amounts of siltstones;
5. greenish black shales with a sandstone–siltstone zone.

Further drilling in the Mesopotamian Basin indicated that these subdivisions are of only local importance, and the final reports of other wells drilled in Rumaila, Nahr Umr, and other fields are filled with alternative schemes of subdivisions. Farther to the north, Ali and Aziz (1993) described the Zubair Formation in East Baghdad Field as consisting of deltaic lobe cycles, in which the uppermost parts are marked by facies indicating wave and tidal action and overlain by shale that ends the cycle. Each cycle contains four different facies of channels, marsh, chenier, and shelf facies. They suggested that the cyclic nature of the formation was a result of regional fluctuations of the sea level due to the Alpine Orogeny.

The Zubair Formation covers most parts of the Mesopotamian Basin and is considered identical in both age and lithology variation to the Biyadh Formation of Saudi Arabia (Powers et al., 1966; Buday, 1980). The formation shows wide vertical and horizontal variation. These are reflected in the number of the cycles, in their facies changes, and in the sandstone/shale ratio. The formation even grades partially into carbonates in some of the eastern fields, such as Halfayia, and into marly limestones of the Sarmord Formation in northern Iraq.

The depositional regime of the Zubair Formation is not fully understood. Although the formation may have been deposited in a fluvial–deltaic complex, a marine influence in some parts of the basin is indicated by the presence of thick shale units and the regional distribution of sandstone bodies of relatively regular thickness. The dominance of the clastic sediments could be attributed to the humid climatic conditions, but other tectonic or local factors may also have contributed to the final configuration of the Zubair basin. The Zubair Formation is the main reservoir in Rumaila North, Rumaila South, Zubair, and many other giant oilfields in southern Iraq. The reservoir consists of three main sand units AB, DI, and LN, with an average porosity of 20% for each unit and average permeabilities of 600, 1000, and 850 md, respectively. The reservoir units are separated by two shale permeability barriers. The thickness of the net pay is around 100 m, mostly of well-sorted sand containing oil of 34° API (Al-Sakini, 1992).

The Ratawi Formation also is a reservoir rock in other Iraqi oilfields, such as Rumaila North and Zubair. The reservoir units are formed of partially leached oolitic and skeletal limestones. According to Al-Sakini (1992), the formation has porosity of 16–18%, which improves significantly in the crestal parts of these fields.

### *Barremian–Middle Aptian Cycle*

The carbonate system recovered from the clastic invasion with the deposition of carbonates of the Lower Qamchuqa Formation in northern Iraq. These deposits consist of neritic shelf shoals with large benthonic foraminifera. In the Mesopotamian Basin, the Shuaiba Formation consists of such shelf carbonates. This formation name is widely used elsewhere in the Arabian Gulf (Alsharhan, 1995).

The Shuaiba Formation was introduced by Rabanit (1951, unpublished report). Owen and Nasr (1958) described it as dolomitic limestone, coarsely crystalline with some neomorphosed rudists and *Orbitolina* spp. and *Choffatella decipiens* Schlumberger. Bellen et al. (1959) amended this and redescribed the formation from the Zubair-3 well to be composed of the following major units, in upward order:

1. pseudo-oolitic limestone with angular sand grains;
2. argillaceous, aphanitic limestone;
3. chalky limestone;
4. recrystallized globigerinid limestone with glauconite;
5. shaly limestone and shale.

The sediments of the Shuaiba Formation represent a transgressive depositional system that was abruptly terminated by an unconformity. In this study, the Shuaiba Formation is considered to be a complete cycle, because it is bounded by a transgressive surface at the base and an unconformity at the top.

The upper contact of the Shuaiba Formation is unconformable with the overlying Nahr Umr Formation, and signals the termination of the Lower Cretaceous major sedimentary cycle (i.e., the Thamama Group), and the beginning of a new major sedimentary cycle (Fig. 6).

Although it is one of the main rudist-bearing reservoirs in onshore and offshore regions of the northern Arabian Gulf (Hamdan and Alsharhan, 1991; Alsharhan, 1995), the Shuaiba is not a major reservoir in the Mesopotamian Basin.

### *Middle Aptian–Albian Cycle*

The base of this cycle includes the "second clastic front". It contains less sand but extends over nearly the same regions of the first clastic front represented by the Lower Cretaceous Ratawi and Zubair Formations. Stratigraphically the clastic sediments of this cycle belong to the Nahr Umr Formation. The deep-water Balambo Formation was still being deposited in the eastern basinal parts of the basin, while in the north a carbonate system was reestablished with the deposition of the Mauddud Formation (previously called the Upper Qamchuqa Formation). Basin-margin lagoons and sabkhas of the Jawan Formation occupied the northwestern parts of the country (Sadooni, 1978, 1995a).

The Nahr Umr Formation was described from the Nahr Umr-2 well in southern Iraq by Owen and Nasr (1958) as consisting of black shale interbedded with medium- to fine-grained sandstones with lignite, amber, and pyrite. For correlation and production purposes, the formation was divided into two members, the Third and the Fourth Sands. These two major sand members are separated by a major shale unit. Hart and Chatton (1960) divided the Nahr Umr Formation into two units, the Nahr Umr Shale and the Nahr Umr Sand. This is because the sand component of the formation practically disappears in northern Iraq. In the upper third of the formation, a limestone unit is designated as the Dair Limestone Member. The fossil content of this unit is very similar to that of the overlying Mauddud Formation. This is one of the justifications to include the Nahr Umr and Mauddud formations in a single cycle.

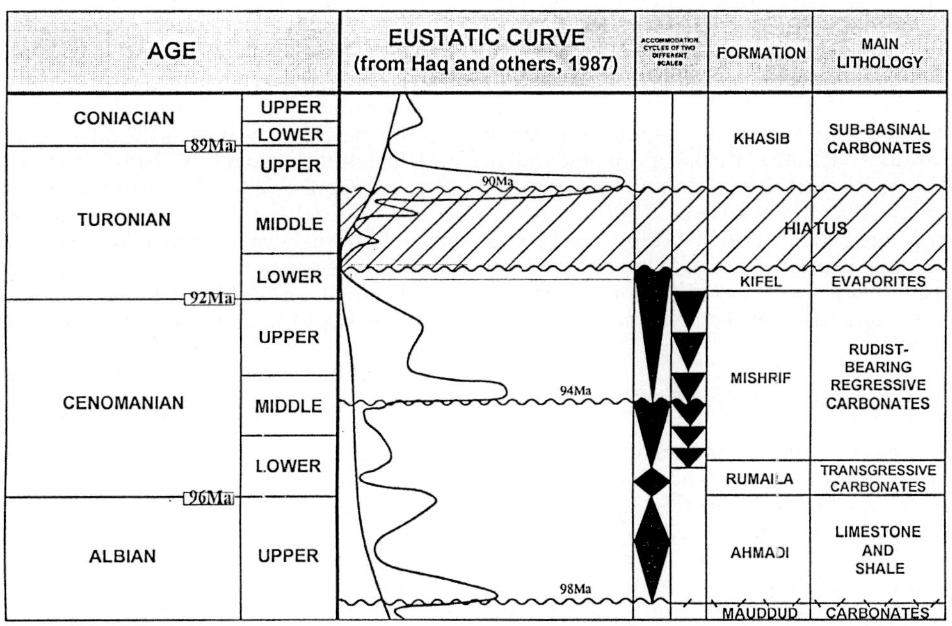


FIG. 7.—The Upper Albian–Coniacian formations and main lithologies of the Mesopotamian Basin as compared to Haq et al. (1987) cycles.

Ibrahim (1981, 1983) investigated the Albian depositional regime, and in particular the Nahr Umr Formation, in some detail in southern Iraq. He suggested that the formation was deposited in a continental setting that ranged from flood plain to delta with probable marine and eolian influence. Alsharhan (1991) studied the stratigraphic status and sedimentological setting of the Nahr Umr Formation in the United Arab Emirates and noted that the deposition of this formation follows the sea-level curves of Haq et al. (1987). Reviewing the sedimentological settings and petroleum potential of the Albian clastics in the western parts of the Arabian Gulf region, including southern Iraq, Alsharhan (1994) noted that the Albian clastics (with different local names) were deposited in a complex setting of alluvial plain to lower coastal plain and shallow marine shelf.

The contact between the Nahr Umr and the Mauddud formations is gradational, as the clastic component decreases upward. The Mauddud is among the few unit names that have been imported to the stratigraphy of Iraq. The formation was first described by Henson (1940, unpublished report) from the Dukhan-1 well in Qatar, and later was amended by Owen and Nasr (1958). The formation has been found also in Burgan-10 well in Kuwait. Al-Shamlan (1975) studied the formation in detail from some of the Kuwaiti oilfields.

According to Owen and Nasr (1958), the formation consists of 110–168 m of "organic, detrital, locally pseudo-oolitic limestones broken by local green or bluish shale". In most of the southern Iraqi oilfields, the Mauddud Formation consists of alternating orbitolinid-bearing limestone with dolomitic limestone and dolomite. Extensive dolomitization of the formation has significantly obliterated its original depositional textures and makes interpretation of the depositional setting difficult (Al-Shididi et al., 1995).

Because of its high shale content the Nahr Umr Formation is not a major oil reservoir in the Mesopotamian basin, as its equivalent Burgan Formation is in Kuwait. Oil, however, was tested in some of the giant fields such as Nahr Umr, Ratawi, Rachi, and Majnoon (Alsharhan, 1994). This oil may be generated from indigenous source beds from the relatively thick shale units or from the underlying shales of the Zubair Formation. The full potential of this formation is not fully investigated, because most of the exploration in the Mesopotamian Basin is still directed toward major structures, whereas the probability of stratigraphic entrapment in the sandstone units of the formation and/or fractured reservoirs in the shales is high.

Although the Mauddud Formation is a good reservoir in northern Iraq (e.g., Kirkuk and Kabbaz fields), its oil potential in the Mesopotamian Basin is relatively low and restricted at present to the Ratawi Field and possibly to the small field of Badra on the Iraq–Iran border. In Ratawi Field, the Mauddud consists of foraminiferal–algal limestones and dolomites. Two reservoir units are separated by massive limestone. The average porosity of each reservoir is around 18%, and some of the pores are filled with heavy oil (Al-Sakini, 1992).

### *Cenomanian–Lower Turonian Cycle*

Activation of the basement faults and differential movement of fault blocks led to prominent horizontal variation in sedimentary deposits in the Mesopotamian Basin (Fig. 2). The upper boundary of this sequence is a well-developed unconformity in the Lower Turonian, and the lower contact shows a visible change in facies from the Mauddud carbonates to the Ahmadi black fissile shales. During the same time, the "Third Clastic Front" was represented by the Rutbah Sandstone Formation in western Iraq and the Ahmadi Shale in the Mesopotamian Basin. This front is, however, of a smaller scale than the previous ones represented by the Zubair and Nahr Umr Formations. A sequence stratigraphic scheme for the Upper Albian–Lower Turonian is here proposed (Fig. 8).

The Mishrif Formation rudist-bearing carbonates were deposited together with the equivalent Gir Bir Formation in the north, following deposition of the sub-basinal, transgressive Rumaila Formation chalky and marly limestones in the Mesopotamian Basin and the basinal carbonates of the Dokan Formation in the north. The latter were deposited in the deeper eastern and intrabasinal parts of the same basin. The Balambo Formation was still being deposited in the bathyal parts of the two basins. The intertonguing of the Mishrif and Rumaila facies in the unstable central parts of the Mesopotamian Basin led some earlier workers to rename their equivalents in these areas the Mahilban, Fahad, and Maotsi formations (Aqrawi, 1983; Sherwani, 1983; Aqrawi and Khaiwka, 1986, 1989; Sherwani and Aqrawi, 1987; Aqrawi et al., 1998). Chatton and Hart (1961) proposed a preliminary model for the deposition of the whole Cenomanian–Lower Turonian cycle (Fig. 8), which was later modified by Aqrawi and Khaiwka (1986).

The Mishrif Formation consists of two large-scale regressive cycles separated by an intraformational unconformity (Reulet, 1982; Aqrawi et al., 1998), particularly on the eastern side of the Mesopotamian Basin near the Iranian border, such as in Amarah Field (Fig. 9). This unconformity can be correlated with the intraformational shale beds of the Mishrif in the southwestern parts of the basin, such as in the Nasiriyya (Kareem et al., 1988) (Fig. 10) and Gharraf (Aqrawi et al., 1998) fields. Each of the two main regressive cycles of the Mishrif Formation can be subdivided into three medium-scale cycles or sequences, particularly in the eastern side of the basin, such as in the Amarah, Halfaya, and Rafidain fields. Almost all the later coarsening-up cycles consist of one reservoir pay zone represented by rudstone and/or rudistid grainstone/packstone facies.

The rudist-bearing carbonates of the Mishrif Formation are the principal reservoir facies in the Mesopotamian Basin (Al-Khersan, 1975; Aqrawi et al., 1998). Dominated by these coarser-grained facies, the Mishrif Formation provides better reservoir zones in the eastern fields, particularly those located in the shallow, uplifted areas, which are also characterized by high subsidence rates along the Amarah paleo-high (Aqrawi et al., 1998). The hydrocarbons in the Mishrif could be sourced partly from the underlying and basinal equivalent Rumaila Formation marly limestones, as well as the deeper carbonates of the Dokan and Balambo Formations. Tight muddy layers in the Mishrif Formation provide local seals for these reservoir units and the overlying Kifel Formation evaporites where present. The sub-basinal Khasib Formation provides the regional seal. The Mishrif petroleum system is one of the unique systems in the whole Arabian Gulf region, but particularly in the Mesopotamian Basin from southern Iraq to the Hemreen Mountains (Fig. 1A). Reservoirs are characterized by high porosities (20% average) and permeabilities (average of 200 md, but may reach up to > 1 d). They contain about 40% of the Cretaceous and up to 30% of the total Iraq oil reserves, which are characterized by 26–28° API (Al-Sakini, 1992).

### *Upper Turonian–Lower Campanian Cycle*

The three formations constituting this cycle, the Khasib, Tanuma, and Sa'di formations, are bounded by two recognizable unconformities in the Mesopotamian Basin, and they may represent a single sequence in the Upper Cretaceous Aruma Group (Aqrawi, 1996). The sub-basinal *Oligostegina* facies of the Khasib Formation carbonates formed in the central and southern parts of

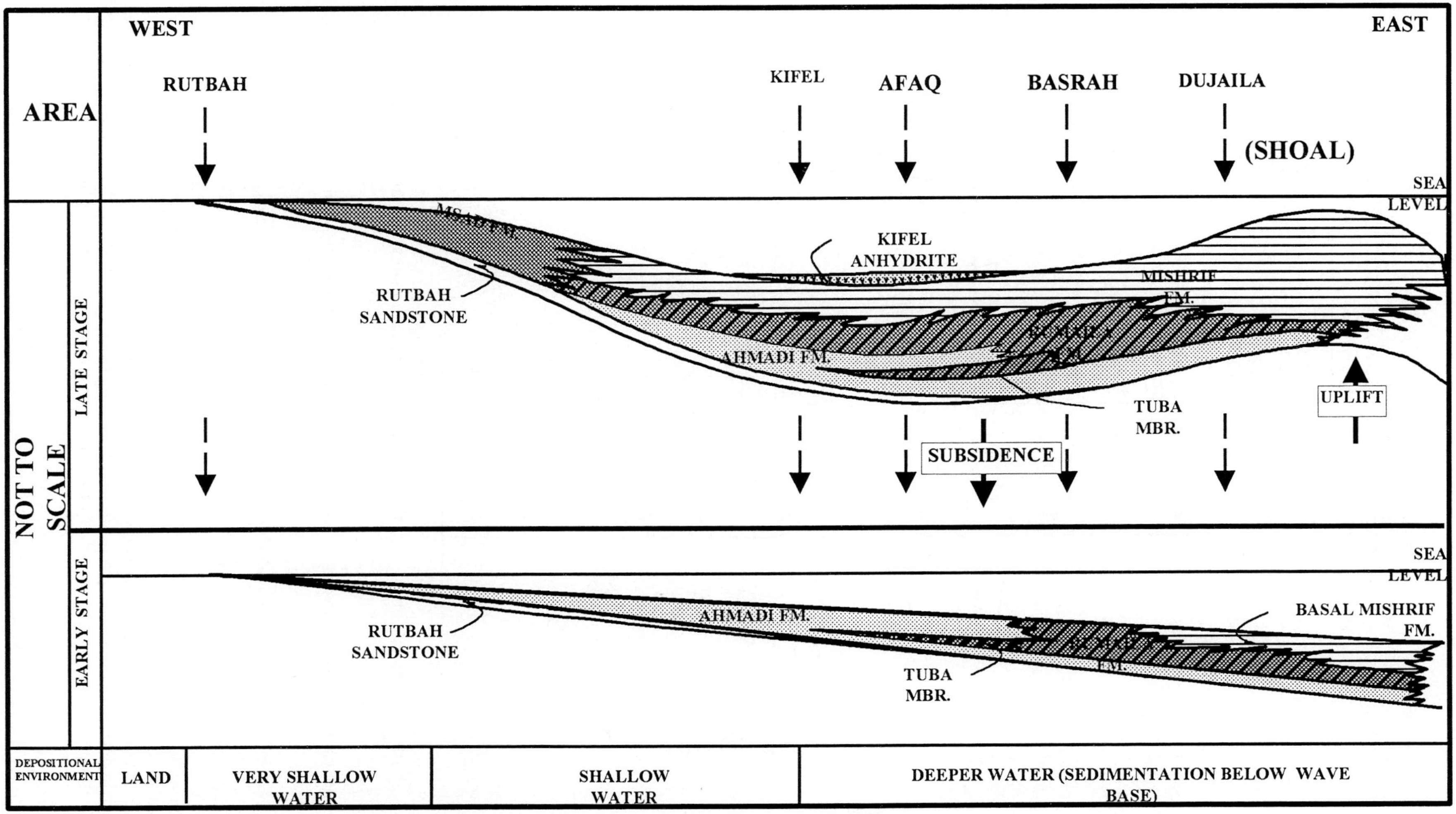


Fɪɢ. 8.—The stratigraphic relations of the Cenomanian–Lower Turonian cycle in the Mesopotamian Basin (after Aqrawi and Khaiwka, 1986, modified after Chatton and Hart, 1961).

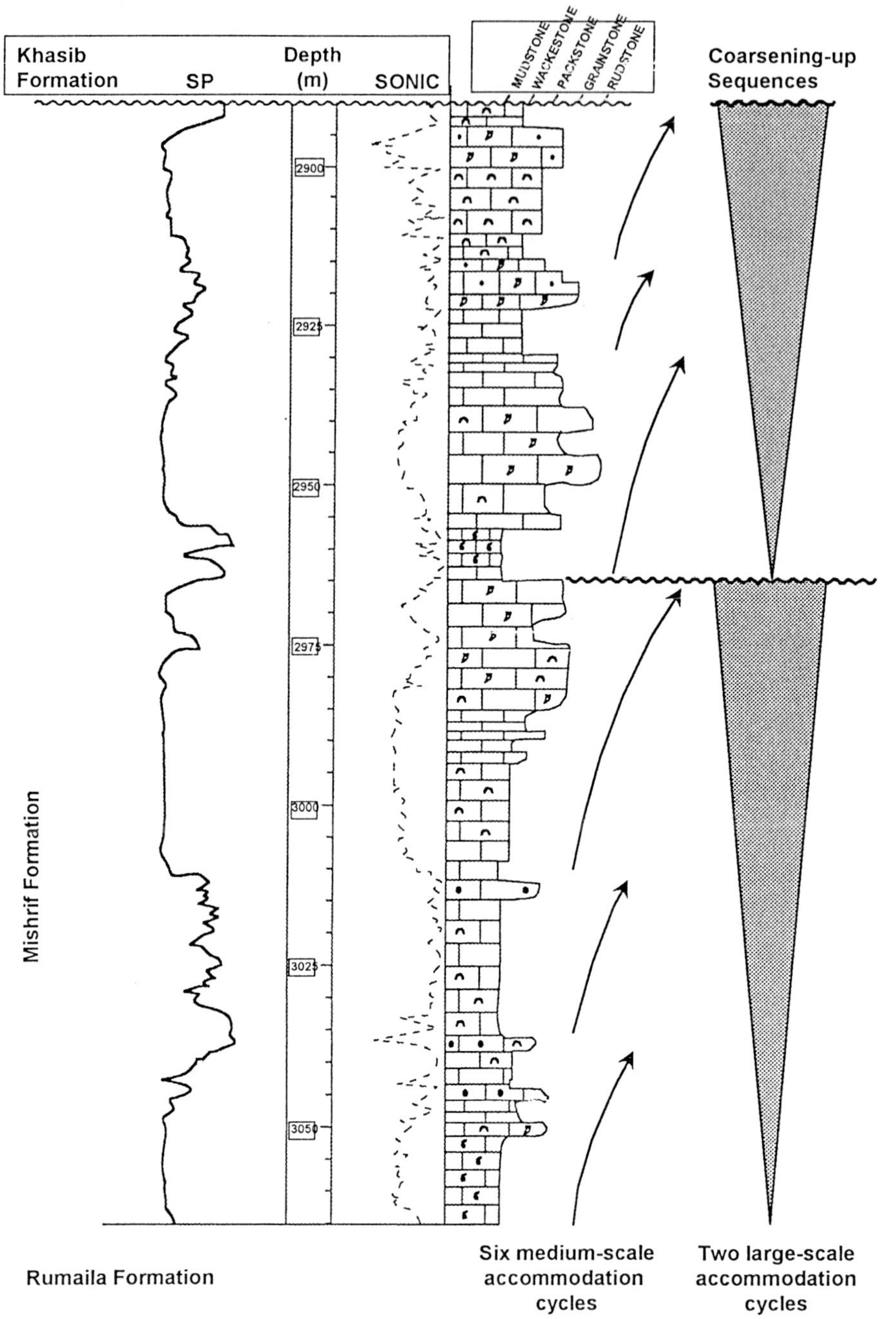


FIG. 9.—Two different-scale sedimentary cycles of the Mishrif Formation in the well Amarah-1 (well Am-1 in Fig. 1).

Iraq during the Turonian after a considerable phase of nondeposition or erosion in early Turonian. A thin calcareous shaly unit of the Tanuma Formation was followed by oolitic grainstone reservoirs (Aqrawi, 1996). A thick Sa'di Formation dominated by wackestone and packstone facies overlies the shaly Tanuma throughout the basin.

The major cycle consists of a mixed siliciclastic–carbonate depositional system in a tectonically unstable middle-shelf environment. The basin configuration was subjected to various changes during deposition of this sequence, which are reflected in the isopach maps of the three formations (Aqrawi, 1996). At least five medium-scale and three large-scale coarsening-up accommodation cycles can be recognized in this major cycle (Fig. 11). Some of the smaller cycles are continuous through adjacent formations. The presence of only detrital illite and kaolinite in the three formations may indicate a tropical or subtropical climate during their deposition (Aqrawi, 1996). This major sequence consists of several carbonate reservoir zones represented by porous oolitic grainstone and skeletal packstone facies, particularly in the Tanuma and Khasib formations, and mainly occurs in the oilfields of Central Iraq, such as the East Baghdad Field. The Tanuma also provides reservoirs in a few southern oilfields. The Khasib and Tanuma carbonate reservoirs account for up to 14% of the Cretaceous (and about 10% of

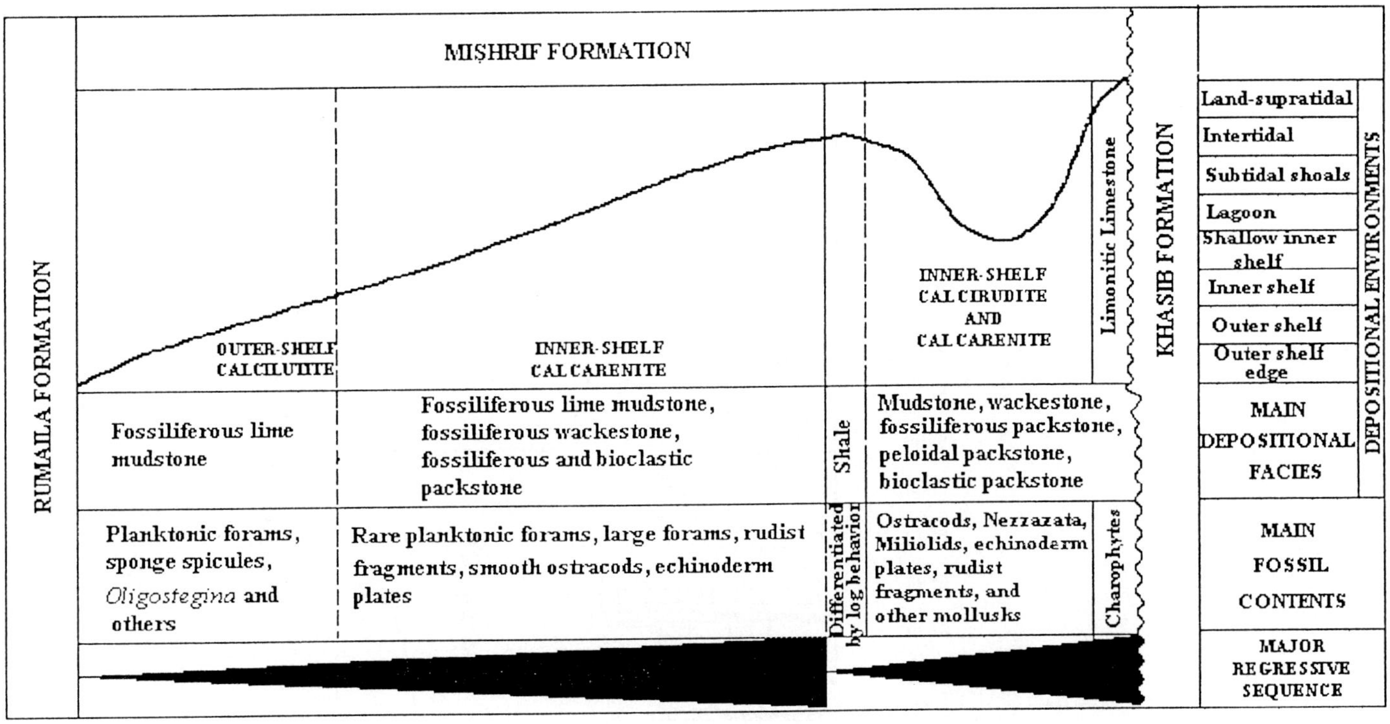


FIG. 10.—Depositional environments of the Mishrif Formation in the Nasiriyya area, referred as well Ns-1 in Fig. 1 (modified after Kareem et al., 1988).

the whole Iraq) oil reserves, and have porosities of up to 30% (Al-Sakini, 1992). The interbedded shale units provide the local seals, but they are not proved as source rocks. The extensive shaly Tanuma provides a regional seal for the Khasib reservoirs, and the tight basinal carbonate facies of the lower parts of the Sa'di Formation may provide a regional seal for the Tanuma reservoirs in addition to the Tanuma shaly beds.

### *Campanian–Maastrichtian Cycle*

The Campanian–Maastrichtian sequence is bounded above by the upper Maastrichtian unconformity, which separates the Cretaceous from the Tertiary in most parts of Iraq. Block faulting of the shelf, which started during the deposition of the previous two cycles (i.e., cycles 5 and 6), was well advanced during this cycle. Rudist patch reefs of the Hartha Formation in the Mesopotamian Basin and the Aqra and Bekhme formations in the northern parts of Iraq developed on the horsts, while deep water carbonates represented by the Shiranish Formation developed in between (Figs. 12 and 13).

The Hartha Formation consists of two major informal sequences, the upper and lower Hartha. Sadooni (1996) divided the Hartha in central Iraq into eight petrographic and petrophysical units. The formation consists of shallow-water foraminiferal shoal and rudist–biostromal deposits, separated by deeper-water muddy carbonates. The Hartha is underlain in some areas by the Safawi Formation evaporites (Al-Naqib, 1967) and overlain by the deep-water sediments of the Shiranish Formation. This, in turn, is overlain by the shelf carbonates of the Tayarat Formation. The sequence terminated after the deposition of the Tayarat, which represents the last phase of Cretaceous deposition before the unconformity separating the Upper Cretaceous Aruma Group from the Tertiary Umm er Radhuma Formation.

According to Al-Sakini (1992), the sediments of this sequence contain less than 2% of the Cretaceous oil reserves in Iraq. The Hartha Formation produced around 700 barrels of oil from the Merjan-1 well in northwest Iraq, with encouraging oil shows in the northern extensions of the East Baghdad and Ahdab fields. This oil may be generated from the deep-water marly limestone of the Shiranish Formation, particularly from the deeper parts of the basin eastward, where these sediments matured enough to generate oil.

## POTENTIAL NEW CRETACEOUS PLAYS

Iraq's largest hydrocarbon reserves are hosted in the Cretaceous sediments, particularly in the Mesopotamian Basin (Al-Sakini, 1992; Ibrahim, 1996). The Cretaceous formations in the Mesopotamian Basin are a unique petroleum system, and are among the richest in the world. Although clastic and carbonate reservoirs, together with their regional and local seals, are found throughout the Cretaceous, the source rocks in many cases are either Lower Cretaceous or Upper Jurassic shale and marly formations (Al-Haba and Abdulla, 1989).

Although the Cretaceous hydrocarbon reserves of Iraq could be increased dramatically by updating the usage of development techniques (Al-Gailani, 1996), exploration of new plays in the Cretaceous system remains high, particularly if attention is diverted from the major structures and structural traps and focused on new geographic areas and stratigraphic traps.

The Yamama Formation is perhaps one of the most promising carbonate reservoirs, because of its wide geographic distribution over most parts of southern Iraq and the presence of a relatively thick section of porous oolitic and skeletal limestones. During the last two decades, the Yamama has provided substantial additions to Iraqi reserves in some southern fields

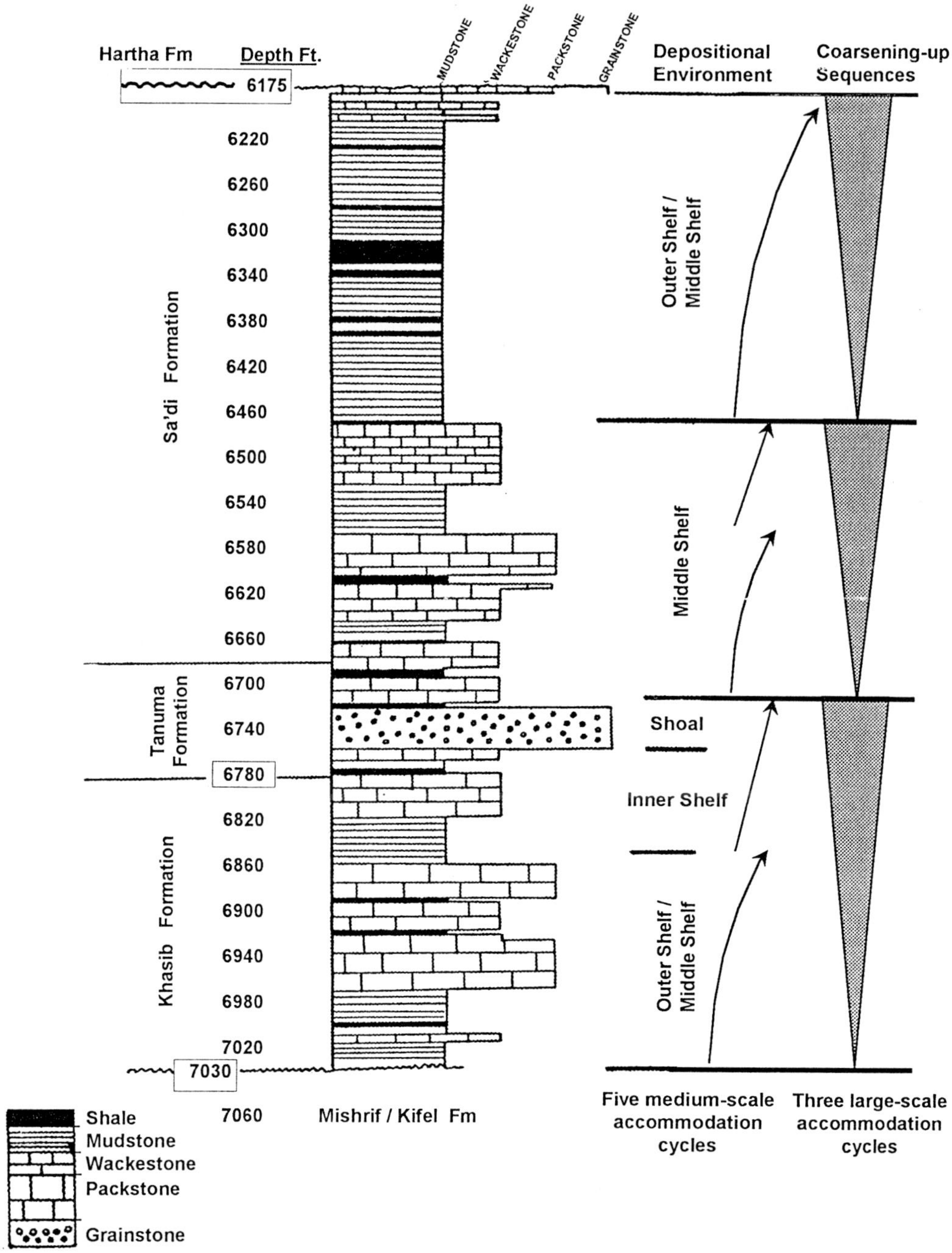


FIG. 11.—Sedimentary cycles of the Turonian–lower Campanian sequence in the Mesopotamian Basin.

where production was formerly from the younger Mishrif Formation, such as West Qurna, or from new fields, such as Nasiriyya.

Both of the major clastic reservoir formations, the Zubair and Nahr Umr Formations, were deposited in complex continental systems with marine influence. A better understanding of these systems requires high-resolution seismic surveys to help define their geometry and architecture. Such surveys will probably reveal the presence of potential reservoirs of stratigraphic entrapment that contain large quantities of oil away from the normal large structures already producing or those explored earlier.

Possible stratigraphic traps can be found within the rudist buildups and grainstones of the Mishrif Formation in areas outside the usual structures. Delineation of such porous units requires high-resolution seismic surveys accompanied by detailed sedimentological work on some of the exploration wells to determine the extension of the reefal bodies extending beyond the structures.

During the last few years, the Sa'di, Tanuma, and Khasib formations have produced oil in central parts of Iraq such as in the East Baghdad Field. The Hartha Formation produced oil from West Iraq that is located outside the usual oil province in the central and the southern parts of the country. The Mauddud Formation also has a high oil potential in some of the southeastern

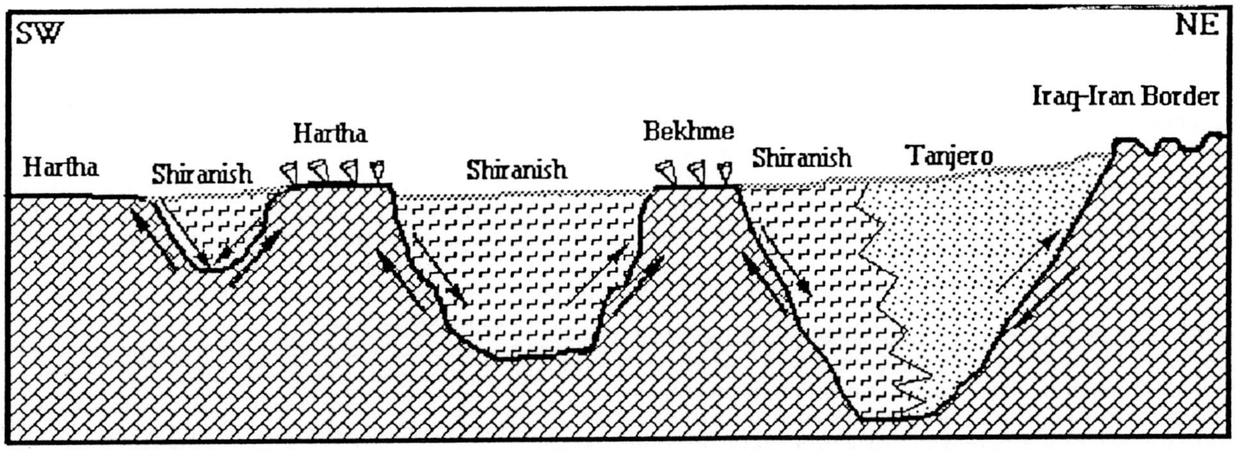


FIG. 12.—Structurally controlled facies development during the Campanian–Maastrichtian. The rudist reefs of the Hartha, Aqra, and Bekhme formations grow on the horsts, whereas the muddy carbonate facies of the Shiranish Formation accumulated in deep grabens, after an intensive period of block-faulting movements started in the Cenomanian.

structures on the Iraq–Iran border. More studies are needed in these areas to explore such potential fully.

Finally, some non-reservoir formations may become reservoirs in other unexplored areas, such as the Ahmadi limestone member in the restricted Iraqi offshore territory, as is the case in Kuwaiti–Saudi Arabian offshore. Extension of the exploration efforts to such areas may succeed in finding new plays.

## SUMMARY AND CONCLUSIONS

Cretaceous sedimentation in the Mesopotamian Basin of Central and Southern Iraq was controlled by local tectonic activity (particularly basement-block displacement and diapirism), and other regional tectonics such as the culmination of Neo-Tethys rifting, in addition to global eustatic sea-level changes. Although tropical and subtropical climate dominated most of the Cretaceous, some recognizable dry periods are well reflected by the deposition of evaporites such as the Kifel and Safawi formations.

This preliminary Cretaceous lithostratigraphic study shows that the Cretaceous formations in the Mesopotamian Basin can be divided into the same three main groups (Thamama, Wasia, and Aruma of Lower, middle and Upper Cretaceous, respectively) used in the whole Arabian Gulf region. In addition, the formations can be further grouped into seven major cycles named according to their time interval of deposition. Some of these major cycles are further subdivided into large- and medium-scale accommodation cycles where additional detailed sedimentological and paleontological data are available.

The Sulaiy and Yamama Formations were deposited on a carbonate ramp and formed the first Cretaceous major cycle. From the middle Valanginian, the ramp underwent flooding of clastics from the uplifted areas in the southwest. Initial stages of this are represented by the thin siltstone and sandstone units of the Ratawi Formation. Clastic deposition reached its peak, creating a major clastic-dominated cycle, with the deposition of the Zubair Formation during the Hautervian–Barremian. These clastics extended as far as the Kirkuk Field in northeastern Iraq.

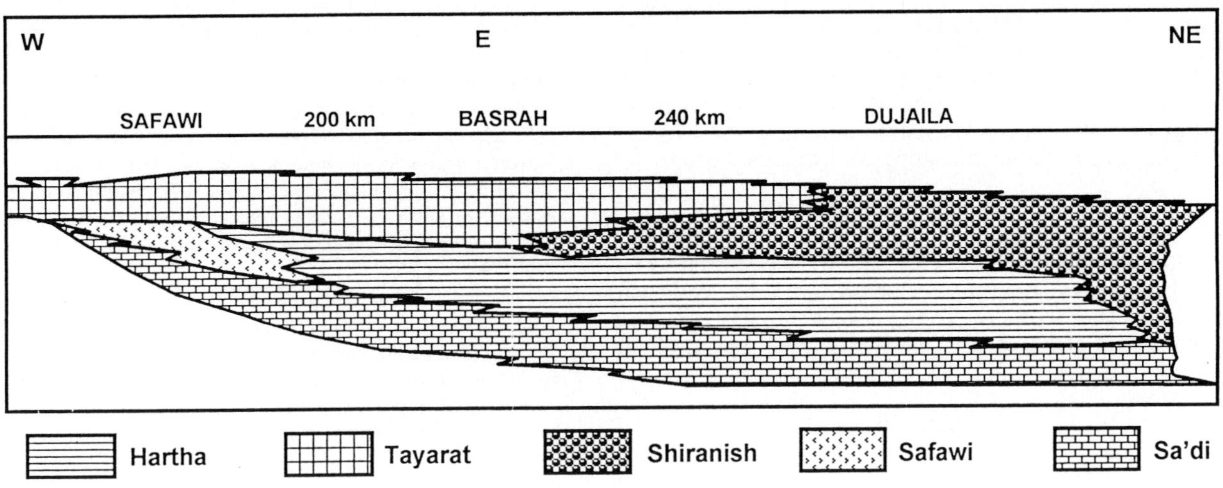


FIG. 13.—Stratigraphic relationships of the main upper Campanian–Maastrichtian carbonate facies in the Mesopotamian Basin (modified from Chatton and Hart, 1961).

The regional clastic influx, which covered most parts of the Arabian Plate, was terminated with the establishment of a new carbonate shelf in the Aptian, when the Shuaiba Formation was deposited, creating the third and final major sequence of the Lower Cretaceous.

After a period of a nondeposition or erosion represented by a well recognized, regional unconformity over large parts of the Middle East, the shelf was once again subjected to a significant clastic influx represented by the Albian Nahr Umr Formation, followed by deposition of Mauddud carbonates. During the Cenomanian–early Turonian after deposition of the transgressive shales and chalky and marly limestones of the Ahmadi and Rumaila Formations, respectively, another major sequence characterized by rudist reefs and related buildups of the Mishrif Formation dominated the shelf areas. Deposition of this major cycle ended in arid climatic conditions, as seen by the Kifel evaporites, followed by a period of erosion or nondeposition during early Turonian. Local tectonic influence is clearly reflected in the large- and medium-size accommodation cycles.

Deposition of the Khasib, Tanuma, and Sa'di formations, as a complete carbonate–clastic sequence, began during the Turonian and lasted until the Campanian. The climate changed again to tropical. Strong tectonic activity was initiated during late middle Cretaceous (i.e., during deposition of the Mishrif Formation) and led to the development of a block-faulted shelf during the Campanian–Maastrichtian. Rudist patch reefs of the Hartha Formation and its equivalents flourished on highs, while deep-water marly limestones of the Shiranish Formation occupied the low regions of the basin. Cretaceous deposition was terminated during the Maastrichtian by regional erosion after the deposition of the Tayarat Formation carbonates. This uppermost unconformity separates the Cretaceous from the Tertiary sequences.

The Cretaceous petroleum system in the Mesopotamian Basin is world ranking in terms of size. Stacked reservoirs are common, created by both local and regional seals, and sourced mainly by hydrocarbons generated in the Lower Cretaceous or Upper Jurassic. Although the Cretaceous system is relatively well explored by traditional methods in the Mesopotamian Basin, it still provides new hydrocarbon prospect potential, both outside the basin and in nonstructural plays that have not yet been proven.

New plays may be found in the Yamama Formation in other old fields that are producing now from the middle and Upper Cretaceous or from new fields especially in the Basrah–Amarah–Naseriyya triangle. Stratigraphic traps are expected to be found in the carbonate system of Mishrif as well as in the clastic systems of Zubair and Nahr Umr. Formations such as the Ahmadi, Khasib, Mauddud, and Hartha may prove to be good reservoirs in areas outside the present "conventional" exploration zones, as demonstrated by the Hartha Formation in central and western Iraq or the Mauddud Formation along the Iraq–Iran border.

## ACKNOWLEDGMENTS

The authors would like to acknowledge the two reviewers, Drs. A.J. Lomando (Chevron Overseas) and G.M. Walkden (University of Aberdeen), for their valuable suggestions and notes. Thanks are also due to Anne Keth Nyvoll (of Smedvig Technologies, Stavanger) for redrafting the figures.

## REFERENCES

ALA, M.A., 1974, Salt diapirism in Southern Iran: American Association of Petroleum Geologists, Bulletin, v. 58, p. 1758–1770.

AL-GAILANI, M., 1996, Iraq's significant hydrocarbon potential remains relatively undeveloped: Oil and Gas Journal, July 29 1996, p. 108–111.

AL-HABA, Y.Q., AND ABDULLAH, M., 1989, A geochemical study of hydrocarbon source rocks in northern Iraq: Oil and Arabic Co-operation Journal, v. 15, p. 11–15 (in Arabic).

ALI, A.J., AND AZIZ, Z.R., 1993, The Zubair Formation, East Baghdad Oilfield, central Iraq: Journal of Petroleum Geology, v. 16, p. 353–364.

AL-KHERSAN, H., 1975, Depositional environments and geological history of the Mishrif Formation in southern Iraq: Ninth Arab Petroleum Congress, Dubai, United Arab Emirates, paper no. 121 (B-3), p. 1–18.

AL-NAQIB, K.M., 1967, Southwest Iraq: U.S. Geological Survey, Professional Paper 560-G, 54 p.

AL-SAKINI, J., 1992, Summary of the petroleum geology of Iraq and the Middle East: Naft-Al-shamal Press, Kirkuk, 180 p. (in Arabic).

AL-SHAMLAN, A.A., 1975, Petrographic and microfacies analysis of the Mauddud Formation in Kuwait: Ninth Arab Petroleum Congress, Dubai, United Arab Emirates, paper 126 (B-3), 34 p.

ALSHARHAN, A.S., 1991, Sedimentological interpretation of the Albian Nahr Umr Formation in the United Arab Emirates: Sedimentary Geology, v. 73, p. 317–327.

ALSHARHAN, A.S., 1994, Albian clastics in the western Arabian Gulf region: a sedimentological and petroleum-geological interpretation: Journal of Petroleum Geology, v. 17, p. 279–300.

ALSHARHAN, A.S., 1995, Facies variation, diagenesis, and exploration potential of the Cretaceous rudist-bearing carbonates of the Arabian Gulf: American Association of Petroleum Geologists, Bulletin, v. 79, p. 531–550.

ALSHARHAN, A.S., AND KENDALL, C.G.ST.C., 1991, Cretaceous chronostratigraphy, unconformities and eustatic sea-level changes in the sediments of Abu Dhabi, United Arab Emirates: Cretaceous Research, v. 12, p. 379–401.

ALSHARHAN, A.S., AND KENDALL, C.G.ST.C., 1995, Facies variation, depositional setting and hydrocarbon potential of the Upper Cretaceous rocks in the United Arab Emirates: Cretaceous Research, v. 16, p. 435–449.

ALSHARHAN, A.S., AND NAIRN, A.E.M., 1986, A review of the Cretaceous formations in the Arabian Peninsula and Gulf: Part I. Lower Cretaceous (Thamama Group), stratigraphy and paleogeography: Journal of Petroleum Geology, v. 9, p. 365–392.

ALSHARHAN, A.S., AND NAIRN, A.E.M., 1988, A review of the Cretaceous formations in the Arabian Peninsula and Gulf: Part II. Mid-Cretaceous (Wasia Group), stratigraphy and paleogeography: Journal of Petroleum Geology, v. 11, p. 89–112.

ALSHARHAN, A.S., AND NAIRN, A.E.M., 1990, A review of the Cretaceous formations in the Arabian Peninsula and Gulf: Part III. Upper Cretaceous (Aruma Group) stratigraphy and paleogeography: Journal of Petroleum Geology, v. 13, p. 247–266.

ALSHARHAN, A.S., AND NAIRN, A.E.M., 1993, Carbonate platform models of Arabian Cretaceous reservoirs, *in* Simo, J.A.T., Scott, R.W., and Masse, J.-P., eds, Cretaceous Carbonate Platforms: American Association of Petroleum Geologists, Memoir 56, p. 173–184.

AL-SHIDIDI, S., THOMAS, G., AND DELFAUD, J., 1995, Sedimentology, diagenesis, and oil habitat of Lower Cretaceous Qamchuqa Group, Northern Iraq: American Association of Petroleum Geologists, Bulletin, v. 79, p. 763–779.

AL-SIDDIKI, A.A., 1978a, Subsurface geology of southeastern Iraq: Tenth Arab Petroleum Congress, Tripoli, Libya, 47 p.

AL-SIDDIKI, A.A., 1978b, Yamama's oil: a big exploration project executed by national staffs: Iraqi National Oil Company, First Science Conference, Baghdad, Iraq, 23 p. (in Arabic).

AQRAWI, A.A.M., 1983, Depositional environment and stratigraphy of the Rumaila Formation and equivalents in selected boreholes, middle and southern Iraq: unpublished M. Sc. thesis, University of Baghdad, 115 p. (in Arabic).

AQRAWI, A.A.M., 1996, Carbonate–siliciclastic sediments of the Upper Cretaceous (Khasib, Tanuma and Sa'di Formations) of the Mesopotamian Basin: Marine and Petroleum Geology, v. 13, p. 781–790.

AQRAWI, A.A.M., KAREEM, B.A., AND AL-REKABI, Y.R., 1988, Diagenetic environments of a shelf carbonate sequence (Mishrif Formation) of middle Cretaceous, southern Iraq: General Management For Reservoir Development, Ministry of Oil, Baghdad, Iraq (unpublished report).

AQRAWI, A.A.M., AND KHAIWKA, M.H., 1986, Depositional environment of Rumaila Formation (Cenomanian) in selected boreholes, central and southern Iraq: Geological Society of Iraq, Journal, v. 19, p. 77–95.

AQRAWI, A.A.M., AND KHAIWKA, M.H., 1989, Microfacies analysis of Rumaila Formation and equivalents (Cenomanian) in Mesopotamian basin, a statistical approach: Journal of the University of Kuwait (Science), v. 16, p. 143–153.

AQRAWI, A.A.M., THEHNI, G.A., SHERWANI, G.H., AND KAREEM, B.M.A., 1998, Mid-Cretaceous rudist-bearing carbonates of the Mishrif Formation: an important reservoir sequence in the Mesopotamian Basin, Iraq: Journal of Petroleum Geology, v. 21, p. 57–82.

BEYDOUN, Z.R., 1991, Arabian plate hydrocarbon geology and potential—a plate tectonic approach: American Association of Petroleum Geologists, Studies in Geology no. 33, 77 p.

BEYDOUN, Z.R., 1993, Evolution of the Northeastern Arabian Plate margin and shelf: Hydrocarbon habitat and conceptual future potential: Institue Français du Pétrole, Revue, v. 48, p. 311–345.

BUDAY, T., 1980, The Regional Geology of Iraq; Vol. 1, Stratigraphy and Palaeogeography: Mosul, Iraq, Dar Al-Kutub Publication House, 445 p.

BUDAY, T., AND JASSIM, S.Z., 1987, The Regional Geology of Iraq. Vol. 2, Tectonism, Magmatism and Metamorphism: Geological Survey and Mineral Investigation, Baghdad, Iraq, 351 p.

BURCHETTE, T.P., 1993, Mishrif Formation (Cenomanian–Turonian), southern Arabian Gulf: carbonate platform growth along a cratonic basin margin, *in* Simo, J.A., Scott, R.W., and Masse, J.-P., eds, Cretaceous Carbonate Platforms: American Association of Petroleum Geologists, Memoir 56, p. 185–199.

CHATTON, M., AND HART, E., 1960, Revision of the Tithonian-to-Albian stratigraphy of Iraq: Iraq Petroleum Company unpublished report, Oil Exploration Company archive, Baghdad, Iraq.

CHATTON, M., AND HART, E., 1961, Review of the Cenomanian to Maastrichtian stratigraphy in Iraq, The Cenomanian Cycle: Iraq Petroleum Company unpublished report, Oil Exploration Company archive, Baghdad, Iraq.

CHATTON, M., AND HART, E., 1962a, Announcement of a rock unit redefined, Ratawi Formation: Iraq Petroleum Company unpublished report, Oil Exploration Company archive, Baghdad, Iraq.

CHATTON, M., AND HART, E., 1962b, Review of the Cenomanian to Maastrichtian stratigraphy in Iraq, the Upper Campanian–Maastrichtian cycle: Iraq Petroleum Company unpublished report, Oil Exploration Company archive, Baghdad, Iraq.

CHATTON, M., AND HART, E., 1962c, Announcement of a new rock unit, Kifel Formation: Iraq Petroleum Company unpublished report, Oil Exploration Company archive, Baghdad, Iraq.

DUNNINGTON, H.V., 1967, Stratigraphic distribution of oilfields in the Iraq–Iran–Arabia Basin: Institute of Petroleum, Journal, v. 53, p. 129–161.

FULORIA, R.C., 1976, Petroleum prospects analysis of southern Iraq, with particular reference to Yamama Formation: Iraq National Oil Company unpublished report, Southern Petroleum Organization, Geological Laboratories, Basrah, Iraq.

GADDO, J., 1961, Announcement of a new rock unit, Batiwah Formation: Iraq Petroleum Company unpublished report, Oil Exploration Company archive, Baghdad, Iraq.

GADDO, J., 1971, The Mishrif Formation paleoenvironment in the Rumaila/Tuba/Zubair region of south Iraq: Geological Society of Iraq, Journal, v. 4, p. 1–12.

HAMDAN, A.R.A., AND ALSHARHAN, A.S., 1991, Palaeoenvironments and palaeoecology of the rudists in the Shuaiba Formation (Aptian) United Arab Emirates: Journal of African Earth Sciences, v. 12, p. 569–581.

HAQ, B.U., HARDENBOL, J., AND VAIL, P.R., 1987, Chronology of fluctuating sea levels since the Triassic (250 million years to present): Science, v. 235, p. 1156–1167.

HARRIS, P.M., FROST, S.H., SEIGLIE, G.A., AND SCHNEIDERMANN, N., 1984, Regional unconformities and depositional cycles, Cretaceous of the Arabian Peninsula, *in* Schlee, J.S., ed., Interregional Unconformities and Hydrocarbon Accumulation: American Association of Petroleum Geologists, Memoir 36, p. 67–80.

IBRAHIM, M.W., 1981, Lithostratigraphy and subsurface geology of the Albian rocks of southern Iraq: Journal of Petroleum Geology, v. 4, p. 147–162.

IBRAHIM, M.W., 1983, Petroleum geology of Iraq: American Association of Petroleum Geologists, Bulletin, v. 67, p. 97–130.

IBRAHIM, M.W., 1984, Geothermal gradients and geothermal oil generation in southern Iraq: A preliminary investigation: Journal of Petroleum Geology, v. 7, p. 77–86.

IBRAHIM, M.W., 1996, Study sizes up Iraq's reserves, exploration status, production potential: Oil and Gas Journal, June 24 1996, p. 53–55.

JASSIM, S.Z., KARIM, S., BASI, M., AL-MUBARAK, M., AND AL-MUNIR, M., 1984, The final report on the regional geological survey of Iraq, 3, stratigraphy: Geological Survey and Mineral Investigation, unpublished report, Oil Exploration Company archive, Iraq.

KAREEM, B.M.A., AQRAWI, A.A.M., AND SHUBBAR, B.A., 1988, Sedimentological and environmental study of the Mishrif Formation in Well Ns-2, Nasiriyya area, southern Iraq: General Management For Reservoir Development, Ministry of Oil, Baghdad, Iraq (unpublished report in Arabic).

KENDALL, C.G.ST.C., BOWEN, B., ALSHARHAN, A., CHEONG, D.K., AND STOUDT, D., 1991, Eustatic controls on carbonate facies in reservoirs, and seals associated with Mesozoic hydrocarbon fields of the Arabian Gulf and the Gulf of Mexico: Marine Geology, v. 102, p. 215–238.

LOUCKS, R.G., AND SARG, J.F., eds., 1993, Carbonate Sequence Stratigraphy: Recent Developments and Applications: American Association of Petroleum Geologists, Memoir 57, 545 p.

MURRIS, R.J., 1980, Middle East: Stratigraphic evolution and oil habitat: American Association of Petroleum Geologists, Bulletin, v. 64, p. 597–618.

OWEN, R.M.S., AND NASR, S.N., 1958, Stratigraphy of the Kuwait–Basrah areas, *in* Weeks, L.G., ed., Habitat of Oil, A Symposium: American Association of Petroleum Geologists, p. 1252–1278.

POWERS, R.W., 1968, Saudi Arabia: Lexique Stratigraphique International, III, Asie, Fasc. 10a., Centre National de la Recherché Scientifique, 177 p.

POWERS, R.W., RAMIREZ, L.F., REDMOND, C.D., AND ELBERG, E.L., JR., 1966, Sedimentary geology of the Saudi Arabia: U.S. Geological Survey, Professional Paper 560-D, 147 p.

REDMOND, C.D., 1964, Lituolid foraminifera from the Jurassic and Cretaceous of Saudi Arabia: Micropaleontology, v. 10, p. 405–414.

REULET, J., 1982, Carbonate reservoir in a marine shelf sequence, Mishrif Formation, Cretaceous of the Middle East, *in* Reeckman, A., and Friedman, G.M., eds., Exploration for Carbonate Platform Reservoirs: New York, John Wiley & Sons, p. 165–173.

SADOONI, F.N., 1978, Sedimentology and petroleum prospects of the Lower Cretaceous Qamchuqa Group, northern Iraq: unpublished Ph.D. thesis, University of Bristol, UK, 376 p.

Sadooni, F.N., 1993, Stratigraphic sequence, microfacies and petroleum prospects of the Yamama Formation, Lower Cretaceous, southern Iraq: American Association of Petroleum Geologists, Bulletin, v. 77, p. 1971–1988.

Sadooni, F.N., 1995a, Diagenetic features of some subsurface Tertiary–Cretaceous evaporites from northern Iraq: Carbonates and Evaporites, v. 10, p. 45–53.

Sadooni, F.N., 1995b, Applications of borehole logs in depositional and diagenetic facies characterization of carbonate reservoirs: First Indian Conference on Petroleum Geophysics, Dehradun, India, Proceedings, 17 p.

Sadooni, F.N., 1996, Stratigraphic and lithological characteristics of Upper Cretaceous carbonates in central Iraq: Journal of Petroleum Geology, v. 19, p. 271–288.

Sadooni, F.N., 1997, Stratigraphy and petroleum prospect of Upper Jurassic carbonates in Iraq: Petroleum Geoscience, v. 3, p. 233–243.

Sadooni, F.N., 1998, Cretaceous rudist-bearing carbonate belt of the Arabian Platform: a probable indicator of a regional climatic front: (abstract) ENERGEX'98, Bahrain.

Sherwani, G.H., 1983, Depositional environments and stratigraphic relationships of the Mishrif Formation in selected boreholes, middle and southern Iraq: unpublished MSc thesis, University of Baghdad, Baghdad (in Arabic).

Sherwani, G.H., and Aqrawi, A.A.M., 1987, Lithostratigraphy and environmental considerations of Cenomanian–early Turonian shelf carbonates (Rumaila and Mishrif Formations) of Mesopotamian basin, central and southern Iraq (abstract): American Association of Petroleum Geologists, Bulletin, v. 71, p. 614.

Stocklin, J., 1972, Iran: Lexique Stratigraphique International, F. 9b, Asie, Centre National Recherche Scientifique, Paris, 376 p.

Sugden, W., and Standring, J.A., 1975, Qatar Peninsula: Lexique Stratigraphique International, F 10b3, Asie, Centre National Recherche Scientifique, Paris, 127 p.

Vail, P.R., Mitchum, R.M., Jr, Todd, R.G., Widmier, J.M., Thompson, S., III, Sangree, J.B., Bubb, J.N., and Hatlelid, W.G., 1977, Seismic stratigraphy and global changes of sea level, *in* Mitchum, R.M., Vail, P.R., and Sangree, J.B., eds, Seismic Stratigraphy—Applications to Hydrocarbon Exploration: American Association of Petroleum Geologists, Memoir 26, p. 49–212.

Van Bellen, R.C., Dunnington, H.V., Wetzel, R., and Morton, D.M., 1959, Iraq: Lexique Stratigraphique International, F 10a, Asie, Centre National Recherche Scientifique, Paris, 333 p.

Van Wagoner J.C., Mitchum, R.M., Campion, K.M., and Rahmanian, V.D., 1990, Siliciclastic sequence stratigraphy in well logs, cores, and outcrops: concepts for high resolution correlation of time and facies: American Association of Petroleum Geologists, Methods in Exploration Series 7, 55 p.

# HYDROCARBON POTENTIAL OF MESOZOIC CARBONATE PLATFORM–BASIN SYSTEMS, U.A.E.

ABDULRAHMAN S. ALSHARHAN
*Faculty of Science, United Arab Emirates University, P.O. Box 17551, Al Ain, U.A.E.*

AND

ROBERT W. SCOTT
*Precision Stratigraphy Associates, RR3 Box 103-3, Cleveland, OK 74020, U.S.A.*

ABSTRACT: The United Arab Emirates (U.A.E.) is located on the eastern Arabian shelf, bounded on the northwest by the Qatar–South Fars Arch and on the east and northeast by the foreland basin and the adjacent foreland fold-and-thrust belt of Oman. Two passive-plate-margin basins were developed in the area, the northern Rub Al Khali basin and the Ras Al Khaimah basin. Since the Permian, a variety of sub-basins, separated by depositional and tectonic highs, have formed, filled, and been buried, but the style of sedimentation has been maintained. The Upper Permian to Holocene rocks and sediments of the U.A.E. consist mainly of epeiric shelf carbonates, associated with minor evaporites and clastics, reflecting major cycles of transgression and regression. These sediments were deposited on the eastern margin of the Arabian Shield, which lay along the southern margin of the Tethys Ocean during the Mesozoic and Cenozoic eras. Sedimentation patterns were controlled by prominent regional structural features, epeirogenic movements, and/or sea-level fluctuations. Abundant oil and gas reserves have been proved in the Mesozoic rocks of the U.A.E., and are contained in the shallow-water carbonates of the Jurassic (Araej and Arab formations) and the Cretaceous (Thamama Group, Mishrif, and Simsima formations). Reservoirs are sourced by deep-water bituminous shale and argillaceous limestones and sealed by evaporites, dense limestones, and shales. The distinctive sedimentologic and structural style, together with the development of source–reservoir–seal, has made the U.A.E. one of the world's richest Mesozoic oil habitats. Detailed studies of the occurrences and distributions of the factors controlling hydrocarbon distribution are important in the understanding of local oil fields and future plays.

## INTRODUCTION

The U.A.E. is located in the eastern part of the Arabian Peninsula between latitudes 22° 40' N and 26° 00' N and longitudes 51° 00' E and 56° 00' E (Fig. 1). In terms of tectonic framework, the U.A.E. is situated within the interior platform of the Arabian shelf, bounded on the northwest by the Qatar–South Fars Arch and on the east and northwest by the foreland basin and adjacent foreland fold-and-thrust belt of Oman. Modern sediments of the Arabian Gulf foreland basin are like the sediments that have accumulated here in the subsurface. Since the Permian, a variety of sub-basins, separated by depositional and tectonic highs, have formed, filled, and been buried, but the overall style of sedimentation has been maintained throughout.

These Upper Permian to Holocene rocks and sediments of the U.A.E. consist mainly of epeiric shelf carbonates, associated with minor evaporites and clastics, reflecting major cycles of transgression and regression. They were deposited on the eastern margin of the Arabian Shield, which lay along the southern margin of the Tethys Ocean during the Mesozoic and Cenozoic eras. Sedimentation patterns were controlled by prominent regional structural features and relatively gentle tectonic subsidence punctuated by eustatic variations in sea level. The resulting facies variations are used to subdivide the stratigraphic section of the U.A.E. into formations (Fig. 2) (Alsharhan, 1989; Alsharhan and Nairn, 1986, 1988; Alsharhan and Kendall, 1991; Kendall et al., 1991a). Indeed, it is possible to correlate most major unconformities with coastal onlap events of Haq et al. (1988), though local tectonic events mask some of the large cycle breaks (Alsharhan and Kendall, 1991; Kendall et al., 1991a). The objective of this paper is to provide a geological background to the sedimentary facies and settings, stratigraphic framework, and structural evolution that controlled the source, migration, and accumulation of hydrocarbons in the U.A.E. (Table 1).

## TECTONIC SETTING

### *Tectonic History*

The tectonic history of the U.A.E. in the Mesozoic and Cenozoic eras is connected with the Permian opening (Metour et al., 1995) and Late Cretaceous–Paleogene closure of the southern Neo-Tethys ocean. Glennie (this volume) divides the NeoTethys into Neo-Tethys-1, which was created in Late Permian by the halving of a microcontinent along the northeast margin of Arabia, and Neo-Tethys-2, which occurred in the Late Triassic and is more recognizable in Iran than in Oman and northeastern U.A.E. Subduction probably began in Neo-Tethys-2 during the Aptian–Albian opening of the South Atlantic. Thus, from Permian to about middle Cretaceous time, the northern part of the U.A.E. was a tectonically passive carbonate shelf. Deposition continued into the Early Turonian in Oman, where diagnostic ammonites date the uppermost part of the Natih Formation (Kennedy and Simmons, 1991). Throughout the rest of the Turonian, regional uplift and the onset of folding of the Oman Mountains occurred in the U.A.E. (Patton and O'Connor, 1988; Robertson and Searle, 1990). This major compressional tectonic event continued until the Maastrichtian, when the plates associated with central Iran and the Arabian Peninsula collided. The result was the rapid emplacement of ocean-floor sediments onto the eastern edge of the Arabian craton, while the adjacent Omani foredeep was filled with argillaceous limestone, chert, and shales. These events increased the amount of clastic sediments that accumulated in the northern U.A.E. area, particularly from the east, though carbonate and evaporite deposition still continued, particularly to the west. During the early Tertiary in the northern Oman Mountains of the U.A.E., the edge of the carbonate platform sequences was thrust over the foredeep. Further thrusting was then probably reinitiated in the latest Miocene–Pliocene. In the middle Tertiary to the late Tertiary, doming

Middle East Models of Jurassic/Cretaceous Carbonate Systems
SEPM Special Publication No. 69, Copyright © 2000
SEPM (Society for Sedimentary Geology), ISBN 1-56576-075-1, p. 335–358.

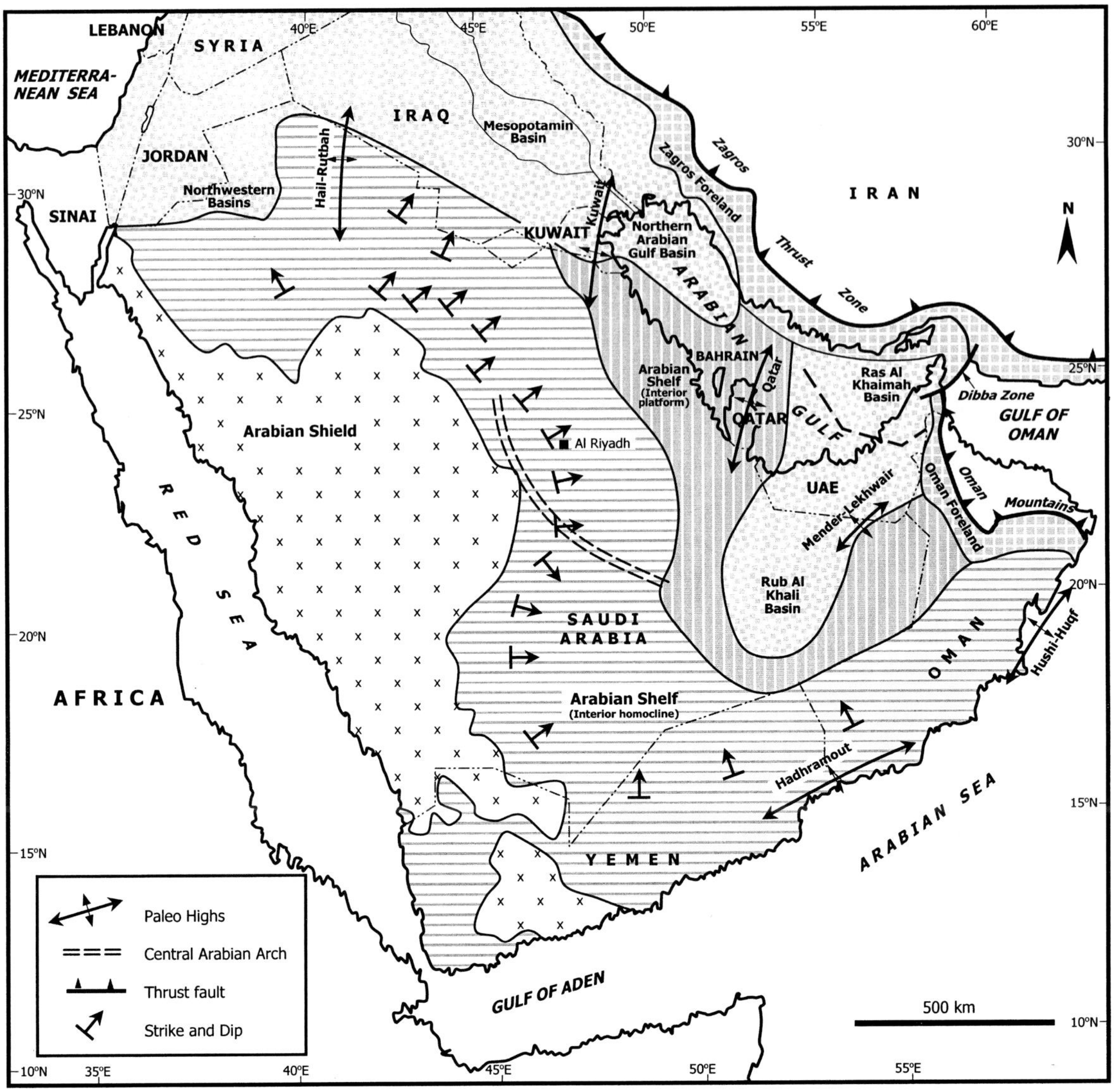


FIG. 1.—Location map of the Arabian Peninsula showing the main tectonic elements (modified from Powers et al., 1966, and Alsharhan and Nairn, 1986; Alsharhan and Kendall, 1986.

and differential uplift occurred in northern U.A.E. over the central foredeep area (Patton and O'Connor, 1988). Broadly, there are two main tectonic provinces in the U.A.E.: (1) a passive continental margin covering Abu Dhabi and part of onshore Dubai, and (2) an eastern thrust belt and associated foreland in onshore and offshore areas of northern U.A.E. These are known as the northern Rub Al Khali basin and the Ras Al Khaimah Basin, respectively (Alsharhan, 1989).

## *Structure*

The passive-continental-margin province is characterized by gentle, simple folds of varying shapes and sizes related to differential regional subsidence or uplift along the deep-seated basement faults or to diapiric movement of Eocambrian salt. The foreland basin is characterized by tensional block-faulted structures, which were related to crustal extension that developed during the Cenomanian–Turonian. Their trends coincide with the hinge line that defines the western margin of the foreland basin described by Alsharhan and Nairn (1997). Large, low-amplitude folds are common in U.A.E., with flank dips of less than 5°. The main fold axes trend N–S with lesser ones that trend NW–SE, NE–SW, or E–W, depending on location (Fig. 3). In western Abu Dhabi, N–S folds are the most common and have been related to deep-seated basement tectonics. They are probably tied to the structural fabric of the underlying Precambrian basement and the superimposed Hercynian folding event, and are similar to the structure of some fields in Eastern Arabia described by Powers et al. (1966) and Murris (1980). Southeast of Abu Dhabi, east of the Falaha syncline, the fold axes trend NE–SW at the Shah, Asab, and Sahil fields. The northeast–southwest trend appears to be related to the late Tertiary overthrusting that occurred in Oman when the Zagros Mountains were folded. In the northeast Abu Dhabi, towards the Jarn Yaphour Field, the trend is NW–SE (Schlumberger, 1981; Alsharhan, 1989). In the central offshore Abu Dhabi, the E–W aligned folds of the Zakum and Ghasha fields formed when the Zagros Mountain fold system developed. They are superimposed on the older N–S trending "Arabian folds" of Schlumberger (1981). Salt diapirism

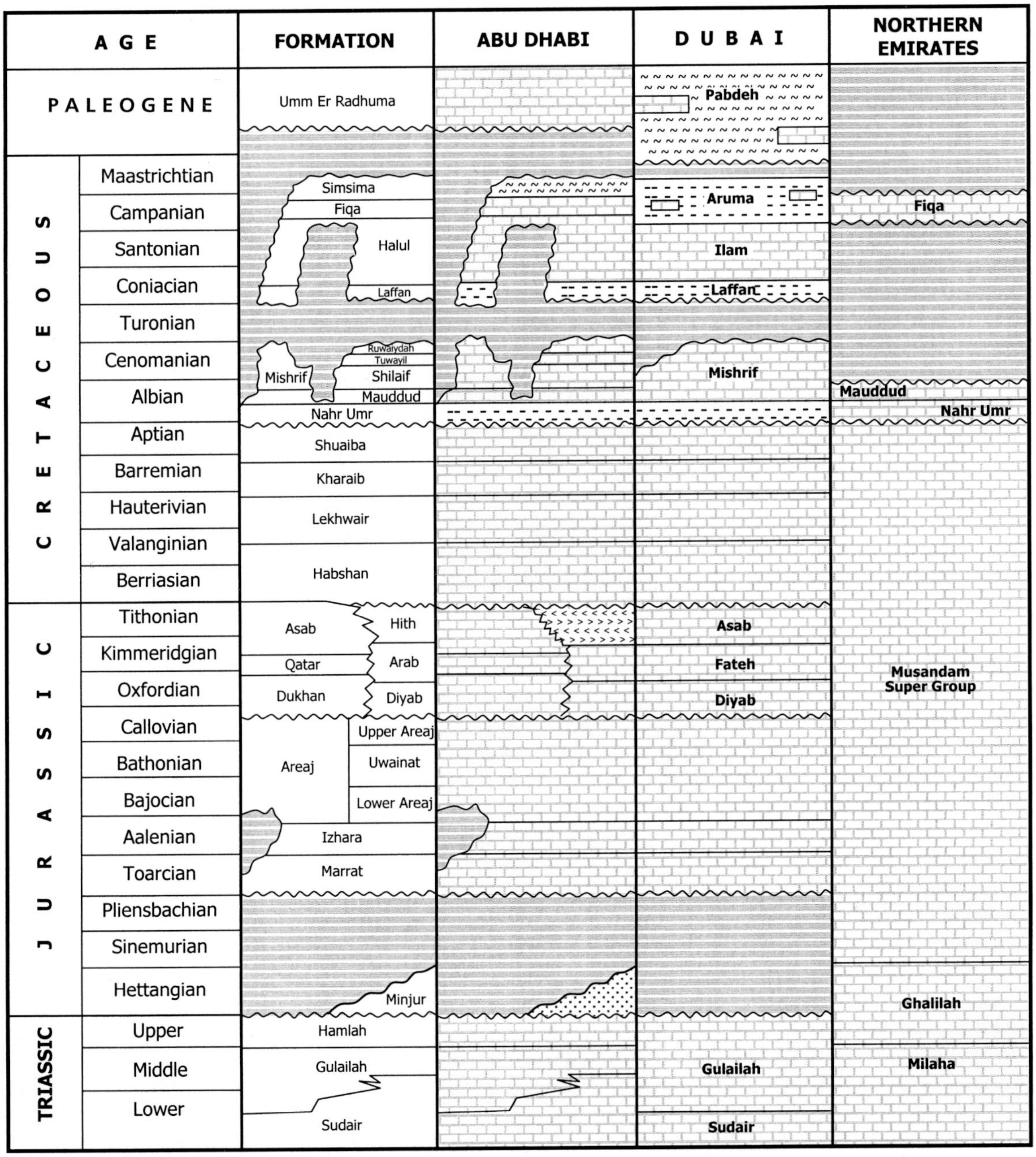


FIG. 2.—Lithostratigraphy of the Mesozoic section in the United Arab Emirates.

has further modified the structures. Indeed, prominent structural anomalies are related to the periods of upward movement of Eocambrian salt deposits, responding in part to readjustments in the configuration of the basement complex. This movement occurred throughout the history of the basin but was particularly common in the Tertiary. Salt diapirs underlie most of the structural highs in Abu Dhabi and offshore Dubai.

## TRIASSIC STRATIGRAPHY AND DEPOSITIONAL SETTING

In the U.A.E. subsurface, the Triassic to lower Lower Jurassic rocks can be divided into four formations (from bottom to top): Sudair, Gulailah, Hamlah, and Minjur (Fig. 2). The Sudair Formation (Scythian) comprises mixed carbonate and terrigenous limestones deposited as three cycles. The lower cycle is formed by limestones interbedded with terrigenous mudstones and minor dolomites deposited in variable-energy shallow marine conditions. The middle cycle is dominated by anhydritic dolomitic mudstones deposited in a supratidal to subtidal setting. The upper cycle is composed of a shallow marine to sabkha facies consisting mainly of interbedded terrigenous mudstones and dolomite with minor oolitic–peloidal packstone/grainstone. In offshore Dubai the percentage of shale increases toward the upper part of the formation, and the top of the formation is marked by thin salt streaks and light green to gray shale and siltstone.

The Gulailah Formation (Carnian) (also known as Jilh Formation in onshore Abu Dhabi) consists predominantly of anhydritic,

TABLE 1.—Depositional settings of Arabian platform and their distinguishing characteristics in the United Arab Emirates.

| **Name** | **Arabian carbonate shelf** |
|---|---|
| Present Location | From 22° to 28° north latitude and 51° to 57° west longitude |
| Geologic time interval | Early Triassic–Late Cretaceous, (Scythian–Maastrichtian) |
| Tectono-sedimentary setting | Passive margin changing to back arc basin |
| Basin type | Epeiric shelf changing to foreland basin |
| Paleoclimate | Generally tropical (ranged from arid, semiarid to humid) |
| Platform type | Rimmed, epeiric shelf changing to distally steepened ramp, shoalΠrudist complex and high-energy shoal rimming an intrashelf basins, high-energy shelf deposited as carbonate sand sheets |
| Platform geometry | Wide to moderately wide, at least 1700 m thick; more than 300 km wide and more than 700 km long |
| Circulation | Open to partially restricted |
| Facies and fossils | Fine- to coarse-grained carbonates (ooliticΠpeloidalΠskeletal sand), thin intervals of shale and sandstone; pelagic and benthic foraminifera, algae, mollusks, sponges, echinoderms, and corals |
| System tracts | Transgressive and progradational highstand |
| Stacking patterns | Buildups, carbonate sand sheets, and nearshore cycles |
| Main Reservoirs (Formation/Age) | Simsima (Maastrichtian), Mishrif (Cenomanian), Shuaiba (Aptian), Kharaib (Barremian), Lekhwair (Hauterivian), Arab (TithonianΠKimmeridgian), Araej (BathonianΠCallovian) |
| Main Source Rocks (Formation/Age) | Diyab Formation (Oxfordian), Shilaif Formation (Cenomanian) |
| Main Seal Rocks (Formation/Age) | Hith anhydrite (Tithonian), Nahr Umr shale (Albian), Laffan shale (Coniacian), basal Umm Er Radhuma shale (Paleocene) |

dolomitic mudstones and fine terrigenous sediments, with minor intraclastic wackestones and oolitic, peloidal packstone/grainstone. The top is marked by gray calcareous shales overlying thin salt and shale beds. The formation reflects a continuation of the Sudair sabkha evaporite, and the presence of terrigenous sediments reveals deposition under shallow-marine sabkha conditions in an area that bordered a very restricted basin. Minor bioclastic and intraclastic sediments represent small incursions of more normal marine conditions.

The Hamlah Formation is upper Carnian–Norian in age, on the basis of palynomorphs and stratigraphic correlation. Its lower part consists of peloidal–molluscan wackestones, overlain by molluscan wackestones and packstones interbedded with bioclastic lime mudstones. These are overlain by highly bioturbated coarser molluscan packstones. The depositional setting suggested is a nearshore marine environment, and the bioturbated limestone facies with marine fauna indicates deposition under quiet-water but well oxygenated conditions. The depositional environment of the formation deepens toward offshore Dubai and the northern Emirates. There it consists of dark gray to brown dolomite overlain by a unit with thin, black to dark gray shale streaks, and it conformably overlies the Middle Triassic (Fig. 4). An abrupt gamma-ray shift mark the unconformable contact with the Lower Jurassic Izhara Formation.

The upper Norian to Hettangian(?) Minjur Formation is well developed in onshore U.A.E. only and is characterized by three coarsening-upward cycles. The lower cycle consists of interbedded argillaceous quartz sandstones and mudstones with thin coal seams near the base. The middle cycle is composed of interbedded shales and sandstones, and the upper cycle consists of thin sands and argillaceous/dolomitic limestones. The formation was deposited in a fluvio-deltaic setting, with the coarsening-upward character of the cycles representing progradation of delta lobes into interdistributary bay areas. The presence of coal horizons reflects a humid continental to marginal marine setting.

In the outcrops of the northern Emirates, the Middle Triassic to Lower Jurassic succession is called the Elphinstone Group and consists of two formations (Figs. 2, 5): the Milaha and the Ghalilah (Searle et al., 1983; Alsharhan and Kendall, 1986). The Milaha Formation (Middle to Upper Triassic) consists of nonresistant, mostly argillaceous, partly dolomitized lime mudstones, which were divided by Hudson (1960) into three members as follows. The lower member consists of thin-bedded dolomitized limestones alternating with siltstones and silty marls and argillaceous dolomites. The middle member consists of well-bedded peloidal, bioclastic, light brown to buff dolomite, and the upper member comprises fossiliferous wackestones and grainstones (mainly bivalves, ostracodes, bryozoans, and stromatolites).

The Milaha Formation was deposited in a sheltered lagoonal, inner-shelf setting, which received occasional influxes of argillaceous sediments. In the northern and central Musandam Peninsula, the Milaha Formation grades from shales and tidal-flat dolomites to muddy, open-shelf, bioclastic, *Megalodon*-rich limestones and mollusc-rich biostromes (Ricateau and Riché, 1980; Searle et al., 1983). In contrast, to the south, along the edge of the Dibba Zone outer shelf, skeletal and peloidal lime-sand shoals and coral–algal reefs accumulated.

The overlying Ghalilah Formation (Upper Triassic–lower Lower Jurassic) represents the culmination of platform-wide clastic deposition and the occurrence of a marine regression. It consists of three units. The lower unit comprises quartz sand-

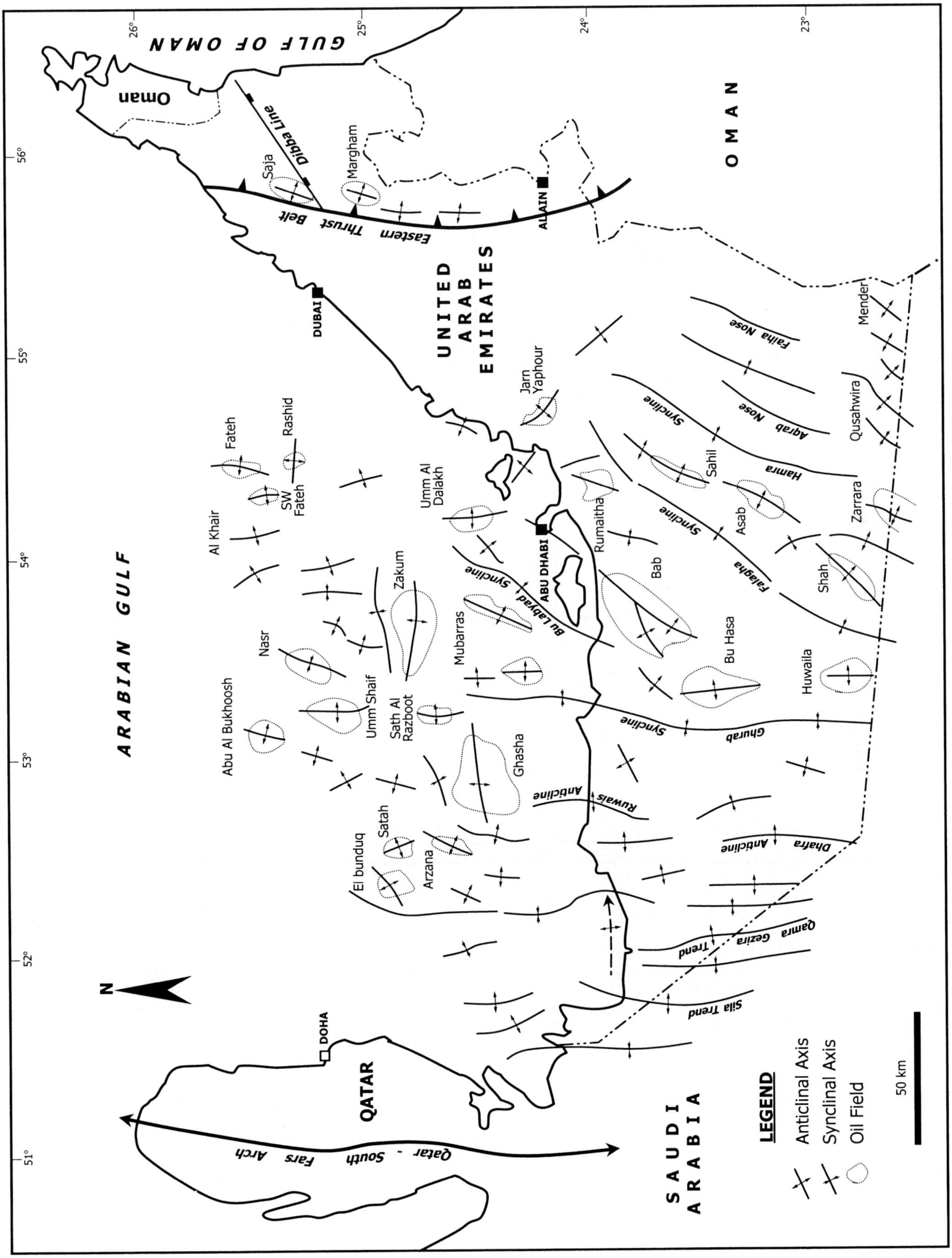


FIG. 3.—Major anticlines and synclines associated with oil and gas fields in the United Arab Emirates (after Alsharhan, 1989; Alsharhan and Nairn, 1997).

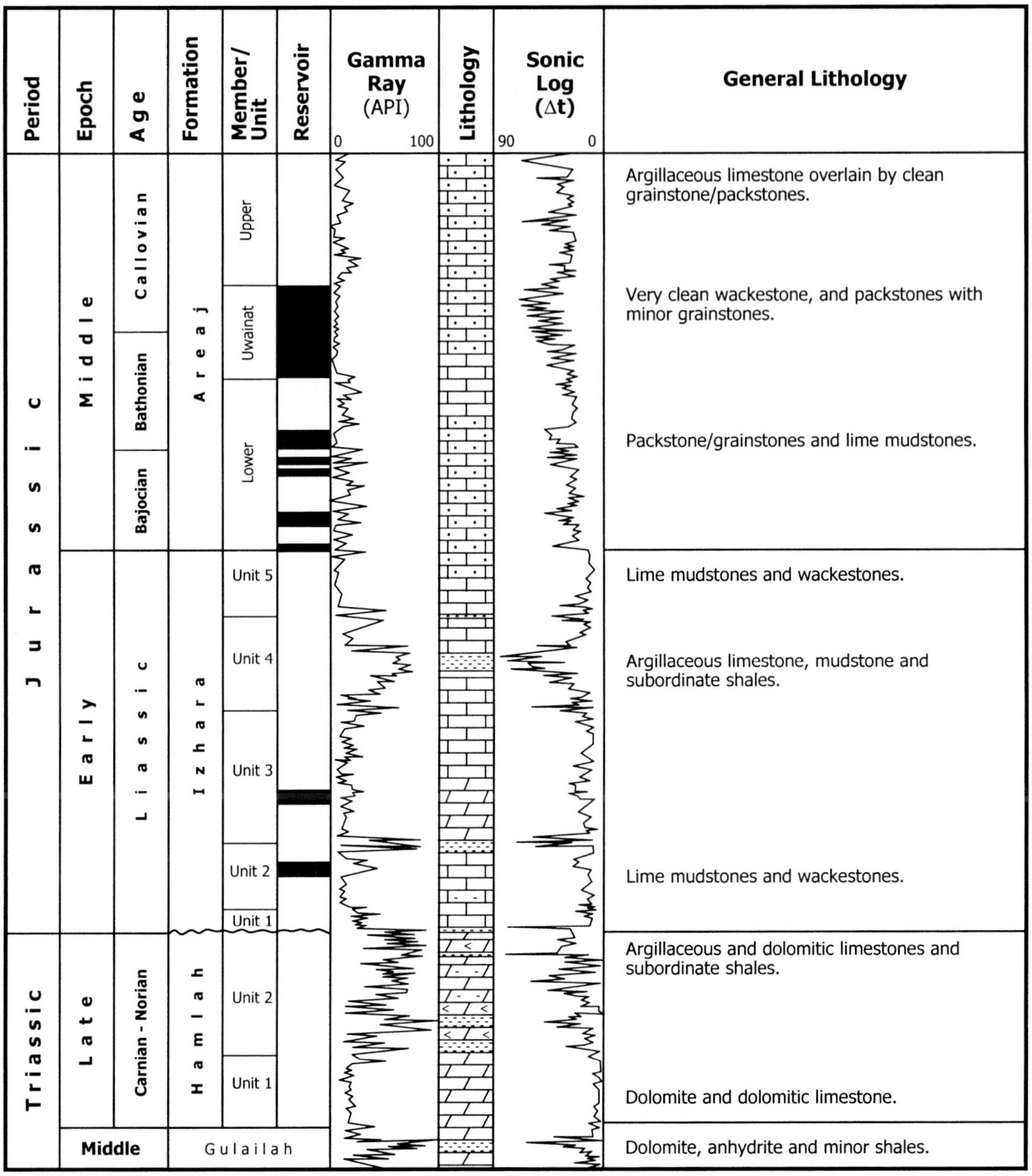


FIG. 4.—Triassic to Middle Jurassic subsurface lithostratigraphy in Abu Dhabi.

stones and marls. The middle unit consists of flaggy grainstones interbedded with marls, and the upper unit is of ferruginous quartz sandstones alternating with marls and calcareous shales. The formation was deposited in a continental to intertidal and shallow subtidal setting (Glennie et al., 1974; Ricateau and Riché, 1980; Alsharhan and Kendall, 1986).

## JURASSIC STRATIGRAPHY AND DEPOSITIONAL SETTING

### *Lower Jurassic*

At the end of Triassic time there was a regional relative drop in sea level caused by a eustatic lowering of sea level or uplift of the basement. In some areas this resulted in widespread progradation of fluvial and coastline clastics and in other areas, erosion or nondeposition (Hassan, 1989; Al-Husseini, 1997). In the subsurface of the U.A.E., the Triassic–Jurassic contact is an erosional sequence boundary marked by a regional erosional truncation (de Matos et al., 1994), but no sedimentary break is recognized in outcrop.

In the subsurface, the Toarcian Marrat Formation (Fig. 2) is found in onshore areas, but it is not present in offshore Abu Dhabi and Dubai. This may be due either to tectonic uplift preventing deposition, or to erosion of this formation. The Marrat is composed of mixed terrigenous clastics and carbonates and is divided into two units. The lower unit consists of interbedded bioclastic packstones/wackestones, intraclastic–peloidal and oolitic packstone/ grainstones, very fine dolomitic lime mudstones, and quartzose sandstones and silty claystones. The upper unit consists of three facies. The lower and upper facies contain mainly dolomitic, oolitic grainstones, micritic

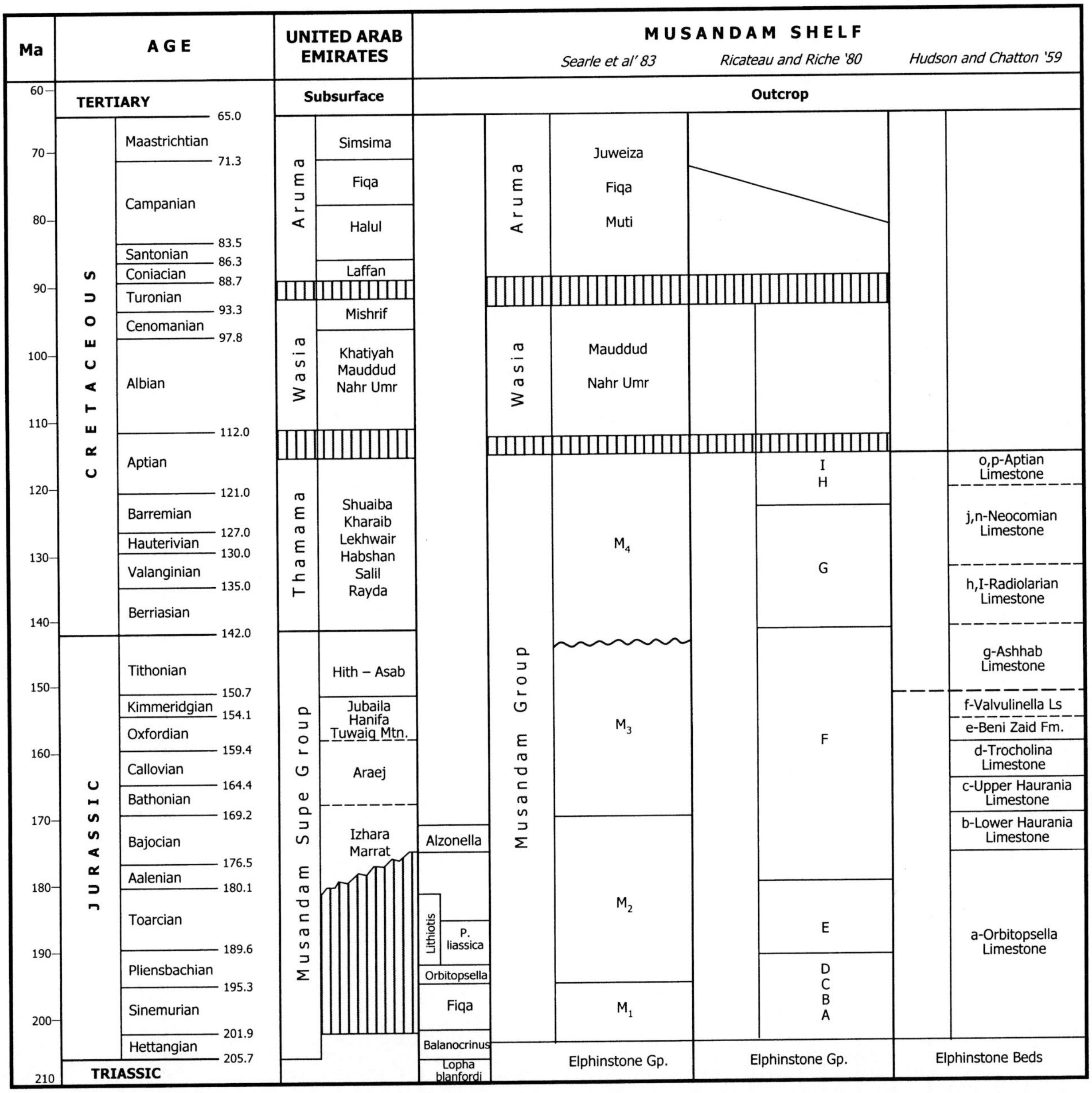


FIG. 5.—Stratigraphic classification of Jurassic–Lower Cretaceous strata in the Musandam Penninsula and subsurface, U.A.E. and Oman. The Cretaceous time scale is based on Obradovich (1993) except for the Albian–Cenomanian boundary as dated by Scott and Stein (1996). The Jurassic time scale is from Gradstein et al. (1995). The classification in the U.A.E. is from Alsharhan and Nairn (1986, 1988, 1990) and de Matos et al. (1994). The outcrop biozonation is from de Matos et al. (1994). Note that the foraminifera *Valvulinella* is now named *Kurnubia*.

sandstones, and sandy limestones. The middle facies comprises a complexly mixed terrigenous–carbonate succession of mudstones and dolomitic, quartzose sandstones. The Marrat Formation represents a regressive sedimentary succession deposited in a marginal marine setting. The lower unit represents an offshore facies and was deposited under conditions of variable energy. The upper unit was probably deposited on a flood plain and on tidal flats.

In the Musandam Mountains of the northern Emirates, the Jurassic to Early Cretaceous sedimentary rocks are called the Musandam Group (Figs. 2, 5, 6). The term "Musandam Limestone" was introduced by Lees in 1928. In 1959, Hudson and Chatton described the Musandam Limestone at its type locality in the wadis Hagil and Milaha. They divided it into a series of units, generally of formational rank and lettered A through P. In 1960, Hudson raised this stratigraphical succession to group status. Glennie et al. (1974) concluded that the Musandam Group generally shows a gradual change in facies from a predominantly deep-water, massive-bedded mudstone facies with subordinate grainstones in the north Wadi Sham and Jebel Hagab area, to a more shallow-water, predominantly skeletal packstone and grainstone facies in the southern mountains of the Peninsula, southeast of Al

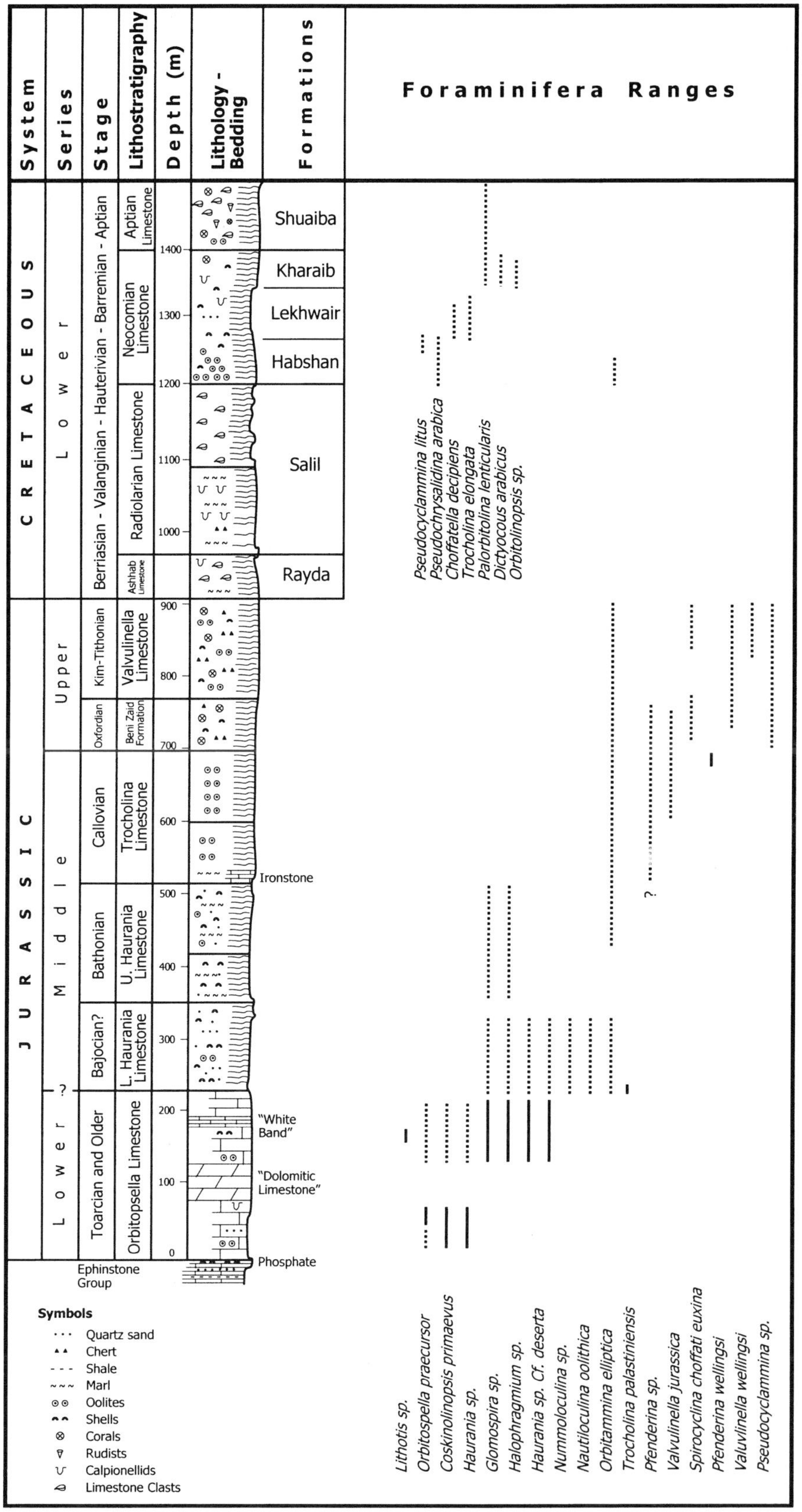


FIG. 6.—Stratigraphy of the Musandam Group, Jabal Hagab between Wadis Hagil and Milaha, Musandam Peninsula, northern Oman Mountains. Foraminifera ranges after Hudson and Chatton (1959).

Khatt village. Further, the Musandam Group was divided into four lithologic units (1–4) by Lippard et al. (1982) and Searle et al. (1983). For more detailed information about these units see Alsharhan and Nairn (1997).

The Jurassic section cropping out in the northern United Arab Emirates provides a detailed record of relative sea-level change, and yields vital clues to the biostratigraphy and sequence stratigraphy of the Jurassic (de Matos et al., 1994; Toland et al., 1993). The basal Liassic beds are locally condensed intervals with crinoids, abundant phosphatic particles, fish debris, and bivalves. However, no clear, visible break in sedimentation is evident between the Triassic and Jurassic in the carbonates of Wadi Naqab (Walkden and de Matos, this volume). If there was an interruption in sedimentation, the hiatus was certainly minor. In excellent outcrops at Wadi Naqab the Liassic carbonates begin with a transgressive, coarse cross-bedded ooid grainstone that is overlain by cyclic shallow carbonate sediments composed of multiple peritidal cycles. The base of each cycle is commonly bioturbated and rich in lituolids, algae, and oncoids, representing a lagoonal subtidal environment. The tops of the cycles generally show disrupted supratidal to intertidal lamination locally displaying birdseye structures. Cyclicity is on the meter scale and, in the Lower and Middle Liassic, cycle tops are commonly marked by paleo-exposure surfaces (paleokarst) with dissolution features, calcretes, dolomitization, and metoric cements, indicating sea-level oscillations (de Matos et al., 1994; Walkden and de Matos, this volume).

### *Middle Jurassic*

Following the development of a significant unconformity over an extensive area in the Middle Jurassic, the Aalenian hiatus, normal marine shelf-carbonate sedimentation was established in the Arabian Gulf (Figs. 2, 5); Al-Husseini, 1997). In western U.A.E., the Middle Jurassic is divided into the lower Izhara Formation and the upper Araej Formation (de Matos and Walkden, this volume) (Figs. 2, 4).

The Izhara Formation in Abu Dhabi is composed largely of argillaceous lime mudstones with some bioclastic, peloidal packstones/grainstones and partly dolomitized and silty quartzose, bioclastic lime mudstones. The formation was deposited in a quiet-water shelf setting, as suggested by the argillaceous lime mudstones. In eastern offshore Abu Dhabi and Dubai the Izhara Formation consists of gray to gray-brown dolomite with anhydrite streaks and inclusions of some pyrite with minor gray to gray-green shales (Fig. 4). It grades upward to dolomitic packstones/wackestones with gray to black shale. Minor high-energy fluctuations in the depositional regime occurred, as evidenced by the grain-supported sediments.

The Araej Formation in Abu Dhabi and Dubai is divided into three members (Fig. 4). The Lower Araej Member comprises bioclastic–intraclastic and peloidal lime mudstones to packstones and minor grainstones, and contains significant amounts of clay and pyrite. The Uwainat Member consists of four different facies: peloidal packstones grading to mudstones; intraclastic wackestones and packstones; mottled bioclastic lime mudstones with traces of carbonaceous material grading into bioclastic–intraclastic packstones; and bioclastic, peloidal packstones and grainstones. The Upper Araej Member consists of dense and argillaceous lime mudstones in the lower part and by bioclastic–peloidal packstones and oolitic grainstones containing sparse bivalve fragments, foraminifera, and echinoderm and molluscan debris in the upper part (Alsharhan and Whittle, 1995).

The Araej Formation represents the continuation of a quiet-water, marine-shelf setting. Intraclasts and peloidal grains suggest higher-energy conditions, whereas the sections corresponding to muddy sediments suggest a slightly deeper, lower-energy setting.

In the northern Emirates at Wadi Hagil the Middle Jurassic Bathonian–Callovian–Bajocian outcrops consist of mudstones, grainstones, and packstones containing bioclasts, mainly algae, gastropods, echinoderms, and bivalves, with intraclasts of ooliths, pellets, and oncoliths (Figs. 5, 6) (Walkden and de Matos, this volume). This succession was deposited in a shallow-water lagoonal to subtidal environment, which extended westward into the Sinai (Holail, this volume).

### *Upper Jurassic*

In western Abu Dhabi, the Upper Jurassic sequence consists of three formations (in ascending order): Diyab, Arab, and Hith (Figs. 2, 7). The Diyab Formation is a thick succession of relatively deep-water, argillaceous lime mudstones and wackestones (Al-Suwaidi et al., this volume). These sediments grade eastward, into relatively clean, shallow-water limestones and dolomites. The Arab Formation reflects a period of marine regression with a shallowing-upward trend. The formation is subdivided into four regressive cycles, referred to as the Arab A to D Members (Fig. 7). The Arab D, the lowermost member, consists of shoal grainstones and lagoonal wackestones/packstones. The Arab A to C Members are characterized by cyclic carbonates and evaporites that were deposited in a peritidal setting over a relatively wide sabkha flat. In onshore western Abu Dhabi, the Arab A to C Members are known as the Qatar Formation, which consists of dense dolomudstones and anhydrites interbedded with dolomitized shallow subtidal to intertidal carbonates. The Hith Anhydrite represents the final regressive stage of the major depositional cycle (Alsharhan, 1989; Alsharhan and Nairn, 1997).

In central and eastern Abu Dhabi, where both the intrashelf basinal limestones and the capping Hith Anhydrite are absent, the Upper Jurassic sequence consists of three formations: the Dukhan, the Qatar, and the Asab (Fig. 2). Toward eastern Abu Dhabi and Dubai, the Upper Jurassic sequence comprises the Fateh and Asab Formations. The Fateh Formation consists of dolomitic packstones and dolomites and is a lateral equivalent to the Dukhan Formation in Abu Dhabi. The Asab Formation is the lateral equivalent of both the Hith and the Arab Formations and consists of lime mudstones and dolomites with traces of anhydrite in the lower part, followed by clean oolitic packstones/grainstones (Ayoub and Ennadi, this volume). It becomes progressively less dolomitic and more limy eastward across Dubai into the northern U.A.E. offshore area (Alsharhan, 1989). Another Arabian example of dolomitic Jurassic carbonates is in Yemen (Al-Thour, this volume).

The Upper Jurassic Arab carbonate platform and/or shelf reservoirs are commonly sealed by supratidal Arab and Hith evaporites. The evaporites of the Hith Formation (Tithonian) represent the final regressive, supratidal stage of the last major Jurassic depositional cycle in the Arabian Gulf region. At the base of the formation, transgressive units occur as local minor lagoonal and intertidal thin, porous carbonate horizons (Alsharhan and Kendall, 1994).

The massive Hith Formation in the subsurface of central and eastern Saudi Arabia, with its nodular and laminated character, gradually thins eastward into Qatar and Abu Dhabi. The eastern margin of this massive anhydritic development trends north–south across central Abu Dhabi. This lateral transition is marked by a change from anhydrites to supratidal/lagoonal and intertidal dolomites. Above the base of the Hith, the anhydrite is progressively replaced by dolomite. Some of the dolomite in the eastern Emirates is probably the product of later diagenesis associated with brines migrating upwards from lower in the Jurassic section in the same way as the Hith dolomites in Saudi Arabia described by Wilson (1985) and Broomhall and Allan (1985).

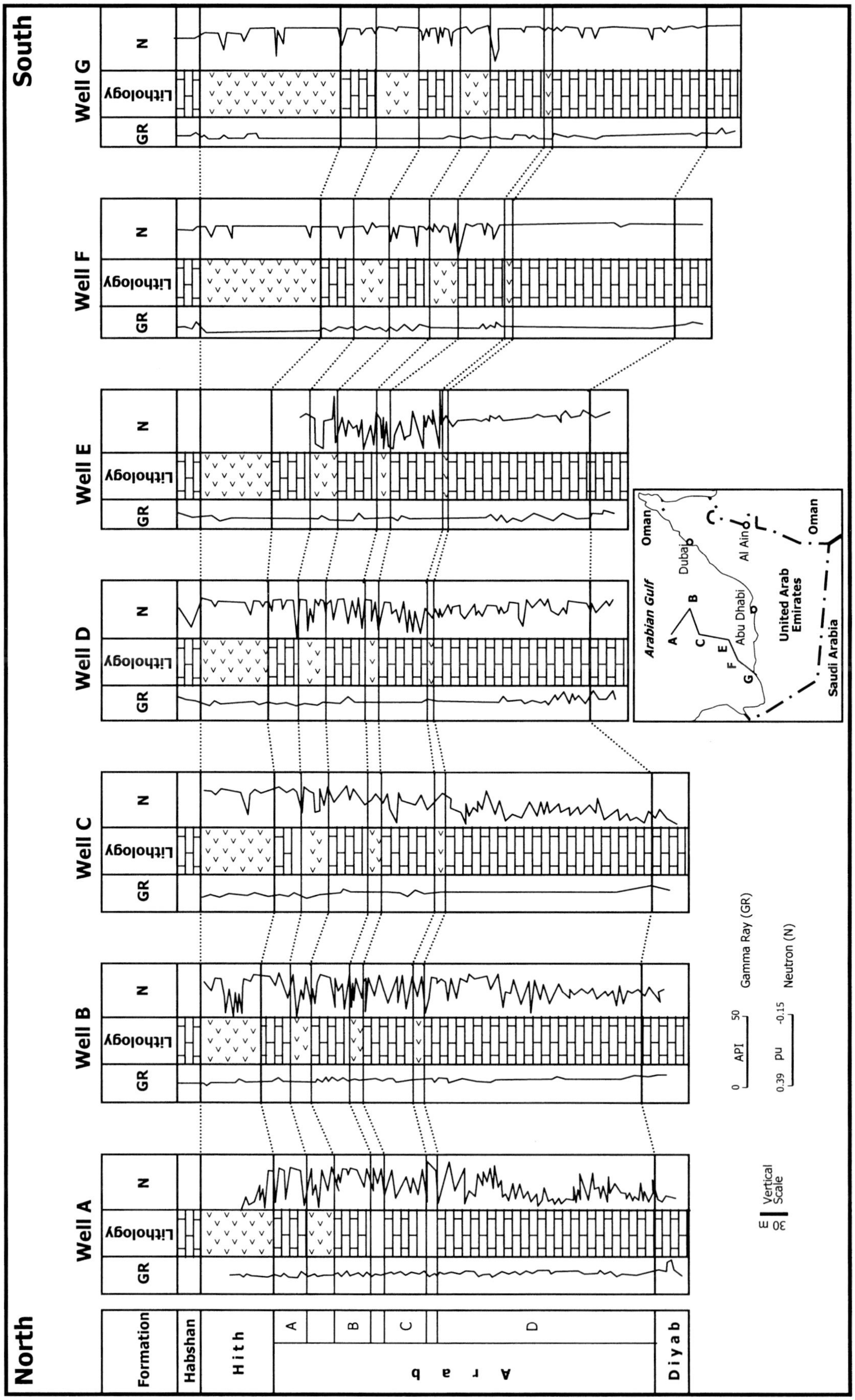


FIG. 7.—Upper Jurassic Arab Formation lithostratigraphic correlation in Abu Dhabi (modified from Al-Silwadi et al., 1996; Alsharhan and Nairn, 1997).

In west-central Abu Dhabi, the Hith Formation consists of massive anhydrites and thinner dolomites. From core observations, the anhydrite here is composed of the chicken-wire variety, and the interbedded dolomites are light to dark brown, and sucrosic. The lower beds are formed by light brown to dark brown peloidal packstones and wackestones instead of dolomites. Towards central Abu Dhabi, where it pinches out into carbonate, the Hith Formation appears to have accumulated locally in a sabkha setting. This is evidenced by the interbedded relationship of the Hith anhydrites with carbonates and the local predominance of horizontally flattened nodules and enterolithic layers of the anhydrite. These sequences match some of the characteristic fabrics found in the Holocene coastal sabkhas of the United Arab Emirates. In Abu Dhabi, the Hith consists of gray to dark brown dolomite, which is sucrosic and medium to coarse-grained and contains minor oolitic grainstones. Anhydrite is dispersed throughout the section, but in very minor quantities. It forms a clear, transparent cement in the dolomites. The Hith Formation thins farther to the east in Abu Dhabi, and to the southeast, toward Oman, dolomite is replaced by limestone (Alsharhan, 1989). The Asab oolite and carbonate grainstones are lateral equivalents of the Arab and Hith formations in eastern Abu Dhabi. De Matos (1994) shows that the Asab Formation oolite is stratigraphically equivalent to the Hith Formation and the Arab-A Member. Al-Silwadi et al. (1996) concluded that correlations are complicated by possible erosion and/or nondeposition of uppermost Jurassic sediments in eastern Abu Dhabi. Therefore, there is no clear evidence of an unconformity beneath the overlying Habshan Formation, and the changes within the Hith appear to be due entirely to lateral facies variations.

In the outcrops of the northern U.A.E., the Upper Jurassic is now known to be no younger than Oxfordian (Toland et al., 1993). The Oxfordian succession at Wadi Hagil is equivalent to the Lower Musandam Limestone unit F of Hudson and Chatton (1959) (Fig. 5) and Alsharhan and Nairn (1997). It is made up of three distinct members. The lower member comprises regressive, thin-bedded, fine to very fine sand, and cortoid packstones, characterized by hummocky cross-stratification, wave-ripple laminae, and local intraclast lag deposits. This is interpreted as a storm-dominated, offshore succession deposited above storm wave base. The middle member comprises regressive, thin- to medium-bedded peloidal packstones and grainstones with commonly partly silicified branching stromatoporoids and dasycladacean algae. The upper member comprises resistant, thick-bedded, peloidal packstones and grainstones with local coral–stromatoporoid floatstone units.

## CRETACEOUS STRATIGRAPHY AND DEPOSITIONAL SETTING

The Cretaceous sediments of the U.A.E. consist mainly of epeiric shelf carbonates and associated minor clastics (Figs. 2, 5, 6). They accumulated during major cycles of transgression and regression, driven by gentle tectonic subsidence punctuated by eustatic sea-level variations (Scott, 1990; Kendall et al., 1991a).

### *Lower Cretaceous*

Overlying the Jurassic Hith Formation of the U.A.E. is the Lower Cretaceous Thamama Group, which consists of the Habshan, Lekhwair, Kharaib, and Shuaiba formations (Alsharhan and Nairn, 1986) (Fig. 8). In the U.A.E. the entire Thamama Group is composed of cycles of porous and nonporous carbonates. The difference in depositional water depth between the Thamama high-energy packstone/grainstone and quiet-water microporous or argillaceous limestones was probably a few tens of meters.

The Habshan Formation is composed mainly of lagoonal carbonates interfingering with minor evaporites. It exhibits one major depositional cycle punctuated by four minor shallowing cycles (Alsharhan, 1989, Alsharhan and Nairn, 1997). The effects of eustatic change are suppressed by tectonic extension and subsidence (Kendall et al., 1991a). The Habshan Formation is generally composed of dense shelf carbonate mudstones and tight grainstones. Only at the Zakum and Bab fields are there porous grainstones in the Habshan reservoir units. Into eastern U.A.E. and in the southern Musandam Peninsula the Habshan grades laterally into and overlies the Rayda and Salil formations (Fig. 6) (Connally and Scott, 1985; Scott, 1990; Hulstrand and Mueller, 1997). Alsharhan and Kendall (1991) and Kendall et al. (1991a) divided this progradational pattern into three cycles of relative sea-level fall, possibly induced by initial collisional of the Arabian and Asian plates. Habshan-equivalent strata crop out in Saudi Arabia as nearshore facies (Shebl and Alsharhan, 1994 and this volume). In the subsurface of Kuwait the equivalent Minagish oolite is an important reservoir (Davies et al., this volume).

The overlying Lekhwair Formation contains five shallowing cycles. The grainstone parts of the cycles commonly form excellent reservoirs sealed between interbedded dense carbonate mudstones. This limestone is interpreted to have been deposited during a gentle sea-level rise with five superimposed low-amplitude, high-frequency sea-level variations (Alsharhan and Kendall, 1991; Kendall et al., 1991a). The Lekhwair is characterized by shale and minor siltstones and extends as far east as central Abu Dhabi. It is probably equivalent to the uppermost extent of the mixed carbonates and clastics of the Buwaib Formation of eastern Saudi Arabia. The Lekhwair is an important reservoir in eastern Abu Dhabi, and many structures in the area contain oil and gas in one or several of its porous members (Alsharhan, 1989).

The overlying Kharaib Formation accumulated under conditions similar to those of the Lekhwair Formation, but it contains more extensive porosity in its shallowing-upward cycles (Alsharhan, 1989; Saotome et al., 1997). The Kharaib Formation is composed of four shallowing-upward carbonate cycles, the grainier parts of which contain depositional and leached-grain porosity (Saotome et al., this volume). These are interpreted to represent deposition during two high-frequency eustatic sea-level changes superimposed on an overall rise (see also Alsharhan and Kendall, 1991; Kendall et al., 1991a).

The Shuaiba Formation is interpreted to represent deposition during one major eustatic rise (Alsharhan and Kendall, 1991; Kendall et al., 1991a; Kendall et al., this volume). It formed as a prograding carbonate shelf around the interior-shelf Bab Basin in central onshore–offshore Abu Dhabi and at the outer shelf margin in the eastern Emirates, and was dominated by rudists (Skelton and Masse, this volume) and calcareous algae. It is most porous in its upper part, where the original grainstones have been leached. In the Bab Basin the Bab Member, a dense, deep shelf to basinal limestone, is overlain by and may be laterally equivalent to the Shuaiba buildup. The Shuaiba is clearly delimited from the dense limestones of the underlying Kharaib Formation and the Nahr Umr Formation above (Fig. 8) (Al-Awar and Humphrey, 1997, this volume; Alsharhan et al., 1997). In the southern part of the basin in Oman the Lower Aptian lower Shuaiba is a highstand aggrading sequence and the Lower–Upper Aptian upper Shuaiba is a lowstand prograding sequence of shallow to deep shelf facies (Witt and Gökdag 1994). Carbon-isotope stratigraphy suggests up to four sea-level changes during deposition of the Shuaiba Formation and its Bab Member (Wagner, 1990; Vahrenkamp 1996; Grötsch et al., 1998; Alsharhan et al., this volume). These events span the entire Aptian. The greatest production of hydrocarbons in the Thamama occurs from leached rudist grainstones associated with rudist buildups in the Aptian Shuaiba Formation.

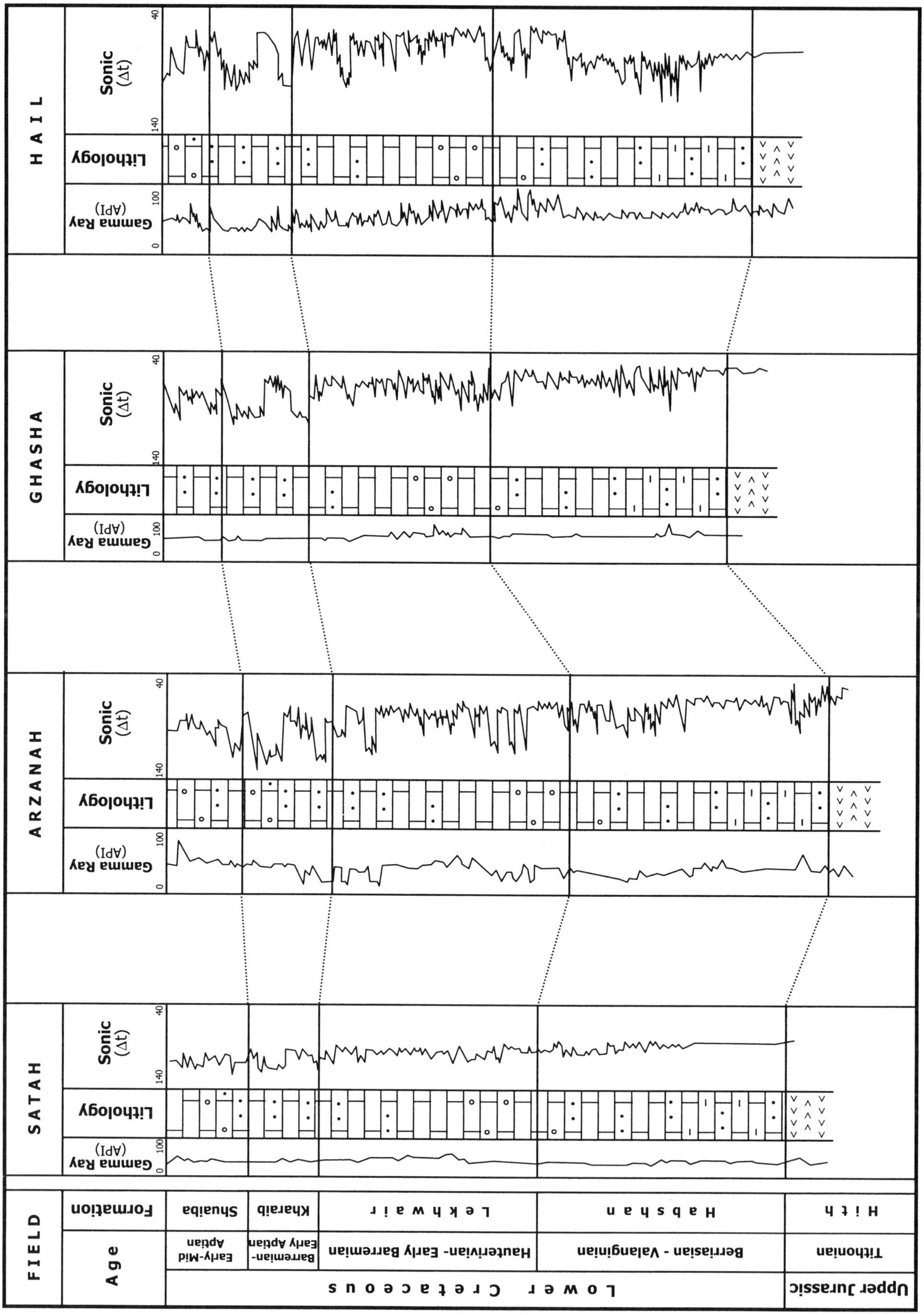


FIG. 8.—Lithostratigraphy and log characteristics of Lower Cretaceous strata in offshore Abu Dhabi.

The occurrence of rudist banks with their off-reef shell debris has produced some of the most prolific fields (as in the Bu Hasa Field of Abu Dhabi). But in many other areas (as in the Sajaa Field of Sharjah, the Idd El Shargi Field of Qatar, and the Yibal and Ghaba North Fields of Oman) the Shuaiba has a microporous fabric, so that reservoir productivity in the unit is highly variable (Moshier et al., 1988; Alsharhan, 1995; Russell and Walkden, 1997). In some fields, stylolite zones divide the Shuaiba into separate reservoirs (Alsharhan and Sadd, this volume).

The basal Thamama Group in the outcrop of northern U.A.E. at Wadi Hagil is the Upper Musandam Limestone unit G (lower part) of Hudson and Chatton (1959) (Fig. 5) and is Berriasian to Valanginian (Toland et al., 1993). There it contains two conglomerate units. The base of the lower conglomerate is a planar erosion surface locally overlain by unstratified, clast-supported conglomerates with well rounded cobbles and small boulders. It is overlain by a thick succession of pelagic lime mudstone with calpionellids and crinoid fragments. The calpionellids and the $^{87}Sr/^{86}Sr$ ratio suggest Early to Middle Berriasian (Toland et al., 1993). The planar bounding surface at the base of the unit marks a significant erosional hiatus spanning the ?Kimmeridgian–earliest Berriasian and defines an abrupt change in depositional environment from shallow platform to platform slope. The basal conglomerate is interpreted to be a transgressive lag deposit, and the lime mudstone interval may represent highstand progradation of a carbonate foreslope into a starved deep-water basin. The base is a prominent downlap surface.

The upper conglomerate fills a channel incised into the lower interval as deep as the Jurassic. In the absence of age-diagnostic fossils Toland et al. (1993) used the $^{87}Sr/^{86}Sr$ ratio in the overlying lime mudstone to suggest a latest Berriasian to Early Valanginian age. This unit is part of the Upper Musandam Limestone unit G (in part) and unit H (in part) of Hudson and Chatton (1959). The conglomerate is characterized by unstratified, clast-supported conglomerates with clasts of pelagic lime mudstone, reworked conglomerate, peloidal grainstone, and chert, among others. These are overlain by pelagic lime mudstone and peloidal grainstone. This boulder bed is interpreted to be a submarine canyon fill. The top of the boulder bed is the onlap surface of the overlying radiolarian lime mudstones, which represent a transgressive system tract. Toland et al. (1993) correlated this channel base with the base of sequence 2-1, cycle LZB-2, at 128.5 Ma (Haq et al., 1988).

### *Middle Cretaceous*

The middle Cretaceous Wasia Group of the Arabian Gulf is bounded below and above by regional unconformities (Figs. 2, 5) (Harris et al., 1984; Alsharhan and Nairn, 1988; Scott, 1990). The Wasia Group appears to drape structures formed by the underlying Thamama Group and was affected by several sea-level variations (Scott, 1990; Alsharhan and Kendall, 1991; Kendall et al., 1991a; Philip et al., 1995; Van Buchem et al., 1996; Immenhauser et al., 1999, this volume; Scott et al., this volume). In Oman the age-equivalent Natih Formation consists of seven members, A–G, that record sea-level events (Philip et al., 1995; Van Buchem et al., 1996). A major sea-level fall occurred in the mid-Cenomanian. This fall overprinted the structural events associated with the onset of the deformation of the Oman Mountains and formed the major unconformity between the E and D members of the Natih, which can be recognized over much of eastern Arabia (Harris et al., 1984; Scott et al., 1988; Alsharhan and Nairn, 1988; Scott, 1990). The top of the Wasia Group is a regional unconformity across the U.A.E. and in Oman. The age of the uppermost strata varies from upper Cenomanian in the U.A.E. to the lowermost Turonian in Oman because the extent of erosion was deeper in the Emirates. The middle Cretaceous stratigraphy of the U.A.E. has been complicated by a history of nomenclature that is still evolving, and also by the fact that the names of formations and lithofacies units have been used interchangeably. In ascending order, the Wasia Group is divided into four formations (Fig. 9) (Alsharhan and Nairn, 1988): The Nahr Umr, Mauddud, Shilaif (or Khatiyah in Dubai), and Mishrif formations (and the age-equivalent Tuwayil and Ruwaydah members).

The shale of the Nahr Umr Formation overlies the Shuaiba Formation and forms the major seal over the Lower Cretaceous reservoirs. The basal Nahr Umr accumulated during a major eustatic sea-level rise beginning in latest Aptian to earliest Albian, which seems to correlate in time with either sequence LZ8-4, 4.2 or UZ Al, 1.1 (Haq et al., 1988) (see also Scott, 1990; Immenhauser et al., 1999; Immenhauser et al., this volume).

Overlying the Nahr Umr Formation are the Mauddud, Shilaif/Khatiyah, and Mishrif formations (Burchette and Britton, 1985; Loutfi et al., 1987, Alsharhan and Nairn, 1988). These formations are separated from each other on the basis of vertical position and lithologies (Fig. 9). This has led to some confusion in the description of the formations and in the understanding of their stratigraphy.

The Mauddud Formation is a dense, micritic shelf limestone that is interpreted to represent a late Albian stillstand. The overlying Shilaif to Mishrif and their equivalent formations contain evidence of six small rises in sea level (Alsharhan and Kendall, 1991; Kendall et al., 1991a). The Shilaif (Khatiyah) Formation is a well-laminated, locally bituminous carbonaceous marl that locally is a source rock. It overlies the Mauddud Formation and locally is laterally equivalent to the Mishrif Formation. The Shilaif Formation is interpreted to have accumulated in a euxinic depression where conditions favored preservation of abundant organic material.

The Mishrif Formation, which forms the progradational rim to the Shilaif Basin, contains most of the hydrocarbons in Middle Cretaceous rocks of the U.A.E. and has the best potential reservoirs. The Mishrif Formation is a bioclastic carbonate that is locally dominated by radiolitid rudists. These reservoirs are sealed by the overlying Coniacian Laffan Formation. The discontinuous distribution of this seal makes it difficult to predict the location of traps in the Mishrif Formation. Where the Wasia Group contains reservoir rocks and is overlain by shale of the Laffan Formation, however, fields may be quite large (Fateh and Umm al Dalkh Fields).

The lower part, however, of the shelfal packstones and wackestones of the Mishrif Formation contain rudist buildups and are equivalent in age to the adjacent Shilaif/Khatiyah basinal, bituminous, pelagic lime mudstones. Much of the upper part of the Mishrif Formation, particularly to the east, has been eroded (as in some parts of the Fateh Field in Dubai). There, the Mauddud, Shilaif, and Mishrif formations are Upper Albian–Cenomanian. Westwards, however, the upper part of the Mishrif (and its basinal equivalent, the Ruwaydah Member, which overlies the Shilaif Formation and the Tuwayil Member) is Upper Cenomanian (Alsharhan and Nairn, 1988). Thus, the shallow-water carbonate Mishrif and Mauddud formations range from Upper Albian to Upper Cenomanian. The Shilaif Formation is Albian to Lower Cenomanian, and the capping basinal facies, the Ruwaydah Member, is Upper Cenomanian (Alsharhan and Nairn, 1988). In Dubai, Pascoe et al. (1995) demonstrated that the bioclastic shoal facies are initially aggradational in local areas in response to an overall rise in sea level in the latest Albian–Early Cenomanian whereas in the Middle to Late Cenomanian they became progressively more progradational into eastern Abu Dhabi. They believe

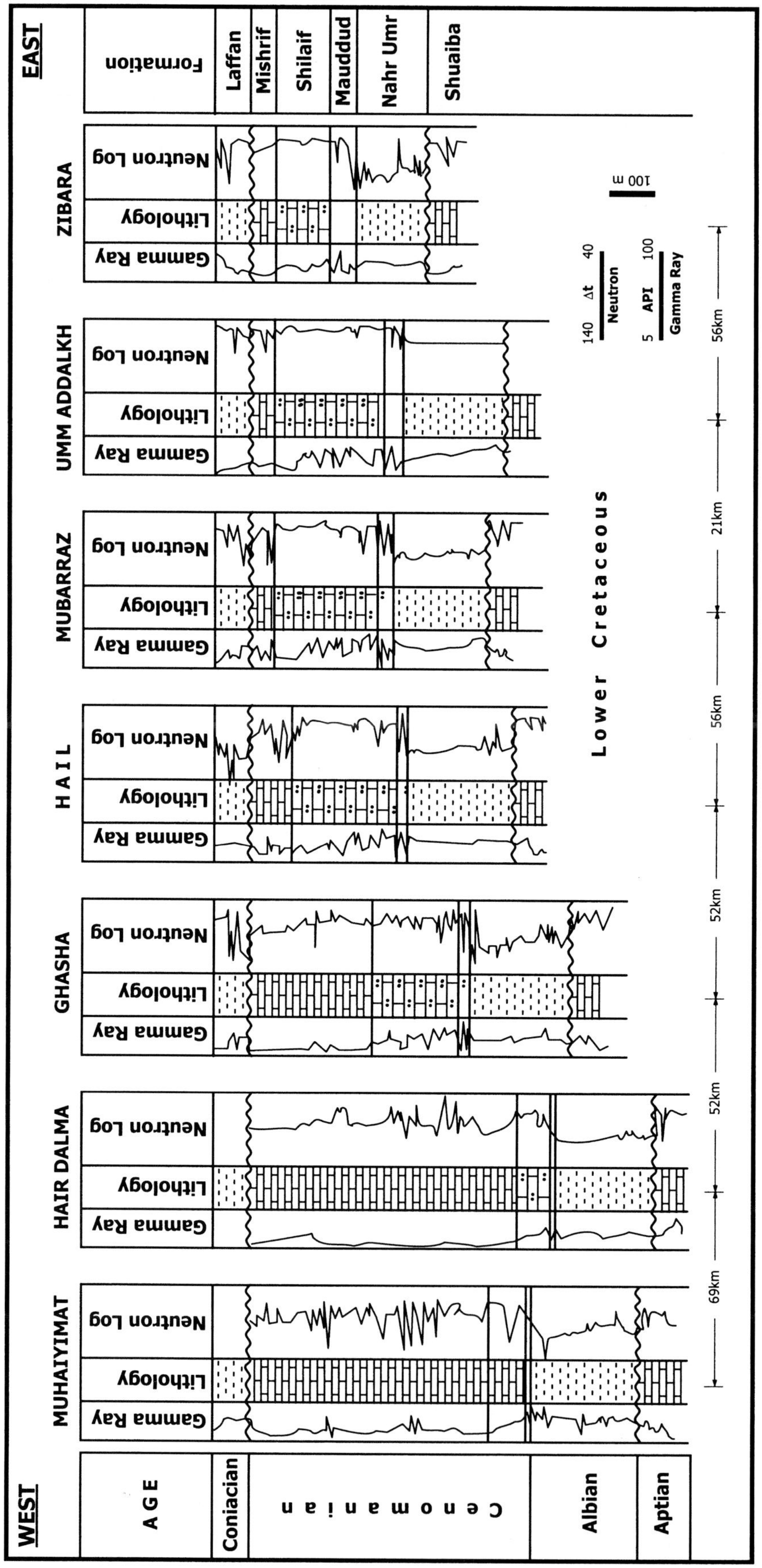


FIG. 9.—Lithostratigraphy and log characteristics of middle Cretaceous strata in offshore Abu Dhabi.

that the latest Mishrif to early Laffan facies are more heterogeneous and are inferred to include equivalents to the Tuwayil and Ruwaydha members of the Mishrif Formation in Abu Dhabi.

### *Upper Cretaceous*

Following the deposition of the Mishrif shelf sediments in the Middle Cretaceous, the region was uplifted and eroded during a coincident Turonian relative fall in sea level. The Aruma foredeep formed while a westward-migrating tectonic high passed through the northern U.A.E. and was exposed as a similarly migrating shelf. These events were the combined result of the subduction at the Arabian continental margin in mid-Cretaceous time and the associated sediment loading of the continental lithosphere (Patton and O'Connor, 1988; Scott, 1990). As this process continued, allochthonous terranes from the Oman area advanced farther onto the continental margin. This caused the area that was upwarped in the Turonian to become part of the floor of the Aruma basin.

Following a period of major erosion and emergence in the Turonian, which lasted as long as 5 My in the northern U.A.E., the Upper Cretaceous Aruma Group (Figs. 2, 5) was deposited during a time span of some 23 My. The Aruma Group sediments have the most complex lithofacies patterns and thickness changes of any of the major Cretaceous rock units in the U.A.E. (Alsharhan and Nairn, 1990). These lateral and vertical changes are the result of the complex interplay of regional subsidence, pronounced eustatic change in sea level, and differential sedimentation rates across the region. The Upper Cretaceous succession can be divided into eight stratigraphic units: the Laffan, Halul, Ilam, Muti, Fiqa (Lower Member, Upper Member), Juweiza, Qahlah, and Simsima formations. The Muti and the Qahlah are known only in outcrop, whereas the Juweiza was reported by Glennie et al. (1974) from the subsurface Juweiza-1 well in Sharjah drilled by Iraq Petroleum Company (IPC) in 1957/1958 at depth of 44–3938 m (145–12,916 ft) with total thickness of at least 2920 m (if interpretation of the well data was correct).

In the northern Emirates part of the Oman Foredeep, sediments of the Upper Cretaceous Aruma Group are deep-water shales and marls, which exceed 3000 m in thickness. On the adjacent shelf, the equivalent section is only 100–150 m thick. Close to the mountain front, the sediments include flysch and transported carbonate blocks eroded from advancing thrust sheets associated with emplacement of the Semail Ophiolite sequence.

The Laffan Formation (Coniacian), which is composed mainly of slightly calcareous shales, lies at the base of the Aruma Group (Fig. 10). As was indicated previously, this shale can locally form a major seal over the Mishrif Formation. It interfingers with and is overlain by interbedded calcareous shales and argillaceous limestone of the Halul Formation. This in turn is overlain by the Campanian Fiqa Formation, which is composed of calcareous shales, marls, and argillaceous limestone with a deep marine fauna. It is followed by the Maastrichtian Simsima Formation, which is composed of fossiliferous and shaly limestone and peloidal–bioclastic packstone and wackestone and dolomitic limestone with coral and rudist buildups (Alsharhan et al., this

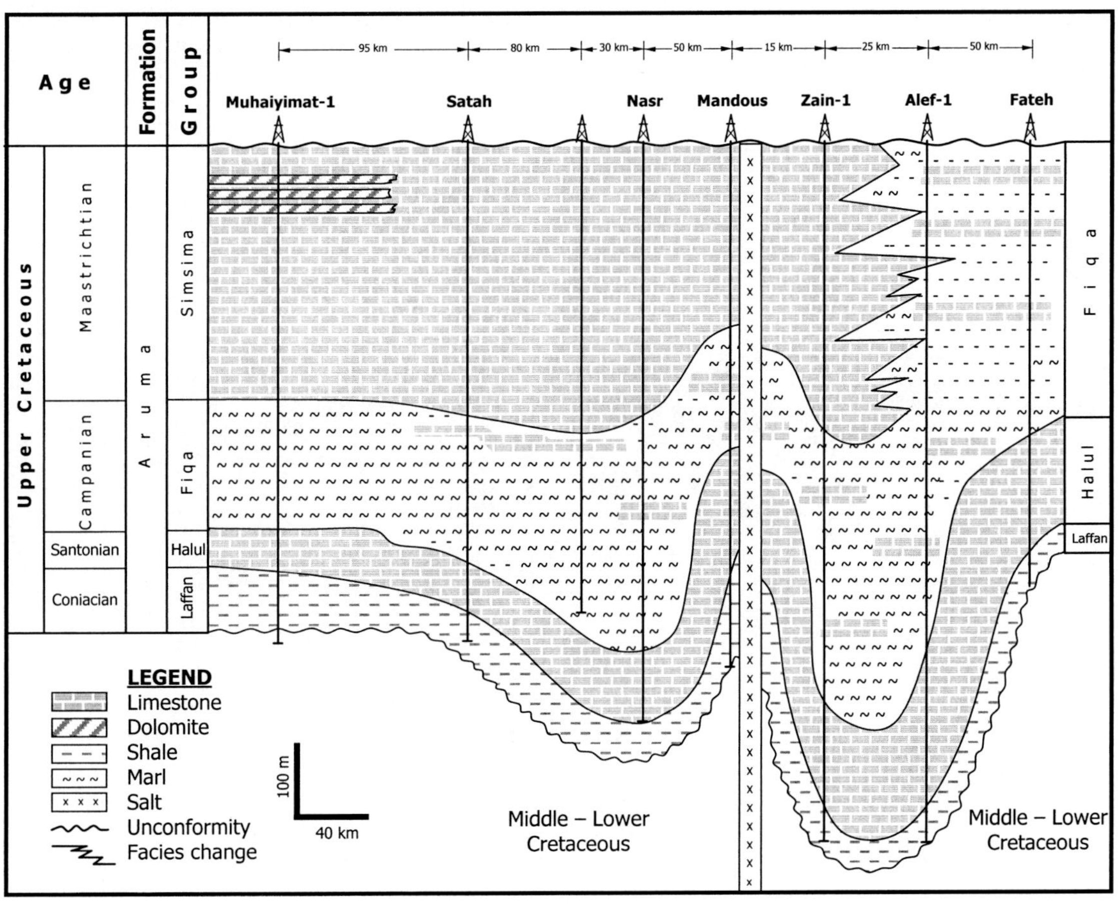


FIG. 10.—Lithostratigraphy of Upper Cretaceous strata in offshore Abu Dhabi–Dubai area.

volume; Schumann, this volume). The Simsima flanks the Oman foredeep (Schlumberger, 1981; Patton and O'Connor 1988, Alsharhan and Nairn, 1990; Smith et al., 1995). In the northern U.A.E., the Aruma Group was deposited in the foreland basin formed in western Oman when west-directed overthrusting occurred in the Oman Mountains. Sediments were then shed from this area farther westward into the Oman trough and across it to the marginal Arabian shelf areas of the U.A.E.

The Upper Cretaceous Aruma Group was apparently affected by epeirogenic movements punctuated by sea-level changes (Scott, 1990). Late Cretaceous deposition began with the Coniacian Laffan shale, a prodeltaic sedimentary rock probably related to either the 89 or 88 Ma sea-level rise of Haq et al. (1988). However, the tectonic overprint of the nappe emplacement of the Oman Mountains masks the sea-level events (Alsharhan and Kendall, 1991, Kendall et al., 1991a) and makes this interpretation difficult. The Halul Formation (or Ilam in Dubai) limestones and marls were deposited in the Santonian to early Campanian and are related to two sea-level rises during this time (see the Haq et al., 1988). As with the rest of the Upper Cretaceous Fiqa, Simsima, and equivalent formations, which were deposited during the Campanian and Maastrichtian, it is much more difficult to recognize the tie to worldwide sea-level events (Alsharhan, 1997).

## POTENTIAL FOR PETROLEUM EXPLORATION

Major hydrocarbon fields are widespread in the U.A.E. (Fig. 11, Table 2). Jurassic oil and gas reservoirs occur in western Abu Dhabi, particularly in the offshore. Lower Cretaceous rocks are productive in central Abu Dhabi from the Zakum Field south through Mubarras, Bab, to the Bu Hasa, Asab, and Sahil fields. The general interpretation for the occurrence of oil in the Jurassic rocks of western Abu Dhabi is that it is related to the presence of the Hith Formation, which traps the hydrocarbons in the underlying Jurassic rocks. In the east, where the anhydrite of the Hith changes facies to dolomites and dolomitic limestones, hydrocarbons have been able to migrate upward into Lower Cretaceous rocks. The presence of the underlying Nahr Umr Formation shale probably prevented much movement of older oil from the Lower Cretaceous reservoirs up through the section. Middle Cretaceous rocks produce from a zone extending south from Umm Al Dalkh Field through the Fateh Field to the Saleh Field in the northern Emirates. Upper Cretaceous rocks of the U.A.E. produce hydrocarbons only at the Shah field in onshore Abu Dhabi, with minor oil and gas condensate occurring in some exploration wells and fields in the northern U.A.E. The oil and gas found in the area are trapped in the structural and/or stratigraphic traps (Fig. 12).

Upper Jurassic evaporitic sulfates and minor chloride strata of Arabia were deposited within a complex of giant salinas and sabkhas (more than $72 \times 10^6$ km$^2$) along the southern margin of the Tethys Ocean (Leeder and Zeidan, 1977; Alsharhan and Nairn, 1997). These evaporites now form an excellent seal to some of the world's most prolific oil reservoirs. The present U.A.E. coastal environments appear to be a reasonable depositional model for some of the evaporite settings of the Hith Formation, but they are lined by a series of vast algal flats (Kendall and Skipwith, 1968; Birdwell et al., 1991). For the Holocene, this association of high concentrations of organic matter interbedded with carbonates and evaporites is unique, but it is also believed to have been common in the geological past. Indeed such associations may represent potential source rocks for ancient carbonate petroleum reservoirs.

Any interpretation of the Upper Jurassic evaporites of the Arabian Gulf region is influenced by the existence of a similar suite of supratidal evaporites and intertidal carbonates in the Holocene tidal flats of the U.A.E. These latter sediments of the

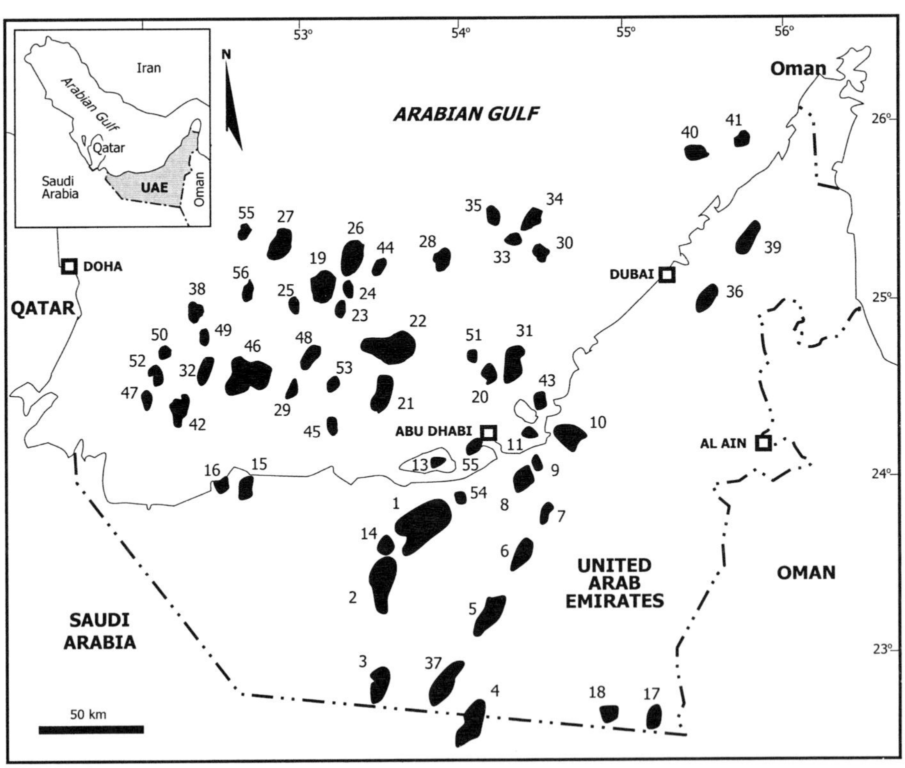


FIG. 11.—Distribution of the oil and gas fields in U.A.E. The names of the fields are shown in Table 2.

Abu Dhabi coastal area are dominated by shallow-water carbonates and associated evaporite sediments (Kirkham, 1997). These include bioclastic grapestone and oolite sands, pelleted aragonitic muds, algal stromatolites, and intertidal and supratidal evaporites (gypsum, anhydrite, halite, and protodolomite). The anhydrite exhibits various structures, including ptygmatic folds, the upturned margins of polygonal saucers, disharmonic folds, and diapiric-like features (Butler et al., 1982).

Shoaling-upward successions of carbonates associated with evaporites are extremely common as hydrocarbon traps. The association of Holocene evaporites, carbonates, and algal-laminated units in Abu Dhabi is an excellent model of this reservoir type. Indeed, Kenig et al. (1990), Birdwell et al. (1991), Kendall et al. (1991b), and Kendall et al. (1994) demonstrate from their studies in Abu Dhabi that this association has some bearing on the occurrence of petroleum in similar sequences in the Middle East and the western United States and Canada. Cycles of ancient sabkha facies have been distinguished in the Jurassic Arab Formation by Wood and Wolf (1969) from Abu Dhabi and the Hith Formation of the eastern Arabian Peninsula. This suggests that other thick repetitions of dolomite and anhydrite in the geological record can also be interpreted in terms of arid, coastal-plain, accretionary processes, forming not only the seal but also the reservoir and the source. In the United States, similar carbonate traps are associated with major oil fields of the Central Basin platform and on the Northwest shelf of the Permian Basin of Texas and New Mexico. In these fields, shelf carbonates interfinger with updip evaporites and clastics (Ward et al., 1986). This association also occurs in fields of the Williston Basin, where evaporite–carbonate couplets of the Madison and Charles Formations trap hydrocarbons and form reservoirs. Similar rocks can also be found in the western Canadian Basin, where shoaling-upward carbonates and evaporites of the Winnepegosis Formation are associated with deposits that are like those seen in Abu Dhabi, and in Lake MacLeod, Australia, described by Alsharhan and Kendall (1994).

Abundant oil and gas reserves have been proved in the Phanerozoic sediments of the U.A.E., with most occurring in the Mesozoic section. Of these, most are contained in Cretaceous sediments (Thamama Group, Mishrif and Simsima formations), but oil and gas also occur in the Jurassic and Lower Cretaceous reservoirs, which are sourced from the Upper Jurassic Diyab Formation (Al-Suwaidi et al., 1997). Thus, the seal of Upper Jurassic Hith Formation traps hydrocarbons in the Jurassic to the west, and in the central U.A.E., where the anhydrite seal is absent, hydrocarbons have migrated upward into the Lower Cretaceous. In contrast, production in the middle and Upper Cretaceous is thought to be sourced from the intrashelf basinal facies of the Albian–Cenomanian Shilaif (Khatiyah) Formation (Fig. 13). The underlying Albian Nahr Umr Formation apparently prevents movement of older oil from the Lower Cretaceous reservoirs upsection. Shales of the Nahr Umr, Laffan, and basal Umm Er Radhuma formations are important seals for the oil and gas accumulations. Secondary seals and barriers in the Lower Cretaceous Thamama Group are mainly dense limestones.

The Lower Cretaceous Thamama Group in the U.A.E. contains many potential structural and stratigraphic traps as in Iraq (Sadooni and Aqrawi, this volume). Most of the major structures have been drilled, but many low-relief features have yet to be explored. Similarly, stratigraphic traps, particularly porosity pinchouts in the cycles of the Thamama carbonates, may prove to be economic. The Thamama contains some thirty such porous units, many of which are thin and pinch out toward the margins of the Rub al Khali basin. Although most of the porous units occur offshore central and western Abu Dhabi, the margins of the Rub al Khali Basin of Abu Dhabi have the greatest undrilled potential in the U.A.E. The rudist bank facies of the Shuaiba Formation is the most productive part of the section. Similarly, rudist facies in the intraplatform shelfal areas mentioned previously also may prove to be important plays.

In central and eastern Arabia, the middle Cretaceous sediments tend to be continental clastics, but in the vicinity of the Arabian Gulf they are carbonate-rich, as in Iraq (Sadooni and Aqrawi, this volume). In the U.A.E., middle Cretaceous Wasia Group rocks reservoir hydrocarbons in many structures that extend from eastern offshore Abu Dhabi to offshore Ras al Khaimah. Traced eastward from Qatar to Oman, the middle Cretaceous section consists of carbonate units locally interfingering with shales. The shales are organically rich, possibly sourcing the oil and gas found in the carbonate reservoir rocks. The presence of rudist, coral-bank, and shell-debris facies is critical to commercial production from middle Cretaceous reservoirs. Successful exploration in the middle Cretaceous interval, both in structural and, more importantly, in stratigraphic traps, depends on mapping porosity trends. The middle Cretaceous of western and eastern Abu Dhabi has great potential. The reservoirs and source exist, but seals and traps will be subtle and a challenge to identify, particularly in western Abu Dhabi.

The Upper Cretaceous Aruma Group has not proved to be as productive as the Lower and middle Cretaceous. This is because reservoir-quality rocks like the Ilam/Halul carbonates and the Juweiza clastics are missing from this part of the geologic section. Until larger hydrocarbon fields are found in the Upper Cretaceous, exploration should concentrate on the rocks beneath it but at locations that can test its potential. Undoubtedly, thicker sections of porous Upper Cretaceous rocks may pinch out along the margins of the Oman Trough and could form large fields. A belt stretching from offshore Ras Al Khaimah to the Oman border parallels the Oman Mountains and provides the trend in which the Upper Cretaceous has the greatest potential.

Most of the Upper Cretaceous sediments deposited in the U.A.E. are regarded as a seal for middle Cretaceous and older reservoirs. Exceptions to this are the Ilam Formation carbonates, near the base of the Upper Cretaceous succession, and the Simsima Formation carbonates, at the top of the Upper Cretaceous interval. Both can be considered attractive exploration targets. The Ilam Formation has a number of oil shows in the offshore Emirates area, primarily from fracture porosity, and the Simsima has production from Shah Field, primarily from moldic and vuggy porosity. The presence of significant matrix porosity and permeability in the Simsima, which is overlain by essentially nonpermeable fine-grained carbonates, makes this unit a potential target over much of the foredeep area and adjacent shelf.

Future potential plays are subtle structural/stratigraphic traps, including: (1) the carbonate buildups of the Shuaiba (Aptian), Mishrif (Cenomanian), and Simsima (Maastrichtian) formations, (2) porosity pinchouts and diagenetic traps of the Habshan, the Lekhwair, and the Kharaib formations and (3) the Cenomanian clastic sediments of western U.A.E., (4) the Diyab Formation (bioclastic sediments) in western offshore Abu Dhabi, and (5) the Arab Formation in western offshore Abu Dhabi, where a well developed Hith Formation anhydrite seal occurs.

## CONCLUSIONS

Abundant oil and gas reserves have been proved in the U.A.E. and are contained in the Jurassic (Araej and Arab Formations) and Cretaceous (Thamama Group, Mishrif and Simsima

TABLE 2.—Oil and gas fields producing from Mesozoic carbonates in the United Arab Emirates. Locations of the fields are shown in Figure 11.

| | | | Reservoir | | | | | | Source rock | | Seal Rock | | | |
|---|---|---|---|---|---|---|---|---|---|---|---|---|---|---|
| No. | Field Name | Dis-covery Date | Forma-tion | Age | Pay Thick-ness (ft) | Depth to top of pay (ft) | Poro-sity (%) | Permea-bility (md) | Forma-tion | Age | Formation | Lithology | Oil Gravity (API°) | Sulfur Content % |
| 1. | Bab | 1953 | Kharaib | Barremian | 160 | 8600 | 1–30 | 16–32 | Diyab | Oxfordian | Kharaib | Dense limestone | 40.6 | 1.007 |
| 2. | Bu Hasa | 1962 | Shuaiba | Aptian | 230 | 7500 | 5–27 | 0.1–120 | Diyab | Oxfordian | Nahr Umr | Shale | 39.2 | 0.95 |
| 3. | Huwailah | 1965 | Shuaiba | Aptian | 250 | 8060 | 8–21 | 5–150 | Diyab | Oxfordian | Nahr Umr | Shale | 38.5 | 0.77 |
| 4. | Zararra | 1970 | Shuaiba | Aptian | 250 | 3700 | 18 | 6–18 | Diyab | Oxfordian | Nahr Umr | Shale | 42.6 | 0.08 |
| 5. | Asab | 1965 | Kharaib | Barremian | 170 | 7700 | 10–30 | 0.1–700 | Diyab | Oxfordian | Kharaib | Dense Limestone | 40.7 | 0.85 |
| 6. | Sahil | 1967 | Kharaib | Barremian | 145 | 8800 | 20–25 | 1–11 | Diyab | Oxfordian | Kharaib | Dense Limestone | 40 | 0.91 |
| 7. | Arjan | 1982 | Kharaib | Barremian | 150 | 9200 | 15–25 | 1–23 | Diyab | Oxfordian | Kharaib | Dense Limestone | 40 | 0.93 |
| 8. | Rumaitha | 1965 | Shuaiba | Aptian | 145 | 9335 | 15–20 | 1–80 | Diyab | Oxfordian | Nahr Umr | Shale | 43 | 0.63 |
| 9. | Shanayal | 1983 | Kharaib | Barremian | 140 | 9200 | | | Diyab | Oxfordian | Kharaib | Dense Limestone | 40 | ? |
| 10. | Jarn Yaphour | 1973 | Shuaiba | Aptian | 200 | 10,500 | 6–18 | 10–40 | Diyab | Oxfordian | Nahr Umr | Shale | 37.2 | 0.97 |
| 11. | Hudairiat | 1974 | Lekhwair | Valanginian | 30 | 10,950 | ? | ? | Diyab | Oxfordia | Lekhwair | Dense Limestone | ? | ? |
| 12. | Zubbaiya | 1969 | Kharaib | Barremian | 150 | 8850 | ? | ? | Diyab | Oxfordian | Kharaib | Dense Limestone | 38 | ? |
| 13. | Bu Labyad | 1985 | Kharaib | Barremian | 140 | 8850 | ? | ? | Diyab | Oxfordian | Kharaib | Dense Limestone | 39–50 | ? |
| 14. | Bida Al Qamzan | 1967 | Kharaib | Barremian | 80 | 7550 | ? | ? | Diyab | Oxfordian | Kharaib | Dense Limestone | 44 | ? |
| 15. | Ruwais | 1968 | Shuaiba | Aptian | 230 | 8350 | ? | ? | Diyab | Oxfordian | Nahr Umr | Shale | 38 | ? |
| 16. | Shuweihat | 1978 | Habshan | Barriasian | 90 | 8700 | ? | ? | Diyab | Oxfordian | Habshan | Dense LImestone | | ? |
| 17. | Mender | 1975 | Shuaiba | Aptian | 80 | 3360 | ? | ? | Diyab | Oxfordian | Nahr Umr | Shale | 39–42 | ? |
| 18. | Qusahwira | 1976 | Shuaiba | Albian | 45 | 5385 | 5–15 | 40 | Diyab | Oxfordian | Nahr Umr | Shale | 35.6 | 0.70 |
| 19. | Umm Shaif | 1958 | Arab | Kimme-ridgian | 575 | 8800 | 8–32 | 60–350 | Diyab | Oxfordian | Hith | Anhdyrite | 39 | 1.9 |
| 20. | Umm Lulu | 1981 | Kharaib | Barremian | 150 | 9250 | ? | ? | Diyab | Oxfordian | Kharaib | Dense Limestone | ? | ? |
| 21. | Mubarraz | 1969 | Kharaib | Barremian | 80–150 | 8800 | 12–25 | 1–20 | Diyab | Oxfordian | Kharaib | Dense Limestone | 35–37 | 1.1 |
| 22. | Zakum | 1964 | Kharaib | Barremian | 145 | 7100 | 15–29 | 15–30 | Diyab | Oxfordian | Kharaib | Dense Limestone | 44.3 | 1.2 |
| 23. | Umm Al Dholou | 1983 | Thamama | Berriasian–Aptian | 150–500 | 5575 | ? | ? | Diyab | Oxfordian | Thamama | Dense Limestone | 31–37 | ? |
| 24. | Belbazem | 1982 | Shuaiba | Aptian | 220 | 5570 | ? | ? | Diyab | Oxfordian | Nahr Umr | Shale | 31 | ? |
| 25. | Umm Al Salsal | 1983 | Kharaib | Barremian | 140 | 5570 | ? | ? | Diyab | Oxfordian | Kharaib | Dense Limestone | 31–47 | ? |
| 26. | Nasr | 1971 | Arab | Kimmer-idgian | 98 | 9300 | 7–17 | 1–45 | Diyab | Oxfordian | Hith | Anhydrite | 30 | ? |
| 27. | Abu Al Bukhoosh | 1969 | Arab | Kimmer-idgian | 28–500 | 8000 | 8–25 | 5–100 | Diyab | Oxfordian | Hith | Anhdyrite | 36 | ? |
| 28. | Mandous | 1967 | Shuaiba | Aptian | 160 | 5850 | 8 | 10 | Bab Member | Aptian | Nahr Umr | Shale | 31.9 | 0.78 |
| 29. | Bu Tini | 1979 | Arab | Kimmer-idgian | 250 | 10,300 | 18 | 20–150 | Diyab | Oxfordian | Hith | Anhdyrite | 39–44 | ? |
| 30. | Rashid | 1973 | Mishrif | Cenomanian | 100 | 9500 | 20 | 80 | Khatiyah | Albian–Cenomanian | Laffan | Shale | 53 | 1.6 |
| 31. | Umm Al Dalkh | 1968 | Mishrif | Cenomanian | 250 | 8000 | 20–25 | 15–50 | Shilaif | Albian–Cenomanian | Laffan | Shale | 30 | 0.08 |

TABLE 2 (continued).—Oil and gas fields producing from Mesozoic carbonates in the United Arab Emirates. Locations of the fields are shown in Figure 11.

| | | | Reservoir | | | | | | Source rock | | Seal Rock | | | |
|---|---|---|---|---|---|---|---|---|---|---|---|---|---|---|
| No. | Field Name | Discovery Date | Formation | Age | Pay Thickness (ft) | Depth to top of pay (ft) | Porosity (%) | Permeability (md) | Formation | Age | Formation | Lithology | Oil Gravity (API°) | Sulfur Content % |
| 32. | Arzanah | 1973 | Arab | Kimmeridgian | 95 | 10,750 | 10–12 | 6–18 | Diyab | Oxfordian | Hith | Anhydrite | 42.6 | 1.3 |
| 33. | SW Fateh | 1970 | Mishrif | Cenomanian | 100 | 7500 | 15–25 | 1–80 | Khatiyah | Albian–Cenomanian | Laffan | Shale | 32 | 1.4 |
| 34. | Fateh | 1966 | Mishrif | Cenomanian | 100 | 8000 | 17–23 | 1–102 | Khatiyah | Albian–Cenomanian | Laffan | Shale | 39 | 1.3 |
| 35. | Fallah | 1976 | Mishrif | Cenomanian | 400 | 8000 | 19 | 16 | Khatiyah | Albian–Cenomanian | Laffan | Shale | 25–47 | ? |
| 36. | Margham | 1982 | Shuaiba | Aptian | 230 | 10,560 | ? | ? | Thamama | Berriasian–Aptian | Nahr Umr | Shale | 50 | 1.4 |
| 37. | Shah | 1965 | Simsima | Maastrichtian | 250 | 4070 | 14–21 | 40–350 | Shilaif | Albian–Cenomanian | Basal Umm Er Radhuma | Shale | 30 | ? |
| 38. | El Bunduq | 1965 | Arab | Kimmeridgian | 280 | 8200 | 9–20 | 6–50 | Diyab | Oxfordian | Hith | Anhydrite | ? | ? |
| 39. | Sajaa | 1980 | Shuaiba | Aptian | 90 | 11,000 | 10 | 0.2–8 | Diyab | Oxfordian | Nahr Umr | Shaly limestone | ? | ? |
| 40. | Mubarek | 1972 | Mishrif | Cenomanian | 60 | 12,940 | ? | ? | Khatiyah | Albian–Cenomanian | Laffan | Shale | 45 | ? |
| 41 | Saleh | 1983 | Mishrif | Cenomanian | 180 | 14,175 | 18 | ? | Shilaif | Albian–Cenomanian | Laffan | Shale | 44 | ? |
| 42. | Hair Dalma | 1968 | Arab | Kimmeridgian | 95 | 9550 | ? | ? | Diyab | Oxfordian | Hith | Anhydrite | 50 | ? |
| 43. | Sadiyat | 1974 | Kharaib | Barremian | 155 | 10,560 | ? | ? | Diyab | Oxfordian | Kharaib | Dense Limestone | ? | ? |
| 44. | Bu Dana | 1983 | Arab | Kimmeridgian | 280 | 10,000 | ? | ? | Diyab | Oxfordian | Hith | Anhydrite | 30 | ? |
| 45. | Hail | 1971 | Arab | Kimmeridgian | 25–100 | 10,500 | ? | ? | Diyab | Oxfordian | Hith | Anhydrite | ? | 0.54–1.5 |
| 46. | Ghasha | 1970 | Arab | Kimmeridgian | 80 | 10,300 | 12–14 | 10–20 | Diyab | Oxfordian | Hith | Anhydrite | 40–47 | ? |
| 47. | Dalma | 1979 | Arab | Kimmeridgian | 13–23 | 10,500 | 9 | 0.4–130 | Diyab | Oxfordian | Hith | Anhydrite | 42–49 | 0.7 |
| 48. | Satah Al Raaz Boot | 1970 | Arab | Kimmeridgian | 100 | 10,000 | 12–15 | 20 | Diyab | Oxfordian | Hith | Anhydrite | 42 | ? |
| 49. | Satah | 1975 | Arab | Kimmeridgian | 25–240 | 9300 | 9–20 | 0.1–130 | Diyab | Oxfordian | Hith | Anhydrite | 37.6 | ? |
| 50. | Jarnain | 1978 | Arab | Kimmeridgian | 20–70 | 10,000 | 8–14 | 0.02–400 | Diyab | Oxfordian | Hith | Anhydrite | 43–49 | ? |
| 51. | Umm Al Anbar | 1982 | Arab | Kimmeridgian | 100–200 | 11,150 | ? | ? | Diyab | Oxfordian | Hith | Anhydrite | 40 | ? |
| 52. | Bu Jufair | 1985 | Arab | Kimmeridgian | 20–210 | 9800–10,500 | ? | ? | Diyab | Oxfordian | Hith | Anhydrite | 43 | ? |
| 53. | Al Bateel | 1982 | Arab | Kimmeridgian | 25–220 | – | ? | ? | Diyab | Oxfordian | Hith | Anhydrite | 40 | ? |
| 54. | Jameelah | 1985 | Kharaib | Barremian | 140 | 8000 | ? | ? | Diyab | Oxfordian | Kharaib | Dense Limestone | 45 | ? |
| 55. | Yaser | 1983 | Araej | Bathonian | 130 | 8950 | ? | ? | Diyab | Oxfordian | Upper Araej | Dense Limestone | 38 | ? |
| 56. | Bu Haseer | 1984 | Arab | Kimmeridgian | 30–230 | 7870 | ? | ? | Diyab | Oxfordian | Hith | Anhydrite | 40 | |

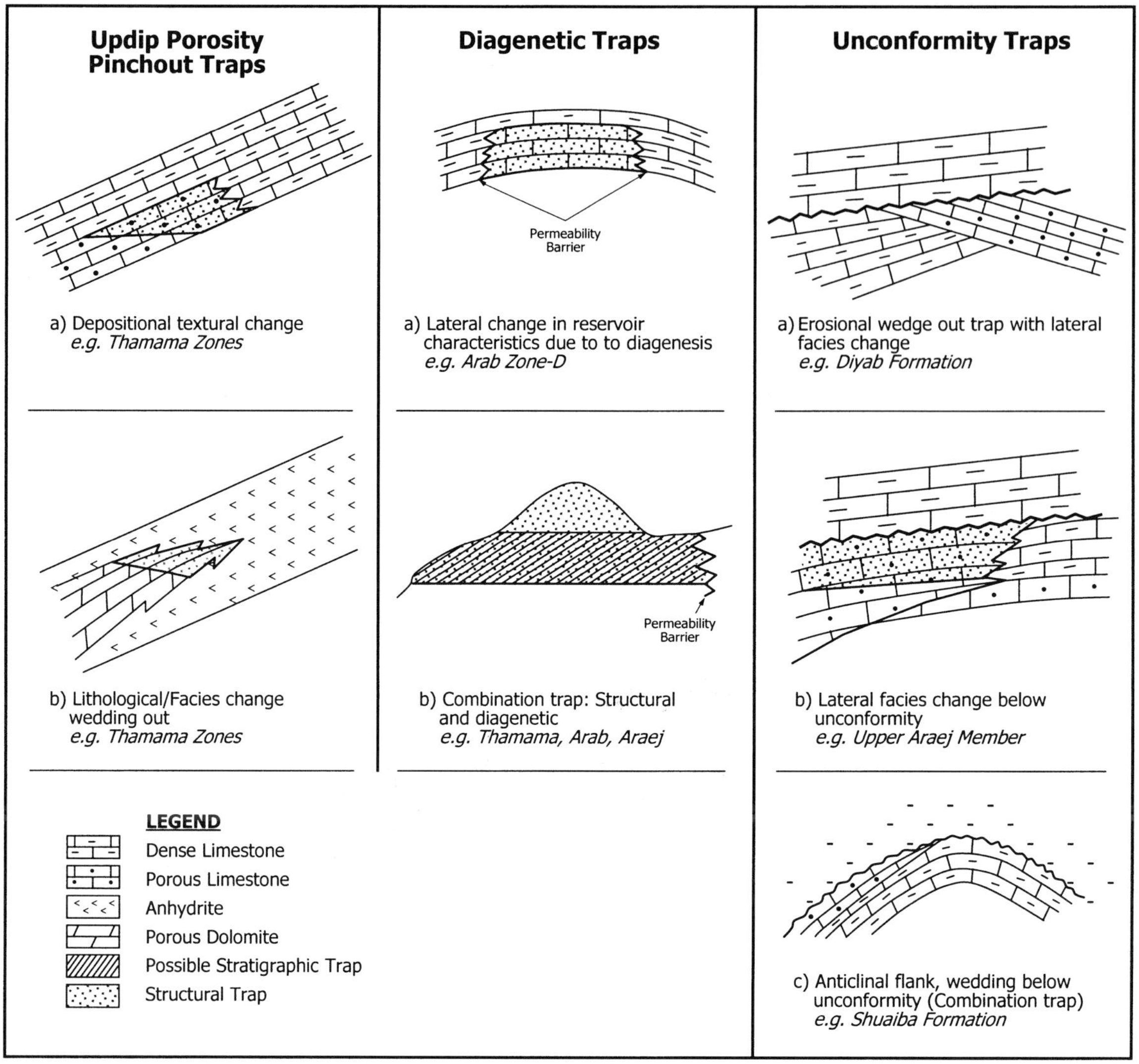


FIG. 12.—The various types of traps occur in United Arab Emirates (from Azer, 1989, reproduced with permission from Society of Petroleum Engineers).

formations) reservoirs. Most of these hydrocarbon accumulations are contained in structural traps, though combined stratigraphic/structural traps exist in some areas. In western Abu Dhabi areas, the bulk of oil and gas is in Jurassic reservoirs whose depositional character is modeled by the Holocene of the U.A.E.; in central areas, the bulk of oil is in Lower Cretaceous shallow-water carbonate reservoirs; and in the eastern offshore Abu Dhabi and Dubai, shallow-water carbonate reservoirs are of middle Cretaceous age. In western Abu Dhabi, structural traps can be shown to have started developing in the Late Jurassic. However, moving eastwards, the initiation of structural development of the traps seems to become progressively younger. In the foreland basin, the Sajaa and Margham structures (which produce gas and condensate) are part of the buried frontal thrust of the Oman Mountains overthrust belt. These structural traps can be shown to have started developing in the Late Cretaceous. In the Miocene they were altered by uplift of the present-day Oman Mountains and reactivation of some of the preexisting faults in the area.

Two main source-rock ages have been identified. One is the Upper Jurassic Diyab Formation, which fed the most prolific Upper Jurassic in the Arab Formation reservoirs and Lower Cretaceous Thamama Group. The other is the middle Cretaceous Shilaif (Khatiyah) Formation, which fed both Mishrif and Simsima reservoirs. Other minor potential source rocks have also been identified in the study area.

There are two principal sealing formations, the Hith Formation anhydrite and the Nahr Umr Formation shale. They are the main seals for the oil and gas accumulations in the underlying Arab Formation and Thamama Group, respectively. Secondary seals and barriers also exist in the stratigraphic sequence.

## ACKNOWLEDGMENTS

The authors would like to express their sincere thanks to Dr. K.W. Glennie of Aberdeen University for his comments and suggestions, which improved this paper.

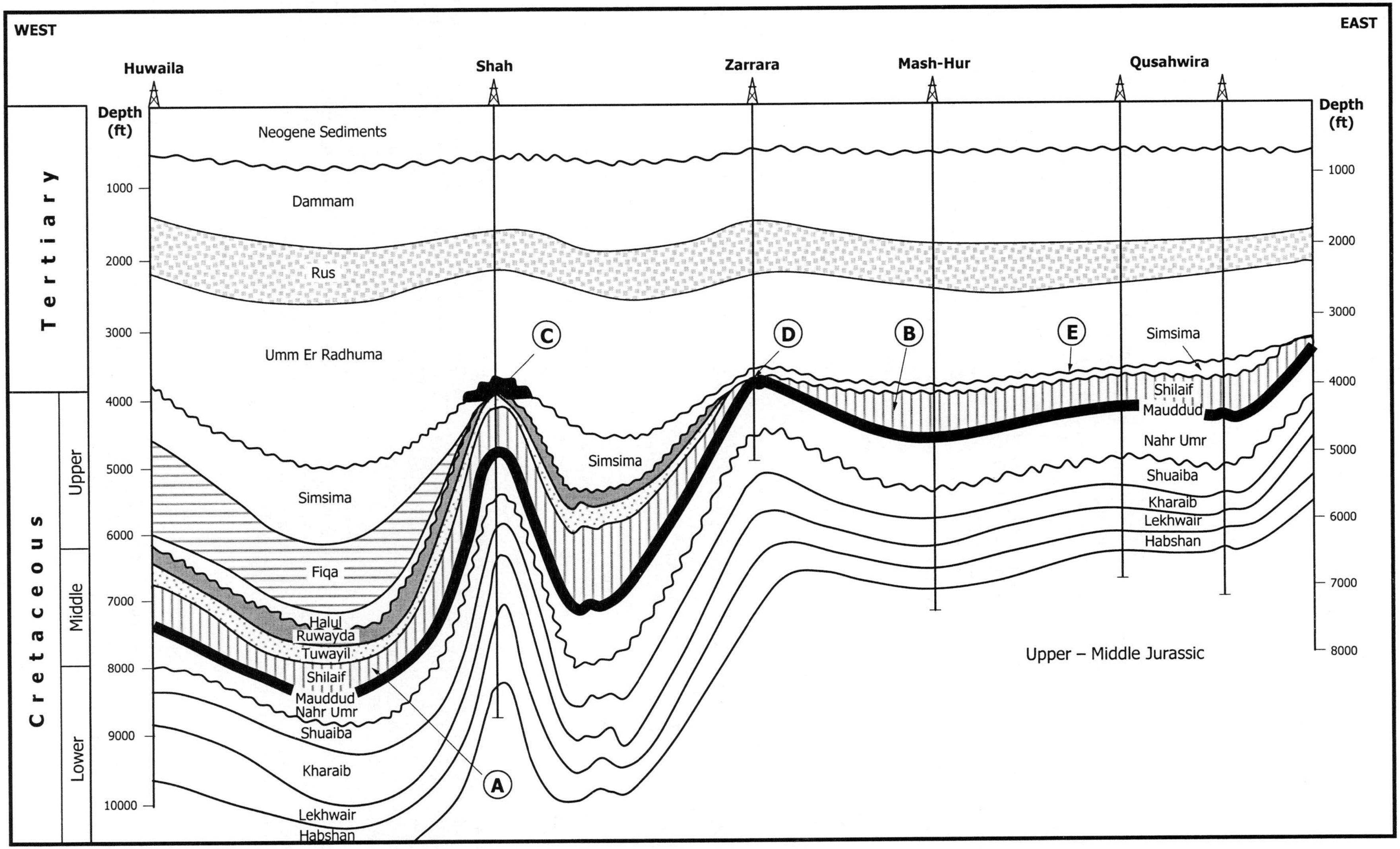


FIG. 13.—West–east cross section of Cretaceous–Tertiary in onshore Abu Dhabi showing prospectivity of middle Cretaceous Reservoirs (modified from Loutfi and El-Bishlawy, 1986). A) Shilaif is mature source rocks and migration is updip toward Shah Field. B) Shilaif is immature for hydrocarbon generation in this area. C) The occurrence of oil in the Simsima Formation here is due to the absence of the Fiqa shale seal. D) Simsima oil is present here only if Shah Field is full to spill point. E) Simsima Formation is non-prospective in southeast Abu Dhabi because of lack of mature source rock.

## REFERENCES


AL-AWAR, A.A., AND HUMPHREY, J.D., 1997, Reservoir characterization of the Shu'aiba Formation, Ghaba North Field, Oman (abstract): GeoArabia, v. 2, p. 476.

AL-HUSSEINI, M.I., 1997, Jurassic sequence stratigraphy of the western and southern Arabian Gulf: GeoArabia, v. 2, p. 361–382.

AL-SILWADI, M.S., KIRKHAM, A., SIMMONS, M.D., AND TWOMBLEY, B.N., 1996, New insights into regional correlations and sedimentology, Arab Formation (Upper Jurassic), offshore Abu Dhabi: GeoArabia, v. 1, p. 6–27.

ALSHARHAN, A.S., 1989, Petroleum geology of the United Arab Emirates: Journal of Petroleum Geology, v. 12, p. 253–288.

ALSHARHAN, A.S., 1995, Facies variation, diagenesis, and exploration potential of the Cretaceous rudist-bearing carbonates of the Arabian Gulf: American Association of Petroleum Geologists, Bulletin, v. 79, p. 531–550.

ALSHARHAN, A.S., 1997, Depositional setting, diagenesis and reservoir properties of the Maastrichtian platform carbonate sequences of the southern Arabian Gulf (abstract): GeoArabia, v. 2, p. 477–478.

ALSHARHAN, A.S., AL-AASM, I.S., AND SALAH, M.G., 1997, Stratigraphy, petrology, chemistry and hydrocarbon potential of the Aptian Shu'aiba Formation, United Arab Emirates (abstract): GeoArabia, v. 2, p. 478–479.

ALSHARHAN, A.S., AND KENDALL, C.G.ST.C., 1986, Precambrian to Jurassic rocks of the Arabian Gulf and adjacent areas: their facies, depositional setting, and hydrocarbon habitat: American Association of Petroleum Geologists, Bulletin, v. 70, p. 977–1002.

ALSHARHAN, A.S., AND KENDALL, C.G.ST.C., 1991, Cretaceous chronostratigraphy, unconformities and eustatic sea-level changes in the sediments of Abu Dhabi U.A.E.: Cretaceous Research, v. 12, p. 379–401.

ALSHARHAN, A.S., AND KENDALL, C.G.ST.C., 1994, Depositional setting of the Upper Jurassic Hith Anhydrite of the Arabian Gulf: An analogue to Holocene Evaporite of the United Arab Emirates and Lake Macleod of western Australia: American Association of Petroleum Geologists, Bulletin, v. 78, p. 1075–1096.

ALSHARHAN, A.S., AND NAIRN, A.E.M., 1986, A review of the Cretaceous formations in the Arabian Peninsula and Gulf: Part I: Lower Cretaceous (Thamama Group) stratigraphy and paleogeography: Journal of Petroleum Geology, v. 9, p. 365–392.

ALSHARHAN, A.S., AND NAIRN, A.E.M., 1988, A review of the Cretaceous formations in the Arabian Peninsula and Gulf: Part II. Mid-Cretaceous (Wasia Group) stratigraphy and paleogeography: Journal of Petroleum Geology, v. 11, p. 89–112.

ALSHARHAN, A.S., AND NAIRN, A.E.M., 1990, A review of the Cretaceous formations in the Arabian Peninsula and Gulf: Part III. Upper Cretaceous (Aruma Group) stratigraphy and paleogeography: Journal of Petroleum Geology, v. 13, p. 247–266.

ALSHARHAN, A.S., AND NAIRN, A.E.M., 1997, Sedimentary Basins and Petroleum Geology of the Middle East: Amsterdam, Elsevier Science, 942 p.

ALSHARHAN, A.S., AND WHITTLE, G.L., 1995, Carbonate–evaporite sequences of the Late Jurassic, southern and southwestern Arabian Gulf: American Association of Petroleum Geologists, Bulletin, v. 79, p. 1608–1630.

AL-SUWAIDI, A.S., TAHER, A.K., ALSHARHAN, A.S., AND SALAH, M., 1997, Geochemical evaluation of the Upper Jurassic Diyab Formation in Abu Dhabi, United Arab Emirates (abstract): GeoArabia, v. 2, p. 479.

AZER, S.R., 1989, Preliminary investigations into possible stratigraphic traps, offshore Abu Dhabi: Proceedings of 5th Middle East Oil Technical Conference and Exhibition, Bahrain, Society of Petroleum Engineers, SPE 17999, p. 739–753.

BIRDWELL, B.A., COHEN, A.D., KENDALL, C.G.ST.C., COLLING, E.L., AND ALSHARHAN, A., 1991, Petrography and petroleum source-rock potential of algal deposits along the coast of Abu Dhabi, U.A.E: Proceedings of 8th Annual Meeting of the Society for Organic Petrology, Sept. 30–Oct. 1st, 1991, Lexington, Kentucky, U.S.A.

BROOMHALL, R.W., AND ALLAN, J.R., 1985, Regional caprock-destroying dolomite on the Middle Jurassic to Early Cretaceous Arabia Shelf: Proceedings of 4th Middle East Oil Show, Bahrain, Society of Petroleum Engineers, SPE 13697, p. 157–164.

BURCHETTE, T.P., AND BRITTON, S.R., 1985, Carbonate facies analysis in the exploration for hydrocarbons: A case-study from the Cretaceous of the Middle East, *in* Brenchley, P.J., and Williams, B.P.J., eds., Sedimentology—Recent Developments and Applied Aspects: Geological Society of London, p. 311–338.

BUTLER, G.P., HARRIS, P.M., AND KENDALL, C.G.ST.C., 1982, Recent evaporites from Abu Dhabi coastal flats, *in* Handford, C.R., Robert, G.L., and Graham, R.D., eds., Depositional and Diagenetic Spectra of Evaporites: SEPM, Core Workshop 3, p. 33–64.

CONNALLY, T.C., AND SCOTT, R.W., 1985, Carbonate sediment-fill of an oceanic shelf, Lower Cretaceous, Arabian Peninsula, *in* Crevello, P.D., and Harris, P.M., eds., Deep-Water Carbonates: Buildups, Turbidites, Debris Flows and Chalks: SEPM, Core Workshop 6, p. 266–302.

DE MATOS, J.E., 1994, Upper Jurassic–Lower Cretaceous stratigraphy; the Arab, Hith and Rayda formations in Abu Dhabi, *in* Simmons, M.D., ed., Micropalaeontology and Hydrocarbon Exploration in the Middle East: London, Chapman & Hall, p. 81–111.

DE MATOS, J.E., HULSTRAND, R., AND WALKDEN, G.M., 1994, The biostratigraphy of the Lower Jurassic of the U.A.E.: Outcrop and subsurface compared: 6th Abu Dhabi International Petroleum Exhibition and Conference, p. 449–459.

GLENNIE, K.W., BOEUF, M.G.A., HUGHES CLARKE, M.W., MOODY-STUART, M., PILAAR, W.F.H., AND REINHARDT, B.M., 1974, Geology of the Oman Mountains: Koninklijk Nederlands Geologisch Mijnbouwkundig Genootstchap, Verhandelingen, 423 p.

GRADSTEIN, F.M., AGTERBERG, F.P., OGG, J.G., HARDENBOL, J., VAN VEEN, P., THIERRY, J., AND HUANG, Z., 1995, A Triassic, Jurassic and Cretaceous time scale, *in* Berggren, W.A., Kent, D.V., Aubry, M.-P., and Hardenbol, J., eds., Geochronology, Time Scales and Global Stratigraphic Correlation: SEPM, Special Publication 54, p. 95–126.

GRÖTSCH, J., BILLING, I., AND VAHRENKAMP, V., 1998, Carbon-isotope stratigraphy in shallow-water carbonates: Implications for Cretaceous black-shale deposition: Sedimentology, v. 45, p. 623–634.

HAQ, B.U., HARDENBOL, J., AND VAIL, P.R., 1988, Mesozoic and Cenozoic chronostratigraphy and cycles of sea-level change, *in* Wilgus, C.K., Hastings, B.S., Kendall, C.G.St.C., Posamenther, H.W., Ross, C.A., and Van Wagoner, J.C., eds., Sea-Level Changes: an Integrated Approach: SEPM, Special Publication 42, p. 71–108.

HARRIS, P.M., FROST, S.H., SEIGLIE, G.A., AND SCHNEIDERMANN, N., 1984, Regional unconformities and depositional cycles, Cretaceous of the Arabian Peninsula, *in* Schlee, J.S., ed., Interregional Unconformities and Hydrocarbon Accumulation: American Association of Petroleum Geologists, Memoir 36, p. 67–79.

HASSAN, T.H., 1989, The Lower and Middle Jurassic in offshore Abu Dhabi: Stratigraphy and hydrocarbon occurrence: Proceedings of 6th Middle East Oil Show, Bahrain, Society of Petroleum Engineers, SPE 18011, p. 847–858.

HUDSON, R.G.S., 1960, The Permian and Trias of the Oman Peninsula, Arabia: Geological Magazine, v. 97, p. 299–308.

HUDSON, R.G.S., AND CHATTON, M., 1959, The Musandam Limestone (Jurassic to Lower Cretaceous) of Oman, Arabia: Notes et Mémoires Moyen-Orient, Paris, v. 7, p. 69–93.

Hulstrand, R.F., and Mueller, H.W., III, 1997, The Early Cretaceous basin-to-shelf sequence exposed in Ras Al Khaimah, U.A.E. (abstract): GeoArabia, v. 2, p. 486.

Immenhauser, A., Schlager, W., Burns, S.J., Scott, R.W., Geel, T., Lehmann, J., van der Gaast, S., and Bolder-Schrijver, L.J.A., 1999, Late Aaptian to Late Albian sea-level fluctuations constrained by geochemical and biological evidence (Nahr Umr Formation, Oman): Journal of Sedimentary Research, v. 69, p. 434–446.

Kendall, C.G.St.C., and Skipwith, P.A., 1968, Recent algal mats of a Persian Gulf Lagoon: Journal of Sedimentary Petrology, v. 38, p. 1040–1058.

Kendall, C.G.St.C., Bowen, B., Alsharhan, A.S., Cheong, D., and Stoudt, D., 1991a, Eustatic controls on carbonate facies in reservoir and seals associated with Mesozoic hydrocarbon fields of the Arabian Gulf and Gulf of Mexico: Marine Geology, v. 102, p. 215–238.

Kendall, C.G.St.C., Cohen, A.S., Birdwell, B., Alsharhan, A.S., Collins, E.L., Rollins, M.S., and Dune, J.R., 1991b, Cyanobacterial mats in strandline evaporite/carbonate deposits along the coastal Abu Dhabi, United Arab Emirates: Potential sources of petroleum (abstract): Dolomieu Conference, Italy.

Kendall, C.G.St.C., Sadd, J.L., and Alsharhan, A.S., 1994, Holocene marine cement coatings on beachrocks of the Abu Dhabi coastline (U.A.E.), an analog for cement fabrics in ancient limestone: Carbonates and Evaporites, v. 9, p. 119–131.

Kenig, F., Huc, A.Y., Purser, B.H., and Oudin, J.L., 1990, Sedimentation, distribution and diagenesis of organic matter in a recent carbonate environment Abu Dhabi, U.A.E.: Organic Geochemistry, v. 16, p. 735–745.

Kennedy, W.J., and Simmons, M.D., 1991, Mid-Cretaceous ammonites and associated microfossils from the Central Oman Mountains, Newsletters on Stratigraphy, v. 25, p. 127–154.

Kirkham, A., 1997, Shoreline evolution, aeolian deflation and anhydrite distribution of the Holocene, Abu Dhabi: GeoArabia, v. 2, p. 403–416.

Leeder, M.R., and Zeidan, R., 1977, Giant Late Jurassic sabkhas of Arabian Tethys: Nature, v. 268, p. 42–44.

Lees, G.M., 1928, The geography and tectonics of Oman and parts of southeastern Arabia: Geological Society of London, Quarterly Journal, v. 84, p. 585–670.

Lippard, S.J., Smewing, J.D., Rothery, D.A., and Browning, P., 1982, The geology of the Dibba Zone, northern Oman Mountains: A preliminary study: Geological Society of London, Journal, v. 139, p. 56–66.

Loutfi, G., and El-Bishlawy, S., 1986, Habitat of hydrocarbon in Abu Dhabi, U.A.E.: Technical papers and case studies presented at the Symposium on "The Hydrocarbon Potential of Intense Thrust Zone", sponsored by Ministry of Petroleum and Mineral Resources, U.A.E. and Organization of Arab Petroleum Exporting Countries, Kuwait, v. 2, p. 63–124.

Loutfi, G., Baslaib, S.M., and Abu Hamd, M., 1987, Cenomanian stratigraphic traps in western Abu Dhabi, U.A.E.: Proceedings of 5th Middle East Oil Show, Bahrain, Society of Petroleum Engineers, SPE 15684, p. 1–8.

Le Metour, J., Bechennee, F., Platel, J.P., and Roger, J., 1995, Late Permian birth of the Neo-Tethys and development of its southern continental margin in Oman, *in* Husseini, M.I., ed., Middle East Petroleum Geosciences, Geo '94, v. 2, p. 643–654.

Moshier, S.O., Handford, C.R., Scott, R.W., and Boutell, R.D., 1988, Giant gas accumulation in a "chalky"-textured micritic limestone, Lower Cretaceous Shuaiba Formation, Eastern United Arab Emirates: SEPM, Core Workshop 12, p. 229–272.

Murris, R.J., 1980, Middle East: stratigraphic evolution and oil habitat: American Association of Petroleum Geologists, Bulletin, v. 64, p. 597–618.

Obradovich, J.D., 1993, A Cretaceous time scale, *in* Caldwell, W.G.E., and Kauffman, E.G., eds., Evolution of the Western Interior Basin: Geological Association of Canada, Special Paper 39, p. 379–396.

Pascoe, R.P., Evans, N.P., and Harland, T.L., 1995, The generation of unconformities within the Mishrif and Laffan formations of Dubai and adjacent areas: applications to exploration and production, *in* Husseini, M.I., ed., Middle East Petroleum Geosciences, Geo '94, v. 2, p. 749–760.

Patton, T.L., and O'Connor, S.J., 1988, Cretaceous flexural history of the northern Oman Mountain Foredeep, U.A.E.: American Association of Petroleum Geologists, Bulletin, v. 72, p. 797–809.

Philip, J., Borgomano, J., and Al-Maskiry, S., 1995, Cenomanian–Early Turonian carbonate platform of northern Oman: Stratigraphy and palaeo-environments: Palaeogeography, Palaeoclimatology, Palaeoceology, v. 119, p. 77–92.

Powers, R.W., Ramirez, L.F., Redmond, C.D., and Elberg, E.L., Jr., 1966, Geology of the Arabian Peninsula: Sedimentary geology of Saudi Arabia: U.S. Geological Survey, Professional Paper 560-D, 127 p.

Ricateau, R., and Riché, P.H., 1980, Geology of the Musandam Peninsula (Sultanate of Oman) and its surroundings: Journal of Petroleum Geology, v. 3, p. 139–152.

Robertson, A.H.F., and Searle, M.P., 1990, The northern Oman Tethyan continental margin: stratigraphy, structure, concepts and controversies, *in* Robertson, A.H.F., Searle, M.P., and Ries, A.C., eds., The Geology and Tectonics of the Oman Region: Geological Society of London, Special Publication 49, p. 3–25.

Russell, S.D., and Walkden, G.M., 1997, Paleohydrological model of fluid flow and differential diagenesis of the Shu'aiba Formation from the subsurface in the U.A.E. to outcrops in Jebel Akhdar, Oman (abstract): GeoArabia, v. 2, p. 491.

Saotome, J., Shibasaki, T., Shikama, T., and Kawaguchi, K., 1997, Geological controls on pore geometry of upper Kharaib reservoir offshore Abu Dhabi, U.A.E. (abstract): GeoArabia, v. 2, p. 492.

Schlumberger Ltd. Staff, 1981, Well evaluation conference—United Arab Emirates/Qatar: Schlumberger Services Technical, Paris, 271 p.

Scott, R.W., 1990, Chronostratigraphy of the Cretaceous carbonate shelf southeastern Arabia, *in* Robertson, A.H.F., Searle, M.P., and Ries, A.C., eds., The Geology and Tectonics of the Oman Region: Geological Society of London, Special Publication 49, p. 89–108.

Scott, R.W., Frost, S.H., and Shaffer, B.L., 1988, Early Cretaceous sea level curves, Gulf Coast and southeastern Arabia, *in* Wilgus, C.K., Posamentier, H., Ross, C.A., and Kendall, C.G.St.C., eds., Sea-Level Changes: An Integrated Approach: SEPM, Special Publication 42, p. 275–284.

Scott, R.W., and Stein, J.A., 1996, Age of the Albian/Cenomanian boundary: a graphic correlation solution: Institut Royal des Sciences Naturelles de Belgique, Brussels, Bulletin, Second International Symposium on Cretaceous Stage Boundaries, Supplement, v. 66, 112 p.

Searle, M.P., James, N.P., Calon, T.J., and Smewing, J.D., 1983, Sedimentological and structural evolution of the Arabian continental margin in the Musandam Mountains and Dibba Zone, United Arab Emirates: Geological Society of America, Bulletin, v. 94, p. 1381–1400.

Shebl, H.T., and Alsharhan, A.S., 1994, Sedimentary facies and hydrocarbon potential of Berriasian–Hauterivian carbonates in central Arabia, *in* Simmons, M.D., ed., Micropalaeontology and Hydrocarbon Exploration in the Middle East: London, Chapman & Hall, p. 159–175.

Smith, A.B., Morris, N.J., Gale, A.S., and Kennedy, W.J., 1995, Late Cretaceous carbonate platform faunas of the United Arab Emirates–Oman border region: Natural History Museum [London], Bulletin, Geology, v. 51, p. 91–119.

Toland, C., Peebles, R.G., and Walkden, G.M., 1993, Upper Jurassic and basal Cretaceous outcrop sequence stratigraphy of Wadi Hagil, Ras al Khaimah: Society of Petroleum Engineers, 8th Middle East Oil Show and Conference, Proceedings, Bahrain, SPE Paper 25581, p. 533–543.

VAHRENKAMP, V.C., 1996, Carbon isotope stratigraphy of the upper Kharaib and Shuaiba Formations: Implications for the Early Cretaceous evolution of the Arabian Gulf Region: American Association of Petroleum Geologists, Bulletin, v. 80, p. 647–662.

VAN BUCHEM, F.S.P., RAZIN, P., HOMEWOOD, P.W., PHILIP, J.M., EBERLI, G.P., PLATEL, J.-P., ROGER, J., ESCHARD, R., DESAUBLIAUX, G.M.J., BOISSEAU, T., LEDUC, J.-P., LABOURDETTE, R., AND CATALOUBE, S., 1996, High resolution sequence stratigraphy of the Natih Formation (Cenomanian/Turonian) in northern Oman: Distribution of source rocks and reservoir facies: GeoArabia, v. 1, p. 65–91.

WAGNER, P.D., 1990, Geochemical stratigraphy and porosity controls in Cretaceous carbonates near the Oman Mountains, *in* Robertson, A.H.F., Searle, M.P., and Ries, A.C., eds., The Geology and Tectonics of the Oman Region: Geological Society of London, Special Publication 49, p. 127–137.

WARD, R.F., KENDALL, C.G.ST.C., AND HARRIS, P.M., 1986, Hydrocarbon occurrence in Guadalupian sediments, the Permian Basin, West Texas and New Mexico: American Association of Petroleum Geologists, Bulletin, v. 70, p. 239–262.

WILSON, A.O., 1985, Depositional and diagenetic facies in Jurassic Arab C and D reservoirs, Qatif field, Saudi Arabia, *in* Roehl, P.O., and Choquette, P.W., eds., Carbonate Petroleum Reservoirs: New York, Springer-Verlag, p. 319–340.

WITT, W., AND GÖKDAG, H., 1994, Orbitolinid biostratigraphy of the Shuaiba Formation (Aptian), Oman—implications for reservoir development, *in* Simmons, M.D., ed., Micropalaeontology and Hydrocarbon Exploration in the Middle East: London, Chapman & Hall, p. 221–241.

WOOD, G.V., AND WOLF, M.H., 1969, Sabkha cycles in the Arab/Darb Formation off Trucial Coast of Arabia: Sedimentology, v. 12, p. 165–192.

# SUBJECT INDEX

# I



# J



# K



# L



# M



# N



# O



# P



# Q



# R

# TAXONOMIC INDEX

## A



## B



## C



## D



## E



## F



## G

# H



# I



# K



# L



# M



# N



# O



# P